# SMALL WASTEWATER SYSTEM OPERATION AND MAINTENANCE

## Volume I

Second Edition

A Field Study Training Program

prepared by

Office of Water Programs
College of Engineering and Computer Science
California State University, Sacramento

Kenneth D. Kerri, Project Director

2012

Cover design: Rich Parkhurst.

In recognition of the need to preserve natural resources, this manual is printed using recycled paper. The text paper and the cover are composed of 10% post-consumer waste. The Office of Water Programs strives to increase its commitment to sustainable printing practices.

Funding for this operator training manual was provided by the Office of Water Programs, California State University, Sacramento. Mention of trade names or commercial products does not constitute endorsement or recommendation for use by the Office of Water Programs or California State University, Sacramento.



First edition published 1997.


Printed in the United States of America

26 25 24 23 22 4 5 6 7 8

ISBN
978-1-59371-062-0

www.owp.csus.edu

## OFFICE OF WATER PROGRAMS

The Office of Water Programs is a nonprofit organization operating under University Enterprises, Inc., California State University, Sacramento, to provide distance learning courses for persons interested in the operation and maintenance of drinking water and wastewater facilities. These training programs were developed by people who explain, through the use of our manuals, how they operate and maintain their facilities. The university, fully accredited by the Western Association of Schools and Colleges, administers and monitors these training programs, under the direction of Dr. Ramzi J. Mahmood.

Our training group develops and implements programs and publishes manuals for operators of water treatment plants, water distribution systems, wastewater collection systems, and municipal and industrial wastewater treatment and reclamation facilities. We also offer programs and materials for pretreatment facility inspectors, environmental compliance inspectors, and utility managers. All training is offered as distance learning, using correspondence, video, or computer-based formats with opportunities for continuing education and contact hours for operators, supervisors, managers, and administrators.

Materials and opportunities available from our office include manuals in print, CD, DVD, or video formats, and enrollments for courses providing CEU (Continuing Education Unit) contact hours. Here is a sample:

- Industrial Waste Treatment, 2 volumes (print, course enrollment)
- Operation of Wastewater Treatment Plants, 2 volumes (print, CD, course enrollment, online courses)
- Advanced Waste Treatment (print, course enrollment)
- Treatment of Metal Wastestreams (print, course enrollment)
- Pretreatment Facility Inspection (print, video, course enrollment)
- Small Wastewater System Operation and Maintenance, 2 volumes (print, course enrollment)
- Operation and Maintenance of Wastewater Collection Systems, 2 volumes (print, course enrollment)
- Collection System Operation and Maintenance Training Videos (video, course enrollment)
- Utility Management (print, course enrollment)
- Manage for Success (print, course enrollment)
- and more

These and other materials may be ordered from the Office of Water Programs:

Office of Water Programs
California State University, Sacramento
6000 J Street
Sacramento, CA 95819-6025
(916) 278-6142 – phone
(916) 278-5959 – FAX

or

visit us on the web at www.owp.csus.edu

## ADDITIONAL VOLUMES OF INTEREST

**Operation of Wastewater Treatment Plants, Volume I**

The Treatment Plant Operator
Why Treat Wastes?
Wastewater Treatment Facilities
Racks, Screens, Comminutors, and Grit Removal
Sedimentation and Flotation
Trickling Filters
Rotating Biological Contactors
Activated Sludge (Package Plants and Oxidation Ditches)
Wastewater Stabilization Ponds
Disinfection Processes

**Operation of Wastewater Treatment Plants, Volume II**

Activated Sludge (Operation of Conventional Activated Sludge Plants)
Sludge Digestion and Solids Handling
Effluent Discharge, Reclamation, and Reuse
Plant Safety
Maintenance
Laboratory Procedures and Chemistry
Applications of Computers for Plant O&M
Analysis and Presentation of Data
Records and Report Writing
Treatment Plant Administration

**Operation and Maintenance of Wastewater Collection Systems, Volume I**

The Wastewater Collection System Operator
Why Collection System Operation and Maintenance?
Wastewater Collection Systems
Safe Procedures
Inspecting and Testing Collection Systems
Pipeline Cleaning and Maintenance Methods
Underground Repair

**Operation and Maintenance of Wastewater Collection Systems, Volume II**

Lift Stations
Equipment Maintenance
Sewer Renewal (Rehabilitation)
Safety/Survival Programs for Collection System Operators
Administration
Organization for System Operation and Maintenance
Capacity Assurance, Management, Operation, and Maintenance (CMOM)

**Small Wastewater System Operation and Maintenance, Volume II**

Wastewater Stabilization Ponds
Activated Sludge (Package Plants and Oxidation Ditches)
Rotating Biological Contactors (RBCs)
Disinfection and Chlorination
Alternative Wastewater Treatment, Discharge, and Reuse Methods
Laboratory Procedures
Management

**Small Water System Operation and Maintenance**

The Small Water System Operator
Water Sources and Treatment
Wells
Small Water Treatment Plants
Disinfection
Safety
Laboratory Procedures
Setting Water Rates and System Security for Small Water Utilities

# PREFACE TO THE SECOND EDITION

This *Small Wastewater System Operation and Maintenance* field study training program serves the following purposes:

1. To develop new qualified small wastewater system operators
2. To expand the abilities of existing operators, permitting better service to both their employers and the public
3. To prepare operators for higher-paying positions by passing civil service and *CERTIFICATION EXAMINATIONS.*[1]

To provide you with the knowledge, skills, and abilities needed to safely operate and maintain small wastewater systems as efficiently and effectively as possible, experienced wastewater system operators prepared the material in each chapter of this manual with an emphasis on what operators need to know to do their jobs.

Small wastewater systems vary from town to town and from region to region. The material in this program is presented to provide you with an understanding of the basic operation and maintenance aspect of your system and with information to help you analyze and solve operation and maintenance problems. This information will help you operate and maintain your system in a safe and efficient manner.

Small wastewater system operators continuously have to learn how to operate and maintain new technologies. They are confronted with having to comply with an ever-expanding array of complex regulations. To stay current in this critical field, operators must have a continuing education program. Joining professional organizations and maintaining contact with regulatory agencies helps operators stay informed so they can perform their jobs and protect the public health.

The project directors are deeply indebted to the many operators and other persons who contributed to this operator training manual. The staff of Orenco Systems, Inc. deserve special recognition and thanks for their cooperation and assistance in helping ensure that this Second Edition contains the most relevant information about current technology in the field of small on-site and decentralized wastewater operation and maintenance systems.

The publications staff of the Office of Water Programs are recognized for their contributions to this training manual: Janet Burton (Publications Manager); Jan Weeks (Senior Editor); Robert Beck (Production Editor); Lea Washington (Technical Editor).

KENNETH D. KERRI
2012

1. *Certification Examination.* An examination administered by a state agency or professional association that operators take to indicate a level of professional competence. In the United States, certification of operators of water treatment plants, wastewater treatment plants, water distribution systems, and small water supply systems is mandatory. In many states, certification of wastewater collection system operators, industrial wastewater treatment plant operators, pretreatment facility inspectors, and small wastewater system operators is voluntary; however, current trends indicate that more states, provinces, and employers will require these operators to be certified in the future. Operator certification is mandatory in the United States for the Chief Operators of water treatment plants, water distribution systems, and wastewater treatment plants.

# OBJECTIVES OF THIS MANUAL

Proper installation, inspection, operation, maintenance, repair, and management of small wastewater systems have a significant impact on the operation and maintenance costs and effectiveness of the systems. The objective of this manual is to provide small wastewater system operators with the knowledge and skills required to operate and maintain these systems effectively, thus eliminating or reducing the following problems:

1. Health hazards created by the discharge of improperly treated wastewater
2. System failures that result from the lack of proper installation, inspection, preventive maintenance, surveillance, and repair programs designed to protect the public's investment in these facilities
3. Fish kills and environmental damage caused by wastewater system problems
4. Corrosion damages to pipes, equipment, tanks, and structures in the wastewater system
5. Complaints from the public or local officials due to the unreliability or failure of the wastewater system to perform as designed
6. Odors caused by insufficient or inadequate wastewater and sludge handling and treatment procedures

# SCOPE OF THIS MANUAL

Operators with the responsibility for septic tanks and on-site treatment and disposal systems, wastewater collection systems, and pumping or lift stations will find this manual very helpful. This manual contains the following information:

1. What small wastewater system operators do
2. Sources of wastewater and damage caused by improperly treated wastewater
3. How to operate and maintain septic tanks and on-site wastewater treatment and disposal systems
4. How to operate and maintain collection systems and pump stations
5. Techniques for recognizing hazards and developing safe procedures and safety programs
6. Procedures for setting up and carrying out an effective maintenance program
7. How to calculate the cost of operating a small wastewater system and develop an appropriate rate structure

Material in this manual furnishes you with information concerning situations encountered by most small wastewater system operators in most areas. These materials provide you with an understanding of the basic operational and maintenance concepts for small wastewater systems and with an ability to analyze and solve problems when they occur. Operation and maintenance programs for small wastewater systems will vary with the age of the system, the extent and effectiveness of previous programs, and local conditions. You will have to adapt the information and procedures in this manual to your particular situation.

Technology is advancing very rapidly in the field of operation and maintenance of small wastewater systems. To keep pace with scientific advances, the material in this program must be periodically revised and updated. This means that you, the system operator, must be aware of new advances and recognize the need for continuous personal training reaching beyond this program. Training opportunities exist in your daily work experience, from your associates, and from attending meetings, workshops, conferences, and classes.

## USES OF THIS MANUAL

This manual was developed to serve the needs of operators in several different situations. The format used was developed to serve as a home-study or self-paced instruction course for operators in remote areas or persons unable to attend formal classes either due to shift work, personal reasons, or the unavailability of suitable classes. This home-study training program uses the concepts of self-paced instruction where you are your own instructor and work at your own speed. In order to certify that a person has successfully completed this program, objective tests and special answer sheets for each chapter are provided when a person enrolls in this course.

Also, this manual can serve effectively as a textbook in the classroom. Many colleges and universities have used this manual as a text in formal classes (often taught by operators). In areas where colleges are not available or are unable to offer classes in the operation of small wastewater systems, operators and utility agencies can join together to offer their own courses using this manual.

Cities or utility agencies can use this manual in several types of on-the-job training programs. In one type of program, a manual is purchased for each operator. A senior operator or a group of operators are designated as instructors. These operators help answer questions when the persons in the training program have questions or need assistance.

This manual was prepared to help operators operate and maintain their small wastewater systems. Please feel free to use the manual in the manner that best fits your training needs and the needs of other operators. We will be happy to work with you to assist you in developing your training program. Please feel free to contact the Project Director:

Project Director
Office of Water Programs
California State University, Sacramento
6000 J Street
Sacramento, CA 95819-6025
(916) 278-6142 – phone
(916) 278-5959 – FAX
wateroffice@csus.edu – e-mail

## TECHNICAL REVIEWERS

Russ Armstrong
Harold Ball
Bill Cagle
David Dauwalder
Steve Desmond
Fred Egger
Ross V. Johnson
Glenn Malliet
Gary Morgan
Mike Myers
Charles Ohr
Frank Paibomsai
Mark S. Richardson
Duane V. Rigge
Richard Solbrig
Jim Van Domelen

## INSTRUCTIONS TO PARTICIPANTS IN HOME-STUDY COURSE

Procedures for reading the lessons and answering the questions in this manual are contained in this section.

To progress steadily through this training program, establish a regular study schedule. Some of the chapters are longer and more difficult than others, so many of them are divided into two or more lessons. The time required to complete a lesson will depend on your background and experience. The important thing is that you understand the material in the lesson before starting the next one.

Each lesson is arranged for you to read a short section, write your answers to the questions at the end of the section, and check your answers against the suggested answers. You can then decide if you understand the material sufficiently to continue or whether you need to read the section again. You may find that this procedure is slower than reading a typical textbook, but you will probably remember much more when you have finished.

Discussion and review questions follow each lesson in the chapters. These questions are provided to help you review the important points covered in the lesson. Write your answers to the discussion and review questions to help yourself retain the material better.

In the appendix at the end of this manual, you will find comprehensive review questions and suggested answers. These questions and answers are provided for you to review how well you remember the material. You will probably need to review the entire manual before attempting to answer these questions. Some of the questions are essay-type questions, which are used by some states for higher-level certification examinations. After you have answered all the questions, compare your answers with those provided and determine the areas in which you might need additional review before your next certification or civil service examination. Please do not send your answers to the Office of Water Programs at California State University, Sacramento.

You are your own teacher in this training program. You could merely look up the suggested answers to the questions at the end of the chapters and in the comprehensive review section, but doing so will not help you understand the material. Consequently, you would not be able to apply the material to the performance of your job or recall it during an examination for certification or a civil service position. You will get out of this program what you put into it, so we encourage you to make the most of the material presented.

## SUMMARY OF PROCEDURE

To complete this program, you need to work through all of the chapters. Study the material in chapter order or in a sequence that works best for you. Many manuals include an arithmetic appendix to give you a quick review of the math concepts found in the manual.

The following is a summary of the suggested procedure for studying the material in this manual:

1. Read what you are expected to learn in each chapter (the Objectives).
2. Read the sections in each lesson or chapter.
3. Write your answers to the questions at the end of each section, just as you would if they were questions on an actual test.
4. Compare your answers with the suggested answers.
5. Decide whether to review the section or to continue on to the next section.
6. Write your answers to the discussion and review questions at the end of each lesson.

*NOTE:* Safety is an important topic in all of the manuals in this operator training series. Because operators daily encounter situations and equipment that can cause serious disabling injury or illness, operators need to be aware of potential dangers and exercise adequate precautions when performing their jobs. As you work through the chapters, pay close attention to the safe procedures that are stressed throughout.

## SMALL WASTEWATER SYSTEM OPERATION AND MAINTENANCE

### Volume I

Second Edition

**Chapters** **Page**

CHAPTER 1

# THE SMALL WASTEWATER SYSTEM OPERATOR

by

Larry Trumbull

and

William Crooks

Revised by

John Brady

# TABLE OF CONTENTS

## Chapter 1. THE SMALL WASTEWATER SYSTEM OPERATOR

# OBJECTIVES

## Chapter 1. THE SMALL WASTEWATER SYSTEM OPERATOR

1. Explain the type of work done by small wastewater system or on-site system operators.
2. Describe where to look for jobs in this profession.
3. Find sources of further information on how to do the jobs performed by small wastewater system or on-site system operators.

# PROJECT PRONUNCIATION KEY

by Warren L. Prentice

The Project Pronunciation Key is designed to aid you in the pronunciation of new words. While this key is based primarily on familiar sounds, it does not attempt to follow any particular pronunciation guide. This key is designed solely to aid operators in this program.

You may find it helpful to refer to other available sources for pronunciation help. Each current standard dictionary contains a guide to its own pronunciation key. Each key will be different from each other and from this key. Examples of the difference between the key used in this program and the *WEBSTER'S NEW WORLD COLLEGE DICTIONARY*[1] key are shown below.

In using this key, you should accent (say louder) the syllable that appears in capital letters. The following chart is presented to give examples of how to pronounce words using the project key.

SYLLABLE

| WORD | 1st | 2nd | 3rd | 4th | 5th |
|---|---|---|---|---|---|
| aerobic | air | O | bick | | |
| bacteria | back | TEER | e | uh | |
| contamination | kun | TAM | uh | NAY | shun |

The first word, *AEROBIC,* has its second syllable accented. The second word, *BACTERIA,* has its second syllable accented. The third word, *CONTAMINATION,* has its second and fourth syllables accented.

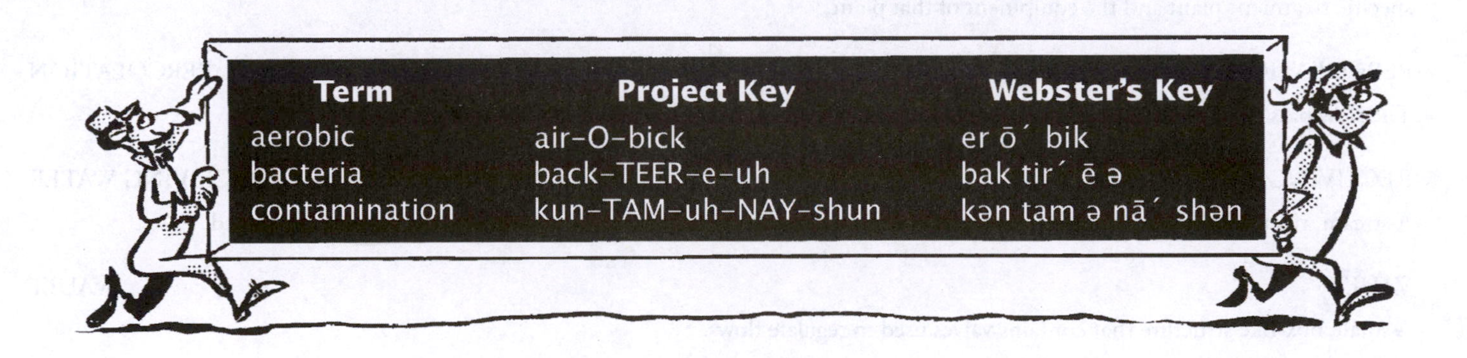

| Term | Project Key | Webster's Key |
|---|---|---|
| aerobic | air-O-bick | er ō´ bik |
| bacteria | back-TEER-e-uh | bak tir´ ē ə |
| contamination | kun-TAM-uh-NAY-shun | kən tam ə nā´ shən |

1. The *WEBSTER'S NEW WORLD COLLEGE DICTIONARY,* Fourth Edition, 1999, was chosen rather than an unabridged dictionary because of its availability to the operator. Other editions may be slightly different.

# WORDS

## Chapter 1. THE SMALL WASTEWATER SYSTEM OPERATOR

AESTHETIC (es-THET-ick) AESTHETIC

Attractive or appealing.

AIR RELEASE AIR RELEASE

A type of valve used to allow air caught in high spots in pipes to escape.

AQUIFER (ACK-wi-fer) AQUIFER

A natural, underground layer of porous, water-bearing materials (sand, gravel) usually capable of yielding a large amount or supply of water.

EFFECTIVE SOIL DEPTH EFFECTIVE SOIL DEPTH

The depth of soil in the leach field trench that provides a satisfactory percolation area for the septic tank effluent.

EFFLUENT (EF-loo-ent) EFFLUENT

Water or other liquid—raw (untreated), partially treated, or completely treated—flowing *FROM* a reservoir, basin, treatment process, or treatment plant.

HOT TAP HOT TAP

Tapping into a sewer line under pressure, such as a force main or a small-diameter sewer under pressure.

O&M MANUAL O&M MANUAL

Operation and Maintenance Manual. A manual that describes detailed procedures for operators to follow to operate and maintain a specific treatment plant and the equipment of that plant.

PERCOLATION (purr-ko-LAY-shun) PERCOLATION

The slow passage of water through a filter medium; or, the gradual penetration of soil and rocks by water.

RECEIVING WATER RECEIVING WATER

A stream, river, lake, ocean, or other surface or groundwaters into which treated or untreated wastewater is discharged.

VAULT VAULT

A small, box-like structure that contains valves used to regulate flows.

WASTEWATER WASTEWATER

A community's used water and water-carried solids (including used water from industrial processes) that flow to a treatment plant. Stormwater, surface water, and groundwater infiltration also may be included in the wastewater that enters a wastewater treatment plant. The term sewage usually refers to household wastes, but this word is being replaced by the term wastewater.

# CHAPTER 1. THE SMALL WASTEWATER SYSTEM OPERATOR

Chapter 1 is prepared especially for new operators or people interested in becoming small *WASTEWATER*[2] system operators. If you are an experienced small wastewater system operator, you may find some new viewpoints in this chapter.

## 1.0 WHY IS THERE A NEED FOR SMALL SYSTEM OPERATORS?

Before people lived clustered together in permanent communities, water was purified in a natural cycle as shown below:

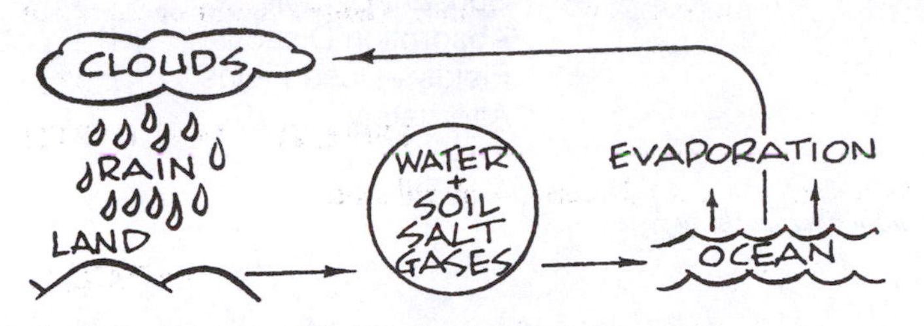


*Simplified natural purification cycle*

But now, with modern society's intensive use of the water resource and the resulting water pollution, it is no longer feasible simply to wait for sun, wind, and time to accomplish the purification of soiled water. Years ago, treatment of human waste in small communities and farmsteads was accomplished through the use of the pit privy. With the advent of indoor plumbing and flush toilets, a better waste treatment method was needed and the septic tank came into extensive use.

The septic tank (see Figure 1.1) principle was first developed in France around 1860 for single family use; later, larger tanks were constructed to provide waste treatment for some small communities both abroad and in the United States.

The conventional on-site system consists of two main components: (1) the septic tank and (2) the subsurface disposal field (also called a leach field or absorption field). The septic tank is designed to remove solids from the wastewater, both floatable solids (rise to surface) and settleable solids (sink to the bottom). The subsurface disposal field disposes of wastewater from the septic tank mainly through *PERCOLATION,*[3] but also by treatment of the wastewater by microorganisms living in the soil. If solids escape the septic tank, the disposal field will clog and fail.

The septic tank was widely used throughout the United States and continues to be used today in rural areas where low population densities permit septic systems to be widely scattered over large areas of land. Many leach field systems failed due to poor or nonexistent maintenance (lack of periodic removal of solids from the septic tank by pumping), inadequate *EFFECTIVE SOIL DEPTH,*[4] but with plenty of land available, new replacement systems could be built. However, leach field failures became a common problem when growth in towns and cities meant that the necessary land area was no longer available for replacement leach fields. Population growth also led to the installation of septic tank leach fields in unsuitable soils or too small an area where, predictably, the leach fields failed. These failures produced unsanitary environments (leach fields clog and untreated wastewater may rise to surface) in communities, polluted streams, and contaminated groundwater *AQUIFERS*[5] with waterborne disease organisms and toxic chemicals. Leach field failures not only affected the human population, but also endangered wildlife and damaged the natural habitat. As population density increased, many septic tank leaching systems were no longer environmentally safe. As a result, towns and cities eventually developed wastewater collection systems and built wastewater treatment facilities.

The 1940s saw a large-scale population shift from rural to metropolitan areas. This shift reduced the use of small septic systems, but in the cities, waterways were being degraded by disposal of greatly increased waste flows. During the 1950s, a program sponsored by the US and state governments mandated expansion and improvement of America's wastewater treatment facilities. By the 1970s, many people seeking relief from the hustle and bustle, noise, crime, and pollution of the big cities migrated back to the rural or semirural countryside. Along with peace and quiet, however, they often found no wastewater collection system, no wastewater treatment plant, and, in many instances, no water systems, phones, or paved streets. Often, the most economical wastewater disposal system was a septic tank and leach field or seepage pit (a 3-foot diameter shallow well used to dispose of wastewater).

---

2. *Wastewater.* A community's used water and water-carried solids (including used water from industrial processes) that flow to a treatment plant. Stormwater, surface water, and groundwater infiltration also may be included in the wastewater that enters a wastewater treatment plant. The term sewage usually refers to household wastes, but this word is being replaced by the term wastewater.
3. *Percolation* (purr-ko-LAY-shun). The slow passage of water through a filter medium; or, the gradual penetration of soil and rocks by water.
4. *Effective Soil Depth.* The depth of soil in the leach field trench that provides a satisfactory percolation area for the septic tank effluent.
5. *Aquifer* (ACK-wi-fer). A natural, underground layer of porous, water-bearing materials (sand, gravel) usually capable of yielding a large amount or supply of water.

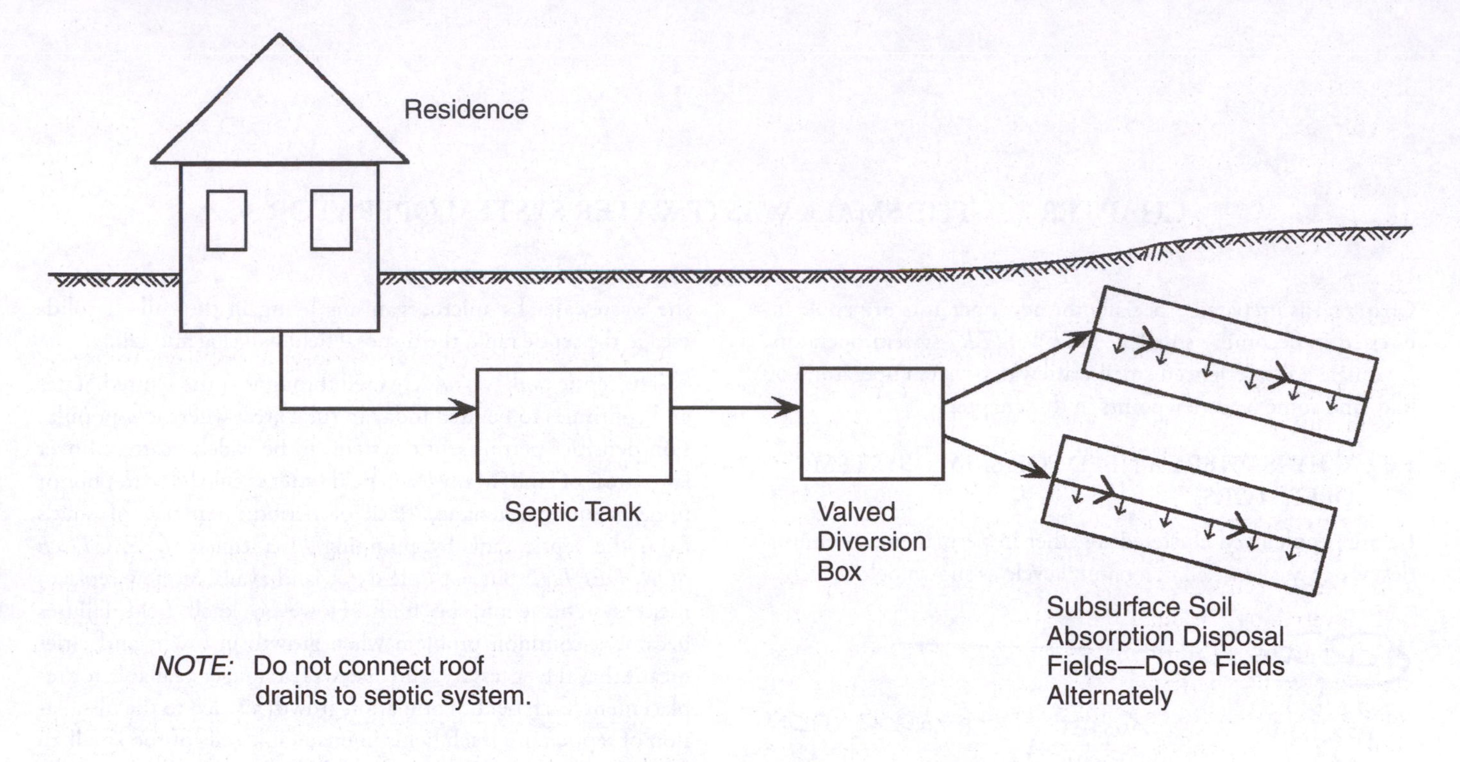


*Fig. 1.1 Septic tank and subsurface leaching system*

The inexperienced homeowner had to operate and maintain a wastewater treatment system. In many cases, no maintenance was performed and, eventually, systems failed because solids that built up in the septic tank were not pumped out before they escaped and plugged the soil disposal system.

In most areas of the United States, local zoning laws and state or federal environmental regulations now govern the installation of new septic systems, making the improper siting (location) of leach fields less likely to occur. In areas where septic system and leach field failures continue to endanger public health, or where other factors rule out the use of individual septic systems (such as poor or wet soil conditions), homeowners sometimes form a community service district, a sanitary district, or a sewer utility agency to develop and provide a community-wide alternative treatment and disposal method to meet environmental requirements. The new wastewater utility agency may serve only four or five homes or could involve several different areas having one, two, or even three hundred homes, stores, or small businesses in each area, all having a common need for a wastewater collection, treatment, and disposal system.

The new system may be a very small community leach field, absorption mounds, a filtering system, or a small wastewater treatment plant such as an oxidation pond, oxidation ditch, activated sludge plant, or rotating biological contactors (RBCs). Treated *EFFLUENT*[6] may be disposed of by using it for irrigation or by discharging it to wetlands or *RECEIVING WATERS*.[7]

The method selected must adequately treat the wastewater and dispose of it in a safe and environmentally sound, economical manner. Whatever method is selected requires competent individuals to maintain the system and ensure that it is functioning properly. This is the job of the small wastewater treatment system operator.

*The operator's duties*

## 1.00 What Does a Small System Operator Do?

The job is not glamorous, often it is downright messy and stinky. It may not be greatly appreciated by others until the system fails, but it is a very important responsibility. As a small wastewater system operator, the goal of your efforts is very much like a doctor's: you must keep your community healthy. You accomplish this by keeping the wastewater system operating properly without exposing the local population to disease or contaminated surface or groundwaters due to pollution caused by accidental sewer overflows, wastewater backups into homes, improper treatment methods, or equipment failures. You are

6. *Effluent* (EF-loo-ent). Water or other liquid—raw (untreated), partially treated, or completely treated—flowing *FROM* a reservoir, basin, treatment process, or treatment plant.

7. *Receiving Water.* A stream, river, lake, ocean, or other surface or groundwaters into which treated or untreated wastewater is discharged.

protecting the environment and the public health of your community by correct treatment and proper disposal or discharge of the wastewater from your community.

Your specific job responsibilities will depend on the types of treatment systems used where you work and local geographic conditions such as terrain and weather. In some instances, the job may even be influenced by regional politics. Wherever you are, the job has challenging tasks to be performed every day; sometimes, it also requires work on weekends, holidays, and, occasionally, in the middle of the night.

Since you will be associated with small systems, help may be difficult to locate.[8] There may not be any accurate as-built records for the location, including size and type of underground pipes. An *O&M MANUAL*[9] may have been prepared and lost, destroyed, or may have never existed. For some older small systems, much of the work may be in locating what has already been installed and evaluating the condition of existing components. You will most likely have to do everything yourself. "Everything" includes tasks too numerous to describe in this chapter, but the duties of a small wastewater system operator will keep life interesting. Depending on the type of system you have and your responsibilities, you may be required to perform many or most of the tasks, mainly outdoors, in the following list:

1. Digging to uncover tanks, *VAULTS,*[10] valves, and lines
2. Installing new pipelines, manhole covers, valves, *AIR RELEASE*[11] valves, and making *HOT TAPS*[12]
3. Cleaning tanks
4. Replacing and repairing pumps
5. Washing screens
6. Reading meters
7. Collecting samples
8. Measuring depths of scum blankets and sludge in septic tanks
9. Adjusting valves
10. Checking electrical control systems
11. Replacing control units
12. Operating or contracting for backhoes
13. Driving trucks
14. Talking to people about their problems or complaints
15. Locating underground system components and protecting them from construction activities

There are inside activities also:

1. Recordkeeping
2. Conducting laboratory analyses
3. Keeping an inventory of spare parts and equipment
4. Preparing, sending out, and collecting bills
5. Preparing budgets
6. Preparing emergency plans
7. Planning for the future
8. Writing reports to your employer and the environmental control agencies
9. Ordering supplies and replacement parts
10. Talking to people about their problems or complaints

In some small systems, there are housekeeping jobs including sweeping and mopping floors, washing windows, waxing floors, setting up chairs, preparing agendas for the next board meeting, bookkeeping, preparing and presenting financial status reports, paying bills, planning for safety, and training programs, not only for you but for your customers as well.

You will have to maintain system maps so you can locate your buried pipelines and your electrical systems in order to show other utilities where not to dig and so that you know where to dig to install a new connection to the system. You may be required to inspect new construction to ensure installations are made to your agency's standards. Or, you may have to construct new facilities yourself.

You may have to go out in the middle of the night to restore electrical power to the system or start a standby generator engine. Clearing a blockage in the collection system could be your Thanksgiving afternoon job. Every system is different, but there are similarities in like systems. You can gain knowledge from other operators and you can share your discoveries with others to help them do a better job of protecting the environment.

### 1.01 Who Does the Small System Operator Work For?

There are many opportunities to work as a small wastewater system operator; some are paid positions, others are not. For example, you could be one of several neighbors who have had to join together to create a small (three or more homes) residential system and you have the privilege of keeping the system operating. There may be no financial compensation because there are no funds to pay for a maintenance person or operator, but the system still must be operated to meet local and state requirements.

8. Help may be available from the National Environmental Services Center (NESC) at (800) 624-8301 in West Virginia. Also, on-site or small flows associations are being formed in many states. The National Onsite Wastewater Recycling Association, 601 Wythe Street, Alexandria, VA 22314, phone (800) 966-2942, is a good source of helpful information, www.nowra.org.
9. *O&M Manual.* Operation and Maintenance Manual. A manual that describes detailed procedures for operators to follow to operate and maintain a specific treatment plant and the equipment of that plant.
10. *Vault.* A small, box-like structure that contains valves used to regulate flows.
11. *Air Release.* A type of valve used to allow air caught in high spots in pipes to escape.
12. *Hot Tap.* Tapping into a sewer line under pressure, such as a force main or a small-diameter sewer under pressure.

Maintaining and operating the wastewater treatment system may be an add-on to your present job. Examples of employees who have become operators include school district maintenance superintendents who have inherited a school's wastewater treatment system; or a maintenance supervisor for a shopping mall; or a large interstate truck stop operator who has had to install a treatment system. Owners of remote resorts, mobile home and RV parks, and apartment complexes all have been required to install and operate their own wastewater treatment systems either because they were not close to an existing local municipal wastewater system or because connection to an existing system was restricted to protect collection or treatment system capacity.

You may be employed by federal, state, or local agencies that are required to operate their own facilities. Federal and state parks and correctional institutions may own and operate wastewater treatment plants. State agencies such as social health services, transportation, and fish and game agencies also may own and operate wastewater treatment plants. Federal military bases, forts, and remote outposts operate small wastewater collection and treatment systems. Also, Native Americans on reservations frequently own and operate small wastewater collection and treatment systems.

You may be employed by a small community service or utility agency to operate and maintain a system or by a privately owned operation and maintenance organization that contracts to operate and maintain wastewater treatment systems or domestic water supply systems. Private contractors such as these remove the burden on a small community or city of having to recruit, interview, and hire personnel who are appropriately trained and state-certified for various competency levels to meet state environmental regulatory agency requirements for operating and maintaining small wastewater systems. Employment opportunities with these specialized private contractors are increasing.

### 1.02 Where Does the Small System Operator Work?

Obviously, the operator works with a small wastewater system or treatment facility. But the different types and locations of systems or treatment plants offer a wide range of working conditions. From the mountains to the sea, wherever people congregate into communities, wastewater systems and treatment plants are found. You may be the sole operator of a small town facility, you may work with several operators in a slightly larger wastewater system, or you may even manage a community's wastewater collection and treatment facilities. You can select your own special place in a wastewater system or treatment plant operation.

### 1.03 What Pay Can a Small System Operator Expect?

In dollars? Prestige? Job satisfaction? Community service? In opportunities for advancement? By whatever scale you use, returns are what you make them. You could choose to work in an affluent subdivision where the pay is good. Choose a small town and pay may not be as good, but job satisfaction, freedom from time-clock hours, community service, and prestige may well add up to a more desirable personal goal. Total reward depends on you.

### 1.04 What Does It Take to Be a Small System Operator?

Desire is what it takes to become a small wastewater system operator. First, you must choose to enter this profession. You can do it with a grammar school, a high school, or a college education. The amount of education needed depends largely on the type of treatment facility, but also on local certification and entry-level job requirements. While some jobs will always exist for manual labor, the real and expanding need is for trained operators. New techniques, advanced equipment, and increasing use of complex instrumentation require a new breed of operator, one who is willing to learn today, and gain tomorrow, for surely the facility will move toward newer and more effective operating procedures and treatment processes. Indeed, the truly service-minded operator assists in adding to and improving the facility's performance on a continuing basis.

*Tomorrow's forgotten operator stopped learning yesterday*

You can be an operator tomorrow by beginning your learning today; or you can be a better operator, ready for advancement, by accelerating your learning today. This manual explains how various wastewater collection and treatment processes and systems function, and it can be used as a guide to assist in operating and maintaining them. It should also be a tool to increase your knowledge so that you may better communicate with local and state environmental agencies. You will learn the terminology needed to communicate effectively with operators of similar systems or larger municipal treatment systems. This manual should also provide you with the opportunity to achieve state certification for the level of competency required by your system. Certification provides the state agency and the public with the assurance that the person operating the system is knowledgeable and capable of successfully treating wastes and can recognize a problem and request assistance before harm is done to the public health, the environment, or the equipment.

This training course, then, is your start toward a better tomorrow, both for you and for the public whose health and environment will benefit.

## QUESTIONS

Place an "X" by the correct answer or answers. After you have answered all the questions, check your answers with those given at the end of the chapter on page 14. Reread any sections you did not understand and then proceed to the next section. You are your own teacher in this training program, and *YOU* should decide when you understand the material and are ready to continue with new material.

**EXAMPLE:**

This is a training course on

________ 1. Accounting
________ 2. Engineering
___X___ 3. Wastewater system operation
________ 4. Salesmanship

1.0A Wastewater is the same thing as

________ 1. Rain
________ 2. Soil
________ 3. Sewage
________ 4. Condensation

1.0B What does an operator do?

________ 1. Collect samples
________ 2. Lubricate equipment
________ 3. Record data

1.0C Who employs small wastewater system operators?

________ 1. Cities
________ 2. Sanitation districts
________ 3. Industries

Check your answers on page 14.

## 1.1 YOUR PERSONAL TRAINING COURSE

Beginning on this page, you are embarking on a training course that has been carefully prepared to help you improve your knowledge of, and ability to operate, wastewater facilities. You will be able to proceed at your own pace; you will have an opportunity to learn a little or a lot about each topic. The course has been prepared this way to fit the various needs of operators, depending on the kind of facility you have or how much you need to learn about it. To study for certification examinations, you will have to cover all the material. You will never know everything about your facility or about the wastewater that flows through it, but you can begin to answer some very important questions about how and when certain things happen in the facility. You can also learn how to manipulate your facility so that it operates at maximum efficiency.

## 1.2 WHAT DO YOU ALREADY KNOW?

If you already have some experience operating a wastewater facility, you may use the first two chapters for a review. If you are relatively new to the wastewater field, these chapters will provide you with the background information necessary to understand the later chapters. The remainder of this introductory chapter describes your role as a protector of water quality, your qualifications to do your job, a little about staffing needs in the wastewater treatment field, and some information on other training opportunities.

## 1.3 THE WATER QUALITY PROTECTOR: *YOU*

Historically, Americans have shown a general lack of interest in the protection of their water resources. We have been content to think that "the solution to pollution is dilution." For years, we were able to dump our wastes with little or no treatment back into the nearest receiving waters. As long as there was enough water to dilute the waste material, nature took care of our disposal problems for us. As more and more towns and industries sprang up, waste loads increased until the natural purification processes could no longer do the job. Many waterways became polluted with wastewater. Unfortunately, for many areas this did not signal the beginning of a cleanup campaign. It merely increased the frequency of the cry: "We do not have the money for a treatment plant," or the ever-popular, "If we make industries treat their wastes, they will move to another state." Thus, the pollution of our waters increased.

*Water quality protector*

Within the last 30 years, we have seen many changes in this depressing picture. We now realize that we must give nature a hand by treating wastes before they are discharged. Today, federal laws and the US Environmental Protection Agency require that each industry must meet the same minimum treatment requirements (effluent standards) in every state.

Adequate treatment of wastes will not only protect our health and that of our downstream neighbors, it will also increase property values, allow game fishing and various recreational uses to be enjoyed, and attract water-using industries to the area. Today, we are seeing massive efforts being undertaken to control water pollution and improve water quality throughout the nation. This includes the effort of your own community, county and state, as well as the federal government.

Great sums of public and private funds are now being invested in complex municipal and industrial wastewater treatment facilities to overcome water pollution, and you, the small system operator, will play a key role in the battle. Without efficient operation of your facility, much of the research, planning, and building that has been done and will be done to accomplish the goals of water quality control in your area will be wasted. Your knowledge and skills could make the difference between a polluted, unusable water supply and safe, clean water. You are, in fact, a water quality protector on the front line of the water pollution battle.

The receiving water quality standards and waste discharge requirements that your facility has been built to meet have been formulated to protect the water users downstream from your facility. The downstream water may be used as the water supply for drinking water treatment plants; for industrial and agricultural uses; to water stock and wildlife; to propagate fish and other aquatic and marine life; for shellfish culture; for recreation, such as swimming, boating, fishing, and simple *AESTHETIC*[13] enjoyment; to generate hydroelectric power; for navigation; and for other purposes. Therefore, you have an obligation to the users of the water downstream as well as to the people of your district or municipality. You are the key water quality protector and must realize that you are in a responsible position to protect the public health and the environment.

## QUESTIONS

Please write your answers to the following questions and compare them with those on page 14.

1.3A How did many receiving waters become polluted?

1.3B Why must municipal and industrial wastewaters receive adequate treatment?

1.3C Why is there a need for small system operators?

## 1.4 YOUR QUALIFICATIONS

The skill and ability required for your job depend to a large degree on the size and type of facility where you are employed. You may operate a small plant serving fewer than a thousand people. You may be the only operator at the plant or, at best, have only one or two additional employees. If this is the case, you must be a jack-of-all-trades because of the diversity of your tasks.

### 1.40 Your Job

To describe the operator's duties, let us start at the beginning. Let us say that the need for a new or improved wastewater treatment plant has long been recognized by the community. The community has voted to issue the necessary bonds to finance the project, and the consulting engineers have submitted plans and specifications. It is in the best interests of the community and the consulting engineer that you be in on the ground floor planning. If it is a new plant, you should be present or at least available during the construction period in order to become completely familiar with the entire plant, including the equipment and machinery and their operation. This will provide you with the opportunity to relate your plant drawings to actual facilities. You also may be the construction inspector during construction.

You and the engineer should discuss how the treatment plant should best be run and the means of operation the designer had in mind when the plant was designed. If it is an old plant being remodeled, you are in a position to offer excellent advice to the consulting engineer. Your experience provides valuable technical knowledge concerning the characteristics of the wastewater, its sources, and the limitations of the present facilities. Together with the consultant, you are a member of an expert team able to advise the district or town.

Once the plant is operating, you become an administrator. In a small plant your duties may not include supervision of personnel, but you are still in charge of records. You are responsible for operating the plant as efficiently as possible, keeping in mind that the primary objective is to protect water quality by continuous and efficient plant performance. Without adequate, reliable records of every phase of operation, the effectiveness of your operation has not been documented (recorded).

You may also be the budget administrator. Most certainly you are in the best position to give advice on budget requirements, management problems, and future planning. You should be aware of the necessity for additional expenditures, including funds for plant enlargement, equipment replacement, and laboratory requirements. You should recognize and define such needs in sufficient time to inform the proper community officials to enable them to accomplish early planning and budgeting.

You are in the field of public relations and must be able to explain the purpose and operation of your plant to visitors, civic organizations, school classes, representatives of news media, and even to the city council or directors of your district. Public interest in water quality is increasing, and you should be prepared to conduct tours that will contribute to public acceptance and support. A well-guided tour for officials of regulatory agencies or other operators may provide these people with sufficient understanding of your plant to allow them to suggest helpful solutions to operational problems.

13. *Aesthetic* (es-THET-ick). Attractive or appealing.

*Special care and safety must be practiced when visitors are taken through your treatment plant. An accident could spoil all of your public relations efforts.*

The appearance of your plant indicates to the visitor the type of operation you maintain. If the plant is dirty and rundown with flies and other insects swarming about, you will be unable to convince your visitors that the plant is doing a good job. Your records showing a high-quality effluent will mean nothing to these visiting citizens unless your plant appears clean and well maintained and the effluent looks good.

Another aspect of your public relations duties involves dealing with the downstream water user. Unfortunately, the operator is often considered by the downstream user, anglers, and environmental groups as a polluter rather than a water quality protector. Through a good public information program, backed by facts supported by reliable data, you can correct the unfavorable impression held by the downstream user and others and establish "good neighbor" relations. This is indeed a challenge. Again, you must understand that you hold a very responsible position and be aware that the main purpose of the operation of your plant is to protect the downstream user, be that user a private property owner, another city or district, an industry, or an angler.

Odor control is another critical part of public relations. If odors from your treatment plant offend neighbors, you and your plant will not be popular with the public. Good housekeeping around your plant helps control odors and convinces the public you are doing a good job.

You are required to understand certain laboratory procedures in order to conduct various tests on samples of wastewater and receiving waters. On the basis of the data obtained from these tests, you may have to adjust the operation of the treatment plant to meet receiving stream standards or discharge requirements.

As an operator, you must have a knowledge of the complicated mechanical principles involved in many treatment mechanisms. In order to measure and control the wastewater flowing through the plant, you must have some understanding of hydraulics. Practical knowledge of electrical motors, circuitry, and controls is also essential.

Safety is a very important operator responsibility. Unfortunately, too many operators take safety for granted. This is one reason why the wastewater treatment industry requires everyone to be safety conscious. You have the responsibility to be sure that your collection, treatment, and disposal system is a safe place to work and visit. Everyone must follow safe procedures and understand why safe procedures must be followed at all times. All operators must be aware of the safety hazards in and around collection, treatment, and disposal systems. You should plan or be a part of an active safety program. Chief operators frequently have the responsibility of training new operators and must encourage all operators to work safely. Especially challenging for small system operators is the fact that you may be the only person on the job and you are the only person responsible for your safety.

Clearly then, the modern small wastewater treatment plant operator must possess a broad range of qualifications.

## QUESTIONS

Please write your answers to the following questions and compare them with those on page 14.

1.4A Why is it important that the operator be present during the construction of a new plant?

1.4B How does the operator become involved in public relations?

## 1.5 STAFFING NEEDS AND FUTURE JOB OPPORTUNITIES

The wastewater treatment field, like so many others, is growing rapidly. New plants are being constructed and old plants are being modified and enlarged to handle the wastewater from our growing population and to treat the new chemicals being produced by advanced technology. New cost-effective small systems are being developed and installed that have never been operated and maintained. You may have the opportunity to work with the design engineer to learn how to safely make the facility operate properly. Operators, maintenance personnel, supervisors, managers, instrumentation experts, and laboratory technicians are sorely needed.

A look at past records and future predictions indicates that water and wastewater treatment is a rapidly growing field. According to the US Bureau of Labor Statistics (BLS),[14] water and wastewater treatment plant operators held about 113,400 jobs

14. Refer to the Bureau of Labor Statistics (BLS) website at www.bls.gov for additional information about the types of jobs available in the water and wastewater industries, working conditions, earnings potential, and the job outlook.

in 2008. The majority of operators worked for local governments but the largest growth in jobs in the future is expected to be in privately owned facilities. The BLS estimates that the number of jobs in the water and wastewater treatment industries will increase by an average of 20 percent in the period 2008–2018. Factors contributing to the increase include population growth, retirement of many current operators, regulatory requirements, more sophisticated treatment, and operator certification. The need for *trained* operators is increasing rapidly and is expected to continue to grow in the future.

## 1.6 TRAINING YOURSELF TO MEET THE NEEDS

This training course and *SMALL WASTEWATER SYSTEM OPERATION AND MAINTENANCE,* Volume II, are not the only courses available to help you improve your abilities. The states offer various types of long- and short-term operator training through their health departments and state environmental training centers. Water pollution control associations and rural water associations also provide training classes conducted by association members, largely on a volunteer basis. The National Environmental Services Center has developed operator training materials. State and local colleges provide valuable training under their own sponsorship or in partnership with others. Many state, local, and private agencies conduct both long- and short-term training as well as interesting and informative seminars. Your state Onsite Wastewater Association may have training materials and helpful seminars, workshops and conferences. The National Onsite Wastewater Recycling Association, 601 Wythe Street, Alexandria, VA 22314, phone (800) 966-2942, is an excellent source of helpful information.

Listed below are several very good references in the field of wastewater treatment plant operation that are frequently referred to throughout this course. The name in quotes represents the term usually used by operators when they mention the reference. Prices listed are those available when this manual was published and will probably increase in the future.

1. *OPERATION OF WASTEWATER TREATMENT PLANTS,* Volume I. Obtain from Office of Water Programs, California State University, Sacramento, 6000 J Street, Sacramento, CA 95819-6025. Price, $49.00.
2. *DESIGN MANUAL: ONSITE WASTEWATER TREATMENT AND DISPOSAL SYSTEMS,* US Environmental Protection Agency. EPA No. 625-1-80-012. Obtain from National Technical Information Service (NTIS), 5301 Shawnee Road, Alexandria, VA 22312. Order No. PB83-219907. Price, $99.00, plus $6.00 shipping and handling.
3. *ONSITE WASTEWATER TREATMENT FOR SMALL COMMUNITIES AND RURAL AREAS* (poster). Obtain from the National Environmental Services Center (NESC), West Virginia University, PO Box 6064, Morgantown, WV 26506-6064, phone (800) 624-8301. Order No. WWPSPE02. Price, $1.25, plus shipping and handling.
4. *"NEW YORK MANUAL." MANUAL OF INSTRUCTION FOR WASTEWATER TREATMENT PLANT OPERATORS* (two-volume set), published by Health Education Services. No longer in print.

These publications cover the broad field of treatment plant operation. At the end of many of the chapters yet to come, lists of other references will be provided.

## SUGGESTED ANSWERS

### Chapter 1. THE SMALL WASTEWATER SYSTEM OPERATOR

You are not expected to have the exact answer suggested for questions requiring written answers, but you should have the correct idea. The numbering of the questions refers to the section in the manual where you can find the information to answer the questions. For example, answers to questions numbered 1.0 can be found in Section 1.0, "Why Is There a Need for Small System Operators?"

Answers to questions on page 11.

1.0A 3

1.0B 1, 2, 3

1.0C 1, 2, 3

Answers to questions on page 12.

1.3A Receiving waters became polluted because of a lack of public concern for the impact of waste discharges and because wastewater was discharged into receiving waters and groundwaters in amounts beyond their natural purification capacity.

1.3B Municipal and industrial wastewaters must receive adequate treatment to protect receiving water users.

1.3C Small system operators are needed to operate their facilities correctly for proper disposal or discharge of wastewater to protect the public health and the environment.

Answers to questions on page 13.

1.4A The operator should be present during the construction of a new system in order to become familiar with the system before the operator begins operating it.

1.4B The operator becomes involved in public relations by explaining the purpose of the treatment facility, how it works, and the impact of the effluent quality discharged to the receiving waters to visitors, civic organizations, news reporters, and city or district representatives.

CHAPTER 2

# SMALL COLLECTION, TREATMENT, AND DISCHARGE SYSTEMS

by

John Brady

and

William Crooks

# TABLE OF CONTENTS

## Chapter 2. SMALL COLLECTION, TREATMENT, AND DISCHARGE SYSTEMS

LESSON 2

# OBJECTIVES

## Chapter 2. SMALL COLLECTION, TREATMENT, AND DISCHARGE SYSTEMS

1. Identify the major classifications of waste materials.
2. List the different types of solids in wastewater.
3. Explain the effects of waste discharges on humans and the environment.
4. Explain the purposes of on-site systems and small wastewater collection, treatment, and discharge systems.
5. Identify and describe the types of collection systems commonly used in small wastewater treatment systems.
6. Describe the typical wastewater flow pattern through conventional treatment processes.
7. Recognize and describe various types of wastewater package treatment plants.
8. Explain how wastewater can be disinfected.
9. Recall and list several common effluent and solids disposal methods.

# WORDS

## Chapter 2. SMALL COLLECTION, TREATMENT, AND DISCHARGE SYSTEMS

ACTIVATED SLUDGE PROCESS — ACTIVATED SLUDGE PROCESS

A biological wastewater treatment process that speeds up the decomposition of wastes in the wastewater being treated. Activated sludge is added to wastewater and the mixture (mixed liquor) is aerated and agitated. After some time in the aeration tank, the activated sludge is allowed to settle out by sedimentation and is disposed of (wasted) or reused (returned to the aeration tank) as needed. The remaining wastewater then undergoes more treatment.

AERATION (air-A-shun) TANK — AERATION TANK

The tank where raw or settled wastewater is mixed with return sludge and aerated. The same as aeration bay, aerator, or reactor.

AEROBIC BACTERIA (air-O-bick back-TEER-e-uh) — AEROBIC BACTERIA

Bacteria that will live and reproduce only in an environment containing oxygen that is available for their respiration (breathing), namely atmospheric oxygen or oxygen dissolved in water. Oxygen combined chemically, such as in water molecules ($H_2O$), cannot be used for respiration by aerobic bacteria.

ALGAE (AL-jee) — ALGAE

Microscopic plants containing chlorophyll that live floating or suspended in water. They also may be attached to structures, rocks, or other submerged surfaces. Excess algal growths can impart tastes and odors to potable water. Algae produce oxygen during sunlight hours and use oxygen during the night hours. Their biological activities appreciably affect the pH, alkalinity, and dissolved oxygen of the water.

ANAEROBIC BACTERIA (AN-air-O-bick back-TEER-e-uh) — ANAEROBIC BACTERIA

Bacteria that live and reproduce in an environment containing no free or dissolved oxygen. Anaerobic bacteria obtain their oxygen supply by breaking down chemical compounds that contain oxygen, such as sulfate ($SO_4^{2-}$).

BOD (pronounce as separate letters) — BOD

Biochemical Oxygen Demand. The rate at which organisms use the oxygen in water or wastewater while stabilizing decomposable organic matter under aerobic conditions. In decomposition, organic matter serves as food for the bacteria and energy results from its oxidation. BOD measurements are used as a surrogate measure of the organic strength of wastes in water.

BIOCHEMICAL OXYGEN DEMAND (BOD) — BIOCHEMICAL OXYGEN DEMAND (BOD)

See BOD.

BIOMASS (BUY-o-mass) — BIOMASS

A mass or clump of organic material consisting of living organisms feeding on the wastes in wastewater, dead organisms, and other debris. Also see ZOOGLEAL MASS.

COLIFORM (KOAL-i-form) — COLIFORM

A group of bacteria found in the intestines of warm-blooded animals (including humans) and also in plants, soil, air, and water. The presence of coliform bacteria is an indication that the water is polluted and may contain pathogenic (disease-causing) organisms. Fecal coliforms are those coliforms found in the feces of various warm-blooded animals, whereas the term coliform also includes other environmental sources.

COMMINUTION (kom-mih-NEW-shun) COMMINUTION

A mechanical treatment process that cuts large pieces of wastes into smaller pieces so they will not plug pipes or damage equipment. Comminution and shredding usually mean the same thing.

DECOMPOSITION or DECAY DECOMPOSITION or DECAY

The conversion of chemically unstable materials to more stable forms by chemical or biological action.

DETENTION TIME DETENTION TIME

The time required to fill a tank at a given flow or the theoretical time required for a given flow of wastewater to pass through a tank. In septic tanks, this detention time will decrease as the volumes of sludge and scum increase.

DEWATER DEWATER

(1) To remove or separate a portion of the water present in a sludge or slurry. To dry sludge so it can be handled and disposed of.

(2) To remove or drain the water from a tank or a trench. A structure may be dewatered so that it can be inspected or repaired.

DISINFECTION (dis-in-FECT-shun) DISINFECTION

The process designed to kill or inactivate most microorganisms in water or wastewater, including essentially all pathogenic (disease-causing) bacteria. There are several ways to disinfect, with chlorination being the most frequently used in water and wastewater treatment plants. Compare with STERILIZATION.

EFFLUENT (EF-loo-ent) EFFLUENT

Water or other liquid—raw (untreated), partially treated, or completely treated—flowing *FROM* a reservoir, basin, treatment process, or treatment plant.

ENTRAIN ENTRAIN

To trap bubbles in water either mechanically through turbulence or chemically through a reaction.

EVAPOTRANSPIRATION (ee-VAP-o-TRANS-purr-A-shun) EVAPOTRANSPIRATION

(1) The process by which water vapor is released to the atmosphere by living plants. This process is similar to people sweating. Also called transpiration.

(2) The total water removed from an area by transpiration (plants) and by evaporation from soil, snow, and water surfaces.

FACULTATIVE (FACK-ul-tay-tive) POND FACULTATIVE POND

The most common type of pond in current use. The upper portion (supernatant) is aerobic, while the bottom layer is anaerobic. Algae supply most of the oxygen to the supernatant.

FLOW LINE FLOW LINE

(1) The top of the wetted line, the water surface, or the hydraulic grade line of water flowing in an open channel or partially full conduit.

(2) The lowest point of the channel inside a pipe, conduit, canal, or manhole. This term is used by some contractors, however, the preferred term for this usage is invert.

GROUNDWATER GROUNDWATER

Subsurface water in the saturation zone from which wells and springs are fed. In a strict sense the term applies only to water below the water table. Also called phreatic water and plerotic water.

IMHOFF CONE IMHOFF CONE

A clear, cone-shaped container marked with graduations. The cone is used to measure the volume of settleable solids in a specific volume (usually one liter) of water or wastewater.

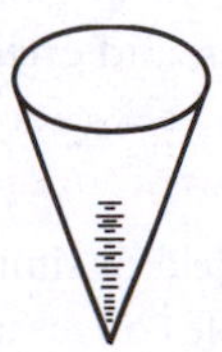

INFILTRATION (in-fill-TRAY-shun) INFILTRATION

The seepage of groundwater into a sewer system, including service connections. Seepage frequently occurs through defective or cracked pipes, pipe joints and connections, interceptor access risers and covers, or manhole walls.

INFLOW INFLOW

Water discharged into a sewer system and service connections from sources other than regular connections. This includes flow from yard drains, foundations, and around access and manhole covers. Inflow differs from infiltration in that it is a direct discharge into the sewer rather than a leak in the sewer itself.

INFLUENT INFLUENT

Water or other liquid—raw (untreated) or partially treated—flowing *INTO* a reservoir, basin, treatment process, or treatment plant.

INORGANIC WASTE INORGANIC WASTE

Waste material such as sand, salt, iron, calcium, and other mineral materials that are only slightly affected by the action of organisms. Inorganic wastes are chemical substances of mineral origin; whereas organic wastes are chemical substances usually of animal or plant origin. Also see NONVOLATILE MATTER, ORGANIC WASTE, and VOLATILE SOLIDS.

INTERCEPTOR INTERCEPTOR

A septic tank or other holding tank that serves as a temporary wastewater storage reservoir for a septic tank effluent pump (STEP) system. Also see SEPTIC TANK.

INTERCEPTOR (INTERCEPTING) SEWER INTERCEPTOR (INTERCEPTING) SEWER

A large sewer that receives flow from a number of sewers and conducts the wastewater to a treatment plant. Often called an interceptor. The term interceptor is sometimes used in small communities to describe a septic tank or other holding tank that serves as a temporary wastewater storage reservoir for a septic tank effluent pump (STEP) system.

INVERT (IN-vert) INVERT

The lowest point of the channel inside a pipe, conduit, canal, or manhole. Also called flow line by some contractors, however, the preferred term is invert.

INVERTED SIPHON INVERTED SIPHON

A pressure pipeline used to carry wastewater flowing in a gravity collection system under a depression, such as a valley or roadway, or under a structure, such as a building. Also called a depressed sewer.

MILLIGRAMS PER LITER, mg/L MILLIGRAMS PER LITER, mg/L

A measure of the concentration by weight of a substance per unit volume. For practical purposes, one mg/L of a substance in water is equal to one part per million parts (ppm). Thus, a liter of water with a specific gravity of 1.0 weighs one million milligrams. If one liter of water contains 10 milligrams of dissolved oxygen, the concentration is 10 milligrams per million milligrams, or 10 milligrams per liter (10 mg/L), or 10 parts of oxygen per million parts of water, or 10 parts per million (10 ppm), or 10 pounds dissolved oxygen in 1 million pounds of water (10 ppm).

MIXED LIQUOR SUSPENDED SOLIDS (MLSS) MIXED LIQUOR SUSPENDED SOLIDS (MLSS)

The amount (mg/L) of suspended solids in the mixed liquor of an aeration tank.

NATURAL CYCLES NATURAL CYCLES

Cycles that take place in nature, such as the water or hydrologic cycle where water is transformed or changed from one form to another until the water has returned to the original form, thus completing the cycle. Other natural cycles include the life cycles of aquatic organisms and plants, nutrient cycles, and cycles of self- or natural purification.

NONPOINT SOURCE NONPOINT SOURCE

A runoff or discharge from a field or similar source, in contrast to a point source, which refers to a discharge that comes out the end of a pipe or other clearly identifiable conveyance. Also see POINT SOURCE.

NONVOLATILE MATTER | NONVOLATILE MATTER

Material such as sand, salt, iron, calcium, and other mineral materials that are only slightly affected by the actions of organisms and are not lost on ignition of the dry solids at 550°C (1,022°F). Volatile materials are chemical substances usually of animal or plant origin. Also see INORGANIC WASTE and VOLATILE SOLIDS.

NUTRIENT | NUTRIENT

Any substance that is assimilated (taken in) by organisms and promotes growth. Nitrogen and phosphorus are nutrients that promote the growth of algae. There are other essential and trace elements that are also considered nutrients. Also see NUTRIENT CYCLE.

NUTRIENT CYCLE | NUTRIENT CYCLE

The transformation or change of a nutrient from one form to another until the nutrient has returned to the original form, thus completing the cycle. The cycle may take place under either aerobic or anaerobic conditions.

ORGANIC WASTE | ORGANIC WASTE

Waste material that may come from animal or plant sources. Natural organic wastes generally can be consumed by bacteria and other small organisms. Manufactured or synthetic organic wastes from metal finishing, chemical manufacturing, and petroleum industries may not normally be consumed by bacteria and other organisms. Also see INORGANIC WASTE and VOLATILE SOLIDS.

OUTFALL | OUTFALL

(1) The point, location, or structure where wastewater or drainage discharges from a sewer, drain, or other conduit.

(2) The conduit leading to the final discharge point or area. Also see OUTFALL SEWER.

OUTFALL SEWER | OUTFALL SEWER

A sewer that receives wastewater from a collection system or from a wastewater treatment plant and carries it to a point of ultimate or final discharge in the environment. Also see OUTFALL.

OXIDATION | OXIDATION

Oxidation is the addition of oxygen, removal of hydrogen, or the removal of electrons from an element or compound; in the environment and in wastewater treatment processes, organic matter is oxidized to more stable substances. The opposite of REDUCTION.

OXIDATION DITCH | OXIDATION DITCH

The oxidation ditch is a modified form of the activated sludge process. The ditch consists of two channels placed side by side and connected at the ends to produce one continuous loop of wastewater flow and a brush rotator assembly placed across the channel to provide aeration and circulation.

PATHOGENIC (path-o-JEN-ick) ORGANISMS | PATHOGENIC ORGANISMS

Bacteria, viruses, protozoa, or internal parasites that can cause disease (such as giardiasis, cryptosporidiosis, typhoid fever, cholera, or infectious hepatitis) in a host (such as a person). There are many types of organisms that do not cause disease and are not called pathogenic. Many beneficial bacteria are found in wastewater treatment processes actively cleaning up organic wastes.

PERCOLATION (purr-ko-LAY-shun) | PERCOLATION

The slow passage of water through a filter medium; or, the gradual penetration of soil and rocks by water.

pH (pronounce as separate letters) | pH

pH is an expression of the intensity of the basic or acidic condition of a liquid. Mathematically, pH is the logarithm (base 10) of the reciprocal of the hydrogen ion activity.

$$pH = Log \frac{1}{\{H^+\}}$$

If $\{H^+\} = 10^{-6.5}$, then pH = 6.5. The pH may range from 0 to 14, where 0 is most acidic, 14 most basic, and 7 neutral.

PHOTOSYNTHESIS (foe-toe-SIN-thuh-sis) PHOTOSYNTHESIS

A process in which organisms, with the aid of chlorophyll, convert carbon dioxide and inorganic substances into oxygen and additional plant material, using sunlight for energy. All green plants grow by this process.

POINT SOURCE POINT SOURCE

A discharge that comes out the end of a pipe or other clearly identifiable conveyance. Examples of point source conveyances from which pollutants may be discharged include: ditches, channels, tunnels, conduits, wells, containers, rolling stock, concentrated animal feeding operations, landfill leachate collection systems, vessels, or other floating craft. A NONPOINT SOURCE refers to runoff or a discharge from a field or similar source.

PRIMARY TREATMENT PRIMARY TREATMENT

A wastewater treatment process that takes place in a rectangular or circular tank and allows those substances in wastewater that readily settle or float to be separated from the wastewater being treated. A septic tank is also considered primary treatment.

PROPORTIONAL WEIR (WEER) PROPORTIONAL WEIR

A specially shaped weir in which the flow through the weir is directly proportional to the head.

PUTREFACTION (PYOO-truh-FACK-shun) PUTREFACTION

Biological decomposition of organic matter, with the production of foul-smelling and -tasting products, associated with anaerobic (no oxygen present) conditions.

REDUCTION (re-DUCK-shun) REDUCTION

Reduction is the addition of hydrogen, removal of oxygen, or the addition of electrons to an element or compound. Under anaerobic conditions (no dissolved oxygen present), sulfur compounds are reduced to odor-producing hydrogen sulfide ($H_2S$) and other compounds. In the treatment of metal finishing wastewaters, hexavalent chromium ($Cr^{6+}$) is reduced to the trivalent form ($Cr^{3+}$). The opposite of OXIDATION.

SCUM SCUM

A layer or film of foreign matter (such as grease, oil) that has risen to the surface of water or wastewater.

SECONDARY TREATMENT SECONDARY TREATMENT

A wastewater treatment process used to convert dissolved or suspended materials into a form more readily separated from the water being treated. Usually, the process follows primary treatment by sedimentation. The process commonly is a type of biological treatment followed by secondary clarifiers that allow the solids to settle out from the water being treated.

SEPTIC (SEP-tick) SEPTIC

A condition produced by anaerobic bacteria. If severe, the sludge produces hydrogen sulfide, turns black, gives off foul odors, contains little or no dissolved oxygen, and the wastewater has a high oxygen demand.

SEPTIC TANK SEPTIC TANK

A system sometimes used where wastewater collection systems and treatment plants are not available. The system is a settling tank in which settled sludge and floatable scum are in intimate contact with the wastewater flowing through the tank and the organic solids are decomposed by anaerobic bacterial action. Used to treat wastewater and produce an effluent that is usually discharged to subsurface leaching. Also referred to as an interceptor; however, the preferred term is septic tank.

SEPTIC TANK EFFLUENT FILTER (STEF) SYSTEM SEPTIC TANK EFFLUENT FILTER (STEF) SYSTEM

A facility in which effluent flows from a septic tank into a gravity flow collection system that flows into a gravity sewer, treatment plant, or subsurface leaching system. The gravity flow pipeline is called an effluent drain.

SEPTIC TANK EFFLUENT PUMP (STEP) SYSTEM SEPTIC TANK EFFLUENT PUMP (STEP) SYSTEM

A facility in which effluent is pumped from a septic tank into a pressurized collection system that may flow into a gravity sewer, treatment plant, or subsurface leaching system.

SHORT-CIRCUITING SHORT-CIRCUITING

A condition that occurs in tanks or basins when some of the flowing water entering a tank or basin flows along a nearly direct pathway from the inlet to the outlet. This is usually undesirable because it may result in shorter contact, reaction, or settling times in comparison with the theoretical (calculated) or presumed detention times.

SHREDDING SHREDDING

A mechanical treatment process that cuts large pieces of wastes into smaller pieces so they will not plug pipes or damage equipment. Shredding and comminution usually mean the same thing.

SLUDGE (SLUJ) SLUDGE

(1) The settleable solids separated from liquids during processing.

(2) The deposits of foreign materials on the bottoms of streams or other bodies of water or on the bottoms and edges of wastewater collection lines and appurtenances.

SOLUBLE BOD SOLUBLE BOD

Soluble BOD is the BOD of water that has been filtered in the standard suspended solids test. The soluble BOD is a measure of food for microorganisms that is dissolved in the water being treated.

SPECIFIC GRAVITY SPECIFIC GRAVITY

(1) Weight of a particle, substance, or chemical solution in relation to the weight of an equal volume of water. Water has a specific gravity of 1.000 at 4°C (39°F). Wastewater particles or substances usually have a specific gravity of 0.5 to 2.5. Particulates with specific gravity less than 1.0 float to the surface and particulates with specific gravity greater than 1.0 sink.

(2) Weight of a particular gas in relation to the weight of an equal volume of air at the same temperature and pressure (air has a specific gravity of 1.0). Chlorine gas has a specific gravity of 2.5.

STABILIZATION STABILIZATION

Conversion to a form that resists change. Organic material is stabilized by bacteria that convert the material to gases and other relatively inert substances. Stabilized organic material generally will not give off obnoxious odors.

STANDARD METHODS STANDARD METHODS

*STANDARD METHODS FOR THE EXAMINATION OF WATER AND WASTEWATER*, 21st Edition. A joint publication of the American Public Health Association (APHA), American Water Works Association (AWWA), and the Water Environment Federation (WEF) that outlines the accepted laboratory procedures used to analyze the impurities in water and wastewater. Available from: American Water Works Association, Bookstore, 6666 West Quincy Avenue, Denver, CO 80235. Order No. 10084. Price to members, $198.50; nonmembers, $266.00; price includes cost of shipping and handling. Also available from Water Environment Federation, Publications Order Department, PO Box 18044, Merrifield, VA 22118-0045. Order No. S82011. Price to members, $203.00; nonmembers, $268.00; price includes cost of shipping and handling.

STERILIZATION (STAIR-uh-luh-ZAY-shun) STERILIZATION

The removal or destruction of all microorganisms, including pathogens and other bacteria, vegetative forms, and spores. Compare with DISINFECTION.

SUBSURFACE LEACHING SYSTEM SUBSURFACE LEACHING SYSTEM

A method of treatment and discharge of septic tank effluent, sand filter effluent, or other treated wastewater. The effluent is applied to soil below the ground surface through open-jointed pipes or drains or through perforated pipes (holes in the pipes). The effluent is treated as it passes through porous soil or rock strata (layers). Newer subsurface leaching systems include chamber and gravelless systems, and also gravel trenches without pipe the full length of the trench.

SUSPENDED SOLIDS SUSPENDED SOLIDS

(1) Solids that either float on the surface or are suspended in water, wastewater, or other liquids, and that are largely removable by laboratory filtering.

(2) The quantity of material removed from water or wastewater in a laboratory test, as prescribed in *STANDARD METHODS FOR THE EXAMINATION OF WATER AND WASTEWATER,* and referred to as Total Suspended Solids Dried at 103–105°C.

TOPOGRAPHY (toe-PAH-gruh-fee) TOPOGRAPHY

The arrangement of hills and valleys in a geographic area.

TRANSPIRATION (TRAN-spur-RAY-shun) TRANSPIRATION

The process by which water vapor is released to the atmosphere by living plants. This process is similar to people sweating. Also called evapotranspiration.

TRICKLING FILTER TRICKLING FILTER

A treatment process in which wastewater trickling over media enables the formation of slimes or biomass, which contain organisms that feed upon and remove wastes from the water being treated.

VOLATILE (VOL-uh-tull) VOLATILE

(1) A volatile substance is one that is capable of being evaporated or changed to a vapor at relatively low temperatures. Volatile substances can be partially removed from water or wastewater by the air stripping process.

(2) In terms of solids analysis, volatile refers to materials lost (including most organic matter) upon ignition in a muffle furnace for 60 minutes at 550°C (1,022°F). Natural volatile materials are chemical substances usually of animal or plant origin. Manufactured or synthetic volatile materials, such as plastics, ether, acetone, and carbon tetrachloride, are highly volatile and not of plant or animal origin. Also see NONVOLATILE MATTER.

VOLATILE SOLIDS VOLATILE SOLIDS

Those solids in water, wastewater, or other liquids that are lost on ignition of the dry solids at 550°C (1,022°F). Also called organic solids and volatile matter.

WASTEWATER WASTEWATER

A community's used water and water-carried solids (including used water from industrial processes) that flow to a treatment plant. Stormwater, surface water, and groundwater infiltration also may be included in the wastewater that enters a wastewater treatment plant. The term sewage usually refers to household wastes, but this word is being replaced by the term wastewater.

WEIR (WEER) WEIR

(1) A wall or plate placed in an open channel and used to measure the flow of water. The depth of the flow over the weir can be used to calculate the flow rate, or a chart or conversion table may be used to convert depth to flow. Also see PROPORTIONAL WEIR.

(2) A wall or obstruction used to control flow (from settling tanks and clarifiers) to ensure a uniform flow rate and avoid short-circuiting.

ZOOGLEAL (ZOE-uh-glee-ul) MASS ZOOGLEAL MASS

Jelly-like masses of bacteria found in both the trickling filter and activated sludge processes. These masses may be formed for or function as the protection against predators and for storage of food supplies. Also see BIOMASS.

# CHAPTER 2. SMALL COLLECTION, TREATMENT, AND DISCHARGE SYSTEMS

(Lesson 1 of 2 Lessons)

## 2.0 NATURE OF WASTEWATER

### 2.00 Types of Waste Discharges

Wastewater from homes, businesses, and industries contains varying amounts of four major types of waste material: *ORGANIC WASTE,*[1] *INORGANIC WASTE,*[2] thermal waste, and radioactive waste. Domestic wastewater and the wastewater from many types of small businesses contains organic material and smaller amounts of inorganic material. The nature of the wastewater discharged from industrial facilities depends on the type of industry; for example, substantial amounts of organic material can be found in the wastewater from industries such as vegetable and fruit packing; dairy processing; meat packing; tanning; and processing of poultry, oil, paper, and fiber (wood). Many other industries discharge primarily inorganic wastes; for instance, wastewater discharges from water softening processes contain sodium and chloride (salt brine) and discharges from metal finishing plants or electroplating shops might contain chromium or copper. Industries such as gravel washing plants discharge varying amounts of soil, sand, or grit, which also may be classified as inorganic wastes.

In addition to organic and inorganic wastes, there are two other major classifications of waste materials: heated (thermal) wastes and radioactive wastes. High-temperature wastewaters may come from cooling processes used by industry and from thermal power stations generating electricity. Radioactive wastes are usually controlled at their source, but could come from hospitals, research laboratories, and nuclear power plants.

### 2.01 Types of Solids in Wastewater

Since the main purpose of wastewater treatment is to remove solids from the wastewater, a detailed discussion of the types of solids is in order. Figure 2.1 and the following paragraphs will help you understand the terms used to describe different types of wastewater solids.

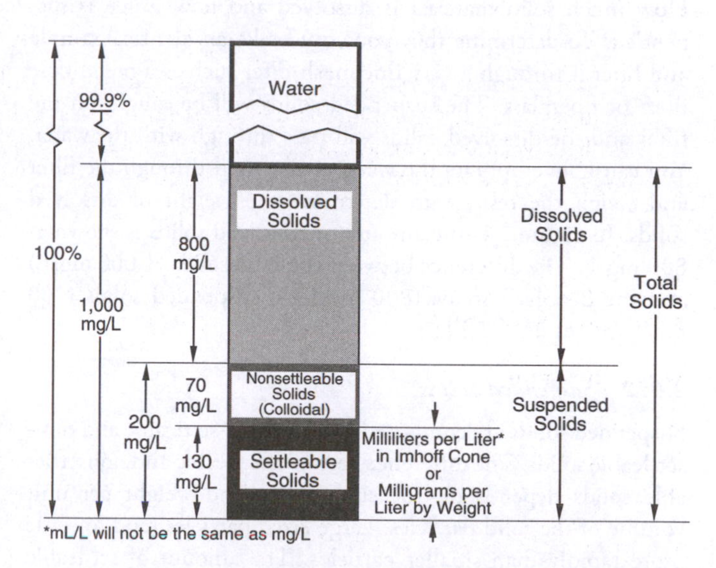


*Fig. 2.1 Typical composition of solids in raw wastewater (floatable solids not shown)*

For normal municipal wastewater, which contains domestic wastewater as well as some industrial and commercial wastes, the goals of the treatment plant designer and operator usually are to remove the organic and inorganic *SUSPENDED SOLIDS,*[3] to remove the dissolved organic solids, and to kill or inactivate the *PATHOGENIC ORGANISMS*[4] by disinfection. The treatment plant does little to remove dissolved inorganic solids; however, some plants are designed to remove nutrients such as nitrogen

---

1. *Organic Waste.* Waste material that may come from animal or plant sources. Natural organic wastes generally can be consumed by bacteria and other small organisms. Manufactured or synthetic organic wastes from metal finishing, chemical manufacturing, and petroleum industries may not normally be consumed by bacteria and other organisms. Also see INORGANIC WASTE and VOLATILE SOLIDS.
2. *Inorganic Waste.* Waste material such as sand, salt, iron, calcium, and other mineral materials that are only slightly affected by the action of organisms. Inorganic wastes are chemical substances of mineral origin; whereas organic wastes are chemical substances usually of animal or plant origin. Also see NONVOLATILE MATTER, ORGANIC WASTE, and VOLATILE SOLIDS.
3. *Suspended Solids.* (1) Solids that either float on the surface or are suspended in water, wastewater, or other liquids, and that are largely removable by laboratory filtering. (2) The quantity of material removed from water or wastewater in a laboratory test, as prescribed in *STANDARD METHODS FOR THE EXAMINATION OF WATER AND WASTEWATER,* and referred to as Total Suspended Solids Dried at 103–105°C.
4. *Pathogenic* (path-o-JEN-ick) *Organisms.* Bacteria, viruses, protozoa, or internal parasites that can cause disease (such as giardiasis, cryptosporidiosis, typhoid fever, cholera, or infectious hepatitis) in a host (such as a person). There are many types of organisms that do not cause disease and are not called pathogenic. Many beneficial bacteria are found in wastewater treatment processes actively cleaning up organic wastes.

and phosphorus. Thermal and radioactive wastes require special treatment processes before discharge to the environment.

### 2.010 *Total Solids*

For discussion purposes, assume that you obtain a one-liter sample of raw wastewater entering the treatment plant. Heat this sample enough to evaporate all the water and weigh all the solid material left (residue); it weighs 1,000 milligrams. Thus, the total solids concentration in the sample is 1,000 milligrams per liter (mg/L).[5] This weight includes both dissolved and suspended solids.

### 2.011 *Dissolved Solids*

How much solid material is dissolved and how much is suspended? To determine this, you could take an identical sample and filter it through a very fine mesh filter such as a membrane filter or fiberglass. The suspended solids will be caught on the filter and the dissolved solids will pass through with the water. You can now evaporate the water that passed through the filter and weigh the residue to determine the weight of dissolved solids. In Figure 2.1, the amount of dissolved solids is shown as 800 mg/L. The difference between the total solids (1,000 mg/L) and the dissolved solids (800 mg/L) is suspended solids (200 mg/L caught on the filter).

### 2.012 *Suspended Solids*

Suspended solids are composed of two parts: settleable and nonsettleable solids. The difference between settleable and nonsettleable solids depends on the size, shape, and weight per unit volume of the solid particles. Large-sized particles tend to settle more rapidly than smaller particles. The amount of settleable solids in the raw wastewater should be estimated in order to design settling basins (primary units), sludge pumps, and sludge handling facilities. Also, measuring the amount of settleable solids entering and leaving the settling basin allows you to calculate the efficiency of the basin for removing the settleable solids. The amount of settleable solids will determine how fast solids will build up in a septic tank.

Settleable solids are determined by using a glass laboratory device called an *IMHOFF CONE.*[6] The Imhoff cone is used to determine the volume of settleable solids in milliliters per liter, mL/L, not by weight, mg/L. (Figure 2.1 shows that the settleable solids weigh 130 mg/L.) A time limit of one hour is all that is provided for the settleability test. Those solids that settle in the one-hour period are classified as settleable solids. The portion of suspended solids that did not have the weight to settle in the one-hour time frame are considered the nonsettleable solids, but are still a portion of the suspended solids. It is important to remember that a volume measurement such as mL/L is not equivalent to a weight measurement of mg/L. The weight measurements in the left of Figure 2.1 are obtained by precise laboratory procedures and sampling with repeatable results when identical tests are made from the same sample, whereas volume measurements are not as precise for settleable solids tests on large-volume samples.

### 2.013 *Floatable Solids*

There is no standard method for the measurement and evaluation of floatable solids. Floatable solids include all types of fats, oils, and greases (FOG). Floatable solids are undesirable in the plant effluent because they plug leach fields and the sight of floatables in receiving waters indicates the presence of inadequately treated wastewater.

### 2.014 *Organic and Inorganic Solids*

For total solids or for any separate type of solids, such as dissolved, settleable, or nonsettleable, the relative amounts of organic and inorganic matter can be determined. Procedures for measuring the amounts of organic and inorganic solids in wastewater are described in laboratory procedures manuals. This information is important for estimating solids handling capacities and for designing treatment processes to remove the organic portion in wastes, which can be very harmful to receiving waters.

## QUESTIONS

Please write your answers to the following questions and compare them with those on page 61.

2.0A What are the four major classifications of waste material in wastewater?

2.0B Domestic wastewater contains mostly (a) organic material, or (b) inorganic material?

2.0C What physical characteristics make some solids settleable and others nonsettleable?

5. *Milligrams per Liter, mg/L.* A measure of the concentration by weight of a substance per unit volume. For practical purposes, one mg/L of a substance in water is equal to one part per million parts (ppm). Thus, a liter of water with a specific gravity of 1.0 weighs one million milligrams. If one liter of water contains 10 milligrams of dissolved oxygen, the concentration is 10 milligrams per million milligrams, or 10 milligrams per liter (10 mg/L), or 10 parts of oxygen per million parts of water, or 10 parts per million (10 ppm), or 10 pounds dissolved oxygen in 1 million pounds of water (10 ppm).
6. *Imhoff Cone.* A clear, cone-shaped container marked with graduations. The cone is used to measure the volume of settleable solids in a specific volume (usually one liter) of water or wastewater.

## 2.1 EFFECTS OF WASTE DISCHARGES

When the treated wastewater from a plant is discharged into receiving waters such as streams, rivers, or lakes, *NATURAL CYCLES*[7] in the aquatic (water) environment may become upset. Whether any problems are caused in the receiving waters depends on the following factors:

1. Type or degree of treatment
2. Size of flow from the treatment plant
3. Characteristics of the wastewater from the treatment plant
4. Amount of flow in the receiving stream or volume of receiving lake that can be used for dilution
5. Quality of the receiving waters
6. Amount of mixing between the treatment plant effluent and the receiving waters
7. Uses of the receiving waters

Natural cycles of interest in wastewater treatment include the natural purification cycles such as the cycle of water from evaporation or *TRANSPIRATION*[8] to condensation to precipitation to runoff and back to evaporation; the life cycles of aquatic organisms; and the cycles of nutrients. These cycles are occurring continuously in wastewater treatment plants and in receiving waters, although the cycles may occur at different rates depending on environmental conditions. Treatment plant operators control and speed up these cycles to work for their benefit in treatment plants and in receiving waters. By controlling the natural cycles, operators also prevent plant operational problems and potentially harmful effects on downstream water users.

### 2.10 Nutrients[9]

*NUTRIENT CYCLES*[10] are an especially important type of natural cycle because of the sensitivity of some receiving waters to nutrients. Treated wastewaters contain nutrients capable of encouraging excessive *ALGAL*[11] growth in receiving waters. These growths hamper domestic, industrial, and recreational uses. Important nutrients include carbon, hydrogen, oxygen, sulfur, nitrogen, and phosphorus. (Conventional wastewater treatment plants do not remove a major portion of the nitrogen and phosphorus nutrients.) All of the nutrients have their own cycles, yet each cycle is influenced by the other cycles. These nutrient cycles are very complex and involve chemical changes in living organisms.

To illustrate the concept of nutrient cycles, a simplified version of the nitrogen cycle will be used as an example (Figure 2.2). A wastewater treatment facility discharges nitrogen in the

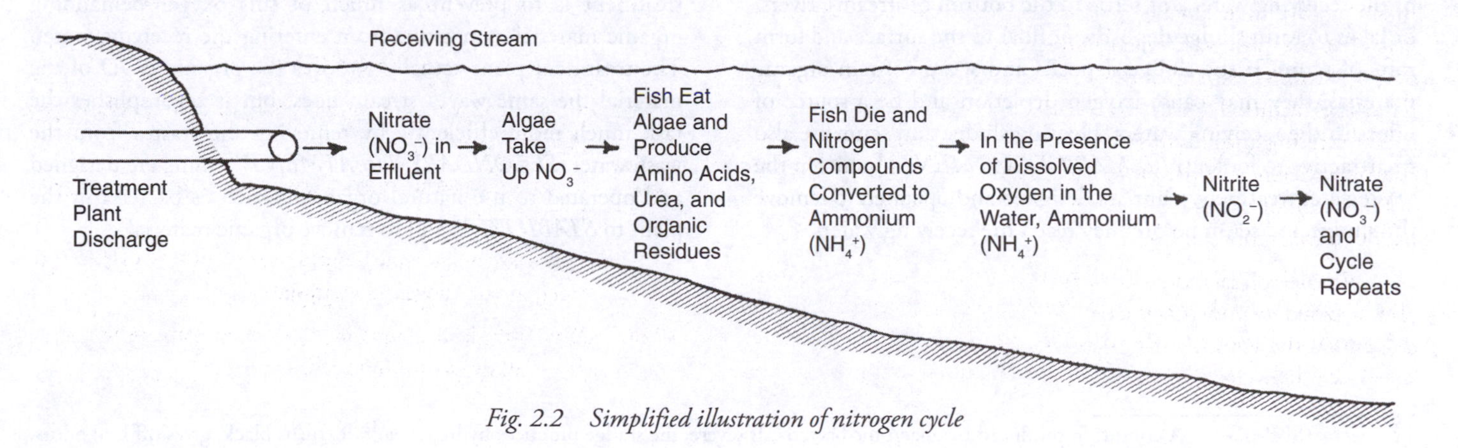


*Fig. 2.2 Simplified illustration of nitrogen cycle*

7. *Natural Cycles.* Cycles that take place in nature, such as the water or hydrologic cycle where water is transformed or changed from one form to another until the water has returned to the original form, thus completing the cycle. Other natural cycles include the life cycles of aquatic organisms and plants, nutrient cycles, and cycles of self- or natural purification.
8. *Transpiration* (TRAN-spur-RAY-shun). The process by which water vapor is released to the atmosphere by living plants. This process is similar to people sweating. Also called evapotranspiration.
9. *Nutrient.* Any substance that is assimilated (taken in) by organisms and promotes growth. Nitrogen and phosphorus are nutrients that promote the growth of algae. There are other essential and trace elements that are also considered nutrients. Also see NUTRIENT CYCLE.
10. *Nutrient Cycle.* The transformation or change of a nutrient from one form to another until the nutrient has returned to the original form, thus completing the cycle. The cycle may take place under either aerobic or anaerobic conditions.
11. *Algae* (AL-jee). Microscopic plants containing chlorophyll that live floating or suspended in water. They also may be attached to structures, rocks, or other submerged surfaces. Excess algal growths can impart tastes and odors to potable water. Algae produce oxygen during sunlight hours and use oxygen during the night hours. Their biological activities appreciably affect the pH, alkalinity, and dissolved oxygen of the water.

form of nitrate ($NO_3^-$) in the plant effluent to the receiving waters. Algae take up the nitrate and produce more algae. The algae are eaten by fish, which convert the nitrogen to amino acids, urea, and organic residues. If the fish die and sink to the bottom, these nitrogen compounds can be converted to ammonium ($NH_4^+$). In the presence of dissolved oxygen and special bacteria, the ammonium is converted to nitrite ($NO_2^-$), then to nitrate ($NO_3^-$), and finally the algae can take up the nitrate and start the cycle all over again.

If too much nitrogen is discharged to receiving waters, too many algae (algal bloom) could be produced. Water with excessive algae can be unsightly. In addition, bacteria decomposing dead algae from occasional die-offs can use up all the dissolved oxygen and cause a fish kill. Thus, the nitrogen cycle has been disrupted, as well as the other nutrient cycles. If no dissolved oxygen is present in the water, the nitrogen compounds are converted to ammonium ($NH_4^+$), the nitrate ($NO_3^-$) compounds are converted to nitrogen gas, the carbon compounds to methane ($CH_4$), and the sulfur compounds to hydrogen sulfide ($H_2S$). Ammonia ($NH_3$) and hydrogen sulfide are odorous gases. Under these conditions, the receiving waters are *SEPTIC,*[12] they stink, and they look terrible. Throughout this manual, you will be provided information on how to control these nutrient cycles in your treatment plant in order to treat wastes and to control odors, as well as to protect receiving waters.

### 2.11 Sludge and Scum[13]

If certain wastes (including domestic wastewater) do not receive adequate treatment, large amounts of solids may accumulate and plug leach fields and filter systems; or accumulate on the banks of the receiving waters; or settle to the bottom of streams, rivers, or lakes to form sludge deposits; or float to the surface and form rafts of scum. If the sludge deposits and scum contain organic material, they may cause oxygen depletion and be a source of odors in the receiving waters. Floating sludge and scum are also unattractive to look at. *PRIMARY TREATMENT*[14] units in the wastewater treatment plant are designed and operated to remove the sludge and scum before they reach the receiving waters.

### 2.12 Oxygen Depletion

Most living creatures, including fish and other aquatic life, need oxygen to survive. Although most streams and other surface waters contain less than 0.001 percent dissolved oxygen (10 milligrams of oxygen per liter of water, or 10 mg/L), most fish can thrive if the water contains at least 5 mg/L and other conditions are favorable.

*Oxygen depletion*

When *OXIDIZABLE*[15] wastes are discharged to a stream, bacteria begin to feed on the waste and decompose or break down the complex substances in the waste into simple chemical compounds. The bacteria, called *AEROBIC BACTERIA,*[16] also use dissolved oxygen (similar to human respiration or breathing) from the water. As more organic waste is added, the bacteria reproduce rapidly. As their population increases, so does their use of oxygen. Where waste flows are high, the population of bacteria may grow large enough to use the entire supply of oxygen from the stream faster than it can be replenished by natural diffusion from the atmosphere. When this happens, fish and most other living things in the stream that require dissolved oxygen die. Therefore, one of the principal objectives of wastewater treatment is to prevent as much of this oxygen-demanding organic material as possible from entering the receiving water. The treatment plant actually removes the organic BOD of the material the same way a stream does, but it accomplishes the task much more efficiently by removing the wastes from the wastewater. *SECONDARY TREATMENT*[17] units are designed and operated to use natural organisms such as bacteria in the plant to *STABILIZE*[18] and to remove organic material.

12. *Septic* (SEP-tick). A condition produced by anaerobic bacteria. If severe, the sludge produces hydrogen sulfide, turns black, gives off foul odors, contains little or no dissolved oxygen, and the wastewater has a high oxygen demand.
13. *Scum.* A layer or film of foreign matter (such as grease, oil) that has risen to the surface of water or wastewater.
14. *Primary Treatment.* A wastewater treatment process that takes place in a rectangular or circular tank and allows those substances in wastewater that readily settle or float to be separated from the wastewater being treated. A septic tank is also considered primary treatment.
15. *Oxidation.* Oxidation is the addition of oxygen, removal of hydrogen, or the removal of electrons from an element or compound; in the environment and in wastewater treatment processes, organic matter is oxidized to more stable substances. The opposite of REDUCTION.
16. *Aerobic Bacteria* (air-O-bick back-TEER-e-uh). Bacteria that will live and reproduce only in an environment containing oxygen that is available for their respiration (breathing), namely atmospheric oxygen or oxygen dissolved in water. Oxygen combined chemically, such as in water molecules ($H_2O$), cannot be used for respiration by aerobic bacteria.
17. *Secondary Treatment.* A wastewater treatment process used to convert dissolved or suspended materials into a form more readily separated from the water being treated. Usually, the process follows primary treatment by sedimentation. The process commonly is a type of biological treatment followed by secondary clarifiers that allow the solids to settle out from the water being treated.
18. *Stabilization.* Conversion to a form that resists change. Organic material is stabilized by bacteria that convert the material to gases and other relatively inert substances. Stabilized organic material generally will not give off obnoxious odors.

Another effect of oxygen depletion, in addition to the killing of fish and other aquatic life, is the problem of odors. When all the dissolved oxygen has been removed, *ANAEROBIC BACTERIA*[19] begin to use the oxygen that is combined chemically with other elements in the form of chemical compounds (such as sulfate (sulfur and oxygen)), which are also dissolved in the water. When anaerobic bacteria remove the oxygen from sulfur compounds, hydrogen sulfide ($H_2S$) gas is produced; hydrogen sulfide has a strong and very offensive "rotten egg" odor. This gas erodes concrete and can discolor and remove paint from homes and structures. Hydrogen sulfide may form explosive mixtures with air and is capable of paralyzing a person's respiratory system. Hydrogen sulfide is a very toxic gas. Other products of anaerobic decomposition (*PUTREFACTION*[20]) also can be objectionable. For example, methane gas is odorless and colorless, but flammable and explosive.

## QUESTIONS

Please write your answers to the following questions and compare them with those on page 61.

2.1A List three natural cycles that are of interest to wastewater treatment facility operators.

2.1B What causes oxygen depletion when organic wastes are discharged to receiving waters?

2.1C What type of bacteria causes hydrogen sulfide gas to be released?

### 2.13 Human Health

Up to now, we have discussed the physical or chemical effects that a waste discharge may have on the uses of water. More important, however, may be the effect on human health through the spread of disease-causing bacteria and viruses. Initial efforts to control human wastes evolved from the need to prevent the spread of diseases. Although untreated wastewater contains many billions of bacteria per gallon, most of these are not harmful to humans, and some are even helpful in wastewater treatment processes. However, humans who have a disease that is caused by bacteria or viruses may discharge some of these harmful organisms in their body wastes. Many serious outbreaks of communicable diseases have been traced to direct contamination of drinking water or food supplies by the body wastes from a human disease carrier.

Some known examples of diseases that may be spread through wastewater discharges are giardiasis (jee-are-DYE-uh-sis), cryptosporidiosis, typhoid, cholera, dysentery, polio, and hepatitis.

Fortunately, the bacteria that grow in the intestinal tract of diseased humans are not likely to find the environment in the wastewater treatment plant or receiving waters favorable for their growth and reproduction. To date, no one working in the wastewater collection or treatment fields is known to have become infected by the HIV virus (AIDS) due to conditions encountered while working on the job. Although many pathogenic organisms are removed by natural die-off during the normal treatment processes, sufficient numbers of some types of organisms can remain to cause a threat to any downstream use involving human contact or consumption. If such uses exist downstream, the treatment plant must also include a *DISINFECTION*[21] process.

The disinfection process most widely used to disinfect wastewater treatment plant effluents is the addition of chlorine. In most cases, proper chlorination of a well-treated waste will result in essentially a complete kill or inactivation of the pathogenic organisms. Operators must realize, however, that the breakdown or malfunction of equipment could result in the discharge at any time of an effluent that contains pathogenic organisms.

In addition to pathogenic organisms, waste treatment chemical by-products can cause illness. In septic tank leach fields, nitrate ($NO_3^-$) is a natural result of aerobic bacteria breaking down ammonia. Nitrate dissolves easily (is very soluble) in water and can contaminate groundwater wells. High nitrate concentration in drinking water interferes with the oxygen uptake in the blood of infants (blue baby syndrome), which can be fatal.

### 2.14 Other Effects

Some wastes adversely affect the clarity and color of the receiving waters making them unsightly and unpopular for recreation.

Many industrial wastes are highly acidic or alkaline (basic), and either condition can interfere with aquatic life, domestic use, and other uses. An accepted measurement of a waste's acidic or basic condition is its *pH.*[22] Before wastes are discharged, they should have a pH similar to that of the receiving water.

Waste discharges may contain toxic substances, such as heavy metals (lead, mercury, cadmium, and chromium) or cyanide, which may affect the use of the receiving water for domestic

19. *Anaerobic Bacteria* (AN-air-O-bick back-TEER-e-uh). Bacteria that live and reproduce in an environment containing no free or dissolved oxygen. Anaerobic bacteria obtain their oxygen supply by breaking down chemical compounds that contain oxygen, such as sulfate ($SO_4^{2-}$).
20. *Putrefaction* (PYOO-truh-FACK-shun). Biological decomposition of organic matter, with the production of foul-smelling and -tasting products, associated with anaerobic (no oxygen present) conditions.
21. *Disinfection* (dis-in-FECT-shun). The process designed to kill or inactivate most microorganisms in water or wastewater, including essentially all pathogenic (disease-causing) bacteria. There are several ways to disinfect, with chlorination being the most frequently used in water and wastewater treatment plants. Compare with STERILIZATION.
22. *pH* (pronounce as separate letters). pH is an expression of the intensity of the basic or acidic condition of a liquid. Mathematically, pH is the logarithm (base 10) of the reciprocal of the hydrogen ion activity.

$$pH = \text{Log}\frac{1}{\{H^+\}}$$

If $\{H^+\} = 10^{-6.5}$, then pH = 6.5. The pH may range from 0 to 14, where 0 is most acidic, 14 most basic, and 7 neutral.

purposes or for aquatic life. Treatment plant effluents chlorinated for disinfection purposes may have to be dechlorinated to protect waters from the toxic effects of residual chlorine.

Taste- and odor-producing substances in industrial waste discharges may reach levels in the receiving waters that are readily detectable in drinking water or in the flesh of fish.

## QUESTIONS

Please write your answers to the following questions and compare them with those on page 61.

2.1D Where do the disease-causing bacteria in wastewater come from?

2.1E What is the most frequently used means of disinfecting treated wastewater?

2.1F What is the term that means "disease-causing"?

2.1G Are all bacteria harmful?

### 2.2 NPDES PERMITS (Figure 2.3)

NPDES stands for National Pollutant Discharge Elimination System. NPDES permits are required by the federal Water Pollution Control Act Amendments of 1972, which were written with the intent of making the nation's waters safe for swimming and for fish and wildlife. The permits regulate discharges into navigable waters from all *POINT SOURCES* [23] of pollution, including industries, municipal wastewater treatment plants, sanitary landfills, large agricultural feedlots, and return irrigation flows. An industry discharging into municipal collection and treatment systems need not obtain a permit but must meet certain specified pretreatment standards. NPDES permits may outline a schedule of compliance for a wastewater treatment facility such as dates for the completion of plant design, engineering, construction, or treatment process changes. Instructions for completing NPDES reporting forms and the necessary forms are available from the regulatory agency issuing the permit.

Your main concern as the operator of a small wastewater treatment system will be maintaining your treatment plant's effluent (discharge) characteristics within the NPDES permit limits for your plant. The permit may specify weekly or monthly average and maximum levels of settleable solids, suspended solids (nonfilterable residue), *BIOCHEMICAL OXYGEN DEMAND (BOD),* [24] and the most probable number (MPN) of *COLIFORM* [25] group bacteria. Larger plants must report effluent temperatures because of the impact of temperature changes on natural cycles. Also, average and maximum flows may be identified as well as an acceptable range of pH values. Almost all effluents are expected to contain virtually no substances that would be toxic to organisms in the receiving waters. The NPDES permit will specify the frequency of collecting samples and the methods of reporting the results. Details on how to comply with NPDES permits will be provided throughout this manual.

Many states issue operating permits for wastewater treatment systems that do not discharge to surface waters (do not require NPDES permits). These permits are issued by state agencies to control wastewater discharges. They may specify the quantity and quality of effluent that may be applied to underground and subsurface absorption systems. The purpose of these state permits is to protect groundwaters, the public health, and the environment.

The NPDES permit is typically a license for a facility to discharge a specified amount of a pollutant into a receiving water under certain conditions. Permits may also authorize facilities to process, incinerate, landfill, or beneficially use sewage sludge. The two basic types of NPDES permits that can be issued are *individual* and *general* permits.

An *INDIVIDUAL PERMIT* is a permit developed for an individual facility. After a facility submits the appropriate applications, the permitting authority develops a permit for that particular facility based on the information contained in the permit application (for example, type of activity, nature of discharge, receiving water quality, and beneficial uses).

---

23. *Point Source.* A discharge that comes out the end of a pipe or other clearly identifiable conveyance. Examples of point source conveyances from which pollutants may be discharged include: ditches, channels, tunnels, conduits, wells, containers, rolling stock, concentrated animal feeding operations, landfill leachate collection systems, vessels, or other floating craft. A NONPOINT SOURCE refers to runoff or a discharge from a field or similar source.

24. *BOD* (pronounce as separate letters). Biochemical Oxygen Demand. The rate at which organisms use the oxygen in water or wastewater while stabilizing decomposable organic matter under aerobic conditions. In decomposition, organic matter serves as food for the bacteria and energy results from its oxidation. BOD measurements are used as a surrogate measure of the organic strength of wastes in water.

25. *Coliform* (KOAL-i-form). A group of bacteria found in the intestines of warm-blooded animals (including humans) and also in plants, soil, air, and water. The presence of coliform bacteria is an indication that the water is polluted and may contain pathogenic (disease-causing) organisms. Fecal coliforms are those coliforms found in the feces of various warm-blooded animals, whereas the term coliform also includes other environmental sources.

MONTH ______, 20____

DIS. NO. ______

PERMIT NUMBER ______

NAME OF FACILITY ______

**MONTHLY OPERATION REPORT OF WASTEWATER TREATMENT FACILITY**
**STABILIZATION PONDS**

______ NAME OF FACILITY

______ PHONE

Send to: MINNESOTA POLLUTION CONTROL AGENCY
1935 WEST COUNTY ROAD B 2
ROSEVILLE, MINNESOTA 55113
ATTN: COMPLIANCE AND ENFORCEMENT SECTION

| TYPE OF POND (PRIMARY, SECONDARY, AERATED, ETC.) | ______ POND | | | | ______ ACRES | ______ POND | | | | ______ ACRES | ______ POND | | | | ______ ACRES |
|---|---|---|---|---|---|---|---|---|---|---|---|---|---|---|---|
| WEEK OF THE MONTH | 1 st | 2 nd | 3 rd | 4 th | 5 th | 1 st | 2 nd | 3 rd | 4 th | 5 th | 1 st | 2 nd | 3 rd | 4 th | 5 th |
| 1. DATE OF TEST OR OBSERVATION | | | | | | | | | | | | | | | |
| 2. ODOR (YES OR NO) | | | | | | | | | | | | | | | |
| 3. AQUATIC PLANTS (% OF COVERAGE, TYPE) | | | | | | | | | | | | | | | |
| 4. FLOATING MATS (% OF COVERAGE, TYPE) | | | | | | | | | | | | | | | |
| 5. POND WATER DEPTH (NEAREST INCH) | | | | | | | | | | | | | | | |
| 6. MUSKRATS, RODENTS ETC. (YES OR NO) | | | | | | | | | | | | | | | |
| 7. DIKE CONDITION (EROSION ETC.) | | | | | | | | | | | | | | | |
| 8. ICE COVER (% OF COVERAGE) | | | | | | | | | | | | | | | |

| GENERAL | | | ANALYSIS - POND INFLUENT, EFFLUENT: INFLUENT | | | | EFFLUENT DISCHARGE | | | | | | | | | | | | RECEIVING WATERS DATA: ABOVE OUTFALL ______ MILES | | | BELOW OUTFALL ______ MILES | | | 33 PRESENT CONDITION | | | | | 34 |
|---|---|---|---|---|---|---|---|---|---|---|---|---|---|---|---|---|---|---|---|---|---|---|---|---|---|---|---|---|---|---|
| DATE | 9 DAY OF WEEK | 10 PRECIPITATION (INCHES) | 11 FLOW (MGD) | 12 SAMPLE TYPE | 13 BOD (mg/L) | 14 TSS (mg/L) | 15 FLOW (MGD) | 16 TIME SAMPLED | 17 BOD (mg/L) | 18 TSS (mg/L) | 19 TURBIDITY (JTU) | 20 FECAL COLIFORM (MPN/100mL) | 21 D.O. (mg/L) | 22 pH (standard units) | 23 TOTAL PHOSPHORUS AS P (mg/L) | 24 KJELDAHL NITROGEN (mg/L) | 25 CHLORINE USED (LBS) | 26 CHLORINE RESIDUAL (mg/L) | 27 TIME OF DAY | 28 WATER TEMP °F | 29 D.O. (mg/L) | 30 TIME OF DAY | 31 WATER TEMP °F | 32 D.O. (mg/L) | DEPTH (ft.) | WIDTH (ft.) | VELOCITY (ft./sec) | NORMAL +/- | DATE | REMARKS INDEX |
| 1 | | | | | | | | | | | | | | | | | | | | | | | | | | | | | 1 | |
| 2 | | | | | | | | | | | | | | | | | | | | | | | | | | | | | 2 | |
| 3 | | | | | | | | | | | | | | | | | | | | | | | | | | | | | 3 | |
| 4 | | | | | | | | | | | | | | | | | | | | | | | | | | | | | 4 | |
| 5 | | | | | | | | | | | | | | | | | | | | | | | | | | | | | 5 | |
| 6 | | | | | | | | | | | | | | | | | | | | | | | | | | | | | 6 | |
| 7 | | | | | | | | | | | | | | | | | | | | | | | | | | | | | 7 | |
| 8 | | | | | | | | | | | | | | | | | | | | | | | | | | | | | 8 | |
| 9 | | | | | | | | | | | | | | | | | | | | | | | | | | | | | 9 | |
| 10 | | | | | | | | | | | | | | | | | | | | | | | | | | | | | 10 | |
| 11 | | | | | | | | | | | | | | | | | | | | | | | | | | | | | 11 | |
| 12 | | | | | | | | | | | | | | | | | | | | | | | | | | | | | 12 | |
| 13 | | | | | | | | | | | | | | | | | | | | | | | | | | | | | 13 | |
| 14 | | | | | | | | | | | | | | | | | | | | | | | | | | | | | 14 | |
| 15 | | | | | | | | | | | | | | | | | | | | | | | | | | | | | 15 | |
| 16 | | | | | | | | | | | | | | | | | | | | | | | | | | | | | 16 | |
| 17 | | | | | | | | | | | | | | | | | | | | | | | | | | | | | 17 | |
| 18 | | | | | | | | | | | | | | | | | | | | | | | | | | | | | 18 | |
| 19 | | | | | | | | | | | | | | | | | | | | | | | | | | | | | 19 | |
| 20 | | | | | | | | | | | | | | | | | | | | | | | | | | | | | 20 | |
| 21 | | | | | | | | | | | | | | | | | | | | | | | | | | | | | 21 | |
| 22 | | | | | | | | | | | | | | | | | | | | | | | | | | | | | 22 | |
| 23 | | | | | | | | | | | | | | | | | | | | | | | | | | | | | 23 | |
| 24 | | | | | | | | | | | | | | | | | | | | | | | | | | | | | 24 | |
| 25 | | | | | | | | | | | | | | | | | | | | | | | | | | | | | 25 | |
| 26 | | | | | | | | | | | | | | | | | | | | | | | | | | | | | 26 | |
| 27 | | | | | | | | | | | | | | | | | | | | | | | | | | | | | 27 | |
| 28 | | | | | | | | | | | | | | | | | | | | | | | | | | | | | 28 | |
| 29 | | | | | | | | | | | | | | | | | | | | | | | | | | | | | 29 | |
| 30 | | | | | | | | | | | | | | | | | | | | | | | | | | | | | 30 | |
| 31 | | | | | | | | | | | | | | | | | | | | | | | | | | | | | 31 | |
| TOTAL | | | | | | | | | | | | | | | | | | | | | | | | | | | | | | |
| MIN. REPORTED VALUE | | | | | | | | | | | | | | | | | | | | | | | | | | | | | | |
| MIN. PERMIT CONDITION | | | | | | | | | | | | | | | | | | | | | | | | | | | | | | |
| AVG. REPORTED VALUE | | | | | | | | | | | | | | | | | | | | | | | | | | | | | | |
| AVG. PERMIT CONDITION | | | | | | | | | | | | | | | | | | | | | | | | | | | | | | |
| MAX. REPORTED VALUE | | | | | | | | | | | | | | | | | | | | | | | | | | | | | | |
| MAX. PERMIT CONDITION | | | | | | | | | | | | | | | | | | | | | | | | | | | | | | |
| REPORTED FREQUENCY OF ANALYSIS | | | | | | | | | | | | | | | | | | | | | | | | | | | | | | |
| PERMIT CONDITION FREQUENCY OF ANALYSIS | | | | | | | | | | | | | | | | | | | | | | | | | | | | | | |
| REPORTED SAMPLE TYPE | | | | | | | | | | | | | | | | | | | | | | | | | | | | | | |
| PERMIT CONDITION SAMPLE TYPE | | | | | | | | | | | | | | | | | | | | | | | | | | | | | | |

35. REMARKS: INCLUDE BYPASS AND OVERFLOW OCCURRENCES, UNUSUAL SEWAGE OR STREAM FLOW PROBLEMS, OPERATIONAL PROBLEMS, OTHER REQUIRED PARAMETER DATA, COMPLAINTS, ETC. IN SPACES ON BACK OF THIS FORM.

*I certify that I am familiar with the information contained in this report and that to the best of my knowledge and belief such information is true, complete, and accurate.*

______ SIGNATURE OF PRINCIPAL EXECUTIVE OFFICER OR AUTHORIZED AGENT

______ DATE

MPCA FORM 704

*Fig. 2.3 Typical NPDES permit reporting form*

A *GENERAL PERMIT* is developed and issued by a permitting authority to cover multiple facilities within a specific category. General permits may offer a cost-effective option for utility agencies because of the large number of facilities that can be covered under a single permit. General permits may be written to cover categories of point sources having common elements:

- Stormwater point sources
- Facilities that involve the same or substantially similar types of operations
- Facilities that discharge the same types of wastes or engage in the same types of sludge use or disposal

NPDES permit components may include best management practices (BMPs), stormwater, combined sewer overflows (CSOs), and sanitary sewer overflows (SSOs).

For additional information, see the EPA's *NPDES PERMIT WRITERS' MANUAL* (EPA-833-K-10-001) at www.epa.gov; the manual is also available for download at http://www.epa.gov/npdes/pubs/pwm_2010.pdf.

## QUESTIONS

Please write your answers to the following questions and compare them with those on page 61.

2.2A What does NPDES stand for?

2.2B What is the main goal of the federal Water Pollution Control Act Amendments of 1972?

## 2.3 COLLECTION, TREATMENT, AND DISCHARGE OF WASTEWATER

Facilities for handling wastewater usually have three major functions: collection, treatment, and discharge (see Figure 2.4). For a community or municipality, these components make up the wastewater facilities. For an individual industry that handles its own wastewater, the same three components are necessary. This is also true for the individual home or on-site wastewater treatment system. This operator training course is designed primarily to train small wastewater treatment system operators, so the discussion in this and later chapters will be related to small wastewater systems. Operator training courses dealing with

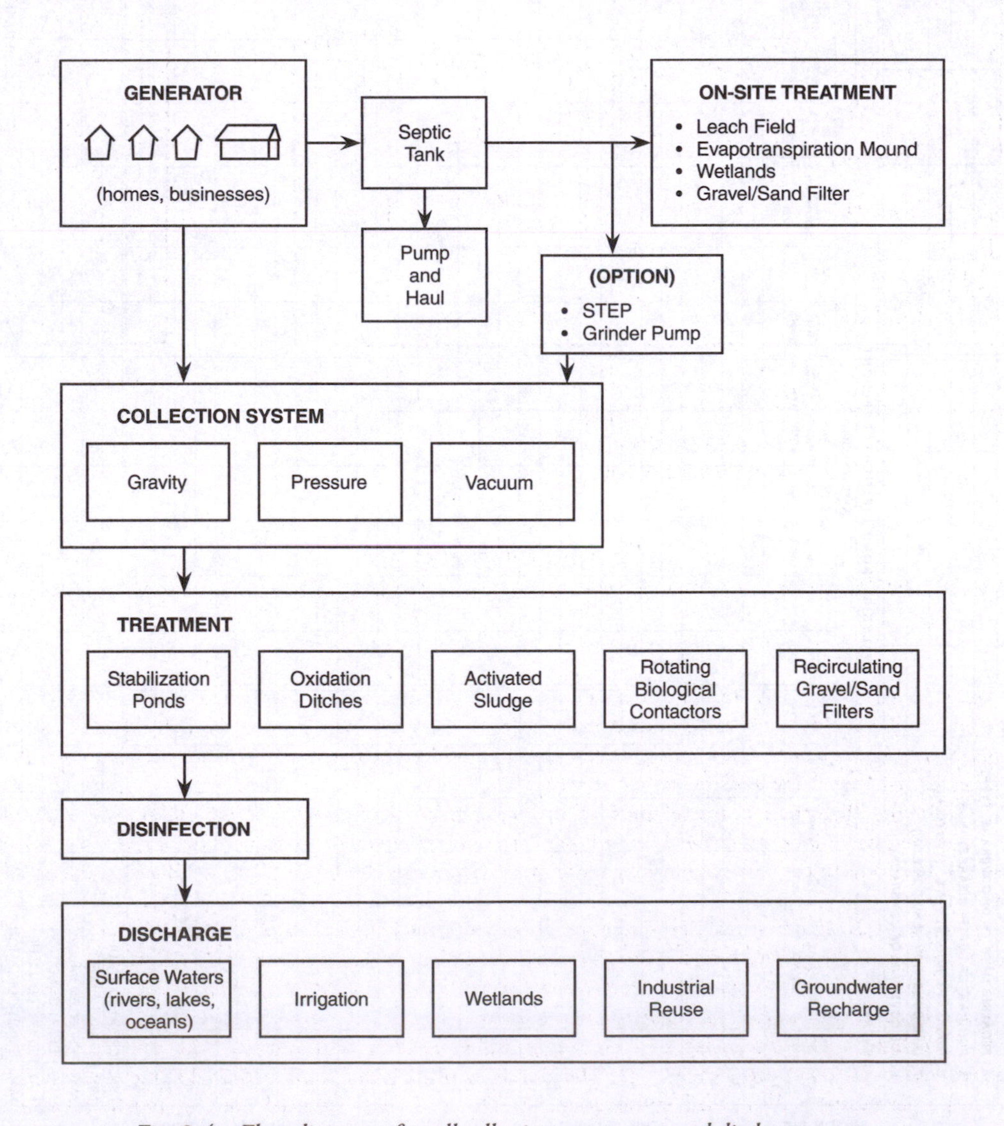


*Fig. 2.4 Flow diagram of small collection, treatment, and discharge systems*

larger municipal wastewater treatment facilities, industrial waste treatment, and the operation and maintenance of collection systems are also available in this series of operator training courses.

The smallest wastewater system is the individual home with its own on-site septic tank and subsurface leaching (discharge) system. In this case, the house wastewater service (sewer) or building sewer line is the collection system. The sewer lines convey wastewater from the house plumbing fixtures to the treatment facility, the *SEPTIC TANK*[26] (Figures 2.5 and 2.6).

The treatment provided by the septic tank is the *REDUCTION*[27] and stabilization of solids. The safe discharge of the septic tank effluent (partially treated wastewater) is usually accomplished in *SUBSURFACE LEACH FIELDS*[28] (Figure 2.7) by the treatment processes of filtering, *PERCOLATION*,[29] soil absorption, and *EVAPOTRANSPIRATION*.[30]

In some areas, the individual home system may become a part of a multiple home or commercial system. This may be accomplished by collecting and conveying the wastewater directly

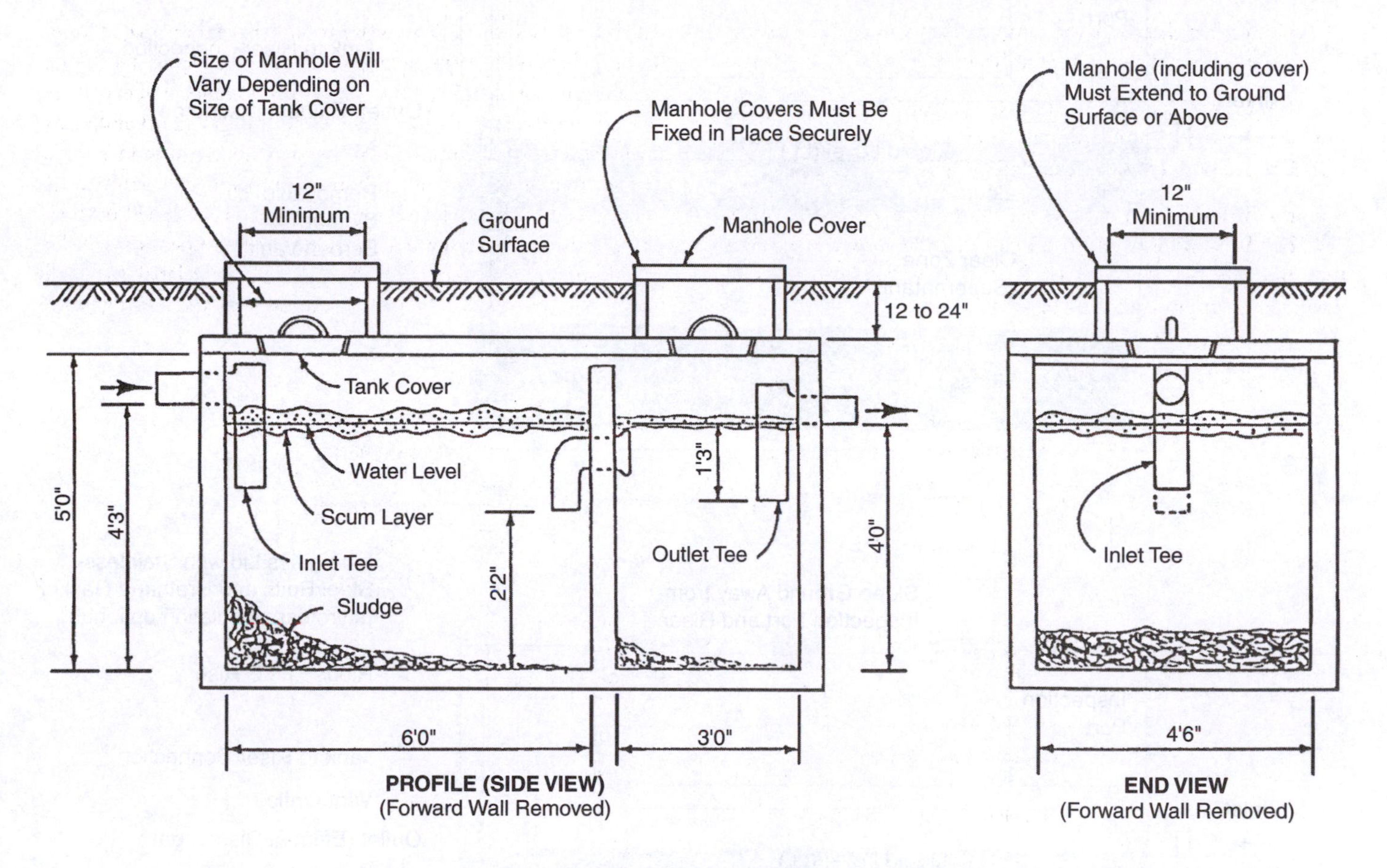


*Fig. 2.5 Typical septic tank*

(Source: *HOMEOWNERS' AND USERS' GUIDE FOR ON-SITE WASTEWATER DISPOSAL SYSTEMS*, California State Water Resources Control Board)

---

26. *Septic Tank.* A system sometimes used where wastewater collection systems and treatment plants are not available. The system is a settling tank in which settled sludge and floatable scum are in intimate contact with the wastewater flowing through the tank and the organic solids are decomposed by anaerobic bacterial action. Used to treat wastewater and produce an effluent that is usually discharged to subsurface leaching. Also referred to as an interceptor; however, the preferred term is septic tank.
27. *Reduction* (re-DUCK-shun). Reduction is the addition of hydrogen, removal of oxygen, or the addition of electrons to an element or compound. Under anaerobic conditions (no dissolved oxygen present), sulfur compounds are reduced to odor-producing hydrogen sulfide ($H_2S$) and other compounds. In the treatment of metal finishing wastewaters, hexavalent chromium ($Cr^{6+}$) is reduced to the trivalent form ($Cr^{3+}$). The opposite of OXIDATION.
28. *Subsurface Leaching System.* A method of treatment and discharge of septic tank effluent, sand filter effluent, or other treated wastewater. The effluent is applied to soil below the ground surface through open-jointed pipes or drains or through perforated pipes (holes in the pipes). The effluent is treated as it passes through porous soil or rock strata (layers). Newer subsurface leaching systems include chamber and gravelless systems, and also gravel trenches without pipe the full length of the trench.
29. *Percolation* (purr-ko-LAY-shun). The slow passage of water through a filter medium; or, the gradual penetration of soil and rocks by water.
30. *Evapotranspiration* (ee-VAP-o-TRANS-purr-A-shun). (1) The process by which water vapor is released to the atmosphere by living plants. This process is similar to people sweating. Also called transpiration. (2) The total water removed from an area by transpiration (plants) and by evaporation from soil, snow, and water surfaces.

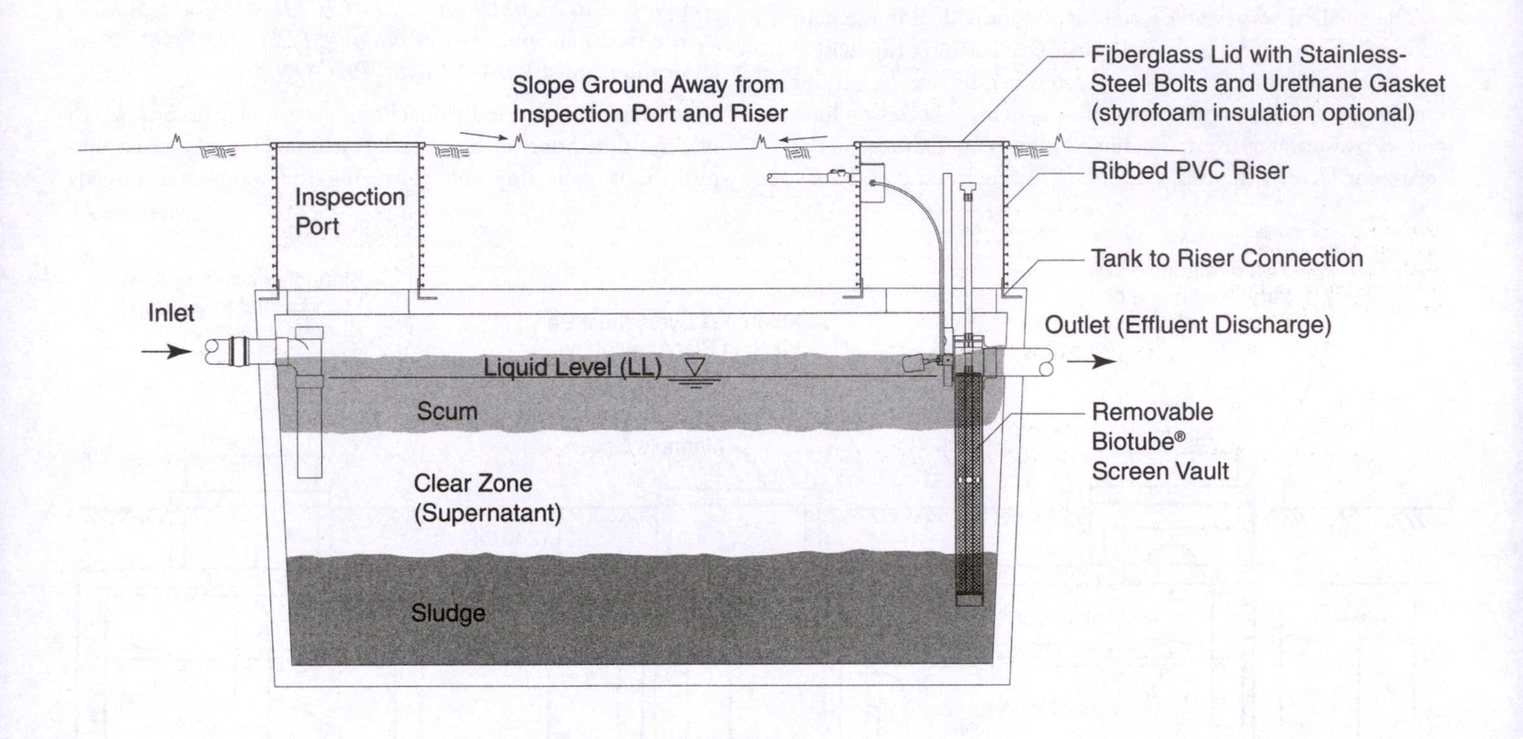


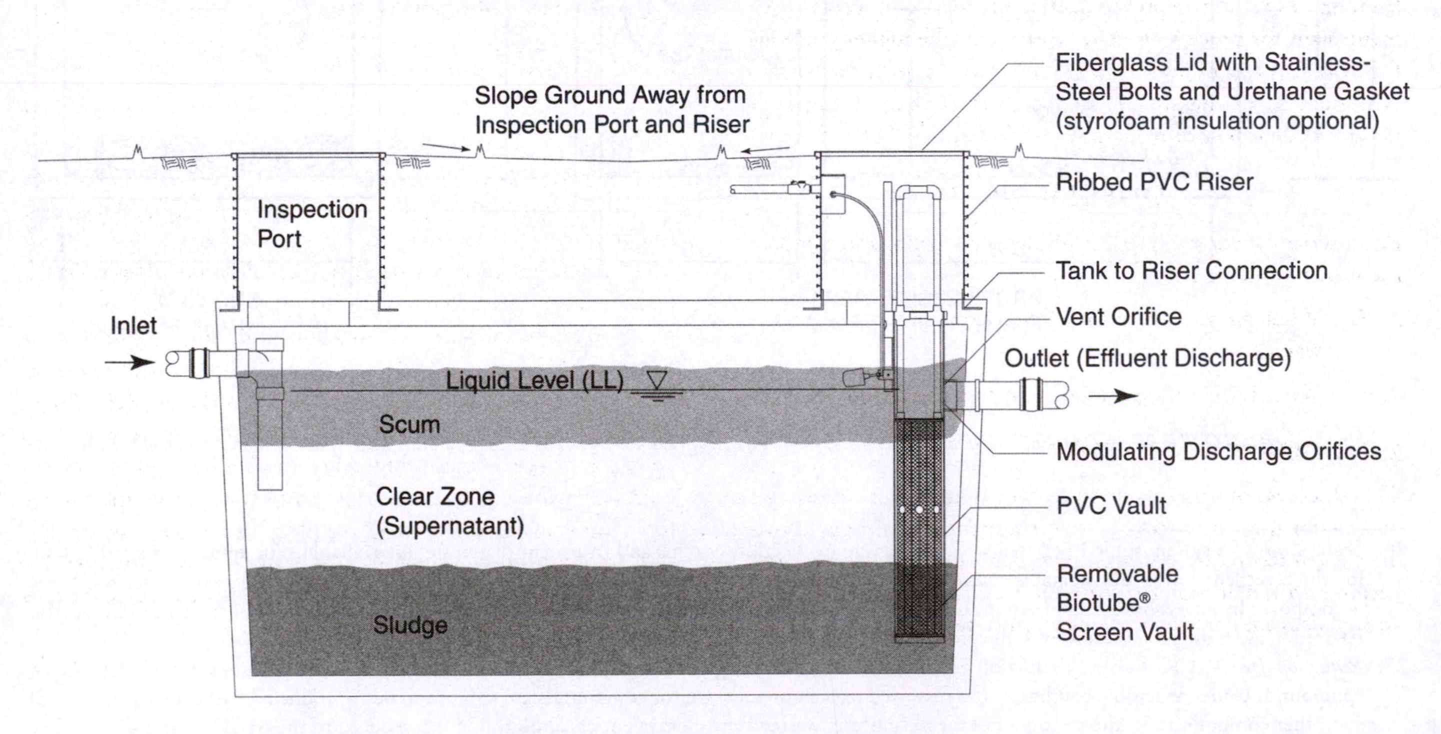


*Fig. 2.6 Newer types of septic tank with effluent filter*

(Courtesy of Orenco Systems, Inc., www.orenco.com)

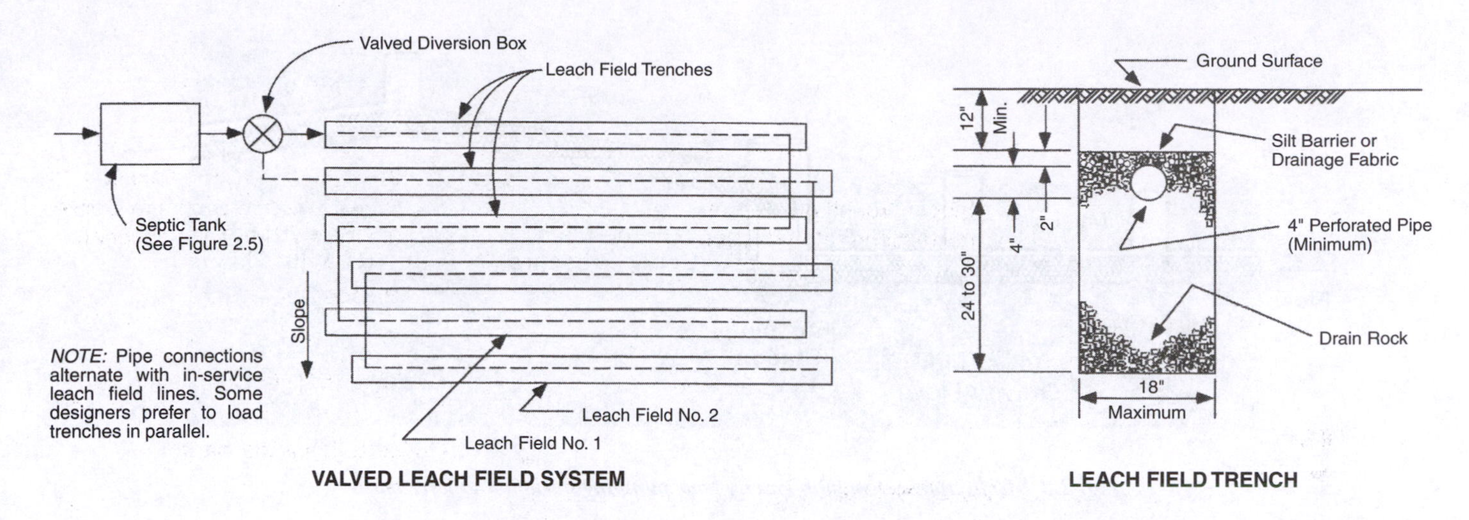


*Fig. 2.7 Typical subsurface leach field system and leach field trench*

(Source: *HOMEOWNERS' AND USERS' GUIDE FOR ON-SITE WASTEWATER DISPOSAL SYSTEMS,* California State Water Resources Control Board)

from homes or from septic tank effluents to a community treatment plant for processing. After treatment, the higher grade effluent is disposed of in the environment by any of several methods, such as discharging it to receiving streams, using land treatment methods, or recharging *GROUNDWATER*[31] basins.

## QUESTIONS

Please write your answers to the following questions and compare them with those on page 61.

2.3A What are the three major functions of a small wastewater system?

2.3B What is the simplest type of on-site wastewater system?

## 2.4 COLLECTION SYSTEMS

The objective of wastewater collection systems is to move the wastewater from the source to the treatment site without allowing wastewater to escape (overflows, backups, leaks, breaks) and keeping unwanted water (*INFLOW*[32] and *INFILTRATION*[33]) and solids from entering the collection system.

### 2.40 Gravity Collection Systems

#### *2.400 Conventional Gravity Flow*

Wastewater in a collection system is commonly conveyed by gravity using the natural slope of the land. Thus, the location and design of the components of a gravity collection system are strongly influenced by the *TOPOGRAPHY*[34] (surface contours) of the service area. In some service areas, the cost to trench becomes excessive because the collection pipeline must be buried very deep to maintain the pipe slope or fall, or due to rock, high groundwater, or unstable soils. In such a case, a system of wastewater pumps is provided to lift the wastewater to a higher elevation for a return to gravity flow (Figure 2.8). These pumping facilities are referred to as lift stations or pumping plants. However, pumps and other mechanical equipment require extensive maintenance and costly energy to operate, so their use is avoided whenever possible. In locations where the topography or ground conditions are unfavorable for a gravity collection system, a pressure or vacuum collection system is often used.

31. *Groundwater.* Subsurface water in the saturation zone from which wells and springs are fed. In a strict sense the term applies only to water below the water table. Also called phreatic water and plerotic water.
32. *Inflow.* Water discharged into a sewer system and service connections from sources other than regular connections. This includes flow from yard drains, foundations, and around access and manhole covers. Inflow differs from infiltration in that it is a direct discharge into the sewer rather than a leak in the sewer itself.
33. *Infiltration* (in-fill-TRAY-shun). The seepage of groundwater into a sewer system, including service connections. Seepage frequently occurs through defective or cracked pipes, pipe joints and connections, interceptor access risers and covers, or manhole walls.
34. *Topography* (toe-PAH-gruh-fee). The arrangement of hills and valleys in a geographic area.

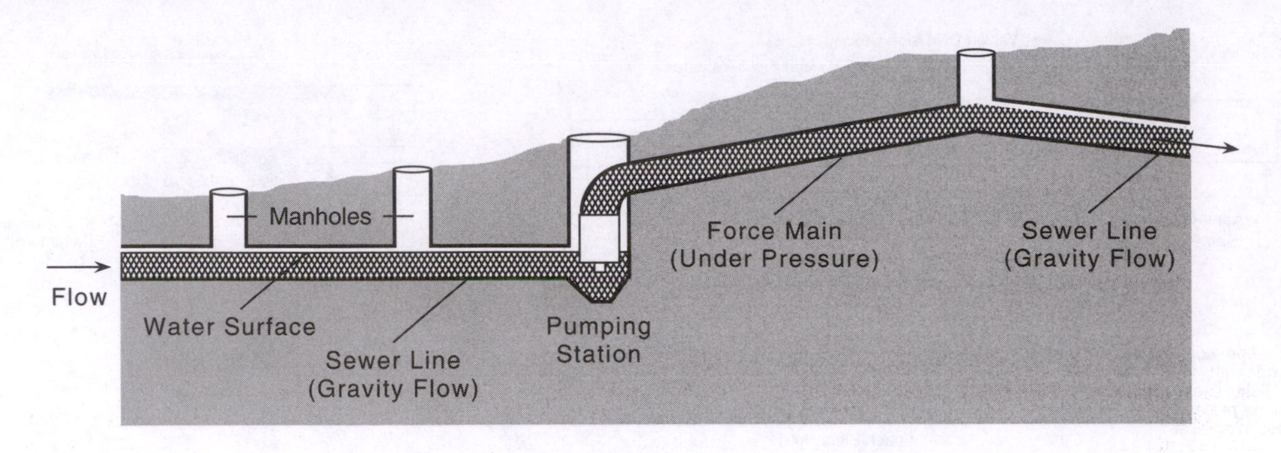


*Fig. 2.8 Collection system with gravity flow, pumping station, and force main*

### *2.401 Effluent Drains*

Small-diameter gravity effluent drains from septic tanks may be feasible (septic tank effluent gravity (STEG) sewers). These effluent drains can follow the ground surface contours as long as the drain pipe does not rise above the *INVERT*[35] of the septic tank outlet. This is possible because the solids have been removed by the septic tank.

## QUESTIONS

Please write your answers to the following questions and compare them with those on page 62.

2.4A How is wastewater commonly conveyed in a collection system?

2.4B Under what conditions would installing a gravity collection system be impractical?

### 2.41 Pressure Collection Systems

Where the topography and ground conditions of an area are not suitable for a conventional gravity collection system due to flat terrain, rocky soils, or extremely high groundwater, pressure collection systems are now becoming a practical alternative. Compared with a gravity collection system, pressure systems offer the following advantages:

1. Deep trenching to maintain flow is unnecessary.
2. *INVERTED SIPHONS*[36] are not needed to cross roads and rivers.
3. Smaller diameter pipe may be used for two reasons: (a) the wastewater moves under pressure, and (b) there are no leaks through joints or access manholes so there is no need to allow for infiltration and inflow volumes.
4. Pressurized systems experience fewer stoppages.
5. Root intrusion problems are eliminated.

Limitations of pressure collection systems compared with gravity systems depend on the topography of the area. Energy requirements and capital cost comparisons depend on whether flow is by gravity or if lift stations are required. Technical expertise and maintenance for each system depend on the design and installed equipment.

The principal components of a pressure collection system (Figure 2.9) include gravity building sewers, septic tanks or pump vaults, pumps and controls, pressure mains, air release valves, and valves for system isolation. If there is not a septic tank, a grinder pump is required (Figure 2.10). A grinder pump is a submersible centrifugal pump with a comminuting (cutting and shredding) blade installed on the pump as the first stage to prevent clogging of the pump impeller with large solids. Where a septic tank is used, effluent pumping from the tank is generally accomplished by a single or multistage submersible centrifugal pump protected by a basket screen.

Gravity building sewers connect a building's wastewater plumbing system to a buried holding tank located on the homeowner's lot, as illustrated in Figures 2.11 and 2.12. A holding tank is commonly a septic tank; in some areas, this tank is referred to as an *INTERCEPTOR.*[37] A septic tank has two

35. *Invert* (IN-vert). The lowest point of the channel inside a pipe, conduit, canal, or manhole. Also called flow line by some contractors, however, the preferred term is invert.
36. *Inverted Siphon.* A pressure pipeline used to carry wastewater flowing in a gravity collection system under a depression, such as a valley or roadway, or under a structure, such as a building. Also called a depressed sewer.
37. *Interceptor.* A septic tank or other holding tank that serves as a temporary wastewater storage reservoir for a septic tank effluent pump (STEP) system. Also see SEPTIC TANK.

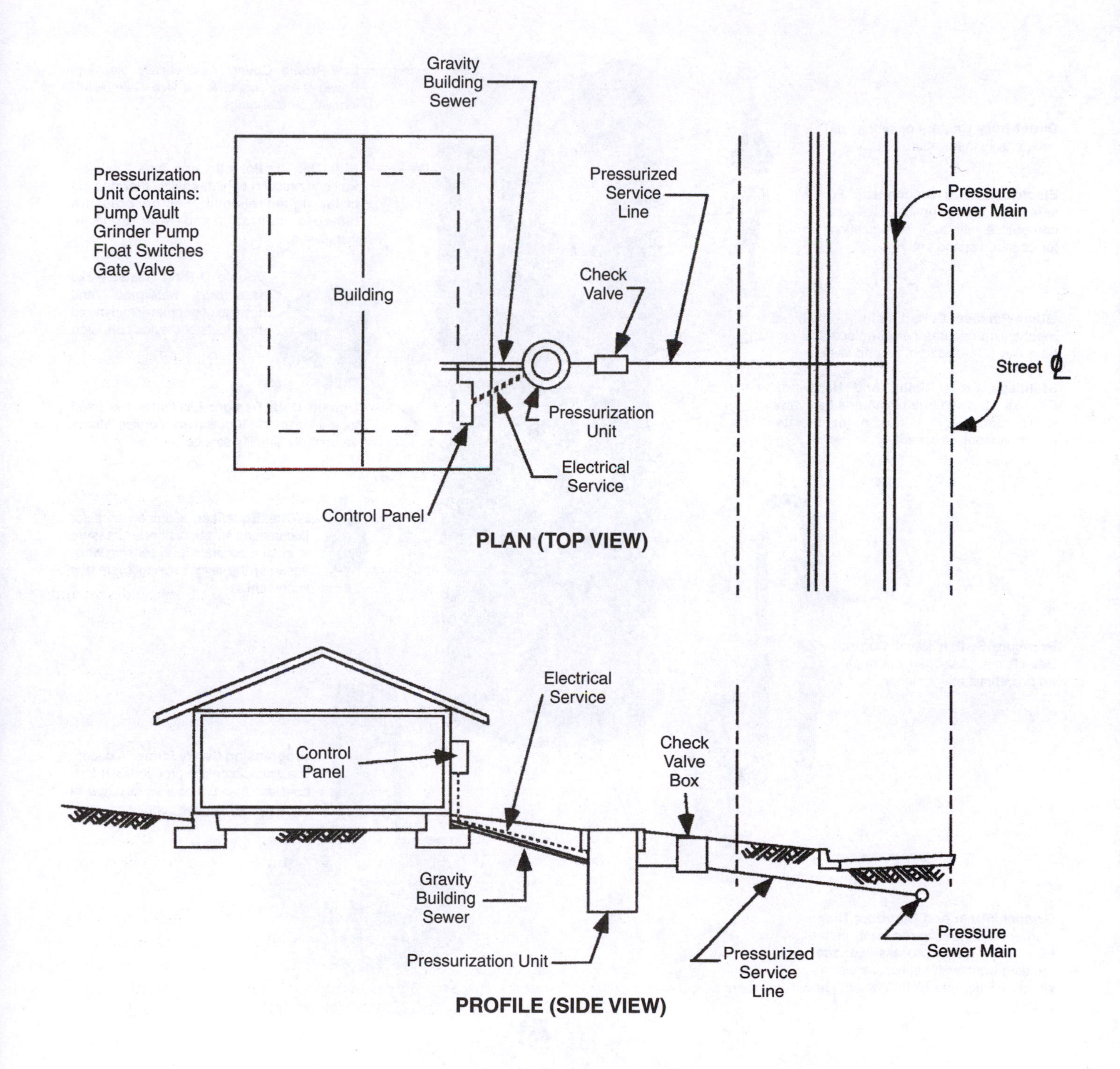


*Fig. 2.9 Principal components of a typical pressure collection system*

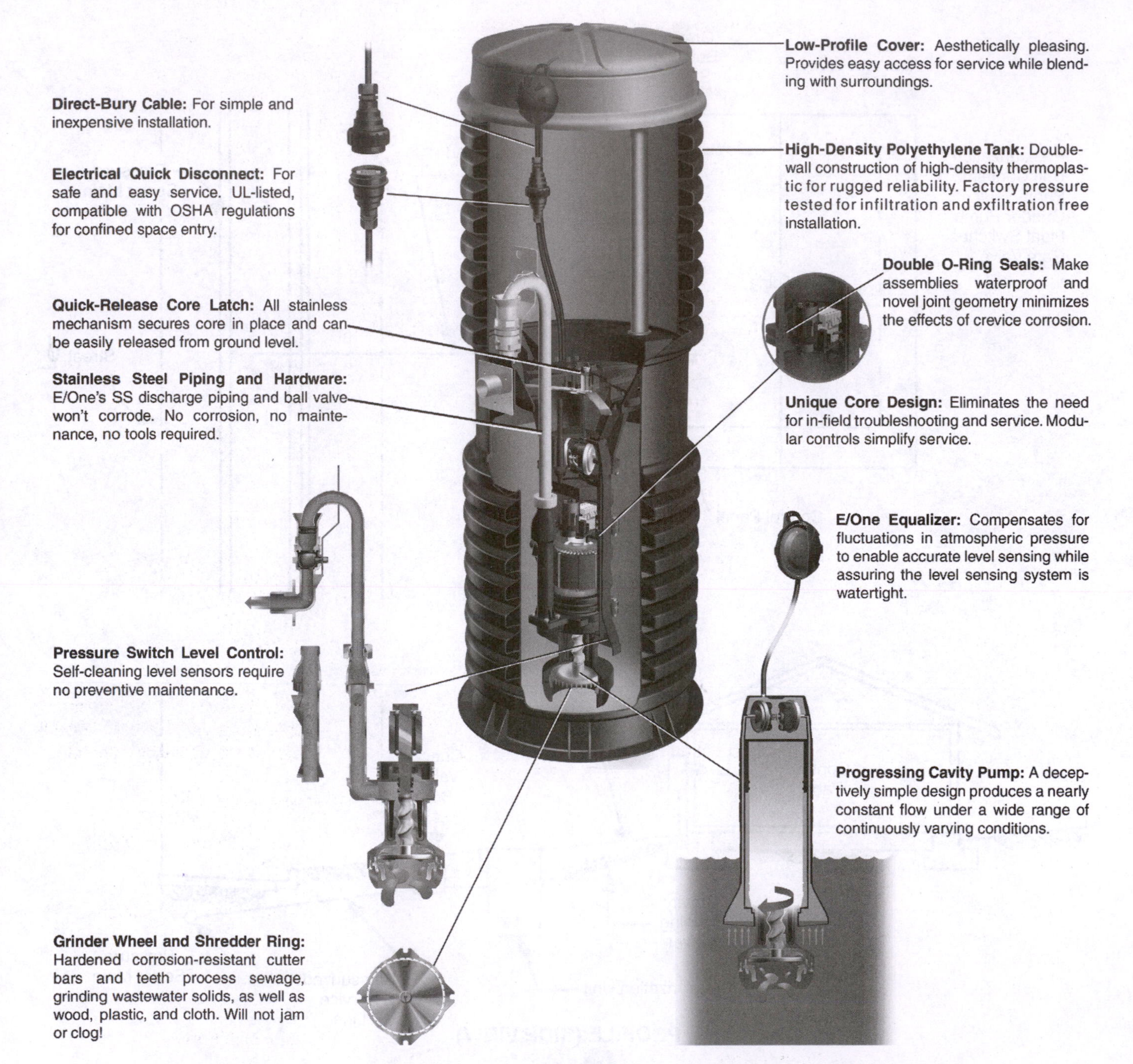

Low-Profile Cover: Aesthetically pleasing. Provides easy access for service while blending with surroundings.
Direct-Bury Cable: For simple and inexpensive installation.
High-Density Polyethylene Tank: Double-wall construction of high-density thermoplastic for rugged reliability. Factory pressure tested for infiltration and exfiltration free installation.
Electrical Quick Disconnect: For safe and easy service. UL-listed, compatible with OSHA regulations for confined space entry.
Double O-Ring Seals: Make assemblies waterproof and novel joint geometry minimizes the effects of crevice corrosion.
Quick-Release Core Latch: All stainless mechanism secures core in place and can be easily released from ground level.
Stainless Steel Piping and Hardware: E/One's SS discharge piping and ball valve won't corrode. No corrosion, no maintenance, no tools required.
Unique Core Design: Eliminates the need for in-field troubleshooting and service. Modular controls simplify service.
E/One Equalizer: Compensates for fluctuations in atmospheric pressure to enable accurate level sensing while assuring the level sensing system is watertight.
Pressure Switch Level Control: Self-cleaning level sensors require no preventive maintenance.
Progressing Cavity Pump: A deceptively simple design produces a nearly constant flow under a wide range of continuously varying conditions.
Grinder Wheel and Shredder Ring: Hardened corrosion-resistant cutter bars and teeth process sewage, grinding wastewater solids, as well as wood, plastic, and cloth. Will not jam or clog!


*Fig. 2.10 Grinder Pump*

(Courtesy of Environment One Corporation)

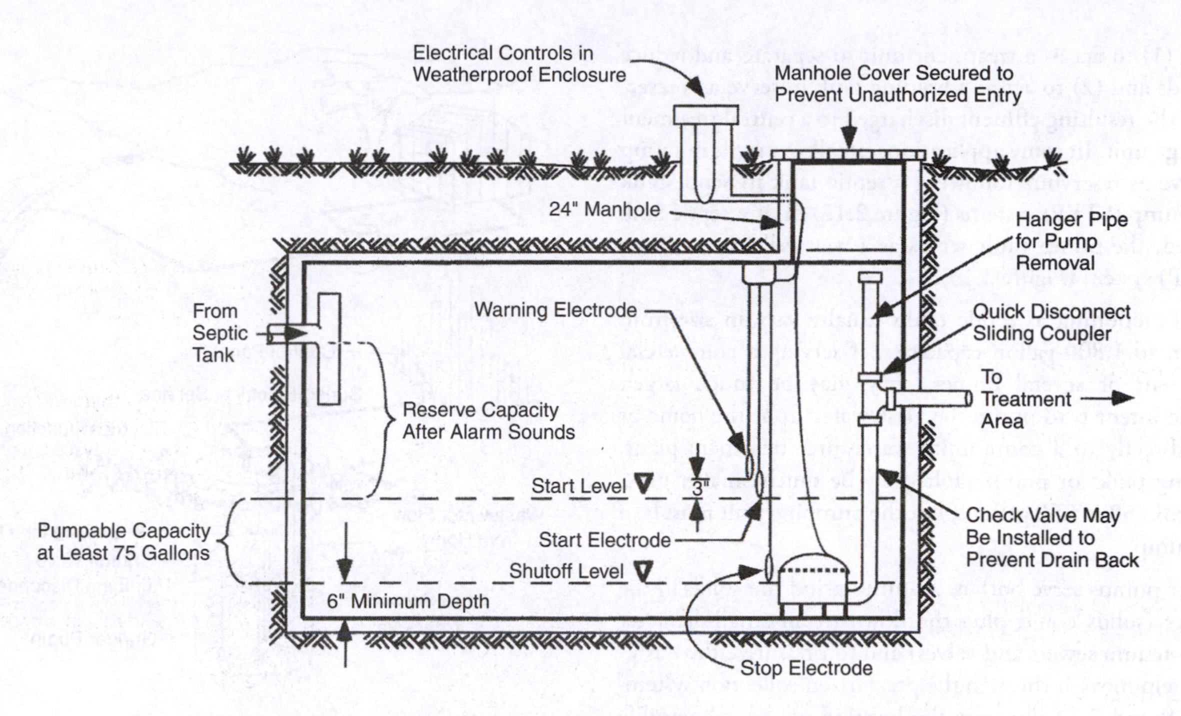


*Fig. 2.11 Typical buried interceptor tank and pump for a single-family home*
(*NOTE:* See Figure 2.12 for a newer design.)
(Source: *RURAL WASTEWATER MANAGEMENT,* California State Water Resources Control Board)

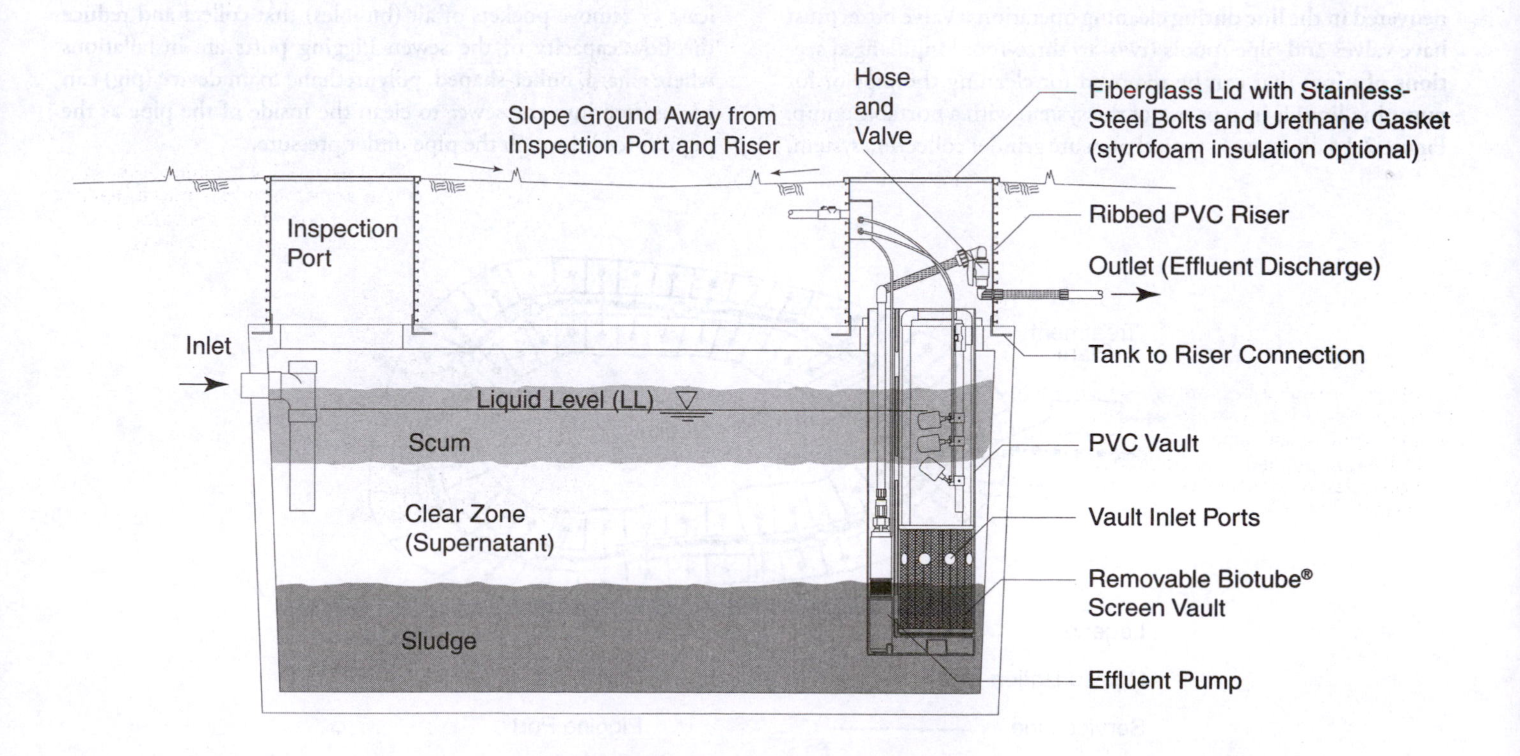


*Fig. 2.12 Septic tank effluent pump (STEP) system*
(Courtesy of Orenco Systems, Inc., www.orenco.com)

purposes: (1) to act as a treatment unit to separate and reduce waste solids and (2) to act as a holding tank to serve as a reservoir, with the resulting effluent discharged to a central treatment or discharge unit. In some applications, small wet wells or pump vaults serve as reservoirs following a septic tank in septic tank effluent pump (STEP) systems (Figure 2.12) or, if a septic tank is not used, the pump vault serves as a wet well for a grinder pump (GP) system (Figure 2.13).

Tanks functioning as septic tanks usually vary in size from 750-gallon to 1,800-gallon capacities. If serving a commercial establishment or several homes, they may be much larger. Where the intent is to pump the wastewater from the home or business directly to a community wastewater treatment plant, the holding tanks or pump vaults may be much smaller units ranging from 50 to 250 gallons, but the pumping unit must be a grinder pump.

Grinder pumps serve both as a unit to grind the solids in the wastewater (solids could plug the downstream small-diameter pressure/vacuum sewers and valves) and to pressurize the wastewater to help move it through the pressurized collection system. Figures 2.9 and 2.13 illustrate the location of the submersible grinder pump in the holding tank (pressurization unit).

Pressure mains are the "arteries" of the pressure collection system; they convey the pressurized wastewater to a treatment plant. Since the wastewater is pushed by pressure, the mains are not dependent on a slope to create a gravity flow and can be laid at a uniform depth following the natural slope of the land along their routes. Like all collection systems, pressure collection systems must be accessible for maintenance. This means line access must be available where a cleaning device can be safely inserted and maneuvered in the line during cleaning operations. Valve boxes must have valves and pipe spools (two- to three-foot long flanged sections of pipe) that can be removed for cleaning the pipe or for pumping liquids into or out of the system with a portable pump. Figure 2.14 illustrates a typical pressure grinder collection system.

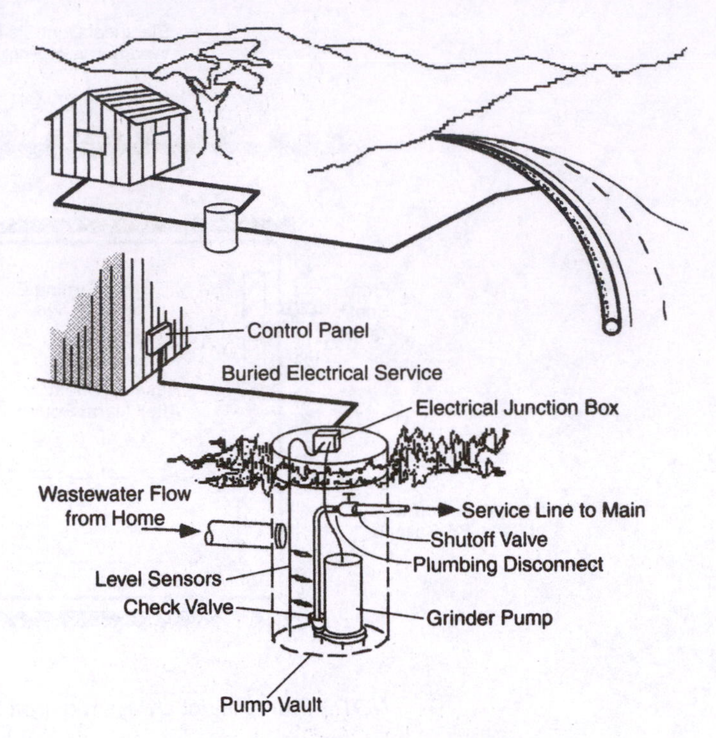


*Fig. 2.13 Grinder pump (GP) system*
(Source: *ALTERNATIVE WASTEWATER COLLECTION SYSTEMS*, US EPA Technology Transfer Manual)

Important parts of a pressure collection system include manual or automatic air release assemblies (Figures 2.15 and 2.16) and pigging ports (Figure 2.17). Air release assemblies are located in the high points of pressure collection systems to release or remove pockets of air (bubbles) that collect and reduce the flow capacity of the sewer. Pigging ports are installations where a hard, bullet-shaped, polyurethane foam device (pig) can be inserted into the sewer to clean the inside of the pipe as the pig is forced through the pipe under pressure.

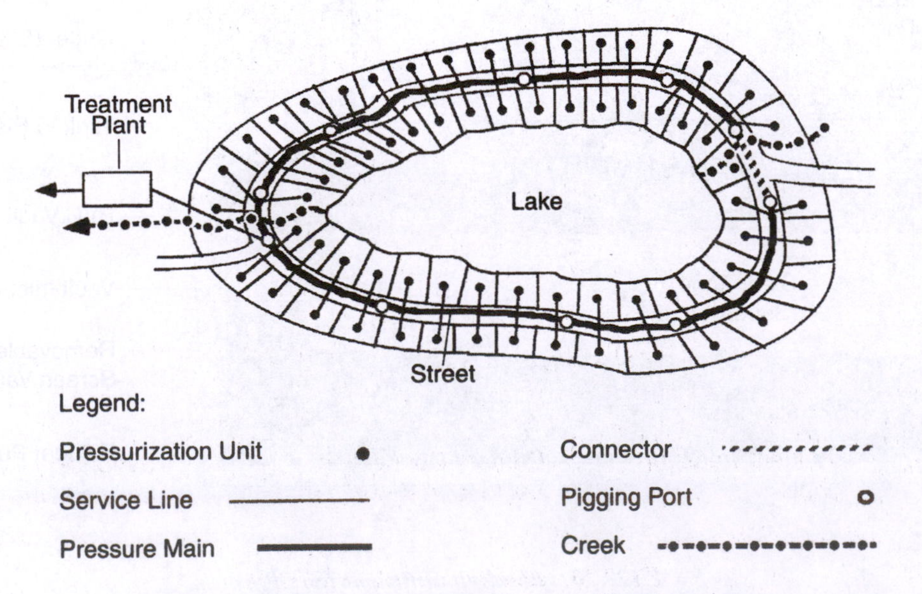


*Fig. 2.14 Schematic of a typical pressure grinder collection system*

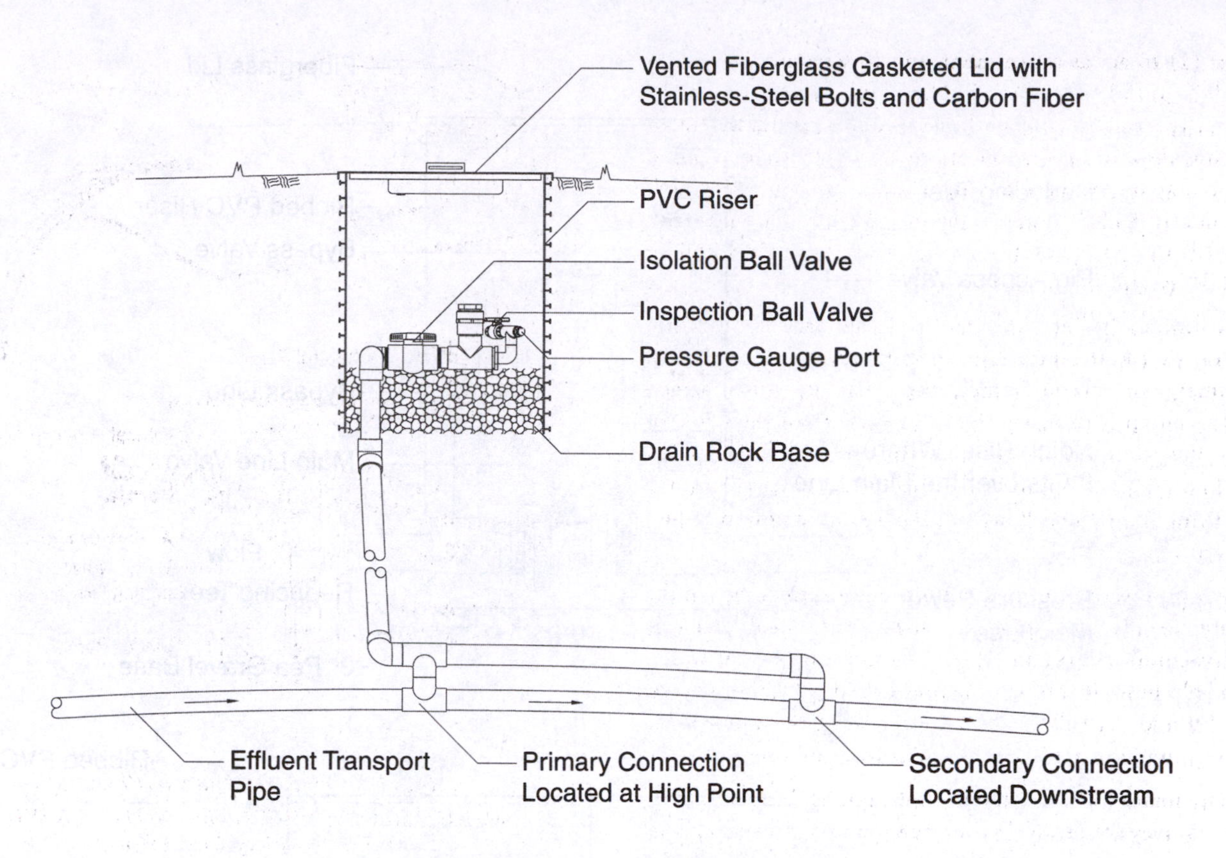


*Fig. 2.15 Manual air release assembly*

(Courtesy of Orenco Systems, Inc., www.orenco.com)

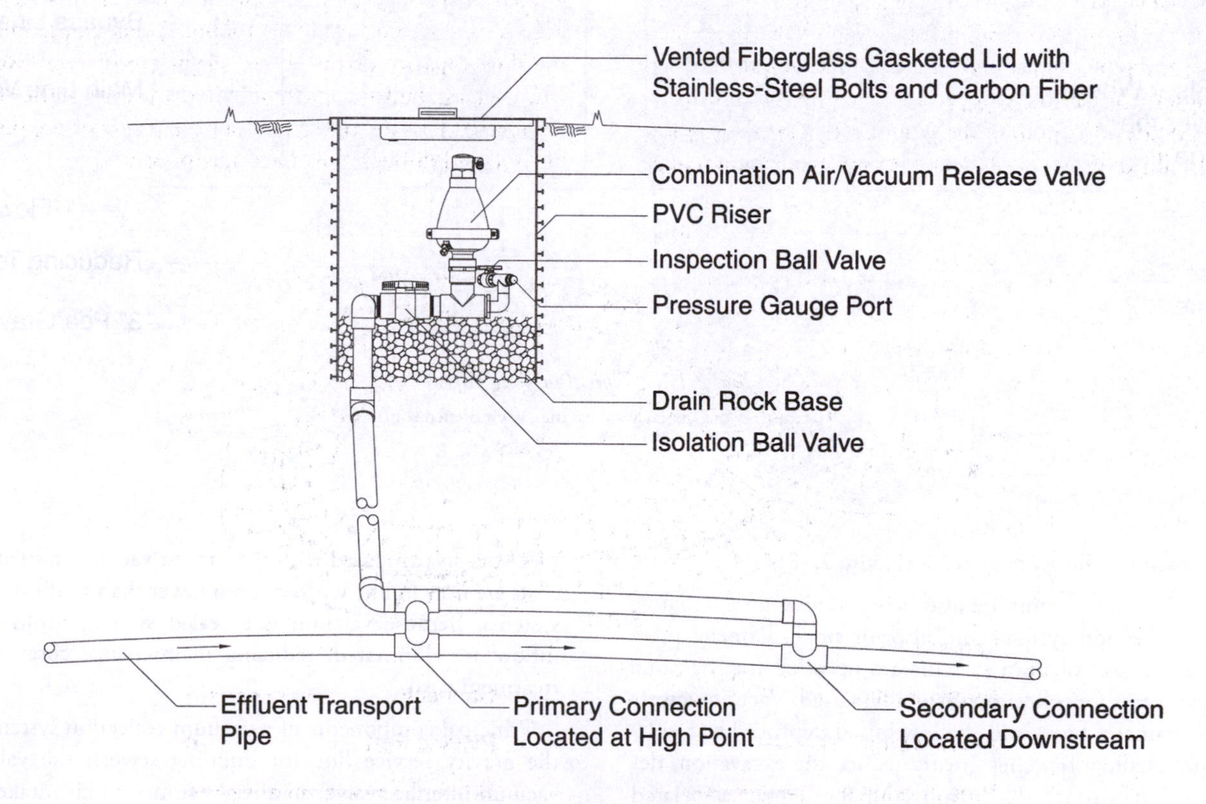


*Fig. 2.16 Automatic air release assembly*

(Courtesy of Orenco Systems, Inc., www.orenco.com)

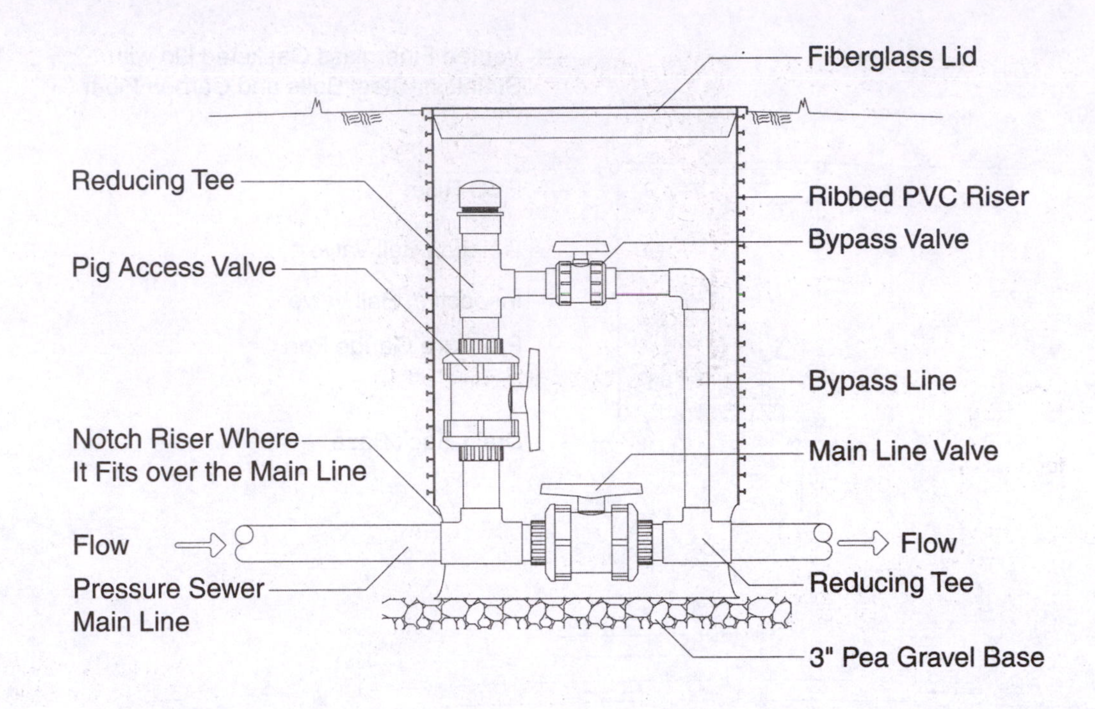


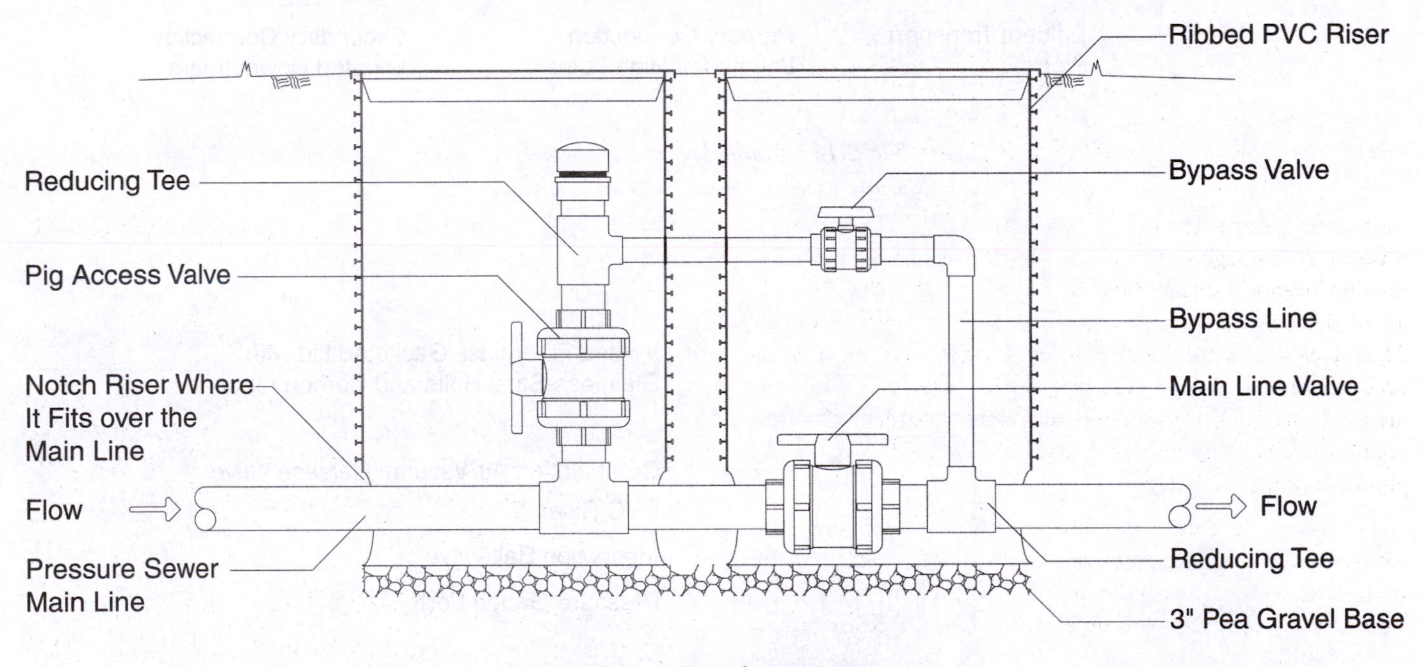


*Fig. 2.17 Typical pigging ports*
(Courtesy of Orenco Systems, Inc., www.orenco.com)

## 2.42 Vacuum Collection Systems (Figure 2.18)

Vacuum collection systems are also being used as an alternative to gravity collection systems and also for small-diameter pressure sewers. These vacuum systems are installed for the same reasons that pressure collection systems are used. Vacuum sewers use small-diameter pipes and shallow burial depths. The resulting narrow, shallow trenches greatly reduce the excavation, dewatering effort, surface disruption, and the danger associated with larger, deeper trenches. The continuing improvement in vacuum technology has resulted in significantly decreased O&M costs compared with the earliest vacuum systems. O&M costs are now in line with, or even lower than traditional gravity systems. Because vacuum is a sealed system, infiltration and inflow are eliminated, reducing maintenance costs as well as treatment costs.

Principal components of a vacuum collection system include the gravity service line (or building sewer); the valve pit; a vacuum interface valve; an airtight sump; an air intake; the service lateral; the vacuum service main; and a central vacuum station, which includes vacuum pumps, a collection tank, and

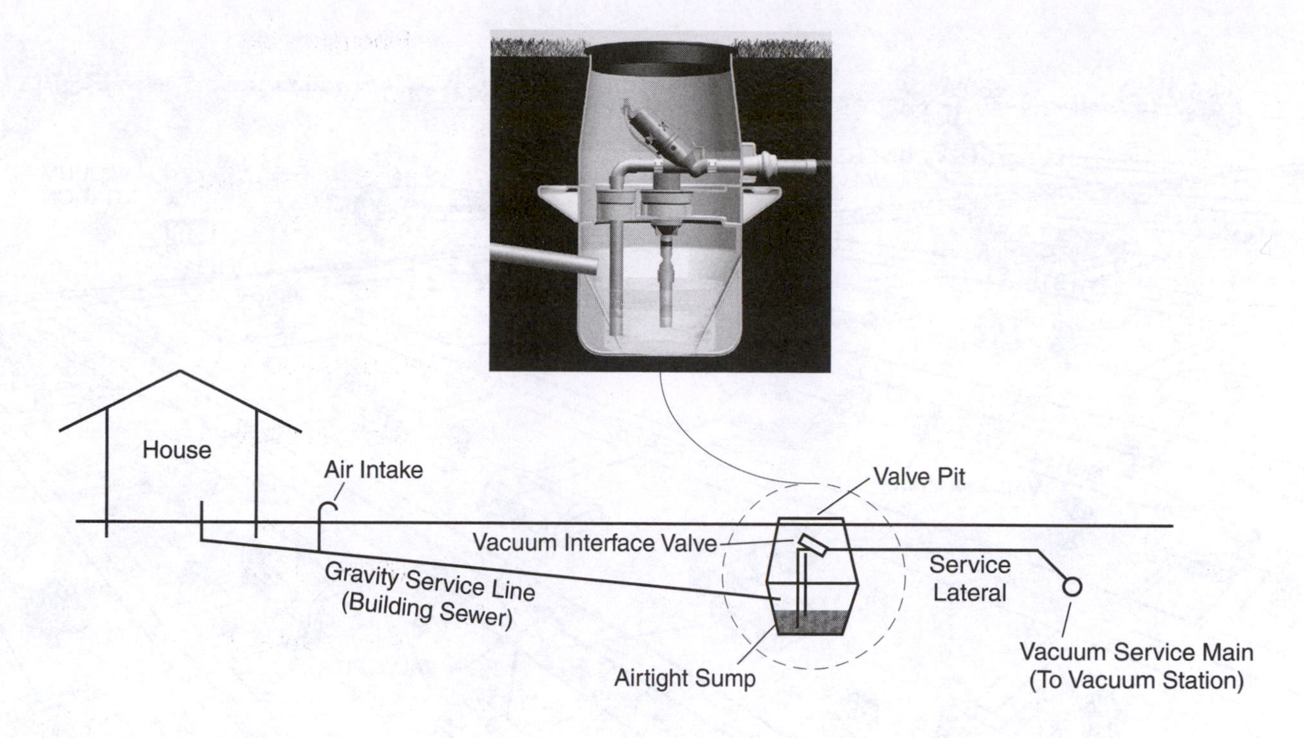


*Fig. 2.18 Principal components of a typical vacuum collection system*
(Courtesy of AIRVAC)

wastewater pumps. The gravity service line connects a building's wastewater drainage system to the valve pit. Vacuum created by vacuum pumps located at the central vacuum station is transferred through the vacuum service mains and to the valve pit. The valve pit is where the interface between gravity and vacuum occurs. Housed in the top chamber of the valve pit is a vacuum interface valve. This valve is normally closed in order to seal the vacuum mains. This ensures that vacuum is maintained in the piping network at all times.

The lower chamber of the valve pit is an airtight sump that receives the wastewater from the house. When 10 gallons of wastewater accumulates in the sump, the vacuum interface valve automatically opens. This is done without any electrical power being required. The valve opens and in 3 to 4 seconds, the contents of the sump are evacuated. The valve stays open for another 2 to 3 seconds to allow for atmospheric air to enter the system. This air comes from the air intake located by the house and forces the wastewater into a service lateral to the vacuum service main, where it is conveyed to the central vacuum station.

Vacuum service mains act as arteries of the vacuum collection system and convey the wastewater to the central vacuum station and then to a treatment plant. Because the wastewater is sucked by vacuum, the vacuum service mains do not have to be laid at a slope to produce a gravity pull on the wastewater; the mains are sloped with steps that can follow the natural slope of their routes.

Vacuum pumps are installed at the central vacuum station, usually next to a treatment plant, to maintain the appropriate vacuum in the main collection system. Wastewater enters the collection tank and when the tank fills to a predetermined level, wastewater pumps transfer the contents to the treatment plant by way of a force main.

Vacuum stations provide a clean, safe environment for maintenance personnel as all wastewater is completely contained within the collection tank. Figure 2.19 illustrates a typical street portion of a vacuum collection system and an overview of the entire collection system. A complete description of the operation and maintenance of vacuum collection systems is presented in Chapter 6, "Collection Systems."

## QUESTIONS

Please write your answers to the following questions and compare them with those on page 62.

2.4C How is wastewater conveyed where the topography is unfavorable for a gravity collection system?

2.4D List the principal components of a pressure collection system.

2.4E List the principal components of a vacuum collection system.

**END OF LESSON 1 OF 2 LESSONS**

**on**

**SMALL COLLECTION, TREATMENT, AND DISCHARGE SYSTEMS**

Please answer the discussion and review questions next.

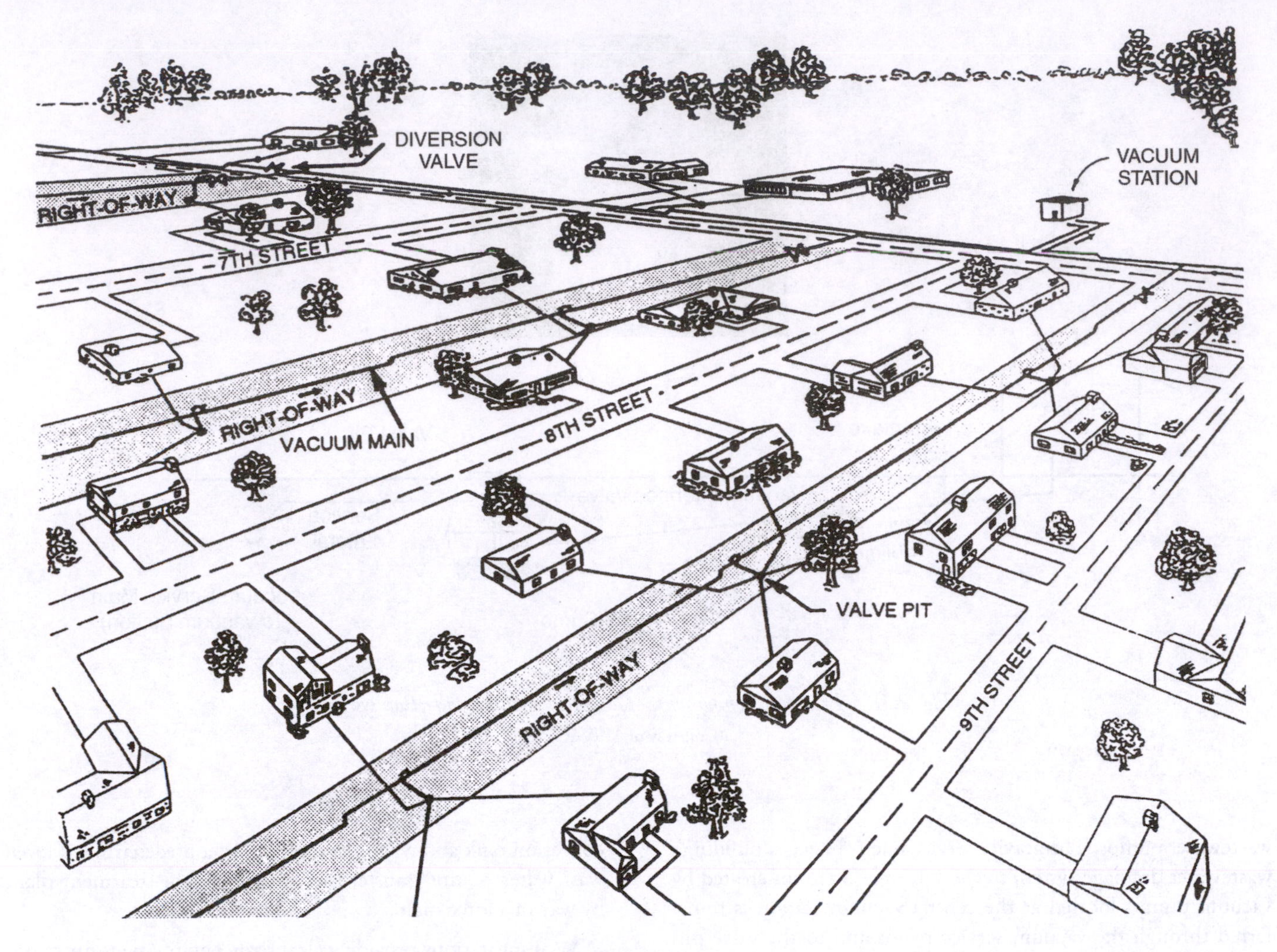


*Fig. 2.19 Typical layout of a vacuum collection system*
(Permission of AIRVAC)

## DISCUSSION AND REVIEW QUESTIONS

### Chapter 2. SMALL COLLECTION, TREATMENT, AND DISCHARGE SYSTEMS

(Lesson 1 of 2 Lessons)

At the end of each lesson in this chapter, you will find discussion and review questions. Please write your answers to these questions to determine how well you understand the material in the lesson.

1. Why should an operator have an understanding of natural cycles?
2. What can happen when nutrient cycles are disrupted and there is no dissolved oxygen in the receiving water?
3. How is the concentration of total solids in a sample of wastewater measured?
4. When treated wastewater from a plant is discharged into receiving waters, what factors influence whether the natural cycles in the aquatic environment become upset? (List as many factors as you can recall.)
5. What problems could be caused in receiving waters by discharging wastewater containing too much nitrogen?
6. What is the purpose of disinfecting wastewater treatment plant effluent?
7. What is the purpose of lift stations or pumping plants in a gravity collection system?
8. Why are vacuum collection systems more difficult to operate and maintain than pressure systems?
9. What are the advantages and limitations of a pressure collection system as compared with a gravity collection system?
10. When can a small-diameter gravity effluent drain from a septic tank be used?

# CHAPTER 2. SMALL COLLECTION, TREATMENT, AND DISCHARGE SYSTEMS

(Lesson 2 of 2 Lessons)

## 2.5 TREATMENT ALTERNATIVES

After wastewater has been conveyed by a gravity wastewater collection system, a pressure system, or a vacuum system, the wastewater reaches a treatment plant. Conventional wastewater treatment usually consists of the following stages:

1. Preliminary treatment or pretreatment—removal of large solids
2. Primary treatment—removal of settleable and floatable solids
3. Secondary treatment—removal of suspended and dissolved solids using biological processes
4. Disinfection—killing or inactivating disease-causing microorganisms

The objective of wastewater treatment is to keep the physical, chemical, and biological processes in balance and in top performance condition.

### 2.50 Physical Facilities

When a new wastewater treatment plant is built by a medium to large city or municipal agency, the facilities are usually designed by engineers and constructed at the site where the plant will operate. The tanks and equipment required to treat large volumes of wastewater are themselves very large; therefore, on-site construction is usually the most cost-effective method of building a treatment plant. For smaller communities, however, purchasing a small package treatment plant may be a more cost-effective solution to treating wastewater.

Package wastewater treatment plants are small treatment plants, which are often fabricated at the manufacturer's factory, transported to the site, and installed as one facility. The package operates as a small conventional treatment plant providing preliminary treatment, primary and secondary treatment, and disinfection. Small package plants are available in several configurations. They operate in the same manner as larger plants of the same types and they achieve comparable levels of waste removal, but package plants are specifically designed to handle smaller flows. The five key waste removal processes described in Section 2.52 are readily available as package plants. These include waste treatment ponds, rotating biological contactors (RBCs), activated sludge, oxidation ditches, and recirculating gravel/sand filters.

### 2.51 Typical Flow Pattern

Upon reaching a wastewater treatment facility, the wastewater flows through a series of treatment processes (Figure 2.20), which remove the wastes from the water and reduce its threat to the public health and the environment before it is discharged from the facility. The number of treatment processes and the degree of treatment usually depend on discharge requirements, which depend on the uses of the receiving waters. Treated wastewaters discharged into a small stream used for a domestic water supply and swimming will require considerably more treatment than wastewater discharged into receiving waters used only for navigation.

Although not all treatment plants are alike, there are certain typical flow patterns that are similar from one plant to another. In the first phase of treatment, wastewater usually flows through a series of pretreatment or preliminary treatment processes: screening, shredding, and grit removal. These processes remove the coarse material from the wastewater. Flow measuring devices are installed after pretreatment processes to record the flow rates and volumes of wastewater treated by the plant. Pre-aeration adds air to freshen the wastewater and to help remove oils and greases.

Next, the wastewater will generally receive primary treatment. During primary treatment, some of the solid matter carried by the wastewater will settle out or float to the water surface where it can be separated from the wastewater being treated.

Secondary treatment processes follow primary treatment and commonly consist of biological processes. This means that organisms living in the controlled environment of the process are used to partially stabilize (oxidize) organic matter not removed by previous treatment processes. In stabilizing the organic material, the organisms convert it into a form that is easier to remove from the wastewater.

Primary and secondary treatment processes remove large quantities of solid material from the wastewater. All of this

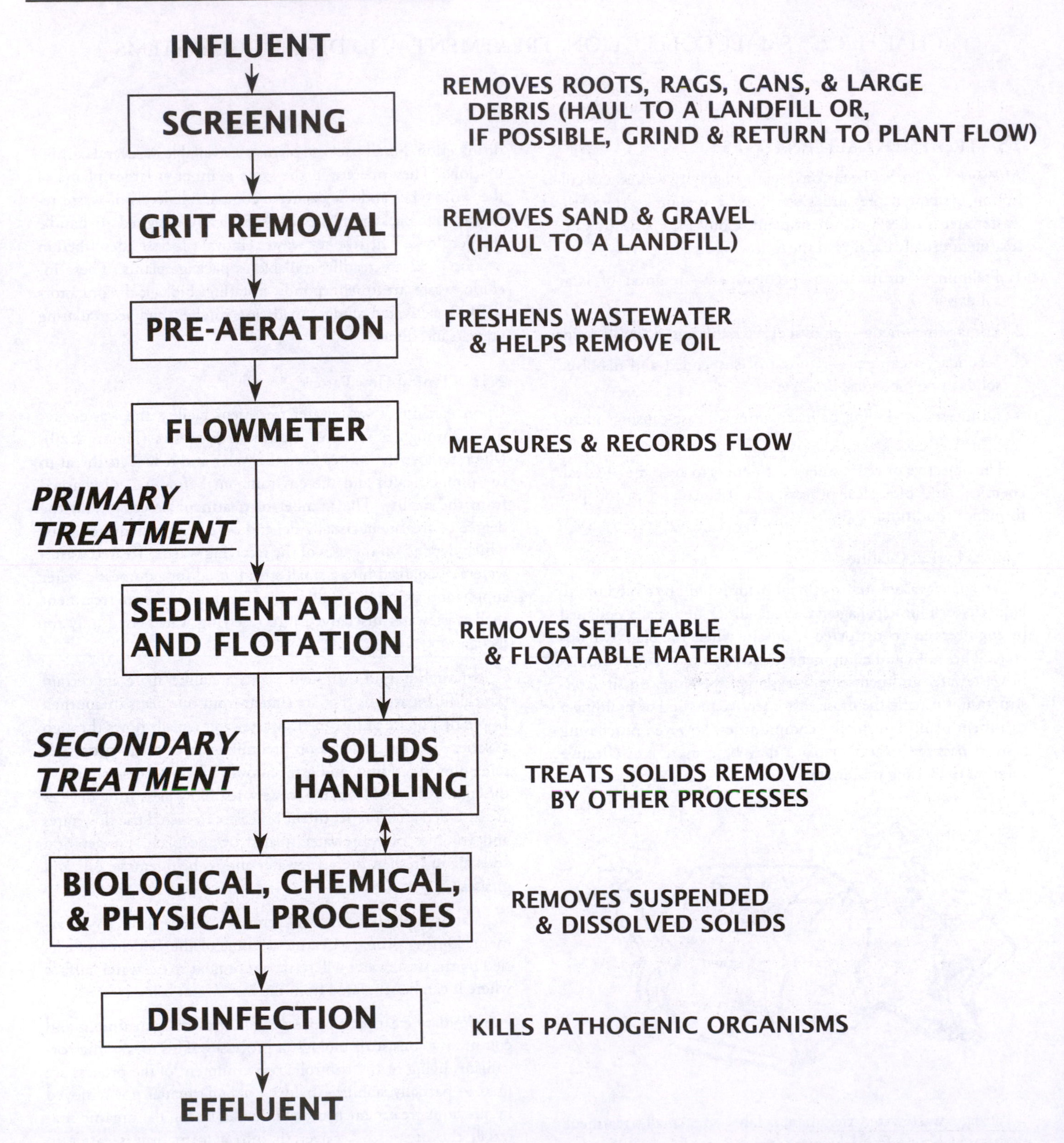


*Fig. 2.20 Flow diagram of wastewater treatment plant processes*

sludge and scum is routed to the solids handling facilities for additional treatment or ultimate disposal.

Waste treatment ponds may be used to treat wastes remaining in wastewater after preliminary treatment, primary treatment, or secondary treatment. Ponds are frequently constructed in rural areas where there is sufficient available land.

Before treated wastewater can be discharged to the receiving waters or routed to other (tertiary) treatment processes, it should be disinfected to prevent the spread of disease. Chlorine is commonly added for disinfection purposes. Depending on the intended use of the treated wastewater or the next step in treatment, it also may be necessary to neutralize the chlorine and thus detoxify the effluent. This can be accomplished by adding sulfur dioxide ($SO_2$) before discharging the effluent from the treatment plant.

## QUESTIONS

Please write your answers to the following questions and compare them with those on page 62.

2.5A What is a package treatment plant?

2.5B What is the purpose of a wastewater treatment facility?

### 2.52 Key Waste Removal Processes

#### *2.520 Waste Treatment Ponds*

The waste treatment pond (Figure 2.21), or stabilization pond, is a special method of biological treatment that is especially suitable for small towns.

The types of treatment processes and the location of ponds are determined by the design engineer on the basis of economics and the degree of treatment required to meet the water quality standards of the receiving waters. In some treatment plants, wastewater being treated may flow through a coarse screen and flowmeter before it flows through a series of ponds. In other types of plants, ponds are located after primary treatment or after *TRICKLING FILTERS.*[38] When used with pressure collection systems, the septic tank (interceptor) provides the pretreatment of solids and scum removal. Figure 2.22 illustrates the typical flow pattern through a pond treatment plant, and Chapter 9 in Volume II describes the process in detail.

When wastewater is discharged to a typical pond, the settleable solids fall to the bottom just as they do in a primary clarifier. The solids begin to decompose and aerobic bacteria soon use up all the dissolved oxygen in the nearby water. A population of anaerobic bacteria then continues the decomposition, much the same as in a septic tank or an anaerobic digester. As the organic matter is destroyed, methane and carbon dioxide are released. When the carbon dioxide rises to the surface of the pond, some of it is used by algae, which convert it to oxygen by the process of *PHOTOSYNTHESIS.*[39] This is the same process used by all green plants. Aerobic bacteria, algae, and other microorganisms feed on the dissolved solids in the upper layer of the pond much the same way they do in a trickling filter or aeration tank. During daylight hours, algae produce oxygen for the other organisms to use.

Some shallow ponds (3 to 6 feet deep) contain dissolved oxygen throughout their entire depth. These ponds are called

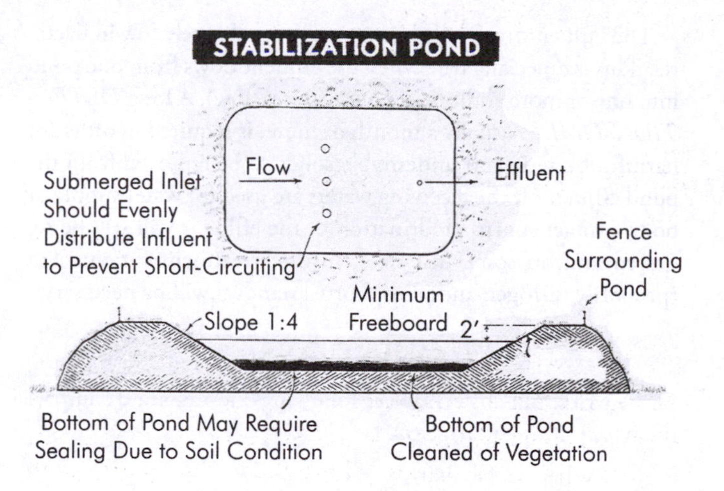


(Courtesy Water Environment Federation)

(Courtesy Water and Sewage Works Magazine)

*Fig. 2.21 Ponds*

38. *Trickling Filter.* A treatment process in which wastewater trickling over media enables the formation of slimes or biomass, which contain organisms that feed upon and remove wastes from the water being treated.

39. *Photosynthesis* (foe-toe-SIN-thuh-sis). A process in which organisms, with the aid of chlorophyll, convert carbon dioxide and inorganic substances into oxygen and additional plant material, using sunlight for energy. All green plants grow by this process.

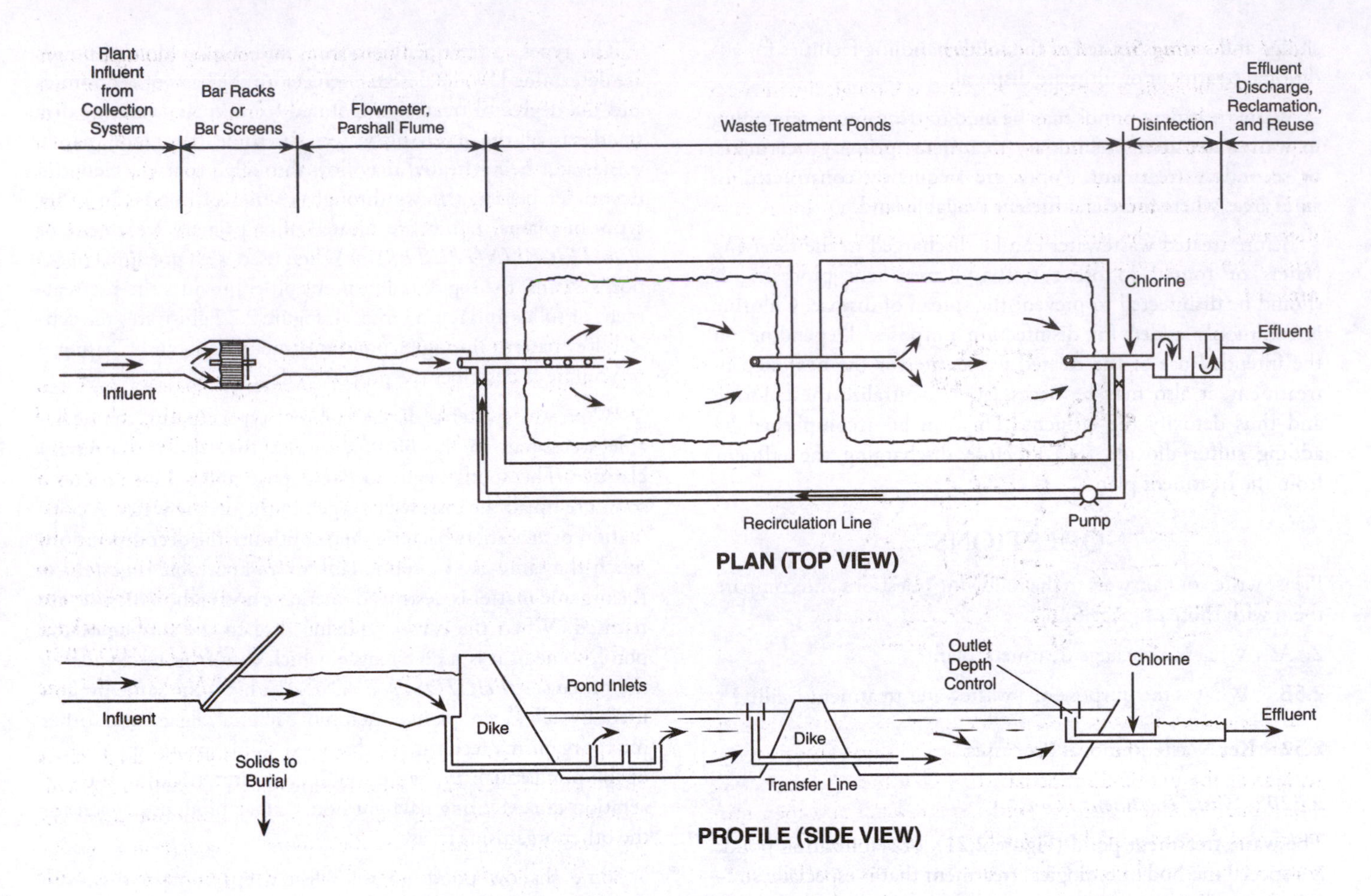


*Fig. 2.22 Possible flow pattern through a pond treatment plant*

aerobic ponds. They usually have a mechanical device (surface aerator) adding oxygen as well as their oxygen supply from algae. In another type of aerated pond, oxygen is delivered by a diffused air system similar to the system used in activated sludge plants.

Deep (8 to 12 feet), heavily loaded ponds may be without oxygen throughout their depth. These ponds are called anaerobic ponds. At times these ponds can be quite odorous and for this reason they are used only in sparsely populated areas.

Ponds that contain an aerobic top layer and an anaerobic bottom layer are called *FACULTATIVE PONDS.*[40] These are the ponds normally seen in most areas. If they are properly designed and operated, they are virtually odor-free and produce a well-oxidized (low BOD) effluent.

Occasionally, ponds are used after a primary treatment unit. In this case, they are usually called "oxidation ponds." When they are used to treat raw wastewater, they are called "raw wastewater lagoons" or "waste stabilization ponds."

The effluent from ponds is normally moderately low in bacteria. This is especially true when the effluent flows from one pond into one or more additional ponds (series flow). A long *DETENTION TIME,*[41] usually a month or more, is required in order for harmful bacteria and undesirable solids to be removed from the pond effluent. If the receiving waters are used for water supply or body contact sports, chlorination of the effluent will still be required and, in some instances, further treatment for nutrient (primarily nitrogen and phosphorus) removal will be necessary.

## QUESTIONS

Please write your answers to the following questions and compare them with those on page 62.

2.5C What is a facultative pond?

2.5D How are facultative ponds similar to the following:

1. a clarifier?
2. a digester?
3. an aeration tank?

40. *Facultative* (FACK-ul-tay-tive) *Pond.* The most common type of pond in current use. The upper portion (supernatant) is aerobic, while the bottom layer is anaerobic. Algae supply most of the oxygen to the supernatant.
41. *Detention Time.* The time required to fill a tank at a given flow or the theoretical time required for a given flow of wastewater to pass through a tank. In septic tanks, this detention time will decrease as the volumes of sludge and scum increase.

### 2.521 *Rotating Biological Contactors*

A rotating biological contactor (RBC) is a biological treatment device in which a population of microorganisms is alternately exposed to wastewater and then air (oxygen) to promote growth of the organisms as they oxidize the waste solids. As illustrated in Figure 2.23, wastewater usually flows through a primary clarifier and then to the RBC unit. However, when treating septic tank effluent wastewater, the primary clarifier is sometimes omitted.

Biological contactors have a rotating shaft surrounded by plastic discs called the "media." A biological slime grows on the surface of the plastic-disc media when conditions are suitable. The slime is rotated into the settled wastewater and then into the atmosphere to provide oxygen for the organisms. The wastewater being treated usually flows parallel to the rotating shaft, but may flow perpendicular to the shaft as it flows from stage to stage or tank to tank. Effluent from the rotating biological contactors flows through secondary clarifiers for removal of suspended solids and dead slime growths. Additional treatment to disinfect or further polish the effluent may be required before it can be discharged from the treatment plant into the environment. A complete description of rotating biological contactors, including their operation and maintenance, is presented in *SMALL WASTEWATER SYSTEM OPERATION AND MAINTENANCE*, Volume II, Chapter 11.

### 2.522 *Activated Sludge Process* [42]

The activated sludge process is another biological treatment process that serves the same functions as a rotating biological contactor. Aerobic bacteria oxidize and thus stabilize the organic wastes in wastewater as they feed on the wastes. This process is usually classified as secondary treatment. There are many variations of the activated sludge process but they all involve the same basic principles. When *AERATION TANKS* [43] are used to grow and maintain an aerobic biological culture, the treatment plant is called an activated sludge plant. Operation of a package activated sludge plant is described fully in *SMALL WASTEWATER SYSTEM OPERATION AND MAINTENANCE,* Volume II, Chapter 10.

The activated sludge process is widely used by large cities and communities where land is expensive and where large volumes of wastewater must be highly treated, economically, without creating a nuisance to neighbors. The activated sludge plant is a common biological treatment process being built today for larger installations (Figure 2.24) or small package

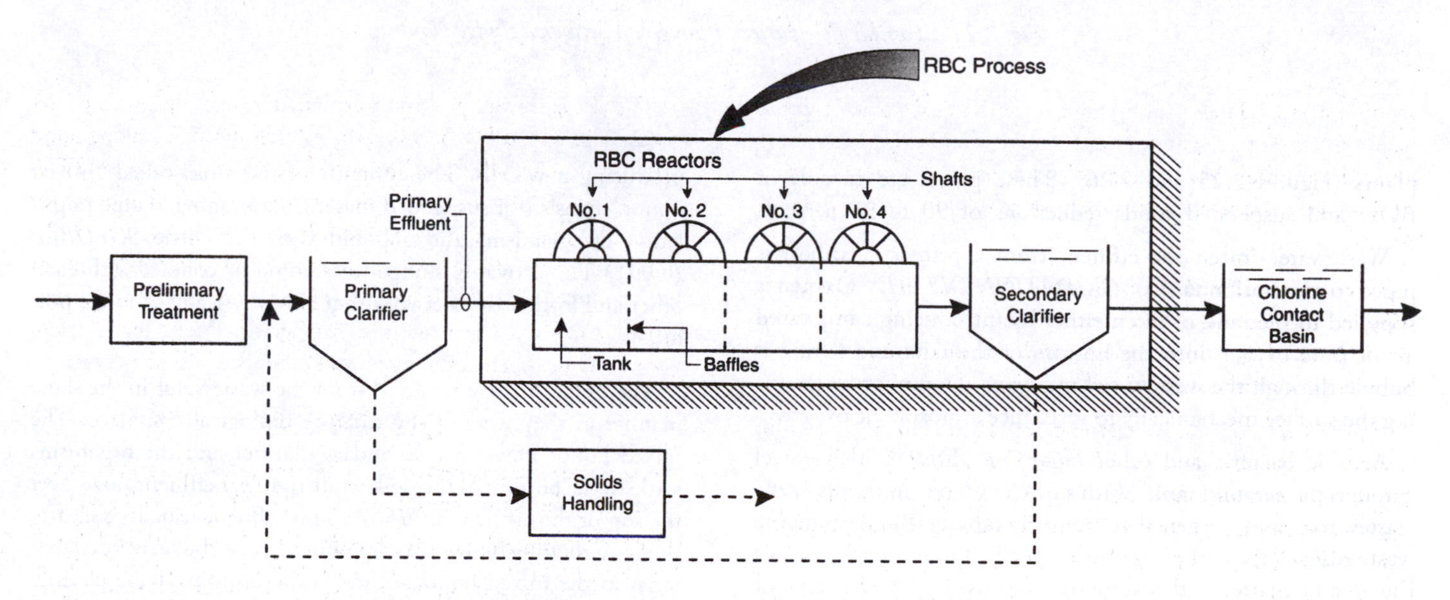


*Fig. 2.23 Typical rotating biological contactor (reactor) treatment plant*

42. *Activated Sludge Process.* A biological wastewater treatment process that speeds up the decomposition of wastes in the wastewater being treated. Activated sludge is added to wastewater and the mixture (mixed liquor) is aerated and agitated. After some time in the aeration tank, the activated sludge is allowed to settle out by sedimentation and is disposed of (wasted) or reused (returned to the aeration tank) as needed. The remaining wastewater then undergoes more treatment.

43. *Aeration* (air-A-shun) *Tank.* The tank where raw or settled wastewater is mixed with return sludge and aerated. The same as aeration bay, aerator, or reactor.

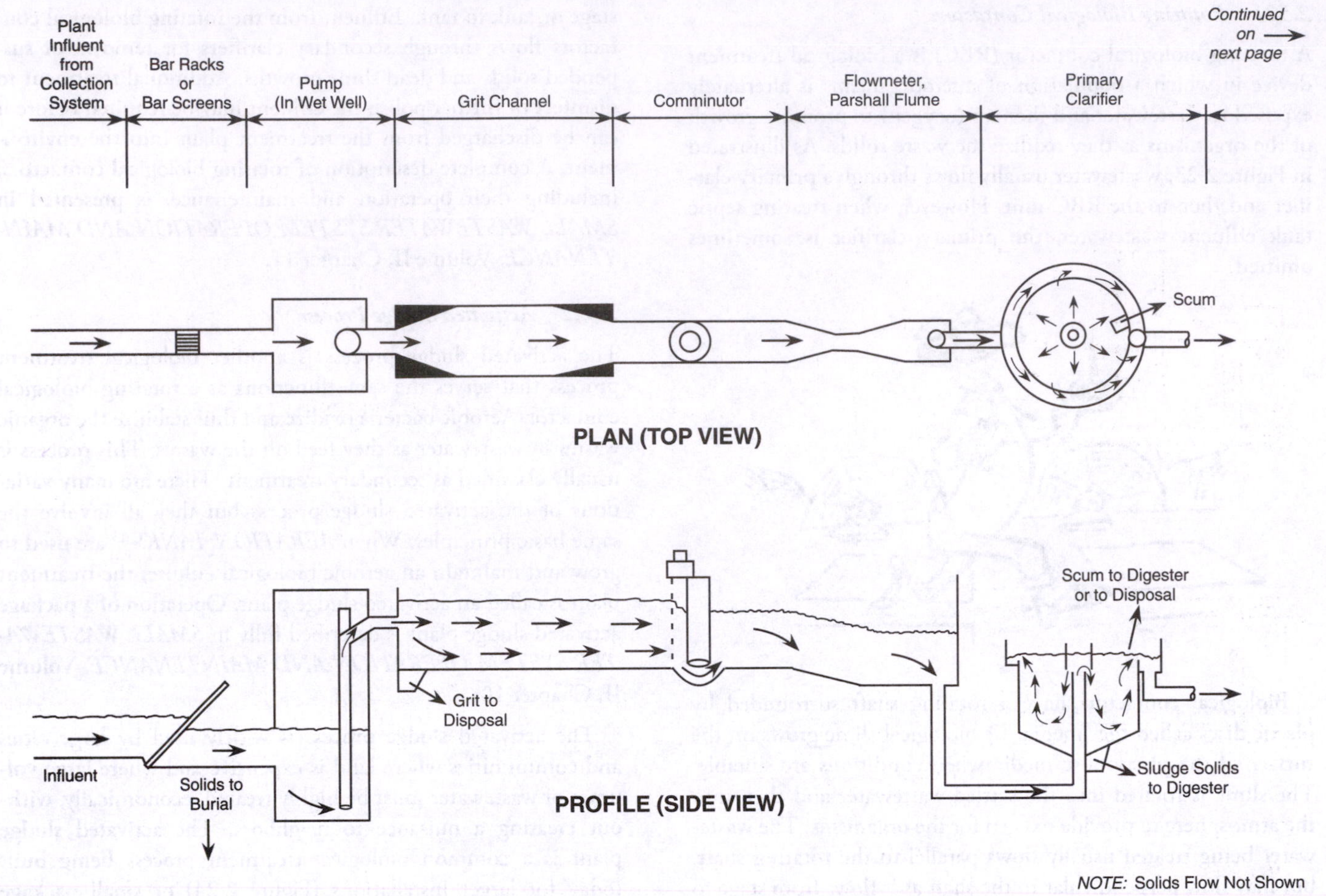

*Fig. 2.24 Possible flow pattern through an activated sludge plant*

plants (Figures 2.25 and 2.26). These plants are capable of BOD and suspended solids reductions of 90 to 98 percent.

Wastewater (often the effluent from a primary clarifier) is piped to an aeration tank or *OXIDATION DITCH.*[44] Oxygen is supplied to the tank or ditch either by introducing compressed air or pure oxygen into the bottom of the tank and letting it bubble through the wastewater and up to the top, or by churning the surface mechanically to introduce atmospheric oxygen.

Aerobic bacteria and other organisms thrive as they travel through the aeration tank. With sufficient food (nutrients in the wastewater) and oxygen they multiply rapidly. By the time the waste reaches the end of the tank (usually 4 to 8 hours), most of the organic matter in the waste has been used by the bacteria for producing new cells. The contents of the tank, called "mixed liquor," consists of suspended material containing a large population of organisms and a liquid with very little *SOLUBLE BOD.*[45] The activated sludge flocs come in contact with each other and form a lacy network that also physically captures pollutants.

The organisms are removed from the wastewater in the same manner as they were in the rotating biological contactor. The mixed liquor flows to a secondary clarifier and the organisms settle to the bottom of the tank while the clear effluent flows over the top of the effluent *WEIRS.*[46] This effluent usually is clearer than in a rotating biological contactor because the suspended material in the mixed liquor settles to the bottom of the clarifier

44. *Oxidation Ditch.* The oxidation ditch is a modified form of the activated sludge process. The ditch consists of two channels placed side by side and connected at the ends to produce one continuous loop of wastewater flow and a brush rotator assembly placed across the channel to provide aeration and circulation.
45. *Soluble Bod.* Soluble BOD is the BOD of water that has been filtered in the standard suspended solids test. The soluble BOD is a measure of food for microorganisms that is dissolved in the water being treated.
46. *Weir* (WEER). (1) A wall or plate placed in an open channel and used to measure the flow of water. The depth of the flow over the weir can be used to calculate the flow rate, or a chart or conversion table may be used to convert depth to flow. Also see PROPORTIONAL WEIR. (2) A wall or obstruction used to control flow (from settling tanks and clarifiers) to ensure a uniform flow rate and avoid short-circuiting.

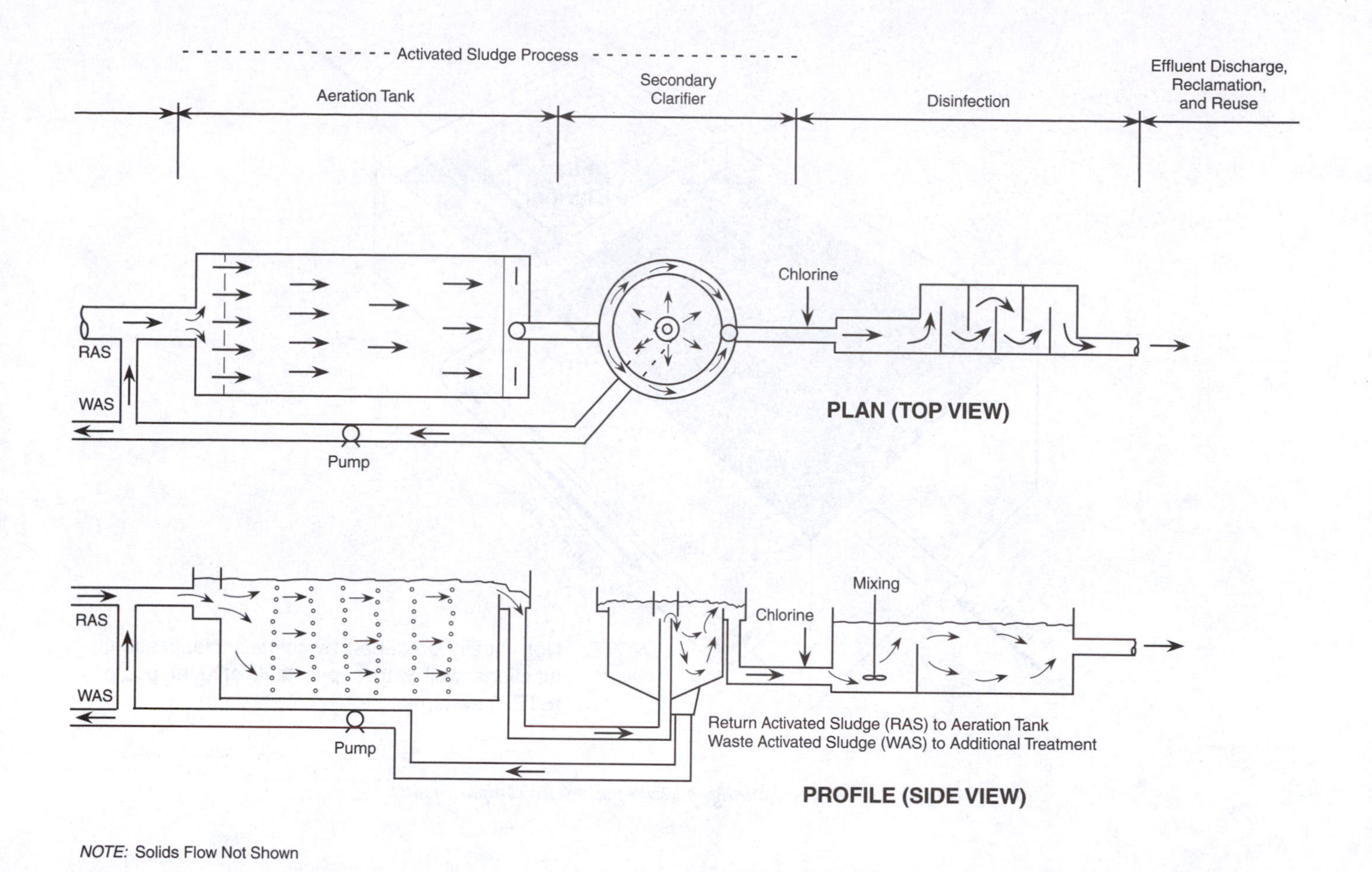


*Fig. 2.24 Possible flow pattern through an activated sludge plant (continued)*

more readily than the material from a rotating biological contactor. The settled organisms are known as "activated sludge." They are extremely valuable to the treatment process. If they are removed quickly from the secondary clarifier, they will be in good condition and hungry for more food (organic wastes). A portion of them are therefore pumped back (recirculated) to the influent end of the aeration tank where they are mixed with the incoming wastewater. Here they begin to feed again on the organic material in the waste, decomposing it and creating new organisms.

Left uncontrolled, the number of organisms would eventually be too great; they would exhaust the food and oxygen supplies and die off. Therefore some organisms must be removed, or "wasted," from the process periodically. This is accomplished by pumping a small amount of the waste activated sludge to the primary clarifier or directly to the sludge handling facilities, which could be an aerobic or anaerobic digester. If the organisms are pumped to the clarifier, they settle along with the raw sludge and then are removed to the sludge handling facilities.

As with the effluents from other biological treatment processes, effluent from the activated sludge process may require additional treatment or disinfection before discharge to the environment.

*Hungry organisms ready for more food*

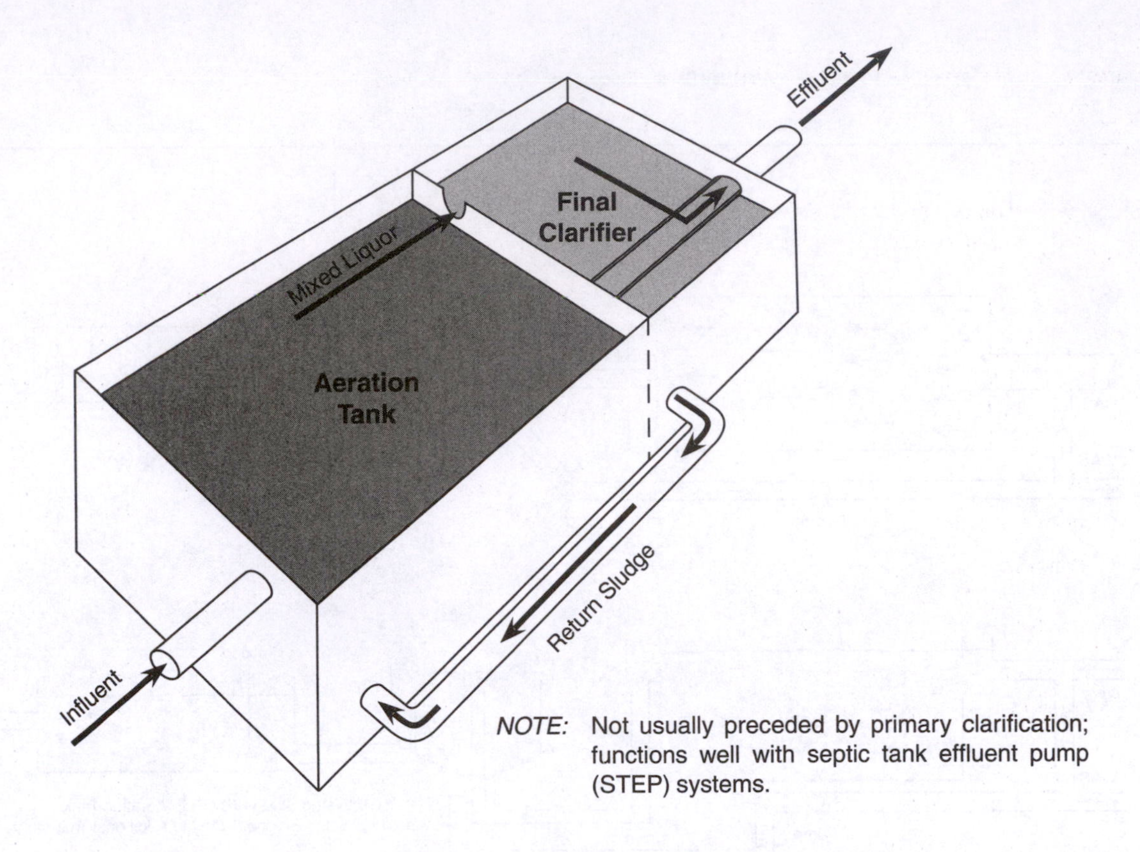


*Fig. 2.25 Activated sludge package plant (two compartments)*

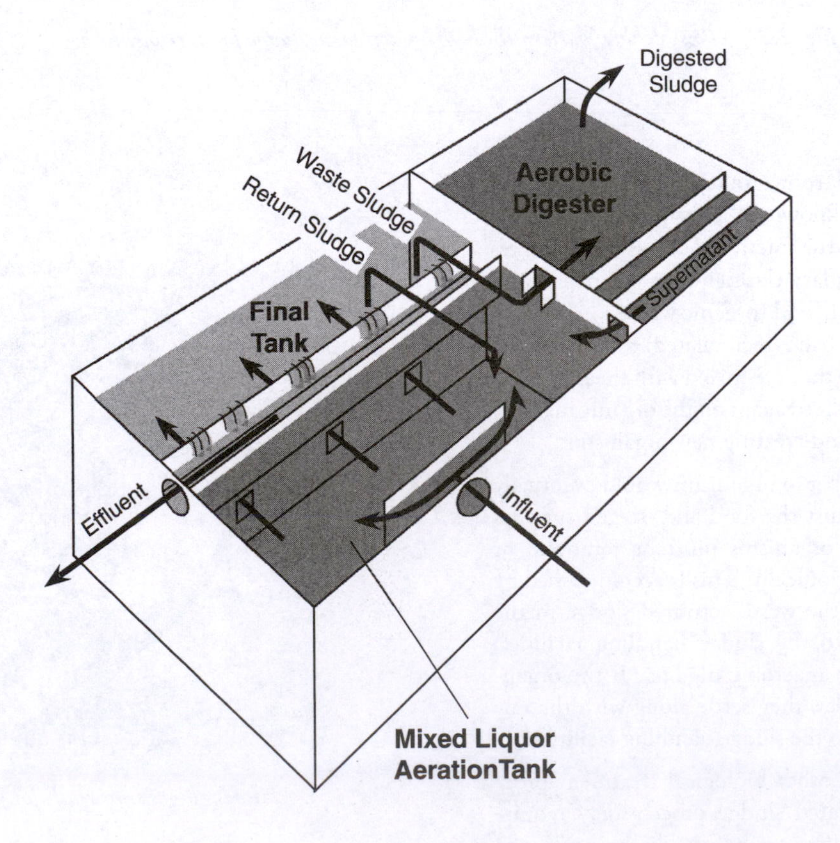


*Fig. 2.26 Activated sludge package plant (three compartments)*

## QUESTIONS

Please write your answers to the following questions and compare them with those on page 62.

2.5E What is a rotating biological contactor?

2.5F What types of wastes are treated (removed) by the activated sludge process?

### *2.523 Oxidation Ditches*

The oxidation ditch (Figure 2.27) is a modified form of the activated sludge process. The main parts of an oxidation ditch treatment unit are the aeration basin (which generally consists of two channels placed side by side and connected at the ends to produce one continuous loop of wastewater flow), a brush rotor assembly, settling tank, return sludge pump, and excess sludge handling facilities. Usually, there is no primary settling tank or grit removal system; inorganic solids such as sand, silt, and cinders are captured in the oxidation ditch and removed during sludge wasting or cleaning operations.

When an oxidation ditch is used with either a gravity or a vacuum collection system, the raw wastewater passes directly through a bar rack or bar screen to the ditch. The bar rack is necessary for the protection of the mechanical equipment such as rotors and pumps. Devices for shredding large solids may be installed after the bar rack or instead of a bar rack. These devices are called comminutors (kom-mih-NEW-ters) or barminutors (bar-mih-NEW-ters).

The oxidation ditch forms the aeration basin; here the raw wastewater is mixed with previously grown active organisms. The rotor is a brush-like aeration device that *ENTRAINS*[47] the necessary oxygen into the liquid for microbial life and keeps the contents of the ditch mixed and moving to prevent settling of solids. The ends of the ditch are well rounded to prevent eddying and dead areas and the outside edges of the curves are given erosion protection.

Oxidation ditches use the extended aeration modification of the activated sludge process. Extended aeration is similar to conventional activated sludge except that the microorganisms are held in the aeration tank (oxidation ditch) longer and do not get as much food. The microorganisms get less food because there are more of them to feed.

The mixed liquor flows from the ditch to a clarifier for separation of the remaining solids from the liquid. The clarified water passes over the effluent weir and is chlorinated. Plant effluent is discharged to either a receiving stream, polishing treatment units, percolation ditches, or a subsurface disposal or leaching system. The settled sludge is removed from the bottom of the clarifier by a pump. The majority of the sludge is returned to the ditch and a small portion is disposed of (wasted). Scum that floats to the surface of the secondary clarifier is removed and either returned to the oxidation ditch for further treatment or disposed of by digestion or burial.

Because the oxidation ditch is operated as a closed system, the amount of *VOLATILE*[48] suspended solids will gradually increase

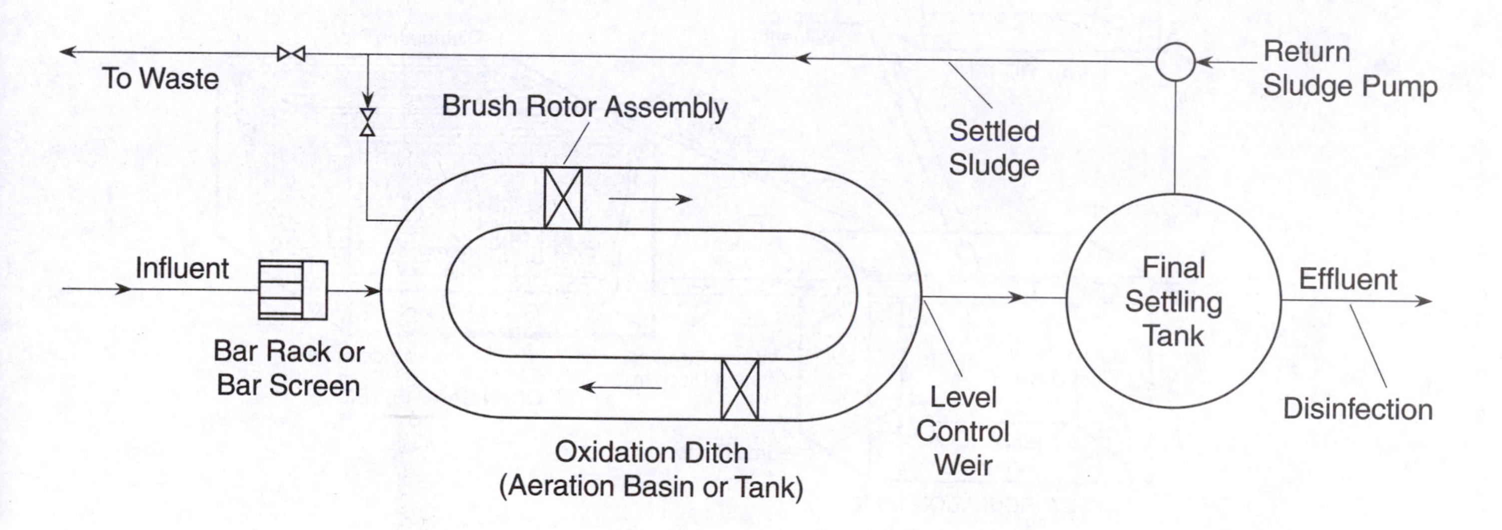


*Fig. 2.27 Oxidation ditch plant*

(Source: "Oxidation Ditch," prepared by William L. Berk for the New England Regional Wastewater Institute, South Portland, Maine 04106, August 1970)

47. *Entrain.* To trap bubbles in water either mechanically through turbulence or chemically through a reaction.
48. *Volatile* (VOL-uh-tull). (1) A volatile substance is one that is capable of being evaporated or changed to a vapor at relatively low temperatures. Volatile substances can be partially removed from water or wastewater by the air stripping process. (2) In terms of solids analysis, volatile refers to materials lost (including most organic matter) upon ignition in a muffle furnace for 60 minutes at 550°C (1,022°F). Natural volatile materials are chemical substances usually of animal or plant origin. Manufactured or synthetic volatile materials, such as plastics, ether, acetone, and carbon tetrachloride, are highly volatile and not of plant or animal origin. Also see NONVOLATILE MATTER.

due to the reproduction of the microorganisms treating the wastes. It will periodically become necessary to remove (or waste) some sludge from the process. Wasting of sludge lowers the *MIXED LIQUOR SUSPENDED SOLIDS (MLSS)*[49] concentration in the ditch (or in any activated sludge process) and keeps the microorganisms more active. It is this control of the activated sludge concentration and wasting of excess activated sludge that make high BOD reductions (90 to 98 percent) possible by this process. Excess activated sludge may be dried directly on sludge drying beds or stored in a holding tank or in sludge lagoons for later transfer to larger treatment plants or disposal in approved sanitary landfills.

The basic process design of extended aeration results in simple, easy operation (as described fully in *SMALL WASTEWATER SYSTEM OPERATION AND MAINTENANCE,* Volume II, Chapter 10). A high mixed liquor suspended solids (MLSS) concentration is carried in the aeration tanks or oxidation ditch and therefore the plant may be capable of handling shock and peak loads without upsetting plant operation. There is no white foam problem after solids buildup, as experienced with other types of activated sludge plants, but thick, leathery brown froth may form. Cold weather operation has less effect on oxidation ditch efficiency than on other activated sludge processes because of the large number of microorganisms in the reactors (aeration tanks).

Because of their high treatment efficiency and relatively simple operation, oxidation ditch package treatment plants are very cost-effective for small communities.

### *2.524 Recirculating Gravel/Sand Filters*

Recirculating gravel/sand filters treat the effluent from septic tanks and package treatment plants by both physical and biological processes. A good effluent is achieved by continuously recirculating the effluent through the gravel/sand filter until an acceptable degree of treatment is achieved. As the wastewater passes through the gravel or sand, microorganisms consume a portion of the waste material. Another portion of the suspended solids is simply filtered out in the sand or gravel. After several passes through the filter, effluent is proportionally discharged to a disposal system, commonly a leach field.

Figures 2.28 and 2.29 show the layout of a recirculating gravel/sand filter package plant for a small wastewater treatment system, such as one that would be used by an individual homeowner or a

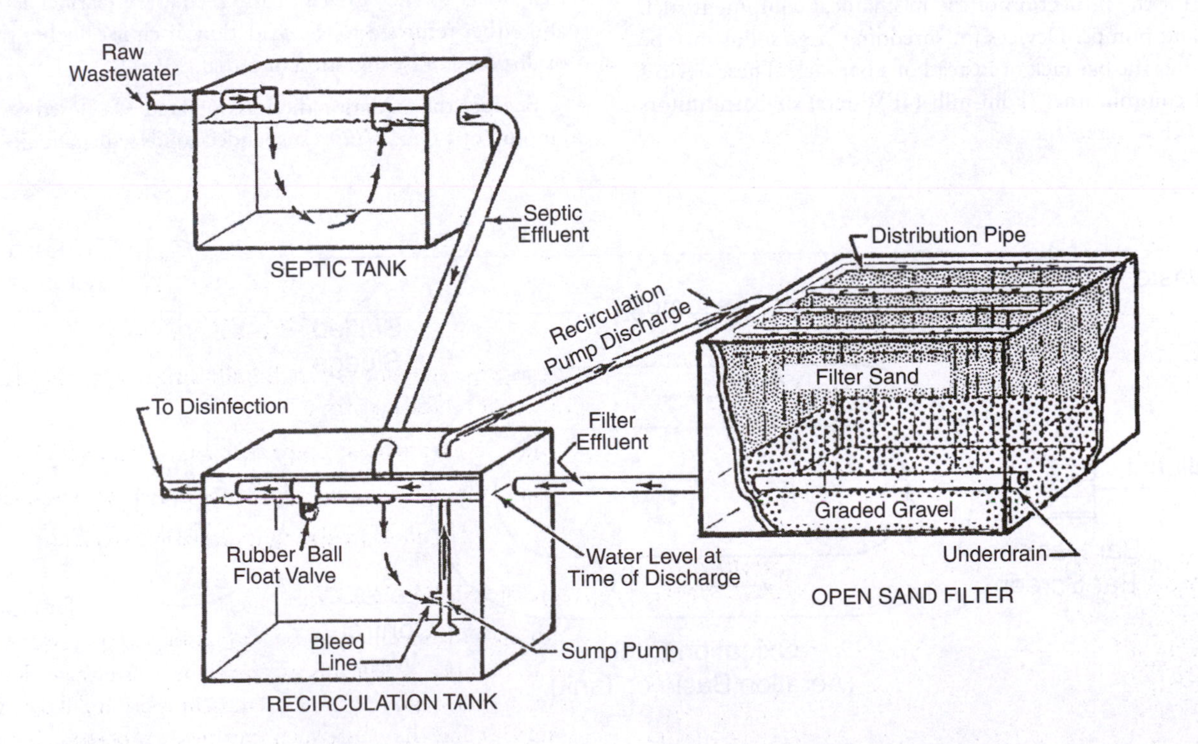


*NOTES:*
1. Water flows from septic tank to recirculation tank.
2. Water is pumped from recirculation tank to open sand filter.
3. Water flows from open sand filter to disinfection. If water is low in recirculation tank, rubber ball float valve drops and filter effluent flows into recirculation tank until water surface rises and rubber ball float valve stops flow into recirculation tank.

*Fig. 2.28 Recirculating sand filter system*

(Source: *RURAL WASTEWATER MANAGEMENT,* California State Water Resources Control Board)

49. *Mixed Liquor Suspended Solids* (MLSS). The amount (mg/L) of suspended solids in the mixed liquor of an aeration tank.

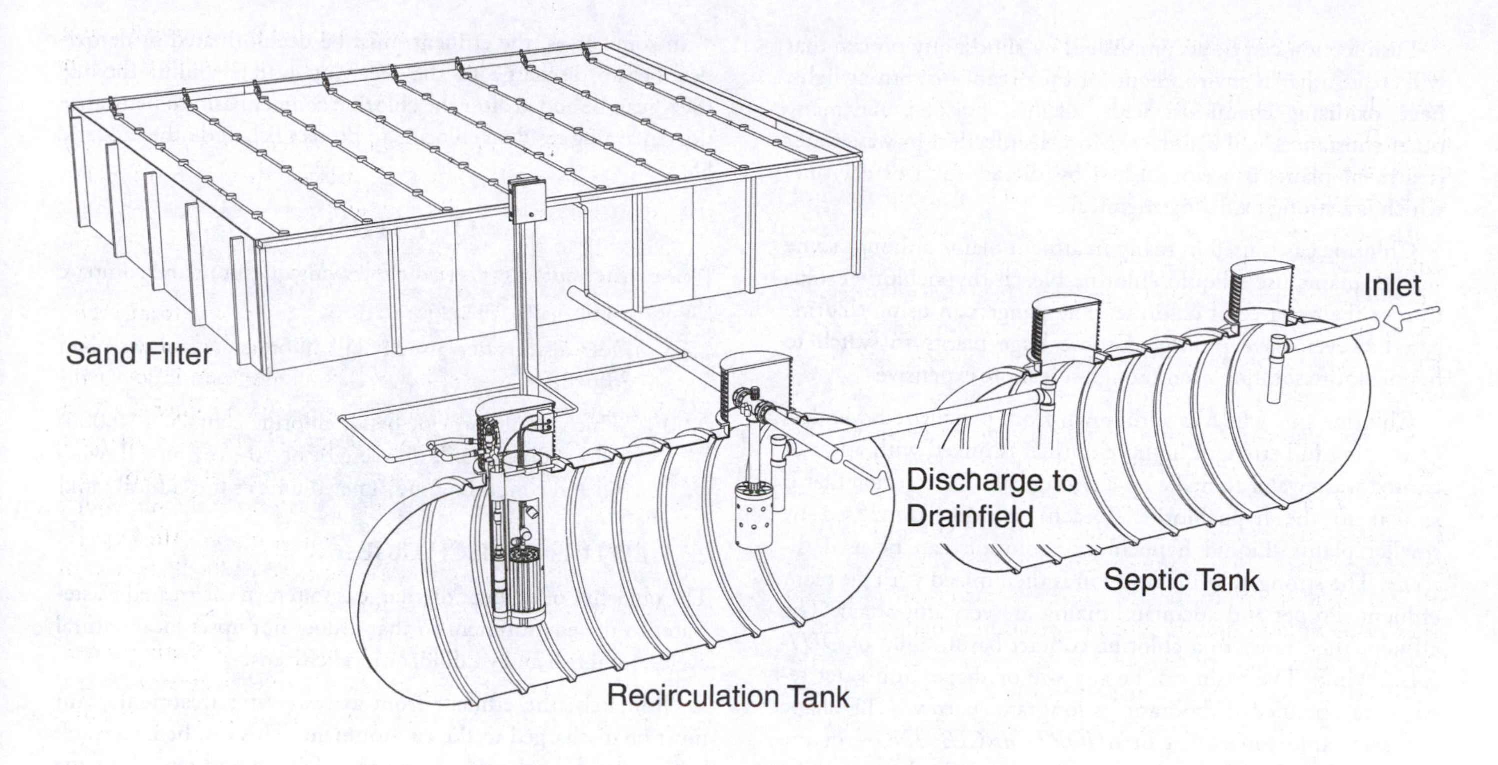


*Fig. 2.29 Newer recirculating sand filter system*
(Courtesy of Orenco Systems, Inc., www.orenco.com)

small community. Units of various sizes are readily available. Figure 2.30 shows the gravel bed of a recirculating gravel/sand filter suitable for a larger community. Complete details of the construction, operation and maintenance of recirculating filters will be presented in Chapter 5.

*Fig. 2.30 Recirculating gravel/sand filter system for a larger community*

## QUESTIONS

Please write your answers to the following questions and compare them with those on page 62.

2.5G What are the main parts of an oxidation ditch?

2.5H Why are the ends of oxidation ditches well rounded?

2.5I What do recirculating gravel/sand filters treat?

## 2.6 DISINFECTION

Although the settling process and biological processes remove a great number of organisms from the wastewater flow, there remain many thousands of bacteria in every milliliter of wastewater leaving the wastewater treatment processes. If there are human wastes in the water, it is possible that some of the bacteria are pathogenic, that is, harmful to humans. Therefore, if the treated wastewater is discharged to a receiving water that is used for a drinking water supply or for swimming or wading, the water pollution control agency or health department will require disinfection of the effluent before it is discharged.

Disinfection is usually defined as the killing or inactivation of pathogenic organisms. The killing of all organisms is called sterilization. Sterilization is not accomplished in treatment plants; the final effluent always contains some living organisms after disinfection due to the inefficiency of the killing process.

Disinfection can be accomplished by almost any process that will create a harsh environment for the organisms. Strong light, heat, oxidizing chemicals, acids, alkalies, poisons, and many other substances will disinfect. Most disinfection in wastewater treatment plants is accomplished by the addition of chlorine, which is a strong oxidizing chemical.

Chlorine gas is used in many treatment plants although some smaller plants use a liquid chlorine bleach (hypochlorite) solution as their source of chlorine. The dangers in using chlorine gas, however, have prompted some large plants to switch to hypochlorite solution even though it is more expensive.

Chlorine gas, which is withdrawn from pressurized cylinders containing full-strength liquid chlorine, is mixed with water or treated wastewater to make up a strong chlorine solution that is similar to the hypochlorite (bleach) solution purchased by smaller plants. Liquid hypochlorite solution can be used directly. The strong chlorine solution is then mixed with the plant effluent. Proper and adequate mixing are very important. The effluent then flows to a chlorine contact basin, tank, or *OUTFALL*[50] line. The basin can be any size or shape, but better results are obtained if the basin is long and narrow. This shape prevents rapid movement or *SHORT-CIRCUITING*[51] of the effluent through the basin. Square or rectangular basins can be baffled to achieve this effect (Figure 2.31). Basins are usually designed to provide approximately 20 to 30 minutes' theoretical contact time, although the trend is to longer times. If the plant's outfall line is of sufficient length, it may function as an excellent contact chamber since short-circuiting will not occur.

In some areas, the effluent must be dechlorinated or detoxified before discharge to the receiving waters. Sulfur dioxide ($SO_2$) can be added after the chlorine contact basin to neutralize the remaining residual chlorine to protect fish and other aquatic life.

## QUESTIONS

Please write your answers to the following questions and compare them with those on page 62.

2.6A Does disinfection usually kill all organisms in the plant effluent?

2.6B Which would provide better chlorine contact, a 10,000-gallon cubical tank or a length of 10-inch pipe flowing full and containing the same volume as the cubical tank?

## 2.7 EFFLUENT DISCHARGE

The objective of effluent discharge is to return the treated wastewater to the environment so that it does not upset local natural cycles, local fish and wildlife, or local citizens.

Ultimately, the effluent from a wastewater treatment plant must be discharged to the environment. This can be into water or onto land, or the water can be reclaimed and reused, or the water can be used for groundwater recharge. Effluents from most wastewater treatment facilities are discharged into receiving waters such as streams, rivers, and lakes. With water becoming scarcer due to increased demands and with higher degrees of treatment being required, plant effluent is becoming a valuable

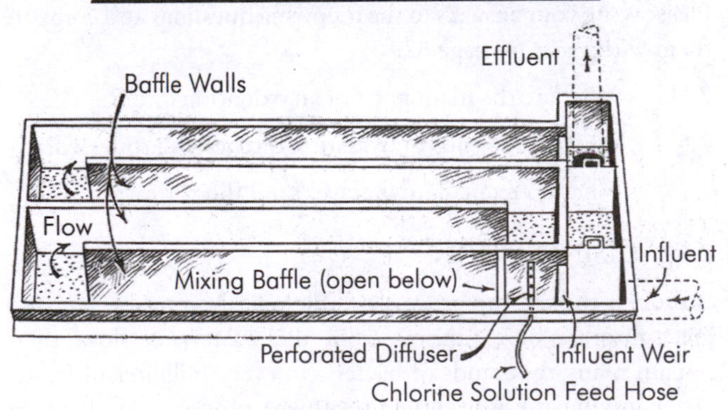


(Courtesy Water Environment Federation)

*Fig. 2.31 Chlorine contact basin*

50. *Outfall.* (1) The point, location, or structure where wastewater or drainage discharges from a sewer, drain, or other conduit. (2) The conduit leading to the final discharge point or area. Also see OUTFALL SEWER.
51. *Short-Circuiting.* A condition that occurs in tanks or basins when some of the flowing water entering a tank or basin flows along a nearly direct pathway from the inlet to the outlet. This is usually undesirable because it may result in shorter contact, reaction, or settling times in comparison with the theoretical (calculated) or presumed detention times.

resource. Both industry and agriculture are discovering that treated effluent may be the most economical source of additional water.

Land treatment is another method of ultimate discharge and can be a means of recharging groundwater basins or storing water for future use. Evaporation ponds are used to dispose of effluents to the atmosphere. Wetlands (Figures 2.32 and 2.33) and aquatic plant systems provide the opportunity to remove nutrients such as nitrogen and phosphorus by the application of natural processes. Regardless of the method of ultimate effluent discharge, operators must carefully operate wastewater treatment plants so that the plant effluent will not cause any adverse impacts on the method of ultimate discharge or on the environment.

## 2.8 SOLIDS DISPOSAL

Final solids disposal is one of the major problems facing many operators today. Solids removed from wastewater by pretreatment processes such as bar racks, screens, and grit removal systems may be disposed of by *DEWATERING*[52] and then direct burial in an approved sanitary landfill or incineration with the remaining ash disposed of in a landfill. Grease and scum from primary and secondary treatment processes and scum and septage (sludge) from interceptors are usually pumped to anaerobic digesters or disposed of in incinerators or in sanitary landfills.

Both aerobic and anaerobic sludge digestion processes produce stabilized or digested solids that ultimately must be disposed of in the environment. Disposal methods include composting with another material such as leaves, farm land application as a soil conditioner or fertilizer, burial in a sanitary landfill, or incineration with ash disposal in a landfill.

## QUESTIONS

Please write your answers to the following questions and compare them with those on page 62.

2.7A Discuss the methods by which the effluent from a wastewater treatment plant could be discharged to the environment.

2.8A How are stabilized or digested solids disposed of?

## 2.9 REVIEW

Operators of wastewater treatment plants provide the best possible treatment of wastes to protect the receiving waters, downstream users, and neighbors. They accomplish these objectives in the following ways:

1. Removing wastes from the wastewater to protect the receiving waters
2. Meeting NPDES or state operating permit requirements
3. Minimizing odors to avoid nuisance complaints
4. Minimizing costs
5. Minimizing energy consumption
6. Maintaining an effective preventive maintenance program
7. Doing their jobs safely

In this chapter, you have read why it is necessary to treat wastewater, something about the types of waste discharges and their effects, and a brief description of the different kinds of solids in wastewater. You also have been given a brief overview of the types of collection, treatment, and discharge systems commonly used in smaller communities. Later chapters will describe

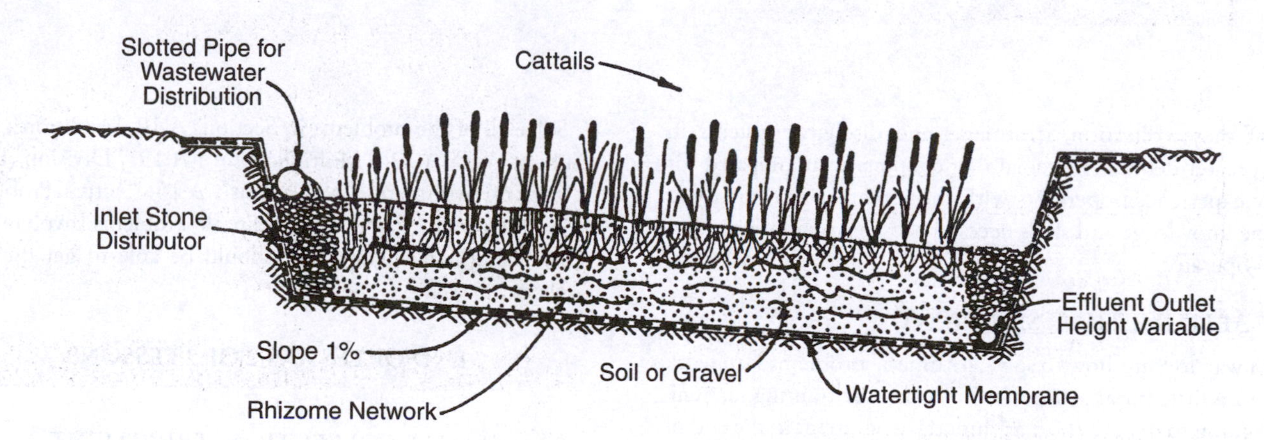


*Fig. 2.32 Wetland subsurface flow system*

(Source: *CONSTRUCTED WETLANDS AND AQUATIC PLANT SYSTEMS FOR MUNICIPAL WASTEWATER TREATMENT,* US EPA Technology Transfer Design Manual)

52. *Dewater.* (1) To remove or separate a portion of the water present in a sludge or slurry. To dry sludge so it can be handled and disposed of. (2) To remove or drain the water from a tank or a trench. A structure may be dewatered so that it can be inspected or repaired.

Multiple inlet port to first treatment cell

Second treatment cell. Cypress trees and ferns growing in the natural wetlands

*Fig. 2.33 Cannon Beach, Oregon, wetlands*

each of these collection, treatment, and discharge systems in much greater detail. The remaining chapters were prepared for you by experienced operators with the intent of providing you with the knowledge and skills necessary to be a small wastewater system operator.

## 2.10 ARITHMETIC ASSIGNMENT

A good way to learn how to solve arithmetic problems is to work on them a little bit at a time. In this operator training manual, we are going to make a short arithmetic assignment at the end of some chapters. If you will work these assignments, you can easily learn how to solve wastewater treatment arithmetic problems.

Turn to the Appendix, "How to Solve Wastewater System Arithmetic Problems," at the back of this manual and read the following sections:

1. "Objectives"
2. A.0, "How to Study This Appendix"
3. A.1, "Basic Arithmetic"

Solve all of the problems in Sections A.10, "Addition"; A.11, "Subtraction"; A.12, "Multiplication"; A.13, "Division"; A.14, "Order of Arithmetic Calculations"; A.15, "Actual Problems"; A.16, "Percentage"; and A.17, "Sample Problems Involving Percent," using a calculator. You should be able to get the same answers.

**END OF LESSON 2 OF 2 LESSONS**

**on**

**SMALL COLLECTION, TREATMENT, AND DISCHARGE SYSTEMS**

Please answer the discussion and review questions next.

# DISCUSSION AND REVIEW QUESTIONS

## Chapter 2. SMALL COLLECTION, TREATMENT, AND DISCHARGE SYSTEMS

(Lesson 2 of 2 Lessons)

Please write your answers to the following questions to determine how well you understand the material in the lesson. The question numbering continues from Lesson 1.

11. List the stages of treatment in a conventional wastewater treatment facility.
12. What is a package wastewater treatment plant?
13. What type of stabilization pond is normally used in most areas?
14. Why is it necessary to periodically remove some of the organisms from the activated sludge process?
15. What is the purpose of the rotor in an oxidation ditch?
16. What chemical means of disinfection is used in most wastewater treatment plants?
17. How can the solids removed from wastewater by pretreatment processes be disposed of?

# SUGGESTED ANSWERS

## Chapter 2. SMALL COLLECTION, TREATMENT, AND DISCHARGE SYSTEMS

### ANSWERS TO QUESTIONS IN LESSON 1

Answers to questions on page 28.

2.0A The four major classifications of waste material in wastewater are organic, inorganic, thermal, and radioactive.

2.0B Domestic wastewater contains mostly organic material.

2.0C The size, shape, and weight per unit volume of solid particles make some solids settleable and others nonsettleable. Large particles tend to settle more rapidly than smaller particles.

Answers to questions on page 31.

2.1A Three natural cycles that are of interest to wastewater treatment facility operators are (1) the water cycle, (2) the life cycles of aquatic organisms, and (3) nutrient cycles.

2.1B Organic wastes in water provide food for the bacteria. These bacteria require oxygen to survive and consequently deplete the oxygen in the water in a way similar to the way oxygen is removed from air when people breathe.

2.1C Hydrogen sulfide gas is produced by anaerobic bacteria.

Answers to questions on page 32.

2.1D Disease-causing bacteria in wastewater come from the body wastes of humans who have a disease.

2.1E Chlorination is the most frequently used means of disinfecting treated wastewater.

2.1F Pathogenic means disease-causing.

2.1G Not all bacteria are harmful. Many bacteria help us treat and purify wastes.

Answers to questions on page 34.

2.2A NPDES stands for National Pollutant Discharge Elimination System.

2.2B The main goal of the federal Water Pollution Control Act Amendments of 1972 is to make the nation's waters safe for swimming and for fish and wildlife.

Answers to questions on page 37.

2.3A The three major functions of a small wastewater system are (1) collection, (2) treatment, and (3) discharge of wastewater.

2.3B The simplest type of on-site wastewater system is the individual home with its own on-site septic tank and subsurface leaching (discharge) system.

Answers to questions on page 38.

2.4A Wastewater in a collection system is commonly conveyed by gravity.

2.4B Gravity collection systems are not used in areas where deep trenching is necessary to maintain the pipe slope or fall. High trenching costs make the gravity system impractical in such areas.

Answers to questions on page 45.

2.4C Where the topography is unfavorable for a gravity collection system, wastewater may be conveyed by a pressure or vacuum collection system.

2.4D Principal components of a pressure collection system include gravity building sewers, septic tanks or pump vaults, pumps and controls, pressure mains, air release valves, and valves for system isolation.

2.4E Principal components of a vacuum collection system include the gravity service line (or building sewer); the valve pit; a vacuum interface valve; an airtight sump; an air intake; the service lateral; the vacuum service main; and a central vacuum station, which includes vacuum pumps, a collection tank, and wastewater pumps.

## ANSWERS TO QUESTIONS IN LESSON 2

Answers to questions on page 49.

2.5A A package treatment plant is a small wastewater treatment plant that is fabricated at the manufacturer's factory, hauled to the site, and installed as one facility. The package plant usually provides conventional treatment including pretreatment or preliminary treatment, primary treatment, secondary treatment, and disinfection.

2.5B The purpose of a wastewater treatment facility is to remove the wastes from the water and reduce its threat to the public health and the environment before it is discharged from the facility.

Answers to questions on page 50.

2.5C A facultative pond is a pond that contains an aerobic top layer and an anaerobic bottom layer.

2.5D A facultative pond acts like a clarifier by allowing solids to settle to the bottom, like a digester because solids on the bottom are decomposed by anaerobic bacteria, and like an aeration tank because of the action of aerobic bacteria in the upper layer of the pond.

Answers to questions on page 55.

2.5E A rotating biological contactor (RBC) is a biological treatment device that has a rotating shaft surrounded by plastic discs called the "media." A biological slime that treats the wastewater grows on the media when conditions are suitable.

2.5F The activated sludge process treats organic wastes and removes BOD and suspended solids; the process can achieve reductions of 90 to 98 percent.

Answers to questions on page 57.

2.5G The main parts of an oxidation ditch are the aeration basin, a brush rotor assembly, settling tank, return sludge pump, and excess sludge handling facilities.

2.5H The ends of oxidation ditches are well rounded to prevent eddying and dead areas.

2.5I Recirculating gravel/sand filters treat the effluent from septic tanks and package treatment plants.

Answers to questions on page 58.

2.6A No, some organisms remain alive due to inefficiencies in the process.

2.6B The pipe would provide better chlorine contact because water cannot short-circuit (take a short route) through a pipe, while it might not move evenly through a tank and thus some of the water would have a shorter contact time.

Answers to questions on page 59.

2.7A The effluent from a wastewater treatment plant must be discharged to the environment. This can be into water or onto land, or the water can be reclaimed and reused, or the water can be used for groundwater recharge. Effluents from most wastewater treatment facilities are discharged into receiving waters such as streams, rivers, and lakes. With water becoming scarcer, plant effluent is becoming a valuable resource for both industry and agriculture as an economical source of additional water. Land treatment is another method of ultimate discharge and can be a means of recharging groundwater basins or storing water for future use. Evaporation ponds are used to dispose of effluents to the atmosphere. Wetlands and aquatic plant systems also provide the opportunity to remove nutrients by means of natural processes.

2.8A Methods for disposing of stabilized or digested solids include composting with another material such as leaves, farm land application as a soil conditioner or fertilizer, burial in a sanitary landfill, or incineration with ash disposal in a landfill.

CHAPTER 3

# SAFETY

by

John Brady

# TABLE OF CONTENTS

## Chapter 3. SAFETY

# OBJECTIVES

## Chapter 3. SAFETY

1. Identify the types of hazards you may encounter operating an on-site or a small wastewater system.
2. Recognize unsafe conditions and correct them whenever they develop.
3. Avoid physical injuries, infections, and diseases.
4. Identify confined space hazards.
5. Take necessary precautions prior to entering a confined space.
6. Inspect safety features of vehicles and equipment.
7. Drive vehicles defensively and safely.
8. Route traffic around a job site.
9. Protect yourself from electrical hazards.
10. Safely handle hazardous chemicals.
11. Use suitable methods to prevent excavation cave-ins.
12. Extinguish fires.
13. Protect yourself from excessive noise, dusts, fumes, mists, gases, and vapors.
14. Develop the habit of always thinking safety and working safely.

# WORDS

## Chapter 3. SAFETY

ACUTE HEALTH EFFECT — ACUTE HEALTH EFFECT

An adverse effect on a human or animal body, with symptoms developing rapidly.

AMPERE (AM-peer) — AMPERE

The unit used to measure current strength. The current produced by an electromotive force of one volt acting through a resistance of one ohm.

ANAEROBIC (AN-air-O-bick) — ANAEROBIC

A condition in which atmospheric or dissolved oxygen (DO) is *NOT* present in the aquatic (water) environment.

CERCLA (SIRK-la) — CERCLA

Comprehensive Environmental Response, Compensation, and Liability Act of 1980. This act was passed primarily to correct past mistakes in industrial waste management. The focus of the act is to locate hazardous waste disposal sites that are creating problems through pollution of the environment and, by proper funding and implementation of study and corrective activities, eliminate the problem from these sites. Current users of CERCLA-identified substances must report releases of these substances to the environment when they take place (not just historic ones). This act is also called the Superfund Act. Also see SARA.

COMMUNITY RIGHT-TO-KNOW — COMMUNITY RIGHT-TO-KNOW

The Superfund Amendments and Reauthorization Act (SARA) of 1986 provides statutory authority for communities to develop right-to-know laws. The act establishes a state and local emergency planning structure, emergency notification procedures, and reporting requirements for facilities. Also see RIGHT-TO-KNOW LAWS and SARA.

CONFINED SPACE — CONFINED SPACE

Confined space means a space that:

(1) Is large enough and so configured that an employee can bodily enter and perform assigned work; and

(2) Has limited or restricted means for entry or exit (for example, manholes, tanks, vessels, silos, storage bins, hoppers, vaults, and pits are spaces that may have limited means of entry); and

(3) Is not designed for continuous employee occupancy.

Also see DANGEROUS AIR CONTAMINATION and OXYGEN DEFICIENCY.

CONFINED SPACE, PERMIT-REQUIRED (PERMIT SPACE) — CONFINED SPACE, PERMIT-REQUIRED (PERMIT SPACE)

A confined space that has one or more of the following characteristics:

(1) Contains or has a potential to contain a hazardous atmosphere,

(2) Contains a material that has the potential for engulfing an entrant,

(3) Has an internal configuration such that an entrant could be trapped or asphyxiated by inwardly converging walls or by a floor that slopes downward and tapers to a smaller cross section, or

(4) Contains any other recognized serious safety or health hazard.

CURRENT CURRENT

A movement or flow of electricity. Electric current is measured by the number of coulombs per second flowing past a certain point in a conductor. A coulomb is equal to about $6.25 \times 10^{18}$ electrons (6,250,000,000,000,000,000 electrons). A flow of one coulomb per second is called one ampere, the unit of the rate of flow of current.

DANGEROUS AIR CONTAMINATION DANGEROUS AIR CONTAMINATION

An atmosphere presenting a threat of causing death, injury, acute illness, or disablement due to the presence of flammable or explosive, toxic, or otherwise injurious or incapacitating substances.

(1) Dangerous air contamination due to the flammability of a gas, vapor, or mist is defined as an atmosphere containing the gas, vapor, or mist at a concentration greater than 10 percent of its lower explosive (lower flammable) limit (LEL).

(2) Dangerous air contamination due to a combustible particulate is defined as a concentration that meets or exceeds the particulate's lower explosive limit (LEL).

(3) Dangerous air contamination due to the toxicity of a substance is defined as the atmospheric concentration that could result in employee exposure in excess of the substance's permissible exposure limit (PEL).

*NOTE:* A dangerous situation also occurs when the oxygen level is less than 19.5 percent by volume (OXYGEN DEFICIENCY) or more than 23.5 percent by volume (OXYGEN ENRICHMENT).

DECIBEL (DES-uh-bull) DECIBEL

A unit for expressing the relative intensity of sounds on a scale from zero for the average least perceptible sound to about 130 for the average level at which sound causes pain to humans. Abbreviated dB.

ELECTROMOTIVE FORCE (EMF) ELECTROMOTIVE FORCE (EMF)

The electrical pressure available to cause a flow of current (amperage) when an electric circuit is closed. Also called voltage.

ELECTRON ELECTRON

(1) A very small, negatively charged particle that is practically weightless. According to the electron theory, all electrical and electronic effects are caused either by the movement of electrons from place to place or because there is an excess or lack of electrons at a particular place.

(2) The part of an atom that determines its chemical properties.

GROUND GROUND

An expression representing an electrical connection to earth or a large conductor that is at the earth's potential or neutral voltage.

HYDROGEN SULFIDE GAS ($H_2S$) HYDROGEN SULFIDE GAS ($H_2S$)

Hydrogen sulfide is a gas with a rotten egg odor, produced under anaerobic conditions. Hydrogen sulfide gas is particularly dangerous because it dulls the sense of smell, becoming unnoticeable after you have been around it for a while; in high concentrations, it is only noticeable for a very short time before it dulls the sense of smell. The gas is very poisonous to the respiratory system, explosive, flammable, colorless, and heavier than air.

INSECTICIDE INSECTICIDE

Any substance or chemical formulated to kill or control insects.

LOADING LOADING

Quantity of material applied to a device at one time.

LOWER EXPLOSIVE LIMIT (LEL) LOWER EXPLOSIVE LIMIT (LEL)

The lowest concentration of a gas or vapor (percent by volume in air) that explodes if an ignition source is present at ambient temperature. At temperatures above 250°F (121°C) the LEL decreases because explosibility increases with higher temperature.

MATERIAL SAFETY DATA SHEET (MSDS) MATERIAL SAFETY DATA SHEET (MSDS)

A document that provides pertinent information and a profile of a particular hazardous substance or mixture. An MSDS is normally developed by the manufacturer or formulator of the hazardous substance or mixture. The MSDS is required to be made available to employees and operators or inspectors whenever there is the likelihood of the hazardous substance or mixture being introduced into the workplace. Some manufacturers are preparing MSDSs for products that are not considered to be hazardous to show that the product or substance is not hazardous.

OSHA (O-shuh) OSHA

The Williams-Steiger Occupational Safety and Health Act of 1970 (OSHA) is a federal law designed to protect the health and safety of workers, including collection system and treatment plant operators. The Act regulates the design, construction, operation, and maintenance of industrial plants and wastewater collection systems and treatment plants. The Act does not apply directly to municipalities, *except* in those states that have approved plans and have asserted jurisdiction under Section 18 of the OSHA Act. *However, contract operators and private facilities do have to comply with OSHA requirements.* Wastewater treatment plants have come under stricter regulation in all phases of activity as a result of OSHA standards. OSHA also refers to the federal and state agencies that administer the OSHA regulations.

OHM OHM

The unit of electrical resistance. The resistance of a conductor in which one volt produces a current of one ampere.

OXYGEN DEFICIENCY OXYGEN DEFICIENCY

An atmosphere containing oxygen at a concentration of less than 19.5 percent by volume.

OXYGEN ENRICHMENT OXYGEN ENRICHMENT

An atmosphere containing oxygen at a concentration of more than 23.5 percent by volume.

PATHOGENIC (path-o-JEN-ick) ORGANISMS PATHOGENIC ORGANISMS

Bacteria, viruses, protozoa, or internal parasites that can cause disease (such as giardiasis, cryptosporidiosis, typhoid fever, cholera, or infectious hepatitis) in a host (such as a person). There are many types of organisms that do not cause disease and are not called pathogenic. Many beneficial bacteria are found in wastewater treatment processes actively cleaning up organic wastes.

PERMISSIBLE EXPOSURE LIMIT (PEL) PERMISSIBLE EXPOSURE LIMIT (PEL)

The legal limit in the United States for exposure of a worker to a hazardous substance (such as chemicals, dusts, fumes, mists, gases, or vapors) or agents (such as occupational noise). OSHA sets enforceable permissible exposure limits (PELs) to protect workers against the health effects of excessive exposure. OSHA PELs are based on an 8-hour time-weighted average (TWA) exposure. Permissible exposure limits are listed in the Code of Federal Regulations (CFR) Title 29 Part 1910, Subparts G and Z. Also see TIME-WEIGHTED AVERAGE (TWA).

PERMIT-REQUIRED CONFINED SPACE (PERMIT SPACE) PERMIT-REQUIRED CONFINED SPACE (PERMIT SPACE)

See CONFINED SPACE, PERMIT-REQUIRED (PERMIT SPACE).

RESISTANCE RESISTANCE

That property of a conductor or wire that opposes the passage of a current, thus causing electric energy to be transformed into heat.

RIGHT-TO-KNOW LAWS RIGHT-TO-KNOW LAWS

Employee Right-To-Know legislation requires employers to inform employees of the possible health effects resulting from contact with hazardous substances. At locations where this legislation is in force, employers must provide employees with information regarding any hazardous substances they might be exposed to under normal work conditions or reasonably foreseeable emergency conditions resulting from workplace conditions. OSHA's Hazard Communication Standard (HCS) (Title 29 CFR Part 1910.1200) is the federal regulation and state statutes are called Worker Right-To-Know laws. Also see COMMUNITY RIGHT-TO-KNOW and SARA.

SARA SARA

Superfund Amendments and Reauthorization Act of 1986. The Comprehensive Environmental Response, Compensation, and Liability Act (CERCLA), commonly known as the Superfund Act, was enacted in 1980. The 1986 amendments increase CERCLA revenues to $8.5 billion and strengthen the EPA's authority to conduct short-term (removal), long-term (remedial), and enforcement actions. The amendments also strengthen state involvements in the cleanup process and the agency's commitments to research and development, training, health assessments, and public participation. A number of new statutory authorities, such as Community Right-To-Know, were also established. Also see CERCLA.

SPECIFIC GRAVITY SPECIFIC GRAVITY

(1) Weight of a particle, substance, or chemical solution in relation to the weight of an equal volume of water. Water has a specific gravity of 1.000 at 4°C (39°F). Wastewater particles or substances usually have a specific gravity of 0.5 to 2.5. Particulates with specific gravity less than 1.0 float to the surface and particulates with specific gravity greater than 1.0 sink.

(2) Weight of a particular gas in relation to the weight of an equal volume of air at the same temperature and pressure (air has a specific gravity of 1.0). Chlorine gas has a specific gravity of 2.5.

TIME-WEIGHTED AVERAGE (TWA) TIME-WEIGHTED AVERAGE (TWA)

A time-weighted average is used to calculate a worker's daily exposure to a hazardous substance (such as chemicals, dusts, fumes, mists, gases, or vapors) or agent (such as occupational noise), averaged to an 8-hour workday, taking into account the average levels of the substance or agent and the time spent in the area. This is the guideline OSHA uses to determine permissible exposure limits (PELs) and is essential in assessing a worker's exposure and determining what protective measures should be taken. A time-weighted average is equal to the sum of the portion of each time period (as a decimal, such as 0.25 hour) multiplied by the levels of the substance or agent during the time period divided by the hours in the workday (usually 8 hours). Also see PERMISSIBLE EXPOSURE LIMIT (PEL).

VOLTAGE VOLTAGE

The electrical pressure available to cause a flow of current (amperage) when an electric circuit is closed. Also called electromotive force (EMF).

WATT WATT

A unit of power equal to one joule per second. The power of a current of one ampere flowing across a potential difference of one volt.

# CHAPTER 3. SAFETY

## 3.0 WHO IS RESPONSIBLE FOR SAFETY?

Operators of small wastewater collection and treatment systems have a unique and very challenging job because it includes a wide range of responsibilities. In this job, more than in many others, an operator's knowledge, skills, and common sense relate directly to safety/survival. A trained professional operator demonstrates an awareness of hazards and a commitment to accomplish every task in a safe manner.

It is not uncommon in small communities for there to be only one or two persons available for operation and maintenance. Often, an operator must work alone. If you are that person, then you and only you are responsible for your safety. The community may supply the necessary tools and safety devices to allow you to work without endangering the public or yourself, but you must know how to recognize dangerous situations and take the precautions necessary to ensure your safety. The purpose of the following section is to acquaint you with some of the common hazards associated with wastewater systems.

## 3.1 TYPES OF HAZARDS

You are equally exposed to accidents whether you are working on the collection system or working in a treatment plant. As an operator, you may be exposed to the following hazards and dangerous conditions:

1. Physical injuries
2. Infections and infectious diseases
3. Insects, bugs, rodents, and snakes
4. Confined spaces
5. Oxygen deficiency or enrichment
6. Explosive, toxic or suffocating gases, vapors, or dusts
7. Vehicles, driving, and traffic
8. Electric shock
9. Toxic or harmful chemicals
10. Cave-ins
11. Fires
12. Noise
13. Dusts, fumes, mists, gases, and vapors

One study of accidents revealed that seventy-one percent of work-related injuries involved these five body parts:

- Backs are by far the most frequently injured part of the body, accounting for 24 percent of the total injuries. Back injuries also cost more, in this case 31 percent of the workers' compensation paid.
- Legs: 13 percent.
- Fingers: 11 percent.
- Arms: 12 percent.
- Trunk: 11 percent.

Obviously, accidents have a significant economic impact as well as taking a toll in human suffering. For example, workers' compensation rates, liability insurance, disability insurance, and other similar costs are directly related to our performance in an agency or company. In recent years, all of these costs have skyrocketed, imposing an economic hardship on many agencies and communities. These are budgetary resources that are no longer available for use in other areas such as operations, maintenance, training, equipment, and salaries.

The probability of your having an accident is directly related to the years of work experience you have. If you have fewer than 2 years of experience, your odds of having an accident are 1 in 7. If you have 15 or more years' experience, the odds of having an accident are 1 in 15, or about half as likely.

There is also a relationship between the risk of injury and certification. In collection systems, there is a 1 in 9 chance of being injured if you are not certified and a 1 in 12 chance if you are certified. For treatment plants, the risk is 1 in 11 for uncertified operators and 1 in 16 for certified operators.

With the seriousness and number of hazards a wastewater treatment operator may be exposed to, attention to safety/survival is critical. The economic cost of accidents is great, but the cost in human suffering is far worse. Consider the effect, for example, on you and your family if you had to spend the rest of your life with only one eye because you failed to wear safety glasses while working around a pump and the coupling happened to disintegrate while you were inspecting it. Or what if a wrench drops out of your co-worker's pocket while the operator is leaning over the entrance tube to the lift station? The wrench strikes you in the head at the bottom of the entrance tube, traveling at about 75 miles per hour. How much pain and anguish could you have spared yourself and your family by taking the time to put on a hard hat?

## 3.2 PHYSICAL INJURIES

### 3.20 Hazards

The most common physical injuries are cuts, bruises, strains, and sprains. Injuries can be caused by moving machinery, improper lifting techniques, or slippery surfaces. Falls from or into tanks, wet wells, catwalks, or conveyors can be disabling. Most of these injuries can be avoided by the proper use of ladders, hand tools, and safety equipment; by following established safety procedures; and by always thinking safety. Strains and sprains are probably the most frequent injuries in wastewater collection systems and treatment plants.

### 3.21 Safety Precautions and Safe Procedures

*OSHA*[1] standards require that all equipment that could unexpectedly start up or release stored energy must be locked out and tagged whenever it is being worked on. Common forms of stored energy are electric energy, spring-loaded equipment, hydraulic pressure, and compressed gases that are stored under pressure.

The operator who will be performing the work on the equipment is the person who installs a lock (either key or combination) on the energy isolating device (switch, valve) to positively lock the switch or valve in a safe position, preventing the equipment from starting or moving. At the same time, a tag such as the one shown in Figure 3.1 is installed at the lock indicating why the equipment is locked out and who is working on it. No other person is authorized to remove the tag or lockout device unless the employer has provided specific procedures and training for removal by others.

Even after equipment is locked out and tagged, it still may not be safe to work on. Stored energy in gas, air, water, steam, and hydraulic systems (such as water/wastewater pumping or collection systems) must be drained or bled down to release the stored energy. Elevated machine members, flywheels, and springs should be positioned or secured in place to prevent movement.

Before bleeding down pressurized systems, think about the pressures involved and what will be discharged to the atmosphere and work area (toxic or explosive gases, corrosive chemicals, sludges). Will other safety precautions have to be taken? What volume of bleed-down material will there be and where will it go? If dealing with sludges, wastewaters, greases, or scums, how can they be cleaned up or contained? Many pumping stations have been flooded when an operator removed a pump volute cleanout only to find that the discharge check valve was not seated or one of the isolation valves was not fully seated. Once the pump was open, however, the force main or pump intake structure drained into the pump room through the opened pump.

All rotating mechanical equipment must have guards installed to protect operators from becoming entangled or caught up in belts, pulleys, drive lines, flywheels, and couplings. Keep the guards installed even when no one is working on the equipment.

DANGER

OPERATOR WORKING ON LINE

DO NOT CLOSE THIS SWITCH WHILE THIS TAG IS DISPLAYED

TIME OFF: ______________________

DATE: ______________________

SIGNATURE: ______________________

This is the ONLY person authorized to remove this tag.

INDUSTRIAL INDEMNITY/INDUSTRIAL UNDERWRITERS/
INSURANCE COMPANIES

4E210–R66

*Fig. 3.1 Typical lockout warning tag*
(Source: Industrial Indemnity/Industrial Underwriters/Insurance Companies)

1. *OSHA* (O-shuh). The Williams-Steiger Occupational Safety and Health Act of 1970 (OSHA) is a federal law designed to protect the health and safety of workers, including collection system and treatment plant operators. The Act regulates the design, construction, operation, and maintenance of industrial plants and wastewater collection systems and treatment plants. The Act does not apply directly to municipalities, *except* in those states that have approved plans and have asserted jurisdiction under Section 18 of the OSHA Act. *However, contract operators and private facilities do have to comply with OSHA requirements.* Wastewater treatment plants have come under stricter regulation in all phases of activity as a result of OSHA standards. OSHA also refers to the federal and state agencies that administer the OSHA regulations.

Various pieces of machinery are equipped with travel limit switches, pressure sensors, pressure reliefs, shear pins, and stall or torque switches to ensure proper and safe operation of the equipment. Never disconnect a device, or install larger shear pins than those specified on the original design, or modify pressure or temperature settings.

### 3.22 Basic Lockout/Tagout Procedure

1. Notify all affected employees that a lockout or tagout system is going to be used and the reason why. The authorized employee is responsible for knowing the type and magnitude of energy that the equipment uses and must understand the potential hazards.
2. If the equipment is operating, shut it down by the normal stopping procedure.
3. Operate the switch, valve, or other energy isolating device(s) so that the equipment is isolated from its energy source(s). Stored energy such as that in springs, elevated machine parts, rotating flywheels, hydraulic systems, and air, gas, steam, or water pressure must be released or restrained by methods such as repositioning, blocking, or bleeding down.
4. Lock out and tag the energy isolating device with your assigned individual lock or tag.
5. After ensuring that no personnel are exposed, and as a check that the energy source is disconnected, operate the push button or other normal operating controls to make certain the equipment will not operate. A common problem when motor control centers (MCCs) contain many breakers is to lock out the wrong equipment (locking pump #3 and thinking it is #2). Always confirm the dead circuit. *CAUTION: Return operating controls to the neutral or OFF position after the test.*
6. The equipment is now locked out or tagged out and work on the equipment may begin.
7. After the work on the equipment is complete, all tools have been removed, guards have been reinstalled, and employees are in the clear, remove all lockout and tagout devices. Operate the energy isolating devices to restore energy to the equipment.
8. Notify affected employees that the lockout and tagout devices have been removed before starting the equipment.

## QUESTIONS

Please write your answers to the following questions and compare them with those on page 106.

3.0A Who is responsible for your safety?

3.1A List at least six types of hazards that the operator of a small wastewater system might be exposed to.

3.2A Whose responsibility is it to lock out and tag equipment before work is performed on it?

3.2B Why is it necessary to bleed down pressurized systems before making repairs?

## 3.3 INFECTIONS AND INFECTIOUS DISEASES

### 3.30 Hazards

Wastewater system operators often come in physical contact with raw wastewater in the course of their daily activities. Even when direct physical contact is avoided, an operator may handle objects that are contaminated. Every disease, parasite, infection, virus, and illness of a community can end up in the wastewater system. The organisms that carry diseases are called *PATHOGENIC ORGANISMS.*[2] Some of the diseases that may be transmitted by wastewater include giardiasis (jee-are-DYE-uh-sis), cryptosporidiosis, anthrax, tuberculosis, paratyphoid fever, cholera, polio, and hepatitis. Tapeworms and the organisms associated with food poisoning may also be present. The possibility that AIDS, which is caused by a virus, can be contracted from exposure to raw wastewater has been discounted by researchers. Although the HIV virus is present in the wastes from AIDS victims, raw wastewater is a harsh environment and there has been no evidence that AIDS can be transmitted by wastewater.

There are three basic routes that may lead to infection:

- Ingestion through splashes, contaminated food, or cigarettes
- Inhalation of infectious agents or aerosols
- Direct contact through an unprotected cut or abrasion

Examples of the major routes of infection are listed below:

- Ingestion: Eating, drinking, or accidentally swallowing pathogenic organisms (for example, hepatitis A)
- Inhalation: Breathing spray or mist containing pathogenic organisms (for example, the common cold)
- Direct contact: Entry of pathogenic organisms into the body through a cut or break in the skin (for example, tetanus)

Ingestion is generally the major route of wastewater operator infection. The common practice of touching the mouth with the hand will contribute to the possibility of infection. Operators who eat or smoke without washing their hands have a much higher risk of infection. Most surfaces near wastewater equipment are likely to be covered with bacteria or viruses. These potentially infectious agents may be deposited on surfaces in the form of an aerosol or may come from direct contact with the wastewater.

2. *Pathogenic* (path-o-JEN-ick) *Organisms.* Bacteria, viruses, protozoa, or internal parasites that can cause disease (such as giardiasis, cryptosporidiosis, typhoid fever, cholera, or infectious hepatitis) in a host (such as a person). There are many types of organisms that do not cause disease and are not called pathogenic. Many beneficial bacteria are found in wastewater treatment processes actively cleaning up organic wastes.

### 3.31 Safety Precautions and Safe Procedures

Good personal hygiene is your best protection against infections and infectious diseases such as giardiasis, cryptosporidiosis, typhoid fever, dysentery, hepatitis, and tetanus. Make it a habit to thoroughly wash your hands before eating or smoking, as well as before and after using the restroom. Always wear proper protective gloves when the possibility of contact with wastewater or sludge exists. Cuts and abrasions, including those that are minor, should be cared for properly. Open wounds invite infection from many of the viruses and bacteria present in wastewater. Bandages covering wounds should be changed frequently. A good rule of thumb is to never touch yourself above the neck whenever there is contact with wastewater. Use the methods below to prevent ingestion of pathogenic organisms:

- Wash hands frequently and never eat, drink, or use tobacco products before washing hands.
- Avoid touching face, mouth, eyes, or nose before washing hands.
- Wash hands immediately after any contact with wastewater.

Immunization shots for protection against tetanus, polio, and hepatitis A and B are essential and are often available free of charge from your local health department. Although tetanus is not a waterborne disease, the organism is extremely widespread in nature and can be found at treatment plants, in manholes, and in the soil. Once a person catches the disease, it is very difficult to treat. Vaccination for tetanus is recommended and the frequency of booster shots should be verified with your physician.

Hepatitis is a viral infection that affects the liver and sometimes produces a symptom known as jaundice, a yellowing of the skin. Hepatitis A and hepatitis B are two strains of the virus. Vaccines are available to protect against hepatitis A and B.

To protect yourself from infections and diseases, make it a habit to thoroughly wash your hands before eating or smoking, as well as before and after using the restroom. Always wear proper protective gloves when the possibility of contact with wastewater or sludge exists. Bandages covering wounds should be changed frequently.

Do not wear your work clothes home because diseases may be transmitted to your family. If possible, provisions should be made in your plant or office for a locker room where each employee has a locker. Work clothes should be placed or hung in these lockers and not thrown on the floor.

Your work clothes should be cleaned as often as necessary. If your employer does not supply you with uniforms and laundry service, investigate the availability of disposable clothing for dirty jobs. If you must take your work clothes home, launder them separately from your regular family wash.

All of these precautions will reduce the possibility of you and your family becoming ill because of your contact with wastewater or sludge.

## 3.4 INSECTS, BUGS, RODENTS, AND SNAKES

### 3.40 Hazards

Many persons have observed a variety of flies, beetles, earthworms, salamanders, frogs, snakes, mice, scorpions, and spiders on top of or in the scum mat of septic tanks. Insect, bug, rodent, and snake bites, though somewhat uncommon, are a double hazard to operators. The insects themselves may be poisonous to humans, such as the black widow spider and the violin spider. Many other types of insect, bug, rodent, and snake bites can lead to infection or serious illness. Rat bites, for example, can transmit rabies and mosquito bites sometimes transmit malaria.

Following is a list of insects and bugs that have been found in wastewater structures such as pump and valve vaults and manholes:

| | | |
|---|---|---|
| wasps | fleas | cockroaches |
| mud daubers | lice | spiders |
| bees | mosquitos | centipedes |
| ticks | blowflies | scorpions |

Always inspect a manhole for insects, bugs, rodents, and snakes before entry.

### 3.41 Safety Precautions and Safe Procedures

Wear protective clothing (such as gloves) and be alert to the possible danger to your health. If you are bitten, gently wash the area with soapy water. Get prompt medical attention if you develop any signs of redness or swelling, or if you develop a fever or an allergic reaction.

Where insects have been a problem, and where rats or other vermin may occupy a collection system or where epidemics of insect-borne diseases may be present, spraying a manhole with an *INSECTICIDE* [3] is suggested. The solution should be water-soluble and leave a toxic residue to be effective against the next hatch of any insects present and breeding in the septic tanks and manholes. Contact your local health agency to determine the appropriate insecticide, if in doubt, or call (800) 858-7378 (an EPA-sponsored information service) to speak to a representative regarding the selection and use of insecticides. Spraying should be conducted at the time of entry if insects or other problems are observed. Ventilate the manhole so insecticide will not be inhaled by operators. If a manhole is especially filthy or odorous from an insecticide or hydrogen sulfide, wash down the manhole with a high-velocity stream of water.

---

3. *Insecticide.* Any substance or chemical formulated to kill or control insects.

## QUESTIONS

Please write your answers to the following questions and compare them with those on page 107.

3.3A What are pathogenic organisms?

3.3B Do all bacteria cause infection in a host?

3.3C How can an operator avoid getting infections and infectious diseases from exposure to wastewater?

3.4A When should an operator seek medical attention for an insect, bug, rodent, or snake bite?

## 3.5 CONFINED SPACES[4]

### 3.50 Definition

A confined space may be defined as an area that has limited means for entry and exit and is not designed for continuous employee occupancy. One easy way to identify a confined space is by whether or not you can enter it by simply walking while standing fully upright. In general, if you must duck, crawl, climb, or squeeze into the space, it is considered a confined space. Examples of confined spaces found in wastewater systems are septic tanks, manholes, valve boxes, pump vaults or pits, channels, sewers, enclosed recirculation structures, anaerobic digesters, and pressure filters.

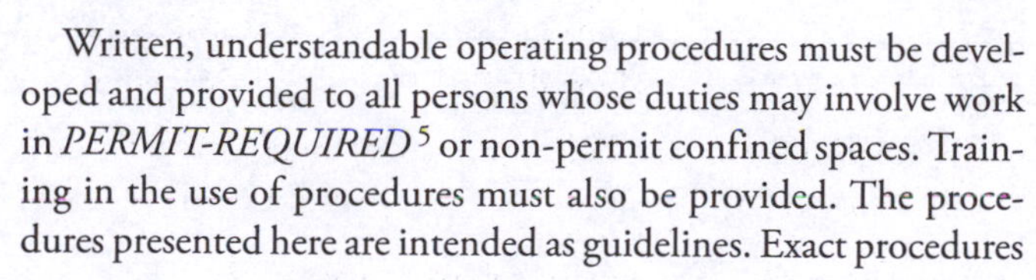

Written, understandable operating procedures must be developed and provided to all persons whose duties may involve work in *PERMIT-REQUIRED*[5] or non-permit confined spaces. Training in the use of procedures must also be provided. The procedures presented here are intended as guidelines. Exact procedures for work in confined spaces may vary with different agencies and geographical locations and must be confirmed with the appropriate regulatory safety agency.

### 3.51 Hazards

The major hazards in confined spaces are oxygen deficiency (less than 19.5 percent oxygen by volume), oxygen enrichment (more than 23.5 percent oxygen by volume), and exposure to *DANGEROUS AIR CONTAMINATION*.[6] Dangerous air contamination presents a threat of causing death, injury, *ACUTE*[7] illness, or disablement due to flammable or explosive, toxic, or otherwise injurious substances. In wastewater treatment, we are concerned primarily with oxygen deficiency/enrichment, methane (explosive), *HYDROGEN SULFIDE*[8] (toxic), carbon monoxide (toxic), and other gases as identified in Table 3.1. A major concern in confined spaces is whether the existing ventilation is able to remove dangerous air contamination or oxygen deficiency that may exist or develop.

#### *3.510 Oxygen Deficiency or Enrichment*

Low oxygen levels may exist in any poorly ventilated, low-lying structure where gases such as hydrogen sulfide, gasoline vapor, carbon dioxide, or chlorine may be produced or may accumulate. Also, low levels of oxygen may develop in confined spaces where operators are working and ventilation is insufficient to replace the oxygen taken up by operator respiration (breathing). Oxygen in a concentration above 23.5 percent (oxygen enrichment) also can be dangerous because it speeds up combustion.

Oxygen deficiency is most likely to occur when structures or channels are installed below grade (ground level). Several gases (including hydrogen sulfide and chlorine) have a tendency to

---

4. *Confined Space.* Confined space means a space that: (1) Is large enough and so configured that an employee can bodily enter and perform assigned work; and (2) Has limited or restricted means for entry or exit (for example, manholes, tanks, vessels, silos, storage bins, hoppers, vaults, and pits are spaces that may have limited means of entry); and (3) Is not designed for continuous employee occupancy. Also see DANGEROUS AIR CONTAMINATION and OXYGEN DEFICIENCY.
5. *Confined Space, Permit-Required (Permit Space).* A confined space that has one or more of the following characteristics: (1) Contains or has a potential to contain a hazardous atmosphere, (2) Contains a material that has the potential for engulfing an entrant, (3) Has an internal configuration such that an entrant could be trapped or asphyxiated by inwardly converging walls or by a floor that slopes downward and tapers to a smaller cross section, or (4) Contains any other recognized serious safety or health hazard.
6. *Dangerous Air Contamination.* An atmosphere presenting a threat of causing death, injury, acute illness, or disablement due to the presence of flammable or explosive, toxic, or otherwise injurious or incapacitating substances. (1) Dangerous air contamination due to the flammability of a gas, vapor, or mist is defined as an atmosphere containing the gas, vapor, or mist at a concentration greater than 10 percent of its lower explosive (lower flammable) limit (LEL). (2) Dangerous air contamination due to a combustible particulate is defined as a concentration that meets or exceeds the particulate's lower explosive limit (LEL). (3) Dangerous air contamination due to the toxicity of a substance is defined as the atmospheric concentration that could result in employee exposure in excess of the substance's permissible exposure limit (PEL). *NOTE:* A dangerous situation also occurs when the oxygen level is less than 19.5 percent by volume (OXYGEN DEFICIENCY) or more than 23.5 percent by volume (OXYGEN ENRICHMENT).
7. *Acute Health Effect.* An adverse effect on a human or animal body, with symptoms developing rapidly.
8. *Hydrogen Sulfide Gas ($H_2S$).* Hydrogen sulfide is a gas with a rotten egg odor, produced under anaerobic conditions. Hydrogen sulfide gas is particularly dangerous because it dulls the sense of smell, becoming unnoticeable after you have been around it for a while; in high concentrations, it is only noticeable for a very short time before it dulls the sense of smell. The gas is very poisonous to the respiratory system, explosive, flammable, colorless, and heavier than air.

## TABLE 3.1 COMMON DANGEROUS GASES ENCOUNTERED IN WASTEWATER COLLECTION SYSTEMS AND AT WASTEWATER TREATMENT PLANTS[a]

| Name of Gas and Chemical Formula | 8TWA PEL[b] | Specific Gravity or Vapor Density[c] (Air = 1) | Explosive Range (% in air by volume) | | Common Properties (% in air by volume) | Physiological Effects (% in air by volume) | Most Common Sources in Sewers | Simplest and Cheapest Safe Method of Testing[d] |
|---|---|---|---|---|---|---|---|---|
| | | | Lower Limit | Upper Limit | | | | |
| Oxygen, $O_2$ (in Air) | | 1.11 | Not flammable | | Colorless, odorless, tasteless, nonpoisonous gas. Supports combustion. | Normal air contains 20.93% of $O_2$. If $O_2$ is less than 19.5%, do not enter space without respiratory protection. | Oxygen depletion from poor ventilation and absorption or chemical consumption of available $O_2$. | Oxygen deficiency indicator. |
| Gasoline Vapor, $C_5H_{12}$ to $C_9H_{20}$ | 300 | 3.0 to 4.0 | 1.3 | 7.0 | Colorless, odor noticeable in 0.03%. Flammable. Explosive. | Anesthetic effects when inhaled. 2.43% rapidly fatal. 1.1% to 2.2% dangerous for even short exposure. | Leaking storage tanks, discharges from garages, and commercial or home dry-cleaning operations. | 1. Combustible gas indicator. 2. Oxygen deficiency indicator. |
| Carbon Monoxide, CO | 50 | 0.97 | 12.5 | 74.2 | Colorless, odorless, nonirritating, tasteless. Flammable. Explosive. | Hemoglobin of blood has strong affinity for gas, causing oxygen starvation. 0.2% to 0.25% causes unconsciousness in 30 minutes. | Manufactured fuel gas. | CO ampoules. |
| Hydrogen, $H_2$ | | 0.07 | 4.0 | 74.2 | Colorless, odorless, tasteless, nonpoisonous, flammable. Explosive. Propagates flame rapidly; very dangerous. | Acts mechanically to deprive tissues of oxygen. Does not support life. A simple asphyxiant. | Manufactured fuel gas. | Combustible gas indicator. |
| Methane, $CH_4$ | | 0.55 | 5.0 | 15.0 | Colorless, odorless, tasteless, nonpoisonous. Flammable. Explosive. | See Hydrogen. | Natural gas, marsh gas, manufactured fuel gas, gas found in sewers. | 1. Combustible gas indicator. 2. Oxygen deficiency indicator. |
| Hydrogen Sulfide, $H_2S$ | 10 | 1.19 | 4.3 | 46.0 | Rotten egg odor in small concentrations, but sense of smell rapidly impaired. Odor not evident at high concentrations. Colorless. Flammable. Explosive. Poisonous. | Death in a few minutes at 0.2%. Paralyzes respiratory center. | Petroleum fumes, from blasting, gas found in sewers. | 1. $H_2S$ analyzer. 2. $H_2S$ ampoules. |
| Carbon Dioxide, $CO_2$ | 5,000 | 1.53 | Not flammable | | Colorless, odorless, nonflammable. Not generally present in dangerous amounts unless there is already a deficiency of oxygen. | 10% cannot be endured for more than a few minutes. Acts on nerves of respiration. | Issues from carbonaceous strata. Gas found in sewers. | Oxygen deficiency indicator. |
| Ethane, $C_2H_4$ | | 1.05 | 3.1 | 15.0 | Colorless, tasteless, odorless, nonpoisonous. Flammable. Explosive. | See Hydrogen. | Natural gas. | Combustible gas indicator. |
| Chlorine, $Cl_2$ | 0.5 | 2.5 | Not flammable Not explosive | | Greenish yellow gas, or amber color liquid under pressure. Highly irritating and penetrating odor. Highly corrosive in presence of moisture. | Respiratory irritant, irritating to eyes and mucous membranes. 30 ppm causes coughing. 40–60 ppm dangerous in 30 minutes. 1,000 ppm apt to be fatal in a few breaths. | Leaking pipe connections. Overdosage. | Chlorine detector. Odor. Strong ammonia on swab gives off white fumes. |
| Sulfur Dioxide, $SO_2$ | 2 | 2.3 | Not flammable Not explosive | | Colorless compressed liquefied gas with a pungent odor. Highly corrosive in presence of moisture. | Respiratory irritant, irritating to eyes, skin, and mucous membranes. Only slightly less toxic than chlorine. | Leaking pipes and connections. | 1. Sulfur dioxide detector. Odor. 2. Strong ammonia on swab gives off white fumes. |

a Originally printed in Water and Sewage Works, August 1953. Adapted from "Manual of Instruction for Sewage Treatment Plant Operators," State of New York.

b 8TWA PEL is the time-weighted average permissible exposure limit, in parts per million, for a normal 8-hour workday and a 40-hour workweek to which nearly all workers may be repeatedly exposed, day after day, without adverse effect.

c Gases with a specific gravity less than 1.0 are lighter than air; those more than 1.0 are heavier than air.

d The first method given is the preferable testing procedure.

collect in low places because they are heavier than air. The *SPECIFIC GRAVITY*[9] of a gas indicates its weight as compared to an equal volume of air. Since air has a specific gravity of exactly 1.0, any gas with a specific gravity greater than 1.0 may sink to low-lying areas and displace the air from that area or structure. (On the other hand, methane may rise to the top or out of a manhole because it has a specific gravity of less than 1.0, which means that it is lighter than air.) Be aware that air movement and temperature differences can affect the concentration and location of gases.

### 3.511 *Toxic or Suffocating Gases or Vapors*

Toxic or suffocating gases may come from industrial waste discharges, process chemicals, or from the decomposition of domestic wastewater. You must become familiar with the waste discharges into your system. Table 3.1, "Common Dangerous Gases Encountered in Wastewater Collection Systems and at Wastewater Treatment Plants," contains information on methods of testing for several gases.

Gases from septic tank sludges are measured and reported as the percent of a specific gas in a volume of all gases. However, atmospheric monitors for confined spaces also monitor gases as percent *LOWER EXPLOSIVE LIMIT (LEL)*[10] and parts per million (ppm). Gas production in the septic tank varies from home to home, but an average range for the following gases is fairly typical: hydrogen sulfide ($H_2S$) = 0.8 percent; methane ($CH_4$) = 12.0 percent; nitrogen ($N_2$) = 13.5 percent; argon (A) = 0.25 percent; oxygen ($O_2$) = 0.65 percent; and hydrogen ($H_2$) less than 0.5 percent. Total yield or production of gases in an individual septic tank through anaerobic digestion by bacteria normally is about 0.8 cubic foot of gas/person/day. Septic tanks are usually vented through the house plumbing vents, so a large accumulation of gases may not be found in septic tanks.

### 3.512 *Explosive Gas Mixtures*

Explosive gas mixtures may develop in many areas of a treatment plant from mixtures of air and methane, natural gas, manufactured fuel gas, hydrogen, or gasoline vapors. Table 3.1 lists the common dangerous gases that may be encountered in a collection system or treatment plant and identifies their explosive range where appropriate. The upper explosive limit (UEL) and lower explosive limit (LEL) indicate the range of concentrations at which combustible or explosive gases will explode when an ignition source is present. No explosion occurs when the concentration is outside these ranges.

Explosive ranges can be measured by using a combustible gas detector calibrated for the gas of concern. Avoid explosions by eliminating all sources of ignition in areas potentially capable of developing explosive mixtures. Only explosion-proof electrical equipment and fixtures should be used in these areas (influent/bar screen rooms, gas compressor areas, battery charging stations). Provide adequate ventilation in all areas that have the potential to develop an explosive atmosphere.[11]

Do not rely on your nose to detect gases. The sense of smell is absolutely unreliable for evaluating the presence of dangerous gases. Some gases have no smell and hydrogen sulfide paralyzes the sense of smell.

## 3.52 Safety Precautions and Safe Procedures

According to OSHA standards, each entry into a permit-required confined space requires a confined space entry permit (Figure 3.2). The confined space entry permit is an authorization and approval in writing that specifies the location and type of work to be done, certifies that all existing hazards have been evaluated by the qualified person, and certifies that necessary protective measures have been taken to ensure the safety of each worker. The permit is renewed each time an operator leaves and re-enters the space after a break of more than 30 minutes.

The qualified person is a person designated in writing as capable (by education or specialized training) of anticipating, recognizing, and evaluating employee exposure to hazardous substances or other unsafe conditions in a confined space. This person must be capable of specifying the control procedures and protective actions necessary to ensure worker safety.

The potential for buildup of toxic or explosive gas mixtures or oxygen deficiency/enrichment exists in all confined spaces.

---

9. *Specific Gravity.* (1) Weight of a particle, substance, or chemical solution in relation to the weight of an equal volume of water. Water has a specific gravity of 1.000 at 4°C (39°F). Wastewater particles or substances usually have a specific gravity of 0.5 to 2.5. Particulates with specific gravity less than 1.0 float to the surface and particulates with specific gravity greater than 1.0 sink. (2) Weight of a particular gas in relation to the weight of an equal volume of air at the same temperature and pressure (air has a specific gravity of 1.0). Chlorine gas has a specific gravity of 2.5.
10. *Lower Explosive Limit (LEL).* The lowest concentration of a gas or vapor (percent by volume in air) that explodes if an ignition source is present at ambient temperature. At temperatures above 250°F (121°C) the LEL decreases because explosibility increases with higher temperature.
11. National Fire Protection Association (NFPA) Standard 820 lists the electrical and ventilation requirements for wastewater facilities within the United States. The National Electrical Code (NEC) must also be considered.

Date and Time Issued: ______ Date and Time Expires: ______ Job Site/Space I.D.: ______

Job Supervisor: ______ Equipment to be worked on: ______ Work to be performed: ______

Standby personnel: ______ ______ ______

1. Atmospheric Checks: Time ______ Oxygen ______ % Toxic ______ ppm
Explosive ______ % LEL Carbon Monoxide ______ ppm

2. Tester's signature: ______

3. Source isolation: (No Entry) N/A Yes No

| | N/A | Yes | No |
|---|---|---|---|
| Pumps or lines blinded, disconnected, or blocked | ( ) | ( ) | ( ) |

4. Ventilation Modification:

| | N/A | Yes | No |
|---|---|---|---|
| Mechanical | ( ) | ( ) | ( ) |
| Natural ventilation only | ( ) | ( ) | ( ) |

5. Atmospheric check after isolation and ventilation: Time ______
Oxygen ______ % > 19.5% < 23.5% Toxic ______ ppm < 10 ppm $H_2S$
Explosive ______ % LEL < 10% Carbon Monoxide ______ ppm < 35 ppm CO

Tester's signature: ______

6. Communication procedures: ______

7. Rescue procedures: ______

8. Entry, standby, and backup persons

| | Yes | No |
|---|---|---|
| Successfully completed required training? | ( ) | ( ) |
| Is training current? | ( ) | ( ) |

9. Equipment:

| | N/A | Yes | No |
|---|---|---|---|
| Direct reading gas monitor tested | ( ) | ( ) | ( ) |
| Safety harnesses and lifelines for entry and standby persons | ( ) | ( ) | ( ) |
| Hoisting equipment | ( ) | ( ) | ( ) |
| Powered communications | ( ) | ( ) | ( ) |
| SCBAs for entry and standby persons | ( ) | ( ) | ( ) |
| Protective clothing | ( ) | ( ) | ( ) |
| All electric equipment listed for Class I, Division I, Groups A, B, C, and D, and nonsparking tools | ( ) | ( ) | ( ) |

10. Periodic atmospheric tests:

Oxygen: ____% Time ____; ____% Time ____; ____% Time ____; ____% Time ____;
Explosive: ____% Time ____; ____% Time ____; ____% Time ____; ____% Time ____;
Toxic: ____ppm Time ____; ____ppm Time ____; ____ppm Time ____; ____ppm Time ____;
Carbon Monoxide: ____ppm Time ____; ____ppm Time ____; ____ppm Time ____; ____ppm Time ____;

We have reviewed the work authorized by this permit and the information contained herein. Written instructions and safety procedures have been received and are understood. Entry cannot be approved if any brackets ( ) are marked in the "No" column. This permit is not valid unless all appropriate items are completed.

Permit Prepared By: (Supervisor) ______ Approved By: (Unit Supervisor) ______

Reviewed By: (CS Operations Personnel) ______
(Entrant) (Attendant) (Entry Supervisor)

This permit to be kept at job site. Return job site copy to Safety Office following job completion.

*Fig. 3.2 Confined space pre-entry checklist/permit*

The atmosphere must be checked with reliable, calibrated instruments (Figure 3.3) prior to every entry. The normal oxygen concentration in the air we breathe is 20.9 percent. The atmosphere in the confined space must not fall below 19.5 percent oxygen. Engineering controls are required to prevent low oxygen levels. However, personal protective equipment is necessary if engineering controls are not possible. In atmospheres where the oxygen content is less than 19.5 percent, a self-contained breathing apparatus (SCBA) (Figure 3.4) is required. SCBAs are sometimes referred to as scuba gear because they look and work much like the oxygen tanks used by divers, but they are not waterproof. For individuals to safely and legally use respiratory protective devices, they should have medical approval, pass a fitness test, be properly trained, and conform to facial hair prohibition requirements.

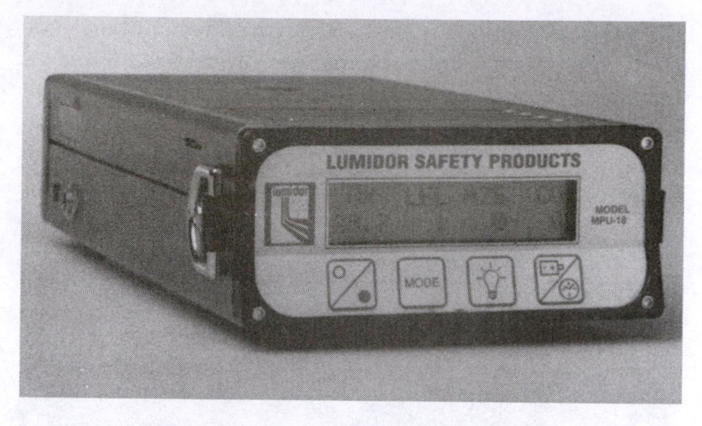


*Fig. 3.3 Portable atmospheric alarm unit*
(Permission of Lumidor Safety Products)

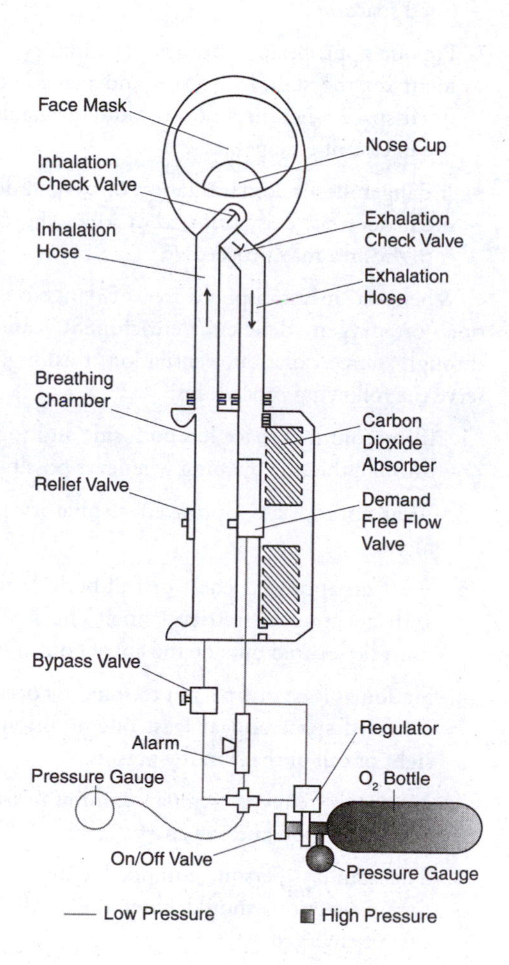


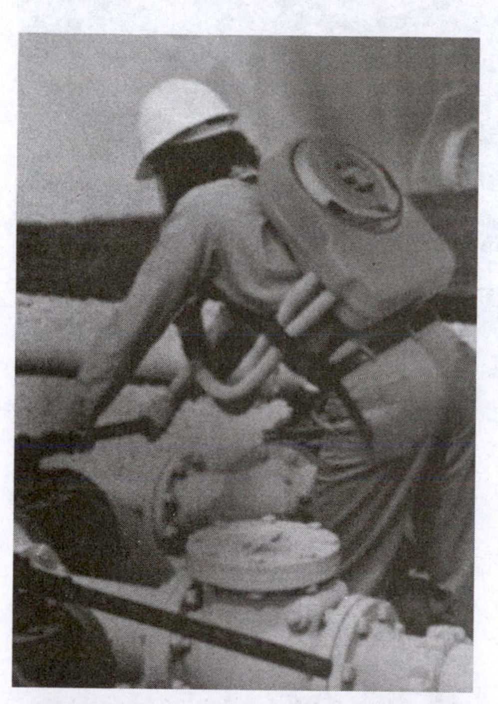

*Fig. 3.4 Self-contained breathing apparatus*
(Permission of BioMarine Industries, Inc.)

Entry into confined spaces is never permitted until the space has been properly ventilated using specially designed forced-air ventilators (Figure 3.5). These blowers force all the existing air out of the space, replacing it with fresh air from outside. This crucial step must always be taken even if gas detection and oxygen deficiency instruments show the atmosphere to be safe. Because some of the gases likely to be encountered in a confined space may be combustible or explosive, the blowers must be electrically rated (explosion proof) so that the blower itself will not create a spark that could cause an explosion.

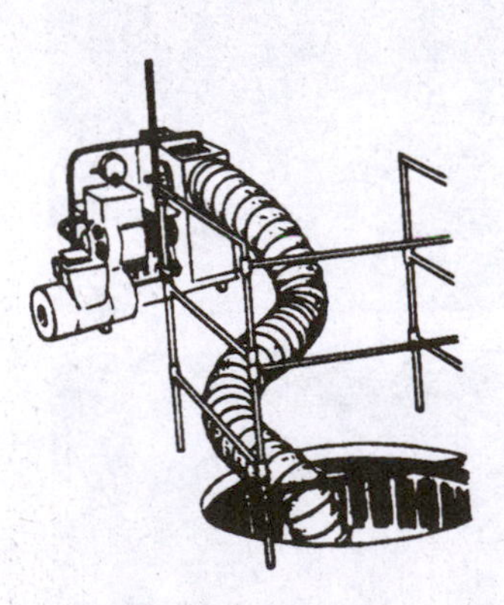

Ventilation of a manhole

*NOTE:* Air inlet should be at least two feet above the street level so street trash will not be picked up by the flowing air. Position the hose inlet in such a way that it will not pick up and blow exhaust gases from work vehicles or traffic into the manhole.

*Fig. 3.5 Ventilation blower with hose*

The following steps are recommended prior to entry into a confined space.

1. Identify and close off or reroute any sewer lines that may convey harmful substances to, or through, the work area.
2. Empty, flush, or purge the space of any harmful substances to the extent possible.
3. Monitor the atmosphere at the work site and within the space to determine if dangerous air contamination or oxygen deficiency/enrichment exists. Test first for oxygen, then for combustible gases and vapors, and then for toxic gases and vapors.
4. Record the atmospheric test results and keep them at the site throughout the work period.
5. If the space is interconnected with another space, test each space. The most hazardous conditions found should govern subsequent steps for entry into the space.
6. If an atmospheric hazard is noted, use portable blowers to further ventilate the area; retest the atmosphere after a suitable period of time. Do not place the blowers inside the confined space.
7. Provide appropriate, approved respiratory protective equipment for the standby person and place it outside the confined space where it will be readily available for immediate use in case of emergency.
8. If dangerous air contamination or oxygen deficiency/enrichment does not exist prior to or following ventilation, entry into the area may proceed.

Whenever an atmosphere free of dangerous air contamination or oxygen deficiency/enrichment cannot be ensured through source isolation, ventilation, flushing, or purging, observe the following procedures:

1. If the confined space has both side and top openings, enter through the side opening whenever possible.
2. Wear appropriate, approved respiratory protective equipment.
3. Wear an approved chest or full-body harness (Figure 3.6) with an attached retrieval line. The free end of the line must be secured outside the entry point.
4. Station at least one person to stand by on the outside of the confined space and at least one additional person within sight or call of the standby person.
5. Maintain frequent, regular communication between the standby person and the entry person.
6. The standby person, equipped with appropriate respiratory protection, should enter the confined space only in case of emergency.
7. If the entry is made through a top opening, use a hoisting device with a harness that suspends a person in an upright position.
8. If the space contains, or is likely to develop, flammable or explosive atmospheric conditions, do not use any tools or equipment (including electrical) that may provide a source of ignition.

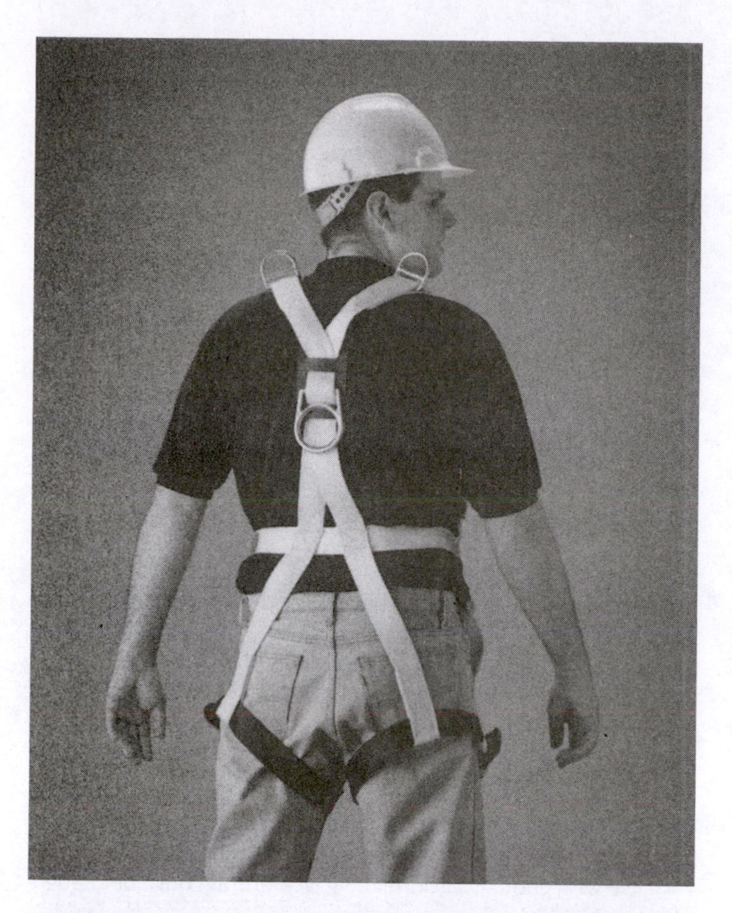

*Fig. 3.6 Full-body harnesses*

(Photos courtesy of Miller® Equipment)

9. Wear appropriate protective clothing when entering a confined space that contains corrosive substances or other substances harmful to the skin.
10. At least one person holding current certification in first aid and cardiopulmonary resuscitation (CPR) must be immediately available during any confined space job.

Confined space work can present serious hazards if you are uninformed or untrained. The procedures presented here are only guidelines and exact requirements for confined space work for your locale may vary. Contact your local regulatory safety agency for specific requirements in your area. Never enter a confined space with an atmosphere containing an explosive condition. If ventilation does not remove the explosive condition, evacuate the area and request assistance from an expert from your local natural gas company.

## QUESTIONS

Please write your answers to the following questions and compare them with those on page 107.

3.5A What is an easy way to identify a confined space?

3.5B What are the major hazards an operator could encounter in confined spaces?

3.5C What are the common properties of hydrogen sulfide gas?

3.5D Can you rely on your nose to detect dangerous gases?

3.5E When and why is it necessary to get an approved confined space entry permit?

3.5F Is it necessary to use blowers to blow all the air out of a confined space when gas detection instruments show the atmosphere to be safe?

3.5G How many persons are needed at the site when it is necessary for someone to enter a confined space?

## 3.6 VEHICLES, DRIVING, AND TRAFFIC

### 3.60 Hazards

Often, collection system work must be performed in a street or right-of-way where passing vehicles present a constant hazard. Also, most of our emergency situations occur when conditions are less than ideal. Darkness, rain, wind, fog, snow, ice, or other weather conditions affect our ability to drive safely. Even when conditions are ideal, the equipment and traffic conditions frequently require extraordinary driving skills just to get to the job site.

To reach the job site, an operator usually drives a utility vehicle of some sort. This vehicle may be a pickup truck or a combination of vehicle and towed equipment. When driving a vehicle day in and day out, it is easy to lose sight of the fact that this equipment needs to be inspected and maintained regularly for our own safety as well as to avoid the inconvenience of a breakdown.

### 3.61 Safety Precautions and Safe Procedures

#### *3.610 Vehicle Inspections*

As part of the routine procedure before heading out to a job site, your work vehicle and any towed equipment should undergo a thorough mechanical/safety item inspection of the following items:

**Mechanical Condition**

1. Windshield wipers.
2. Horn.
3. Seat belts.
4. Mirrors and windows.
5. Lighting system, including backup lights, turn signals, and brake lights. Also, check the towed vehicle lighting system.
6. Brakes on the vehicle and trailer, if so equipped.
7. Tire tread and inflation.
8. Wheel attachment (vandals may have loosened lug nuts).
9. Trailer hitch and tongue.
10. Auxiliary equipment such as winches or hoists.
11. Safety chain. After the trailer is coupled, the safety chain should be securely attached to the frame of the towing vehicle with enough slack to allow jackknifing, but not enough to drag on the ground.

**Safety Equipment**

1. Flashing lights or rotating beacons on work vehicles
2. Flashlight or spotlight
3. First-aid kit
4. Fire extinguisher
5. Road flares
6. Traffic cones
7. Street barricades
8. Traffic control signs and flags
9. Personal equipment such as hard hat, boots, gloves, and rain gear
10. Disposable towels and hand cleansers
11. Ventilation blowers
12. Atmospheric monitors

Inspect each piece of equipment you will be using to ensure it is working. Before an atmospheric testing device is taken out to a job site, for example, it should be inspected for calibration and proper function. If there is any doubt about the functional capacity of an item, use replacement equipment while the defective equipment is repaired or further tested.

### *3.611 Defensive Driving*

Having inspected your work vehicle and equipment, the next hazard you face is getting to the job site safely, frequently during rush hour and with heavy vehicle traffic conditions. Obey all traffic laws and remember to fasten your seat belts. Many states now have seat belt laws, and most agencies have seat belt policies. Also remember it is your responsibility to be sure that any passengers are using their seat belts.

Because collection systems are spread out, operators drive more miles annually on the job than the average person's total miles driven during a year. Therefore, defensive driving is an important part of our daily routine.

Defensive driving is not only a mature attitude toward driving, but a strategy as well. You must always be on the alert for the large number of new drivers, inattentive or sleepy drivers, people driving under the influence of alcohol or drugs, and those drivers who are just plain incompetent. In addition, weather, construction, and the type of equipment you are driving can also present hazards. Key elements of defensive driving include the following:

1. Always be aware of what is going on ahead of you, behind you, and on both sides of you.
2. Always have your vehicle under control.
3. Be willing to surrender your legal right-of-way if it might prevent an accident.
4. Know the limitations of the vehicle and towed equipment you are operating (stopping distance, impaired road vision, distance to change lanes safely, acceleration rate).
5. Take into consideration weather or other unusual conditions.
6. Develop a defensive driving attitude, that is, be aware of where you are, be alert for potential dangers, and be prepared with strategies to avoid accidents. For example, in residential areas, you should assume that a child could run out into the street at any time, so always be thinking of what type of evasive action you could take. Be especially alert driving through construction areas during rush hour.
7. Always maintain a safe stopping distance behind the vehicle in front of you. Rear end collisions, either due to driver inattentiveness or tailgating, are one of the most frequent types of vehicle accidents. A safe stopping distance is the total minimum distance your vehicle will travel under ideal conditions, with good brakes, and at various speeds, as you perceive a road hazard, react to the hazard, apply the brakes, and come to a complete stop. Traveling at 55 miles per hour (88 km/hr), a minimum safe stopping distance could range from 290 to 300 feet (88 to 91 meters) or more, depending on vehicle mass or weight, type of brakes (air or hydraulic), weather conditions, road conditions, time of day, road incline, and the driver's visual acuity (clearness of vision) and reaction time.
8. Keep in mind that you are driving a vehicle that is highly visible to the public. It may be necessary to swallow your pride when confronted by discourteous or unsafe drivers. Even when in the wrong, those types of drivers are quick to alert your agency, and thus your supervisor, to real or imagined violations that you committed while driving an agency vehicle.

### *3.612 Job Site Protection*

Working in a roadway represents a significant hazard to a small wastewater system operator as well as pedestrians and drivers. Motor vehicle drivers may be distracted while driving and may not be focused on safe driving. At any given time of the night or day, a certain percentage of drivers could be driving while under the influence of drugs or alcohol. Given the amount of time small wastewater system operators work in traffic while cleaning pipes, performing inspections, doing rehabilitation work, and making repairs, the control of traffic is necessary to reduce the risk of injury or death while working in this hazardous area. Consequently, traffic movement and street or utility repair work must be regulated to provide optimum safety and convenience for all.

Any time traffic may be affected by road work activities, you must notify appropriate authorities in your area before work begins. These could be state, county, or local authorities depending on whether it is a state, county, or local street. Frequently, a permit must be issued by the authority that has jurisdiction before traffic can be diverted or disrupted. In some cases, when traffic diversion or disruption may obstruct access by emergency response agencies such as fire and police departments, these agencies must be notified before traffic is diverted or disrupted as well. In most cases, you will need to plan ahead to secure permits and notify authorities. This could involve a phone call or two or it could mean several days' or weeks' advance planning if you need to make extensive traffic control arrangements.

The primary function of temporary traffic control (TTC) is to provide for the safe and efficient movement of vehicles, bicyclists, and pedestrians (including persons with disabilities in accordance with the Americans with Disabilities Act of 1990 (ADA)) through or around TTC zones while reasonably protecting workers and equipment. This can be accomplished by appropriate and consistent use of TTC devices. Most TTC zones are divided into four areas: the advance warning area, the transition area, the activity area, and the termination area. A TTC device is defined as a sign, signal, marking, or other device used to regulate, warn, or guide traffic, placed on, over, or adjacent to a street, highway, or pedestrian facility, or shared-use path by authority of a public agency having jurisdiction. Figures 3.7 and 3.8 show examples of TTC devices used to warn and guide traffic.

Part 6 of the US Department of Transportation's *MANUAL ON UNIFORM TRAFFIC CONTROL DEVICES* (MUTCD)[12] is the national standard for all TTC devices used during construction, maintenance, utility activities, and incident management. The MUTCD contains basic principles, a description of the standard traffic control devices used in work areas, guidelines for the application of the devices, and typical application diagrams. Information concerning proper flagging is also presented.

*Fig. 3.7 Signs warning traffic in the advance warning area*

12. *MANUAL ON UNIFORM TRAFFIC CONTROL DEVICES* (MUTCD), Federal Highway Administration (FHWA). The current edition is available as a PDF version at http://mutcd.fhwa.dot.gov/.

*Fig. 3.8 Portable arrows guiding traffic around a work activity area*
(Permission of Safety Tech, Inc.)

While not all of the examples will meet the specific requirements of the laws in your geographical area, they should serve to make you aware of various aspects of traffic control. Check your local and state standards for TTC guidelines in specific circumstances and incorporate these control requirements into your agency's procedures.

Each person whose actions affect TTC zone safety, from upper-level management to field workers, should receive training appropriate to the job decisions each individual is required to make. Only those individuals who are trained in proper TTC practices and have a basic understanding of the principles (established by applicable standards and guidelines, including those of the *MANUAL ON UNIFORM TRAFFIC CONTROL DEVICES* (MUTCD)) should supervise the selection, placement, and maintenance of TTC devices used for TTC zones and for incident management. Individuals become qualified to control traffic by gaining the following knowledge and experience:

- A basic understanding of the principles of TTC in work zones
- Knowledge of the standards and guidelines governing TTC
- Adequate training in safe TTC practices
- Experience in applying TTC in work zones

Any work in public streets or highways must be regulated to ensure proper coordination of the work, thus protecting the public's interest. To accomplish this, any person, firm, corporation, or agency must obtain permission from the governing road authority before starting work within the right-of-way of any street or highway. Regulatory TTC devices or signs must be approved by the governing road authority before they are installed on any street or highway.

The governing road authority may determine or define the times when work may be performed. During peak traffic periods, construction work may be restricted or not permitted. Peak periods of traffic movement may vary in different areas.

Good public relations should be maintained at all times and all TTC devices must be removed as soon as practical when they are no longer needed.

## QUESTIONS

Write your answers in a notebook and then compare your answers with those on page 107.

3.6A List the items on a work vehicle and towed equipment that should undergo a thorough mechanical inspection as part of the routine procedure before heading out to a job site.

3.6B What are some of the variables that affect actual vehicle stopping distances versus theoretical total stopping distance?

3.6C What is the primary function of temporary traffic control (TTC) on streets and highways?

3.6D Who must be notified any time traffic may be affected by road work activities?

3.6E What publication is the national standard for all temporary traffic control (TTC) devices used during construction, maintenance, utility activities, and incident management?

## 3.7 ELECTRICITY

Wastewater collection and treatment system operators are routinely exposed to a variety of hazardous field conditions related to electrical equipment. The first rule that applies to electrical systems is: If you are not qualified, do not, under any circumstances, attempt to work on electrical systems. The objective of this section is not to make you an instant electrical expert qualified to work on electrical equipment, but it is to make you more knowledgeable so that you have an awareness of the hazards associated with electrical systems.

To simplify our understanding of electrical systems, we can compare the similarities of a hydraulic system, in which a pump supplies water and pressure, to an electrical system, in which a power company supplies current and voltage (Figure 3.9).

1. The energy or pressure from the pump is similar to the energy or voltage from the power company. In both systems, an energy source is needed to move a mass of physical particles.
2. The movement of a mass of water (molecules) through a pipe can be compared to the movement of a mass of current (electrons) through a conductor such as wire. In both systems, physical particles (molecules or electrons) are moved through a conduit or channel.
3. A variable valve such as a gate valve on a water pipe is like a variable resistor on an electrical wire. Both create a blockage

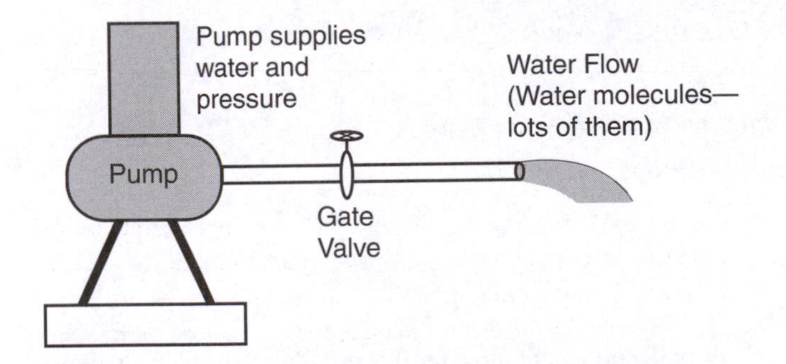


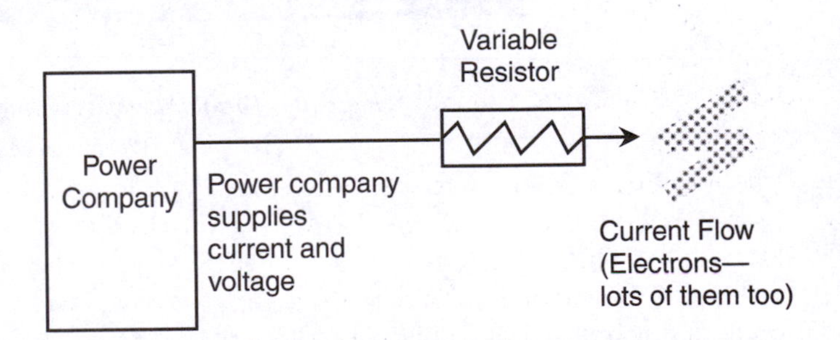


Maximum resistance (valve closed) = no water flow

Minimum resistance (valve open) = maximum water flow

Maximum resistance (open circuit) = no current flow

Minimum resistance (short circuit) = maximum current flow

| | Energy (Potential) | Mass (Particles) | Restriction (To Flow) |
|---|---|---|---|
| Hydraulic System | Pressure | Water (Molecules) | Variable Valve |
| Electrical System | Voltage | Current (Electrons) | Variable Resistor |

*Fig. 3.9 Electrical system and hydraulic system similarities*

or resistance to the movement of flow (water or current). (The term "variable" simply means the valve or resistor is capable of adjustment, similar to a volume knob on a radio.) In both systems, flow of the moving mass of particles can be adjusted from no flow to maximum flow to levels of flow in between, by adjusting the valve or resistor. In a hydraulic system, a closed valve (maximum resistance) means no water will flow, whereas an open valve means maximum water will flow. In an electrical system, an open circuit (maximum resistance) means no current will flow, whereas a closed or short circuit means maximum current will flow. In electrical systems, resistance is referred to as the "load," and the variable resistor may take the form of motor windings in an electric drill or a filament in a light bulb. A variable resistor can also take the form of a human body.

In the case of the electrical system, the power company or utility can be compared to the pump. It supplies the electric energy just as the pump supplies hydraulic energy to the pipeline. Other similarities exist between hydraulic systems and electrical systems. Both of these systems want to become a "closed loop." In other words, in an electrical system, the electrons (current measured in amperes (amps), milliamps, or microamps) want to return to the source, which is commonly referred to as "ground" (at a zero voltage state). In the case of the hydraulic system, the pressure in the pipe also wants to return to a zero pressure state. When you open a closed, pressurized pipeline, the pressure will want to drop to zero if the pump is off.

The significance of this, in small wastewater system work, is that if the human body accidentally becomes the load or inadvertently creates a blockage or resistance to the electric current, then current will flow through the body in an attempt to return to its ground state. The amount of current that flows through the human body has a direct bearing on the amount of damage that is done to the body, ranging from a painful shock to irreparable damage to the body tissue.

Many misconceptions exist about electricity. Among the more common is that low voltage (120 volts) is less dangerous than a higher voltage (240 volts, 3 phase, or 480 volts, 3 phase). In fact, it is the current flow (electrons) through the body that causes the problems. The amount of current flowing through the body depends on the voltage (volts) and the resistance (ohms) and can be expressed in the following formula:

Amps are equal to voltage divided by resistance.

$$\text{Current, amps} = \frac{\text{Voltage, volts}}{\text{Resistance, ohms}} \quad \text{or} \quad \frac{\text{Watts}}{\text{Volts}} = \text{Amps}$$

For example, a 100-watt light bulb in your house that is in a 120-volt (AC) circuit has a resistance in the filament of approximately 144 ohms.

$$\text{Current, amps} = \frac{120 \text{ volts}}{144 \text{ ohms}} = 0.83 \text{ amp} \quad \text{or} \quad \frac{100 \text{ W}}{120 \text{ V}} = 0.83 \text{ amp}$$

Similarly, a 7.5-watt bulb in a 120-volt circuit results in a current flow of 0.06 amp. This appears to be a very small amount of current, but let us take a look at the effects of current in the human body.

In this case, we will use milliamps, which is a thousandth (0.001) of an amp; 0.06 amp becomes 60 milliamps and 0.83 amp becomes 830 milliamps.

- 1 milliamp or less, no sensation, not felt.
- More than 5 milliamps, painful shock. (This is 5 thousandths of an amp, 0.005 amp.)
- More than 10 milliamps, local muscle contractions sufficient to cause freezing of the muscles for 2.5 percent of the human population.
- More than 15 milliamps, local muscle contractions sufficient to cause freezing to 50 percent of the population.
- More than 30 milliamps, breathing is difficult, can cause unconsciousness.
- 50 to 100 milliamps, possible ventricular fibrillation of the heart (uncontrolled, rapid beating of the heart muscle).
- Over 200 milliamps, severe burns and muscular contractions, heart more apt to suffer stoppage rather than only fibrillation. (This is only two-tenths of an amp.)

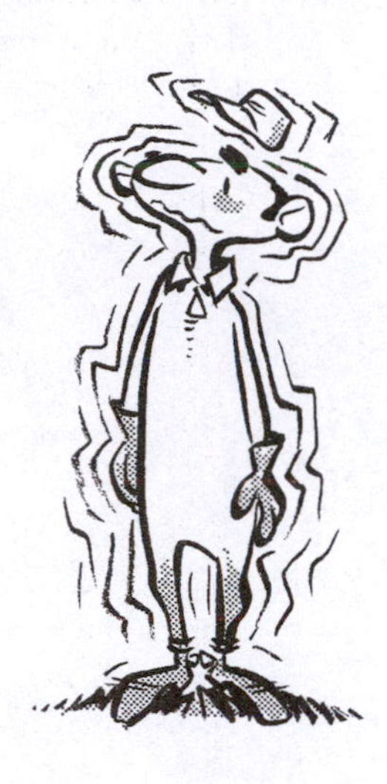

- Over a few amperes, irreparable damage to the body tissue.

So, the current flowing in the 120-volt circuit that lights a 7.5-watt light bulb is enough to cause severe muscular contractions of the heart. Again, the controlling factor that determines the extent of injury is the amount of resistance to the flow of current. Figure 3.10 illustrates some typical body resistances and the resulting current flow.

- If your head happens to become the load in an electric circuit, the current flows through the brain from ear to ear. At 120 volts, that could allow 1.2 amps of current to flow, which is 1,200 milliamps. Chances for permanent damage or death are very high, since the head offers only 100 ohms of resistance.

$$\frac{\text{Volts}}{\text{Ohms}} = \text{Amps} \qquad \frac{120 \text{ volts}}{100 \text{ ohms}} = 1.2 \text{ amps}$$

- If the current flows from the hand to the foot, passing through the heart, this represents a resistance of about 500 ohms, which will allow 240 milliamps of current to flow through this vital organ. Severe burns and muscular contractions, heart stoppage, and death are likely results.
- Resistance to current flow changes significantly if the skin is wet. Dry skin will have a resistance of 100,000 to 600,000 ohms while wet skin will have a resistance of only 1,000 to 6,000 ohms. When dealing with higher voltages such as 220 volts, or 480 volts, or even higher, wet skin allows more current to flow through the body.

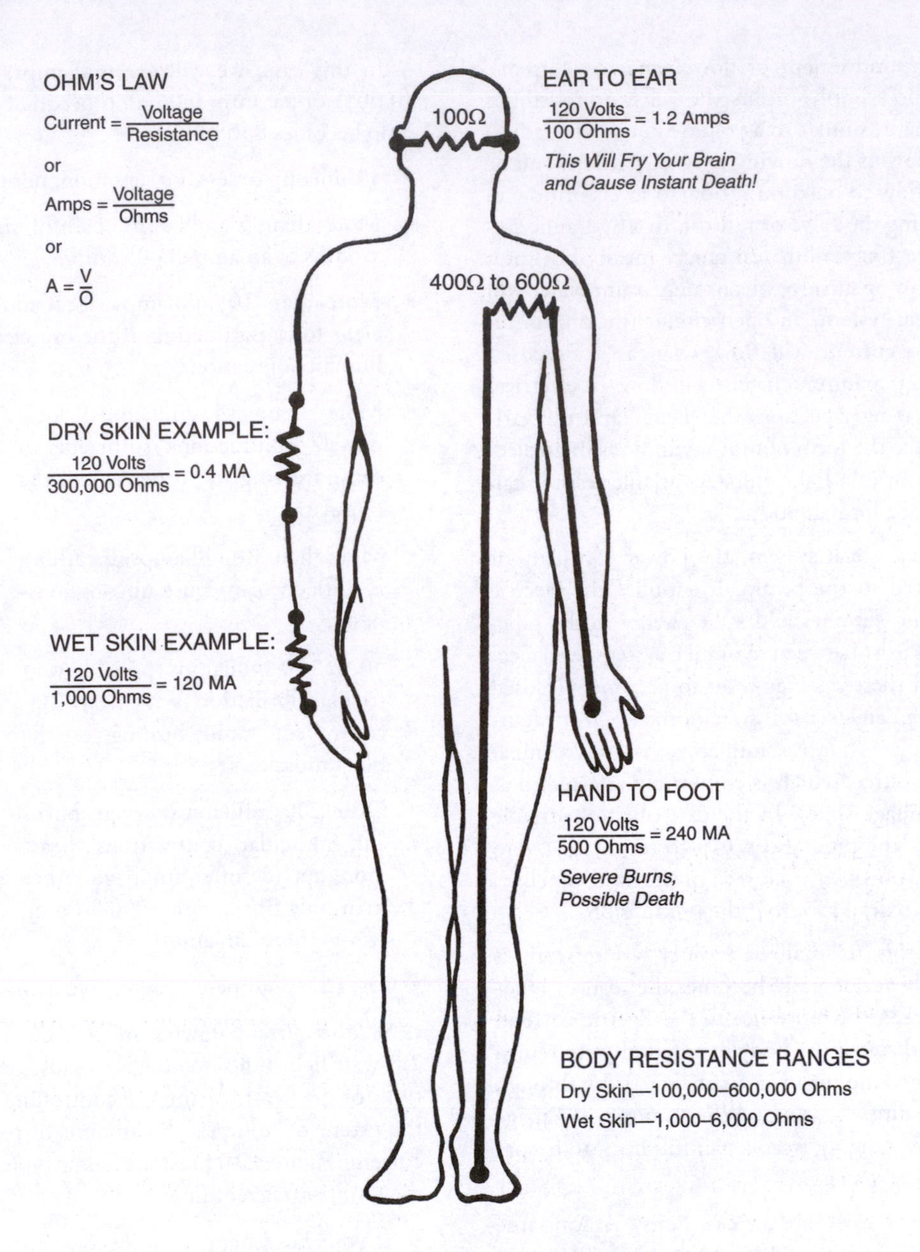


*Fig. 3.10 Typical body resistances and resulting current flow*

(Permission of General Electric)

### 3.70 Hazards

Let us take a look at some of the types of equipment and work situations that increase our exposure to electrical hazards:

1. Lift stations have a variety of electrical systems, single phase, three phase, alternating current (AC), direct current (DC), low voltage, and medium voltage.
2. Flow metering stations, which are usually wet locations, may require explosion-proof wiring methods.
3. Telemetry systems.
4. Standby/emergency power equipment.
5. Voltage ranges from low to high, voltages range from 24 volts to 4,160 volts, and as high as 13,800 volts (typical exposure is in the 120-volt to 480-volt range).
6. Much of our equipment is located in wet or damp locations.
7. We frequently have to work under difficult conditions such as during thunderstorms when downed power lines are frequent hazards.
8. Failures occur frequently at night when it is dark.
9. We may be required to provide standby power through the use of portable generators and temporary wiring.
10. When working under emergency conditions, time is usually very limited. In order to prevent bypassing of raw wastewater or backups into private homes, we are under a great deal of pressure to rig pumps, generators, or other equipment.
11. Construction repair projects increase the hazard of encountering live underground power lines.

These hazards increase the possibility of serious problems:

- Electrocution resulting in slight to serious burns or even death
- Physical injury due to failure to lock out or disconnect electrical systems; for example, someone starts a pump while you have your hand in the impeller pulling out rags
- Explosions resulting from electrical equipment operating in explosive atmospheres such as wet wells and underground vaults
- Protective equipment such as circuit breakers exploding due to high short-circuit currents
- Temporary or permanent blindness and severe burns due to flashing or arcing

### 3.71 Safety Precautions and Safe Procedures

Once again, if you are not qualified you have no business working in or around electrical systems. Not only are you creating a hazard to yourself, but to your colleagues and the public as well. Electrical maintenance/installation is controlled by rigid federal and state regulations including the National Electrical Code (NEC) and local building codes.

If you are qualified and are working on electrical equipment, observe the basic safety rules listed in the following paragraphs.

Disconnect and lock out all equipment prior to working on it. Always lock out and tag electrical equipment being serviced. This is very important in Septic Tank Effluent Pump (STEP) systems where electrical power is taken from the power panel of the customer. If the control circuit breaker is not locked or tagged, a homeowner could accidentally energize the system while you are working on it. (See Section 3.22, "Basic Lockout/Tagout Procedure.")

Observe common-sense procedures that will prevent or reduce the possibility of electrocution through the accidental contact with a live circuit. Use rubber mats, rubber boots, leather gloves, or some other form of isolation between you and ground and use only grounded power tools, especially when working in wet conditions.

Avoid wearing eyeglasses with metal frames, watches, jewelry, or metal belt buckles, which can increase the possibility of severe burns and electrocution if they come in contact with electric circuits.

Always maintain adequate clearance of body parts with live circuits. Professional electricians will always try to work with one hand only when working around live electric circuits. This, coupled with adequate isolation from the ground through rubber mats and gloves, will minimize the possibility of completing a circuit.

Electrical codes in many states specify that when working on certain levels of voltage in live circuits, you should not be alone. Your partner and colleague should be trained in CPR or other first-responder first-aid measures in case of accidental electrocution.

Wear safety glasses when working on live equipment. Tinted glasses are preferred. Flashing or arcing can damage the eyes

through the instantaneous generation and explosion of molten metal and high-intensity arcing from short circuits.

When resetting tripped circuit breakers, always investigate the cause of the circuit breaker tripping. Although circuit breakers are prone to nuisance type trips, they are also protective devices indicating a possible short circuit or other fault condition downstream. There are cases where even qualified people have neglected to check out the cause of the tripping and have tried to reset the circuit breaker, which then exploded. Enclosures should always be closed. For example, shut panel doors when reenergizing a circuit. The doors are a protective device designed to protect you against the explosion of circuit breakers or other electrical components within the enclosure.

Another possibility for electrocution occurs when we are working in damp and wet places and using electric hand tools (drills, saws) for routine maintenance work. Solid-state electronics and the recognition that very small amounts of current flowing through the human body can have fatal results prompted the development of the ground-fault circuit interrupter (GFCI). Today, most of us have this type of device in our homes because the 1975 National Electrical Code (NEC) requires them in bathrooms and for outdoor receptacles. Hotels and other commercial facilities have installed them as well. When working in damp or wet locations, protect yourself from shock by using a ground-fault circuit interrupter when you are using portable power tools. Again, STEP systems are a major hazard because you may be working on pump systems in a septic tank.

### 3.72 Emergency Procedures

In the event of someone suffering from electric shock, the following steps should be taken:

1. Survey the scene and see if it is safe to enter.
2. If necessary, free the victim from a live power source by shutting power off at a nearby disconnect, or by using a dry stick or some other nonconducting object to move the victim.
3. Request help by calling 911 or the emergency number in your community.
4. Check for breathing and pulse. If the person is unresponsive, begin CPR (cardiopulmonary resuscitation) immediately. Position the person on his or her back. Use an automated external defibrillator (a device used to correct a dangerously abnormal heart rhythm), if available. Or, begin hands-only

CPR with straight arms and forceful compressions of the chest (press down about 2 inches) at about 100 a minute. Lift hands slightly after each compression to allow the chest to recoil. Take turns with a bystander until emergency medical services arrive. Keep CPR interruptions to a minimum.

Remember, only trained and qualified individuals working as teams of two or more should be allowed to service, repair, or troubleshoot electrical equipment and systems.

## QUESTIONS

Please write your answers to the following questions and compare them with those on page 107.

3.7A What is the first rule that applies to electrical systems?

3.7B What is the "closed loop" similarity between hydraulic and electrical systems?

3.7C When working with electric circuits, how can you provide isolation between you and the ground?

## 3.8 HAZARDOUS CHEMICALS

Wastewater operators may come in contact with a wide variety of chemicals. Chlorine gas or chlorine compounds, which are used for wastewater disinfection or sanitizing, are among the most frequently used hazardous materials in wastewater systems. Plant laboratories require a variety of acids and bases along with other chemicals to perform sample analyses. Commonly used materials such as engine fuels, paints and thinners, and cleaning agents also require safe storage, use, and precautions to prevent personal injury.

If you are employed in a wastewater system that has several operators, the supervisor should have a listing of all the chemicals used by personnel in the system. If you are the lone operator or lead operator, then it is your responsibility to establish a list with descriptions of proper handling and safety procedures for each chemical used. The list should start with the chemical that is used in the greatest volume by you or the operators in the system. The federal worker *RIGHT-TO-KNOW LAW*[13] states that a person using a chemical has the right to know the hazards associated with that chemical.

The best source of information about an individual chemical is the *MATERIAL SAFETY DATA SHEET (MSDS)*.[14] The Material Safety Data Sheet with the required information can be obtained from the manufacturer or vendor of the chemical or material. The MSDS tells of the types of hazards the chemical presents and what to do in case of an emergency. Figure 3.11 shows the MSDS form recommended by OSHA and an actual MSDS for hydrogen sulfide is shown on the two pages following the OSHA form.

Another reliable source of information about specific chemicals is the label on the material container. There are a number of different labeling systems in use, including one designed by the National Fire Protection Association (NFPA) (Figure 3.12). In this design, four diamond-shaped divisions are typically color-coded, with blue indicating level of health hazard, red indicating level of fire hazard (flammability), yellow indicating (chemical) reactivity, and white containing special codes for specific hazards. Each division of health, flammability, and reactivity is rated on a scale from 0 (no hazard or ordinary hazard) to 4 (severe hazard). Private labeling systems such as the one illustrated in Figure 3.13 produced by J. T. Baker Chemical Company are also widely available.

All operators using the chemicals must have immediate access to this information and must be familiar with its proper use and response to emergencies through training and personal study.

If you have an emergency and need technical help, call *CHEMTREC* at (800) 424-9300.

Systems that use chlorine gas should have leak repair kits for the size of chlorine containers used, along with access to self-contained breathing units (local fire departments normally have several of these units). Both fire department personnel and wastewater operators should be well trained in the use of the repair kits and how to handle various chlorine leaks. The chlorine supplier can provide this training. Once trained, refresher practice sessions should be held routinely with the operators and fire department personnel.

## QUESTIONS

Please write your answers to the following questions and compare them with those on page 108.

3.8A What is the purpose of the MSDS?

3.8B Where could an operator obtain the MSDS for a particular chemical?

---

13. *Right-To-Know Laws.* Employee Right-To-Know legislation requires employers to inform employees of the possible health effects resulting from contact with hazardous substances. At locations where this legislation is in force, employers must provide employees with information regarding any hazardous substances they might be exposed to under normal work conditions or reasonably foreseeable emergency conditions resulting from workplace conditions. OSHA's Hazard Communication Standard (HCS) (Title 29 CFR Part 1910.1200) is the federal regulation and state statutes are called Worker Right-To-Know laws. Also see COMMUNITY RIGHT-TO-KNOW and SARA.

14. *Material Safety Data Sheet (MSDS).* A document that provides pertinent information and a profile of a particular hazardous substance or mixture. An MSDS is normally developed by the manufacturer or formulator of the hazardous substance or mixture. The MSDS is required to be made available to employees and operators or inspectors whenever there is the likelihood of the hazardous substance or mixture being introduced into the workplace. Some manufacturers are preparing MSDSs for products that are not considered to be hazardous to show that the product or substance is not hazardous.

Material Safety Data Sheet
May be used to comply with
OSHA's Hazard Communication Standard,
29 CFR 1910.1200 Standard must be
consulted for specific requirements.

**U.S. Department of Labor**
Occupational Safety and Health Administration
(Non-Mandatory Form)
Form Approved
OMB No. 1218-0072

| **IDENTITY** *(As Used on Label and List)* | Note: Blank spaces are not permitted. If any item is not applicable, or no information is available, the space must be marked to indicate that. |
|---|---|

**Section I**

| | |
|---|---|
| Manufacturer's Name | Emergency Telephone Number |
| Address *(Number, Street, City, State, and ZIP Code)* | Telephone Number for Information |
| | Date Prepared |
| | Signature of Preparer *(optional)* |

**Section II — Hazardous Ingredients/Identity Information**

| Hazardous Components (Specific Chemical Identity; Common Name(s)) | OSHA PEL | ACGIH TLV | Other Limits Recommended | % *(optional)* |
|---|---|---|---|---|
| | | | | |
| | | | | |
| | | | | |
| | | | | |
| | | | | |
| | | | | |
| | | | | |
| | | | | |

**Section III — Physical/Chemical Characteristics**

| | | | |
|---|---|---|---|
| Boiling Point | | Specific Gravity ($H_2O$ = 1) | |
| Vapor Pressure (mm Hg) | | Melting Point | |
| Vapor Density (AIR = 1) | | Evaporation Rate (Butyl Acetate = 1) | |
| Solubility in Water | | | |
| Appearance and Odor | | | |

**Section IV — Fire and Explosion Hazard Data**

| | | | |
|---|---|---|---|
| Flash Point (Method Used) | Flammable Limits | LEL | UEL |
| Extinguishing Media | | | |
| Special Fire Fighting Procedures | | | |
| Unusual Fire and Explosion Hazards | | | |

(Reproduce locally)

OSHA 174, Sept. 1985

*Fig. 3.11 Material Safety Data Sheet*

## Section V — Reactivity Data

| Stability | Unstable | | Conditions to Avoid |
|---|---|---|---|
| | Stable | | |

Incompatibility *(Materials to Avoid)*

Hazardous Decomposition or Byproducts

| Hazardous Polymerization | May Occur | | Conditions to Avoid |
|---|---|---|---|
| | Will Not Occur | | |

## Section VI — Health Hazard Data

| Route(s) of Entry: | Inhalation? | Skin? | Ingestion? |
|---|---|---|---|

Health Hazards *(Acute and Chronic)*

| Carcinogenicity: | NTP? | IARC Monographs? | OSHA Regulated? |
|---|---|---|---|

Signs and Symptoms of Exposure

Medical Conditions
Generally Aggravated by Exposure

Emergency and First Aid Procedures

## Section VII — Precautions for Safe Handling and Use

Steps to Be Taken in Case Material is Released or Spilled

Waste Disposal Method

Precautions to Be Taken in Handling and Storing

Other Precautions

## Section VIII — Control Measures

Respiratory Protection *(Specify Type)*

| Ventilation | Local Exhaust | Special |
|---|---|---|
| | Mechanical *(General)* | Other |

| Protective Gloves | Eye Protection |
|---|---|

Other Protective Clothing or Equipment

Work/Hygienic Practices

Page 2

* U.S.G.P.O.:1986-491-529/45775

*Fig. 3.11 Material Safety Data Sheet (continued)*

**Genium Publishing Corporation**
One Genium Plaza
Schenectady, NY 12304-4690 USA
(518) 377-8854

***Material Safety Data Sheets Collection:***

**Sheet No. 52**
**Hydrogen Sulfide**

Issued: 7/79 Revision: B, 9/92

## Section 1. Material Identification

**39**

**Hydrogen Sulfide ($H_2S$) Description:** Formed as a byproduct of many industrial processes (breweries, tanneries, slaughter houses), around oil wells, where petroleum products are used, in decaying organic matter, and naturally occurring in coal, natural gas, oil, volcanic gases, and sulfur springs. Derived commercially by reacting iron sulfide with dilute sulfuric or hydrochloric acid, or by reacting hydrogen with vaporized sulfur. Used in the production of various inorganic sulfides and sulfuric acid, in agriculture as a disinfectant, in the manufacture of heavy water, in precipitating sulfides of metals; as a source of hydrogen and sulfur, and as an analytical reagent.

**Other Designations:** CAS No. 7783-06-4, dihydrogen monosulfide, hydrosulfuric acid, sewer gas, stink damp, sulfuretted hydrogen, sulfur hydride.

**Manufacturer:** Contact your supplier or distributor. Consult latest *Chemical Week Buyers' Guide*[73] for a suppliers list.

**Cautions:** Hydrogen sulfide is a highly flammable gas and reacts vigorously with oxidizing materials. It is highly toxic and can be instantly fatal if inhaled at concentrations of 1000 ppm or greater. Be aware that the sense of smell becomes rapidly fatigued at 50 to 150 ppm, and that its strong rotten-egg odor is not noticeable even at very high concentrations.

R 2
I 4
S 3
K 3

**NFPA**
4 (top), 3 (left), 0 (right), – (bottom)

HMIS
H 3
F 4
R 0
PPE*
* Sec. 8

## Section 2. Ingredients and Occupational Exposure Limits

Hydrogen sulfide: 98.5% *technical*, 99.5% *purified*, and CP *(chemically pure grade)*

**1991 OSHA PELs**
8-hr TWA: 10 ppm (14 mg/m$^3$)
15-min STEL: 15 ppm (21 mg/m$^3$)

**1990 IDLH Level**
300 ppm

**1990 NIOSH REL**
10-min Ceiling: 10 ppm (15 mg/m$^3$)

**1992-93 ACGIH TLVs**
TWA: 10 ppm (14 mg/m$^3$)
STEL: 15 ppm (21 mg/m$^3$)

**1990 DFG (Germany) MAK**
TWA: 10 ppm (15 mg/m$^3$)
Category V: Substances having intense odor
Peak exposure limit 20 ppm, 10 min momentary value, 4/shift

**1985-86 Toxicity Data***
Human, inhalation, $LC_{Lo}$: 600 ppm/30 min; toxic effects not yet reviewed
Man, inhalation, $LD_{Lo}$: 5700 µg/kg caused coma and pulmonary edema or congestion.
Rat, intravenous, $LD_{50}$: 270 µg/kg; no toxic effect noted

* See NIOSH, *RTECS* (MX1225000), for additional toxicity data.

## Section 3. Physical Data

**Boiling Point:** -76 °F (-60 °C)
**Freezing Point:** -122 °F (-86 °C)
**Vapor Pressure:** 18.5 atm at 68 °F (20 °C)
**Vapor Density (Air = 1):** 1.175
**pH:** 4.5 (freshly prepared saturated aqueous solution)
**Viscosity:** 0.01166 cP at 32 °F/0 °C and 1 atm
**Liquid Surface Tension (est):** 30 dyne/cm at -77.8 °F/-61 °C
**Molecular Weight:** 34.1
**Density:** 1.54 g/L at 32 °F (0 °C)
**Water Solubility:** Soluble*; 1g/187 mL (50 °F/10 °C), 1g/242 mL (68 °F/20 °C), 1g/ 314 mL (86 °F/30 °C)
**Other Solubilities:** Soluble in ethyl alcohol, gasoline, kerosine, crude oil, and ethylene glycol.
**Odor threshold:** 0.06 to 1.0 ppm†

**Appearance and Odor:** Colorless gas with a rotten-egg smell.

* $H_2S$ solutions are not stable. Absorbed oxygen causes turbidity and precipitation of sulfur. In a 50:50 mixture of water and glycerol, $H_2S$ is stable.
† Sense of smell becomes rapidly fatigued and can not be relied upon to warn of continuous $H_2S$ presence.

## Section 4. Fire and Explosion Data

| **Flash Point:** None reported | **Autoignition Temperature:** 500 °F (260 °C) | **LEL:** 4.3% v/v | **UEL:** 46% v/v |
|---|---|---|---|

**Extinguishing Media:** Let small fires burn unless leak can be stopped immediately. For large fires, use water spray, fog, or regular foam. **Unusual Fire or Explosion Hazards:** $H_2S$ burns with a blue flame giving off sulfur dioxide. Its burning rate is 2.3 mm/min. Gas may travel to a source of ignition and flash back. **Special Fire-fighting Procedures:** Because fire may produce toxic thermal decomposition products, wear a self-contained breathing apparatus (SCBA) with a full facepiece operated in pressure-demand or positive-pressure mode. Structural firefighter's protective clothing is not effective for fires involving $H_2S$. If possible without risk, stop leak. Use unmanned device to cool containers until well after fire is out. Withdraw immediately if you hear a rising sound from venting safety device or notice any tank discoloration due to fire. Do not release runoff from fire control methods to sewers or waterways.

## Section 5. Reactivity Data

**Stability/Polymerization:** $H_2S$ is stable at room temperature in closed containers under normal storage and handling conditions. Hazardous polymerization cannot occur. **Chemical Incompatibilities:** Hydrogen sulfide attacks metals forming sulfides and is incompatible with 1,1-bis(2-azidoethoxy) ethane + ethanol, 4-bromobenzenediazonium chloride, powdered copper + oxygen, metal oxides, finely divided tungsten or copper, nitrogen trichloride, silver fulminate, rust, soda-lime, and all other oxidants. **Conditions to Avoid:** Exposure to heat and contact with incompatibles. **Hazardous Products of Decomposition:** Thermal oxidative decomposition of hydrogen sulfide can produce toxic sulfur dioxide .

## Section 6. Health Hazard Data

**Carcinogenicity:** The IARC,[164] NTP,[169] and OSHA[164] do not list hydrogen sulfide as a carcinogen. **Summary of Risks:** $H_2S$ combines with the alkali present in moist surface tissues to form caustic sodium sulfide, causing irritation of the eyes, nose, and throat at low levels (50 to 100 ppm). Immediate death due to respiratory paralysis occurs at levels greater than 1000 ppm. Heavy exposure has resulted in neurological problems, however recovery is usually complete. $H_2S$ exerts most of it's toxicity on the respiratory system. It inhibits the respiratory enzyme cytochrome oxidase, by binding iron and blocking the necessary oxydo-reduction process. Electrocardiograph changes after over-exposure have suggested direct damage to the cardiac muscle, however some authorities debate this. **Medical Conditions Aggravated by Long-Term Exposure:** Eye and nervous system disorders. **Target Organs:** Eyes, respiratory system and central nervous system. **Primary Entry Routes:** Inhalation, eye and skin contact. **Acute Effects:** Inhalation of low levels can cause headache, dizziness, nausea, cramps, vomiting, diarrhea, sneezing, staggering, excitability, pale

*Continued on next page*

No. 52 Hydrogen Sulfide 9/92

## Section 6. Health Hazard Data, *continued*

complexion, dry cough, muscular weakness, and drowsiness. Prolonged exposure to 50 ppm, can cause rhinitis, bronchitis, pharyngitis, and pneumonia. High level exposure leads to pulmonary edema (after prolonged exposure to 250 ppm), asphyxia, tremors, weakness and numbing of extremeties, convulsions, unconsciousness, and death due to respiratory paralysis. Concentrations near 100 ppm may be odorless due to olfactory fatigue, thus the victim may have no warning. Lactic acidosis may be noted in survivors. The gas does not affect the skin although the liquid (compressed gas) can cause frostbite. The eyes are very susceptible to $H_2S$ keratoconjunctivitis known as 'gas eye' by sewer and sugar workers. This injury is characterized by palpebral edema, bulbar conjunctivitis, mucous-puss secretions, and possible reduction in visible capacity.
**Chronic Effects:** Chronic effects are not well established. Some authorities have reported repeated exposure to cause fatigue, headache, inflammation of the conjunctiva and eyelids, digestive disturbances, weight loss, dizziness, a grayish-green gum line, and irritability. Others say these symptoms result from recurring acute exposures. There is a report of encephalopathy in a 20 month old child after low-level chronic exposure.
**FIRST AID Eyes:** *Do not* allow victim to rub or keep eyes tightly shut. Gently lift eyelids and flush immediately and continuously with flooding amounts of water. Treat with boric acid or isotonic physiological solutions. Serious exposures may require adrenaline drops. Olive oil drops (3 to 4) provides immediate treatment until transported to an emergency medical facility. Consult a physician immediately. **Skin:** *Quickly* remove contaminated clothing and rinse with flooding amounts of water. For frostbite, rewarm in 107.6°F (42 °C) water until skin temperature is normal. *Do not* use dry heat. **Inhalation:** Remove exposed person to fresh air and administer 100% oxygen. Give hyperbaric oxygen if possible. **Ingestion:** Unlikely since $H_2S$ is a gas above -60 °C. **Note to Physicians:** The efficacy of nitrite therapy is unproven. Normal blood contains < 0.05 mg/L $H_2S$; reliable tests need to be taken within 2 hr of exposure.

## Section 7. Spill, Leak, and Disposal Procedures

**Spill/Leak:** Immediately notify safety personnel, isolate and ventilate area, deny entry, and stay upwind. Shut off all ignition sources. Use water spray to cool, dilute, and disperse vapors. Neutralize runoff with crushed limestone, agricultural (slaked) lime, or sodium bicarbonate. If leak can't be stopped in place, remove cylinder to safe, outside area and repair or let empty. Follow applicable OSHA regulations (29 CFR 1910.120).
**Ecotoxicity Values:** Bluegill sunfish, TLm = 0.0448 mg/L/96 hr at 71.6 °F/22 °C; fathead minnow, TLm = 0.0071 to 0.55 mg/L/96 hr at 6 to 24 °C.
**Environmental Degradation:** In air, hydrogen sulfides residency (1 to 40 days) is affected by temperature, humidity, sunshine, and the presence of other pollutants. It does not undergo photolysis but is oxidated by oxygen containing radicals to sulfur dioxide and sulfates. In water, $H_2S$ converts to elemental sulfur. In soil, due to its low boiling point, much of $H_2S$ evaporates quickly if spilled. Although, if soil is moist or precipitation occurs at time of spill, $H_2S$ becomes slightly mobile due to its water solubility. $H_2S$ does not bioaccumulate but is degraded rapidly by certain soil and water bacteria. **Disposal:** Aerate or oxygenate with compressor. For in situ amelioration, carbon removes some $H_2S$. Anion exchanges may also be effective. A potential candidate for rotary kiln incineration (1508 to 2912 °F/820 to 1600 °C) or fluidized bed incineration (842 to 1796 °F/450 to 980 °C). Contact your supplier or a licensed contractor for detailed recommendations. Follow applicable Federal, state, and local regulations.

**EPA Designations**
Listed as a RCRA Hazardous Waste (40 CFR 261.33): No. U135
SARA Toxic Chemical (40 CFR 372.65): Not listed
Listed as a SARA Extremely Hazardous Substance (40 CFR 355), TPQ: 500 lb
Listed as a CERCLA Hazardous Substance* (40 CFR 302.4): Final Reportable Quantity (RQ), 100 lb (45.4 kg) [* per RCRA, Sec. 3001 & CWA, Sec. 311 (b)(4)]

**OSHA Designations**
Listed as an Air Contaminant (29 CFR 1910.1000, Table Z-1-A & Z-2)
Listed as a Process Safety Hazardous Material (29 CFR 1910.119), TQ: 1500 lb

## Section 8. Special Protection Data

**Goggles:** Wear protective eyeglasses or chemical safety goggles, per OSHA eye- and face-protection regulations (29 CFR 1910.133). Because contact lens use in industry is controversial, establish your own policy. **Respirator:** Seek professional advice prior to respirator selection and use. Follow OSHA respirator regulations (29 CFR 1910.134) and, if necessary, wear a MSHA/NIOSH-approved respirator. For < 100 ppm, use a supplied-air respirator (SAR) or SCBA. For < 250 ppm, use a SAR operated in continuous-flow mode. For < 300 ppm, use a SAR or SCBA with a full facepiece. For emergency or nonroutine operations (cleaning spills, reactor vessels, or storage tanks), wear an SCBA. *Warning! Air-purifying respirators do not protect workers in oxygen-deficient atmospheres.* If respirators are used, OSHA requires a respiratory protection program that includes at least: a written program, medical certification, training, fit-testing, periodic environmental monitoring, maintenance, inspection, cleaning, and convenient, sanitary storage areas. **Other:** Wear chemically protective gloves, boots, aprons, and gauntlets to prevent skin contact. Polycarbonate, butyl rubber, polyvinyl chloride, and neoprene are suitable materials for PPE. **Ventilation:** Provide general & local exhaust ventilation systems to maintain airborne concentrations below the OSHA PEL (Sec. 2). Local exhaust ventilation is preferred because it prevents contaminant dispersion into the work area by controlling it at its source.[103] **Safety Stations:** Make available in the work area emergency eyewash stations, safety/quick-drench showers, and washing facilities. **Contaminated Equipment:** Separate contaminated work clothes from street clothes and launder before reuse. Clean PPE. **Comments:** Never eat, drink, or smoke in work areas. Practice good personal hygiene after using this material, especially before eating, drinking, smoking, using the toilet, or applying cosmetics.

## Section 9. Special Precautions and Comments

**Storage Requirements:** Prevent physical damage to containers. Store in steel cylinders in a cool, dry, well-ventilated area away from incompatibles (Sec. 5). Install electrical equipment of Class 1, Group C. Outside or detached storage is preferred. **Engineering Controls:** To reduce potential health hazards, use sufficient dilution or local exhaust ventilation to control airborne contaminants and to keep levels as low as possible. Enclose processes and continuously monitor $H_2S$ levels in the plant air. Keep pipes clear of rust as $H_2S$ can ignite if passed through rusty pipes. Purge and determine $H_2S$ concentration before entering a confined area that may contain $H_2S$. The worker entering the confined space should have a safety belt and life line and be observed by a worker from the outside. Follow applicable OSHA regulations (1910.146) for confined spaces. $H_2S$ can be trapped in sludge in sewers or process vessels and may be released during agitation. Calcium chloride or ferrous sulfate should be added to neutralize process wash water each time $H_2S$ formation occurs. Control $H_2S$ emissions with a wet flare stack/scrubbing tower. **Administrative Controls:** Consider preplacement and periodic medical exams of exposed workers emphasizing the eyes, nervous and respiratory system.

**Transportation Data (49 CFR 172.101)**

**DOT Shipping Name:** Hydrogen sulfide, liquefied
**DOT Hazard Class:** 2.3
**ID No.:** UN1053
**DOT Packaging Group:** --
**DOT Label:** Poison Gas, Flammable Gas
**Special Provisions (172.102):** 2, B9, B14

***Packaging Authorizations***
**Exceptions:** --
**Non-bulk Packaging:** 304
**Bulk Packaging:** 314, 315

***Vessel Stowage Requirements***
**Vessel Stowage:** D
**Other:** 40

***Quantity Limitations***
**Passenger, Aircraft, or Railcar:** Forbidden
**Cargo Aircraft Only:** Forbidden

***MSDS Collection*** **References:** 26, 73, 89, 100, 101, 103, 124, 126, 127, 132, 136, 140, 148, 149, 153, 159, 163, 164, 168, 171, 180
**Prepared by:** M Gannon, BA; **Industrial Hygiene Review:** PA Roy, MPH, CIH; **Medical Review:** AC Darlington, MPH, MD

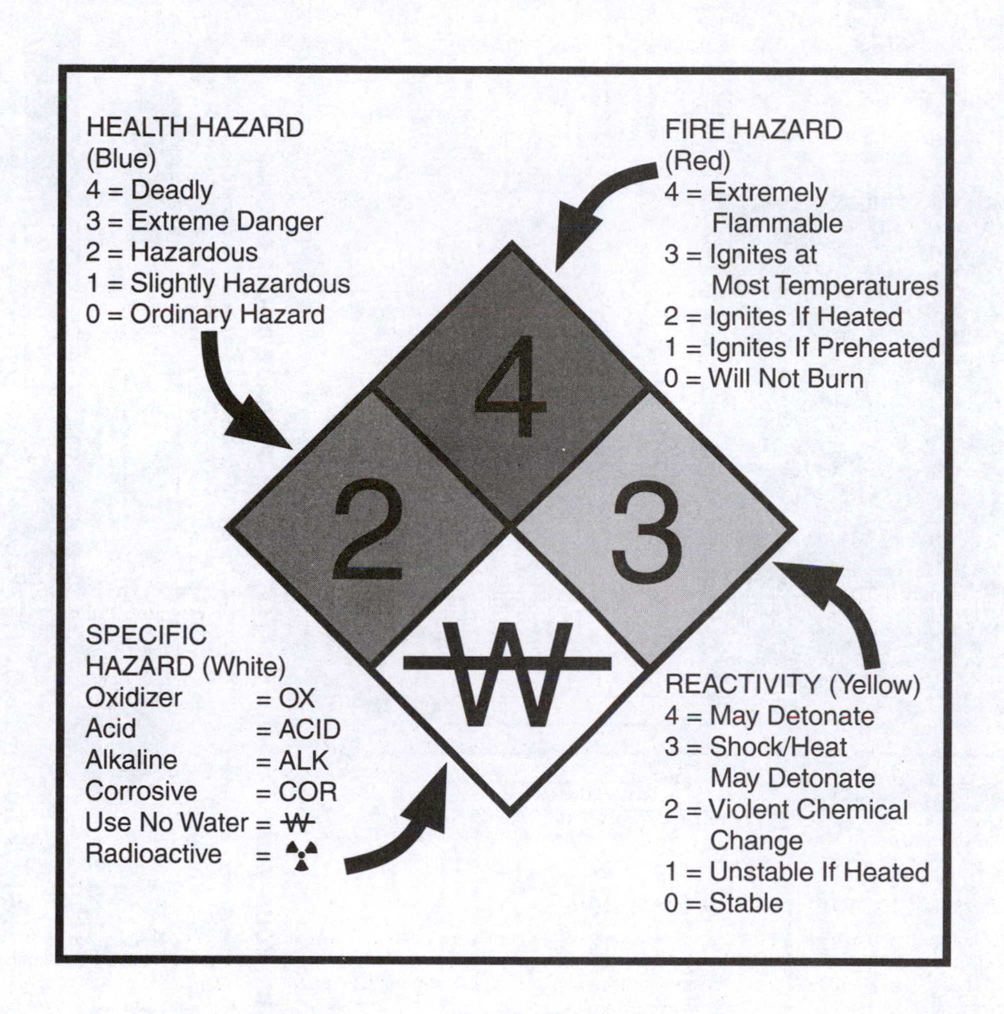


*Fig. 3.12 Typical NFPA hazard warning label*

# BAKER SAF-T-DATA™ Guide

J.T.Baker®

An easy-to-understand hazard classification appears on J. T. Baker labels. It will help increase your awareness of vital occupational health and safety practices. Using the BAKER SAF-T-DATA™ System as a guide you can quickly learn the hazards each substance presents to your health and safety, personal laboratory protective equipment that should be used for handling, and the recommended storage of compatible products by color code.

## A NUMERICAL HAZARD CODE

Substances are rated on a scale of 0 (non-hazardous) to 4 (extremely hazardous) in each of four hazard categories:

- **Health hazard** - the danger or toxic effect a substance presents if inhaled, ingested, or absorbed.
- **Flammable hazard** - the tendency of the substance to burn.
- **Reactivity hazard** - the potential of a substance to explode or react violently with air, water or other substances.
- **Contact hazard** - the danger a substance presents when exposed to skin, eyes, and mucous membranes.

Rating Scale

| 4 | 3 | 2 | 1 | 0 |
|---|---|---|---|---|
| Extreme | Severe | Moderate | Slight | None* |

*No scientific data in the standard references that suggests the substance is hazardous.

## HAZARD SYMBOL

A substance rated 3 or 4 in any hazard category will also display a hazard symbol. These easy-to-understand pictograms emphasize the serious hazards related to a substance:

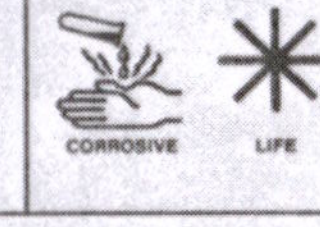
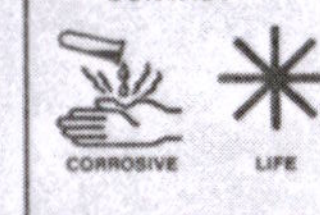

Sample label:

BAKER SAF-T-DATA™ System

| HEALTH | FLAMMABILITY | REACTIVITY | CONTACT |
|---|---|---|---|
| 1 | 3 | 2 | 1 |
| SLIGHT | SEVERE | MODERATE | SLIGHT |

LABORATORY PROTECTIVE EQUIPMENT: SAFETY GLASSES, LAB COAT, VENT HOOD, PROPER GLOVES, EXTINGUISHER

STORAGE COLOR: RED

4 L 9006-3

**Acetone**

$(CH_3)_2CO$ FW 58.1

'BAKER ANALYZED' Reagent

ACTUAL ANALYSIS, LOT 238093 MEETS A.C.S. SPECIFICATIONS

| | | | |
|---|---|---|---|
| Assay $[(CH_3)_2CO]$(by GC, corrected for $H_2O$) | | 99.8 | % |
| Color (APHA) | < | 5 | |
| Density (g/ml) at 25°C | | 0.7852 | |
| Residue after Evaporation | | 0.00004 | % |
| Solubility in Water | | Passes Test | |
| Titrable acid | | 0.0002 | meq/g |
| Titrable base | | 0.00008 | meq/g |
| Aldehyde (as HCHO) | < | 0.002 | % |
| Isopropyl Alcohol $(CH_3CHOHCH_3)$(by GC) | < | 0.01 | % |
| Methanol $(CH_3OH)$(by GC) | < | 0.01 | % |
| Substances Reducing Permanganate | | Passes Test | |
| Water $(H_2O)$(by Karl Fischer titrn) | | 0.2 | % |
| Trace Impurities (in ppm): | | | |
| Copper (Cu) | < | 0.05 | |
| Heavy Metals (as Pb) | | 0.3 | |
| Iron (Fe) | | 0.06 | |
| Nickel (Ni) | < | 0.1 | |

**DANGER!**

**EXTREMELY FLAMMABLE**
**HARMFUL IF SWALLOWED OR INHALED**

Keep away from heat, sparks, flame. Keep in tightly closed container. Avoid breathing vapor. Avoid contact with eyes, skin, clothing. Use with adequate ventilation. Wash thoroughly after handling. In case of fire, use water spray, alcohol foam, dry chemical, $CO_2$. Flush spill area with water spray.
**FIRST AID: If swallowed,** if conscious, immediately induce vomiting and call a physician. **If inhaled,** remove to fresh air. If not breathing, give artificial respiration. If breathing is difficult, give oxygen. Call a physician.
**FLASH POINT:** ~0°F(SCT)
**CAS 67-64-1** 402-2275
**DOT Description:** Ⓡ Acetone UN 1090
**IMO Description:** Ⓡ Acetone
**EPA-HW:** Ignitables
**Solusorb**® solvent absorbent is recommended for spills of Acetone.

NFPA 3 1 0

© J.T. Baker Chemical Co.
Phillipsburg, NJ 08865

## B LABORATORY PROTECTIVE EQUIPMENT

This series of pictograms suggests the personal protective clothing and equipment recommended for use when handling the substance in a laboratory situation. The pictograms relate to the combination of hazards presented by the substance.

SAFETY GLASSES, GOGGLES, GOGGLES & SHIELD, LAB COAT, LAB COAT & APRON, PROPER GLOVES, VENT HOOD, EXTINGUISHER (A), EXTINGUISHER (B), EXTINGUISHER (D)

STOP — The stop sign indicates the substance represents a special extreme hazard and the MSDS and other references should be consulted before handling.

## C STORAGE COLOR CODING

The SAF-T-DATA label suggests a unique method for setting up your chemical storage area. Compatible products are labelled with the same color. Simply group these colors together and follow the recommendations for appropriate storage:

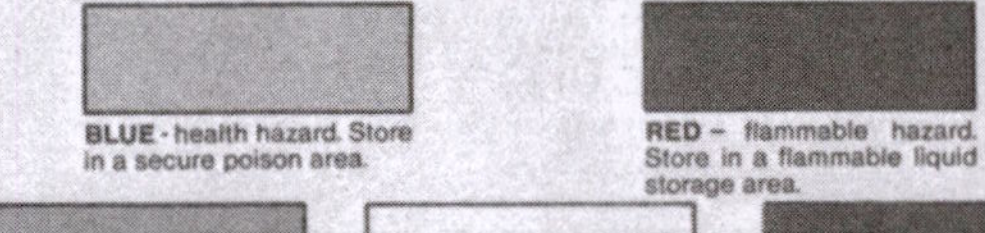

**BLUE** - health hazard. Store in a secure poison area.

**RED** – flammable hazard. Store in a flammable liquid storage area.

**YELLOW** - reactivity hazard. Store separately and away from flammable or combustible materials.

**WHITE** – contact hazard. Store in a corrosion-proof area.

**ORANGE** - substances with no rating higher than 2 in any hazard category. Store in a general chemical storage area.

**STRIPED LABELS** - incompatible materials of the same color class have striped labels. These approximately 40 products should not be stored adjacent to substances with the same colored labels. Proper storage must be individually assessed.

## D SPILL CONTROL CODE

This statement indicates which J.T. Baker spill control kit is recommended for use with the substance.

## E NFPA SYSTEM*

This system was adopted by the NFPA in 1975 to safeguard the lives of firefighters. It is based on the hazards created by a substance in a fire situation. For this reason, the hazard ratings in the Baker SAF-T-DATA™ System, which are based on substances in a laboratory situation, will not always correspond with the NFPA ratings.

*National Fire Protection Association

*Fig. 3.13 J. T. Baker hazardous substance labeling system*

(Permission of J. T. Baker Chemical Company)

## 3.9 EXCAVATIONS

Operators involved in activities such as septic tank installation, collection system pipe laying or repairing, digging to place tanks, pump vaults, or trenching to install or repair pipe in excavations over four feet in depth are exposed to a life-threatening hazard—cave-ins that could bury them in a few moments beneath tons of earth without warning or time to escape. Operators must be aware of the serious hazards in this type of work and must use safe procedures to prevent cave-ins.

There are three recognized methods of excavation protection: sloping, shoring, and shielding. Different soils and conditions dictate the type of protection that may be safely used. It is strongly recommended that adequate cave-in protection be provided when the trench or excavation is four feet deep or deeper. OSHA requires adequate protection if the trench is five feet or more in depth.

### 3.90 **Sloping** (Figure 3.14)

Sloping is only practiced in very stable soils, and usually where the excavation is not required to be very deep. Sloping is simply removing the trench wall itself. Allowable slopes may range from ½:1 (one-half foot back for every one foot of trench depth) to 1½:1, depending on the type of soil. The less stable the soil is, the wider the trench will become at the top, creating a less severe angle on the side walls. In other words, the walls will be at a flatter incline. Due to excavating time and space required to deposit spoil material (excavated material, such as soil, from the trench of a water main or sewer), sloping is rarely used except in very stable soils. Also, material that is returned to the excavation (backfill) must be compacted. Typically, it is in the best interest of the project to disturb as little soil as necessary. This will minimize any future soil settlement or sloughing.

### 3.91 **Shoring**[15]

Shoring is the most frequently used protective system when trenching or excavating. Shoring is a complete framework of wood or metal that is designed to support the walls of a trench (see Figure 3.15). Sheeting is the solid material placed directly against the side of the trench. Either wooden sheets or metal plates might be used. Uprights are used to support the sheeting. They are usually placed vertically along the face of the trench

(1) In lieu of a shoring system, the sides or walls of an excavation or trench may be sloped, provided equivalent protection is thus afforded. Where sloping is a substitute for shoring that would otherwise be needed, the slope shall be at least ¾ horizontal to 1 vertical unless the instability of the soil requires a slope flatter than ¾ to 1.

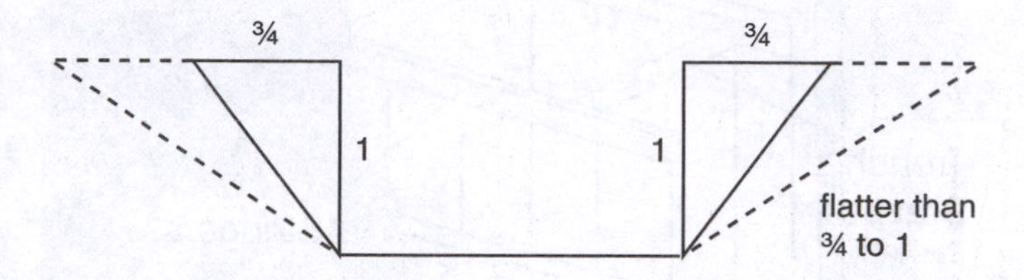


EXCEPTIONS: In hard, compact soil where the depth of the excavation or trench is 8 feet or less, a vertical cut of 3½ feet with sloping of ¾ horizontal to 1 vertical is permitted.

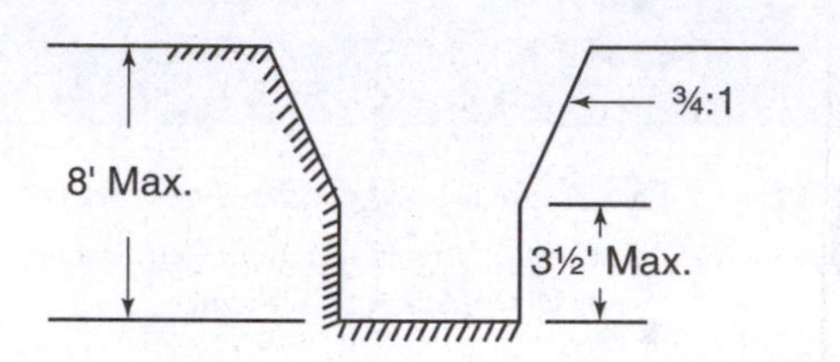


In hard, compact soil where the depth of the excavation or trench is 12 feet or less, a vertical cut of 3½ feet with sloping of 1 horizontal to 1 vertical is permitted.

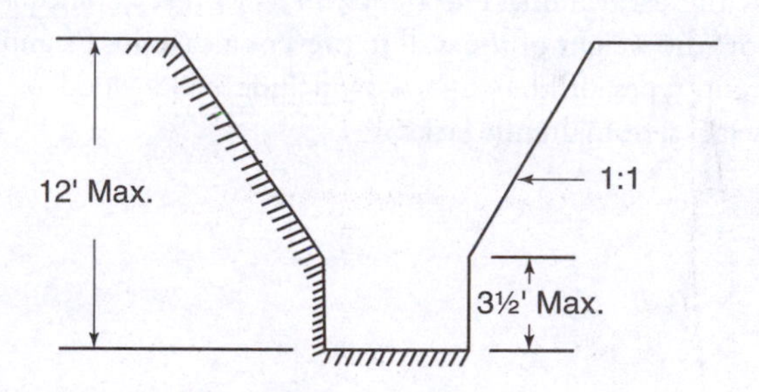


(2) Benching in hard, compact soil is permitted provided that a slope ratio of ¾ horizontal to 1 vertical, or flatter, is used.

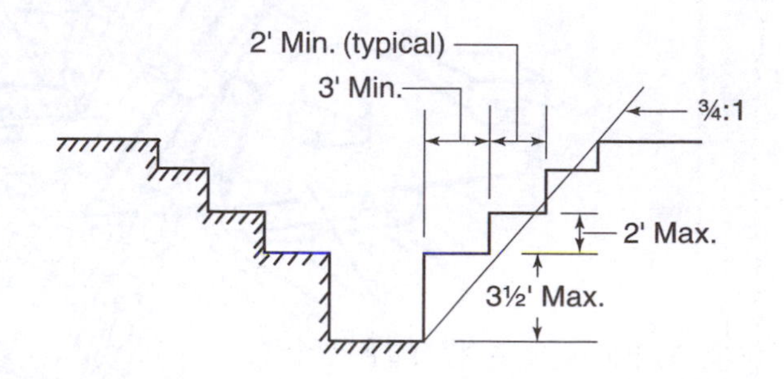


*Fig. 3.14 Sloping or benching systems*

(Reproduced from *CONSTRUCTION SAFETY ORDERS,* Title 8, Section 1541, California Code of Regulations)

15. For additional information about shoring regulations, types of shores, and shoring size requirements, see *OPERATION AND MAINTENANCE OF WASTEWATER COLLECTION SYSTEMS,* Volume I, Chapter 7, Section 7.1, "Shoring," in this series of operator training manuals.

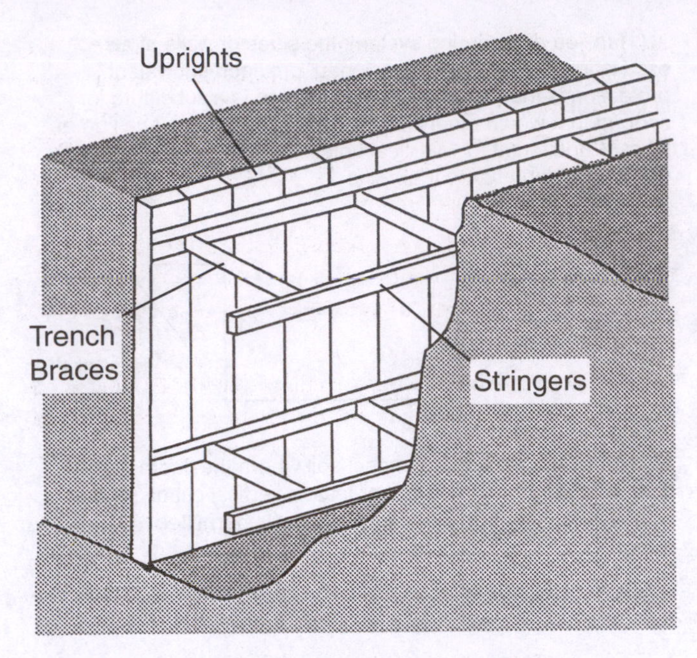


*Fig. 3.15 Close sheeting method of shoring*
(Source: "Wastewater System Operations Certification Study Guide,"State of Oklahoma)

wall. Spacing between the uprights varies depending upon the stability of the soil. Stringers are placed horizontally along the uprights. Trench braces are attached to the stringers and run across the excavation. The trench braces must be adequate to support the weight of the wall to prevent a cave-in. Examples of different types of trench braces include solid wood or steel, screw jacks, or hydraulic jacks.

The space between the shoring and the sides of the excavation should be filled in and compacted in order to prevent a cave-in from getting started. If properly done, shoring may be the operator's best choice for cave-in protection because it actually prevents a cave-in from starting and does not require additional space. Try to visit a construction site to observe various types of shoring.

### 3.92 **Shielding** (Figure 3.16)

Shielding is accomplished by using a cylinder or a two-sided, braced steel box that is open on the top, bottom, and both ends. This "drag shield," as it is sometimes called, is pulled through the excavation as the trench is dug out in front and filled in behind. Operators using a drag shield should always work only within the walls of the shield. If the trench is left open behind or in front of the shield, the temptation could be to wander outside of the shield's protection. Shielding does not actually prevent a cave-in as the space between the trench wall and the drag shield is left open, allowing a cave-in to start. There have been cases where a drag shield was literally crushed by the weight of a collapsing trench wall.

### 3.93 Safety Precautions

Under certain conditions, you may need the help of a soils engineer, a shoring specialist, or a competent person to evaluate the soil and design a safe protective shoring system. Excavations in soil that is unstable, sandy, saturated with water, or recently disturbed may require the expertise of a soils engineer or competent person. Excavations in areas that experience vibration in the soil due to trains, traffic, or heavy equipment may need the help of a shoring specialist. If in doubt, ask for help; do not take a

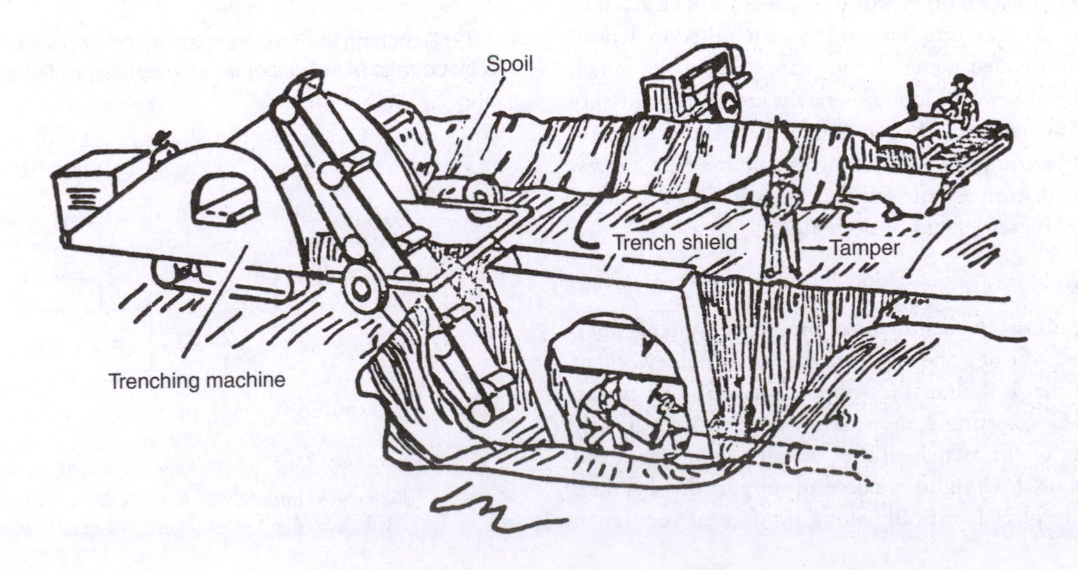


*Fig. 3.16 Trench or drag shields*

chance with your life or someone else's life. Always follow these basic rules concerning excavation work:

1. Spoil should be placed on only one side of the excavation.
2. Spoil must be placed a minimum of two feet away from the edge of the excavation.
3. A stairway, ramp, ladder, or other safe means of exit is required in the trench if it is more than four feet deep.
4. The means of exit must be placed so that one is available every twenty-five feet. If a ladder is used, it must extend at least three feet above the excavation wall.

Accidents at the site of trenching and shoring activities are all too common. In addition to protecting workers from the danger of a cave-in, safety precautions must also be taken to protect them from traffic hazards if the work is performed in a street. It may also be necessary to provide forced-air ventilation and to observe confined space procedures if circumstances require such precautions. Check with your local safety regulatory agency. They can provide you with the appropriate regulations. Do not wait until an emergency arises to obtain the information.

### 3.94 Acknowledgment

Major portions of Section 3.5, "Confined Spaces," Section 3.7, "Electricity," and Section 3.9, "Excavations," were adapted from "Wastewater System Operations Certification Study Guide," developed by the State of Oklahoma. The use of this material is greatly appreciated.

## QUESTIONS

Please write your answers to the following questions and compare them with those on page 108.

3.9A What are the three recognized methods of excavation protection?

3.9B Which excavation protection method can only be used in shallow excavations in very stable soils?

3.9C Why is the space between the shoring and the sides of the excavation filled in and compacted?

3.9D Where should excavated spoil be placed in relation to an excavation?

## 3.10 FIRES

Everyone must know what to do in case of a fire. If proper action is taken quickly, fires are less likely to cause damage and injury. The use of the proper type of extinguisher for each class of fire will give the best control of the situation and avoid compounding the problem. For example, water should not be poured on grease fires, electrical fires, or metal fires because water could increase the hazards, such as splattering of the fire and electric shock. The classes of fires given here are based on the type of material being consumed.

Fire classifications are important for determining the type of fire extinguisher needed to control the fire. Classifications also aid in recordkeeping. Fires are classified as A, B, C, or D fires based on the type of material being consumed: A, ordinary combustibles; B, flammable liquids and vapors; C, energized electrical equipment; and D, combustible metals. Fire extinguishers are also classified as A, B, C, or D to correspond with the class of fire each will extinguish.

**Class A** fires: ordinary combustibles such as wood, paper, cloth, rubber, many plastics, dried grass, hay, and stubble. Use foam, water, soda-acid, carbon dioxide gas, or almost any type of extinguisher.

**Class B** fires: flammable and combustible liquids such as gasoline, oil, grease, tar, oil-based paint, lacquer, and solvents, and also flammable gases. Use foam, carbon dioxide, or dry chemical extinguishers.

**Class C** fires: energized electrical equipment such as starters, breakers, and motors. Use carbon dioxide or dry chemical extinguishers to smother the fire; both types are nonconductors of electricity.

**Class D** fires: combustible metals such as magnesium, sodium, zinc, and potassium. Operators rarely encounter this type of fire. Use a Class D extinguisher or use fine dry soda ash, sand, or graphite to smother the fire. Consult with your local fire department about the best methods to use for specific hazards that exist at your facility.

Multipurpose extinguishers are also available, such as a Class BC carbon dioxide extinguisher that can be used to smother Class B and Class C fires. A multipurpose ABC carbon dioxide extinguisher will handle most laboratory fire situations. (When using carbon dioxide extinguishers, remember that the carbon dioxide can displace oxygen—take appropriate precautions.)

There is no single type of fire extinguisher that is effective for all fires so it is important that you understand the class of fire you are trying to control. You must be trained in the use of the different types of extinguishers, and the proper type should be located near the area where that class of fire may occur.

Mount fire extinguishers in conspicuous locations and install signs or lights nearby to assist people in finding them. Everyone should know how fires are classified and the appropriate extinguisher for each class. Post the phone number of the local fire department in an obvious location by every phone.

Inspect all firefighting equipment and extinguishers on a regular basis and after each use. Fire extinguishers must be visually inspected monthly and must receive an annual maintenance check. Always refill extinguishers promptly after use.

Clearly mark all exits from pump stations and other buildings. Post conspicuous signs identifying flammable storage areas and signs prohibiting smoking where smoking is not allowed. Be sure your fire department is aware of your collection system and possible causes of fires.

The best way to fight fires is to prevent them from occurring. Fires can be prevented by careful extinguishing of cigarettes and good housekeeping of combustibles, flammable liquids, and electrical equipment and circuits.

## 3.11 NOISE

Wastewater treatment facilities contain some equipment that produces high noise levels, intermittently or continuously. Operators (and their employers) must be aware of this and use safeguards such as hearing protectors that eliminate or reduce noise to acceptable levels. In general, if you have to shout or cannot hear someone talking to you in a normal tone of voice, the noise level is excessive. Prolonged or regular daily exposure to high noise levels can produce at least two harmful, measurable effects: hearing damage and masking of desired sounds such as speech or warning signals.

The ideal method of dealing with any high-noise environment is the elimination or reduction of all sources through feasible engineering or administrative controls. This approach is frequently not possible; therefore, employers are required to identify and monitor operators whose normal noise exposure might equal or exceed an 8-hour *TIME-WEIGHTED AVERAGE (TWA)*[16] of 85 *DECIBELS*[17] (A-scale). Hearing protectors are to be used only if feasible administrative or engineering controls are unsuccessful in reducing noise exposures to permissible levels. Table 3.2 lists the typical sound pressure levels for three pieces of equipment, general street traffic, and a residential subdivision.

TABLE 3.2 TYPICAL SOUND PRESSURE LEVELS (IN DECIBELS (dB))

| Equipment | Sound Level |
|---|---|
| Jackhammer | 100 dB |
| Air Compressor | 94 dB |
| Heavy Truck | 85 dB |
| Street Traffic | 70 dB |
| Residential Subdivision | 50 dB |

To ensure that the welfare of operators is not compromised and to comply with federal regulations (29 CFR 1910.95), a comprehensive hearing conservation program should be implemented. All individuals whose normal noise exposure equals or exceeds the 8-hour TWA of 85 dBA must be included in this program. The primary elements of the program are monitoring, audiometric testing, hearing protection, training in the use of protective equipment and procedures, access to noise level information, and recordkeeping. The purpose of the conservation program is to prevent hearing loss that might affect an operator's ability to hear and understand normal speech.

A selection of hearing protection devices (Figures 3.17 and 3.18) must be available to operators; however, a certain degree of confusion can arise concerning the adequacy of a particular protector. To estimate the adequacy of a hearing protector, use the noise reduction rating (NRR) shown on the hearing protector package. Subtract 7 dB from the NRR and subtract the remainder from the individual's A-weighted TWA noise environment to obtain the estimated A-weighted TWA under the ear protector. To provide adequate protection, the value under the ear protector should be 85 dB or less, the lower the better.

It is essential that individuals use properly rated protective devices in high-noise areas or during high-noise activities. Employee training must include the following information: (1) the effects of noise on hearing; (2) the purpose of hearing protectors and the advantages, limitations, and effectiveness of various types, as well as instruction on selection, fitting, use, and care; and (3) the purpose of audiometric testing and an explanation of the test procedures. The training must be repeated annually. Audiometric test records must be retained for as long as the affected employee works at the plant. Contact your local health department or safety regulatory agency for assistance in the development of a hearing protection (conservation) program for your specific treatment plant.

## 3.12 DUSTS, FUMES, MISTS, GASES, AND VAPORS

The ideal way to control occupational diseases caused by breathing air contaminated with harmful dusts, fumes, mists, gases, and vapors is to prevent atmospheric contamination from occurring. This can sometimes be accomplished through engineering control measures. Remember, OSHA requires that engineering controls be implemented whenever feasible to eliminate or reduce operator exposure to a hazard. When effective engineering

16. *Time-Weighted Average (TWA).* A time-weighted average is used to calculate a worker's daily exposure to a hazardous substance (such as chemicals, dusts, fumes, mists, gases, or vapors) or agent (such as occupational noise), averaged to an 8-hour workday, taking into account the average levels of the substance or agent and the time spent in the area. This is the guideline OSHA uses to determine permissible exposure limits (PELs) and is essential in assessing a worker's exposure and determining what protective measures should be taken. A time-weighted average is equal to the sum of the portion of each time period (as a decimal, such as 0.25 hour) multiplied by the levels of the substance or agent during the time period divided by the hours in the workday (usually 8 hours). Also see PERMISSIBLE EXPOSURE LIMIT (PEL).
17. *Decibel* (DES-uh-bull). A unit for expressing the relative intensity of sounds on a scale from zero for the average least perceptible sound to about 130 for the average level at which sound causes pain to humans. Abbreviated dB.

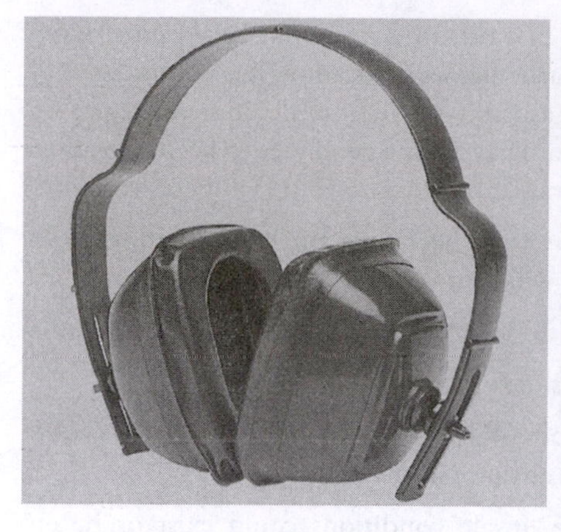

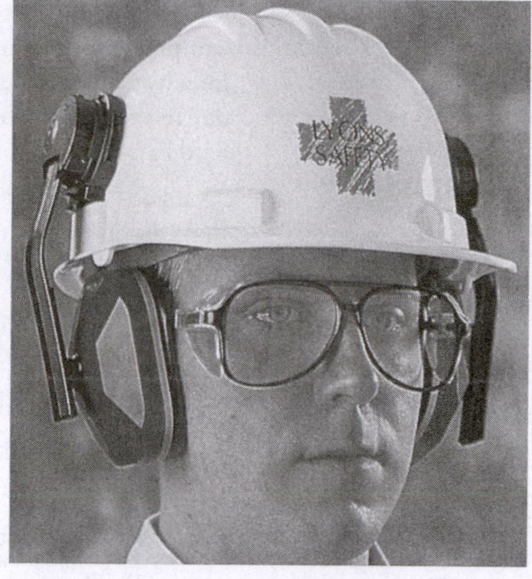


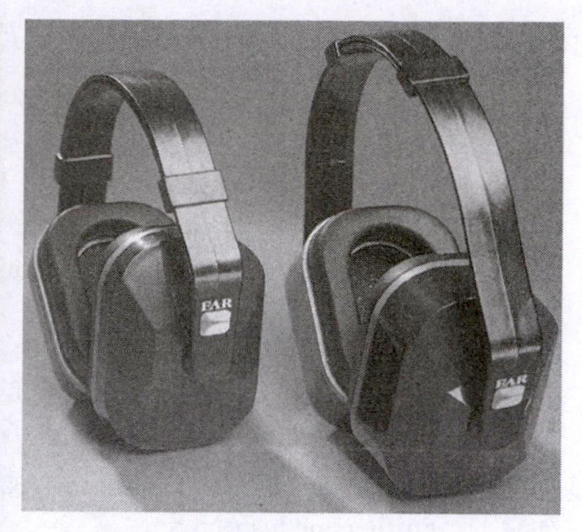


*Fig. 3.17 Muff-type hearing protection*

(Permission of Lyons Safety)

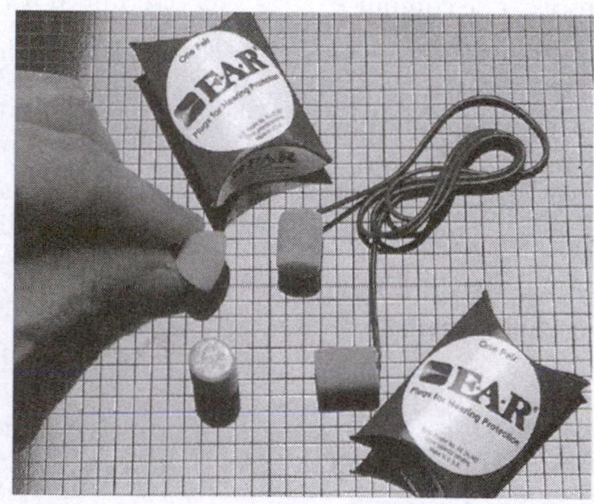


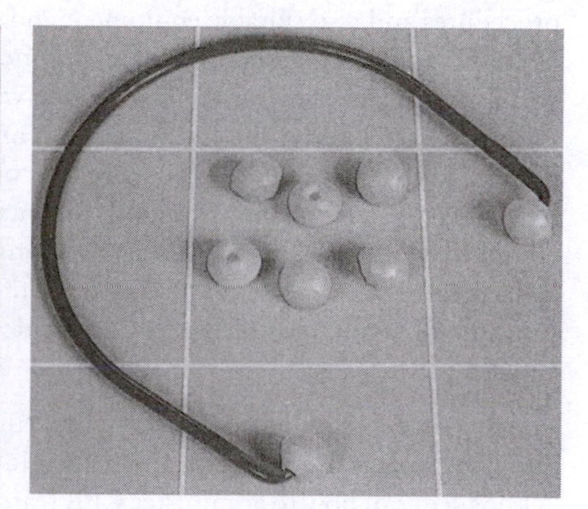

*Fig. 3.18 Earplug insertion-type hearing protection*

(Permission of Lyons Safety)

controls are not feasible, however, appropriate respirators must be used.

Respirators must be provided by the employer when they are necessary to protect the health of the operator. The respirators must be appropriate and suitable for the purpose intended. Listed below are the four most common types of respirators:

1. Self-Contained Breathing Apparatus (SCBA)
2. Supplied-Air Respirators (SAR)
3. Powered Air-Purifying Respirators (PAPR)
4. Air-Purifying Respirators (APR)

Remember, you, the operator, must use the provided respiratory protection in accordance with instructions and training provided to you.

Employers are also responsible for establishing and maintaining a respiratory protection program, including these basic elements:

1. Written standard operating procedures (SOPs) governing the selection and use of respirators
2. Instruction and training in the proper use of respirators and their limitations (to include annual fit testing)
3. Physical assessment of individuals assigned tasks requiring the use of respirators

These are only a few requirements for the safe use of respiratory protection. Specific requirements for a respiratory protection program for your application must be confirmed with your local regulatory safety agency.

## QUESTIONS

Please write your answers to the following questions and compare them with those on page 108.

3.10A Fires are classified as A, B, C, or D fires based on the type of material being consumed. List the types of materials for each of the fire classifications?

3.10B Fire extinguishers are classified as A, B, C, or D to correspond with the class of fire each will extinguish. What are the methods of extinguishing each classification of fire?

3.10C How often should firefighting equipment and extinguishers be inspected?

3.11A What should operators do if feasible administrative or engineering controls are unsuccessful in reducing noise exposures to permissible levels?

3.12A What is the ideal way to control occupational diseases caused by breathing contaminated air?

## 3.13 CREATING A SAFE WORK ENVIRONMENT

Safety on the job requires operators to routinely use safe work procedures and also suitable equipment and tools. The community you work for may not be aware of some of the dangers that you as the operator may encounter in the wastewater system. It is your responsibility to educate yourself and to communicate your needs for training; for the assistance of another person or persons, on occasion; and for certain tools and equipment to accomplish your tasks safely. You are responsible for the proper care, maintenance, and use of that equipment so that it functions properly and is available for use, particularly in emergency situations.

All operators should carry a first-aid kit in their work vehicle and every operator should be trained in basic first-aid procedures. A knowledge of how to administer CPR (cardiopulmonary resuscitation) is also recommended. In most areas, training to acquire and maintain these skills is conducted by the Red Cross or the local fire department. Although first aid generally involves the one-time treatment of minor injuries such as scratches, cuts, and small burns, knowledge of first aid and especially CPR can be a life saver in emergencies. Additionally, where medical personnel are not readily available—this has been interpreted as being three or four minutes away—OSHA requires employers to ensure that an adequate number of employees with a current certificate from a first-aid program (American Red Cross) are available to provide emergency care for injured employees.

You are your own personal protector. If you have other operators working with you, you must also make sure they are working safely and understand the dangers or the hazardous conditions associated with their work to prevent injury to you, them, and the public, and to prevent possible damage to the facility.

It is your duty to stop the work if an unsafe condition or procedure is observed. There have been instances where a local official (mayor) visited a job site to review work and was asked to leave due to the fact that the official did not have proper safety equipment for that particular work location. It is not uncommon for a bystander to be asked to move from the immediate job site due to the lack of a hard hat. (It is a good idea to keep a few extra hard hats on hand for use by visitors to your facility or work site.)

A good practice is to review the task to be performed by thinking about what will be involved.

1. Plan ahead
   a. What does this next job consist of?
   b. What applicable OSHA or state safety standards should be reviewed in advance of the job?
   c. What possible unsafe conditions could exist or be encountered during the job?
      (1) The function
      (2) The site location
      (3) Special considerations
      (4) If there is an accident, how do I get help?
   d. What tools or equipment are required to do this job safely?
   e. Do I need help?
      (1) Knowledge

      Do you have the knowledge and skills to perform the task? You may be capable of driving an automobile, but that does not ensure that you can operate a backhoe to dig an excavation. Or if you can plug in an electric lamp, that does not indicate you are capable of troubleshooting a 220-volt motor control center. Ego or pride can create problems. Ask questions. Learn how to do each task correctly and safely from an experienced person.

      (2) Assistance
         (A) Lifting
         (B) Help in assembling or constructing facilities
         (C) Safety backup
            - Confined space operations
            - Traffic control
            - Chemical handling
      (3) Do not take shortcuts—do the job properly using the correct tools and procedures.
      (4) If you are a single-operator system and occasionally must perform work in remote areas or must respond to trouble calls or complaints in the night, take another person along with you—one who is capable of going for help if there is an accident and you need assistance.
2. Care for Equipment and Tools

   The operator is responsible for the care and maintenance of equipment provided for safely doing each job. Tools should be kept clean and in good working order. Safety devices in particular should be routinely checked for operation, battery replacement, calibration, and recharging of pressurized containers. As examples, an empty fire extinguisher is useless. A chemical eye wash/deluge shower will not function if the water supply valve has been closed, and you are not going to find that valve and open it if you have been sprayed in the face with chlorine or an acid solution. A self-contained breathing apparatus is of no value if the air tank is empty. There are numerous items that if not checked, cleaned, and tested can cause you or someone else serious injury or loss of life.

3. Personal Protection

   Wear protective clothing of gloves and boots when working in or around wastewater. When handling chemicals, protect your eyes with goggles or face shields and wear a rubber apron. Always wash your hands before eating or smoking. Always wear a hard hat in construction zones. If possible, change your work clothes before going home. Keep in mind that new wastewater operators do not have your knowledge and experience to protect themselves, so do not hesitate to correct or instruct a person in correct procedures or safe practices.

## 3.14 THE OPERATOR'S SAFETY RESPONSIBILITIES TO THE PUBLIC

The public is not aware of dangers associated with wastewater systems. While performing your job, you must take precautions to protect people and property. The operator's full attention is needed when dealing with the following situations:

1. Vehicle Operation

   Vehicles and other equipment are commonly used to transport ourselves and tools through the community to perform various tasks. The equipment you use should be well maintained and all components should be working properly. The safe and responsible operation of work vehicles can only be controlled by you.

2. Wastewater Overflows, Spills, and Hazardous Materials

   The public dislikes raw wastewater overflows or spills from plugged or broken systems. Not all wastewater spills are recognized by the public, in particular children. Every effort must be made to contain wastewater discharges by stopping or reducing the flow. The spill and consequent flow area should not be accessible by the public. The spill area should be cleaned up promptly and made safe by removal of the contaminant, or, if permissible, by soil burial, or by applying sanitizing chemicals to the spill area.

   Hazardous materials used in and around wastewater systems, such as chlorine or sulfur dioxide, are occasionally spilled or leaked to the atmosphere due to faulty connections or equipment. Areas where these chemicals are used should be well posted with signs to warn the public of these locations. When a leak or spill occurs, issue an immediate warning to the local authorities to assist in containment of the material and protect local residents or passersby.

   Environmental control agencies of states or cities require immediate notification of a spill or discharge of wastewater or hazardous materials. You should have their phone numbers available in your office and vehicles for immediate notification. Their immediate assistance will help you control an undesirable situation. If you are ever involved in a major hazardous material emergency and are on the scene, a nationwide agency, *CHEMTREC,* at (800) 424-9300, will provide advice.

To work safely in many areas requires special training and equipment. The equipment and tools are useless if you do not know how to use and take care of them properly. For all of the areas discussed in this safety review, additional training material is available through state safety offices, police and fire departments, universities, libraries, wastewater associations, and training courses such as this one.

## 3.15 ADDITIONAL READING


1. *SAFETY AND HEALTH IN WASTEWATER SYSTEMS* (MOP 1). Obtain from Water Environment Federation (WEF), Publications Order Department, 601 Wythe Street, Alexandria, VA 22314-1994. Order No. MO2001. Price to members, $48.00; nonmembers, $65.00; plus shipping and handling.
2. *CHLORINE BASICS,* Seventh Edition. A PDF of the document is available as a free download from the Chlorine Institute, Inc., at www.chlorineinstitute.org.
3. *FISHER SAFETY CATALOG.* Obtain from Literature Fulfillment at the Fisher Scientific website, www.fishersci.com, or phone (800) 772-6733.
4. *STANDARD METHODS FOR THE EXAMINATION OF WATER AND WASTEWATER,* 21st Edition, 2005. Obtain from Water Environment Federation (WEF), Publications Order Department, PO Box 18044, Merrifield, VA 22118-0045. Order No. S82011. Price to members, $203.00; nonmembers, $268.00; price includes cost of shipping and handling.


## 3.16 ARITHMETIC ASSIGNMENT

Turn to the Appendix, "How to Solve Wastewater System Arithmetic Problems," at the back of this manual and read Section A.2, "Areas," and Section A.3, "Volumes."

### QUESTIONS

Please write your answers to the following questions and compare them with those on page 108.

3.13A Whose responsibility is it to stop work if an unsafe condition or procedure is observed?

3.14A What should an operator do to protect the public in the event of a spill of wastewater or hazardous material?

Please answer the discussion and review questions next.

## DISCUSSION AND REVIEW QUESTIONS

### Chapter 3. SAFETY

Please write your answers to the following discussion and review questions to determine how well you understand the material in the chapter.

1. What is the operator's responsibility with regard to safety?
2. How can an operator avoid physical injuries?
3. How can operators be protected from injury by the accidental discharge of stored energy?
4. Immunization shots protect against what infection and infectious disease?
5. What precautions should you take to avoid transmitting disease to your family?
6. What should you do when you discover an area with an oxygen deficiency?
7. What safety equipment on a work vehicle should be inspected before leaving for the job site?
8. Any time traffic may be affected by road work activities, who must you notify before work begins or before traffic is diverted or disrupted?
9. What types of equipment and work situations increase an operator's exposure to electrical hazards?
10. What are the basic safety rules for qualified persons working on electrical equipment?
11. Why should wastewater system operators be familiar with Material Safety Data Sheets?
12. Name the main components of a shoring system.
13. How would you extinguish an electrical fire?
14. How can wastewater system operators protect their hearing from loud noises?
15. What would you look for when inspecting safety devices?
16. Accidents do not just happen—they are ______________!

**S** SAFETY FIRST
**A** ACCIDENTS COST LIVES
**F** FASTER IS NOT ALWAYS BETTER
**E** EXPECT THE UNEXPECTED
**T** THINK BEFORE YOU ACT
**Y** YOU CAN MAKE THE DIFFERENCE

**ACCIDENTS DO NOT JUST HAPPEN...**
**THEY ARE CAUSED!**

## SUGGESTED ANSWERS

### Chapter 3. SAFETY

Answers to questions on page 75.

3.0A You and only you are responsible for your safety.

3.1A Hazards that the operator of a small wastewater system might be exposed to include the following:

1. Physical injuries
2. Infections and infectious diseases
3. Insects, bugs, rodents, and snakes
4. Confined spaces
5. Oxygen deficiency or enrichment
6. Explosive, toxic or suffocating gases, vapors, or dusts
7. Vehicles, driving, and traffic
8. Electric shock
9. Toxic or harmful chemicals
10. Cave-ins
11. Fires
12. Noise
13. Dusts, fumes, mists, gases, and vapors

3.2A The operator who will be working on the equipment is the person who installs a lock (either key or combination) on the energy isolating device (switch, valve) to positively lock the switch or valve in a safe position, preventing the equipment from starting or moving. At the same time, the operator must install a tag at the lock indicating why the equipment is locked out and who is working on it.

3.2B Stored energy must be bled down to prevent its sudden release, which could injure an operator working on the system or equipment.

Answers to questions on page 77.

3.3A Pathogenic organisms are disease-carrying organisms.

3.3B No, not all bacteria cause infection in a host.

3.3C An operator can avoid getting infections and infectious diseases by practicing good personal hygiene, wearing proper protective gloves, and getting recommended immunizations.

3.4A Get prompt medical attention for insect, bug, rodent, or snake bites if you develop any signs of redness or swelling, or if you develop a fever or an allergic reaction.

Answers to questions on page 84.

3.5A One easy way to identify a confined space is by whether or not you can enter it by simply walking while standing fully upright. In general, if you must duck, crawl, climb, or squeeze into the space, it is considered a confined space.

3.5B The major hazards in confined spaces are oxygen deficiency (less than 19.5 percent oxygen by volume), oxygen enrichment (more than 23.5 percent oxygen by volume), and exposure to dangerous air contamination. Another major concern is whether the existing ventilation is able to remove dangerous air contamination or oxygen deficiency that may exist or develop.

3.5C Common properties of hydrogen sulfide gas include a rotten egg odor in small concentrations, but sense of smell is rapidly impaired. Odor is not evident at high concentrations. Hydrogen sulfide gas is colorless, flammable, explosive, poisonous (toxic), and heavier than air.

3.5D No, you cannot rely on your nose to detect dangerous gases.

3.5E An approved confined space entry permit is needed to ensure that all existing hazards in the work area have been evaluated by a qualified person, and that necessary protective measures have been taken to ensure the safety of each worker. Each time an operator leaves the permit-required space for more than 30 minutes, the permit must be renewed.

3.5F Yes, it is always necessary to use blowers to blow all of the air out of a confined space, even when gas detection instruments show the atmosphere to be safe.

3.5G When someone must enter a confined space, station at least one person to stand by on the outside of the confined space and at least one additional person within sight or call of the standby person.

Answers to questions on page 88.

3.6A As part of the routine procedure before heading out to a job site, your work vehicle and towed equipment should undergo a thorough mechanical inspection of the following items:

1. Windshield wipers.
2. Horn.
3. Seat belts.
4. Mirrors and windows.
5. Lighting system, including backup lights, turn signals, and brake lights. Also, check the towed vehicle lighting system.
6. Brakes on the vehicle and trailer, if so equipped.
7. Tire tread and inflation.
8. Wheel attachment (vandals may have loosened lug nuts).
9. Trailer hitch and tongue.
10. Auxiliary equipment such as winches or hoists.
11. Safety chain. After the trailer is coupled, the safety chain should be securely attached to the frame of the towing vehicle with enough slack to allow jackknifing, but not enough to drag on the ground.

3.6B Actual vehicle stopping distances vary depending on vehicle mass or weight, type of brakes (air or hydraulic), weather conditions, road conditions, time of day, road incline, and the driver's visual acuity (clearness of vision) and reaction time.

3.6C The primary function of temporary traffic control (TTC) on streets and highways is to provide for the safe and efficient movement of vehicles, bicyclists, and pedestrians (including persons with disabilities in accordance with the Americans with Disabilities Act of 1990 (ADA)) through or around TTC zones while reasonably protecting workers and equipment.

3.6D Any time traffic may be affected by road work activities, you must notify appropriate authorities in your area before work begins. These could be state, county, or local authorities depending on whether it is a state, county, or local street. In some cases, when traffic diversion or disruption may obstruct access by emergency response agencies such as fire and police departments, these agencies must be notified before traffic is diverted or disrupted as well.

3.6E The US Department of Transportation's *MANUAL ON UNIFORM TRAFFIC CONTROL DEVICES* (MUTCD), Part 6, is the national standard for all temporary traffic control (TTC) devices used during construction, maintenance, utility activities, and incident management.

Answers to questions on page 92.

3.7A This is the first rule that applies to electrical systems: If you are not qualified, do not, under any circumstances, attempt to work on electrical systems.

3.7B The "closed loop" similarity between hydraulic and electrical systems refers to the fact that in electrical systems, the electrons want to return to the source, which is commonly referred to as "ground." In hydraulic systems, the pressure in the pipe also wants to return to a zero pressure state.

3.7C When working with electric circuits, provide isolation between you and the ground by using rubber mats, rubber boots, or leather gloves.

Answers to questions on page 92.

3.8A The purpose of the MSDS (Material Safety Data Sheet) is to identify the types of hazards a particular chemical presents and to tell what to do in case of an emergency.

3.8B The Material Safety Data Sheet with the required information can be obtained from the manufacturer or vendor of the chemical or material.

Answers to questions on page 101.

3.9A The three recognized methods of excavation protection are sloping, shoring, and shielding.

3.9B Sloping should only be used in shallow excavations in very stable soils.

3.9C The space between the shoring and the sides of the excavation should be filled in and compacted to prevent a cave-in from getting started.

3.9D Spoil should be placed on only one side of an excavation and at least two feet away from the edge of the excavation.

Answers to questions on page 104.

3.10A Fires are classified as A, B, C, or D fires based on the type of material being consumed: A, ordinary combustibles; B, flammable liquids and vapors; C, energized electrical equipment; and D, combustible metals.

3.10B Fire extinguishers are classified as A, B, C, or D to correspond with the class of fire each will extinguish. The methods of extinguishing each classification of fire are summarized as follows:

**Class A** fires: Use foam, water, soda-acid, carbon dioxide gas, or almost any type of extinguisher.

**Class B** fires: Use foam, carbon dioxide, or dry chemical extinguishers.

**Class C** fires: Use carbon dioxide or dry chemical extinguishers to smother the fire; both types are nonconductors of electricity.

**Class D** fires: Use a Class D extinguisher or use fine dry soda ash, sand, or graphite to smother the fire.

3.10C Inspect all firefighting equipment and extinguishers on a regular basis and after each use. Fire extinguishers must be visually inspected monthly and must receive an annual maintenance check. Always refill extinguishers promptly after use.

3.11A If feasible administrative or engineering controls are unsuccessful in reducing noise exposures to permissible levels, hearing protectors are to be used.

3.12A The ideal way to control occupational diseases caused by breathing contaminated air is to prevent atmospheric contamination from occurring.

Answers to questions on page 105.

3.13A It is your duty to stop work if an unsafe condition or procedure is observed.

3.14A Every effort must be made to contain wastewater spills or discharges by stopping or reducing the flow. The spill and consequent flow area should not be accessible by the public. The spill area should be cleaned up promptly and made safe by removal of the contaminant, or, if permissible, by soil burial, or by applying sanitizing chemicals to the spill area.

CHAPTER 4

# SEPTIC TANKS AND PUMPING SYSTEMS

by

John Brady

# TABLE OF CONTENTS

## Chapter 4. SEPTIC TANKS AND PUMPING SYSTEMS

# OBJECTIVES

## Chapter 4. SEPTIC TANKS AND PUMPING SYSTEMS

1. Describe the main components of a septic tank effluent pump (STEP) system.
2. Describe the main components of a grinder pump system.
3. Operate and maintain a STEP system.
4. Operate and maintain grinder pump (GP) systems.
5. Troubleshoot septic tank effluent pump and grinder pump system problems.
6. Develop a maintenance program for your system.
7. Safely perform your duties.

## WORDS

### Chapter 4. SEPTIC TANKS AND PUMPING SYSTEMS

ABSORPTION (ab-SORP-shun) ABSORPTION

The taking in or soaking up of one substance into the body of another by molecular or chemical action (as tree roots absorb dissolved nutrients in the soil).

ADSORPTION (add-SORP-shun) ADSORPTION

The gathering of a gas, liquid, or dissolved substance on the surface or interface zone of another material.

AMPERAGE (AM-purr-age) AMPERAGE

The strength of an electric current measured in amperes. The amount of electric current flow, similar to the flow of water in gallons per minute.

ANAEROBIC (AN-air-O-bick) ANAEROBIC

A condition in which atmospheric or dissolved oxygen (DO) is *NOT* present in the aquatic (water) environment.

ANAEROBIC (AN-air-O-bick) DECOMPOSITION ANAEROBIC DECOMPOSITION

The decay or breaking down of organic material in an environment containing no free or dissolved oxygen.

ANAEROBIC (AN-air-O-bick) DIGESTER ANAEROBIC DIGESTER

A wastewater solids treatment device in which the solids and water (about 5 percent solids, 95 percent water) are placed in a large tank where bacteria decompose the solids in the absence of dissolved oxygen.

APPURTENANCE (uh-PURR-ten-nans) APPURTENANCE

Machinery, appliances, structures, and other parts of the main structure necessary to allow it to operate as intended, but not considered part of the main structure.

ARTIFICIAL GROUNDWATER TABLE ARTIFICIAL GROUNDWATER TABLE

A groundwater table that is changed by artificial means. Examples of activities that artificially raise the level of a groundwater table include agricultural irrigation, dams, and excessive sewer line exfiltration. A groundwater table can be artificially lowered by sewer line infiltration, water wells, and similar drainage methods.

CATHODIC (kath-ODD-ick) PROTECTION CATHODIC PROTECTION

An electrical system for prevention of rust, corrosion, and pitting of metal surfaces that are in contact with water, wastewater, or soil. A low-voltage current is made to flow through a liquid (water) or a soil in contact with the metal in such a manner that the external electromotive force renders the metal structure cathodic. This concentrates corrosion on auxiliary anodic parts, which are deliberately allowed to corrode instead of letting the structure corrode.

CLEAR ZONE CLEAR ZONE

See SUPERNATANT.

DECOMPOSITION or DECAY DECOMPOSITION or DECAY

The conversion of chemically unstable materials to more stable forms by chemical or biological action.

DETENTION TIME

DETENTION TIME

The time required to fill a tank at a given flow or the theoretical time required for a given flow of wastewater to pass through a tank. In septic tanks, this detention time will decrease as the volumes of sludge and scum increase.

DRY WELL

DRY WELL

A dry room or compartment in a lift station, near or below the water level, where the pumps are located, usually next to the wet well.

ENERGY GRADE LINE (EGL)

ENERGY GRADE LINE (EGL)

A line that represents the elevation of energy head (in feet or meters) of water flowing in a pipe, conduit, or channel. The line is drawn above the hydraulic grade line (gradient) a distance equal to the velocity head ($V^2/2g$) of the water flowing at each section or point along the pipe or channel. Also see HYDRAULIC GRADE LINE (HGL).

[SEE DRAWING ON PAGE 116]

EXFILTRATION (EX-fill-TRAY-shun)

EXFILTRATION

Liquid wastes and liquid-carried wastes that unintentionally leak out of a sewer pipe system and into the environment.

FRICTION LOSS

FRICTION LOSS

The head, pressure, or energy (they are the same) lost by water flowing in a pipe or channel as a result of turbulence caused by the velocity of the flowing water and the roughness of the pipe, channel walls, or restrictions caused by fittings. Water flowing in a pipe loses head, pressure, or energy as a result of friction. Also called head loss.

GRINDER PUMP

GRINDER PUMP

A small, submersible, centrifugal pump with an impeller, designed to grind solids into small pieces before they enter the collection system.

GROUNDWATER TABLE

GROUNDWATER TABLE

The average depth or elevation of the groundwater over a selected area. Also see ARTIFICIAL GROUNDWATER TABLE, SEASONAL WATER TABLE, and TEMPORARY GROUNDWATER TABLE.

HEAD LOSS

HEAD LOSS

The head, pressure, or energy (they are the same) lost by water flowing in a pipe or channel as a result of turbulence caused by the velocity of the flowing water and the roughness of the pipe, channel walls, or restrictions caused by fittings. Water flowing in a pipe loses head, pressure, or energy as a result of friction. The head loss through a comminutor is due to friction caused by the cutters or shredders as the water passes through them and by the roughness of the comminutor walls conveying the flow through the comminutor. Also called friction loss.

[SEE DRAWING ON PAGE 117]

HYDRAULIC GRADE LINE (HGL)

HYDRAULIC GRADE LINE (HGL)

The surface or profile of water flowing in an open channel or a pipe flowing partially full. If a pipe is under pressure, the hydraulic grade line is that level water would rise to in a small, vertical tube connected to the pipe. Also see ENERGY GRADE LINE (EGL).

[SEE DRAWING ON PAGE 116]

HYDROGEN SULFIDE GAS ($H_2S$)

HYDROGEN SULFIDE GAS ($H_2S$)

Hydrogen sulfide is a gas with a rotten egg odor, produced under anaerobic conditions. Hydrogen sulfide gas is particularly dangerous because it dulls the sense of smell, becoming unnoticeable after you have been around it for a while; in high concentrations, it is only noticeable for a very short time before it dulls the sense of smell. The gas is very poisonous to the respiratory system, explosive, flammable, colorless, and heavier than air.

IMHOFF CONE

IMHOFF CONE

A clear, cone-shaped container marked with graduations. The cone is used to measure the volume of settleable solids in a specific volume (usually one liter) of water or wastewater.

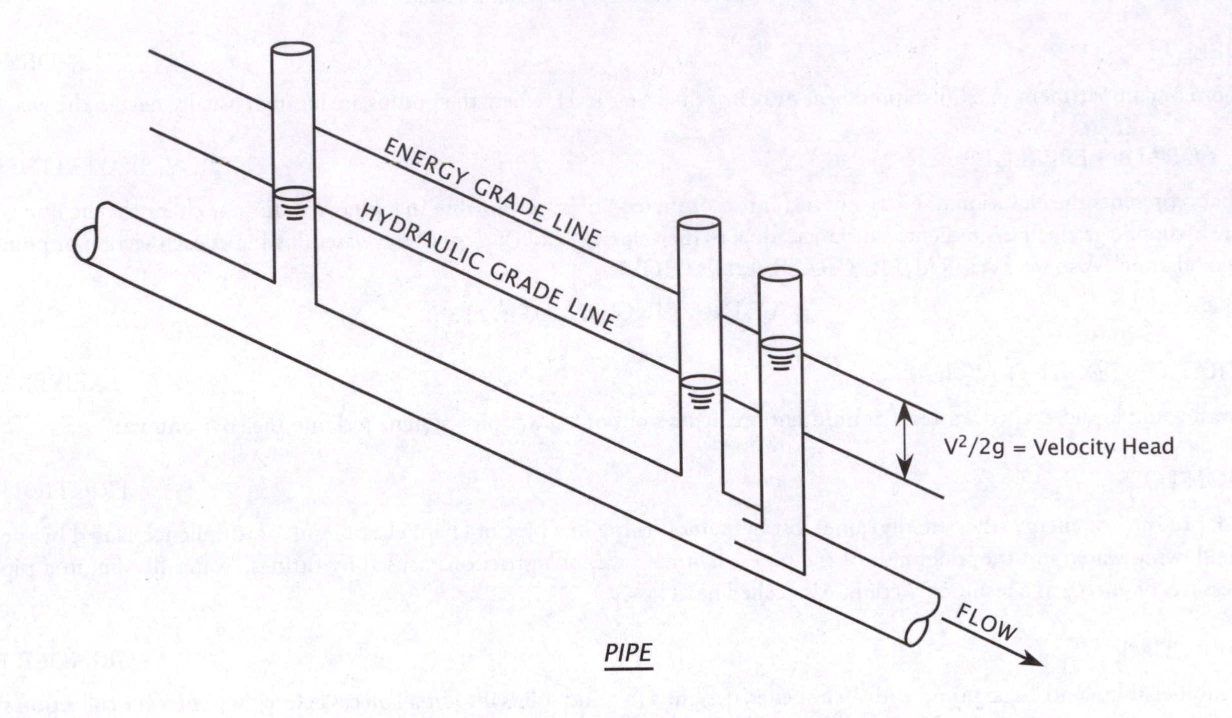


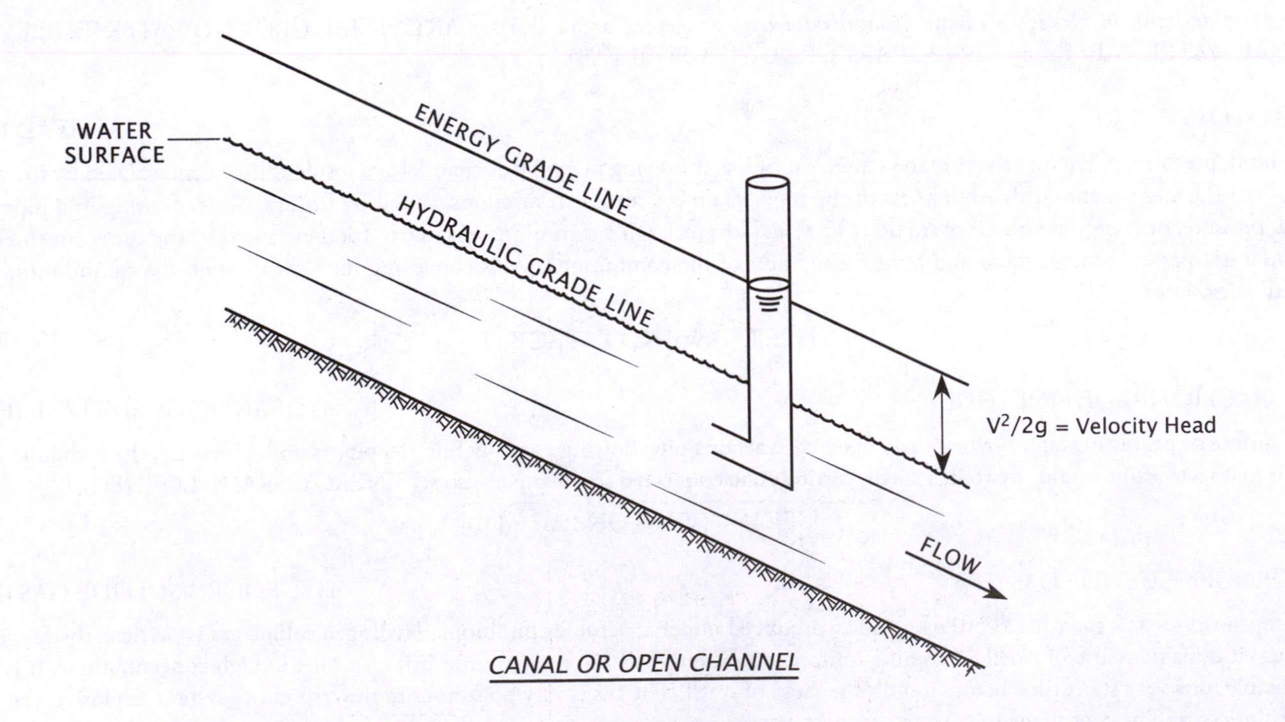


*Energy grade line and hydraulic grade line*

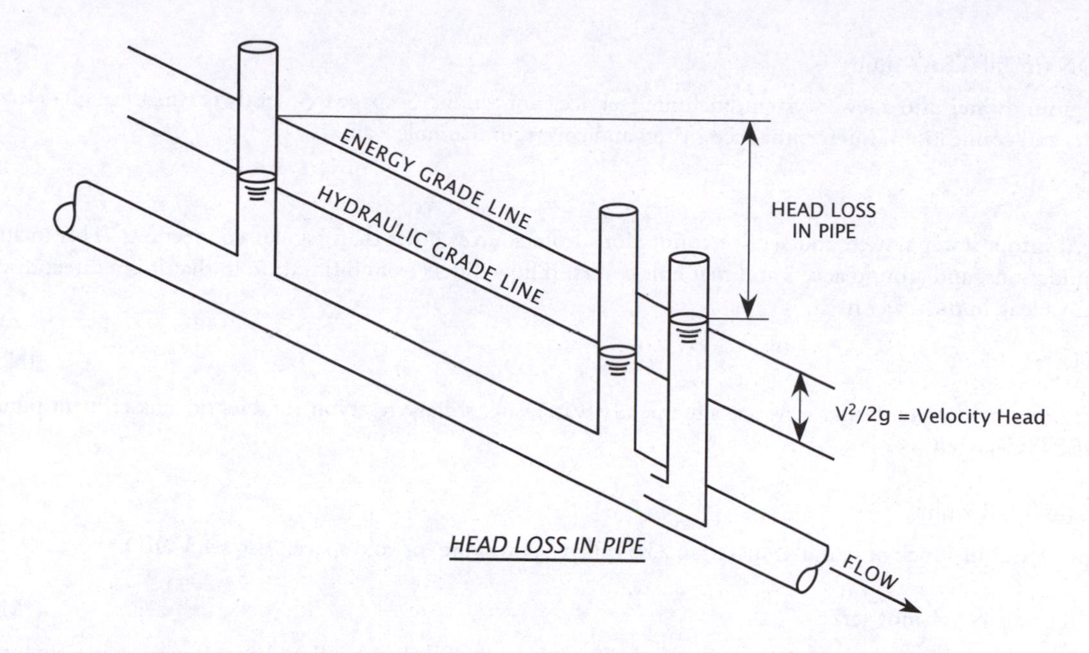
ENERGY GRADE LINE
HYDRAULIC GRADE LINE
HEAD LOSS
IN PIPE
V2/2g = Velocity Head
HEAD LOSS IN PIPE
FLOW

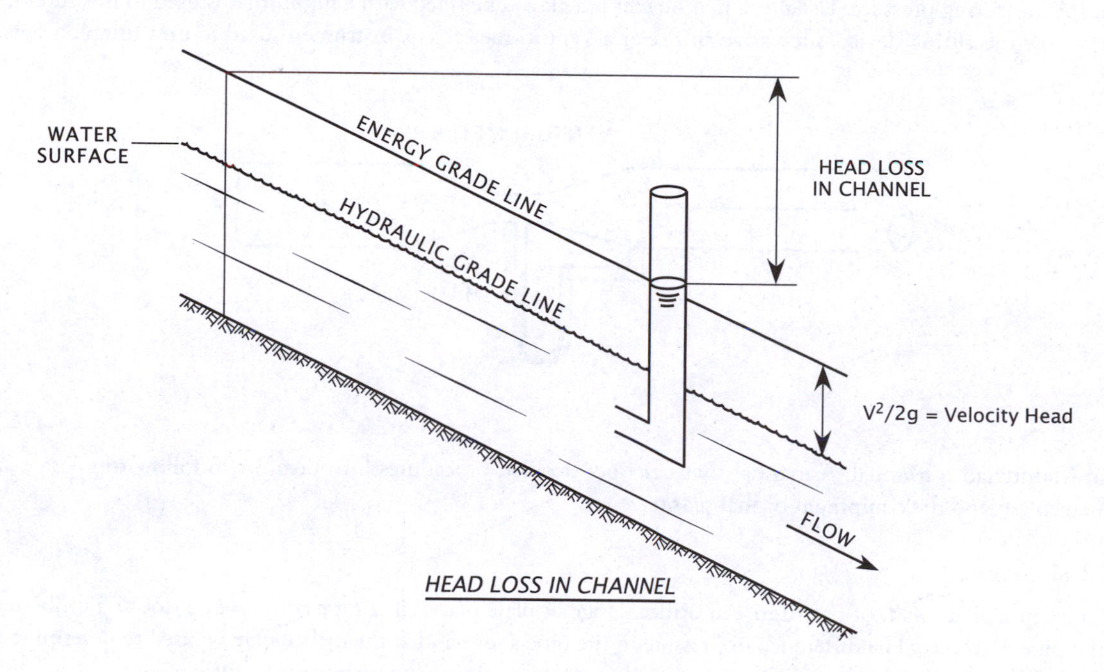
WATER
SURFACE
ENERGY GRADE LINE
HYDRAULIC GRADE LINE
HEAD LOSS
IN CHANNEL
V2/2g = Velocity Head
FLOW
HEAD LOSS IN CHANNEL

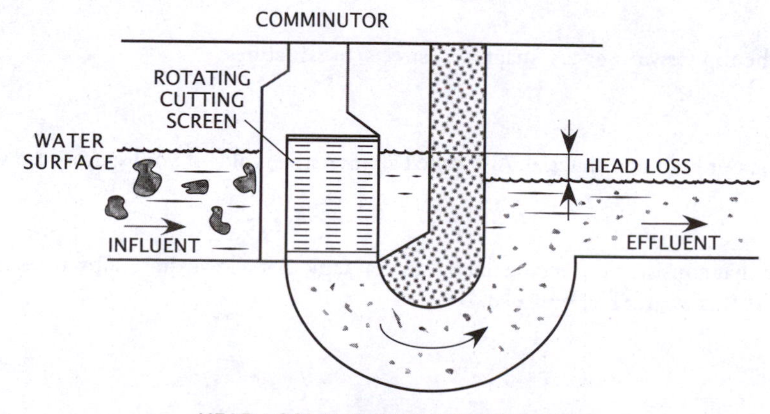
COMMINUTOR
ROTATING
CUTTING
SCREEN
WATER
SURFACE
HEAD LOSS
INFLUENT
EFFLUENT
HEAD LOSS THROUGH COMMINUTOR

*Head loss*

INFILTRATION (in-fill-TRAY-shun) INFILTRATION

The seepage of groundwater into a sewer system, including service connections. Seepage frequently occurs through defective or cracked pipes, pipe joints and connections, interceptor access risers and covers, or manhole walls.

INFLOW INFLOW

Water discharged into a sewer system and service connections from sources other than regular connections. This includes flow from yard drains, foundations, and around access and manhole covers. Inflow differs from infiltration in that it is a direct discharge into the sewer rather than a leak in the sewer itself.

INTERCEPTOR INTERCEPTOR

A septic tank or other holding tank that serves as a temporary wastewater storage reservoir for a septic tank effluent pump (STEP) system. Also see SEPTIC TANK.

INTERSTICE (in-TUR-stuhz) INTERSTICE

A very small open space in a rock or granular material. Also called a pore, void, or void space. Also see VOID.

MANOMETER (man-NAH-mut-ter) MANOMETER

An instrument for measuring pressure. Usually, a manometer is a glass tube filled with a liquid that is used to measure the difference in pressure across a flow measuring device, such as an orifice or a Venturi meter. The instrument used to measure blood pressure is a type of manometer.

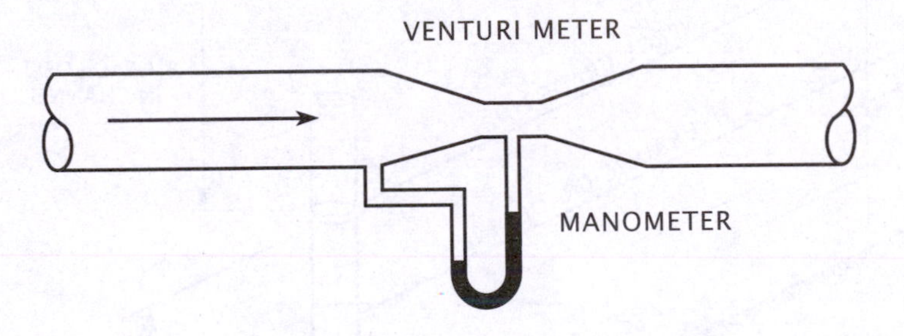


O&M MANUAL O&M MANUAL

Operation and Maintenance Manual. A manual that describes detailed procedures for operators to follow to operate and maintain a specific treatment plant and the equipment of that plant.

ORIFICE (OR-uh-fiss) ORIFICE

An opening (hole) in a plate, wall, or partition. An orifice flange or plate placed in a pipe consists of a slot or a calibrated circular hole smaller than the pipe diameter. The difference in pressure in the pipe above and at the orifice may be used to determine the flow in the pipe. In a trickling filter distributor, the wastewater passes through an orifice to the surface of the filter media.

PLAN or PLAN VIEW PLAN or PLAN VIEW

A drawing or photo showing the top view of sewers, manholes, streets, or structures.

PORE PORE

A very small open space in a rock or granular material. Also called an interstice, void, or void space. Also see VOID.

PRIMARY CLARIFIER PRIMARY CLARIFIER

A wastewater treatment device that consists of a rectangular or circular tank that allows those substances in wastewater that readily settle or float to be separated from the wastewater being treated.

PROFILE — PROFILE

A drawing showing elevation plotted against distance, such as the vertical section or side view of sewers, manholes, or a pipeline.

ROTAMETER (ROTE-uh-ME-ter) — ROTAMETER

A device used to measure the flow rate of gases and liquids. The gas or liquid being measured flows vertically up a tapered, calibrated tube. Inside the tube is a small ball or bullet-shaped float (it may rotate) that rises or falls depending on the flow rate. The flow rate may be read on a scale behind or on the tube by looking at the middle of the ball or at the widest part or top of the float.

SCUM — SCUM

A layer or film of foreign matter (such as grease, oil) that has risen to the surface of water or wastewater.

SEASONAL WATER TABLE — SEASONAL WATER TABLE

A water table that has seasonal changes in depth or elevation.

SECONDARY CLARIFIER — SECONDARY CLARIFIER

A wastewater treatment device consisting of a rectangular or circular tank that allows separation of substances that settle or float not removed by previous treatment processes.

SEPTAGE (SEPT-age) — SEPTAGE

The sludge produced in septic tanks.

SEPTIC (SEP-tick) — SEPTIC

A condition produced by anaerobic bacteria. If severe, the sludge produces hydrogen sulfide, turns black, gives off foul odors, contains little or no dissolved oxygen, and the wastewater has a high oxygen demand.

SEPTIC TANK — SEPTIC TANK

A system sometimes used where wastewater collection systems and treatment plants are not available. The system is a settling tank in which settled sludge and floatable scum are in intimate contact with the wastewater flowing through the tank and the organic solids are decomposed by anaerobic bacterial action. Used to treat wastewater and produce an effluent that is usually discharged to subsurface leaching. Also referred to as an interceptor; however, the preferred term is septic tank.

SEPTIC TANK EFFLUENT PUMP (STEP) SYSTEM — SEPTIC TANK EFFLUENT PUMP (STEP) SYSTEM

A facility in which effluent is pumped from a septic tank into a pressurized collection system that may flow into a gravity sewer, treatment plant, or subsurface leaching system.

SET POINT — SET POINT

The position at which the control or controller is set. This is the same as the desired value of the process variable. For example, a thermostat is set to maintain a desired temperature.

SLUDGE (SLUJ) — SLUDGE

(1) The settleable solids separated from liquids during processing.

(2) The deposits of foreign materials on the bottoms of streams or other bodies of water or on the bottoms and edges of wastewater collection lines and appurtenances.

SLURRY — SLURRY

A watery mixture or suspension of insoluble (not dissolved) matter; a thin, watery mud or any substance resembling it (such as a grit slurry or a lime slurry).

STABILIZATION — STABILIZATION

Conversion to a form that resists change. Organic material is stabilized by bacteria that convert the material to gases and other relatively inert substances. Stabilized organic material generally will not give off obnoxious odors.

SUMP — SUMP

This term refers to a facility or structure that connects an industrial discharger to a public sewer. The sump could be a sample box, a clarifier, or an intercepting sewer.

SUPERNATANT (soo-per-NAY-tent) — SUPERNATANT

The relatively clear water layer between the sludge on the bottom and the scum on the surface of an anaerobic digester or septic tank (interceptor).

(1) From an anaerobic digester, this water is usually returned to the influent wet well or to the primary clarifier.

(2) From a septic tank, this water is discharged by gravity or by a pump to a leaching system or a wastewater collection system.

Also called clear zone.

SUSPENDED SOLIDS — SUSPENDED SOLIDS

(1) Solids that either float on the surface or are suspended in water, wastewater, or other liquids, and that are largely removable by laboratory filtering.

(2) The quantity of material removed from water or wastewater in a laboratory test, as prescribed in *STANDARD METHODS FOR THE EXAMINATION OF WATER AND WASTEWATER*, and referred to as Total Suspended Solids Dried at 103–105°C.

TEMPORARY GROUNDWATER TABLE — TEMPORARY GROUNDWATER TABLE

(1) During and for a period following heavy rainfall or snow melt, the soil is saturated at elevations above the normal, stabilized, or seasonal groundwater table, often from the surface of the soil downward. This is referred to as a temporary condition and thus is a temporary groundwater table.

(2) When a collection system serves agricultural areas in its vicinity, irrigation of these areas can cause a temporary rise in the elevation of the groundwater table.

TOTAL DYNAMIC HEAD (TDH) — TOTAL DYNAMIC HEAD (TDH)

When a pump is lifting or pumping water, the vertical distance (in feet or meters) from the elevation of the energy grade line on the suction side of the pump to the elevation of the energy grade line on the discharge side of the pump. The total dynamic head is the static head plus pipe friction losses.

VOID — VOID

A pore or open space in rock, soil, or other granular material, not occupied by solid matter. The pore or open space may be occupied by air, water, or other gaseous or liquid material. Also called an interstice, pore, or void space.

WET WELL — WET WELL

A compartment or tank in which wastewater is collected. The suction pipe of a pump may be connected to the wet well or a submersible pump may be located in the wet well.

# CHAPTER 4. SEPTIC TANKS AND PUMPING SYSTEMS

(Lesson 1 of 2 Lessons)

*NOTE:* Operation and maintenance programs for small wastewater systems will vary with the age of the system, the extent and effectiveness of previous programs, and local conditions. The purpose of this chapter is to provide an overview of septic tanks and pumping systems and should be used to acquire general knowledge. You will need to adapt the information and procedures in this chapter to specific equipment manuals and site-specific operation and maintenance manuals for your particular situation.

## 4.0 SEPTIC TANKS (Figures 4.1 and 4.2)

On-site wastewater treatment systems usually consist of two major parts: (1) a septic tank,[1] which provides primary treatment; and (2) some type of soil absorption system, usually a leach field, where the treated wastewater is discharged. Many community systems also use individual septic tanks but the clarified effluent is transported off site for discharge to a leach field, seepage pit, mound system, sand filters, or wetlands.

### 4.00 Treatment Process Description

A typical septic tank may be a single-compartment tank (Figure 4.1) or divided into two compartments (Figure 4.2). The first compartment acts as the *PRIMARY CLARIFIER*[2] where the majority of grease, oils, and retained and digested solids are removed. Wastewater is retained in the first compartment for a period *(DETENTION TIME*[3]*)* long enough to allow the settleable solids to separate from the wastewater and be deposited on the bottom of the tank, and for the scum (grease, oils, and other light materials) to float to the surface. The primary compartment also performs the function of an *ANAEROBIC DIGESTER*[4] where bacteria in the tank break down or reduce some of the heavy solids (sludge) that have accumulated on the bottom.

The first compartment is separated from a second compartment by an interior baffle or wall (see side view of Figure 4.2) which is fitted with a port or a pipe tee. The baffle permits the wastewater from the clear water *(SUPERNATANT*[5]*)* space between the sludge and scum layers to flow from the first compartment into the second compartment without carrying solids over from the first compartment. The second compartment acts as another settling chamber and is similar to a *SECONDARY CLARIFIER.*[6]

The sludge volume reduction that occurs in the first compartment is the result of a natural biological treatment process. No chemicals or other commercial additives are needed. This reduction of the solids content in the wastewater entering the second compartment is important because it greatly reduces the quantity of solids entering the leach field or discharge system. Carryover of suspended solids and floatable grease and oil in the effluent (treated wastewater) from the septic tank tends to seal the soil in the leach field preventing the wastewater from soaking into the soil (percolation and absorption), or a large amount of solids can plug the leach field's distribution piping. Either condition may create a wastewater backup of the system and possibly flood the home through the fixtures, building sewer, or illegal storm and floor drain connections.

---

1. Septic tanks are also referred to as "interceptors" by some agencies, including the US Environmental Protection Agency. The term "septic tanks" is more commonly used throughout the wastewater industry, however, and "septic tanks" will be used in this training manual.
2. *Primary Clarifier.* A wastewater treatment device that consists of a rectangular or circular tank that allows those substances in wastewater that readily settle or float to be separated from the wastewater being treated.
3. *Detention Time.* The time required to fill a tank at a given flow or the theoretical time required for a given flow of wastewater to pass through a tank. In septic tanks, this detention time will decrease as the volumes of sludge and scum increase.
4. *Anaerobic* (AN-air-O-bick) *Digester.* A wastewater solids treatment device in which the solids and water (about 5 percent solids, 95 percent water) are placed in a large tank where bacteria decompose the solids in the absence of dissolved oxygen.
5. *Supernatant* (soo-per-NAY-tent). The relatively clear water layer between the sludge on the bottom and the scum on the surface of an anaerobic digester or septic tank (interceptor). (1) From an anaerobic digester, this water is usually returned to the influent wet well or to the primary clarifier. (2) From a septic tank, this water is discharged by gravity or by a pump to a leaching system or a wastewater collection system. Also called clear zone.
6. *Secondary Clarifier.* A wastewater treatment device consisting of a rectangular or circular tank that allows separation of substances that settle or float not removed by previous treatment processes.

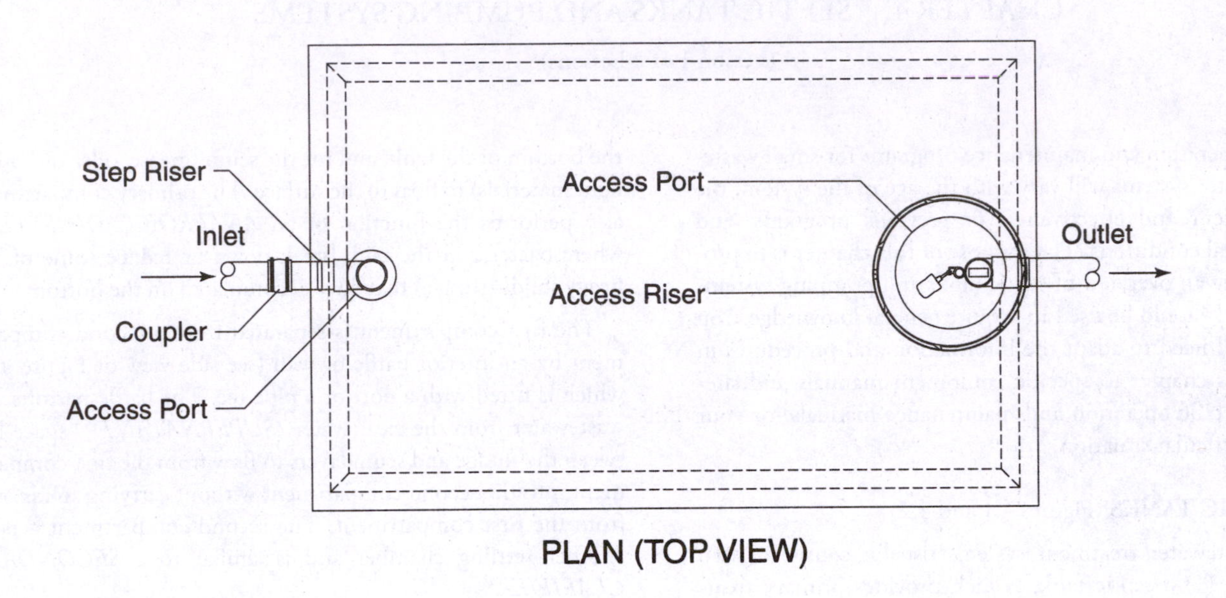


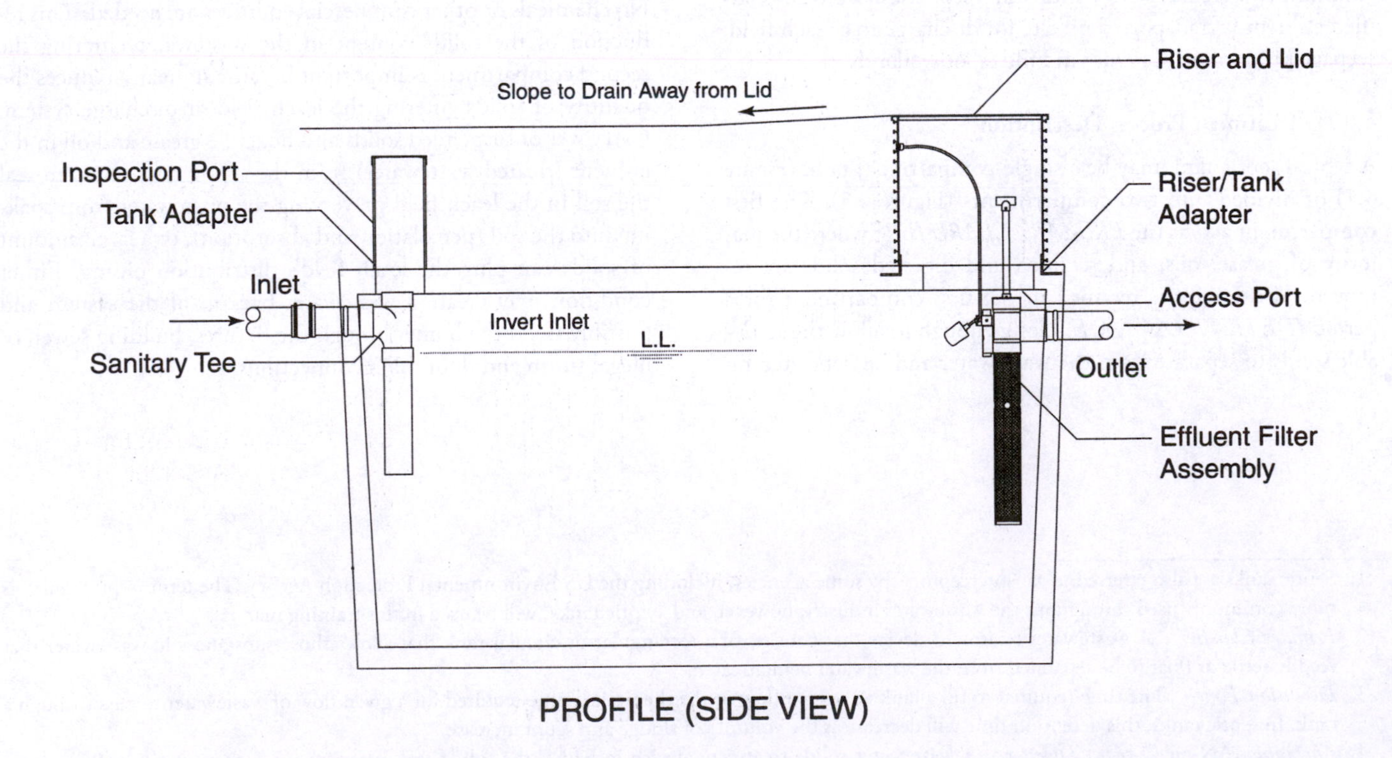


*Fig. 4.1 Typical single-compartment septic tank*

(Courtesy of Orenco Systems, Inc., www.orenco.com)

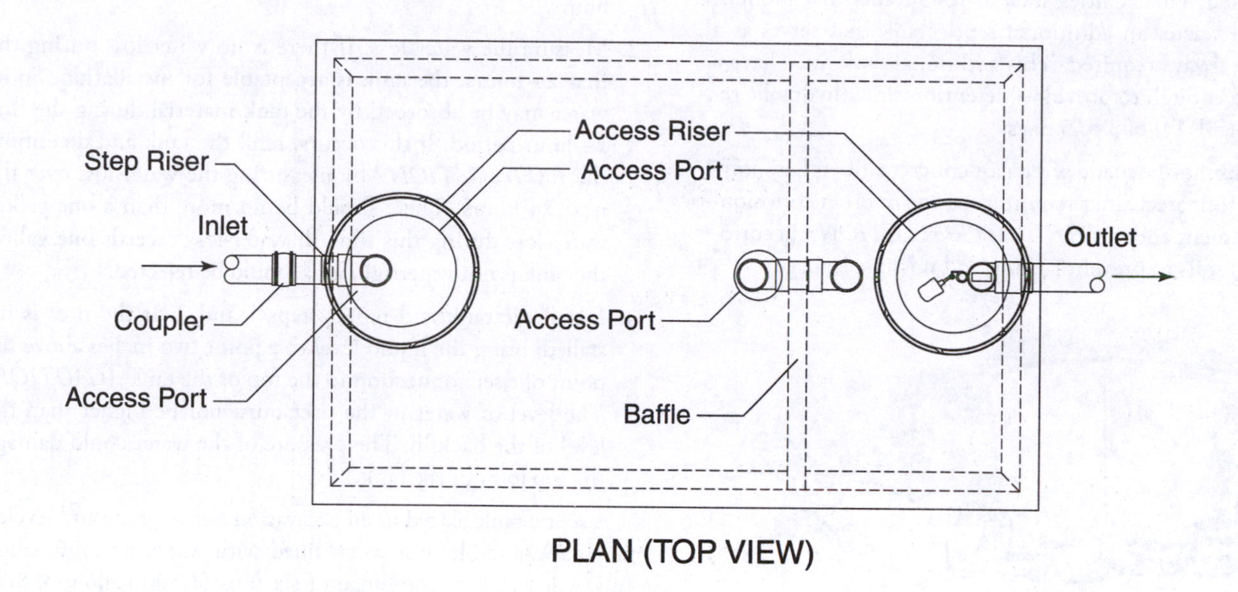


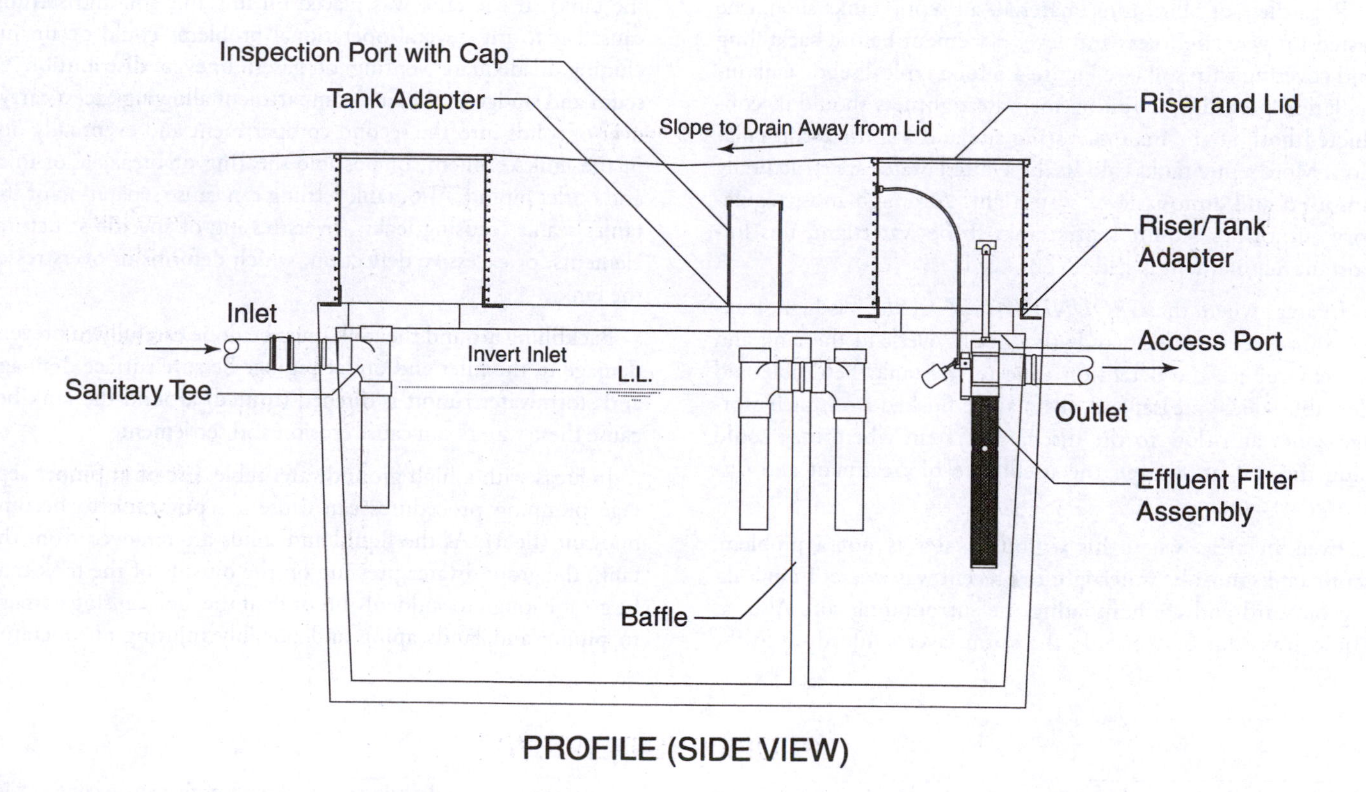


*Fig. 4.2 Typical septic tank with two compartments*

(Courtesy of Orenco Systems, Inc., www.orenco.com)

### 4.01 Tank Construction and Installation

Typical residential septic tanks have a volume of 1,000, 1,500, or 2,000 gallons. Larger homes are using larger tanks. For commercial establishments, the local septic tank codes are used to determine the necessary volume required to properly handle the waste generated. For facilities such as restaurants that produce high-strength waste, an additional septic tank that serves as a grease trap is always required. This additional tank must have a capacity large enough to provide a detention time (hydraulic retention time (HRT)) of 3 to 5 days.

Septic tanks may be made of precast concrete, fiberglass, polyethylene, or steel. Steel tanks require a coating or other corrosion-resistant treatment and *CATHODIC PROTECTION*[7] in corrosive (low pH) soils to prevent rusting and possible leakage.

Regardless of fabricating material, all septic tanks should be tested for watertightness and level placement before backfilling and covering with soil (see Figure 4.3 for a typical septic tank installation procedure). Testing for watertightness should be conducted both at the manufacturing site and again after installation. Most septic tanks sold in the United States are structurally unsound and almost never watertight. Although most regulatory authorities require septic tanks to be watertight, this important requirement is seldom enforced.

In areas where the *GROUNDWATER TABLE*[8] is high, leakage into a septic tank may hydraulically overload the tank and thereby reduce the detention time in the tank. Settleable and floatable solids, grease, and oils may be flushed from their storage zones and flow to the discharge system where they could plug the soil or shorten the useful life of treatment and discharge systems.

Even in areas where high groundwater is not a problem, septic tanks must be watertight to prevent wastewater from leaking outward and contaminating the surrounding soil. Also, as liquid leaks out of the tank, the scum layer could drop to the outlet level where floatable solids, fats, soaps, oils, and greases may be washed through the outlet pipe.

To test the watertightness of a septic tank, use the following procedure:

1. Fill the tank to its brim with water and let it stand for 24 hours.
2. Measure the water loss. If there is no water loss during the first 24 hours, the tank is acceptable for installation. Some water may be absorbed by the tank material during the first 24-hour period. If this occurs, refill the tank and determine the *EXFILTRATION*[9] by measuring the water loss over the next 24 hours. There should be no more than a one-gallon water loss during this time. If water loss exceeds one gallon the tank is not watertight and should be rejected.
3. Install the tank and repeat steps 1 and 2. If the riser is installed, bring the liquid level to a point two inches above the point of riser connection to the top of the tank. (*CAUTION:* The level of water in the riser must not be higher than the level of the backfill. The pressure of the water could damage the top joint of the tank.)

A septic tank placed in an excavation, set to grade and leveled may appear stable, but when filled with water, a 1,500-gallon tank will weigh in the range of six tons (1,500 gallons × 8.34 pounds per gallon of water = 12,510 pounds plus the weight of the tank). If the tank was placed on unstable soil and settling caused it to tilt, several operational problems could occur, including inadequate venting of gases, uneven distribution of scum and sludge in the first compartment allowing excess carry-over of solids into the second compartment and eventually out in the tank's effluent, or possible shearing or breakage of inlet and outlet piping. Also, tank settling can cause separation of the tank's seams (causing leaks), overstressing of specific structural elements, or excessive deflection, which deforms or overstresses the tank.

Backfilling around the tank must be done carefully to prevent damage to the inlet and outlet piping. Be sure surface drainage and stormwater runoff is ditched (routed) around the tank because these waters can cause erosion and settlement.

In areas with a high groundwater table, use of improper septage pumping procedures can cause a septic tank to become buoyant (float). As the liquid and solids are removed from the tank, the groundwater pressure on the outside of the tank may be great enough to suddenly lift or float the tank causing damage to piping and landscaping and possibly injuring an operator.

---

7. *Cathodic* (kath-ODD-ick) *Protection.* An electrical system for prevention of rust, corrosion, and pitting of metal surfaces that are in contact with water, wastewater, or soil. A low-voltage current is made to flow through a liquid (water) or a soil in contact with the metal in such a manner that the external electromotive force renders the metal structure cathodic. This concentrates corrosion on auxiliary anodic parts, which are deliberately allowed to corrode instead of letting the structure corrode.
8. *Groundwater Table.* The average depth or elevation of the groundwater over a selected area. Also see ARTIFICIAL GROUNDWATER TABLE, SEASONAL WATER TABLE, and TEMPORARY GROUNDWATER TABLE.
9. *Exfiltration* (EX-fill-TRAY-shun). Liquid wastes and liquid-carried wastes that unintentionally leak out of a sewer pipe system and into the environment.

Set tank level and to a uniform bearing on a minimum 4″ thick sand or granular bed overlying a firm and uniform base. Tank should not bear directly on large boulders or massive rock edges.

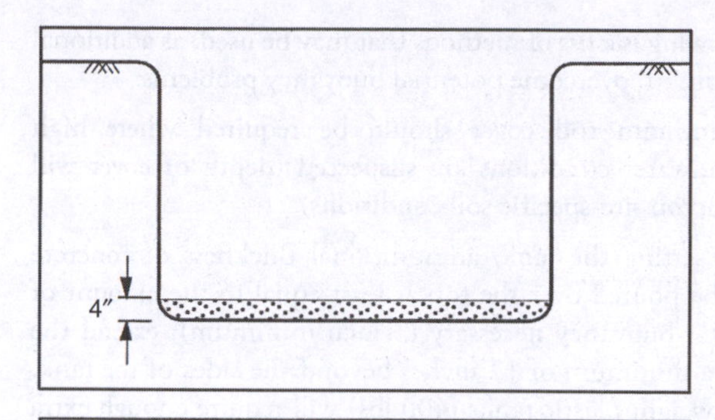


Unstable or wet foundations should be stabilized and cared for by over excavation and backfill with select materials, or other means as required to ensure a stable and uniform bearing foundation for the tank.

Excavation for the tank should extend at least twenty four inches (24″) outside the farthest outer dimension of the tank to allow room for proper compaction.

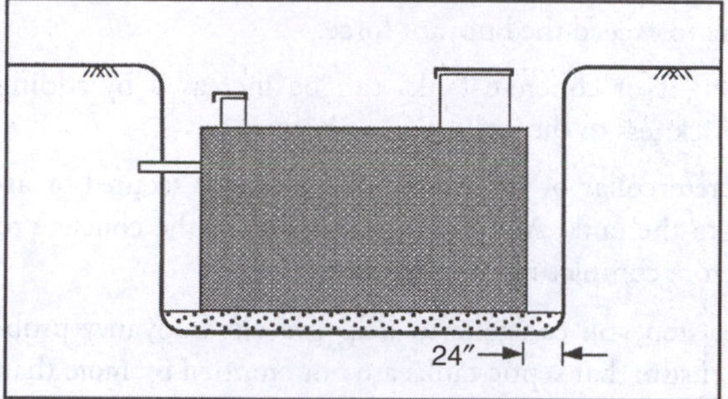


Backfill should be placed in uniform, mechanically compacted layers no greater than 24″ thick and of nearly equal height on each side of the tank to minimize settlement and to provide support for the tank walls. Backfill should be of proper size and gradation (and free of stones over 4″ diameter, and any other deleterious materials).

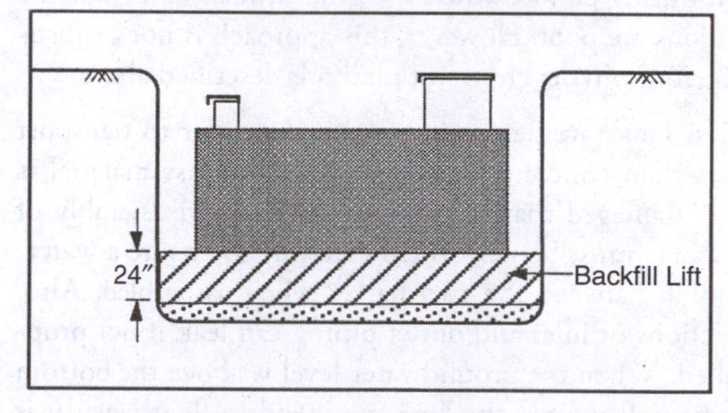


Each layer should be thoroughly tamped, making sure the soil or sand backfill contains sufficient moisture to allow for proper compaction.

*Jetting or flooding should not be used to settle backfill.*

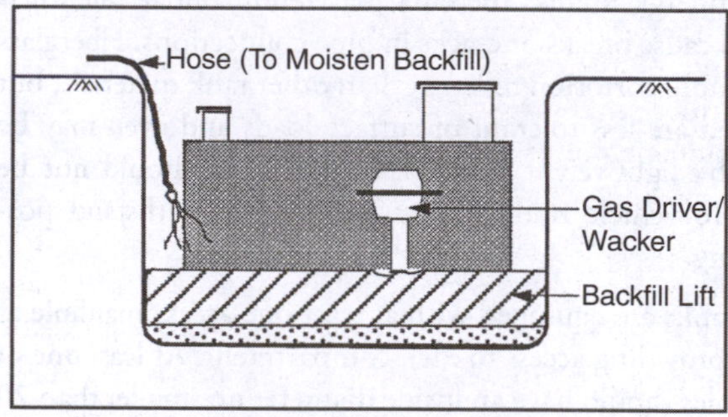


Tanks should be placed at such depth to facilitate a minimum ¼″ per foot slope of the building sewer. In some instances, buildings may have more than one building sewer. Care should be taken to establish the tank elevation such that a minimum ¼″ per foot slope is attainable from any building sewer.

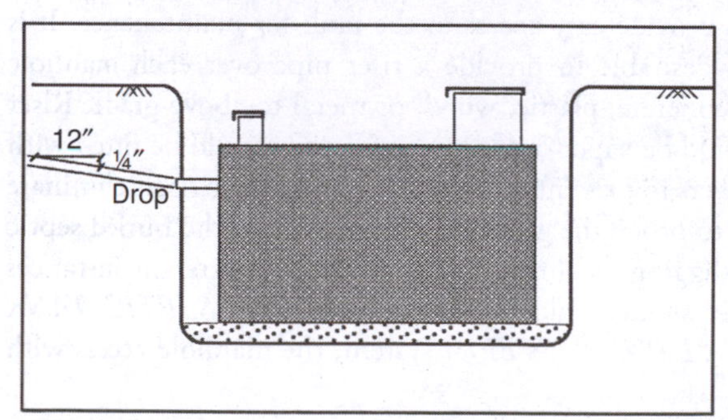


*Fig. 4.3 Septic tank installation procedure*

(Courtesy of Orenco Systems, Inc., www.orenco.com)

The following is a list of methods that may be used, as additional precautions, to overcome potential buoyancy problems.

1. A minimum soil cover should be required where high groundwater conditions are suspected (depth of cover will depend on site-specific soil conditions).
2. After setting the tank, an additional thickness of concrete may be poured over the top at least equal to the amount of counter-buoyancy necessary (3-inch minimum); extend the sides a minimum of 12 inches beyond the sides of the tank. Lightweight plastic tanks (400 lbs) will require enough extra concrete to exceed the buoyant force.
3. The weight of concrete tanks can be increased by adding more thickness to the walls, top, or bottom.
4. A concrete collar or ballast may be properly secured or attached to the tank. Allow sufficient time for the concrete to cure before completing the backfilling.

Another step you can take to help prevent buoyancy problems is to ensure that septic tanks are not emptied by more than 50 percent during periods when the groundwater is high or the soil conditions are poor. However, this approach is not a substitute for the four physical restraint methods described above.

Fiberglass tanks are lighter in weight and easier to transport to job sites than concrete tanks, but the fiberglass material is more easily damaged than concrete or steel. Field assembly of fiberglass tanks must be performed carefully to ensure a watertight bonding between the two halves when assembled. Also, the connections of inlet and outlet piping can leak if not properly installed. When the groundwater level is above the bottom of a light fiberglass tank, the tank may tend to float when it is empty and cause breaks or cracks in pipe connections. Fiberglass tanks are more corrosion resistant than other tank materials, but once buried are less tolerant of surface loads and even may be damaged by lightweight vehicles. Septic tanks should not be subjected to vehicle traffic unless designed to withstand possible loading.

Septic tanks are equipped with at least one access manhole at each end, providing access to each compartment. At least one of the manholes should have an inside diameter no smaller than 20 inches to provide easy access to the tank for maintenance. It is also very desirable to provide a riser pipe over each manhole made of concrete, plastic, wood, or metal to above grade. Riser pipes should be vapor- and watertight and should be fitted with secure covers to prevent unauthorized entry. The risers eliminate the need to probe the ground surface to locate the buried septic tank or digging to uncover the manhole covers. In instances where the septic tanks are being used in a *SEPTIC TANK EFFLUENT PUMP (STEP)*[10] system, the manhole access with riser is required over the second compartment to provide access to the effluent pump and associated equipment. Manholes with risers are also used when measuring sludge blanket depths and scum layer thicknesses to determine the need for pumping the tank for routine cleaning.

Piping and fittings used in the system or tanks are usually ABS or PVC plastic; however, vitrified clay pipe also may be used. Cast-iron pipe has been found not to be suitable for this type of installation because of corrosion.

In some utility districts the homeowner has the responsibility for buying the septic tank and installing it or having it installed by a private contractor. In this situation, the wastewater agency usually provides written instructions (local wastewater discharge ordinance) and specifications for the installation when the connection permit is issued. In other utility districts, the agency itself installs the septic tank and *APPURTENANCES*[11] and the homeowner is responsible only for construction of the building sewer. A "License to Enter upon Property" and other legal documents are required to protect the wastewater agency from liability when utility personnel perform work on private property. See Appendix A at the end of this chapter for some examples of the types of permits and liability release forms that should be completed before working on private property.

## QUESTIONS

Please write your answers to the following questions and compare them with those on page 174.

4.0A List the two major parts of a basic individual on-site wastewater treatment system.

4.0B Why do solids carried by wastewater need to be removed by a septic tank?

4.0C What tests should be performed on a septic tank before the excavation is backfilled and the tank is covered with soil?

4.0D Why should riser pipes be installed over both the inlet and outlet manholes on a septic tank?

### 4.02 Homeowner's Responsibility for Proper Use

The septic tank is designed to accept two loads: wastewater and the solids carried by the wastewater. Floatable solids (scum) such as grease and oils accumulate on the water surface and the settleable solids (sludge) accumulate on the bottom of the tank. Some of the solids are broken down (volume reduced) by microorganisms. The scum and sludge accumulations must be monitored (measured) annually and pumped out of the septic tank before they build up to the extent they escape through the overflow outlet in the first chamber.

---

10. *Septic Tank Effluent Pump (STEP) System.* A facility in which effluent is pumped from a septic tank into a pressurized collection system that may flow into a gravity sewer, treatment plant, or subsurface leaching system.
11. *Appurtenance* (uh-PURR-ten-nans). Machinery, appliances, structures, and other parts of the main structure necessary to allow it to operate as intended, but not considered part of the main structure.

### *4.020 Solids*

Persons using septic tank systems must be encouraged not to use the septic tank system for anything that can be disposed of safely as trash or in some other way such as by composting. The less material put into the septic tank, the less often it will require pumping to remove the accumulated septage (sludge and scum solids).

Keep these common items out of a septic tank:

1. Eggshells, cantaloupe seeds, gum, coffee grounds
2. Grease or cooking oils
3. Tea bags, chewing tobacco, cigarette butts, kitty litter
4. Condoms, dental floss, feminine hygiene products, diapers
5. Paper towels, facial tissue, newspapers, candy wrappers
6. Rags, stringy materials, large amounts of hair
7. "Flushable" wipes, baby wipes, medicated wipes, cleaning wipes
8. Plastics, rubber goods, flammables, acids

Garbage disposal units should be avoided because their use greatly speeds up the accumulation of floating scum in the septic tank. Food scraps should be composted or disposed of with other household solid waste, and grease should be collected in a container rather than poured down the drain.

### *4.021 Liquids*

Water conservation is the easiest and cheapest way to protect the septic system because when less water is used, there is less wastewater to be discharged by the septic tank to the leaching system. Also, when less wastewater enters the septic tank, the detention time in the tank increases and reduction of solids improves. Attempt to educate those using the system to practice the following measures:

1. Do not let water run while brushing teeth or washing hands, vegetables, and dishes.
2. Use water-saving devices in the shower and toilet tanks. Ultra-low-flush toilets use only 1.5 gallons per flush whereas the common older style toilets use 5 to 6 gallons per flush.
3. Repair leaky fixtures. A leaking toilet can contribute a gallon a minute or more to the household wastewater volume. You can check a toilet by adding food dye to the tank and see if it turns up in the bowl without flushing.

4. Use the washing machine or dishwasher only when there is a full load. Avoid doing several loads in one day as it hydraulically overloads the septic tank, thus reducing detention time, which increases solids carryover to the leach field.
5. Do not let rain or surface water runoff enter the septic tank through roof or yard drains. Seal tank risers and manhole covers with epoxy or a good grade of caulking at the tank and cover.
6. Large homes with multiple sinks, bathrooms, and showers should use larger tanks to ensure adequate hydraulic detention time. Tankage is cheap compared to the cost of repairing or replacing a failed drain field.

### *4.022 Toxic Materials*

Ordinary use of household chemicals will not harm the microorganisms in the septic tank and leaching discharge system but do not use excessive amounts. Never dump strong medications, photo chemicals, petroleum oils, solvents, paint thinners, or poisonous chemicals (insecticides or pesticides) down the drain to the septic tank to dispose of them. These toxic substances will kill the beneficial organisms in the septic tank and effluent discharge system that reduce solids and stabilize waste constituents and they could harm the surrounding environment.

### *4.023 Septic Tank Additives*

Septic tank additives purchased as over-the-counter chemicals, enzymes, and bacterial cultures are not recommended. Manufacturers are always trying to improve the performance of their products. Some additives will do more harm by breaking down the scum mat and allowing the grease and oil products to pass through the tank and cause the drain field soils to clog sooner. Additives also can have harmful effects on the organisms decomposing wastes in septic tanks if not applied properly. The use of additives typically does not reduce the need for periodic cleaning of the septic tank by pumping its contents and disposing of the sludge and scum at an approved disposal site for such material. The natural growth of a bacterial culture in a septic tank will accomplish waste solids *DECOMPOSITION*[12] and *STABILIZATION*[13] if the microorganisms are not killed by harmful additives (toxic substances) as described in Section 4.022.

See Appendix B at the end of this chapter for an example of a simple user's guide that could be distributed by a wastewater agency to inform the public about the proper use of a septic tank system.

### 4.03 Septic Tank Safety

Extreme care must be exercised in opening a septic tank or any chamber, vault, or manhole that has been exposed to wastewater, especially where sludge has been deposited. When a large septic tank is being cleaned, do not enter the tank until it has been thoroughly ventilated and gases have been removed to prevent

12. *Decomposition* or *Decay.* The conversion of chemically unstable materials to more stable forms by chemical or biological action.
13. *Stabilization.* Conversion to a form that resists change. Organic material is stabilized by bacteria that convert the material to gases and other relatively inert substances. Stabilized organic material generally will not give off obnoxious odors.

explosion hazards, toxic gas (hydrogen sulfide) hazards, or suffocation of workers due to lack of oxygen. If you have not already studied Chapter 3, "Safety," or are not able to identify the hazards associated with septic tanks and other wastewater structures, be sure to review Chapter 3 before opening or inspecting a septic tank. Pay particular attention to sections describing confined spaces, gas production in septic tanks, oxygen deficiency/enrichment, and infections and infectious diseases.

### 4.04 Measuring Sludge and Scum

If an operating problem is suspected, the tank may be inspected by removing the access manhole at the inlet end of the tank. This is the manhole over the larger compartment that acts as the primary clarifier and the anaerobic digester. The best practice is to measure the thickness of both the scum and the sludge layers near the outlet baffle or pipe. A simple method that works in most instances is to use a shovel to push the scum layer away from the side of the tank so that the thickness of the scum can be estimated or measured. If there is less than two inches between the bottom of the scum layer and the bottom of the tank outlet device (this may be an overflow pipe, baffle, or tee), the scum (not the liquid) should be removed. A thick scum blanket (typically 14 inches or more) approaching the bottom of the overflow indicates that the area designed for scum accumulation is approaching its capacity. By measuring the sludge depth and scum thickness, the need for pumping the tank for routine cleaning can be determined (see Section 4.05, "Sludge Removal and Disposal").

A "sludge judge" can be used to measure the depth of sludge on the bottom of a tank. This device is a long, clear plastic tube. The tube must be long enough to reach the bottom of the tank from ground level. A tube diameter of ½ to 1 inch is sufficient. A larger diameter tube is harder to seal when withdrawing a sample from the tank. Push the tube to the bottom of the tank. Place your hand or a plug over the open end on top to form an airtight seal. Slowly remove the tube from the tank. Measure the height of sludge column in the tube to determine the thickness or depth of the sludge in the tank. The sludge judge causes light particles to be suspended and they do not settle readily, which gives incorrect measurements.

Several devices have been developed that are more accurate, more reliable, and simpler to use than the sludge judge for measuring both scum blanket thickness and the depth that the top of the sludge blanket has reached in a tank. One such device for measuring sludge and scum levels is the "flopper" device shown in Figure 4.4A. To use the flopper scum measuring device, pull the string so the flopper will be in a vertical position along the

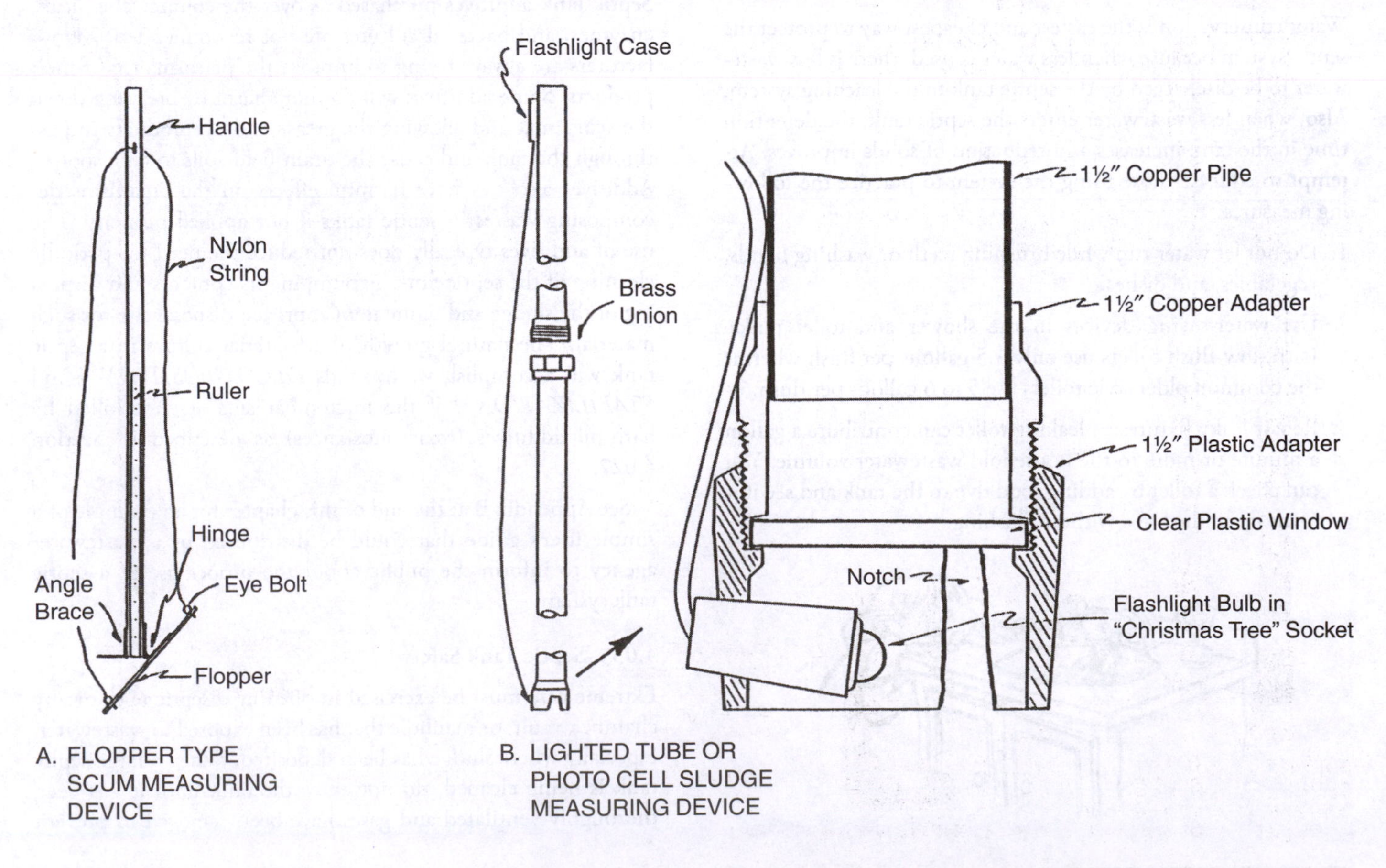


*Fig. 4.4 Devices for measuring the thickness of a scum mat and the depth of sludge in septic tanks*

(Source: *SEPTIC TANK SYSTEMS, ONE CONSULTANT'S TOOL KIT*, by Dr. John T. Winneberger. Reproduced by permission of Dr. John T. Winneberger)

side of the handle. Push the flopper and handle down through the scum mat. Pull the string so the flopper is against the angle brace and at right angles (perpendicular) to the handle. Lift the device until the flopper reaches the bottom of the scum mat (you can feel the flopper touch the bottom of the scum mat) and read the scum mat thickness on the ruler attached to the handle of the device.

After determining the scum blanket thickness with the flopper device, the sludge blanket depth may be measured by slowly lowering the device (with the flopper perpendicular to the handle) until a slight resistance is felt. Resistance indicates that the device is touching the surface of the sludge blanket. Note the measured depth. Pull the string to place the flopper in the vertical position and continue to lower the device until the bottom of the tank is reached. Note the measured depth and add to it the length of the flopper from tip to angle brace. To determine the sludge blanket depth, subtract the first measurement (the top of the sludge blanket) from the second measurement (the bottom of the tank).

To use the lighted tube sludge measuring device (Figure 4.4B), turn the light on and slowly lower the device into the tank. Look down the tube and observe the light until it disappears into the thick sludge. When the light is no longer visible in the tube, the sludge blanket has been reached. This is the distance to the top of the sludge blanket. Distances are easily recorded if the tube has a ruler of distances marked on the side. Subtract this distance from the distance to the bottom of the tank to obtain the sludge thickness.

Sludge Thickness, inches = Distance to Bottom of Tank, inches − Distance to Top of Sludge Blanket, inches

| Sludge Thickness, inches | = Distance to Bottom of Tank, inches | − Distance to Top of Sludge Blanket, inches |
|---|---|---|

Figure 4.5 shows similar sludge and scum measuring gauges produced by another major manufacturer.

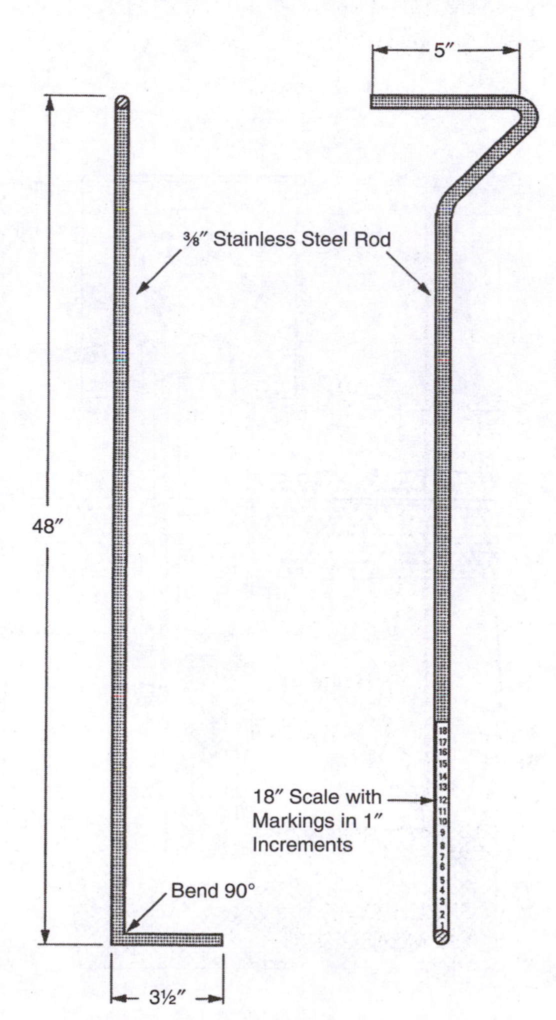


*Fig. 4.5 Gauges for measuring septic tank sludge and scum accumulations*

(Courtesy of Orenco Systems, Inc., www.orenco.com)

Many septic tank manufacturers provide charts (Figure 4.6) showing how to convert a depth of sludge or scum to a volume.

The supernatant (water with low solids content) zone of the tank is the area between the bottom of the scum layer and the top of the sludge blanket. If that zone is reduced by a thick scum blanket and a deep sludge deposit, then there is little space left in the first compartment to provide for a reasonable detention time for incoming wastewater to allow the solids to settle to the bottom and scum to float to the surface. The solids that are not removed in the first compartment must then be removed in the second compartment or they will be discharged with the tank's effluent to the discharge leach field system where the solids will clog the discharge system. When clogging occurs, liquid may break through to the ground surface and the wastewater may back up into the home's plumbing fixtures. For this reason, it is always necessary to pump out a septic tank at regular intervals.

## QUESTIONS

Please write your answers to the following questions and compare them with those on page 174.

4.0E List the two loads septic tanks are designed to accept.

4.0F How can the water load on a septic tank be reduced?

4.0G List the major hazards an operator may be exposed to when inspecting a septic tank.

4.0H What can cause an excessive reduction of detention time in a septic tank?

### 4.05 Sludge Removal and Disposal

Under normal operating conditions, wastewater solids settle to the bottom of the first compartment of the septic tank where they tend to spread out in a flat blanket. This volume is reduced through the process of microorganisms feeding on and breaking

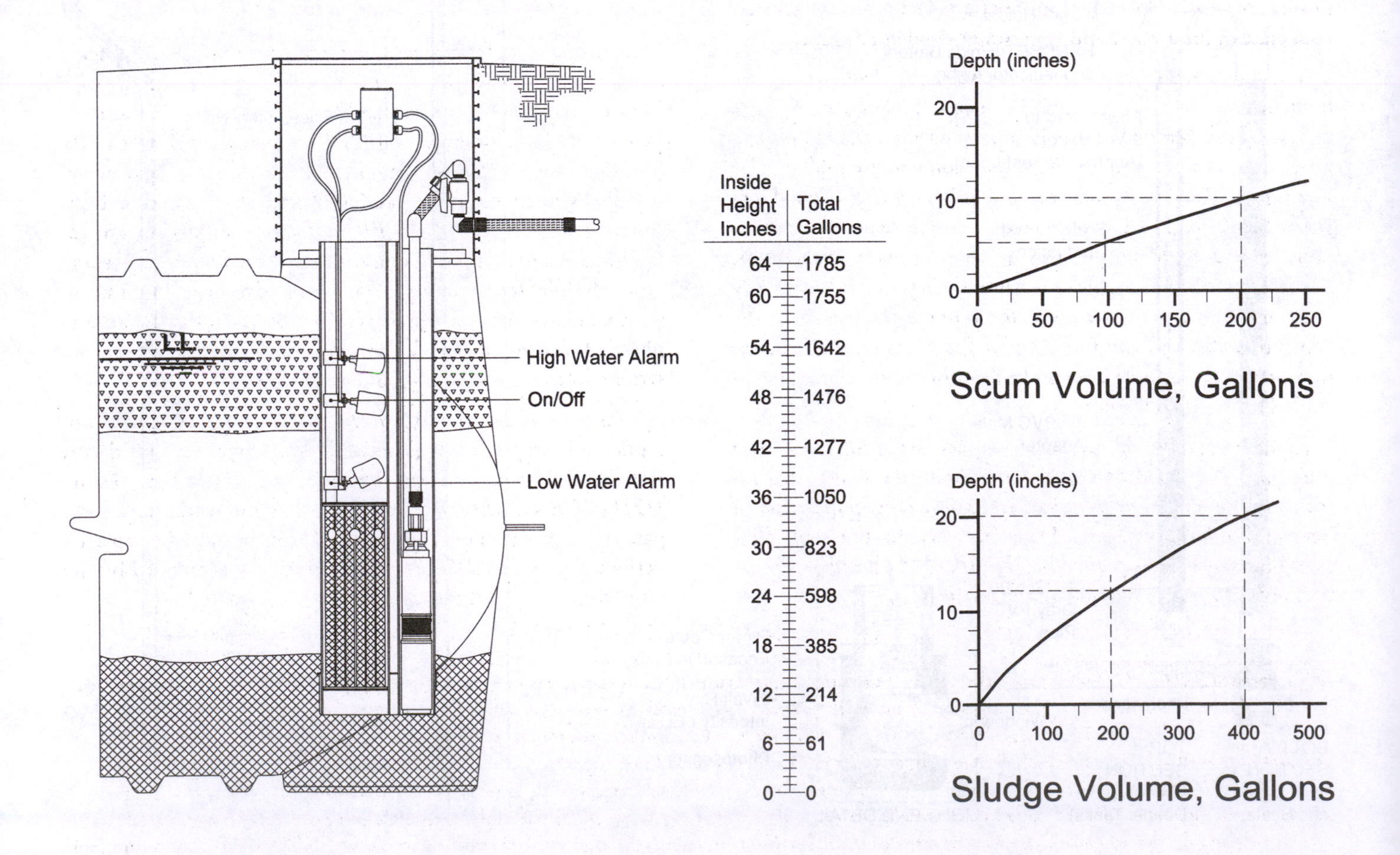


*Fig. 4.6 Conversion of sludge and scum depths to volumes*

(Courtesy of Orenco Systems, Inc., www.orenco.com)

down the organic components of the settled waste material. A properly sized septic tank commonly removes about 65 percent of the *SUSPENDED SOLIDS*[14,15] that pass through the tank. Most of the solids removal occurs in the first compartment. Over time, however, solids and scum accumulate in the tank.

A septic tank should be pumped to remove accumulated solids before they escape to the second compartment; typically, this is when the scum layer is about 12 to 14 inches thick and the sludge blanket reaches a depth of about 24 inches. The volume of sludge and scum pumped from septic tanks can be estimated by measuring the volume pumped into a tank (change in tank contents) on a truck that hauls away the sludge and scum. This information should be recorded for future reference.

Most wastewater agencies hire private companies on annual contracts to pump and dispose of septic tank solids (septage) as well as to respond to calls for emergency pumping. Other utilities own their own pumping trucks and perform this function themselves. In either case, septic tank sludge must be disposed of in an approved manner and in accordance with local and state regulations. The material may be buried with permission of the proper authorities or emptied into a sanitary sewer system. However, solids from septic tanks should never be emptied into storm drains or discharged directly into any stream or waterway. A septage handling and disposal plan must be developed that complies with existing state and federal regulations. If you are unsure about any aspect of the proper disposal of sludge, contact your state or local water and wastewater regulatory agencies.

The frequency for pumping septic tanks depends on the homeowners' or residents' cooking or food preferences, hygiene practices, and lifestyles. Periodic pumping intervals commonly range from two years to ten years, with some households requiring a more frequent pumping and others going beyond ten years. Septic tanks at commercial facilities, especially restaurants, may need to be pumped as often as every six months. Actual inspection of sludge and scum levels is the only sure way to determine when a tank needs to be pumped. Therefore, the depth of solids and the thickness of the scum layer should be measured on an annual basis and this information should be recorded for future reference.

Septic tanks should never be washed or disinfected after pumping. The small amount of wastewater or sludge that remains in the tank after pumping contains a large population of beneficial microorganisms. Their presence in the tank after pumping may speed up the reestablishment of a healthy population of organisms to treat incoming wastewater.

Many operators report that when pumping out and cleaning a septic tank, they normally find a mound of very dense sludge directly under the tank's inlet tee, and find it more difficult to remove this heavier sludge than the rest of the sludge in the tank. This concentration of dense sludge may be caused by the rapid settling of heavier material such as eggshells, coffee grounds, or silt from washing vegetables. If dense sludge is a problem or if sludge and scum levels build up too quickly, talk with the homeowner or resident about the types of solids that should not be disposed of in a septic tank system.

## 4.06 Troubleshooting Septic Tanks

### *4.060 Cracks and Blockages*

Little can go wrong with a septic tank itself. Redwood or fiberglass tanks sometimes suffer structural damage or deteriorate through age. Steel and concrete deteriorate due to hydrogen sulfide corrosion. Problems most often occur in the plumbing or in the disposal system. Blockages in the building sewer between the home and the septic tank can usually be cleared with plumbing tools. If the plumbing backs up suddenly under normal use in dry weather, a blockage is the most probable cause. Some pipe blockages are caused by tree roots entering the pipe or by grease or detergent buildup, which can develop over a period of time.

### *4.061 Odors*

Odor complaints from areas known to have septic tank systems are common and odors are one of the greatest concerns of a homeowner with a septic tank. Odors have often been traced to improperly located house vents drafting the gas downward over the house roof. Odors commonly occur when starting a newly installed septic tank or for a period of time after a tank has been pumped out. Until *ANAEROBIC*[16] microorganisms become established, odors through house vents are common. In some areas, the inlet tee to the septic tank is fitted with a cleanout cap, which isolates the household venting system from the atmosphere of the septic tank to prevent odors. A separate venting system for the septic tank should be provided.

Another method of odor control that may be purchased from septic tank suppliers is an activated carbon filter canister that is installed on the house vents. The activated carbon removes the *HYDROGEN SULFIDE GAS* ($H_2S$)[17] as the septic tank gases pass through the filter.The activated carbon must be replaced periodically as its *ADSORPTION*[18] capacity is reduced by the saturation of $H_2S$ on the carbon.

---

14. *Suspended Solids.* (1) Solids that either float on the surface or are suspended in water, wastewater, or other liquids, and that are largely removable by laboratory filtering. (2) The quantity of material removed from water or wastewater in a laboratory test, as prescribed in *STANDARD METHODS FOR THE EXAMINATION OF WATER AND WASTEWATER,* and referred to as Total Suspended Solids Dried at 103–105°C.
15. You may wish to review the discussion on suspended solids in Section 2.012 in Chapter 2.
16. *Anaerobic* (AN-air-O-bick). A condition in which atmospheric or dissolved oxygen (DO) is *NOT* present in the aquatic (water) environment.
17. *Hydrogen Sulfide Gas* ($H_2S$). Hydrogen sulfide is a gas with a rotten egg odor, produced under anaerobic conditions. Hydrogen sulfide gas is particularly dangerous because it dulls the sense of smell, becoming unnoticeable after you have been around it for a while; in high concentrations, it is only noticeable for a very short time before it dulls the sense of smell. The gas is very poisonous to the respiratory system, explosive, flammable, colorless, and heavier than air.
18. *Adsorption* (add-SORP-shun). The gathering of a gas, liquid, or dissolved substance on the surface or interface zone of another material.

Temperature also may play an important role in odor production from septic tanks. In an actively used tank, the temperature of the wastewater tends to be warmer near the top and cooler near the bottom of the tank. During periods of low flows, the temperature is essentially the same throughout the tank. The warmer the temperature of the wastewater, the greater the rate of gas production.

### 4.062 *Solids Carryover*

Solids carryover in the septic tank effluent is not a direct problem for the septic tank but it can plug downstream wastewater discharge components such as leach fields, mounds, and sand filters. Solids can block or plug distribution piping *ORIFICES*[19] in the discharge system, or solids may seal off and plug the *VOIDS*[20] of the soil or permeable absorption media.

Solids carryover can be controlled by the addition of an effluent filter that is installed in the septic tank in the second compartment or near the outlet (effluent) end of the septic tank. Figure 4.7 shows a septic tank effluent filter (STEF) system. This filter is designed so that it may be removed easily from the septic tank for cleaning, hosed off with water, and then reinstalled in the tank. According to the manufacturer, this cleaning procedure is required about as often as the tank requires pumping.

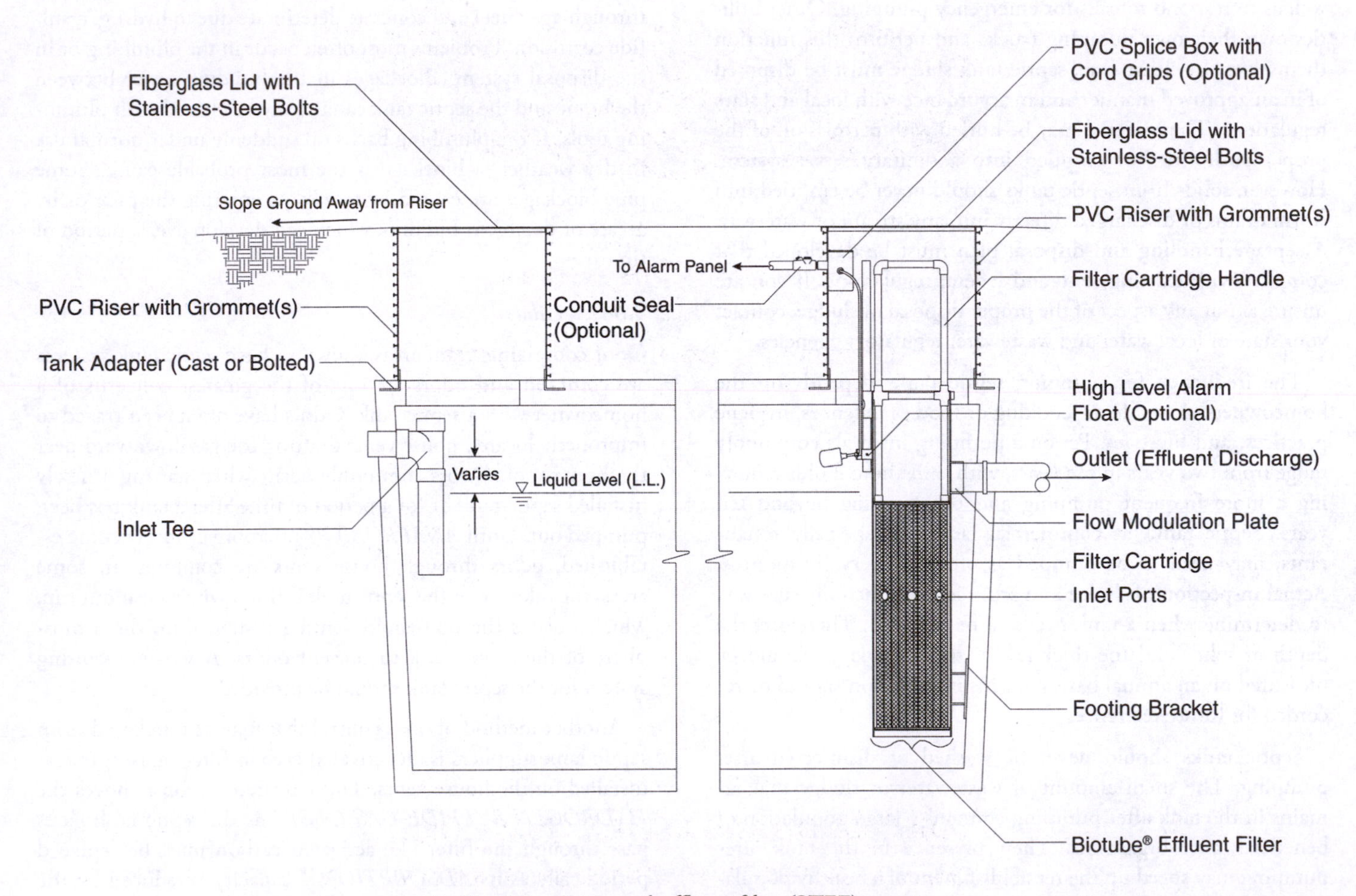


*Fig. 4.7 Septic tank effluent filter (STEF) system*

(Courtesy of Orenco Systems, Inc., www.orenco.com)

19. *Orifice* (OR-uh-fiss). An opening (hole) in a plate, wall, or partition. An orifice flange or plate placed in a pipe consists of a slot or a calibrated circular hole smaller than the pipe diameter. The difference in pressure in the pipe above and at the orifice may be used to determine the flow in the pipe. In a trickling filter distributor, the wastewater passes through an orifice to the surface of the filter media.
20. *Void.* A pore or open space in rock, soil, or other granular material, not occupied by solid matter. The pore or open space may be occupied by air, water, or other gaseous or liquid material. Also called an interstice, pore, or void space.

If an effluent filter clogs and wastewater backs up in the septic tank, the problem may simply be a plugged filter. However, if cleaning the filter does not correct the problem, the backup may mean it is time to pump accumulated solids and scum out of the tank or that the biological system in the tank has been upset. In this case, it will be necessary to investigate further to determine the cause of the backup.

### *4.063 Troubleshooting Guide*

Table 4.1 describes several problems that occur when septic tanks are not functioning properly and suggests steps that can be taken to identify and correct the problem.

## QUESTIONS

Please write your answers to the following questions and compare them with those on page 174.

4.0I When should a septic tank be pumped to remove accumulated solids?

4.0J What can cause pipe blockages in a septic tank system?

4.0K What is a common cause of odor complaints from septic tank owners?

**END OF LESSON 1 OF 2 LESSONS**

**on**

**SEPTIC TANKS AND PUMPING SYSTEMS**

Please answer the discussion and review questions next.

## DISCUSSION AND REVIEW QUESTIONS

**Chapter 4. SEPTIC TANKS AND PUMPING SYSTEMS**

(Lesson 1 of 2 Lessons)

At the end of each lesson in this chapter, you will find discussion and review questions. Please write your answers to these questions to determine how well you understand the material in the lesson.

1. How is the volume of wastewater solids reduced in a septic tank?
2. What is the purpose of the baffle between the first and second compartments of a septic tank?
3. What steps should homeowners be advised to take to protect their septic tank systems?
4. What problems could develop in a septic tank when the scum blanket depth becomes excessive (more than 14 inches)?
5. What are some of the most common septic system problems?

TABLE 4.1 TROUBLESHOOTING—SEPTIC TANKS

| Indicator/Observation | Probable Cause | Check | Solution |
|---|---|---|---|
| Flooded fixtures in home, water will not drain or drains very slowly | 1. Backup in septic tank system | 1a. Liquid level in septic tank effluent end | 1a. If normal, see 1b. |
| | | 1b. Cleanout just prior to septic tank inlet | 1b. Run plumber's snake clearing inlet to septic tank of blockage. If inlet is clear, see 2 for compartmented tanks. |
| | 2. Compartmented tanks—transfer port between first and second compartment is blocked, thus restricting flow between compartments | 2. Liquid levels of first and second compartments | 2. If first compartment higher than second compartment, have tank pumped out and cleaned by septic tank pumper. |
| Septic tank liquid level high, tank flooded and creating backup | 1a. Blocked septic tank outlet | 1a. Outlet/discharge pipe | 1a. Clear outlet pipe of blockage, determine cause of blockage—rags, paper, plastic, scum, or sludge. |
| | 1b. Rags, paper, plastic, greases caused blockage | | 1b. Educate home residents of proper disposal of undesirable material. |
| | 1c. Scum/sludge | 1c. Measure sludge and scum blanket depths | 1c. If sludge and scum levels indicate full tank, have solids (not wastewater) pumped out of tank. |
| | 1d. Drain field saturation (failure) | 1d. Water flooding drain field | 1d. Allow drain field to rest (dry out). |
| | 1e. Tank inlet settled or broken outlet pipe (reverse water gradient) | 1e. Has tank settled or outlet pipe broken | 1e. Install new tank or repair outlet pipe. |
| | 2. Excessive water flow to tank | 2. Review water use in home, toilets or faucets leaking, large clothes wash days | 2. Reduce water load. |
| | 3. Effluent discharge system backup or failure | 3. Flow to and through effluent distribution box to discharge system | 3. Restore flow to discharge system. If unable to restore, have septic tank pumped to provide wastewater storage until discharge system is put back into service. |
| | 4. Broken or restricted effluent discharge line due to septic tank settlement or traffic load | 4. Effluent discharge line; look for moist soil or pooling adjacent to effluent end of septic tank | 4. Excavate and repair pipeline. |
| | 5. Infiltration/inflow causing high water flow rate through septic tank, effluent clearer than usual, low solids content | 5. Flow rate through tank; clarity of effluent | 5. Locate source of incoming water and isolate from tank. |
| | 6. Effluent filter plugged with solids | 6. Head loss across filter, septic tank full of water but surface inside filter is much lower | 6. Remove septic tank effluent filter (STEF) system filter. Clean filter element with brush and water spray, reinstall filter. Check or inspect assembly and discharge hose. |

TABLE 4.1 TROUBLESHOOTING—SEPTIC TANKS *(continued)*

| Indicator/Observation | Probable Cause | Check | Solution |
|---|---|---|---|
| Low liquid in septic tank | 1. Septic tank leaking (exfiltration); could contaminate groundwater supply | 1. Mark existing liquid level and monitor level for several days | 1. If liquid level does not rise during normal use, open both access covers. Have septic tank pumped down, hosed out, ventilate septic tank with portable blower by forcing air in one access and exhausting through other access opening at least 30 minutes. Use appropriate confined space entry procedures. Check for cracks or other damage to interior of tank including pipe connections. Repair or replace tank. |
| | 2. Excessive pumping by effluent pumping units | 2. ON/OFF effluent pump controls | 2. Adjust ON/OFF level controls to operate within desired tank liquid levels. |
| Noxious odors, rotten egg smell (hydrogen sulfide or $H_2S$) | | | |
| 1. Odors in home | 1a. Improper vent system on house plumbing | 1a. Odors in house predominate around sinks, tubs, and shower drains | 1a. Have plumber check home plumbing vent system. Add proper traps and lines to existing plumbing. (Sometimes a ¼-inch diameter hole is drilled in the cleanout cap on the tank inlet, which lets the odor vent to atmosphere at the tank instead of through the house vents.) Be sure this hole does not become a source of stormwater inflow to the system. |
| | 1b. Sink traps not deep enough and empty between uses, thus allowing gases to drift into building | 1b. Depth of sink traps | 1b. Replace with deeper sink traps. |
| 2. Odors around home | 2a. and b. Vents on septic tank or down draft from house vents exhausting on home roof | 2a. Inlet to septic tank | 2a. Install an inlet tee equipped with a cleanout cap or an ell on the inlet isolating household vents from the septic tank atmosphere. |
| | | 2b. House vents improperly located | 2b. Exhaust from the vents drafting downward over the house roof. Most reasonable solution is addition of activated carbon filter to house vents to remove hydrogen sulfide gas. |
| | 2c. Broken inlet or building sewer line | 2c. Tank inlet or building sewer line for breaks | 2c. Repair breaks. |
| | 2d. Cleanout cap left off | 2d. Cleanout caps | 2d. Install cleanout caps. |
| Insects, flies | Breeding in septic tank in area above or on surface of scum blanket escape septic tank through house plumbing vent system | Insects leaving house plumbing vents | Install stainless-steel screens across or inside plumbing vents that open to the atmosphere to prevent passage of insects |

*NOTES:*

Any septic tank sludge, scum, or wastewater that has overflowed the tank or leaked through broken lines should be limed (add lime to disinfect) and covered with soil. Be sure to use procedures approved by the local health agency.

Odor, insect, and house service line problems in septic tank effluent pump (STEP) systems are similar to the troubleshooting guide for septic tanks.

The home resident or customer will be the first to notice a problem of insects, odors, or a service line backup.

Other problems covered in high/low liquid level-related problems in interceptors/pump vaults (septic tanks) will initiate an audio/visual alarm. The alarm should notify the home resident of a high/low liquid level or a problem creating the high/low liquid level condition before a backup of wastewater into the home occurs.

An operator should respond to an alarm situation as quickly as possible.

In some localities, water quality control or environmental control agencies require notification by the operating agency within a prescribed time limit when a wastewater spill or overflow occurs.

# CHAPTER 4. SEPTIC TANKS AND PUMPING SYSTEMS

(Lesson 2 of 2 Lessons)

## 4.1 TYPES OF WASTEWATER PUMPING SYSTEMS

Two types of pumping systems are commonly used in small communities to move the wastewater generated in homes and businesses into a community collection system. The two types of pumping arrangements are septic tank effluent pump (STEP) systems, and grinder pump (GP) systems.

In a STEP system, wastewater from a home or business flows into an on-site septic tank or a *WET WELL*.[21] The wet well may contain a submersible pump or may be connected to the suction side of a pump in a *DRY WELL*.[22] (See Figure 4.8.) The wastewater is then pumped into a pressurized collection system[23] and conveyed to a community treatment or discharge facility. When standard septic tanks are used in a STEP system, they function in the usual manner to reduce and stabilize wastewater solids. A pump installed in the septic tank effluent compartment pumps the clarified liquids into a community collection or discharge system. The collection and discharge systems may be individual, cluster, or community systems.

When the community system provides wastewater treatment as well as discharge services, the septic tank is not required for solids removal and treatment and a smaller wet well or pump vault may be used if gravity flow cannot be maintained. Wastewater and solids are pumped from the vault into a pressurized collection system. Since the small wet well or pump vault is only a holding tank for the wastewater, little if any of the solids content of the wastewater will be removed before pumping. *GRINDER PUMPS*[24] are therefore installed to reduce the size of solids entering the collection system.

## 4.2 COMPONENTS OF PUMPING SYSTEMS

Table 4.2 lists the main components of STEP and GP systems and describes the function or purpose of each part.

### 4.20 Building Sewer

Often, a new building sewer at least three inches in diameter is laid from the house to the septic tank when a new tank is installed. This component is relatively inexpensive. Installing a new one ensures that the system is watertight to prevent *INFILTRATION*[25]/*INFLOW*[26] (I/I), and greatly reduces possible problems associated with older building sewers such as broken pipe, offset joints, root intrusion, and improper grade. It is good practice to install a cleanout to grade in the building sewer line just ahead of where it enters the septic tank. The cleanout provides easy access for the agency personnel to investigate problems in the system and to determine whose responsibility it is to restore service, when necessary. In most systems, the homeowner is responsible for maintaining the building sewer.

In those instances where a septic tank is not installed and a grinder pump is used to pump the wastewater into the collection system, the building sewer ends at the grinder pump vault.

### 4.21 Septic Tank

When a STEP collection system is being constructed to serve homes that have existed for some time with on-site septic systems, many agencies are requiring the installation of a new septic tank at each home and abandonment of the older tanks. Authorities usually require the removal and proper disposal of the contents of the old tank. The old septic tank is then filled with sand or it is crushed in place and the hole is backfilled after a newly installed tank is placed in service. Installation of a new septic tank (and a new building sewer, as noted previously) ensures a watertight system that reduces the possibility of infiltration. Installation of all new tanks and building sewers also means that components throughout the new system will all be similar. This makes for more dependable operation and easier maintenance of the system.

---

21. *Wet Well.* A compartment or tank in which wastewater is collected. The suction pipe of a pump may be connected to the wet well or a submersible pump may be located in the wet well.
22. *Dry Well.* A dry room or compartment in a lift station, near or below the water level, where the pumps are located, usually next to the wet well.
23. Chapter 6 describes the types of gravity, pressure, and vacuum collection systems commonly used by small agencies.
24. *Grinder Pump.* A small, submersible, centrifugal pump with an impeller, designed to grind solids into small pieces before they enter the collection system.
25. *Infiltration* (in-fill-TRAY-shun). The seepage of groundwater into a sewer system, including service connections. Seepage frequently occurs through defective or cracked pipes, pipe joints and connections, interceptor access risers and covers, or manhole walls.
26. *Inflow.* Water discharged into a sewer system and service connections from sources other than regular connections. This includes flow from yard drains, foundations, and around access and manhole covers. Inflow differs from infiltration in that it is a direct discharge into the sewer rather than a leak in the sewer itself.

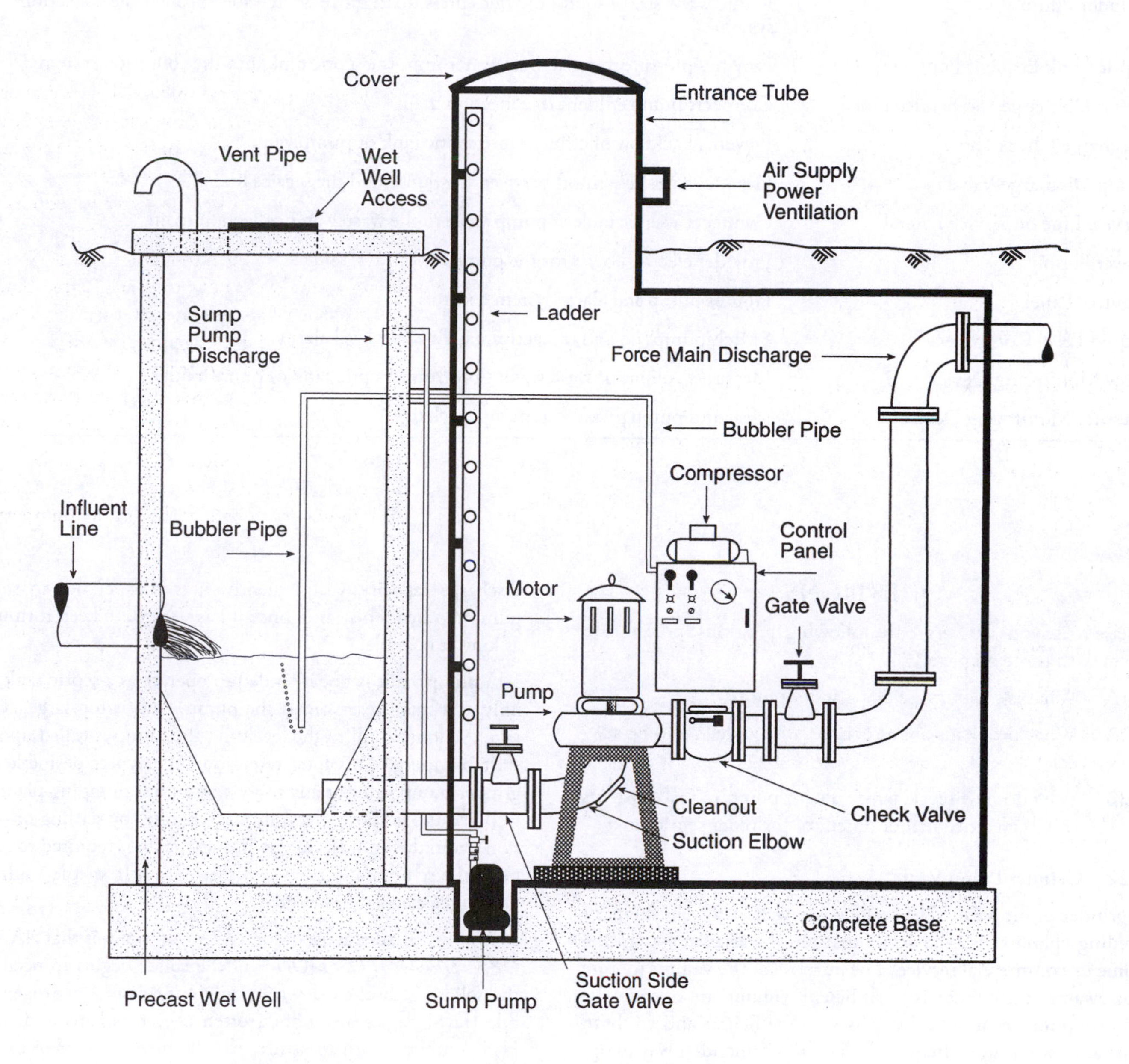


*Fig. 4.8 Prefabricated factory lift station showing wet well (left side) and dry well (right side)*

TABLE 4.2 COMPONENTS OF STEP AND GRINDER PUMP SYSTEMS

| Part | Purpose |
|---|---|
| Building Sewer | Connects home or business plumbing system to septic tank or grinder pump vault |
| Septic Tank | Captures, reduces, and stabilizes solids |
| Grinder Pump Vault | Reservoir where wastewater is stored prior to being pumped into the pressurized collection system |
| Grinder Pump | Reduces the size of solids; provides pressure to move wastewater through the collection system |
| Septic Tank Effluent Pump | Provides pressure to move wastewater from the septic tank into the collection system |
| Pump Discharge and Service Line | Connects pump effluent to collection main |
| Discharge Check Valve | Prevents backflow of effluent into septic tank or pump vault |
| Pump Discharge Valve | Provides a positive cutoff between the pump and the service line |
| Service Line or Service Lateral | Connects a septic tank or pump vault to the wastewater collection main |
| Power Supply | Provides electric power for the pumps |
| Control Panel | Houses pump and alarm system controls |
| Liquid Level Controllers | Switch pumps on and off; activate high water level alarm |
| Flow Monitoring Device | Measures volume of wastewater flow from a septic tank or pump vault |
| Pressure Monitoring Device | Measures pump pressure at pump discharge |

## QUESTIONS

Please write your answers to the following questions and compare them with those on page 174.

4.1A What is the purpose of a grinder pump?

4.2A What problems are associated with older building sewers?

4.2B Why do new septic tank effluent pump (STEP) systems usually require abandonment of the older septic tank?

### 4.22 Grinder Pump Vault

A grinder pump vault is basically a wet well, dry well, sump, or holding chamber for the accumulation of wastewater from the home or commercial service. The purpose of the vault is to store wastewater until there is a sufficient volume to operate the grinder pump economically. It is very inefficient and costly to start and stop pumps frequently. The recommended way to operate a pump that has started is to continue pumping for a reasonable amount of time. A pump that is started and stopped too frequently (five to six times per hour may be acceptable) may suffer damage from overheated windings. Starting a pump requires more *AMPERAGE*[27] than operating the pump at a full discharge capacity. Consequently, it is cheaper not to start a pump too frequently and, once it has started, to keep it running as long as possible.

A pump vault is not intended to operate as a septic tank, but only as a small reservoir. If the pump vault is too large, it will take too long to fill to the level that the pump is called upon to start pumping. This long retention time is not desirable in a grinder pump vault or any other wastewater pumping plant wet well because it allows the solids to separate (by settling or flotation) from the liquids. Energy in some form is required to resuspend the scum and settled solids in the vault, or the vault will need to be cleaned more often.

Another disadvantage of long retention times is that *ANAEROBIC DECOMPOSITION*[28] of the solids begins to occur. As the solids are broken down by microorganisms, hydrogen sulfide ($H_2S$) gas creates odor (rotten egg) problems and, when $H_2S$ combines with moisture, it will corrode concrete or steel structures. For these reasons, grinder pump vaults are generally kept from 24 to 30 inches in diameter, and 6 to 8 feet deep. This size provides adequate storage (140 to 375 gallons total volume) for pump operation and some reserve of storage for power outages or pump failures, but also keeps the wastewater

27. *Amperage* (AM-purr-age). The strength of an electric current measured in amperes. The amount of electric current flow, similar to the flow of water in gallons per minute.

28. *Anaerobic* (AN-air-O-bick) *Decomposition.* The decay or breaking down of organic material in an environment containing no free or dissolved oxygen.

relatively fresh and the vault reasonably free of scum and solids accumulation. This volume permits longer periods of time between routine cleanings of the vault.

Grinder pump vaults are sometimes installed fairly close to the residence. This reduces the cost of the building sewer line and shortens the run of electrical service for the grinder pump controls.

Pump vaults and covers are manufactured of fiberglass or thin-wall fiberglass reinforced polyester (FRP) and are commonly supplied as a complete package containing the pump and associated controls and alarms (Figure 4.9).

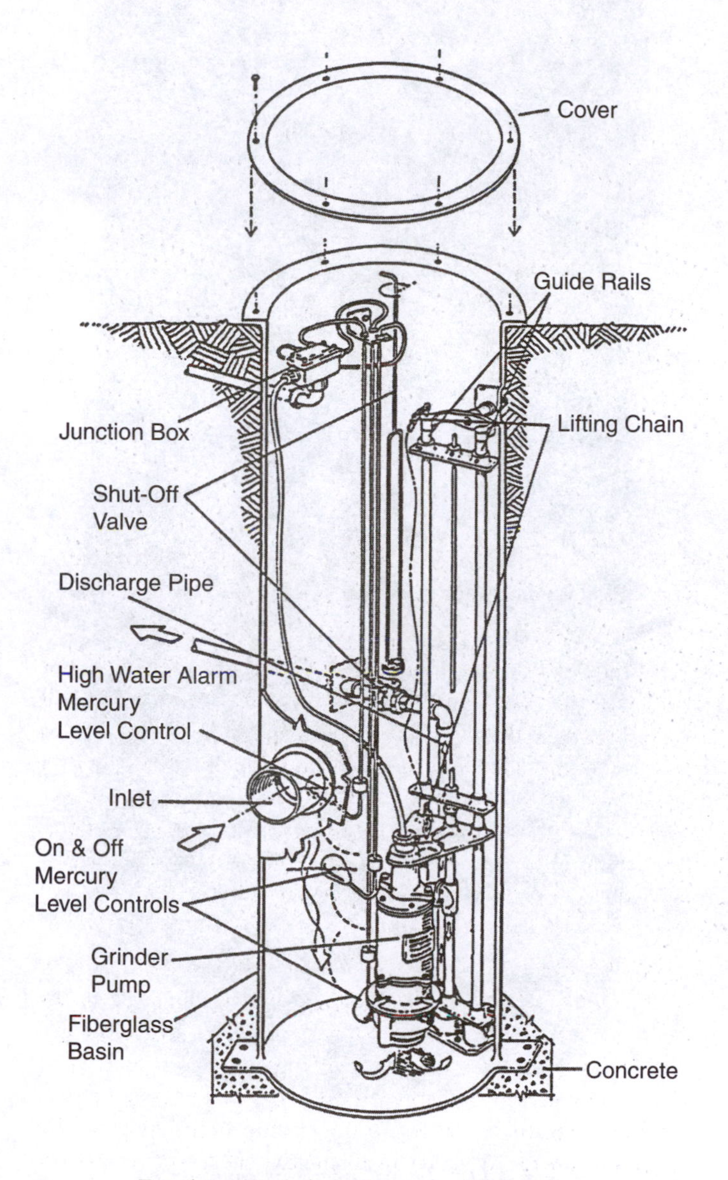


*Fig. 4.9 Typical grinder pump (GP) package using slide-away coupling and guide rails*
(Permission of F. E. Myers Pump Co.)

### 4.23 Grinder Pump

A grinder pump serves two purposes: (1) it creates the pressure that moves wastewater through the collection system and (2) it reduces the size of solids present in the wastewater by grinding the solids into a pumpable *SLURRY.*[29] Reliability of the system requires the cooperation of the user in keeping particularly troublesome solids from being discharged to the pumping unit. Such items include hard solids, plastics, feminine hygiene products, rags, hair, or other stringy material.

Pump manufacturers provide preassembled packages that include the pump vault, the pump, in-vault piping and valves, liquid level sensors, electrical control panel, electrical junction box, and associated equipment. The availability of preassembled packages is a benefit to the agency and the operators because the design of the assembly has usually been extensively field tested and has been refined to provide an economical unit that requires very little service. In addition, the use of identical package units throughout the system contributes to greater reliability and easier maintenance.

A typical grinder pump (GP) unit is shown in Figure 4.9. The pump shown is suspended and aligned by means of a slide-away coupling on the guide rails. A long-handled operating wrench is used when it is necessary to reach the shutoff valve to prevent backflow from the service line. A lifting chain is provided for removing the pump when maintenance is necessary.

Three liquid level float sensors that are equipped with mercury switches are located within the pump vault: pump OFF, pump ON, and high water level alarm. In some instances, an additional OFF float is added to stop the pump and activate a low level alarm. This prevents the pump from operating dry and causing it to overheat or burn up.

When a dual pump unit is installed, additional sensors and switches are necessary to start the lag pump during high flows and to alternate the pump lead/lag start cycles to ensure even wear on the pump units.

Electrical wiring is run from the electrical junction box in the pump vault to an electrical control panel. The control panel is normally mounted on an outside wall of the home close to the grinder pump vault. In some instances, the control panel is mounted on a post or pipe pedestal next to the pump vault. In either case, it is desirable that the control panel be in plain view and close to the vault. This provides maintenance personnel safe and easy access for testing and interruption of power during servicing. Wiring from the control panel is generally laid in conduit to the home power distribution panel. The control panel usually is provided with its own individual circuit breaker.

Figure 4.10 shows a centrifugal grinder pump suspended from the vault cover. This arrangement can be used for installation in the home, often in the basement, but may also be used outdoors.

29. *Slurry.* A watery mixture or suspension of insoluble (not dissolved) matter; a thin, watery mud or any substance resembling it (such as a grit slurry or a lime slurry).

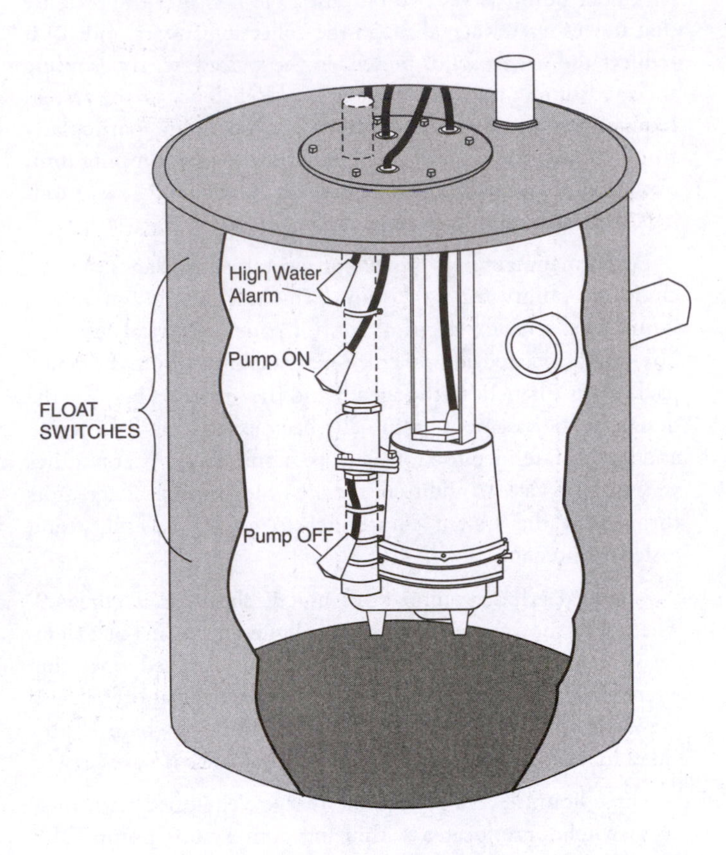


*Fig. 4.10 Typical centrifugal grinder pump (GP) package with pump suspended from basin cover*
(Permission of Barnes Pump Co.)

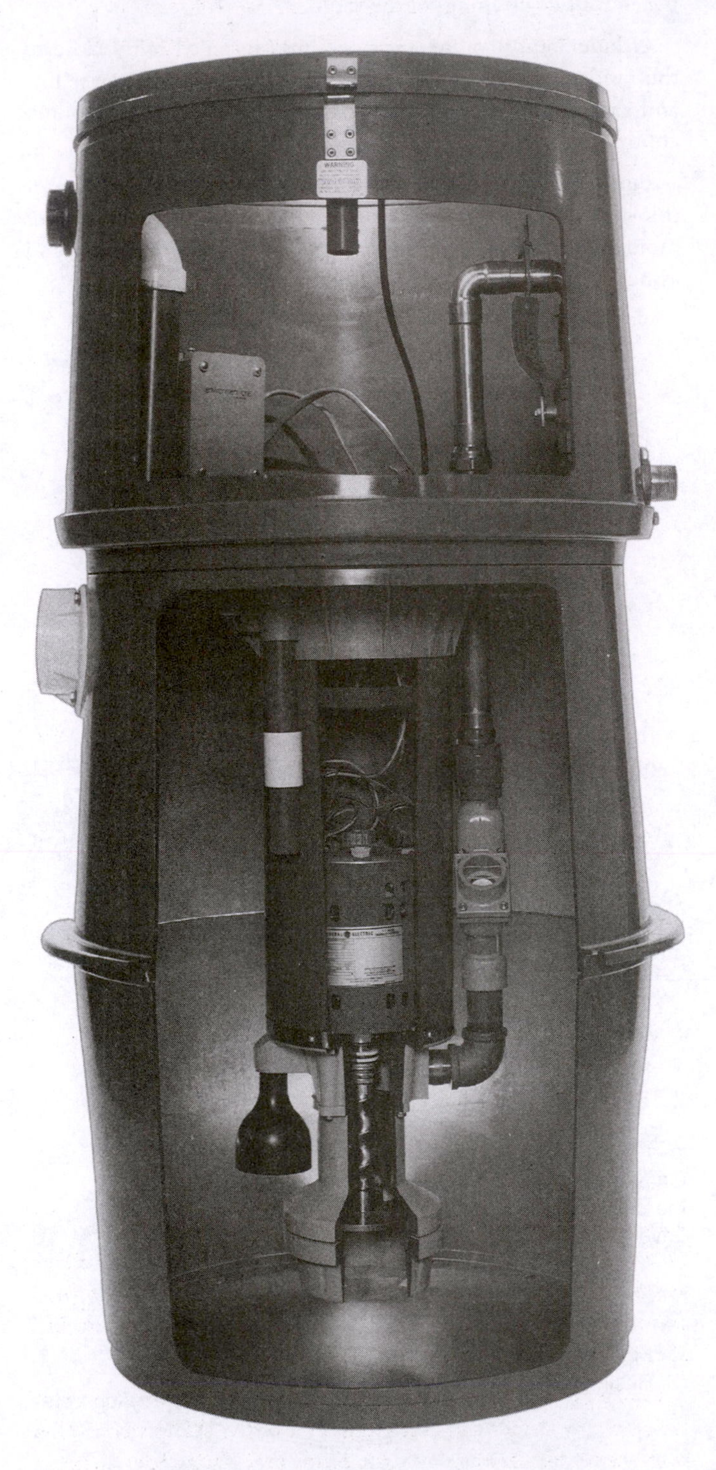

*Fig. 4.11 Typical progressive cavity-type grinder pump (GP) package*
(Courtesy of Environment One Corporation)

Figures 4.11 and 4.12 show a semipositive displacement progressive cavity pump that is suspended in the vault. The core assembly consists of the pump, motor, grinder, piping, valving, and electrical controls. Liquid levels are sensed using a pressure sensor in which the pressure of air trapped in a tube increases as the water depth increases. The trapped air pressure is measured and converted to a water level reading.

Since progressive cavity pumping units are essentially self-contained, the only functions of the control panel are to house a branch circuit disconnect (to disconnect power to the unit during servicing) and to provide for a high water level alarm annunciator (usually a horn or light) that may be seen or heard by the home resident in case there is a problem with the unit. Upon hearing or seeing the alarm, the resident or a neighbor may notify the operating agency of a problem and silence the

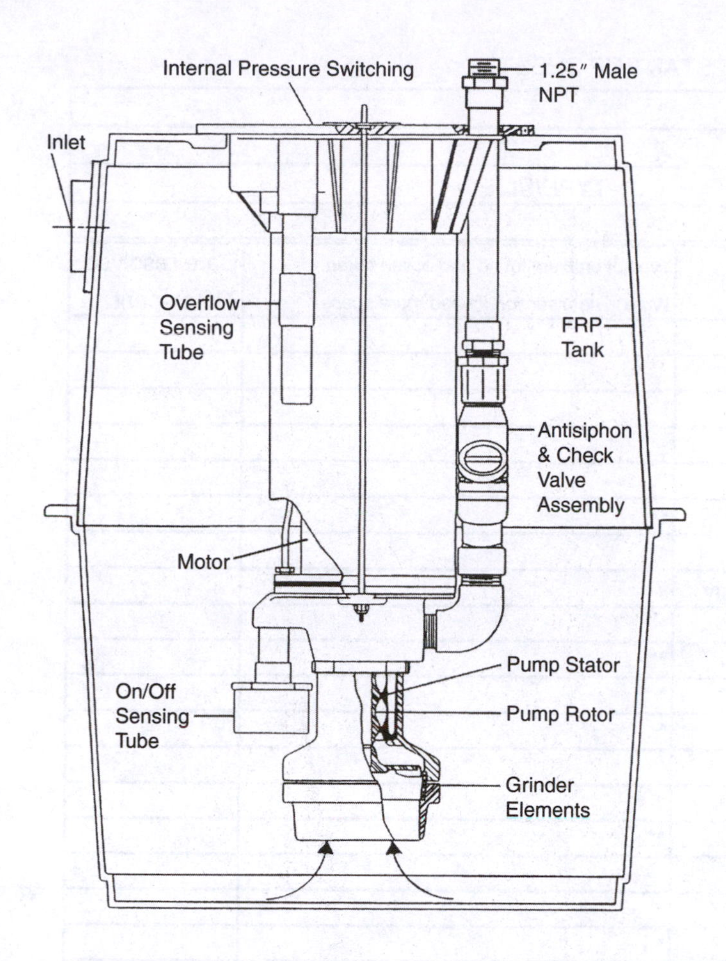


*Fig. 4.12 Basic components of a progressive cavity grinder pump (GP)*
(Courtesy of Environment One Corporation)

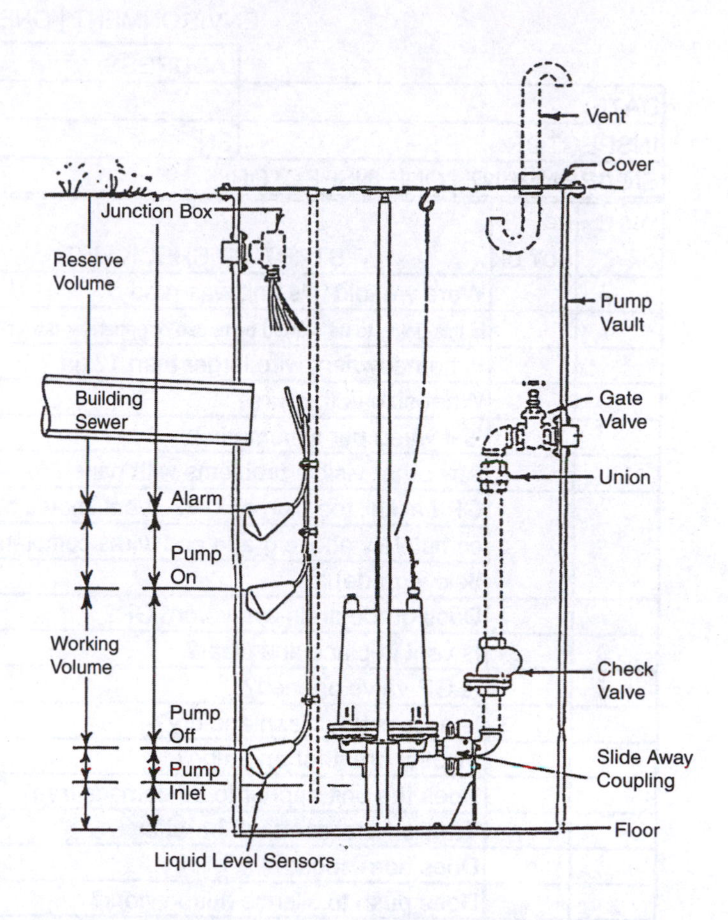


*Fig. 4.13 Alarm and pump ON/OFF levels for a grinder pump (GP) vault*
(Source: *ALTERNATIVE WASTEWATER COLLECTION SYSTEMS,* Technology Transfer Manual, US Environmental Protection Agency)

alarm at the control panel. The operating agency must respond in a reasonable time and correct the problem before a wastewater overflow or flooding has time to occur. In some situations, alarms have been silenced without notification to the operating agency and overflows have occurred.

As the grinder pump illustrations indicate, the pumping units are normally suspended above the floor of the pump vault. This is to ensure free passage of waste solids into the grinder element of the pump. Figure 4.13 provides another view of a grinder pump installation and identifies the various conditions or wastewater levels that are expected to occur in normal and abnormal circumstances.

Key to a successful grinder pump operation is proper initial installation of the system. A detailed start-up checklist and report such as the ones shown in Figures 4.14 and 4.15 will make a tremendous contribution to getting a grinder pump system off to a trouble-free start. The checklist is especially useful to contractor personnel.

## QUESTIONS

Please write your answers to the following questions and compare them with those on page 174.

4.2C What problems might occur if a grinder pump vault is too large?

4.2D What is the purpose of the three liquid level float sensors used in grinder pump vaults?

4.2E Where should the control panel for a grinder pump be located (for an outdoor installation)?

ENVIRONMENT | ONE START UP SHEET

| | ADDRESS: |
|---|---|
| DATE: | |
| INSPECTOR: | SN: TYPE/SIZE: |
| ENVIRONMENT \| ONE INSPECTOR: | |

| YES OK | NO NOT OK | START-UP CHECK LIST | Type of problem found and action taken. Write in note section if need more space. | Date resolved Time spent |
|---|---|---|---|---|
| | | Were we told this unit was ready? | | |
| | | Is this going to be started by house or generator power? | | |
| | | Is homeowners wire larger than 12 ga.? | | |
| | | What size is the wire? | | |
| | | Is it wired per schematic? | * | |
| | | Any other visible problems with panel? | * | |
| | | GP backfill too high or low? (vent should be completely above grade and wires completely below grade). | ** | |
| | | Does grade drain away from GP? | ** | |
| | | Is vent in place and clear? | * | |
| | | Is GP valve opened? | * | |
| | | Is accessway clean and dry? | | |
| | | Inspect breather and tube? | * | |
| | | Does the unit appear to be damage free? | ** | |
| | | **Turn on alarm circuit breaker** | | |
| | | Does horn sound? | * | |
| | | Does push to silence button work? | * | |
| | | Does alarm light work? | * | |
| | | Is alarm light tight? | | |
| | | **Turn on pump circuit breaker** | | |
| | | Volt Amps | * | |
| (MIN-MAX) | | (220-260) (4.4-6) | | |
| | | Does the system appear to be leak free? | * | |
| | | Did the alarm shut off properly? | * | |
| | | **Leave both breakers on if all is ok** | | |
| | | Does accessway cover fit properly? | | |
| | | **Lock both panel and GP** | | |

Notes:

Location note:

SIGN-OFF: This pump is ready for gravity tie-in:

This pump has been approved by Environment | One:

Gravity tie-in was made on: Water meter reading:

Date:

note: * can not pass start up and needs to be repaired before completion of start up.
** If it will effect the pump operation. than it can not pass start up until repaired

*Fig. 4.14 Grinder pump start-up checklist*

(Courtesy of Environment One Corporation)

## GRINDER PUMP SERVICE RECORD

Address:

Special unit or location:

Special Note:

| DATE | TECHNICIAN | SERIAL # | DESCRIBE PROBLEM AND SERVICE PERFORMED |
|---|---|---|---|

*Fig. 4.14 Grinder pump start-up checklist (continued)*

(Courtesy of Environment One Corporation)

**FREEMIRE & ASSOCIATES, INC.**
9160 RUMSEY ROAD, B-8, COLUMBIA, MD 21045
(410) 995-0305 (301) 621-4928
ENVIRONMENT ONE GRINDER PUMP START-UP REPORT

SERIAL # SERVICE # MODEL # DATE #

| **ADDRESS/LOT #** | **NAME OF DEVELOPMENT/PROJECT** |
|---|---|
| | |
| | **TEST & INSPECTION** |
| **HOMEOWNER:** | **Voltage: Pump** ______ **Alarm** ______ |
| **CONTRACTOR/PHONE:** | **Amp Reading** ______ |
| | **Union** ______ **Valve** ______ |
| | **Street Valve** ______ |
| | **Breather** ______ **QD** ______ **Float Vlv** ______ |
| | **Location of Alarm** ______ **Type of Disconnect** ______ |
| | **Accessway Ht.** ______ **Vent** ______ |
| **CONTACT PERSON:** | **Rope Harness** ______ **Wrench** ______ |
| | **Redundant Check Valve** ______ |
| | **E.Q.D.** ______ **Field Joint** ______ |

**WARRANTY:**

**COMMENTS:**

Freemire & Assoc. Representative

Contractor's Representative

*Fig. 4.15 Grinder pump start-up report*
(Courtesy of Environment One Corporation)

## 4.24 Septic Tank Effluent Pump (STEP) Systems (Figures 4.16 and 4.17)

### *4.240 Description of System*

Wastewater from the home flows through the building sewer into the septic tank where the treatment of floatables such as oils, greases, and settleable solids occurs. The treatment provided by the septic tank eliminates the requirement of the grinder element on the pump. In early STEP systems, centrifugal pumps capable of passing small solids were installed. Recently installed systems use a turbine-type centrifugal pump similar to the pumps installed in domestic water wells. These pumps are less tolerant of solids; therefore, a strainer or screen is installed around the pump, shown as the "filter cartridge" in Figure 4.7, page 132. Screens, strainers, and filter cartridges are essentially the same thing and perform the same function.

Screens prevent large solids or stringy material from entering the pump's suction, thus prolonging the life of the pumping unit and reducing the number of service calls. The basket strainers or screens are of sufficient size to permit the pump and control sensors to be placed inside the strainer assembly. This permits the STEP pump to rest on or near the bottom of the tank or vault. However, the screens or baskets must be cleaned periodically to ensure uninterrupted flow of tank effluent to the pump. When the septic tank has only one compartment (no divider wall), the bottom of the screen or basket strainer is mounted above floor level so that it is out of the sludge settling zone.

It is common practice to use a single wet well or pump vault for multiple septic tanks in locations such as mobile home parks or commercial establishments. Figure 4.18 shows such a multiple-unit septic tank and pump assembly. In some instances, a pump vault very similar to the grinder pump vault may be used after the tank for a single family residence. Figure 4.19 shows a septic tank effluent pump as it might be installed in an external vault for a single residence. The pump vaults in both illustrations may be as deep as 12 feet. When this type of arrangement is used, the pump sits on legs on the bottom of the vault. Also, the discharge piping is not rigid piping, but flexible hose equipped with a quick-disconnect coupling. The pump discharge valve is placed at a higher elevation in the vault for easy access. In many installations, lever-operated ball valves are being used in place of the discharge gate valve shown in Figure 4.19. (Lever-operated valves are described in Section 4.252, "Pump Discharge Valve.")

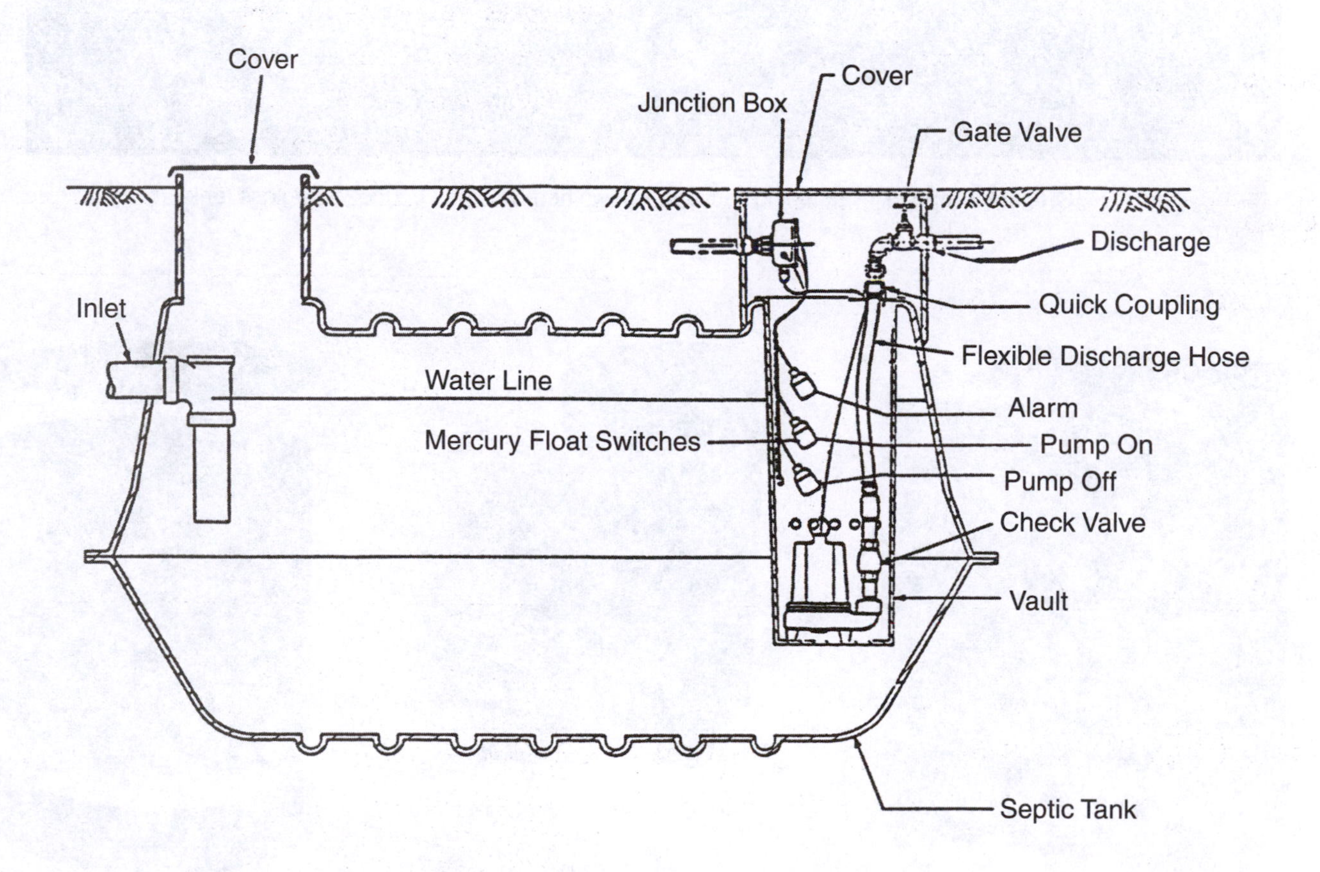


*Fig. 4.16 Septic tank effluent pump (STEP) package with internal pump vault*

(Permission of Barnes Pump Co.)

Orenco® ProSTEP™ pumping package, shown here with an Orenco® fiberglass tank

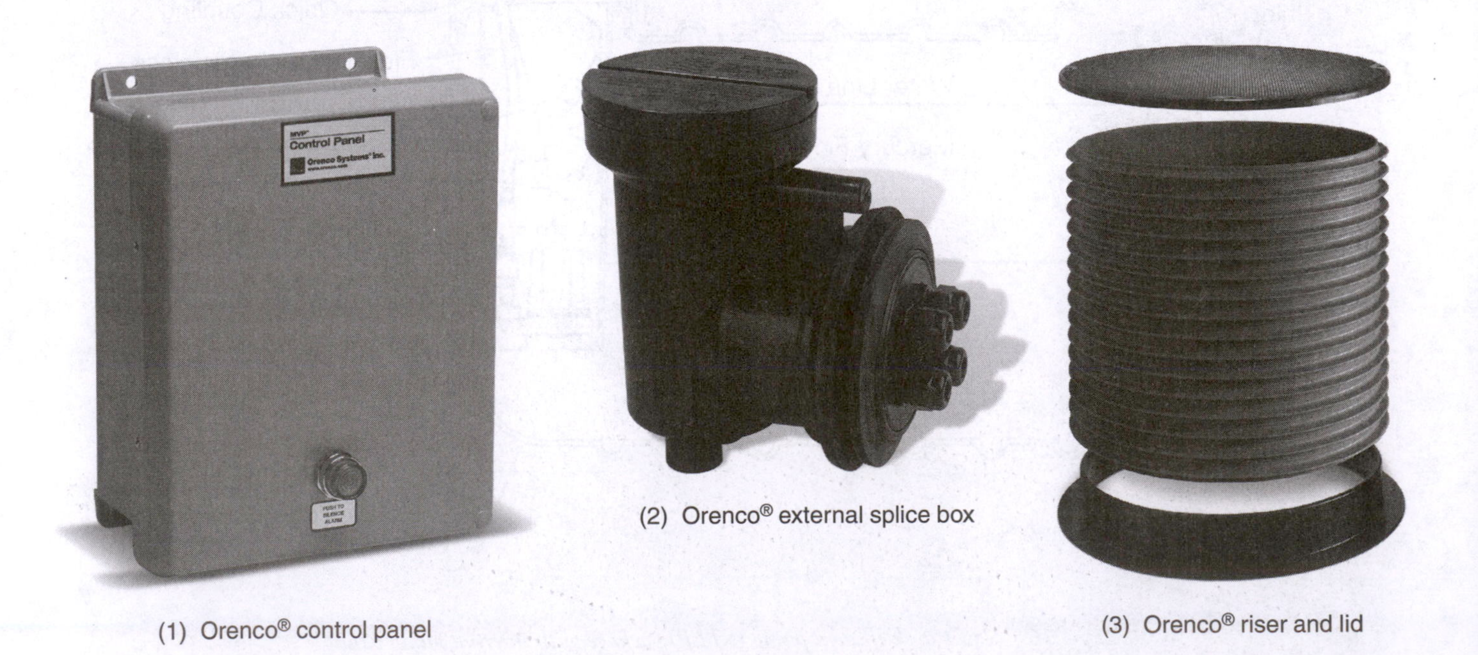


(1) Orenco® control panel

(2) Orenco® external splice box

(3) Orenco® riser and lid

*Fig. 4.17 STEP system components*

(Orenco® ProSTEP™ Pumping System. Courtesy of Orenco Systems, Inc., www.orenco.com)

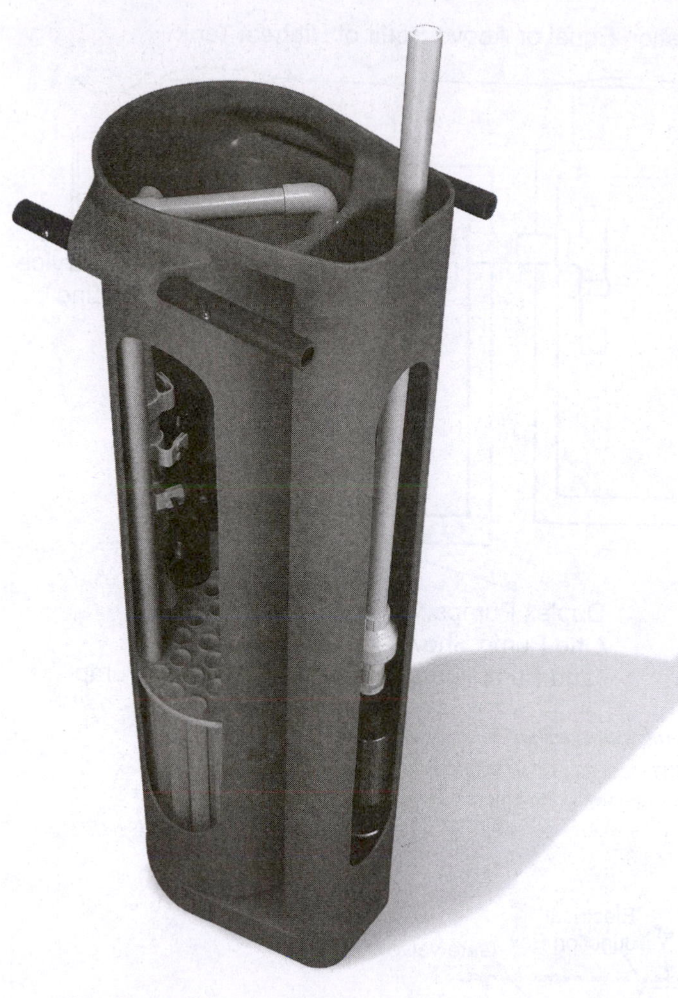

(4) Orenco® Biotube® pump vault

(5) Orenco® effluent pump

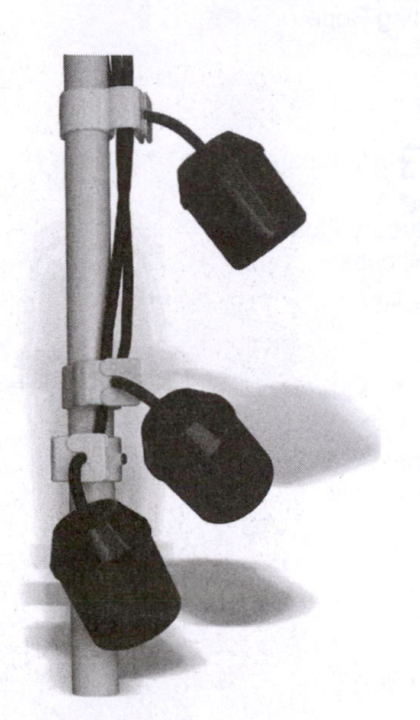

(6) Orenco® float switch assembly

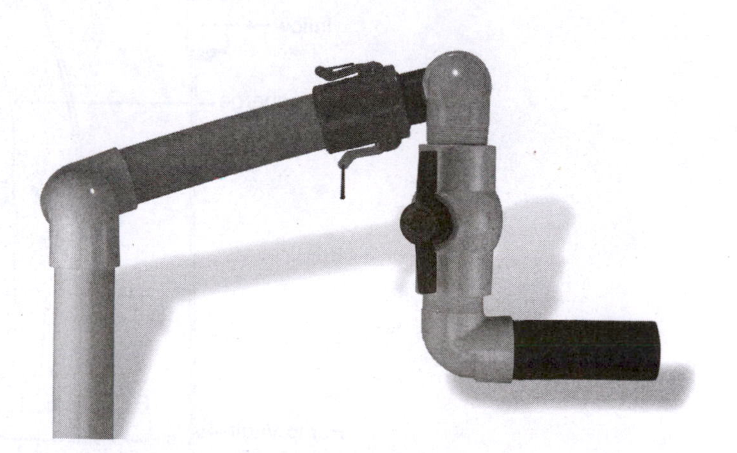

(7) Orenco® discharge assembly

*Fig. 4.17 STEP system components (continued)*

(Orenco® ProSTEP™ Pumping System. Courtesy of Orenco Systems, Inc., www.orenco.com)

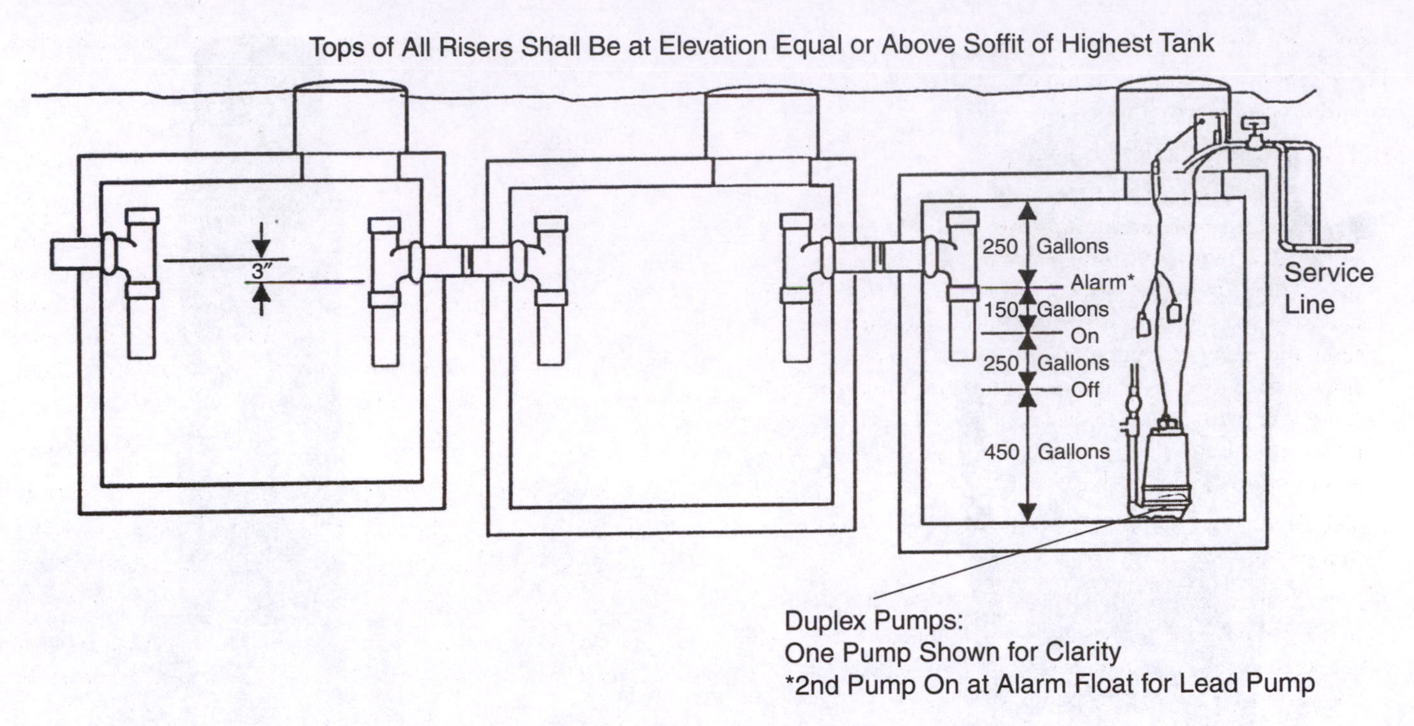


*Fig. 4.18 Multiple-unit septic tank and external pump assembly*

(Source: *ALTERNATIVE WASTEWATER COLLECTION SYSTEMS,* Technology Transfer Manual, US Environmental Protection Agency)

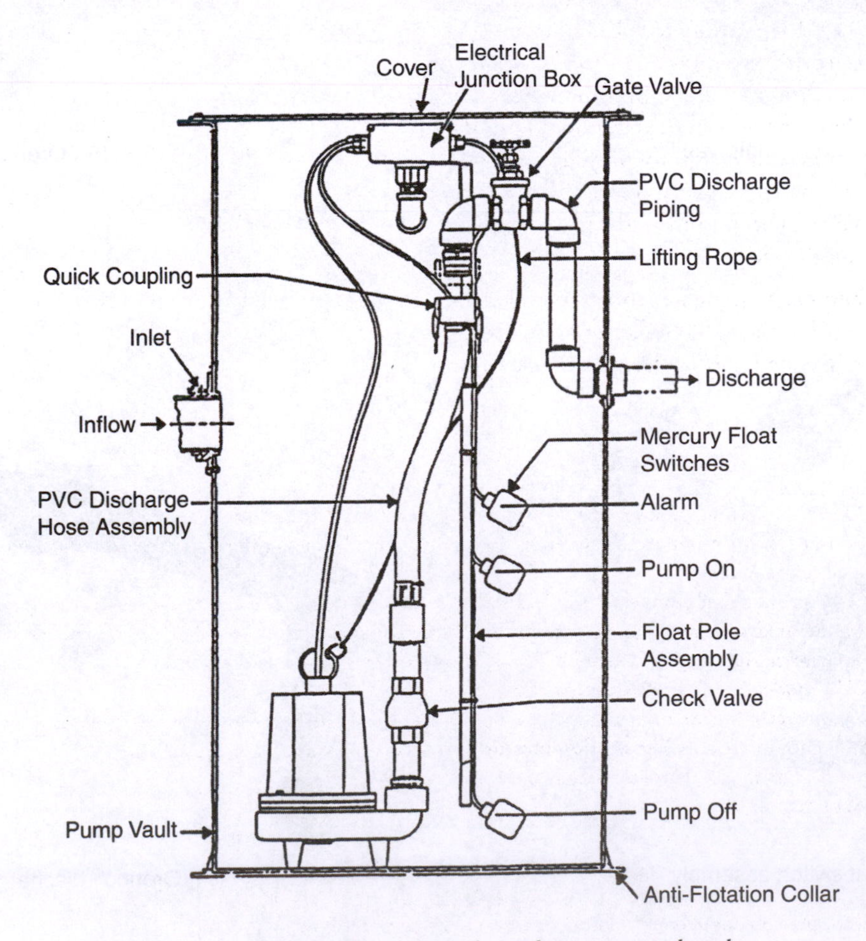


*Fig. 4.19 Septic tank effluent pump located in an external vault*

(Permission of Barnes Pump Co.)

### *4.241 Selection of Effluent Pumps*

The selection of an effluent pump is greatly dependent upon the conditions described in this section.

1. The type of wastewater to be pumped.

   If the wastewater is directly discharged from a grinder pump vault, then the pump must not only grind up the solids but it must also pump larger size solids into the collection system. If the wastewater first passes through a septic tank, then the solids content is reduced and the wastewater to be pumped is relatively free of solids. The pump for this situation requires less horsepower than the grinder pump because it is not required to grind solids. It will have a higher efficiency due to closer clearances on the pump's impeller and the absence of the solids.

2. The *TOTAL DYNAMIC HEAD (TDH)*[30, 31] of the collection system.

   Basically, the term "TDH" means how high the pump has to lift and push the wastewater being pumped. The lift portion is the necessary vacuum the pump must create to draw the wastewater up into the pump. Once the wastewater is in the pump, the push is the amount of energy the pump must apply to the wastewater to push it to wherever it is to be discharged.

   TDH is measured two ways: as feet in elevation, or as pounds per square inch (psi) in pressure, determined either by a pressure gauge or *MANOMETER*.[32] Either expression can be converted to the other because 2.31 feet of water is equivalent to one psi. For example, if a pressure wastewater collection system main is operated in the range of 35 psi, the pump must impart sufficient energy to the wastewater to overcome the 35 psi in the main. In this case, 35 psi × 2.31 ft/psi = 80.85 feet that the pump must push (pressurize) the wastewater just to enter the main.

   The pump must also exert enough energy to compensate for *FRICTION LOSSES*[33] created by the wastewater flowing through the service line and any difference in elevation between the septic tank wastewater level and the collection system main.

*EXAMPLE*

A home with a septic tank is located at the bottom of a hill. The pressure main is laid at the top of the hill, which is 172 feet above the septic tank, and the pressure main operates at a pressure of 26 psi. Assume that the friction losses created by the valving, other fittings, and piping of the service line add an additional 9 feet to the head (push) that the pump must overcome. What is the total dynamic (discharge) head in feet and in pounds per square inch (psi)?

| **Known** | | **Unknown** |
|---|---|---|
| Main Operating Pressure, psi | = 26 psi | 1. Discharge Head, ft |
| Difference in Elevation, ft | = 172 ft | |
| Service Line Friction Losses, ft | = 9 ft | 2. Discharge Head, psi |

1. Calculate the main operating pressure in feet. Recall that 2.31 feet of water is equivalent to 1 pound per square inch.

$$\text{Main Operating Pressure, ft} = 26 \text{ psi} \times 2.31 \text{ ft/psi}$$
$$= 60 \text{ ft}$$

2. Find the discharge head in feet.

$$\text{Discharge Head, ft} = \text{Difference in Elevation, ft} + \text{Friction Losses, ft} + \text{Operating Pressure, ft}$$
$$= 172 \text{ ft} + 9 \text{ ft} + 60 \text{ ft}$$
$$= 241 \text{ ft}$$

3. Find the discharge head in pounds per square inch.

$$\text{Discharge Head, psi} = \frac{\text{Discharge Head, ft}}{2.31 \text{ ft/psi}}$$
$$= \frac{241 \text{ ft}}{2.31 \text{ ft/psi}}$$
$$= 104.3 \text{ psi}$$

30. *Total Dynamic Head (TDH).* When a pump is lifting or pumping water, the vertical distance (in feet or meters) from the elevation of the energy grade line on the suction side of the pump to the elevation of the energy grade line on the discharge side of the pump. The total dynamic head is the static head plus pipe friction losses.
31. *Energy Grade Line (EGL).* A line that represents the elevation of energy head (in feet or meters) of water flowing in a pipe, conduit, or channel. The line is drawn above the hydraulic grade line (gradient) a distance equal to the velocity head ($V^2/2g$) of the water flowing at each section or point along the pipe or channel. Also see HYDRAULIC GRADE LINE (HGL).
32. *Manometer* (man-NAH-mut-ter). An instrument for measuring pressure. Usually, a manometer is a glass tube filled with a liquid that is used to measure the difference in pressure across a flow measuring device, such as an orifice or a Venturi meter. The instrument used to measure blood pressure is a type of manometer.

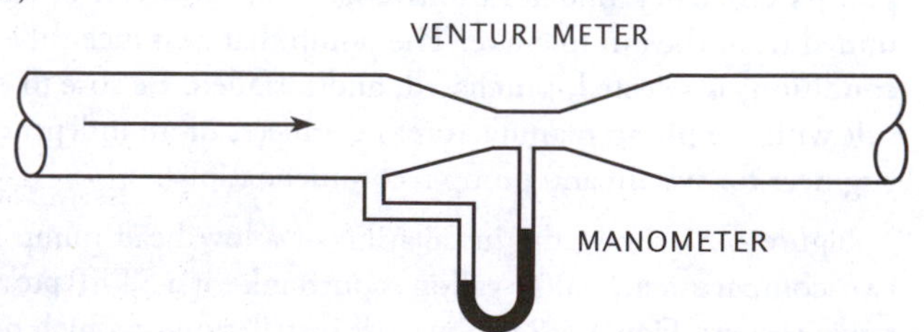


33. *Friction Loss.* The head, pressure, or energy (they are the same) lost by water flowing in a pipe or channel as a result of turbulence caused by the velocity of the flowing water and the roughness of the pipe, channel walls, or restrictions caused by fittings. Water flowing in a pipe loses head, pressure, or energy as a result of friction. Also called head loss.

This example illustrates the discharge conditions that a pump must meet. The pump performance (discharge) curves shown on Figure 4.20 indicate that at a total dynamic head of 200 feet, three of the pumps shown could deliver a net discharge of 2.4 GPM, 3.6 GPM, and 4.8 GPM, depending on the pump model and number of stages.

3. The flow rate to be pumped.

   The flow rate determines the pump capacity in gallons per minute (GPM) that the pump must deliver against the expected total dynamic head. This flow depends on the flow into the septic tank and the size of the septic tank.

#### *4.242 Types of Effluent Pumps*

Most STEP systems limit their selection of pumps to the centrifugal type, but other types of pumps have been used. Centrifugal pumps common to STEP units are classified as "submersibles," and as single-stage (one impeller), or multistage turbines (two or more impellers).

Turbines with stacked impellers (one impeller on top of the other) are designed so that the wastewater being pumped passes through each impeller and each impeller induces (adds) additional energy to the wastewater. This design produces a pump with the capability of developing high discharge heads; that is, it is capable of pushing the water to high elevations of 400 to 500 feet above the pump. The multistage turbine pump usually discharges a lower wastewater flow rate than other types of pumps. The single-stage pump does not have the capability of discharging a head in excess of 50 feet but can deliver a much higher flow at a lower energy input to the pump than the multistage turbine pumps.

For the proper selection or sizing of a pump for a particular application, the site-specific conditions must be determined. In addition to the type of liquid (water), the total dynamic head (TDH), and the flow rate, other factors that must be considered include piping sizes for both pump discharge and suction, if required; types and number of pipe fittings in the suction and discharge lines; length of pipes in the suction and discharge lines; whether the pump discharges into a pressurized vessel or system or has an open, free discharge; physical space limitations; and power availability. Once this information is collected and the conditions determined, then a review of various pump manufacturers' data is used to select a pump. This information includes a pump curve (Figure 4.20). The pump curve shows the characteristics of a particular pump at various operating ranges of flow, TDH, and HP (horsepower). For larger pumps, the pump's efficiency under various conditions also can be determined from the pump curve. The pump that best meets the site conditions is selected, purchased, and installed. Be sure to consult with the pump manufacturer's engineers or an independent engineer for system and pump recommendations.

Figure 4.21 shows the installation of a low head pump in a two-compartment, 1,000-gallon septic tank for a STEP pressure sewer system. Figure 4.22 shows the installation of a high head, low flow pump in a 1,500-gallon, two-compartment septic tank. There are many similarities between the two tanks in the location of strainer assemblies, float switches, discharge piping, and valving. The strainer assembly, pump, and controls in both tanks can all be reached from the surface through the tank riser, and all of the above-mentioned components can be removed from the tank for servicing or repair without dewatering (draining) the tank. Piping connections, valves, and electrical junction boxes are located high in the tank riser so that maintenance is made much easier and there is less chance of them being submerged in wastewater if a power or pump failure occurs.

In most instances, one operator can service the tank and pumping components. However, two operators may be required in the following situations: (1) to remove grinder pumps that are heavier than one operator can safely handle; (2) to remove the strainer assembly if a large buildup of debris is attached to the filter screen unit; or (3) if the drain port at the bottom of the pump assembly is blocked and the liquid cannot drain out as the unit is lifted from the tank.

## QUESTIONS

Please write your answers to the following questions and compare them with those on pages 174 and 175.

4.2F Why are screens or basket strainers installed around turbine-type centrifugal STEP pumps?

4.2G Selection of an effluent pump depends primarily on what three factors?

4.2H What type of pump is used to achieve high discharge heads?

4.2I Why should piping connections, valves, and electrical junction boxes be located high in the septic tank riser?

### 4.25 STEP System Pump Discharge and Service Line

This section describes the piping, valves, and hoses that connect the septic tank effluent pump to the pressurized wastewater collection main.

#### *4.250 Pump Discharge*

STEP discharge lines are commonly made of flexible PVC (polyvinyl chloride) or neoprene hose or pipe and are at least one inch in diameter. The flexible hose is connected directly to the pump discharge.

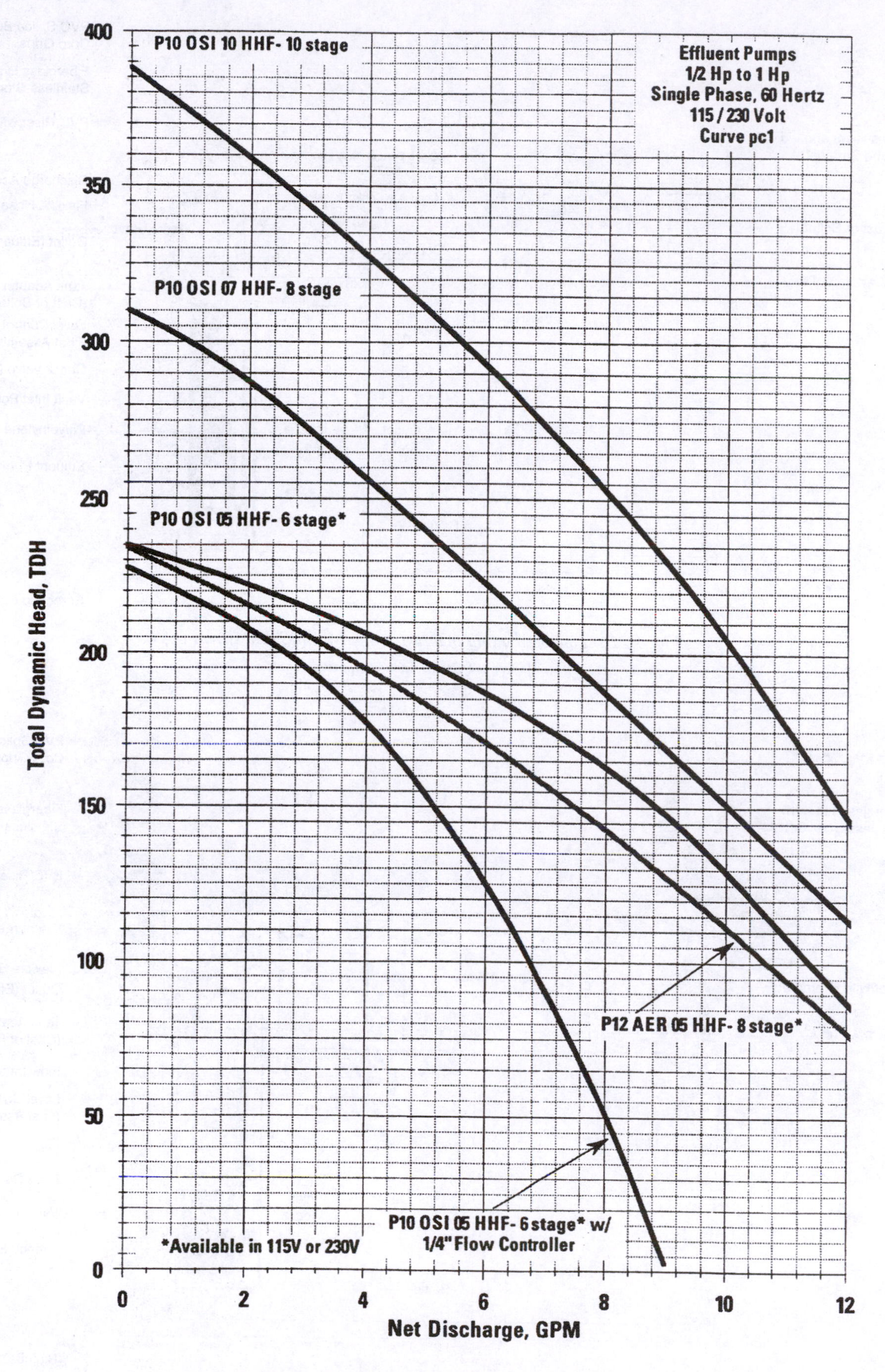


*Fig. 4.20 Pump performance (discharge) curves*

(Courtesy of Orenco Systems, Inc., www.orenco.com)

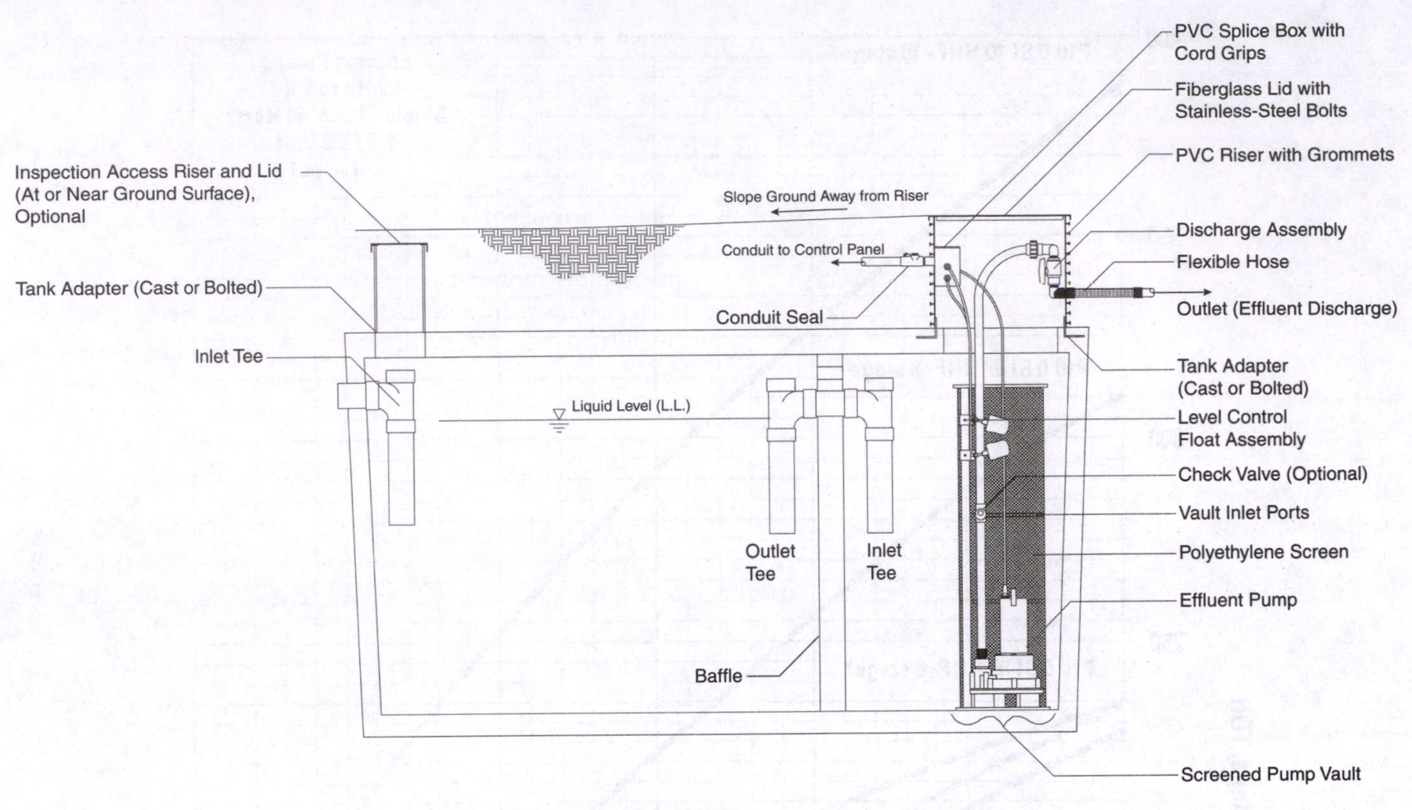


*Fig. 4.21 Low head pump in a two-compartment STEP system*
(Courtesy of Orenco Systems, Inc., www.orenco.com)

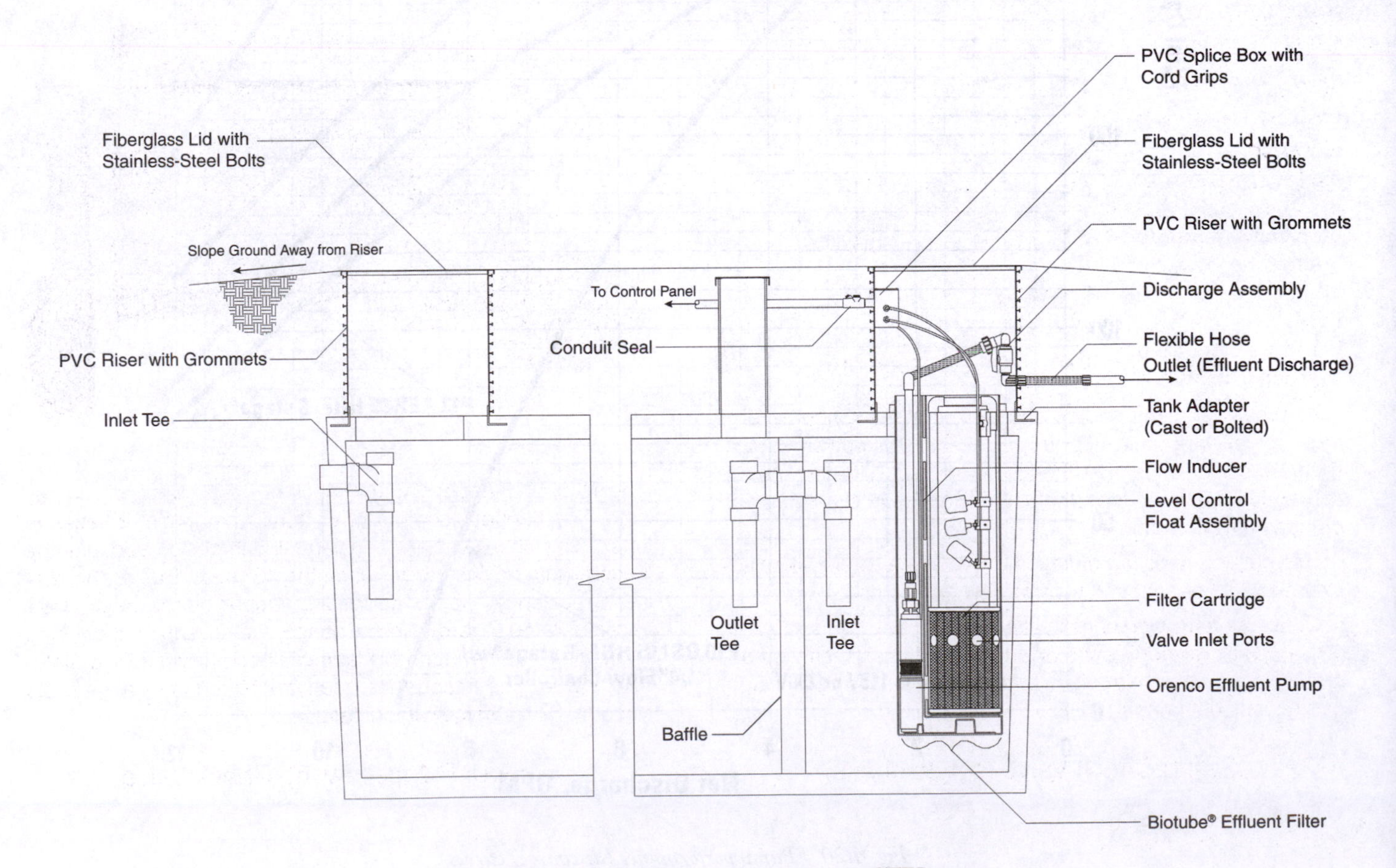


*Fig. 4.22 High head, low flow, two-compartment STEP system*
(Courtesy of Orenco Systems, Inc., www.orenco.com)

### 4.251 *Pump Discharge Check Valve (Figures 4.23 and 4.24)*

A check valve the same diameter as the discharge hose (usually one to two inches) is installed just downstream of the pump discharge. This valve opens when the pump starts and effluent is discharged from the pump. The wastewater level is pumped down until the lower float switch is activated, thus shutting off the pump. When the pump impeller stops rotating, the water pressure drops to zero. The last effluent discharged from the pump prior to shutdown is still in the discharge hose and instantly starts to drain back through the pump into the septic tank. At this time, the reverse flow in the discharge hose closes the check valve, preventing the effluent just pumped from draining back into the tank. If this check valve did not seat (close all the way), the effluent would drain back into the tank, raise the wastewater level, and cause the pump to start again. Thus, the pump would cycle on and off pumping basically the same water over and over again. It is probable the septic tank would not flood, but the pump would be running almost continuously and would keep pumping until the material preventing the check valve from fully closing was flushed out or the valve was replaced.

An additional check valve should be, and usually is, installed in the service line just prior to the service line connection to the collection system main. The main purpose of this check valve is to prevent the wastewater in the main line, which is also under pressure from other septic tank effluent pumps, from flowing back through the service line once the effluent pump shuts off. Installation of a second check valve also protects against spills in the event that the service line is damaged during excavation work. However, if both the pump discharge check valve and the service line check valve fail to seat, backflow would call for the pump to run continuously, or could possibly flood the tank.

Ball valves (Figure 4.25) operate similar to check valves. When the pump starts, the flowing water moves (lifts) the ball and allows the water to flow. When the pump stops and the

Swing Check shutoff depends on gravity (weight of disc) and reverse flow. The pivot point of the swing check is outside the periphery of the disc and the greater the head, the greater the possibility that the fluid will flow back through the valve before the disc can shut off.

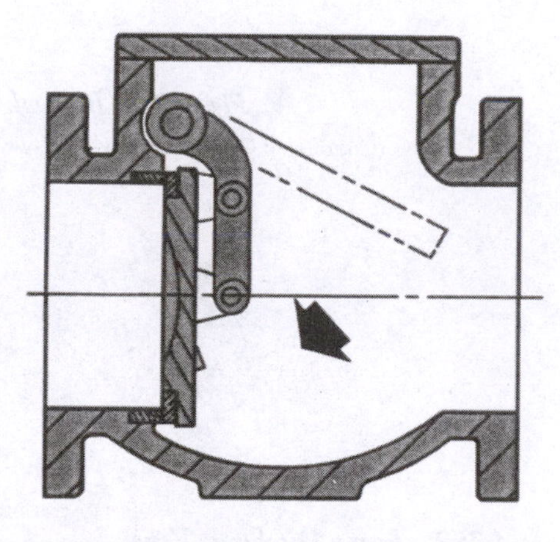

To effect complete shutoff, the disc of a swing check valve must travel through a 90° arc to the valve seat, as shown. Without resistance to slow the disc's downward thrust, and encouraged by reverse flow, the shutoff results in slamming and damaging water hammer.

Use caution applying this type check valve where velocities exceed 3 feet per second reverse flow and 30 psi.

Conventional swing check valve. Disc closes through a long 90° arc with reverse flow started. Shutoff characteristics are greatly improved with addition of outside lever spring or weight or air cushion chamber.

Swing Check Valve (single check)

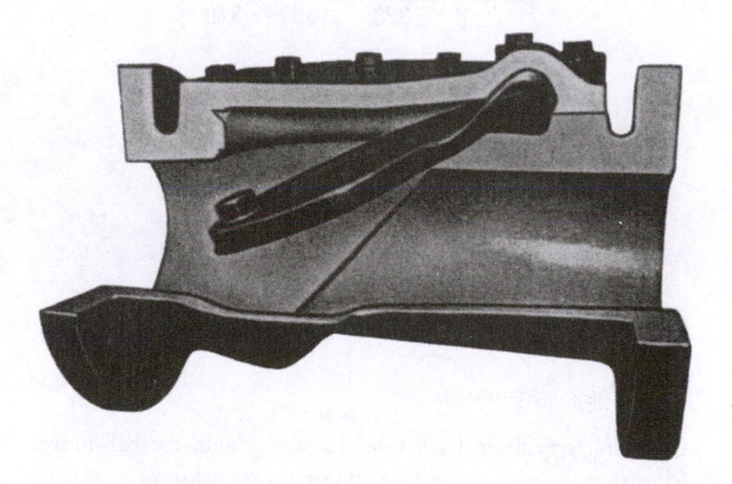

The Rubber Flapper Swing Check is uniquely simple in design. The flapper does not pivot from a hinge pin, it flexes to open. This check valve is nonslamming in design because the flapper only travels 35° to reach the 45° seat. Due to the rubber's resiliency, the spring loaded flapper closes before flow reversal or near zero velocity, hence, the severity of slamming is negligible. The Rubber Flapper Check is virtually maintenance free. The flapper has been flex tested equivalent to 20 years of field service without fatigue.

Rubber Flapper Swing Check Valve

*Fig. 4.23 Pump discharge check valves*

(Permission of APCO/Valve and Primer Corporation)

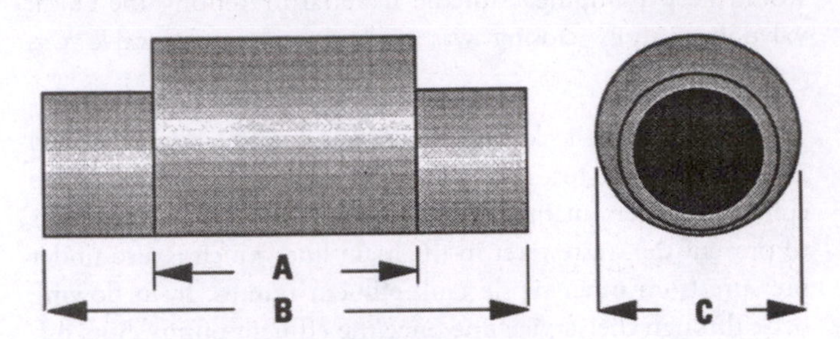


**Check valves** are available in the sizes shown below. Composed of sch. 40 HI-IMPACT PVC Type II per ASTM D-1785. Swing gate composed of EPDM. Only 1/2 lb. back pressure required for complete closure. All sizes available in either slip or threaded connections.

| Dia. Size (in.) | A (in.) | B (in.) | C (in.) |
|---|---|---|---|
| 1/2 | 2.82 | 5.38 | 2.19 |
| 3/4 | 2.82 | 5.38 | 2.19 |
| 1 | 2.82 | 5.13 | 2.19 |
| 1 1/4 | 3.54 | 6.31 | 2.81 |
| 1 1/2 | 3.54 | 6.31 | 2.79 |
| 2 | 3.88 | 6.63 | 3.40 |

*Fig. 4.24 Check valve*

(Courtesy of Orenco Systems, Inc., www.orenco.com)

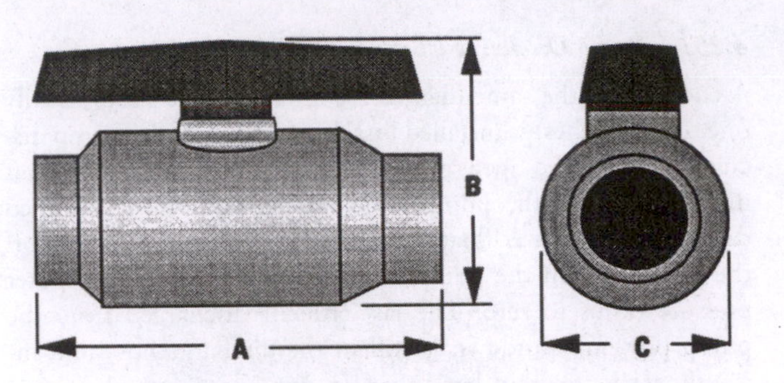


**Ball valves** are available in the sizes shown below. Composed of sch. 40 HI-IMPACT PVC Type II per ASTM D-1785. O-rings composed of EPDM. Stem seal is pre-loaded for longer life. All sizes available in either slip or threaded connections.

| Dia. Size (in.) | A (in.) | B (in.) | C (in.) |
|---|---|---|---|
| 1/2 | 3.22 | 2.57 | 1.60 |
| 3/4 | 4.04 | 3.71 | 2.08 |
| 1 | 4.38 | 3.56 | 2.33 |
| 1 1/4 | 5.30 | 4.50 | 3.15 |
| 1 1/2 | 5.30 | 4.50 | 3.15 |
| 2 | 6.00 | 5.45 | 3.80 |

*Fig. 4.25 Ball valve*

(Courtesy of Orenco Systems, Inc., www.orenco.com)

water starts to drain back into the septic tank, the ball moves (drops) to its original position and prevents backflow.

PVC swing checks using neoprene flappers have performed very well. These tend to have large, clear waterways, open and close easily, and sulfide deposits are not a problem.

Many ball check valves, especially spring-operated ones, have protruding parts that catch stringy material. Other designs have clear openings and are more suitable for wastewater systems. Effluent screens prevent problems with stringy material.

Proven experience with the valve is desired. This can usually be provided when the valve is part of a package supplied by an experienced pressure sewer products manufacturer.

Check valves called "backwater valves" are made to be installed on building sewers, but are rarely used. These are swing check valves with the working parts being removable from the top of the valve; they are available in 3-, 4-, and 6-inch diameters.

### *4.252 Pump Discharge Valve*

This valve is usually the same size as the diameter of the effluent pump discharge hose, and will vary from one to two inches. The purpose of the valve is to provide a positive cutoff between the effluent pump and the service line. If the automatic service line check valve failed to seat at the connection to the pressure main, manually closing the pump discharge valve would eliminate the problem of backflow into the septic tank. The STEP system can be isolated from the service line using the pump discharge valve when the pump discharge hose must be disconnected from the service line to remove the pump, to bleed trapped air or gases from the discharge hose, or to service or replace a faulty pump discharge check valve.

The ball valve is the most common type of pump discharge valve. One type is a lever-operated valve, which is equipped with an operating handle similar to a wrench. A 90-degree turn of the wrench completely opens the valve. The valve is open when the valve handle is in line with the discharge flow and the valve is closed when the handle rests at a 90-degree angle to the discharge flow.

Another type of ball valve is a handle-operated valve (Figure 4.26). This valve is open when the handle is in line with flow and closed when the handle is across the flow line.

### 4.253 *STEP Service Lines or Service Laterals (Figure 4.27)*

The service line carries the pumped effluent from the septic tank across the homeowner's property and discharges into the STEP pressure wastewater collection main or a gravity sewer main. The size (diameter) of the service line depends on the volume of wastewater to be pumped. For a single home the size ranges from one inch to two inches in diameter. Pipe diameters of 1¼ inches and 1½ inches commonly serve single homes. Larger flow from multiple tanks (serving several homes) or a commercial system may range from 1-inch up to 6-inch diameter service lines.

Service lines are buried at very shallow depths, but usually at least below the frost line (minimum of 6 inches). A toning wire or warning tape should be buried just above the service line to aid in locating the line when excavation work must be performed in the line's general area. If the location is known, precautions can be taken to prevent damage to the line and an interruption of service.

At the discharge end of the service line (often next to the property line) a valve box is installed to house the check valve and the isolation valve. The box provides access for the operating staff to service the check valve or to close the isolation valve to prevent backflow from the pressure main during check valve servicing. Easy access is critical in an emergency if damage to the service line could cause wastewater to be discharged back from the main and contaminate the area.

Figure 4.28 shows a typical pressure service line connected to a gravity sewer with the check valve and isolation valve.

When new service lines must be added to the system, use of a pipe saddle (Figure 4.29) permits connections to be made while the main remains under pressure. This type of connection is referred to as a "hot tap" because the line is under pressure. The pipe saddle is a device that is wrapped around a pipe where a new service connection is desired. With the saddle in place, a main can be tapped and the service line connected without taking the pressure main out of service. Many operators recommend that the section of PVC being tapped be valved off during the tap. Problems have been observed of PVC pipe splitting under hot taps for water systems. The connection is leakproof and normally provides service for many years without any problem. The toning wire over the new service lateral is attached to the collection system main toning wire to ensure continuity and positive location of the connecting service.

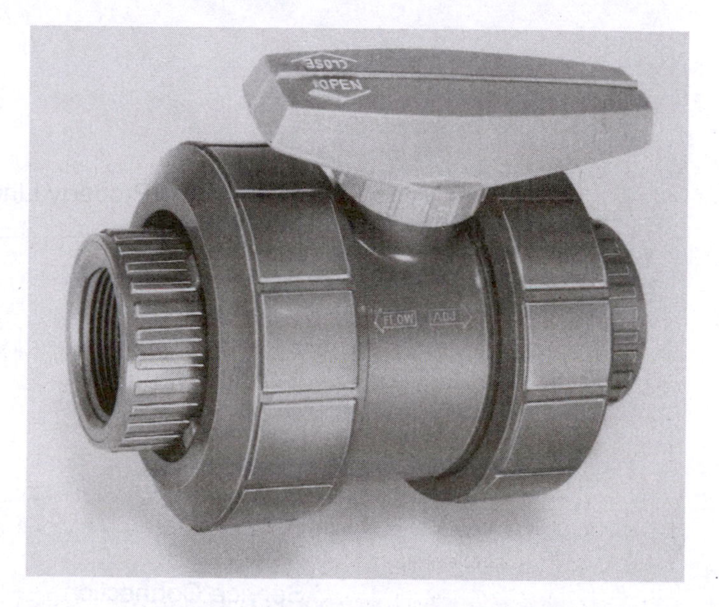


*Fig. 4.26 PVC pump discharge ball valves*

(Courtesy of Orenco Systems, Inc., www.orenco.com)

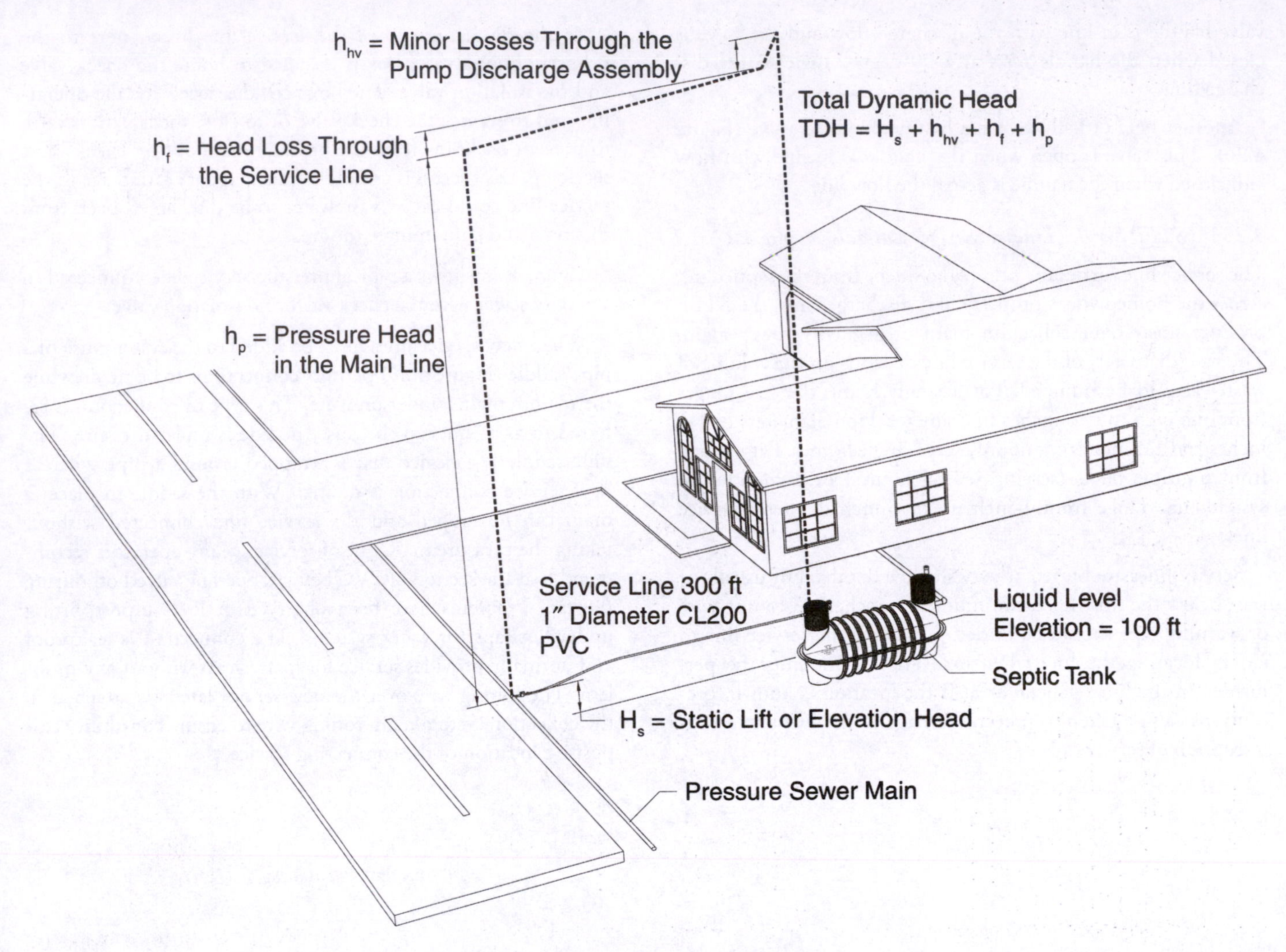


*Fig. 4.27 Typical service line for STEP system*

(Courtesy of Orenco Systems, Inc., www.orenco.com)

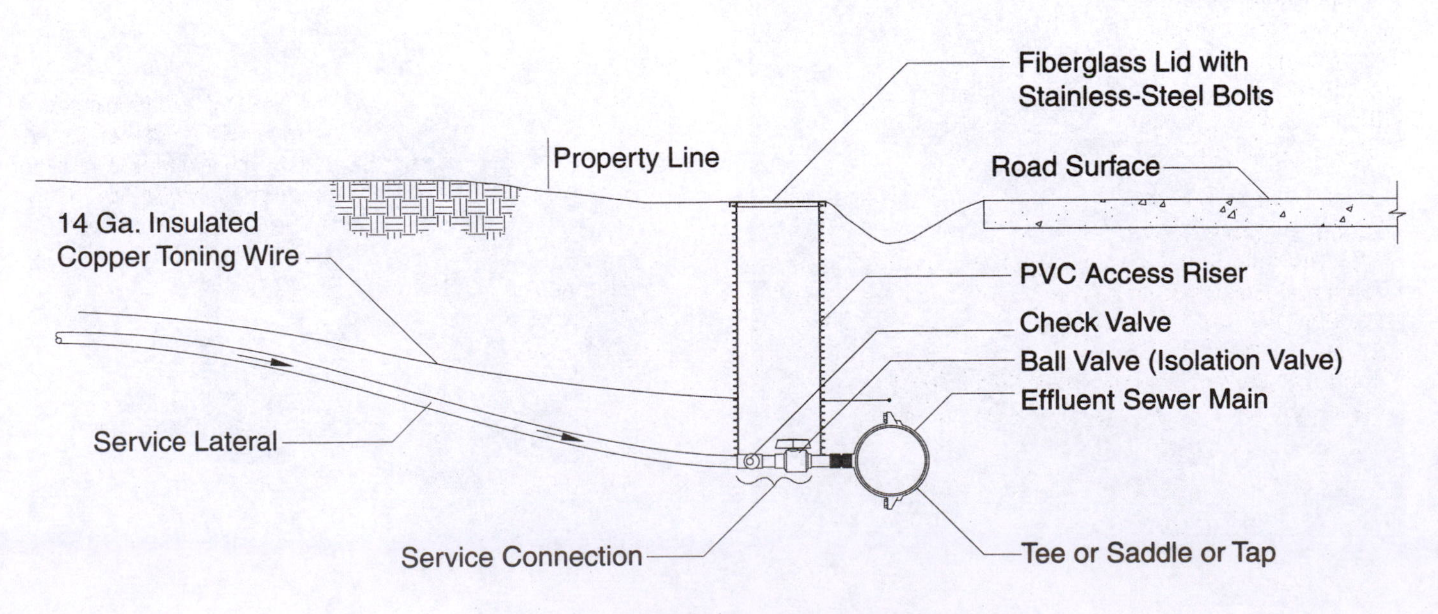


*Fig. 4.28 Typical gravity service connection*

(Courtesy of Orenco Systems, Inc., www.orenco.com)

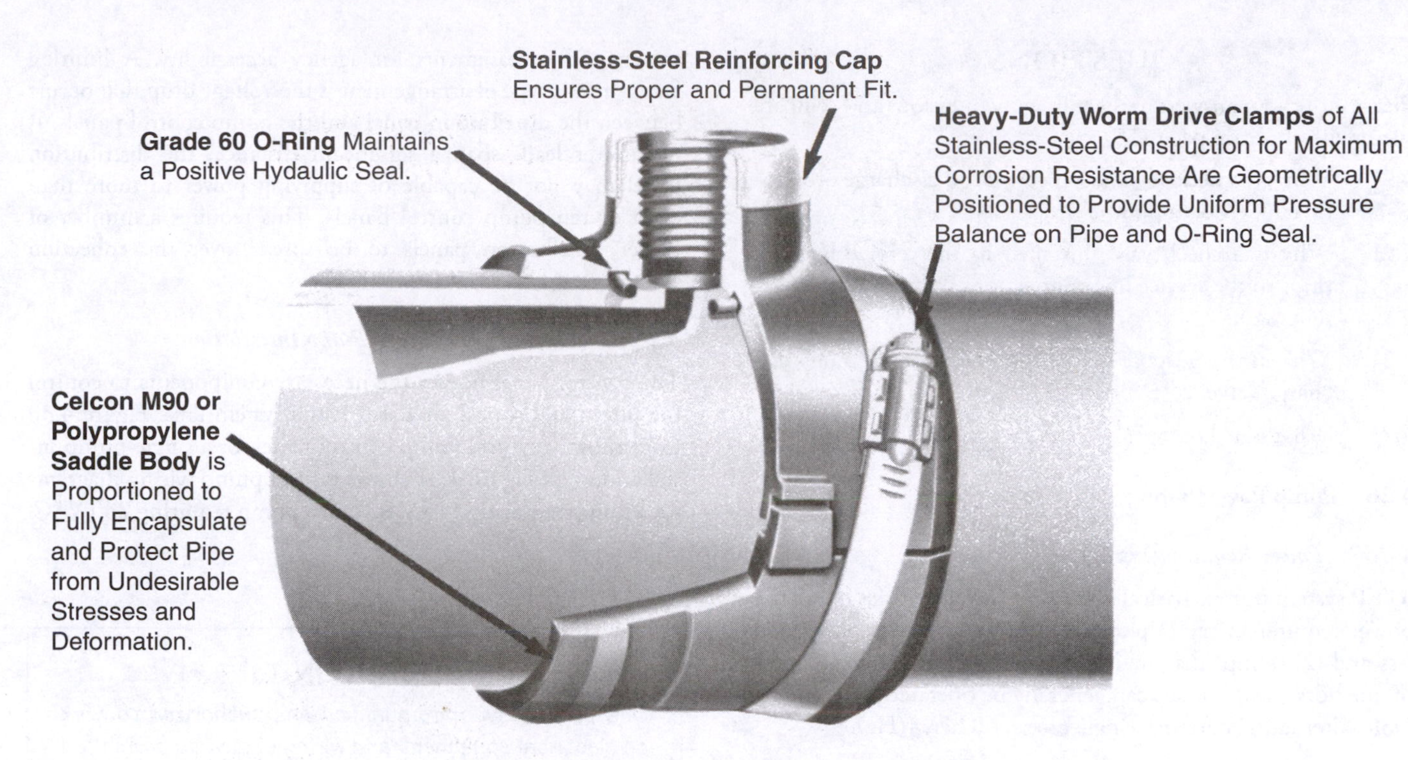
Stainless-Steel Reinforcing Cap Ensures Proper and Permanent Fit.
Grade 60 O-Ring Maintains a Positive Hydaulic Seal.
Heavy-Duty Worm Drive Clamps of All Stainless-Steel Construction for Maximum Corrosion Resistance Are Geometrically Positioned to Provide Uniform Pressure Balance on Pipe and O-Ring Seal.
Celcon M90 or Polypropylene Saddle Body is Proportioned to Fully Encapsulate and Protect Pipe from Undesirable Stresses and Deformation.

Saddles for plastic pipe sizes 1 inch through 4 inch

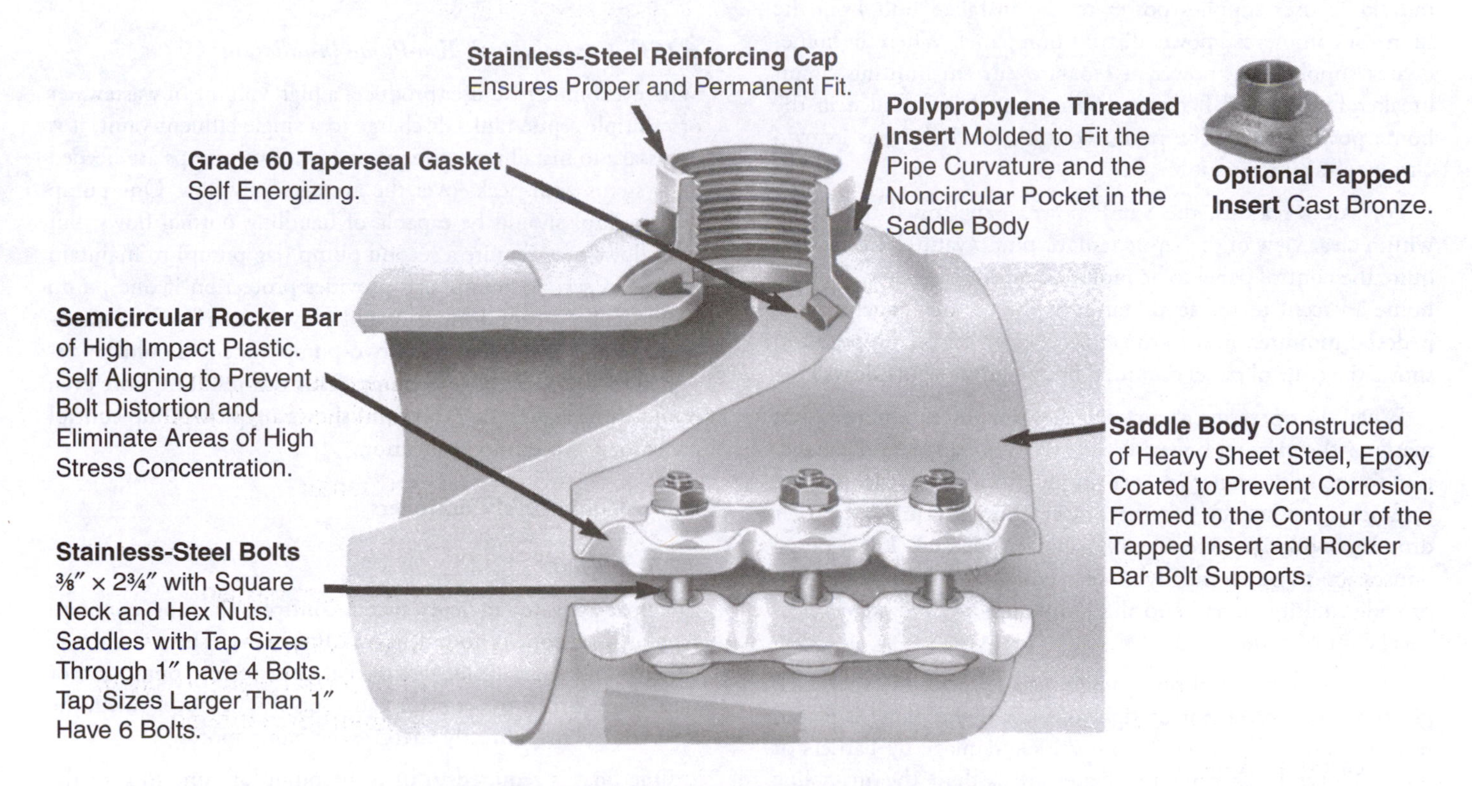
Stainless-Steel Reinforcing Cap Ensures Proper and Permanent Fit.
Polypropylene Threaded Insert Molded to Fit the Pipe Curvature and the Noncircular Pocket in the Saddle Body
Optional Tapped Insert Cast Bronze.
Grade 60 Taperseal Gasket Self Energizing.
Semicircular Rocker Bar of High Impact Plastic. Self Aligning to Prevent Bolt Distortion and Eliminate Areas of High Stress Concentration.
Saddle Body Constructed of Heavy Sheet Steel, Epoxy Coated to Prevent Corrosion. Formed to the Contour of the Tapped Insert and Rocker Bar Bolt Supports.
Stainless-Steel Bolts ⅜″ × 2¾″ with Square Neck and Hex Nuts. Saddles with Tap Sizes Through 1″ have 4 Bolts. Tap Sizes Larger Than 1″ Have 6 Bolts.

Saddles for plastic pipe sizes 6 inch through 12 inch

*Fig. 4.29 Tapping saddles for plastic pipes*

(Courtesy of Orenco Systems, Inc., www.orenco.com)

## QUESTIONS

Please write your answers to the following questions and compare them with those on page 175.

4.2J Why is a check valve installed on the discharge of a septic tank effluent pump?

4.2K Why is a check valve installed in the service line just prior to the service line connection to the collection system main?

4.2L What is the purpose of a septic tank effluent pump discharge valve?

4.2M What is a "hot tap"?

## 4.26 Pump Power Supply

### *4.260 Power Requirements*

STEP system pumps basically fall into two categories of electric power requirements: (1) pumps requiring one horsepower and less and (2) pumps that are larger than one horsepower. Pumps of one horsepower or less can generally be operated by 120 VAC (volts alternating current), single phase, 60 hertz (Hz).

### *4.261 Control Panels, General (Figure 4.30)*

The electric power supply for the STEP system pumps may be purchased by the operating agency. More often, however, the individual user supplies power to the installed unit from the home or commercial power distribution panel. When the homeowner supplies the power, a separate current-limiting circuit breaker for the STEP pump control panel is installed in the home power panel. The pump control panel also has its own current-limiting circuit breaker.

For safety reasons, the STEP power panel must be located within clear view of the septic tank or pump vault. This may require the control panel to be mounted on the outside wall of the home adjacent to the septic tank, or the control panel may be pedestal-mounted next to the tank. (Figure 4.17-1 on page 146 shows the control panel mounted on the home's outside wall.)

When the operating agency supplies the power, it may be that many of the existing house services are inadequate and connection to the homeowner's power supply cannot be made. In these instances, the agency has the power company make a power drop to an off-site metered service power pedestal. The pedestal houses the power company's meter and a distribution panel to provide multiple circuits to the individual STEP systems' electrical control panels.

The metered distribution panel is usually placed in the STEP pressure main easement or right-of-way. If the panel is placed near a road, it is protected from vehicle damage by barriers or posts. Power leads from the panel are laid in the main line trench, down the trenches of the service lines or laterals, and finally into the individual pump control panels, which are usually mounted on pedestals next to the septic tanks. This arrangement keeps power, piping, and toning wire all in the same trench and in easements for agency accessibility. A limiting factor in this type of arrangement is the voltage drop that occurs between the distribution panel and the pump control panels. If the power leads span a significant distance, the distribution panel may not be capable of supplying power to more than eight or ten pump control panels. This requires a number of power distribution panels to be spread over the collection system area.

### *4.262 Control Panel, Single-Pump Installation*

The control panel houses the necessary components to control the pump and sound an alarm if the system fails. Figure 4.30 (top) shows a typical pump control panel for a single-pump installation and Figure 4.31 shows typical pump wiring diagrams for a pump requiring 115 VAC and a pump requiring 230 VAC.

**CAUTION**

Only persons who are qualified and authorized to work on electrical equipment and wiring should be permitted to install or work on electrical control panels.

### *4.263 Control Panel, Two-Pump Installation*

Where a commercial user produces a high volume of wastewater or multiple septic tanks discharge to a single effluent vault, it is necessary to install two effluent pumps. Two pumps are needed for systems with peaks over the average daily flow. One pump (lead pump) should be capable of handling normal flows, but peak flows may require a second pump (lag pump) to maintain service. A second pump also provides protection if one pump should fail. A second pump could prevent overflows of wastewater to the environment. A two-pump system should be designed to alternate the lead pump. This will result in more even pump wear. Figure 4.30 (bottom) shows a typical pump control panel for a two-pump installation.

## 4.27 Liquid Level Controllers

### *4.270 Floats*

The float-activated mercury switch control is the most common type of float control found in STEP systems. The switch may either perform as a pilot device that controls the pump motor start contactor in the control panel, or, if it is motor rated, the float control may directly energize the pump motor.

One float is required to turn the pump off, one to turn the pump on, and one for activating a high water level alarm. If a backup OFF switch is desired, an additional float switch is used. In two-pump installations, one more float switch is added for starting the lag pump.

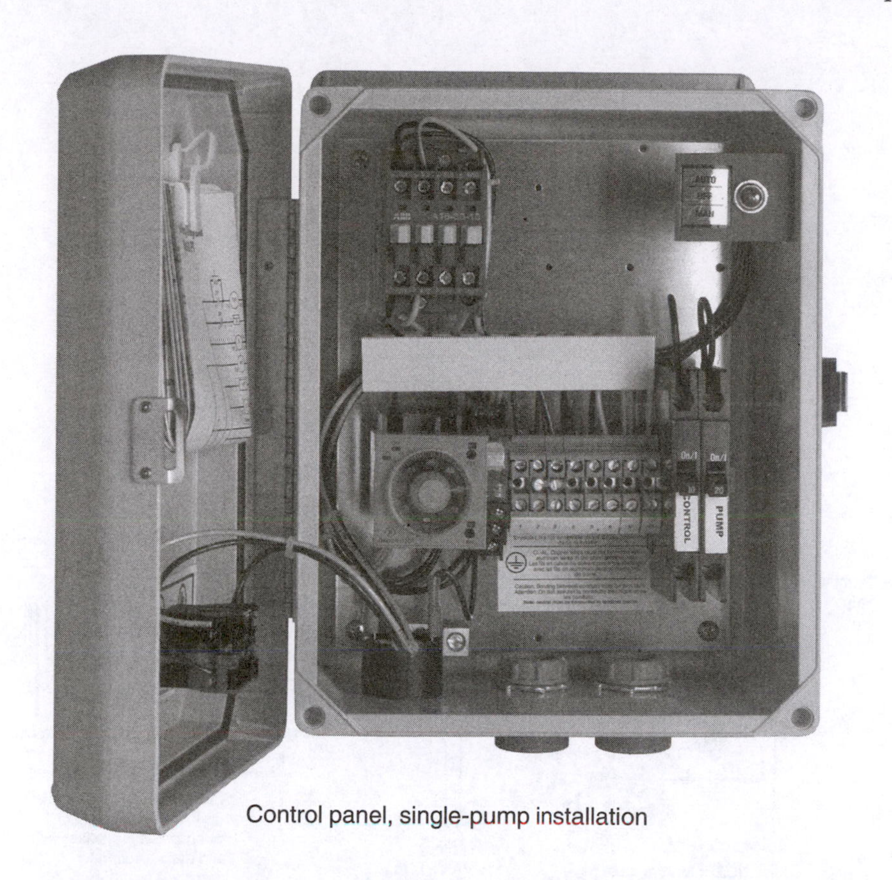


Control panel, single-pump installation

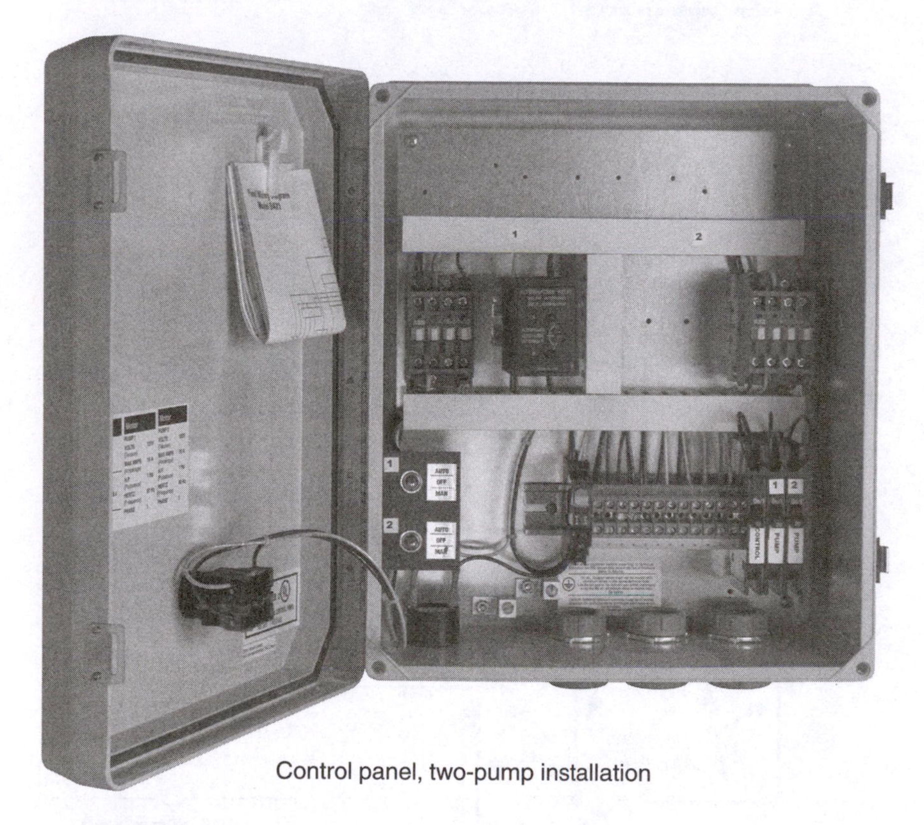


Control panel, two-pump installation

*Fig. 4.30 Typical pump control panels*

(Courtesy of Orenco Systems, Inc., www.orenco.com)

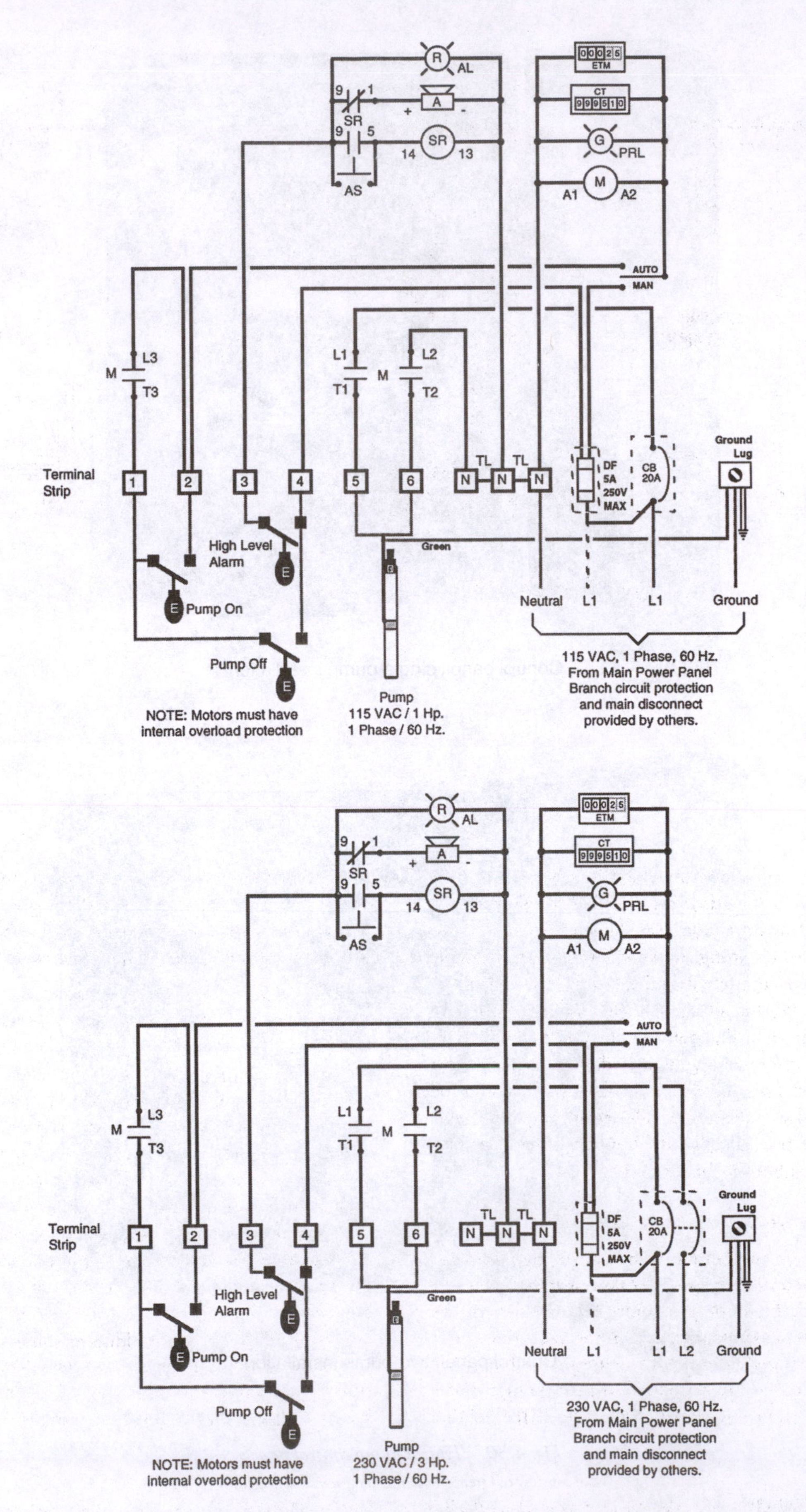


*Fig. 4.31 Typical pump wiring diagrams for a pump requiring 115 VAC and a pump requiring 230 VAC*

(Courtesy of Orenco Systems, Inc., www.orenco.com)

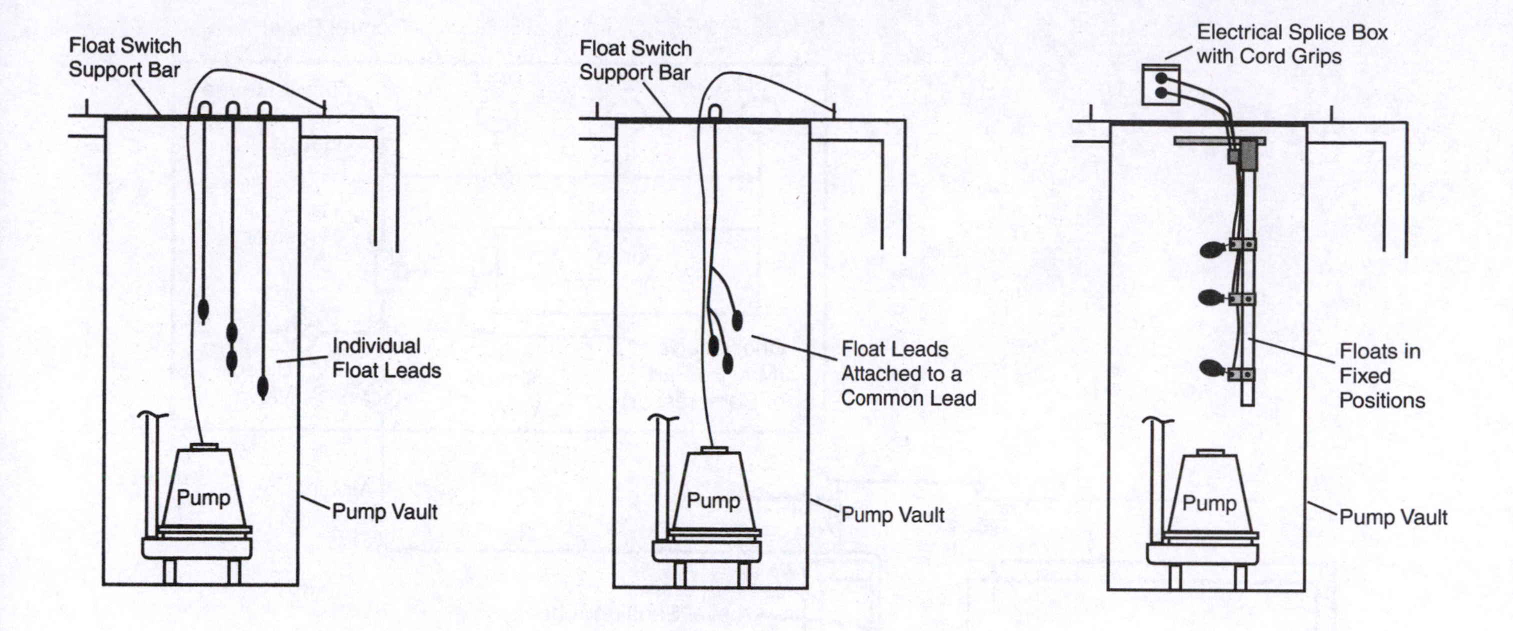


*Fig. 4.32 STEP system liquid level control float switch assemblies*

Figure 4.32 shows three common liquid level control float switch assemblies. In the first drawing, the float switches are suspended from each float's individual lead. This arrangement is not recommended because the rising and falling water tends to tangle the float leads and thus interfere with the desired level control. In the second drawing, shorter, individual float leads are attached to a common lead. This helps maintain the float switches in position but movement still may be obstructed. The third drawing shows the float switch assembly providing fixed positions for the floats within the pump vault strainer. This arrangement prevents entanglement of the individual floats, but permits them to move up or down freely to perform their function.

#### *4.271 Air Pressure (Back Pressure) Sensors*

Air pressure sensors are another means of controlling pump operation. The system consists of a small compressor that blows air down a sensing tube in the pump vault. The pressure in the sensing tube increases as the liquid level in the vault rises. Pressure switches in the pump control panel sense the back pressure of the air and turn the pump on or off, or activate high water level alarms. Figure 4.33 indicates the location of the air line and the placement of the sensing tube in the pump vault for a pressure sensing level control system. In most installations, the initial cost is higher than for float systems because a compressor is needed to supply air to the pump vault level sensing tube.

Some air pressure sensor systems are based on a continuous air flow to an air pipe that emits a constant stream of air bubbles. As water levels rise and fall, the pressure required to pump air bubbles into the vault also changes. A small *ROTAMETER*[34] is installed in the air line between the compressor and the sensing tube to indicate the rate of air flow. The rotameter is adjusted to admit a constant air stream to the sensing tube or bubbler. Air-actuated pressure switches sense the back pressure on the sensing tube from the rise or fall of the liquid levels in the tank. Based on the pressure changes, the pressure switches control the pump on/off operation and activate the high water alarm in the case of a power or pump failure. The continuous air flow system is more reliable than the trapped air level bell system because there is less chance of blockage by a buildup of solids or grease. Fish tank aquarium compressors have been used to provide the air pressure. One supplier of sensor control systems uses a trapped air system similar to the bubbler but without the compressor.

34. *Rotameter* (ROTE-uh-ME-ter). A device used to measure the flow rate of gases and liquids. The gas or liquid being measured flows vertically up a tapered, calibrated tube. Inside the tube is a small ball or bullet-shaped float (it may rotate) that rises or falls depending on the flow rate. The flow rate may be read on a scale behind or on the tube by looking at the middle of the ball or at the widest part or top of the float.

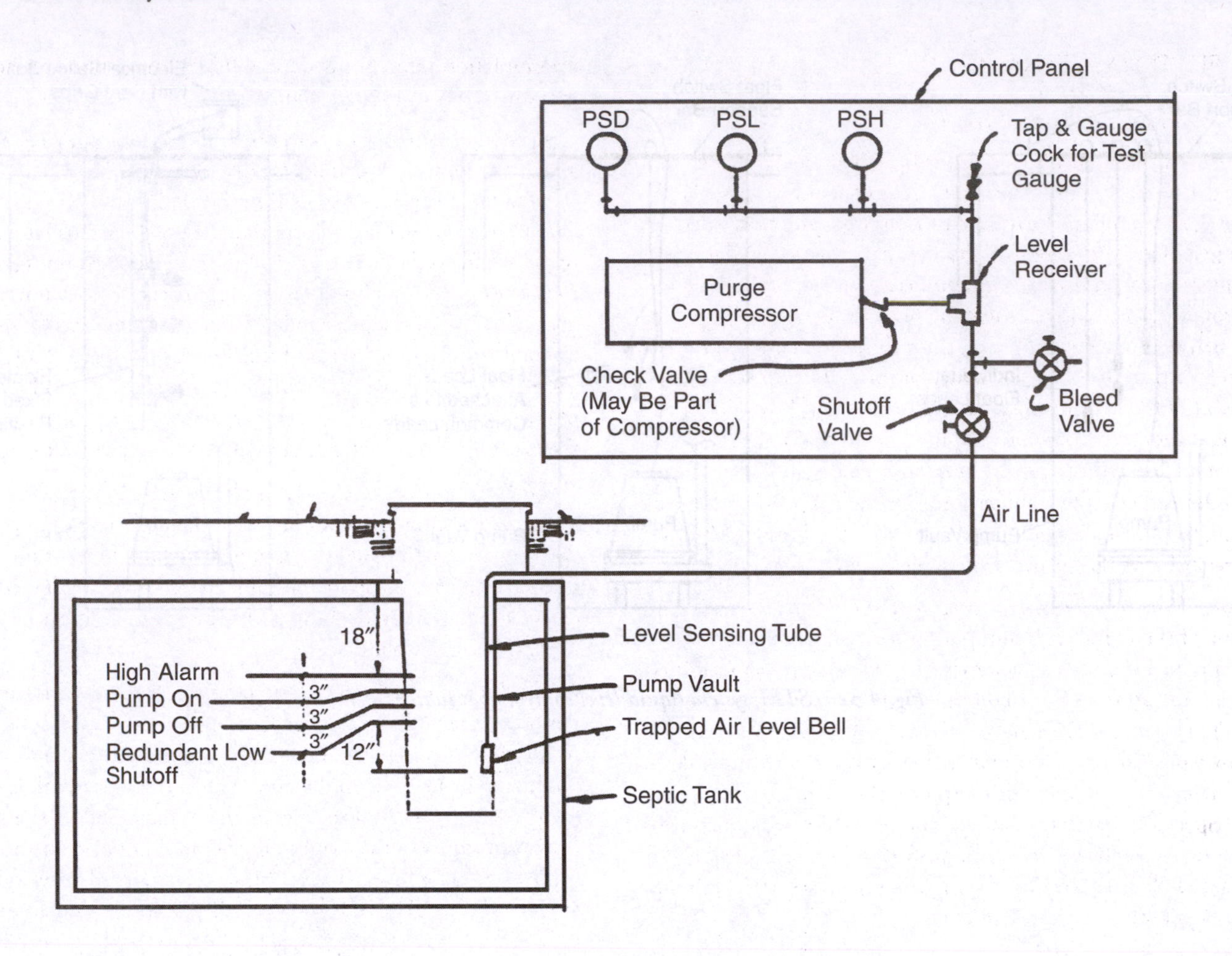


*Fig. 4.33 Air pressure sensor pump control system*

(Source: "Manila Community Service District, Design-Construction Report for the Septic Tank Effluent Pumping Sewerage System," California State Water Resources Control Board)

### *4.272 Other Level Sensor Units*

Numerous other liquid level sensing methods have seen limited use in STEP systems, such as diaphragm switches, reed switches, probes, transducers, and ultrasonics. The mercury float and mechanical float switches continue to be widely used due to their greater reliability and ease of maintenance. The greatest complaint from operational staff is that there is no easy method to repair the floats; either they work or you replace them. In some units, this means replacing the entire float level control system.

### 4.28 Flow and Pressure Monitoring Devices

On some septic tank effluent pump (STEP) units, the operational staff occasionally need to check the pump discharge pressure and, in some instances, the volume of flow generated from a customer's septic tank. The pressure readings may be obtained by installing a gauge or pressure recorder on the pump discharge inside the septic tank riser before the service line leaves the tank.

If a wastewater flow measurement is required, a standard single service water meter is installed in the wastewater service line. The water meter functions very well on the septic tank effluent, particularly if it has passed through a strainer pump vault. Grinder pump systems have been metered but solids do occasionally plug the flowmeter, which then requires servicing. If the agency happens to have responsibility for both the domestic water supply and wastewater discharge, make certain that meters previously used for wastewater are never installed in drinking water systems.

## QUESTIONS

Please write your answers to the following questions and compare them with those on page 175.

4.2N Why should the STEP power panel be located within clear view of the septic tank or pump vault?

4.2O What are the two most common types of effluent level pump controls?

4.2P How can pump discharge pressure be measured on a STEP unit?

4.2Q How can the wastewater flow from a home be measured?

## 4.3 FIELD MAINTENANCE

### 4.30 Working on Private Property

The sewer utility usually is responsible for maintenance of the entire collection system. The utility district assumes responsibility for all components downstream from the septic tank inlet. In some instances, the responsibility for maintenance of the components located on private property has been left to the individual property owner. This avoids the need for utility operators to enter private property. However, since the septic tank or grinder pump is critical to the performance of the collection system, responsibility for maintenance should be retained by the district. If your duties as an operator require you to maintain or repair equipment located on private property, you may wish to read Appendix A, "Legal Considerations," at the end of this chapter. It contains information about an operator's legal responsibilities with regard to entering and protecting private property.

Most of the problems associated with STEP systems occur in the area of the septic tank pumping systems. The operators who maintain STEP systems therefore bear an added responsibility that the operators of most conventional wastewater systems do not. That is, the STEP system operator will often find it necessary to work on private property and interact with individual customers. It is the operator's responsibility to keep the equipment operating sanitarily, safely, and with the least inconvenience to the resident. To accomplish this, an operator's public relations skills have to be of the highest quality. Special training in public relations is a good investment.

### 4.31 Preventive Maintenance

For the most part, routine maintenance of STEP systems and grinder pumps consists of visual inspection and general housekeeping types of activities. Whenever an operator makes a service call, the liquid level in the tank, the sludge blanket depth, and the scum blanket depth should be checked. Check for normal levels, color of the wastewater, and odors. As explained previously, all septic tanks require pumping and cleaning at some time interval. How often the septic tank needs to be pumped for removal of scum and sludge depends largely upon the user. The pumping interval for a residential septic tank typically ranges from three to ten years. Pumping a septic tank too often may interfere with the development of a healthy population of microorganisms and thus interfere with proper treatment of the wastewater solids.

Restaurants and other high-use facilities such as taverns require more frequent pumping than most residences. Common practice is to require a grease trap or an additional septic tank to serve as a grease trap for grease removal and to pump the grease trap or tank serving these high-use facilities every three to six months. During the pumping operations, utility district personnel should be present to record the depth of sludge and thickness of any scum in each tank so that the pumping schedule can be adjusted in the future according to actual sludge and scum accumulation rates. See Section 4.04 for suggestions on how to measure scum and sludge depths.

A good resource for operators, homeowners, and restaurant and other commercial food facility owners is National Onsite Wastewater Recycling Association (NOWRA), whose mission is "to provide leadership and promote the on-site wastewater treatment and recycling industry through education, training, communication, and quality tools to support excellence in performance." NOWRA's helpful guide, "Restaurant Do's and Dont's," is available on their website at http://www.nowra.org. NOWRA's "Septic Locator" tool is useful for locating septic product suppliers, service providers, and on-site wastewater professionals in the user's area. It is available at http://www.septiclocator.com.

Tank inspection is usually performed by agency personnel immediately after the tank has been pumped out to check for cracks, leaks, baffle integrity, and general condition of the tank. Effluent screens used on the tank outlet must be pulled and cleaned regularly by flushing with water if they are to be effective at removing solids.

The liquid level in the septic tank can be easily observed by removing the tank riser access cover. If the wastewater level is up in the riser, this is an indication of a problem with the pump or the hydraulic loading on the tank. When the septic tank is operating normally, the wastewater at the effluent end of the septic tank or pump vault is a light gray-colored liquid with the look of dirty water. Liquid that is darker than normal may indicate that the sludge blanket is too deep. If the pump is enclosed in a strainer vault, check the water level in the vault. A low liquid level inside the strainer may indicate that the sludge or scum has plugged the screen openings and the screen will need to be cleaned. If there are indications that a problem exists or may be developing, review the troubleshooting guide in Section 4.063, "Troubleshooting Guide," and Section 4.36, "Troubleshooting," for possible causes of the problem and some suggested solutions.

On routine maintenance visits, wash down grinder pump vaults, particularly those equipped with float switches, to reduce grease buildup. STEP pump vaults accumulate less grease so there is less need to clean them. Run the pump and controllers through their cycles to see that all components are in working order. Take voltage and amperage readings; high amperage indicates that something is restricting the movement of the rotor, usually a clogged impeller or grinder. The pumps and associated components are not routinely removed from their vaults but removal may be necessary to clear the blockage or if there are motor problems.

Normal operation of a septic tank should not produce unpleasant odors but odor problems may develop if the system is not used for a period of time. Without a fairly constant inflow to circulate the tank contents, the wastewater in the vault may become *SEPTIC*[35] and generate rotten egg odors. If the tank is producing noticeable odors, investigate the cause using Table 4.1, "Troubleshooting—Septic Tanks," on pages 134 and 135.

---

35. *Septic* (SEP-tick). A condition produced by anaerobic bacteria. If severe, the sludge produces hydrogen sulfide, turns black, gives off foul odors, contains little or no dissolved oxygen, and the wastewater has a high oxygen demand.

Another maintenance task that can be attended to during routine visits is checking the motor starter contacts. The starter contacts may occasionally need to be cleaned in areas of heavy insect infestation. Also, keep the panel enclosure in good repair to prevent insects from getting inside.

During the first year or two after installation of a pump vault, the soil under and around the vault tends to settle. Minor earth settlement is common and not a serious problem. However, if the top of the vault drops below ground surface, the problem needs to be corrected to prevent the inflow of surface waters into the vault. Inspect the inlet and outlet pipes for breakage that will allow leakage into the system. Retrofitting the tank with a watertight extension will eliminate the potential inflow problem. However, if the vault settles, there could be damage to the vault and piping, such as cracks.

### 4.32 Maintenance Staff Requirements

A minimum of two people should be trained and available. On some systems, one person can attend to the service calls. Some pumping units, especially grinder pumps, may require two persons (and possibly a lifting device) to remove the pump from the tank or vault due to the weight of the pump. Similarly, two or more operators and a lifting device may be needed if it is necessary to remove pumping units surrounded by vault strainers. If the strainer drain valve fails to open when the unit is lifted, the weight of the retained water adds considerably to the weight that must be lifted.

A fully trained and experienced backup person is needed for times when the lead person is unavailable. The two people do not necessarily have to be full-time employees, but at least one person has to be available on call.

### 4.33 Maintenance Safety

Hazardous gases (hydrogen sulfide and methane) are commonly encountered at STEP septic tanks, at relatively inactive grinder pump (GP) vaults, and points where pressure sewers discharge. The proper practices of pressure sewer maintenance are usually such that the gases are not hazardous due to dilution with surrounding air, but service personnel need to be aware of the potential hazards of these gases. For example, hydrogen sulfide gas (which smells like rotten eggs) is known to deaden a person's ability to smell it after a brief exposure. This may cause the person to wrongly think the gas has drifted away. Many deaths have been attributed to exposure to hydrogen sulfide gas. All operators must be advised of procedures and precautions regarding exposure to hazardous gases. In addition, the entire system should be designed so that no exposure is required.

Extensive electrical connections and equipment are used at each pumping unit, typically including an electrical control panel, mercury float switches, and electrical junction boxes. Operators are exposed to the possibility of electrical burn or shock in their maintenance duties. They may be standing on wet ground while servicing electrical components, and working under adverse conditions such as darkness, rain, or snow. Electrical safety should be emphasized in the *O&M MANUAL*[36] and system designs must minimize the exposure of O&M personnel to live electrical systems.

Personal hygiene should also be discussed in the O&M manual. As with any project involving contact with wastewater, maintenance of septic tank systems exposes operators to the possibility of infections or disease. For additional information about safety hazards associated with septic tank systems and recommended safety precautions, see Chapter 3, "Safety."

### 4.34 Emergency Calls

Most STEP systems are equipped with either audible or visual alarms to notify the homeowner when a high/low liquid level problem exists in the septic tank or pump vault. The alarm will be triggered before a backup of wastewater into the home occurs. Homeowners should be instructed to promptly notify the operating agency by telephone whenever they become aware of an alarm condition. Although the building sewer is the property owner's responsibility, an operator should respond to the emergency as quickly as possible to minimize damage from an overflow or backup and inconvenience to the customer. Prompt operator response is crucial when the pump vaults used in the system have small reserve volumes between the high level alarm *SET POINT*[37] and the top of the basin. A typical reserve volume is 150 gallons. In some cases, maintenance personnel must respond to such calls whenever the call is received, even during late night or early morning hours. However, a 12-hour to 24-hour response time is usually adequate.

Odor complaints occur when a system is poorly designed. As with household plumbing backups, faulty venting in the building plumbing is usually the cause. If improved venting fails to eliminate the odor complaints, the carbon filters can be installed in the septic tank inlet vent or running traps can be placed in the service lateral to prevent the sewer main from venting through the service connection.

During extended power outages, pump basins may fill to capacity. However, water use and corresponding wastewater flows usually are greatly reduced during power outages. Portable standby generators, gasoline-powered pumps, or a septic tank pumper truck should be available to pump down flooded tanks and prevent overflows.

---

36. *O&M Manual.* Operation and Maintenance Manual. A manual that describes detailed procedures for operators to follow to operate and maintain a specific treatment plant and the equipment of that plant.
37. *Set Point.* The position at which the control or controller is set. This is the same as the desired value of the process variable. For example, a thermostat is set to maintain a desired temperature.

As a community grows, the collection system is extended. High level alarms at the individual pumping units are likely to occur more frequently, usually due to the mains having reached capacity, the mains having developed a problem with air-caused head loss, or pumps that are too small.

In some localities, a wastewater treatment operating agency is required to notify the local or state water quality control or environmental control agencies whenever a wastewater spill or overflow occurs. Often, such notification must be made within prescribed time limits. Check with your local regulatory agencies to verify the spill reporting requirements in your area.

### 4.35 Information Resources

If you, the operator, are fortunate enough to be on hand during construction of a new STEP system, learn as much as you can about all of the components, even down to the smallest type of steel pipe band clamp. The design engineer will most likely be more than willing to help you understand how the system is intended to work and the function of each component. The designer wants it to function flawlessly, just as you do. If the design engineer is not available, request the training needed to understand the system.

The operator's most valuable resource for maintenance information is a complete, up-to-date operation and maintenance (O&M) manual. If yours is out of date or has been lost, the design engineer should have copies on hand or you can use the guidelines presented in Appendix C at the end of this chapter to update an old manual or prepare a new one.

Equipment suppliers are an excellent source of information for operators. They want their products to be properly installed and maintained and will often provide literature on the products they manufacture and sell. A supplier may also be aware of methods other operators have used to solve a particular problem or improve the performance of a piece of equipment. Operators who take the time to establish good working relationships with their suppliers can benefit from their expertise and may receive better support service when problems arise. Good working relationships also benefit suppliers and they usually appreciate comments from operators who use their products and have discovered a better way of maintaining them or have suggestions for improving a product.

The increasing use of STEP systems is making them quite common throughout the world, and particularly in the United States. This increases the number of people who, like yourself, are responsible for the proper operation and maintenance of these systems. Visiting or communicating with other system operators often provides ideas or methods that you or they can use to make the job easier and the system perform better and more economically. Be aware, however, that while there may be other wastewater systems similar to yours, the problems and solutions are never identical. Different geographical locations, terrain, products, equipment, and community makeup mean each system is unique. Just as there is no single document that can tell you what must be done and when, there is no other system just like yours. You will have to evaluate other operators' suggestions in the context of your own system and try to find your own best operating strategy.

## QUESTIONS

Please write your answers to the following questions and compare them with those on page 175.

4.3A What items should be checked on a septic tank whenever an operator makes a service call?

4.3B What is the typical pumping interval for a septic tank at (1) a residence and (2) a high-use facility such as a restaurant?

4.3C What preventive maintenance tasks should be performed on grinder pumps during routine maintenance visits?

4.3D What general types of safety hazards could an operator encounter while maintaining a STEP system?

4.3E What types of emergency calls should maintenance personnel be prepared to respond to?

4.3F Where could an operator find information about the operation and maintenance of STEP systems?

### 4.36 Troubleshooting

This section contains tables that can be used to quickly troubleshoot problems that could develop in septic tank systems. A typical troubleshooting sequence in response to a high liquid level alarm might proceed as follows:

- First, confirm that power is being supplied to the control panel and observe the liquid level in the tank to confirm that a high water condition exists. If no reserve volume remains available to receive flow, pump the basin down by running the pump manually (if it will run), or by using a portable pump if normal service cannot be restored.
- Then, turn the power off so work can be performed more safely with the malfunctioning installation. Guided by experience with the pumping units on the project, or a history of performance of the particular pumping unit being serviced, first consider the most likely causes of the problem. (Refer to Table 4.3, "Troubleshooting—STEP Systems: Septic Tanks and Pump Vaults," for a description of pump problems and possible solutions.)

### TABLE 4.3 TROUBLESHOOTING—STEP SYSTEMS: SEPTIC TANKS AND PUMP VAULTS

| Indicator/Observation | Probable Cause | Check | Solution |
|---|---|---|---|
| High liquid level or wastewater backup into home or business | Electrical power off | 1. Pump panel power main breaker | 1. Reset main circuit breaker if turned off accidentally by home resident or tripped out by overload. |
| | | 2. Fuse, if used in pump control panel | 2. Replace if burned out. |
| Audio/visual alarm (service request sent to wastewater operator) | | | |
| 1. Power supply OK—high liquid level | 1. Pump system | 1. Amperage drawn by pump motor | 1a. Manually turn pump on; check amperage drawn. If amperage reading is low, pump is not primed; turn pump off. Disconnect discharge hose or pipe at entrance to service lateral (in riser), bleed off air or gas by filling pump with water. Reconnect discharge, put pump in service. If amperage draw is normal and liquid level in tank is falling, place on AUTO control and observe through a pump-down cycle. |
| | | | 1b. Amperage reading high—pump impeller jammed or pump motor burned out. Turn electrical power to pump control panel to OFF. Follow proper electrical lockout procedures. Disconnect pump discharge piping; remove pump (may have to cut power supply leads to pump if not equipped with quick disconnect). Clear debris from impeller or, if pump motor is burned out, replace the pump with similar type pump, reconnect discharge piping and electrical supply. If power leads were cut for removal, apply heat shrink insulation to new splice (see Figure 4.34, page 171). Prime pump and test run on AUTO through one cycle. |
| | | | 1c. Amperage draw normal. Pump operates, liquid level falling. Stop pump. Should start and continue pumping to low level cutoff. |
| 2. Pump will not run on AUTO control | 2. Controller failure | | |
| | 2a. Trapped air in controller | 2a. Sensing tube or bell | 2a. 1. Restore air to sensing tube. Run pump on MANUAL control, pump septic tank or vault down to the lowest level to admit air into bell of sensing tube.<br>2. Clean sensing line by rodding, flushing, or purging foreign material. If tube was clear, check tube and piping for air leaks. |
| | 2b. Float control | 2b. Float position | 2b. Float control hung up out of position. Untangle. Place in proper position or order. If still inoperative, replace float or float array with new unit. |

TABLE 4.3 TROUBLESHOOTING—STEP SYSTEMS: SEPTIC TANKS AND PUMP VAULTS *(continued)*

| Indicator/Observation | Probable Cause | Check | Solution |
|---|---|---|---|
| 2. Pump will not run on AUTO control *(continued)* | 2c. Air bubbler control | 2c. Compressor and air flow through rotameter | 2c. 1. Restore power to air compressor. If compressor will not run, replace. *NOTE:* If a replacement compressor is not readily available, you may want to use a compressed inert gas (nitrogen or carbon dioxide) as a temporary replacement. Typical air flow for a bubbler is very low and a cylinder can be used for a short time period. Set rotameter to correct air flow rate. If compressor is OK,<br>2. Check sensing tube for leaks and to ensure that it is not broken off. Check that air flows to end of sensing tube. If sensing tube is functional, check pressure switch for open/close while applying back pressure on sensing tube. Replace faulty pressure switch. |
| | 2d. Sonic or other control | 2d. Power supply to sensor and transducer | 2d. Replace sensor units if still inoperative, replace control circuit with new module. *NOTE:* In grinder pump vaults, grease and soap buildups can interfere with all control sensors. Routine cleaning by hosing is necessary. |
| 3. Pump operates on AUTO control but liquid level is still high | 3. Pump filter or screen creating low head, or turbine pump intake is plugged | 3. Check filter or screen for binding or high head loss (restricts liquid flow from one side to the other) | 3. Dewater vault to low cutoff level. Liquid does not flow through screen filling pump area at normal flow rate. Remove filter or screen unit, clean by brushing and hosing off. Make sure no large pieces of scum or other debris are left inside of screen that would plug pump impeller. For turbine pumps, also clean screen over intake port to impeller. Reinstall, check pump operation. *NOTE:* Temporary service can be restored by scraping or hosing exterior surface of screen while in place; this is not a long-term solution. |
| 4. High liquid level, pump operation OK | 4. Blocked discharge | 4a. Pump discharge isolation valves | 4a. Valve in closed position. Open valve. |
| | | 4b. Service lateral isolation valve at main | 4b. Valve in closed position. Open valve. |
| | | 4c. Discharge hose or pipe | 4c. Turn off pump—disconnect discharge line. Turn on pump. Check flow.<br>1. Normal flow—reconnect discharge to service lateral. Check pump discharge pressure with pressure gauge. Low pressure indicates pump is worn out. Replace pump.<br>2. High pressure—blockage in service lateral or collection main. Check collection system main. If normal, blockage is in lateral; clear blockage.<br>3. No flow from disconnected pump discharge pipe or hose.<br>• Blockage between pump and connection to service line. Remove or clear blockage.<br>• Pump impeller clogged, worn, or loose. Pump not pumping. Replace pump. |

TABLE 4.3 TROUBLESHOOTING—STEP SYSTEMS: SEPTIC TANKS AND PUMP VAULTS *(continued)*

| Indicator/Observation | Probable Cause | Check | Solution |
|---|---|---|---|
| 5. Low liquid level | 5a. Pump unit not shutting off at low liquid level set point. Control problem | 5a. Pump control—see 2 above. Controller failure | 5a. Check, clean, clear control devices so they can function as intended—see 3 above. |
| | 5b. Pump not shutting off. Controls OK. Pump starter contactor stuck closed | 5b. Pump starter contactor | 5b. Turn off electric power at main supply panel. Lock out and tag.<br>1. Replaceable panels—remove and replace pump control panel.<br>2. Noninterchangeable panels—check pump motor starter relay for operation, corrosion, and insects. Moisture may have caused one of the contacts to arc, thus welding it closed and keeping the pump energized. If a contact is welded, pitted, burned, or greatly discolored, replace relay, restore power, and check pump operation. |
| | 5c. Pump unit operates normally. Septic tank or vault damaged and leaking | 5c. Interceptor or vault damaged and leaking | 5c. Have septic tank or vault pumped out and cleaned; ventilate and check for structural damage. Follow approved confined space procedures for entry. |
| Pump cycles on-off frequently | 1a. Check valves not seating | 1a. Check valve operation | 1a. Disconnect discharge pipe. Open lateral isolation valve. If liquid flows, check valve at end of service lateral at main is not seating, permitting backflow. If no check valve (at end of service lateral) then pump check valve is not seating. Clean and repair or replace check valve. |
| | 1b. Design flows are being exceeded | 1b. Measure flows and record | 1b. Increase pumping capacity or decrease use (flows). |
| House service line problems | | See Table 4.1, "Troubleshooting—Septic Tanks," pages 134 and 135, and Table 4.4, "Troubleshooting—House Service Lines." | |
| Insects, odors | | See Table 4.1, "Troubleshooting—Septic Tanks," pages 134 and 135. | |

### *4.360 House Service Lines*

See Table 4.4, "Troubleshooting—House Service Lines."

Clogged building sewers represent a maintenance problem even though maintenance of the building sewer is the responsibility of the homeowner. Service line stoppages can be cleared through the home's plumbing cleanouts with standard plumbing snakes or rods. Service lines that have been damaged or broken by construction activities, surface loads from vehicles driving over them, or root intrusion require repair or replacement.

### *4.361 STEP Systems: Septic Tanks and Pump Vaults*

For help in troubleshooting problems with STEP systems, see Table 4.3, "Troubleshooting—STEP Systems: Septic Tanks and Pump Vaults," and Table 4.5, "Diagnosing STEP System Problems." When using Table 4.5, read across the top of the chart to find the symptom and then read down to find the likely cause. For any symptom, the possible causes are numbered in the order most likely to occur. The cause identified by #1 should be investigated first, followed by #2, and so on. (Also, see Section 4.063, Table 4.1, "Troubleshooting—Septic Tanks," pages 134 and 135.)

TABLE 4.4 TROUBLESHOOTING—HOUSE SERVICE LINES

| Indicator/Observation | Probable Cause | Check | Solution |
|---|---|---|---|
| Flooded fixtures in home, water will not drain or drains very slowly | 1a. Backup in septic tank system | 1a. Liquid level in septic tank | 1a. Normal liquid level in septic tank indicates service line problem. |
| | 1b. Service line blockage (grease, detergent, paper, or plastic material) | 1b. Cleanout just prior to septic tank inlet<br>1. if full of water | 1b. 1. Run plumber's snake or rods down cleanout clearing obstruction in septic tank inlet. |
| | | 2. empty, no water standing in cleanout | 2. Run plumber's snake or rods from cleanout at home (beginning of service line) to septic tank inlet cleanout. If stoppage is removed, water will flow, restoring service, or snake will not pass by or through blockage. |
| | 1c. Broken or offset joint in service line, possible root intrusion, crushed pipe, or broken pipe caused by settlement or surface traffic | 1c. Service line | 1c. 1. Push plumber's snake down service line to blockage, try to clear blockage with power snake or hand rods equipped with spearhead blade or root saw.<br>2. If pipe cannot be cleared or is broken, excavate service line at point of blockage, cut service line one foot downstream of the broken/plugged pipe. Open or cut service line slightly above broken or plugged segment. Let wastewater drain through excavation to open downstream service line. After draining, remove rest of broken pipe or material causing blockage. Trim ends of cut pipe. Run snake through service line from house cleanout to septic tank inlet. If clear, replace removed pipe segment with new pipe and band clamps. Run water through the service line and check for leakage before backfilling. |
| | 1d. Blocked or inadequate venting | 1d. Vent | 1d. Clear blockage, provide vent. |

**TABLE 4.5 DIAGNOSING STEP SYSTEM PROBLEMS***

| Causes ↓ / Symptoms → | Alarm Light On —High Level | Alarm Light On—Low Level | Pump Does Not Run | Control Box Breaker Trips | Pump Runs—Does Not Pump | Alarm Sounds Intermittently | Short Pump Cycles | Excessive Pump Counts |
|---|---|---|---|---|---|---|---|---|
| Main Circuit Breaker Tripped Off | 3 | | 3 | | | | | |
| Manual/Off/Auto (MOA) Switch Off | 1 | | 1 | | | | | |
| Pump Circuit Breaker Off | 2 | | 2 | | | | | |
| Poor Electrical Connections | 4 | | 4 | 1 | | 1 | 3 | |
| Tangled or Inoperative Floats | 5 | 1 | 5 | | | 3 | | |
| Water in Junction Box | | | | | | 2 | | |
| Improper Wiring | | | | 3 | | 4 | | |
| Inoperative Pump | 8 | | 6 | 2 | | | | |
| Broken Discharge Plumbing | 11 | | | | 4 | | | |
| Pump Inlet Fouled | 6 | | | | 2 | | 1 | 2 |
| Improper Float Settings | | | | | | 5 | 4 | |
| Worn Pump Impellers | 7 | | | | 3 | | | |
| Valve Shut Off | 9 | | | | 1 | | | |
| Vault Screen in Need of Cleaning | | | | | | | 2 | 3 |
| Siphoning | | 2 | | | | | | |
| Leaky Tank (Exfiltrating) | | 3 | | | | | | |
| Groundwater Infiltration | 10 | | | | | | | 1 |

*Reprinted with permission of Orenco Systems, Inc.

Many problems with septic tanks will be indicated by changes in liquid level, sludge blanket depth, or scum blanket thickness.

1. LIQUID LEVEL. This is quickly determined by removing the septic tank riser access cover and observing the liquid level in the tank. If the wastewater level is up in the riser, either (1) the pump is not operating and discharging to the main; or (2) a high hydraulic load is being applied to the tank from the customer, surface runoff, or infiltration/inflow from high groundwater entering through the building sewer or the septic tank itself. Pump operating and discharge pressure will indicate if the pumping system is operating properly. If the high liquid level is a pumping problem, it can be remedied. If the problem is high flows, then the source must be located and corrected by isolation or repair.

2. DEEP SLUDGE BLANKET. The indication of this problem is that the effluent in the discharge end of the septic tank or pump vault is no longer a light gray-colored dirty water but instead is much darker than normal. To confirm that this is the problem, collect a small sample, 300 to 1,000 milliliters (⅓ to 1 quart), of the tank effluent wastewater in a clear glass jar, beaker, or *IMHOFF CONE.*[38] Let the sample settle for at least 30 minutes. If a black sludge settles to the bottom, this indicates that the sludge blanket is too deep and the tank's septage (sludge solids) needs to be pumped and disposed of at an approved disposal site.

3. THICK SCUM BLANKET. Often, the indication of this problem, particularly if a strainer vault is used in the pump system, is when the septic tank water level is high but there is a low liquid level inside the strainer where the pump is positioned. This indicates that sludge or scum has plugged the screen openings preventing wastewater from reaching the pump. More commonly, this is caused by scum material rather than by sludge. The screen must be cleaned. Occasionally, partial cleaning can be accomplished by hosing the screen from the pump side and attempting to remove the debris from the septic tank side of the screen. This generally is only a partial solution. The screen assembly must be removed for a thorough cleaning. When performing this task, be careful not to contaminate the hose and nozzle with wastewater. Also, never leave the hose hanging in the septic tank (avoid backflow).

   When the screen is sealed with scum in some installations, the wastewater level in the tank can rise and flow over the top of the screened pump vault, thus filling the pump area with grease, hair, fibers, and other scum material and plugging the pump. It will be necessary to remove the pumping system for a complete cleaning and to pump and clean the septic tank.

The intrusion of water into a pump vault should be thoroughly investigated because the excess water can overload the system and cause failures. Possible causes of water intrusion include roof drains discharging too close to the pump vault or careless irrigation practices. In some instances, the pump vault may have been installed too low causing water to puddle around and over the vault cover. Leaking check valves can also permit backflow into the vault.

Other pump vault problems range from accidental damage caused by the home resident to faulty installation during construction or inadequate piping that fails to properly convey the wastewater from the pump vault. If odors become a problem in systems with vault cover vents, check to see if an accumulation of water is blocking the venting of gases to the atmosphere.

38. *Imhoff Cone.* A clear, cone-shaped container marked with graduations. The cone is used to measure the volume of settleable solids in a specific volume (usually one liter) of water or wastewater.

### *4.362 Electrical Equipment*

Problems with electrical control panels are highly diverse, ranging from inoperative components to loose wiring or the intrusion of ants or other insects. Exercise caution when working around electrical control panels and know your limits. Only qualified and authorized persons should attempt to troubleshoot or repair electrical equipment. When troubleshooting electrical equipment, consider the following common sources of problems.

- Always check for blown fuses or tripped circuit breakers. Circuit breakers may have been turned off by the homeowner or may be tripped out due to an overload of the ground-fault circuit interrupter (GFCI) circuit.
- The wiring splices in electrical junction boxes can be a major source of maintenance problems, especially if the splices are not watertight. Junction boxes typically leak, despite claims to the contrary. To reduce the occurrence of this problem, seal the splices by applying heat shrink tubing as shown in Figure 4.34.
- The homeowner's power supply can be at fault for some electrical problems.
- If the area experiences insect infestations, the motor starter contacts may occasionally need to be cleaned.

Also see the Appendix to Chapter 7 for control panel troubleshooting procedures.

1. Remove approximately ⅜″ of insulation from the end of each motor lead and wire to be connected.

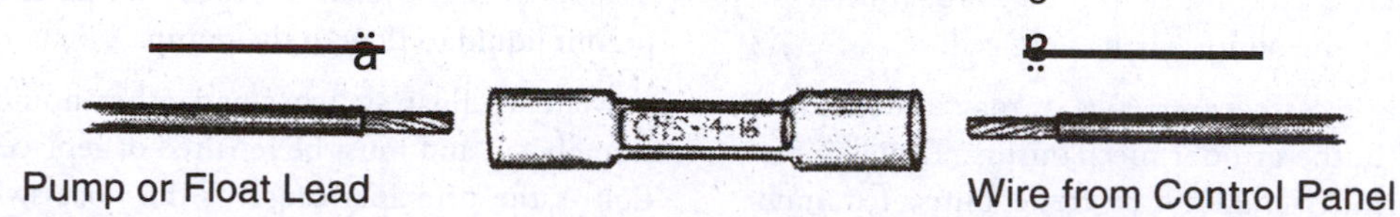


2. Insert bare lead ends into the butt connector and crimp with a crimping tool designed to crimp insulated connectors. Other types of tools can puncture the heat shrink tubing. Once a connection is crimped, tug on the butt connector to check the connection.

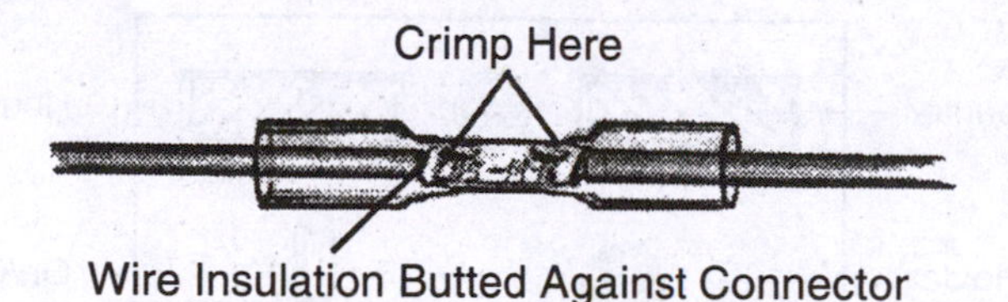


3. To shrink the insulated heat shrink tubing, apply moderate heat with a propane torch or heat gun (or any tool that will provide adequate heat). ***Caution: Keep the torch moving; too much concentrated heat will damage the tubing.***
4. When tubing begins to shrink, increase concentration of heat at the edge of the butt connector. As the tube collapses on the wire, work heat out to each end until entire tube has collapsed tightly around the wire. Enough heat should be applied to melt the sealing glue on the inside of the shrink tube. As the tube collapses around the wire, some sealant should ooze out of the end of the tube providing a watertight seal.

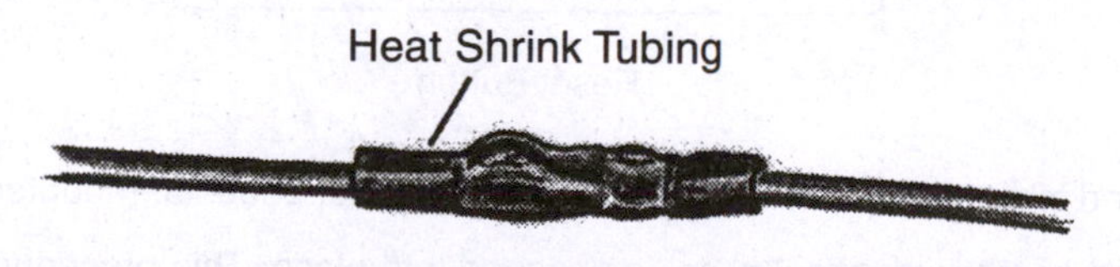


*Fig. 4.34 Splicing with butt connectors and heat shrink tubing*

(Courtesy of Orenco Systems, Inc., www.orenco.com)

### 4.363 *Grinder Pumps*

See Figures 4.14 and 4.15 (pages 142 to 144) for a checklist and start-up report used to ensure proper installation and start-up of grinder pump systems. This checklist and the report form also can be used to troubleshoot grinder pump installations.

Problems an operator may encounter with grinder pumps could relate to the associated electrical equipment, the pump itself, the pump vault, or the piping. Grinder pumps operate in a harsh environment and may require frequent overhaul or replacement. Keep an adequate supply of spare parts and replacement pumps on hand to minimize any inconvenience to customers.

Problems with junction boxes, electrical control panels, and tripped or thrown breakers are described in Section 4.362, "Electrical Equipment," and problems with the vault are described in Section 4.361, "STEP Systems: Septic Tanks and Pump Vaults."

High and low pump motor amperage readings are indications that a pump is not operating properly. Amperage can be measured at the pump control panel using the pair of pump power conductors exposed externally on the face of the panel (Figure 4.35). Each conductor lead has sufficient room between the panel face and the lead to attach a snap-on amprobe meter to measure current draw by the pump when energized.

If a pump is drawing excessive amperage, it may indicate that something is jammed in the grinder mechanisms. Troublesome materials found in grinders include cigarette butts, feminine hygiene products, condoms, rags, kitty litter, and similar items. Remove the obstruction and discuss the problem with the homeowner, emphasizing the importance of their cooperation in properly disposing of bulky or troublesome materials. If a pump cannot be cleared of an obstruction or is inoperative for some other reason, remove it and install a similar type spare pump to restore service. The inoperable pump can be taken to the shop to be repaired and used as a future standby pump. A high amperage reading or current draw may also indicate that some pump motor windings are burned out and require replacement. Usually heaters are provided that shut off the motor if the current is above a certain value.

A low amperage reading indicates the pump is drawing air (rather than water) and needs to be primed. This is accomplished by loosening the discharge hose just prior to the discharge isolation valve and releasing the trapped air or gas. Problems with air binding or gas buildup can be prevented by installing a pump air release valve above the pump discharge check valve in the discharge hose. A one-inch Morrison bleeder valve can be used. If the pump is installed in a screened basket strainer, a low amperage reading may indicate that the screen is plugged with solids. Clean the screen as described in Section 4.361, "STEP Systems: Septic Tanks and Pump Vaults," to permit liquid to flow to the pump.

Mercury float switches and other liquid level sensors occasionally fail and must be repaired or replaced. Grease accumulation is the principal cause of the float switches not operating

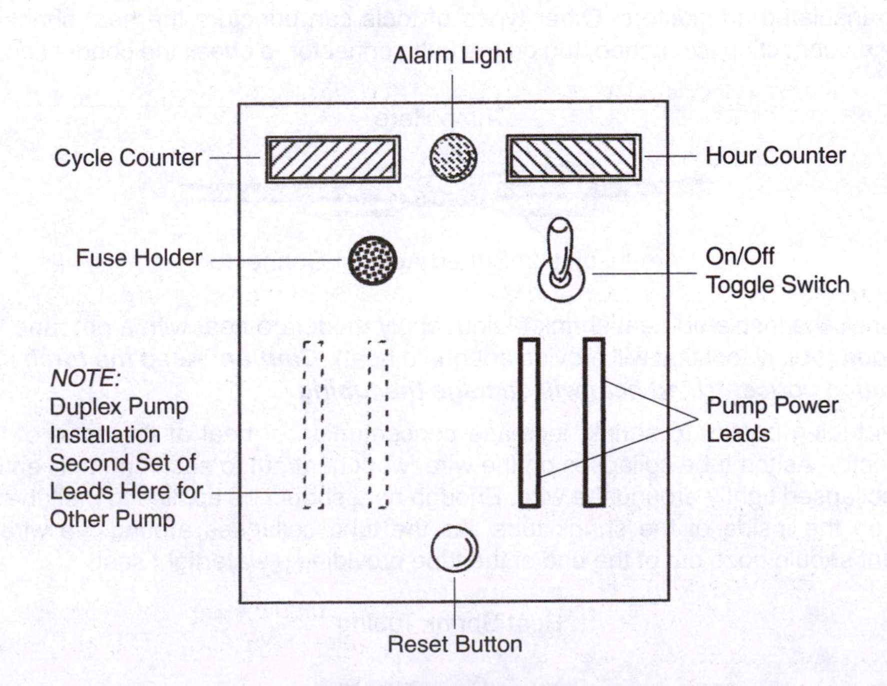


*NOTES:* 1. The panel is mounted in a weather-proof enclosure and is wall- or pedestal-mounted.

2. Panel snaps out and another identical panel is snapped into place. This procedure changes all control components. The removed panel is returned to the shop for repair.

*Fig. 4.35 Pump power panel*

properly. Remove the grease and ask the residents to dispose of cooking grease elsewhere. Other malfunctions include obstruction of the float switches; loose, cracked, stiff cords restricting movement of the float; and simply not working. Adjustment or replacement of pressure switches also is sometimes necessary.

## QUESTIONS

Please write your answers to the following questions and compare them with those on page 175.

4.3G What test can be used to confirm that the sludge blanket in a septic tank is too deep?

4.3H What are some possible sources of water intrusion into a pump vault?

4.3I What is the principal cause of float switches not working properly?

### 4.4 ACKNOWLEDGMENTS

We wish to acknowledge the references used to prepare the material in this chapter.

1. *ALTERNATIVE WASTEWATER COLLECTION SYSTEMS,* Technology Transfer Manual, US Environmental Protection Agency.
2. *MANILA COMMUNITY SERVICES DISTRICT, DESIGN-CONSTRUCTION REPORT FOR THE SEPTIC TANK EFFLUENT PUMPING SEWERAGE SYSTEM,* California State Water Resources Control Board.
3. *RURAL WASTEWATER MANAGEMENT,* California State Water Resources Control Board.
4. *HOMEOWNERS' AND USERS' GUIDE FOR ON-SITE WASTEWATER DISPOSAL SYSTEMS,* California State Water Resources Control Board.
5. *ORENCO SYSTEMS, INC., ENGINEERS' CATALOG,* Orenco Systems, Inc., 814 Airway Avenue, Sutherlin, OR 97479.
6. Glide-Idleyld Park Sewer District Forms and Publications, Douglas County Department of Public Works, Roseburg, OR 97479.

### 4.5 ADDITIONAL INFORMATION

Useful information on check valve selection is available on the APCO Willamette Valve and Primer Corporation website at www.apcovalves.com.

### 4.6 ARITHMETIC ASSIGNMENT

Turn to the Appendix, "How to Solve Wastewater System Arithmetic Problems," at the back of this manual and read Section A.4, "Metric System."

**END OF LESSON 2 OF 2 LESSONS**

**on**

**SEPTIC TANKS AND PUMPING SYSTEMS**

Please answer the discussion and review questions next.

## DISCUSSION AND REVIEW QUESTIONS

**Chapter 4. SEPTIC TANKS AND PUMPING SYSTEMS**

(Lesson 2 of 2 Lessons)

Please write your answers to the following questions to determine how well you understand the material in the lesson. The question numbering continues from Lesson 1.

6. Why are new building sewers and new septic tanks usually installed when a septic tank effluent pump (STEP) system is installed?
7. How does the size of a grinder pump vault affect the cost and efficiency of pump operation?
8. Why is a long retention time not desirable in a grinder pump vault?
9. Why is a grinder pump usually larger (more horsepower) than an effluent pump for a septic tank?
10. How is total dynamic head (TDH) measured?
11. What factors must be considered when calculating the required pump discharge head for a particular location?
12. What types of STEP system maintenance situations require that more than one operator be present?
13. What is the recommended location of the control panel for a septic tank or pump vault?
14. When would it be necessary to install two septic tank effluent pumps at the same location?
15. How can float switches be installed so that they move up or down freely but do not become tangled together?
16. Describe how an air pressure (back pressure) sensor controls pump operation.
17. Routine preventive maintenance of septic tanks and grinder pumps in a STEP system includes what tasks?
18. What are the main types of safety hazards operators of STEP systems could be exposed to?
19. What is an operator's best source of information about a particular STEP system?
20. What do high and low grinder pump motor amperage readings indicate?

# SUGGESTED ANSWERS

## Chapter 4. SEPTIC TANKS AND PUMPING SYSTEMS

### ANSWERS TO QUESTIONS IN LESSON 1

Answers to questions on page 126.

4.0A The two major parts of a basic individual on-site wastewater treatment system are the septic tank and the leach field.

4.0B Solids carried by wastewater need to be removed by a septic tank to prevent the solids from plugging the leach field distribution piping or sealing the soil (loss of absorption or percolation capacity) in the leach field.

4.0C A septic tank should be tested for watertightness at the manufacturing facility. Once the tank has been installed but before the excavation is backfilled and covered with soil, the tank should be tested again for inward or outward leakage and for level placement in the excavation.

4.0D Riser pipes should be installed over both the inlet and outlet manholes on a septic tank to eliminate the need to probe the ground surface to locate the buried septic tank or digging to uncover the manhole covers.

Answers to questions on page 130.

4.0E Septic tanks are designed to accept (1) wastewater and (2) solids carried by the wastewater.

4.0F The water load on a septic tank can be reduced by water conservation practices and by keeping rain or surface water runoff from entering the septic tank.

4.0G When inspecting a septic tank an operator may be exposed to the hazards associated with confined spaces, including hazardous gases and oxygen deficiency/enrichment as well as infections/infectious diseases.

4.0H Detention time in a septic tank can be reduced by too thick a scum layer on the surface or too thick a sludge blanket on the bottom.

Answers to questions on page 133.

4.0I A septic tank should be pumped to remove accumulated solids when the scum layer or the sludge blanket is approaching the opening of the outlet device.

4.0J Pipe blockages in the building sewer between a home and a septic tank can be caused by tree roots entering the pipe or by grease or detergent buildup.

4.0K Odor complaints from septic tank owners have often been traced to improperly located house vents drafting the gas downward over the house roof.

### ANSWERS TO QUESTIONS IN LESSON 2

Answers to questions on page 138.

4.1A The main purpose of a grinder pump is to reduce the size of solids entering the collection system. A grinder pump also provides pressure to move wastewater through the collection system.

4.2A Problems associated with older building sewers include broken pipe, offset joints, root intrusion, and improper grade.

4.2B New septic tank effluent pump (STEP) systems usually require abandonment of the older septic tanks to reduce the possibility of infiltration and to maintain similarity of components throughout the new system for more dependable operation and easier maintenance.

Answers to questions on page 141.

4.2C If a grinder pump vault is too large, it will take too long to fill to the level that the pump is called upon to start pumping. This long retention time allows the solids to separate (by settling or flotation) from the liquids. Energy in some form is required to resuspend the scum and settled solids in the vault, or the vault will need to be cleaned more often. Also, anaerobic decomposition of the solids may produce hydrogen sulfide gas, which can create odor and corrosion problems.

4.2D The three liquid level float sensors in grinder pump vaults turn the pump ON, turn the pump OFF, and turn on a high water level alarm.

4.2E A grinder pump control panel should be mounted on a pedestal or outside wall of the home, near the pump vault, and within plain view of the vault.

Answers to questions on page 150.

4.2F Turbine-type centrifugal pumps are less tolerant of solids than some other types of pumps. Screens or basket strainers are installed to prevent large solids or stringy material from entering the pump's suction, thus prolonging the life of the pumping unit and reducing the number of service calls.

4.2G Selection of an effluent pump depends primarily on the type of wastewater to be pumped, the total dynamic head (TDH) of the collection system, and the flow rate to be pumped.

4.2H High discharge heads are achieved by using multistage turbine pumps.

4.2I Piping connections, valves, and electrical junction boxes are located high in the septic tank riser so that there is less chance of them being submerged in wastewater if a power failure or pump failure occurs.

Answers to questions on page 158.

4.2J A check valve is installed on the discharge of a septic tank effluent pump to prevent the pumped effluent from draining back into the tank, raising the wastewater level, and causing the pump to start again.

4.2K A check valve is installed in the service line just prior to the service line connection to the collection system main to prevent the wastewater in the main line from flowing back through the service line.

4.2L A septic tank effluent pump discharge valve provides a positive cutoff between the effluent pump and the service line. It can be operated manually to prevent backflow into the septic tank and can be used to isolate the system for pump repair or replacement.

4.2M When new service lines must be added to a system, use of a pipe saddle permits connections to be made while the main remains under pressure. This type of connection is referred to as a "hot tap" because the line is under pressure.

Answers to questions on page 162.

4.2N The STEP power panel should be located within clear view of the septic tank or pump vault for the safety of the operator working on the system.

4.2O The two most common types of effluent level pump controls are float-activated mercury switch controls and air pressure (back pressure) sensor systems.

4.2P Pump discharge pressure readings may be obtained by installing a gauge or pressure recorder on the pump discharge inside the septic tank riser before the service line leaves the tank.

4.2Q Wastewater flow from a home can be measured by using a standard single service water meter installed in the wastewater service line.

Answers to questions on page 165.

4.3A Whenever an operator makes a service call, the liquid level in the tank, the sludge blanket depth, and the scum blanket depth should be checked.

4.3B The typical pumping interval for a residential septic tank is three to ten years; for a high-use facility, the typical interval is every three to six months.

4.3C During routine maintenance visits, the operator should wash down grinder pump vaults to remove grease, run the pump and controllers through their cycles, check motor amperage, and clean the motor starter contacts, if necessary.

4.3D Safety hazards an operator might encounter while maintaining a STEP system include electrical burns or shock, infections, diseases, and exposure to hazardous gases such as hydrogen sulfide.

4.3E Maintenance personnel should be prepared to respond to the following types of emergency calls: (1) high/low liquid level alarms, (2) odor complaints, and (3) extended power outages.

4.3F An operator could find information about the operation and maintenance of STEP systems in the plant O&M manual, from equipment suppliers, and from other operators of these systems.

Answers to questions on page 173.

4.3G To confirm that the sludge blanket is too deep, collect a small sample, 300 to 1,000 milliliters (⅓ to 1 quart), of the tank effluent wastewater in a clear glass jar, beaker, or Imhoff cone. Let the sample settle for at least 30 minutes. If a black sludge settles to the bottom, this indicates that the sludge blanket is too deep and the tank's septage (sludge solids) needs to be pumped and disposed of at an approved disposal site.

4.3H Possible sources of water intrusion into a pump vault include roof drains discharging too close to the pump vault, careless irrigation practices, and improper vault installation that causes water to puddle around and over the vault cover.

4.3I Grease accumulations are the principal cause of float switches not working properly.

# APPENDIX A

Legal Considerations

## Easement Provisions, Rights, and Conditions

Any utility that provides a community service must obtain an easement from each property owner where a portion of the system may lie upon, cross through, or where maintenance personnel may have to enter upon those premises. An example of an easement agreement from an existing operating agency is shown in Figure 4.36.

## License to Enter upon Property (Figure 4.37)

A license to enter upon property is primarily an agreement between adjacent property owners where one service line serves two or more homes. In these instances each home is equipped with a septic tank (interceptor) or grinder pump vault, but the discharge from each residence's pumping system uses a common service line to convey the wastewater to the collection system. Often, one line must cross the adjoining property to tie into the service line leading to the main. This agreement allows one or the other property owner to perform work on the other's property to connect to the service line.

## Connection Permit (Figure 4.38)

A connection permit can serve several purposes. By signing the permit, the homeowner agrees (and is legally bound) to pay the charges or fees assessed by the utility for its services. The permit also gives the utility permission to install the facilities required to provide service. Besides giving the utility these legal authorities, a connection permit system also enables the utility to maintain a complete listing of all connections to the system. Complete and up-to-date records greatly simplify the task of locating problems if any should develop within the system.

## Installation Specifications (Installation Guide)

An installation guide for the property owner is often issued with the connection permit. Where the district or sewer authority installs the necessary components, the only specifications supplied to the property owner are those required for the construction of the building sewer. When the property owner also has the responsibility for purchasing and installing the septic tank and associated components, these additional installation requirements and specifications may be provided:

1. Safety
2. Responsibilities of the property owner and the installer or contractors
3. Tank installation
4. Service line installation
5. Electrical/mechanical equipment
6. Special conditions that the district may require

## The Operator's Role

In most instances, the system's wastewater treatment plant operator is assigned the responsibility of scheduling, overseeing, and inspecting the installation of each individual's pressure sewer system. This provides the operator with direct knowledge of the property owner and the installed system, and ensures that all work was properly performed to the district's standards. Although the operator has the authority to enforce the district's requirements, it is important to remember that operators only have permission to perform certain tasks on private property.

In performance of their duties, operators must take care not to disturb or damage anything on that property. Operators do not own the property and only have the right to attend to the components that are associated with the wastewater system. Close gates so that animals may not leave or enter the property. If some form of extra maintenance work is required, notify the property owner if at all possible. If work must proceed for sanitary or safety reasons before the owner can be notified, take all reasonable means to protect the property. For example, if excavation is necessary to repair pipe or electrical systems, cut sod or remove shrubs and plants with minimal damage so they may be replaced when the work is completed. Place excavated soil on ground cloths or tarps during repairs. This provides a neat, easy way to replace the backfill, protects lawn areas, walks, and swimming pool decks from soil staining, and leaves a nice clean job when completed. When sod or plants are replaced, attempt to restore them to their original location and elevation and add a little water from the hose to promote their rapid recovery. The property owner will appreciate your thoughtfulness and will respect the operator and the district for the professional manner in which you perform your duties.

GLIDE-IDLEYLD PARK SEWERAGE SYSTEM EASEMENT

In consideration of the prospective benefits to be derived from the locating, constructing, and maintaining a pressure sanitary sewer by Douglas County, the undersigned hereby grants to Douglas County, a political subdivision of the State of Oregon, an easement for the purpose of constructing, installing, maintaining, and inspecting pressure sewer service lines, interceptor tanks, sump pumps, pump vaults, and facilities incidental thereto on the following described property in Douglas County, Oregon;

Part of Section ______________, Township ______________, Range ______________, Willamette Meridian, as more particularly described in that instrument recorded at Volume ______________, page ______________; deed records, Douglas County, Oregon.

For the installation, construction, maintenance, or inspection of service lines, interceptor tanks, sump pumps, pump vaults, or facilities incidental thereto that serve the undersigned's property:

A. The undersigned releases Douglas County from any and all claims necessarily incident to such installation, construction, maintenance, or inspection and is responsible for repair and maintenance of the sewer line between the interceptor tank and the structure being served.

B. Douglas County will be responsible for routine maintenance and inspection but the undersigned shall pay for repairs caused by abuse or misuse of the system.

Douglas County may use roads upon the above described property for access for all purposes mentioned herein, if such roads exist, otherwise by such route as shall occasion the least damage and inconvenience to the undersigned. The undersigned shall not erect any structure or excavate or substantially add to or diminish the ground cover within 10 feet of any septic/interceptor tank or pump vault facilities or within 5 feet of any service lines installed by Douglas County.

The rights, conditions, and provisions of the easement shall insure to the benefit of and be binding upon the heirs, successors, and assigns of the parties hereto.

DATED this ______________ day of ______________, 20 ______.

______________________________________________

STATE OF OREGON )
)ss
County of Douglas ) (Date) ______________________

Personally appeared the above named ______________________

______________________________________________

and acknowledged the foregoing instrument to be their voluntary act and deed.

______________________

Notary Public for Oregon
My Commission Expires: ______________________

STATE OF OREGON )
)ss
County of Douglas ) (Date) ______________________

Personally appeared the above named ______________________

______________________________________________

and acknowledged the foregoing instrument to be their voluntary act and deed.

______________________

Notary Public for Oregon
My Commission Expires: ______________________

*Fig. 4.36 Sewerage system easement*

(Permission of Oregon/Glide-Idleyld Park Sewer District)

## LICENSE TO ENTER UPON PROPERTY

LICENSE AGREEMENT made on ______________________________ by and between ________________________________________________, herein referred to as Owner, and ______________________________________________________, herein referred to as Licensee.

IT IS HEREBY AGREED BETWEEN THE PARTIES:

1. GRANT: Owner hereby grants to Licensee the right to enter upon Owner's property to install and connect a sewer line to the Glide-Idleyld Park Sewerage System that is owned and operated by Douglas County. Owner's property is located in Section ___________, Township ___________, Range ___________, Willamette Meridian, and is described in instrument no. ___________ which is recorded in Volume ___________ at page ___________, deed records of Douglas County, Oregon. The sewer line will serve Licensee's property located in Section ___________, Township ___________, Range ___________, Willamette Meridian, and described in instrument no. ___________ which is recorded in Volume ___________ at page ___________, deed records of Douglas County, Oregon.

The granting of this license does not give Licensee any interest in Owner's property.

2. INDEMNIFICATION: Licensee shall save, defend, and hold Owner harmless from any claim, suit, or action whatsoever for damage to property, or injury or death to any person or persons due to the negligence of the Licensee, Licensee's agents, employees, or independent contractors, arising out of the performance of any work or project covered by this permit.

3. REVOCATION: Both parties acknowledge and expressly agree that this license is a personal privilege of Licensee to enter upon the property of Owner and this license is revocable at the will of Owner by written notice to Licensee. However, Owner may not revoke this license if Licensee has started work in reliance on the license and is not in default under the terms of this license.

4. ASSIGNMENTS: The rights created by this license shall not be assigned, nor shall the right to enter upon Owner's property be conferred by Licensee on any third person (except agents, employees, or independent contractors engaged by Licensee to perform the work covered by this license on behalf of Licensee) without the prior written consent of Owner. Any attempt to assign or transfer this license shall be a nullity and shall result in immediate cancellation of the license.

5. PERFORMANCE OF WORK: All work covered by this license shall be completed by ______________________________. The work shall be performed by Licensee, or agents, employees, or independent contractors engaged by Licensee, in accordance with plans and specifications approved by Douglas County. The sewer line shall be located as specified in the diagram attached hereto as Exhibit A and incorporated herein. After the sewer line is installed, Licensee shall restore Owner's property to the same condition it was prior to performance of the work. If Owner's property cannot be completely restored, Licensee shall compensate Owner for damage that is incident to performance of the work.

*Fig. 4.37 License to enter upon property*

(Permission of Oregon/Glide-Idleyld Park Sewer District)

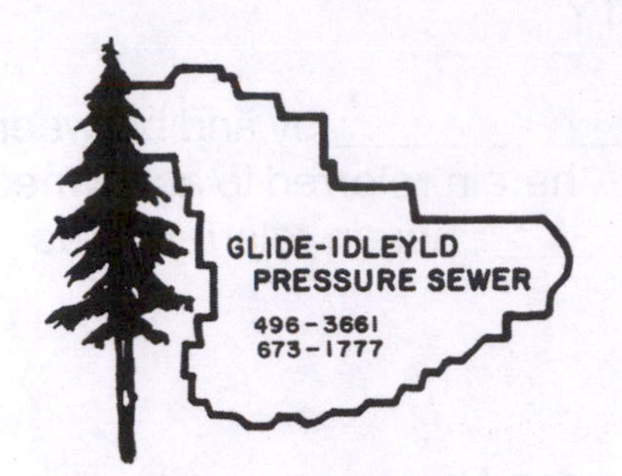


DOUGLAS COUNTY ENGINEERING OFFICE
ROOM 219
DOUGLAS COUNTY COURTHOUSE
ROSEBURG, OREGON 97470
PHONE: 440-4481

# CONNECTION PERMIT

No. ______

PROPERTY OWNER (S) ______

MAILING ADDRESS ______ PHONE ______

STREET ADDRESS ______

TAX ACCOUNT NO. ______

CONNECTION CLASS A B C ______

REMARKS ______

| CONNECTION FEE | AMOUNT | |
|---|---|---|
| SYSTEM COSTS | | ☐ EASEMENT<br>☐ INSTALLATION GUIDE<br>☐ SYSTEM USER MANUAL |
| EQUIVALENT DWELLING UNITS<br>UNITS ______ X FEE ______ | $ ______ | BUILDING / PLACEMENT PERMIT NO.<br># ______ |
| ☐ PRIOR ASSESSMENT | | ______<br>ISSUED BY: |
| ______<br>SIGNATURE DATE | | **NOTICE**<br>I HEREBY CERTIFY THAT I HAVE READ AND EXAMINED THIS FORM AND KNOW THE SAME TO BE TRUE AND CORRECT. ALL PROVISIONS OF LAWS AND ORDINANCES GOVERNING THIS INSTALLATION WILL BE COMPLIED WITH. THE GRANTING OF A PERMIT DOES NOT PRESUME TO GIVE AUTHORITY TO VIOLATE OR CANCEL THE PROVISIONS OF ANY OTHER STATE OR LOCAL LAW REGULATING CONSTRUCTION OF THE INSTALLATION. |
| HOOK UP COSTS | | |
| COMPONENT PACKAGES<br>TYPE ______ UNITS ______ X FEE ______ | $ ______ | |
| TYPE ______ UNITS ______ X FEE ______ | $ ______ | |
| SPECIAL CHARGES<br>SERVICE LINE FEET ______ X FEE ______ | $ ______ | |
| ELECTRICAL LINE FEET ______ X FEE ______ | $ ______ | ______<br>PROPERTY OWNER DATE |
| HOOK UP DATE ______ | | |
| CONNECTION FEE TOTAL | $ | |

*Fig. 4.38 Connection permit*

(Permission of Oregon/Glide-Idleyld Park Sewer District)

# APPENDIX B

Do's and Don'ts for Septic Tank Systems

(Excerpt from *Homeowner's Manual,* courtesy of Orenco Systems®, Inc.)

## Do's and Don'ts for INSIDE the House

There are a number of do's and don'ts that will help ensure a long life and minimal maintenance for your system. As a general rule, nothing should be disposed into any wastewater system that hasn't first been ingested, other than toilet tissue, mild detergents, and wash water. Here are some additional guidelines.

**Don't** flush dangerous and damaging substances into your wastewater treatment system. (Please refer to the "Substitutes for Household Hazardous Waste," on the next panel.) Specifically, do not flush . . .

- Pharmaceuticals
- Excessive amounts of bath or body oils
- Water softener backwash
- Flammable or toxic products
- Household cleaners, especially floor wax and rug cleaners
- Chlorine bleach, chlorides, and pool or spa products
- Pesticides, herbicides, agricultural chemicals, or fertilizers

**Don't** ignore leaky plumbing fixtures; repair them. A leaky toilet can waste up to 2,000 gallons (7500 liters) of water in a single day. That's 10-20 times more water than a household's typical daily usage. Leaky plumbing fixtures increase your water bill, waste natural resources, and overload your system.

**Don't** leave interior faucets on to protect water lines during cold spells. A running faucet can easily increase your wastewater flow by 1,000 to 3,000 gallons (4,000 to 12,000 liters) per day and hydraulically overload your system. Instead, properly insulate or heat your faucets and plumbing.

**Don't** use special additives that are touted to enhance the performance of your tank or system. Additives can cause major damage to other areas in the collection system. The natural microorganisms that grow in your system generate their own enzymes that are sufficient for breaking down and digesting nutrients in the wastewater.

**Do** collect grease in a container and dispose with your trash. And avoid using garbage disposals excessively. Compost scraps or dispose with your trash, also. Food by-products accelerate the need for septage pumping and increase maintenance.

**Do** keep lint out of your wastewater treatment system by cleaning the lint filters on your washing machine and dryer before every load. Installing a supplemental lint filter on your washing machine would be a good precautionary measure. (This normally takes just a few minutes. Lint and other such materials can make a big difference in the frequency and cost of pumping out your primary treatment tank.)

**Do** use your trash can to dispose of substances that cause maintenance problems and/or increase the need for septage pumping. Dispose of the following with your trash:

- Egg shells, cantaloupe seeds, gum, coffee grounds
- Tea bags, chewing tobacco, cigarette butts
- Condoms, dental floss, sanitary napkins, diapers
- Paper towels, newspapers, candy wrappers
- Rags, large amounts of hair
- "Flushable" wipes, baby wipes, medicated wipes, cleaning wipes

"FLUSHABLE" WIPES ARE
NOT
FLUSHABLE.
DISPOSE IN TRASH.


**DON'T** plumb water softener discharge brine into your wastewater system. (The softened WATER is OK, just not the BRINE that's produced during the regeneration cycle.)

**DO** route the brine around your wastewater system so it discharges directly into the soil. This is a cost-effective solution that ensures the long-term performance of your system and the biological processes that occur inside it.

Water softener brine interferes with nitrogen removal. And it degrades treatment by interfering with the settling process inside the tank. Without proper settling, solids, grease, and oils are carried through your system, clogging components. This increases your costs by...

- requiring the tank to be pumped more often (at hundreds of dollars per pumpout)
- requiring filters to be cleaned more often
- fouling drainfields and other downstream equipment

(Excerpt from *Homeowner's Manual,* courtesy of Orenco Systems®, Inc.)

## Do's and Don'ts for INSIDE the House

**Don't** use excessive amounts of water. Using 50 gallons (200 liters) per person per day is typical. If your household does not practice any of the "water conserving tips" below, you may be using too much water.

**Do** conserve water:

- Take shorter showers or take baths with a partially filled tub. Be cautious about excessive use of large soaking tubs.
- Don't let water run unnecessarily while brushing teeth or washing hands, food, dishes, etc.
- Wash dishes and clothes when you have a full load.
- When possible, avoid doing several loads in one day.
- Use water-saving devices on faucets and showerheads.
- When replacing old toilets, buy low-flush models.

**Do** use substitutes for household hazardous waste. Replace the following hazardous products with products that are less environmentally harmful. The hazardous cleaners are listed below, followed by the suggested substitute.

**Ammonia-based cleaners:**

- For surfaces, sprinkle baking soda on a damp sponge.
- Or for windows, use a solution of 2 tbs (30 mL) white vinegar to 1 qt (1 L) water. Pour the mixture into a spray bottle.

**Disinfectants:**
Use borax: 1/2 cup (100 g) in a gallon (4 L) of water; deodorizes also.

**Drain decloggers:**
Use a plunger or metal snake, or remove and clean trap.

**Scouring cleaners & powders:**
Sprinkle baking soda on a damp sponge or add 4 tbs(50 g) baking soda to 1 qt (1 L) warm water. Or use Bon Ami® cleanser; it's cheaper and won't scratch.

**Carpet/upholstery cleaners:**
Sprinkle dry cornstarch or baking soda on, then vacuum. For tougher stains, blot with white vinegar in soapy water.

**Toilet cleaners:**
Sprinkle on baking soda or Bon Ami; then scrub with a toilet brush.

**Furniture/floor polishes:**
To clean, use oil soap and warm water. Dry with soft cloth. Polish with 1 part lemon juice and 2 parts oil (any kind), or use natural products with lemon oil or beeswax in mineral oil.

**Metal cleaners:**

- Brass and copper: scrub with a used half of lemon dipped in salt.
- Stainless steel: use scouring pad and soapy water.
- Silver: rub gently with toothpaste and soft wet cloth.

**Oven cleaners:**
Quickly sprinkle salt on drips; then scrub. Use baking soda and scouring pads on older spills.

**Laundry detergents:**
Choose a liquid detergent (not a powder) that doesn't have chlorine or phosphates.

## At the Control Panel

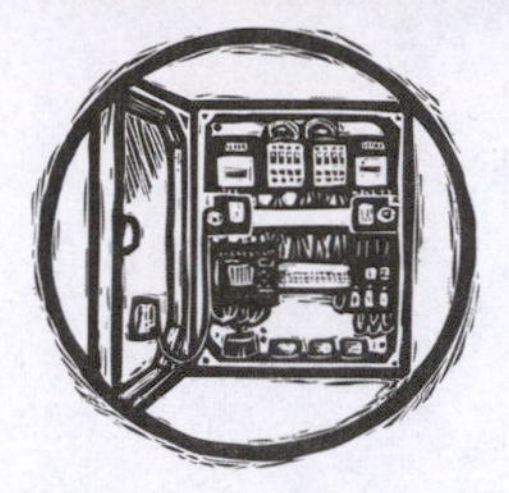

**Do** locate your electrical control panel where it will be protected from potential vandalism and have unobstructed access.

**Do** familiarize yourself with the location of your wastewater system and electrical control panel. Refer to the panel's model and UL number (inside the door panel) when reporting a malfunction in the system.

**Do** take immediate action to correct the problem in the event of an alarm condition. Call your system operator or maintenance company immediately whenever an alarm comes on. (It sounds like a smoke alarm.)

**Do** remember that the audible alarm can be silenced by pushing the lighted button located directly above the "Push to Silence" label on the front of the electrical control panel. With normal use, the tank has a reserve storage capacity good for 24-48 hours.

**Don't** turn off the main circuit breaker to the wastewater pumps when going on vacation. If there is any infiltration or inflow into the system, the pumps will need to handle it.

(Excerpt from *Homeowner's Manual,* courtesy of Orenco Systems®, Inc.)

## Do's and Don'ts for OUTSIDE the House

### IMPORTANT! CAUTION!

Only a qualified electrician or authorized installer/operator should work on your control panel. Before anyone does any work on either the wiring to the level control floats and pumps in the vault or on the control panel itself, it is imperative to first switch the isolation fuse/breaker and the circuit breakers in the panel to the "Off" positions, then switch "Off" the power to the system at the main breaker!

**Don't** enter your tank. Entering an underground tank without the necessary confined space entry training and procedures can result in death from asphyxiation or drowning. Keep children away from tank openings if lids are off or lid bolts are removed.

**Do** keep the tank access lid fastened to the riser at all times with stainless steel lid bolts. If the lid or riser becomes damaged, BLOCK ACCESS TO THE TANK OPENING, IMMEDIATELY. Then call your service provider to repair it. If you or your service provider needs replacement bolts, call Orenco at 800-348-9843 or 541-459-4449.

**Don't** dig without knowing the location of your wastewater system. As much as possible, plan landscaping and permanent outdoor structures before installation. But easily removable items, such as bird baths and picnic tables, are OK to place on top of your system.

**Don't** drive over your tank or any buried components in your system, unless it's been equipped with a special traffic lid. If the system is subject to possible traffic, put up a barricade or a row of shrubs.

**Don't** dump RV waste into your wastewater system. It will increase the frequency of required septage pumping. When dumped directly into the pumping vault, RV waste clogs or fouls equipment, causing undue maintenance and repair costs. (Also, some RV waste may contain chemicals that are toxic or that may retard the biological digestion occurring within the tank.)

**Don't** ever connect rain gutters or storm drains to the sewer or allow surface water to drain into it. And don't discharge hot-tub water into your system. The additional water will increase costs, reduce the capacity of the collection and treatment systems, and flood the drainfield. It can also wash excess solids through the tank.

**Do** make arrangements with a reliable service person to provide regular monitoring and maintenance. Place the service person's phone number on or in your control panel!

**Do** keep a file copy of your service provider's sludge and scum monitoring report and pumpout schedule. This information will be beneficial for real estate transactions or regulatory visits.

**Do** keep an "as built" system diagram in a safe place for reference.

(Excerpt from *Homeowner's Manual,* courtesy of Orenco Systems®, Inc.)

# APPENDIX C

The Operation and Maintenance (O&M) Manual

## WHAT IS AN O&M MANUAL?

An operation and maintenance (O&M) manual is a book, binder, or manual that describes the detailed procedures for operators to follow to operate and maintain a specific wastewater collection, treatment, or pretreatment facility and the equipment of that facility. The engineer who designs the system usually provides an O&M manual as part of the contracted services.

The typical O&M manual also contains a listing of system components and services and typical drawings detailing the design and construction of each component. In addition, the manual should contain a comprehensive maintenance log to document all maintenance performed and any performance problems and the corrective actions taken.

If the O&M manual is kept up to date as the system expands or is modified, it can be an operator's most valuable maintenance guide. Keep in mind, however, that even the designer cannot foresee every detail that will be required to keep the system functioning. If you do not have an O&M manual you should produce your own, whether yours is a new system or one that has been in operation for some time.

## CONTENTS OF AN O&M MANUAL

A good O&M manual should contain, at a minimum, the documents and information described in the following paragraphs.

### Identification of Customers and Flow Inventory

You or your agency will require a listing of your connected customers. You will wish to know the type of each connected customer to determine billing rates, volumes of flows, type of wastewater generated by that customer, and the location of that customer in the service system. Therefore, you will require the following information:

1. Street address, or county assessor's parcel number. If you use the assessor's parcel number you will need a set of the assessor's maps to identify and locate various parcels of property.
2. Type of service: Single family residence

   Multiple family (duplex, apartment building, condominium)

   Commercial (store, restaurant)

   Industrial
3. Volume of flow generated, if not single family residence, in gallons per day.
4. If commercial or industrial type of wastewater, include a special notation if excessive amounts of solids, oils, or greases might be discharged, and list the possible toxic substances that may be discharged with the wastewater. For example, a plating company manufacturing electronic circuit boards could discharge volumes of acids, alkaline solutions, toxic heavy metals, and toxic/explosive cleaning solvents or degreasers. All of these constituents could and often do require some type of pretreatment before being discharged to a wastewater collection system. Any one of these constituents would be especially harmful to the small wastewater treatment or discharge system you are operating.

### System Drawings

No matter how small the system, someone had to map it out to have it built. For four or five homes, this may be a simple freehand sketch. In most instances, a civil or sanitary engineer will design the system. Along with the O&M manual, the designer should provide the agency (you) with a complete set of prints, known as record prints or as-built prints. Two sets of prints are desirable, one to put away in a safe file in case the other set that you will use becomes lost or damaged.

Always keep a clean set of prints up to date. If valves, new services, or any new segments of pipe are added to the system, it is to your benefit to add the changes to the as-built prints. You will often be called upon to locate a portion of your system when other utilities need to repair their system or install a new portion of their system. Accurate records will save you both time and the embarrassment of not being able to locate your buried equipment quickly. Similarly, if a contractor accidentally cuts your pressure main and floods the neighborhood with raw wastewater, you can reduce the volume of the spill provided you can quickly locate the proper isolation valves on the main to stop the flow.

On the as-built prints, the *PLAN*[39] view of the collection system will show the routing of the collection system main and the connection of the home service lines or laterals. The system will be laid out with detailed measurements beginning at the starting point, normally the discharge end of the collection system or pipe. The start point is labeled "station 0.00 feet."

39. *Plan* or *Plan View.* A drawing or photo showing the top view of sewers, manholes, streets, or structures.

The next station or control point is generally 100 feet upstream on the pipe route. The measurements are accurate to one one-hundredth (1/100th) of a foot. If you were looking at a collection system plan, and an air release valve and vault was indicated to be at station 4 + 25.26, this tells you that the air release valve is located 425.26 feet upstream from the end of the pressure main. Every valve, service connection, cleanout, branch sewer, or other component should be indicated on the print and its exact location specified in feet measured from the discharge end of the main to the location of the component.

The *PROFILE*[40] view of the system indicates the depth at which the collection system is laid. In STEP systems, the pressure main follows the surface contour, generally thirty inches below grade (ground surface), so a profile print is not always necessary. However, to prevent possible future interference with other utilities, the depth of the main through river crossings or other necessary grade changes should also be noted on the as-built drawings or plan. Remember, your system includes the septic tank or interceptor so be sure that the tanks and associated components are also included on the as-built prints or a customer's plot plan (Figure 4.39).

"As built" means that not all connections, valves, or other components are always installed where the design engineer anticipated and indicated on the original plans. "As-built plans" are plans corrected to show any changes from the design plans. Therefore, when the contractor or you install or add components, or change pipe size, the inspector or you should mark the exact location on the plan. You may not be capable of measuring to a 1/100 (0.01) of a foot, but you should be able to come within one foot.

**Description of the System**

A description of the system and each of its components should be provided. The component descriptions should include the function of each component, its relation to adjacent components, and typical performance characteristics.

***Septic Tanks or Interceptors***

Your system records should include a data sheet for each type of septic tank that is currently installed or is approved for installation. This data sheet should include the specifications for construction, size, configuration, operating volume, tank limits for sludge deposit and scum blanket layers, and cost. Also, list the supplier's name and address and the time required to order and receive delivery at the site. The same information should be gathered if multiple tanks, separate pump vaults, or duplex pump vaults are installed. The type of unit or units that are installed at each location should be indicated as described above in "Identification of Customers and Flow Inventory."

***Types and Sizes of Effluent Pump Systems***

Many STEP pressure systems use a variety of types and sizes of pumps to meet site-specific needs. For each type of pumping package that is used in the system, prepare data sheets on each component. The data should include type of strainer pump vault, if used; type, size, and manufacturer of the pumping unit; discharge check valve style, size, and manufacturer; and the type of discharge hose or pipe, quick pipe coupling disconnect, and size and type of isolation valve. Also, record the level control method, set points, and wiring arrangements. For example, indicate such items as heat shrink insulation at the splice both for level control and pump power circuits, type of control panel and location or mounting arrangement, and power requirement. This sounds like a lot of information to record, but usually not more than three or four types of pumps are used in a system, and the same components are normally used for any particular type or size of pump. Two or three pages will carry all the data and information for each of the pumping systems.

It is sometimes possible to designate a particular pumping system type using a single number or letter. For example, a Peabody Barnes pump may require a certain screen vault of a given perforation dimension and size along with a standard flexible hose with quick disconnect, and a certain check valve and isolation valve, using a float control with a standard low-voltage power panel with a standard alarm system. This configuration could be called a type (1) pumping system. Similarly, a deep well, multistage turbine type pump may use a different dimension strainer vault and discharge or level control system, but all the turbine pumps of this size in the system should use the same components. Therefore, the multistage turbine pump could be designated a type (2) pumping system. For each type of pumping system that your district uses, you will need to record the replacement cost, where to obtain it, and how long it takes for delivery. If delivery takes a week and the system has thirty-eight similar units in operation, you may want a spare unit of pump, control panel, or level control system in the shop for emergency replacement.

40. *Profile.* A drawing showing elevation plotted against distance, such as the vertical section or side view of sewers, manholes, or a pipeline.

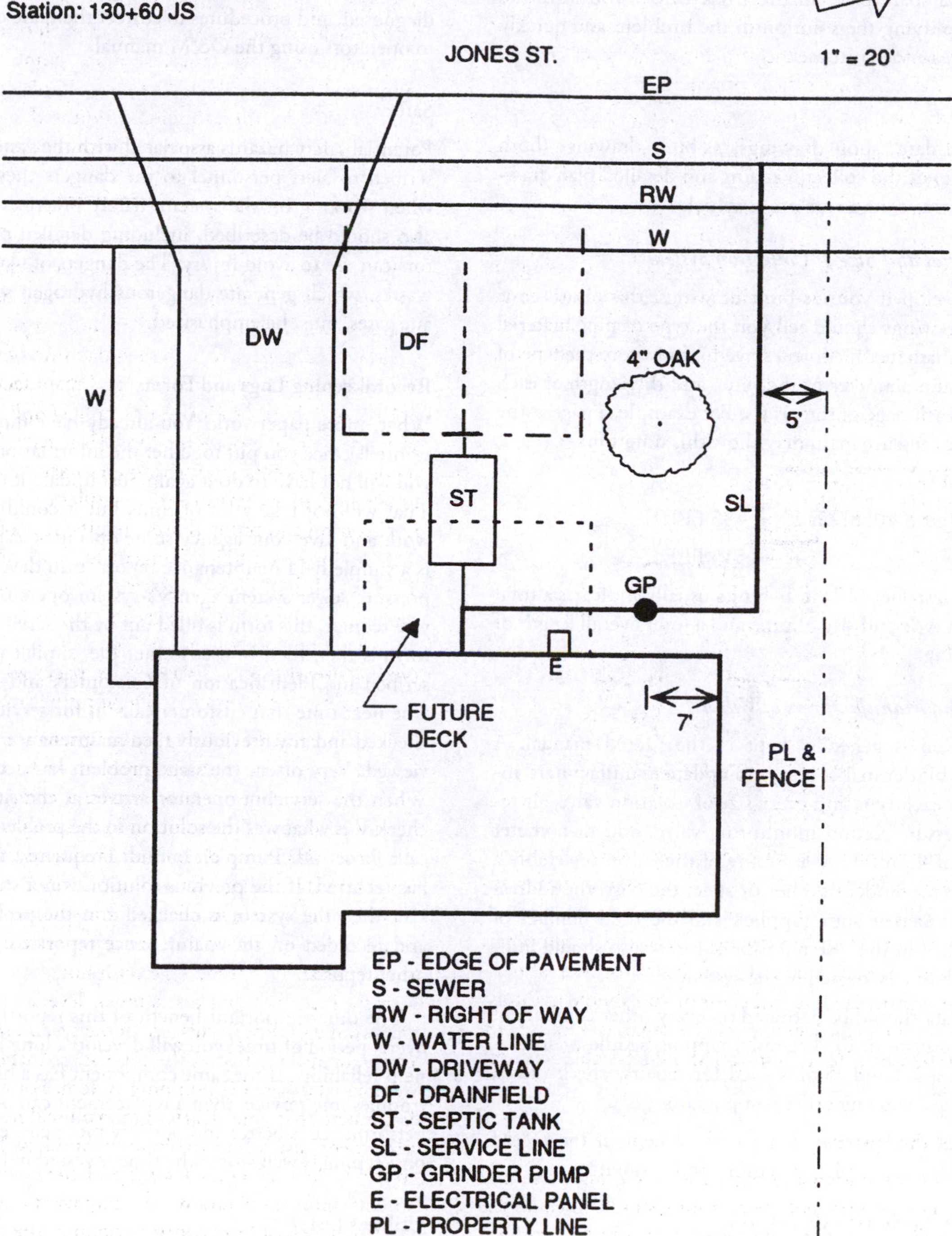

*Fig. 4.39 Typical lot facility plan*

(Source: *ALTERNATIVE WASTEWATER COLLECTION SYSTEMS,* Technology Transfer Manual, US Environmental Protection Agency)

Again, the customer service listing should also indicate the pumping unit that is installed at that location, such as a type (2) or a type (1). If you have this information and a customer calls to report that they have an alarm condition and no service, a quick check of your records will tell you what you will be looking at when you arrive at the site. You may wish to put a couple of the indicated spare parts in the truck before you leave the shop, thus simplifying the solution to the problem and quickly restoring service to the customer.

#### *Collection System*

Specific design data, shop drawings, as-built drawings (both plan and profile) of the collector mains and detailed plan drawings of each service connection are essential.

#### *Pipeline Material and Size of Collection System*

If you have developed your as-built drawings, this task is easy. The as-built drawings should tell you the type of pipe material and size (pipe diameters). All you have to do is note the type of pipe installed, the diameter of the pipe, and the length of each size shown on each page of the prints. For example, a page at the head end of the pressure main may show this notation:

SHEET 6 OF 11;

PVC, SCHEDULE 40, SIZE 2″ = 485 FEET

4″ = 515 FEET

The first or last sheet of the as-builts usually indicates a total footage of each type and size of pipe and a total overall length of the system piping.

#### *Sizes, Types, and Manufacturers of Valves*

This information is generally kept in the O&M manual. A loose-leaf style binder makes it easy to update manufacturers' information. For each type and every size of isolation valve, air release valve, pressure station monitoring valve, and manometer tap, the manual should have a picture of the valve, preferably a cutaway, the exact model number or style, the cost, the address and phone number of your supplier, and the total number of each size installed in the system. Your as-builts also should indicate where each one is located in the system.

The same data should be gathered on every other component installed in the system; for example, tapping saddles, corporation cocks if used, band clamps used for repairs, check valves, dresser couplings, and any other components.

Other useful information that could be kept in the O&M manual includes a spare parts inventory and a tool list.

### Description of the System Operation

Normal operation, emergency operation situations and procedures, and fail-safe features should be described.

### System Testing, Inspection, and Monitoring

The purpose, methods, and schedule of all recommended testing, inspections, and monitoring should be described. Sample recording forms should also be included.

### Prevention Maintenance Procedures and Schedules

A clear description of all preventive maintenance procedures is needed with specific schedules for their performance.

### Troubleshooting

Descriptions of common operating problems, how they may be diagnosed, and procedures to correct them are extremely helpful to operators using the O&M manual.

### Safety

Potential safety hazards associated with the system should be described to alert personnel to the dangers they may encounter when working on the system. Safety practices and precautions also should be described, including detailed procedures operators can use to avoid injury. The dangers of working with septic wastes, which generate dangerous hydrogen sulfide and methane gases, must be emphasized.

### Recordkeeping Logs and Forms

What, more paperwork? You already have more than you ever wanted! Once you put together the information described above you will not have to do it again, just update it to keep it current. That will not take a lot of time, but it could save you a lot of work and save your agency some operating capital. Figure 4.40 is a simple field maintenance report form developed by a STEP pressure sewer system agency's system operators. On every service request, this form is filled out by the servicing operator. The form is then filed in a customer file, similar to the system described in "Identification of Customers and Flow Inventory." The next time that customer calls in for service, the file can be checked and any previously filed maintenance reports can be reviewed. Very often, the same problem has occurred previously. When the servicing operator arrives at the site, the first thing checked is what was the solution to the problem on the previous call? Power off? Pump air bound? Frequently, the same problem has recurred. If the previous solution is not the remedy for this call, then the system is checked and the problem is corrected and recorded on the maintenance report to be filed with the other reports.

The other important benefit of this reporting system is that, over a period of time, you will develop a long history of component reliability. If the same component has a history of failure or troublesome service, then a replacement can be tried to see if its performance is better and possibly more cost effective. This recordkeeping is well worth the time invested in it.

### Utilities List

A list of all other utilities (water, phone, electricity, gas, oil) in the district and a list of contact names, addresses, and phone numbers should be included in the O&M manual. Also, the O&M manual should contain a list of contact names, addresses, and phone numbers for regulatory agencies (health department, water pollution control, safety, or OSHA) and police, fire, hospital, ambulance, and physician.

PERSON CALLING:________________ MAINTAINED BY:________________
DATE CALLED/TIME:________________ DATE RESPONDED/TIME:________________
ADDRESS:________________ FIELD TIME:________________
TELEPHONE:________________ TRAVEL TIME:________________

SYSTEM
☐ SIMPLEX
☐ MULTIPLE
☐ GRAVITY

VAULT
☐ SHALLOW
☐ DEEP
☐ W/SCREEN
☐ W/O SCREEN

PUMP / MODEL
☐ EFFLUENT ______
☐ HIGH HEAD ______
☐ GRINDER ______

CONTROLS / MODEL
☐ PANEL ______
☐ LEVEL ______

METER READINGS:

| | DATE | COUNT | TIME |
|---|---|---|---|
| RESPONSE: | ______ | ______ | ______ |
| PREVIOUS: | ______ | ______ | ______ |

DAYS: ______ CYCLES: ______ ELAPSED TIME: ______
FREQUENCY: ______ CYCLES/DAYS ☐ NORMAL ☐ HIGH ☐ LOW
DURATION: ______ MINUTES/CYCLE ☐ NORMAL ☐ HIGH ☐ LOW

CONDITIONS LEADING TO CALL:
☐ ALARM ☐ SURFACE RUNOFF ☐ NEW INSTALLATION
☐ ODOR ☐ TANK OVER FLOW ☐ OTHER ______
☐ NOISE ☐ SEWAGE BACKUP ______

ALARM
☐ HIGH
☐ LOW
☐ OFF

TANK LL
☐ HIGH
☐ LOW
☐ NORMAL

PUMP
☐ ON
☐ OFF

CIRCUIT BREAKER
☐ ON
☐ OFF

PUMP TEST (MANUAL):
☐ INOPERABLE: ______ AMPS.
☐ OPERABLE: ______ AMPS.

DRAW DOWN:
TIME ______ (SEC.) (MIN.)
DEPTH ______ INCHES

PUMPING HEAD: ______ FT.
SHUTOFF HEAD: ______ FT.

CAUSE OF MALFUNCTION:

MECHANICAL
☐ PUMP
☐ BUILDING SEWER
☐ CHECK VALVE
☐ LEVEL CONTROL
☐ CONTROL PANEL
☐ SERVICE LINE
☐ SCREEN

PHYSICAL OR SYMPTOMATIC
☐ BACK PRESSURE
☐ AIR BOUND
☐ SLUDGE & SCUM
☐ CLOG
☐ POWER
☐ INFILTRATION/INFLOW
☐ EXFILTRATION
☐ OTHER ______

MAINTENANCE:
☐ REPAIR: ______
☐ REPLACE: ______
SHOP TIME: ______
COMMENTS: ______

*Fig. 4.40 Field maintenance report*

(Permission of the Oregon/Glide-Idleyld Park Sewer District)

CHAPTER 5

# WASTEWATER TREATMENT AND EFFLUENT DISCHARGE METHODS

by

John Brady

# TABLE OF CONTENTS

## Chapter 5. WASTEWATER TREATMENT AND EFFLUENT DISCHARGE METHODS

# OBJECTIVES

## Chapter 5. WASTEWATER TREATMENT AND EFFLUENT DISCHARGE METHODS

1. Explain how sand filters work.
2. Safely operate and maintain a sand filter.
3. Safely operate and maintain a recirculating gravel filter.
4. Describe subsurface infiltration systems and how they work.
5. Monitor and control a subsurface leaching system.
6. Monitor and control seepage beds and pits.
7. Operate and maintain absorption mounds.
8. Operate and maintain evapotranspiration systems.
9. Safely perform your duties.

## WORDS

### Chapter 5. WASTEWATER TREATMENT AND EFFLUENT DISCHARGE METHODS

ABSORPTION (ab-SORP-shun) ABSORPTION

The taking in or soaking up of one substance into the body of another by molecular or chemical action (as tree roots absorb dissolved nutrients in the soil).

ADSORPTION (add-SORP-shun) ADSORPTION

The gathering of a gas, liquid, or dissolved substance on the surface or interface zone of another material.

AEROBIC (air-O-bick) AEROBIC

A condition in which atmospheric or dissolved oxygen is present in the aquatic (water) environment.

ALGAE (AL-jee) ALGAE

Microscopic plants containing chlorophyll that live floating or suspended in water. They also may be attached to structures, rocks, or other submerged surfaces. Excess algal growths can impart tastes and odors to potable water. Algae produce oxygen during sunlight hours and use oxygen during the night hours. Their biological activities appreciably affect the pH, alkalinity, and dissolved oxygen of the water.

BOD (pronounce as separate letters) BOD

Biochemical Oxygen Demand. The rate at which organisms use the oxygen in water or wastewater while stabilizing decomposable organic matter under aerobic conditions. In decomposition, organic matter serves as food for the bacteria and energy results from its oxidation. BOD measurements are used as a surrogate measure of the organic strength of wastes in water.

BIOMASS (BUY-o-mass) BIOMASS

A mass or clump of organic material consisting of living organisms feeding on the wastes in wastewater, dead organisms, and other debris. Also see ZOOGLEAL MASS.

CAPILLARY (KAP-uh-larry) ACTION CAPILLARY ACTION

The movement of water through very small spaces due to molecular forces.

DISCHARGE HEAD DISCHARGE HEAD

The pressure (in pounds per square inch (psi) or kilopascals (kPa)) measured at the centerline of a pump discharge and very close to the discharge flange, converted into feet or meters. The pressure is measured from the centerline of the pump to the hydraulic grade line of the water in the discharge pipe.

Discharge Head, ft = (Discharge Pressure, psi)(2.31 ft/psi)

or

Discharge Head, m = (Discharge Pressure, kPa)(1 m/9.8 kPa)

DYNAMIC HEAD DYNAMIC HEAD

When a pump is operating, the vertical distance (in feet or meters) from a point to the energy grade line. Also see ENERGY GRADE LINE (EGL), STATIC HEAD, and TOTAL DYNAMIC HEAD (TDH).

DYNAMIC PRESSURE DYNAMIC PRESSURE

When a pump is operating, pressure resulting from the dynamic head.

Dynamic Pressure, psi = (Dynamic Head, ft)(0.433 psi/ft)

or

Dynamic Pressure, kPa = (Dynamic Head, m)(9.8 kPa/m)

ENERGY GRADE LINE (EGL) ENERGY GRADE LINE (EGL)

A line that represents the elevation of energy head (in feet or meters) of water flowing in a pipe, conduit, or channel. The line is drawn above the hydraulic grade line (gradient) a distance equal to the velocity head ($V^2/2g$) of the water flowing at each section or point along the pipe or channel. Also see HYDRAULIC GRADE LINE (HGL).

[SEE DRAWING ON PAGE 198]

EVAPOTRANSPIRATION (ee-VAP-o-TRANS-purr-A-shun) EVAPOTRANSPIRATION

(1) The process by which water vapor is released to the atmosphere by living plants. This process is similar to people sweating. Also called transpiration.

(2) The total water removed from an area by transpiration (plants) and by evaporation from soil, snow, and water surfaces.

GARNET GARNET

A group of hard, reddish, glassy, mineral sands made up of silicates of base metals (calcium, magnesium, iron, and manganese). Garnet has a higher density than sand.

GRADE GRADE

(1) The elevation of the invert (or bottom) of a pipeline, canal, culvert, sewer, or similar conduit.

(2) The inclination or slope of a pipeline, conduit, stream channel, or natural ground surface; usually expressed in terms of the ratio or percentage of number of units of vertical rise or fall per unit of horizontal distance. A 0.5 percent grade would be a drop of one-half foot per hundred feet (one-half meter per hundred meters) of pipe.

HEAD HEAD

The vertical distance, height, or energy of water above a reference point. A head of water may be measured in either height (feet or meters) or pressure (pounds per square inch or kilograms per square centimeter). Also see DISCHARGE HEAD, DYNAMIC HEAD, STATIC HEAD, SUCTION HEAD, SUCTION LIFT, and VELOCITY HEAD.

HEAD LOSS HEAD LOSS

The head, pressure, or energy (they are the same) lost by water flowing in a pipe or channel as a result of turbulence caused by the velocity of the flowing water and the roughness of the pipe, channel walls, or restrictions caused by fittings. Water flowing in a pipe loses head, pressure, or energy as a result of friction. The head loss through a comminutor is due to friction caused by the cutters or shredders as the water passes through them and by the roughness of the comminutor walls conveying the flow through the comminutor. Also called friction loss.

[SEE DRAWING ON PAGE 199]

HEADER HEADER

A large pipe to which the ends of a series of smaller pipes are connected. Also called a manifold.

HYDRAULIC GRADE LINE (HGL) HYDRAULIC GRADE LINE (HGL)

The surface or profile of water flowing in an open channel or a pipe flowing partially full. If a pipe is under pressure, the hydraulic grade line is that level water would rise to in a small, vertical tube connected to the pipe. Also see ENERGY GRADE LINE (EGL).

[SEE DRAWING ON PAGE 198]

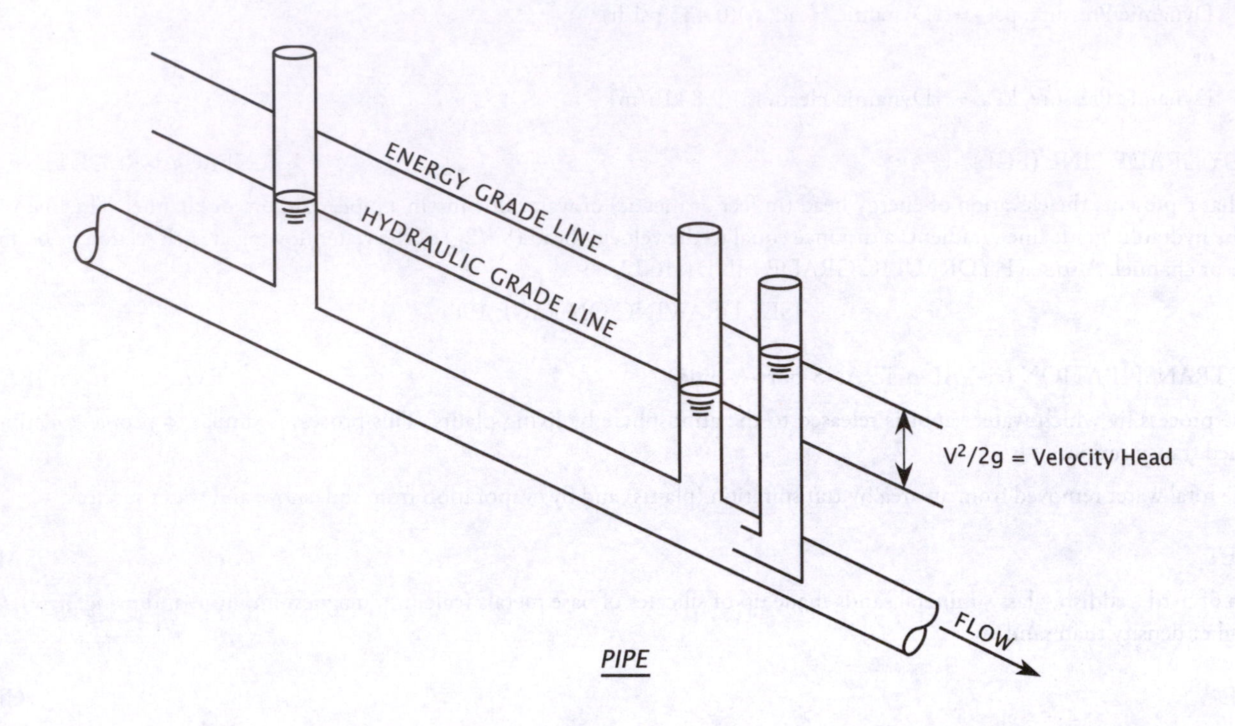


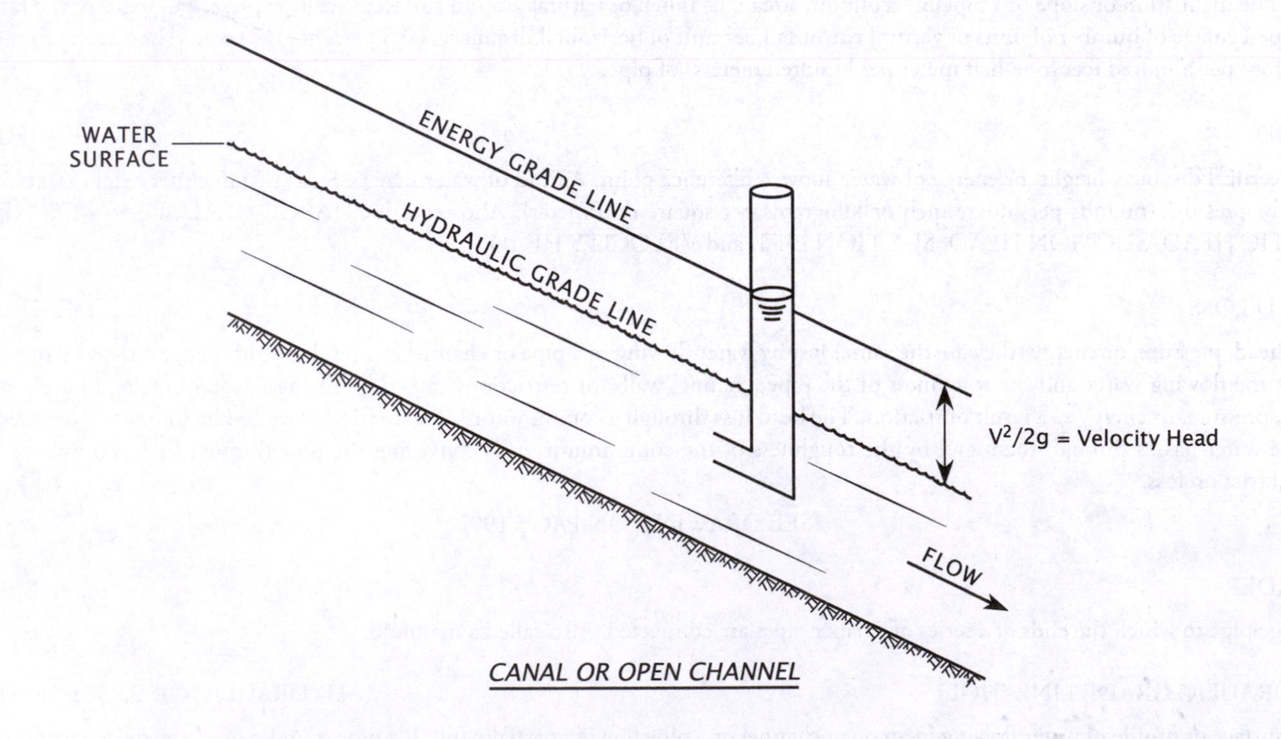


*Energy grade line and hydraulic grade line*

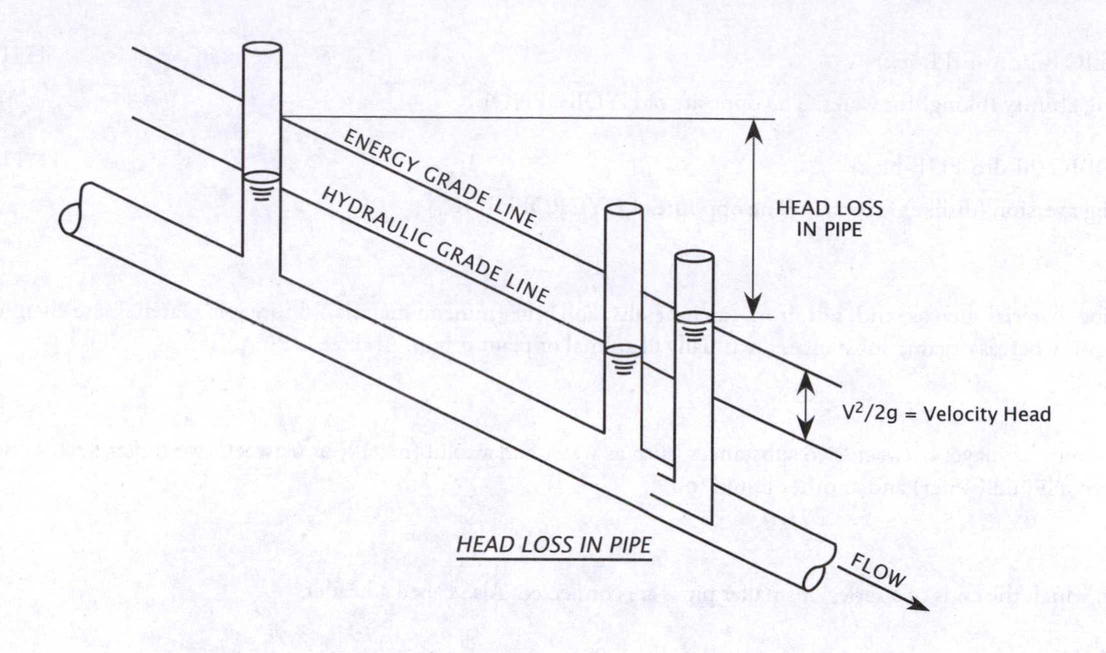
ENERGY GRADE LINE
HYDRAULIC GRADE LINE
HEAD LOSS IN PIPE
V²/2g = Velocity Head
HEAD LOSS IN PIPE
FLOW

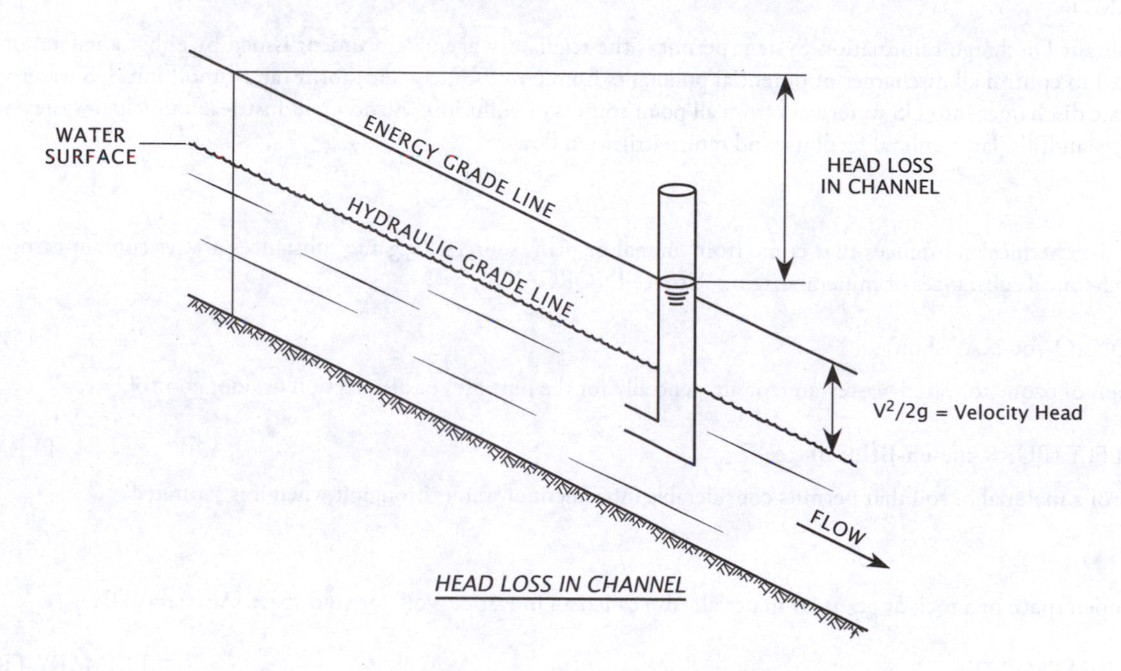
WATER SURFACE
ENERGY GRADE LINE
HYDRAULIC GRADE LINE
HEAD LOSS IN CHANNEL
V²/2g = Velocity Head
FLOW
HEAD LOSS IN CHANNEL

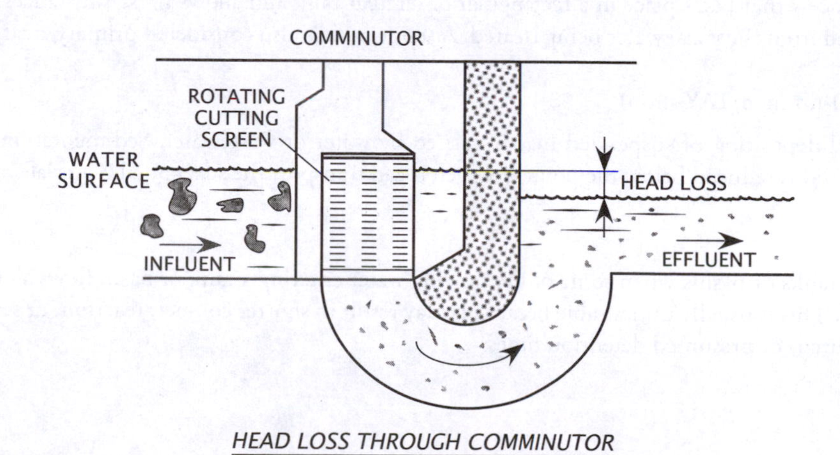
COMMINUTOR
ROTATING CUTTING SCREEN
WATER SURFACE
HEAD LOSS
INFLUENT
EFFLUENT
HEAD LOSS THROUGH COMMINUTOR

*Head loss*

HYDROPHILIC (hi-dro-FILL-ick) HYDROPHILIC

Having a strong affinity (liking) for water. The opposite of HYDROPHOBIC.

HYDROPHOBIC (hi-dro-FOE-bick) HYDROPHOBIC

Having a strong aversion (dislike) for water. The opposite of HYDROPHILIC.

INORGANIC INORGANIC

Used to describe material such as sand, salt, iron, calcium salts, and other mineral materials. Inorganic materials are chemical substances of mineral origin, whereas organic substances are usually of animal or plant origin. Also see ORGANIC.

INTERFACE INTERFACE

The common boundary layer between two substances, such as water and a solid (metal); or between two fluids, such as water and a gas (air); or between a liquid (water) and another liquid (oil).

MANIFOLD MANIFOLD

A large pipe to which the ends of a series of smaller pipes are connected. Also called a header.

NPDES PERMIT NPDES PERMIT

National Pollutant Discharge Elimination System permit is the regulatory agency document issued by either a federal or state agency that is designed to control all discharges of potential pollutants from point sources and stormwater runoff into US waterways. NPDES permits regulate discharges into US waterways from all point sources of pollution, including industries, municipal wastewater treatment plants, sanitary landfills, large animal feedlots, and return irrigation flows.

ORGANIC ORGANIC

Used to describe chemical substances that come from animal or plant sources. Organic substances always contain carbon. (Inorganic materials are chemical substances of mineral origin.) Also see INORGANIC.

OZONATION (O-zoe-NAY-shun) OZONATION

The application of ozone to water, wastewater, or air, generally for the purposes of disinfection or odor control.

PERMEABILITY (PURR-me-uh-BILL-uh-tee) PERMEABILITY

The property of a material or soil that permits considerable movement of water through it when it is saturated.

PORE PORE

A very small open space in a rock or granular material. Also called an interstice, void, or void space. Also see VOID.

PRIMARY TREATMENT PRIMARY TREATMENT

A wastewater treatment process that takes place in a rectangular or circular tank and allows those substances in wastewater that readily settle or float to be separated from the wastewater being treated. A septic tank is also considered primary treatment.

SEDIMENTATION (SED-uh-men-TAY-shun) SEDIMENTATION

The process of settling and depositing of suspended matter carried by water or wastewater. Sedimentation usually occurs by gravity when the velocity of the liquid is reduced below the point at which it can transport the suspended material.

SHORT-CIRCUITING SHORT-CIRCUITING

A condition that occurs in tanks or basins when some of the flowing water entering a tank or basin flows along a nearly direct pathway from the inlet to the outlet. This is usually undesirable because it may result in shorter contact, reaction, or settling times in comparison with the theoretical (calculated) or presumed detention times.

SLOPE — SLOPE

The slope or inclination of a trench bottom or a trench side wall is the ratio of the vertical distance to the horizontal distance or rise over run. Also see GRADE (2).

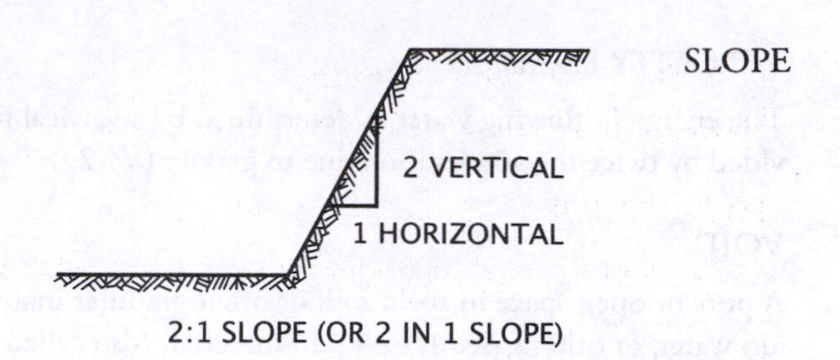


STABILIZATION — STABILIZATION

Conversion to a form that resists change. Organic material is stabilized by bacteria that convert the material to gases and other relatively inert substances. Stabilized organic material generally will not give off obnoxious odors.

STATIC HEAD — STATIC HEAD

When water is not moving, the vertical distance (in feet or meters) from a reference point to the water surface is the static head. Also see DYNAMIC HEAD, DYNAMIC PRESSURE, and STATIC PRESSURE.

STATIC PRESSURE — STATIC PRESSURE

When water is not moving, the vertical distance (in feet or meters) from a specific point to the water surface is the static head. The static pressure in psi (or kPa) is the static head in feet times 0.433 psi/ft (or meters × 9.81 kPa/m). Also see DYNAMIC HEAD, DYNAMIC PRESSURE, and STATIC HEAD.

SUCTION HEAD — SUCTION HEAD

The positive pressure [in feet (meters) of water or pounds per square inch (kilograms per square centimeter) of mercury vacuum] on the suction side of a pump. The pressure can be measured from the centerline of the pump up to the elevation of the hydraulic grade line on the suction side of the pump.

SUCTION LIFT — SUCTION LIFT

The negative pressure [in feet (meters) of water or inches (centimeters) of mercury vacuum] on the suction side of a pump. The pressure can be measured from the centerline of the pump down to (lift) the elevation of the hydraulic grade line on the suction side of the pump.

THRESHOLD ODOR — THRESHOLD ODOR

The minimum odor of a gas or water sample that can just be detected after successive dilutions with odorless gas or water. Also called odor threshold.

TOTAL DYNAMIC HEAD (TDH) — TOTAL DYNAMIC HEAD (TDH)

When a pump is lifting or pumping water, the vertical distance (in feet or meters) from the elevation of the energy grade line on the suction side of the pump to the elevation of the energy grade line on the discharge side of the pump. The total dynamic head is the static head plus pipe friction losses.

TRANSPIRATION (TRAN-spur-RAY-shun) — TRANSPIRATION

The process by which water vapor is released to the atmosphere by living plants. This process is similar to people sweating. Also called evapotranspiration.

UNIFORMITY COEFFICIENT (UC) — UNIFORMITY COEFFICIENT (UC)

The ratio of (1) the diameter of a grain (particle) of a size that is barely too large to pass through a sieve that allows 60 percent of the material (by weight) to pass through, to (2) the diameter of a grain (particle) of a size that is barely too large to pass through a sieve that allows 10 percent of the material (by weight) to pass through. The resulting ratio is a measure of the degree of uniformity in a granular material, such as filter media.

$$\text{Uniformity Coefficient} = \frac{\text{Particle Diameter}_{60\%}}{\text{Particle Diameter}_{10\%}}$$

VELOCITY HEAD | VELOCITY HEAD

The energy in flowing water as determined by a vertical height (in feet or meters) equal to the square of the velocity of flowing water divided by twice the acceleration due to gravity ($V^2/2g$).

VOID | VOID

A pore or open space in rock, soil, or other granular material, not occupied by solid matter. The pore or open space may be occupied by air, water, or other gaseous or liquid material. Also called an interstice, pore, or void space.

ZOOGLEAL (ZOE-uh-glee-ul) MASS | ZOOGLEAL MASS

Jelly-like masses of bacteria found in both the trickling filter and activated sludge processes. These masses may be formed for or function as the protection against predators and for storage of food supplies. Also see BIOMASS.

ZOOGLEAL (ZOE-uh-glee-al) MAT | ZOOGLEAL MAT

A complex population of organisms that form a slime growth on the sand filter media and break down the organic matter in wastewater. These slimes consist of living organisms feeding on the wastes in wastewater, dead organisms, silt, and other debris. On a properly loaded and operating sand filter, these mats are so thin as to be invisible to the naked eye. Slime growth is a more common term.

# CHAPTER 5. WASTEWATER TREATMENT AND EFFLUENT DISCHARGE METHODS

(Lesson 1 of 2 Lessons)

## 5.0 WASTEWATER TREATMENT AND DISCHARGE ALTERNATIVES FOR SMALL SYSTEMS

This lesson describes various types of sand and gravel filters, which are being used by growing numbers of small communities and small wastewater agencies to produce high-quality effluents suitable for discharge. These systems have been successfully designed to treat wastewater flows ranging from 200 gallons per day up to 500,000 gallons per day. Due to their dependable performance and ease of operation, however, these systems are being adapted by larger municipal wastewater agencies to serve as polishing units for other treatment processes such as oxidation lagoons, rotating biological contactors, trickling filters, and activated sludge plants.

Lesson 2 of this chapter describes some wastewater treatment and discharge options that are appropriate for use either as on-site treatment systems or as small community treatment systems.[1] The treatment alternatives include leach fields, seepage beds and pits, absorption mounds, and evapotranspiration systems.

Before wastewater can be discharged using the methods mentioned above, *PRIMARY TREATMENT*[2] of the wastewater is required. This treatment is usually accomplished in septic tanks. A tank may be installed at each home, or one tank may serve a cluster of homes, or the raw wastewater from the community may be collected and treated in a community septic tank or primary treatment plant.

Some of the treatment and discharge methods described in this chapter are not approved by all regulatory agencies (local health departments) for use as on-site treatment systems, or in some states as community treatment systems. Check with your local and state regulatory agencies to find out whether a treatment or discharge method you are considering using is approved for use in your area.

## 5.1 SAND FILTRATION

### 5.10 Types of Filters

Sand filters are widely used at individual homes and in small communities to treat wastewater effluent from septic tanks. Two

types of downflow sand filters are used: (1) buried (covered with soil) intermittent filters and (2) recirculating filters. In recent designs of the recirculating sand filter, the sand is covered with a layer of pea gravel. The bed is still open to the atmosphere through the gravel to maintain *AEROBIC*[3] conditions. While both types of sand filters are somewhat alike in design, they may differ in media, operation, access, and performance.

Buried sand filters (Figures 5.1 and 5.2) are intermittent sand filters used for wastewater treatment primarily at individual residences. These filters are a single-pass type of filter in which the effluent from a septic tank is applied to the filter by intermittent gravity flow or by a dosing siphon (described in Section 5.150). After the wastewater passes through the media, all of the filter effluent is discharged, usually in a leach field system. If the septic tank that the filter serves is properly operated and maintained, a buried sand filter should perform properly for more than 20 years.

Recirculating sand filters (RSFs) (Figure 5.3) are used mainly in small communities, especially those with STEP (septic tank effluent pump) systems or a community septic tank. RSFs may or may not have a gravel cover around and over the distribution system. The gravel keeps the sand surface in place and level. The septic tank effluent is screened to trap larger undesirable solids, thus reducing maintenance of the sand media surface below. Recirculating sand filters produce a high-quality effluent that may be discharged to a leach field, irrigation system, or surface waters after disinfection.

---

1. Volume II of this manual, Chapter 13, "Alternative Wastewater Treatment, Discharge, and Reuse Methods," describes two other effluent discharge methods: wetlands and land treatment systems.
2. *Primary Treatment.* A wastewater treatment process that takes place in a rectangular or circular tank and allows those substances in wastewater that readily settle or float to be separated from the wastewater being treated. A septic tank is also considered primary treatment.
3. *Aerobic* (air-O-bick). A condition in which atmospheric or dissolved oxygen is present in the aquatic (water) environment.

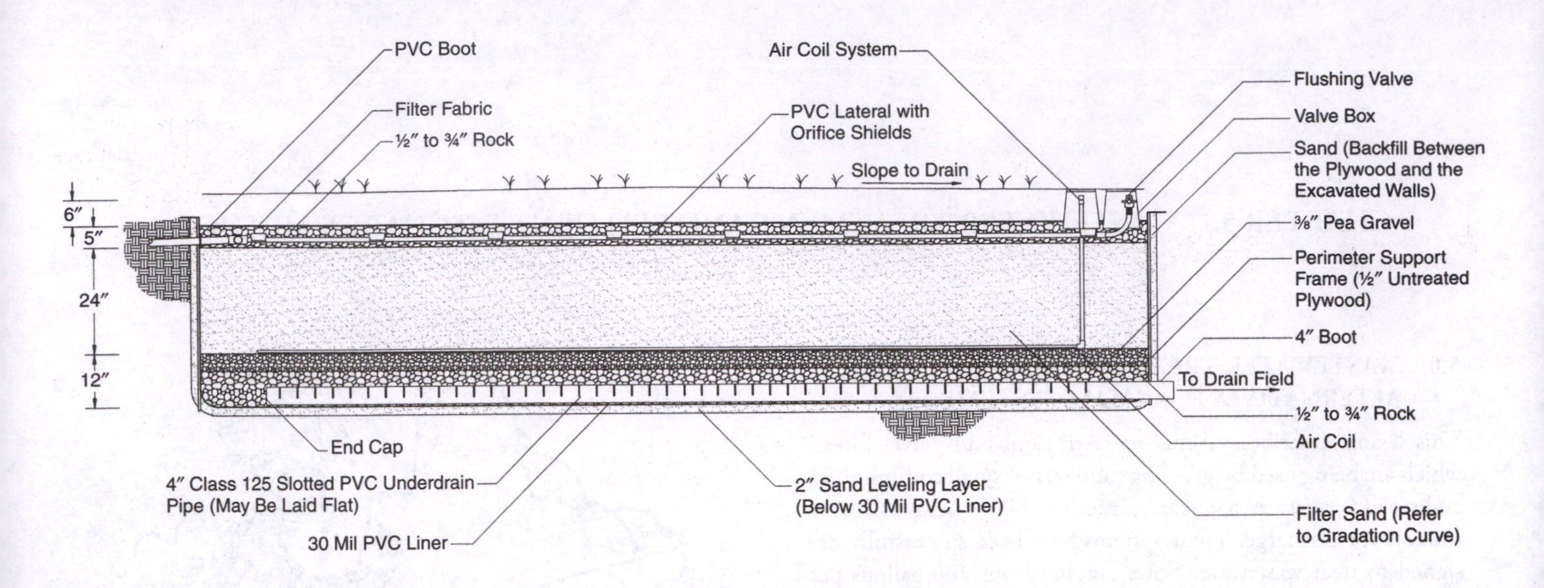


*Fig. 5.1 Gravity discharge sand filter (side view)*

(Courtesy of Orenco Systems, Inc., www.orenco.com)

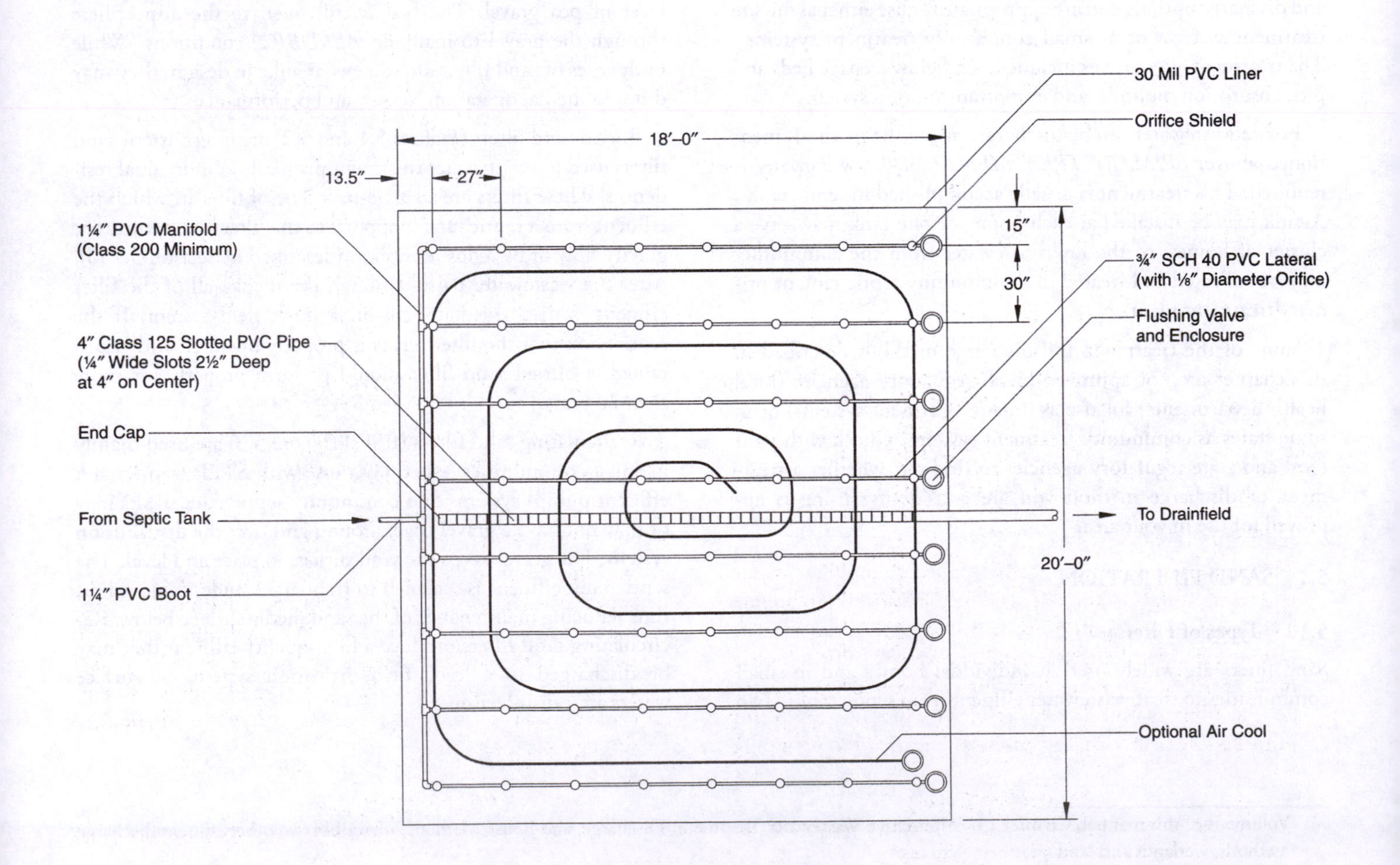


*Fig. 5.2 Gravity discharge sand filter (plan view)*

(Courtesy of Orenco Systems, Inc., www.orenco.com)

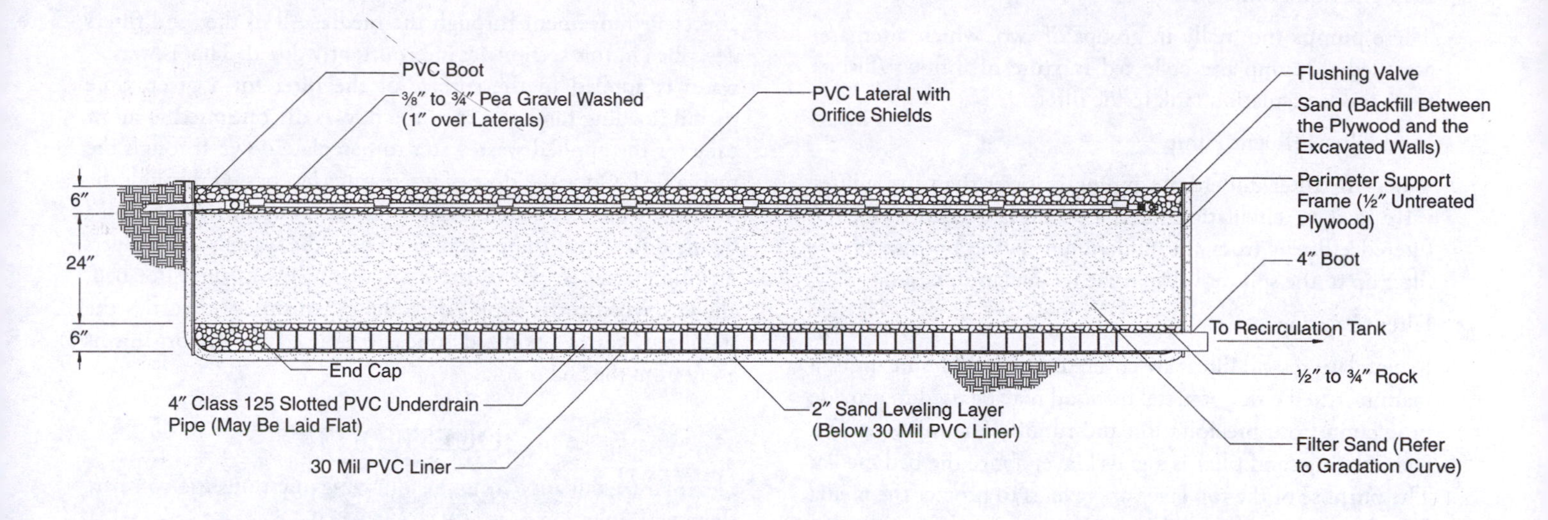


*Fig. 5.3 Recirculating sand filter (side view or elevation)*

(Courtesy of Orenco Systems, Inc., www.orenco.com)

## 5.11 Components

### *5.110 Intermittent Sand Filters*

- Bed

  Sand filters have one or more rectangular beds. The size of the bed depends on the quality and amount of wastewater that will be applied to the filter.

- Media

  The bed holds the filter media, which is usually a washed and graded sand that is placed two to three feet deep. Although sand is the most commonly used media material, other materials such as *GARNET,*[4] anthracite (coal), mineral tailings, crushed glass (recycled), and bottom ash also can be used and may be more readily available than sand in some regions.

- Distribution Piping

  The pretreated wastewater to be applied to the filter is carried to the surface of the filter bed by the distribution piping. The preferred method of treatment is to spread the influent just beneath the filter surface through spaced orifices (holes) in the distribution piping. Spraying or spreading the influent on the filter surface usually results in filtration problems caused by *ALGAE.*[5] On buried filters, the distribution piping is vented to the atmosphere to ensure that wastewater will flow freely and evenly across the surface of the filter media.

- Underdrain or Collection Piping

  A single pipe or a network of pipes is placed on the bottom of the filter bed and covered with eight inches to one foot of washed and graded pea gravel. The underdrains may be vented to the atmosphere on buried filters to collect the filtered wastewater and to maintain aerobic conditions within the filter media.

### *5.111 Recirculating Sand Filters (RSFs)*

Recirculating sand filter components include the parts used in intermittent sand filters as well as the following additional equipment:

- Recirculation Structure

  A recirculation tank is used to store filter influent (wastewater discharged from the home or community septic tank(s)) to be treated by the filter plus a portion (usually 80 percent) of the effluent (filtrate) that has just previously been applied and drained from the filter. The remaining 20 percent portion of the filter treated effluent is discharged through a splitter valve.

---

4. *Garnet.* A group of hard, reddish, glassy, mineral sands made up of silicates of base metals (calcium, magnesium, iron, and manganese). Garnet has a higher density than sand.

5. *Algae* (AL-jee). Microscopic plants containing chlorophyll that live floating or suspended in water. They also may be attached to structures, rocks, or other submerged surfaces. Excess algal growths can impart tastes and odors to potable water. Algae produce oxygen during sunlight hours and use oxygen during the night hours. Their biological activities appreciably affect the pH, alkalinity, and dissolved oxygen of the water.

- Recirculation Pump System

  These pumps (normally in groups of two, which alternate) are used to pump the collected mixture of influent/filtrate from the recirculation tank to the filter.

- Sand Filter Effluent Pump

  When the filter underdrain piping is lower than the splitter valve in the recirculation tank, a pump is needed to lift the filtered effluent from the underdrain in the bottom of the filter up to the splitter valve in the recirculation tank.

- Filter Covering

  Recirculating sand filters are covered. On small home units, a manufactured cover of metal or wood may be used to provide protection from precipitation and runoff. The top layer of a recirculating sand filter is a gravel layer above the bed media. The purpose of the top layer of gravel is to protect the media from being disturbed while permitting easy access to the distribution piping just under the surface of the gravel layer. A shade cloth is used to protect the filter from leaves and debris.

### 5.12 How Sand Filters Work

Sand filters provide treatment to the wastewater using both physical and biological processes. First, suspended solids are filtered out as the wastewater passes down through the media. This is strictly a physical action. To provide effective filtration, it is important for the filter media to be clean and composed of sand particles having basically the same size and shape. The media traps the solids and holds them in the bed as the water passes through.

Once the suspended solids (mostly *ORGANIC*[6] materials) have been trapped in the slime film or mat on the media, they are *STABILIZED*[7] by microorganisms living in the film or mat on the media. As in other aerobic wastewater treatment processes, the organisms in this microbial colony consume the organic material trapped in the slime film or mat.

Most of the organisms in sand filters are aerobic and therefore require oxygen to live and function. The wastewater collects oxygen from the air in the filter bed as it passes through the media. Changes in atmospheric pressure and air temperature, due to changing weather conditions, create air pressure differences that cause air to move through the filter.

The way in which wastewater is applied to the filter bed also affects air movement through the media. All of the sand filters described in this section are intermittently dosed. That is, wastewater is applied to the surface of the filter for a given time period (loading rate) and then the flow is discontinued to allow time for the applied wastewater to percolate down through the media bed. Once the dose of wastewater has passed through the media, air fills any remaining media voids not already occupied by organisms and trapped solids. Thus, the wastewater delivers organic matter (food) to the microorganisms in the filter bed, allows oxygen to be supplied to the organisms, and carries the stabilized wastes (oxidized solids) produced by the organisms away from the bed media.

## QUESTIONS

Please write your answers to the following questions and compare them with those on page 242.

5.0A List three alternative wastewater treatment and discharge systems.

5.1A List two types of sand filters.

5.1B What are the four basic sand filter components?

5.1C How do sand filters treat wastewater?

### 5.13 Operation and Maintenance

Recirculating sand filters are designed to treat a recycled portion of the filter's effluent/filtrate. A proportion of the filtrate is returned from the filter to be mixed at a ratio of from 3 to 1 to 5 to 1 (3:1 to 5:1) with the influent to be applied to the filter. Mixing filtrate with the incoming wastewater dilutes the wastewater being applied to the filter. This procedure improves overall treatment results and reduces odor problems. In addition, the filtrate contains dissolved oxygen, which helps to maintain the filter bed in an aerobic state. Once the pumping cycles have been set and the filtrate proportioning valves have been adjusted to control recirculation or effluent discharge flow, recirculating filters require relatively little operational control.

On systems operating with *NPDES PERMITS,*[8] influent and effluent monitoring of flow rates, suspended solids (SS), *BIOCHEMICAL OXYGEN DEMAND (BOD),*[9] and *E. coli*[10] is necessary.

---

6. *Organic.* Used to describe chemical substances that come from animal or plant sources. Organic substances always contain carbon. (Inorganic materials are chemical substances of mineral origin.) Also see INORGANIC.
7. *Stabilization.* Conversion to a form that resists change. Organic material is stabilized by bacteria that convert the material to gases and other relatively inert substances. Stabilized organic material generally will not give off obnoxious odors.
8. *NPDES Permit.* National Pollutant Discharge Elimination System permit is the regulatory agency document issued by either a federal or state agency that is designed to control all discharges of potential pollutants from point sources and stormwater runoff into US waterways. NPDES permits regulate discharges into US waterways from all point sources of pollution, including industries, municipal wastewater treatment plants, sanitary landfills, large animal feedlots, and return irrigation flows.
9. *BOD* (pronounce as separate letters). Biochemical Oxygen Demand. The rate at which organisms use the oxygen in water or wastewater while stabilizing decomposable organic matter under aerobic conditions. In decomposition, organic matter serves as food for the bacteria and energy results from its oxidation. BOD measurements are used as a surrogate measure of the organic strength of wastes in water.
10. *E. coli* is a type of bacteria found in the wastes of warm-blooded animals. The presence of *E. coli* in water indicates possible contamination by pathogenic (disease-causing) organisms.

Filter maintenance consists of routine cleaning of dosing equipment. Weed growth should be removed along with debris such as leaves, papers, and other foreign material.

As with any system relying on bacteria for treatment, the operator must monitor the health of the bacterial slimes on the surface of the sand. The filter bed needs to stay aerobic with thin bacterial slimes throughout the sand. Never add anything to the wastewater that may be toxic to or upset the filter's bacterial equilibrium (for example, degreasers, enzymes, nonbiodegradable cleansers).

Check the surface of gravel-covered filters for pooling and check the distribution pipe outlet orifices for flows. In spots where minor pooling occurs, the surface of the media has most likely become blocked. Spots of this type usually can be opened up by raking the gravel to break up the material that was creating the pooling. To check distribution pipe orifices, use a garden rake to remove several inches of gravel over each orifice. When the dosing pump is on, a small stream of water will be squirted from the orifice into the air above the filter bed. If no flow is observed, clean the orifice with a wire and replace the orifice shield and gravel over the pipe. If more than one orifice in a lateral is blocked, the lateral should be cleaned by flushing or using a bottle brush or using a high-pressure washer. Weeds may grow in the gravel cover of recirculating filters and should be removed.

## 5.14 Loading Guidelines

Table 5.1 lists typical loading guidelines for two types of sand filters.

**TABLE 5.1 SAND FILTER LOADING GUIDELINES**[a]

| Loading Guideline | Type of Filter | |
|---|---|---|
| | Intermittent (Buried) | Recirculating |
| Pretreatment | minimum of sedimentation | |
| Media | washed and graded sand | |
| Media Depth | | |
| (feet) | 2.0 | 2.0–3.0 |
| (meters) | 0.6–0.9 | 0.6–0.9 |
| Hydraulic Loading | | |
| (gallons/day/sq ft) | 1–1.5 | 3–10 |
| (centimeters/day[b]) | 4–6 | 12–20 |
| Dosing Frequency | 12–24 doses/day | 1–3 min/30 min |
| Recirculation Ratio | N/A[c] | 3:1–5:1 |

a EPA manual, *WASTEWATER TREATMENT AND DISPOSAL FOR SMALL COMMUNITIES*

b 4 cm/day = 0.04 cu m/day/sq m

c N/A = not applicable

The influent to a sand filter must first receive primary treatment, usually *SEDIMENTATION*[11] in a septic tank, to remove a large percentage of the settleable solids and a portion of the biochemical oxygen demand (BOD). The effluent from screened STEP systems typically contains about 150 mg/L BOD and 30 mg/L total suspended solids (TSS).

Sand filter design and operation are based on the hydraulic loading (flow) on the filter. The hydraulic loading is based on the amount of BOD/TSS in the applied wastewater, the type of filter design, and the media makeup. If the applied wastewater has a higher BOD/TSS, then a lower hydraulic loading rate is used. In instances where the BOD/TSS is unusually high, use more septic tanks in series and provide better management of the BOD/TSS source. For example, encourage restaurants to scrape waste food into a garbage can to reduce BOD/TSS greatly. Some type of septic tank effluent filter is needed for proper sand filter operation.

Hydraulic loading rates can range from 0.75 to 10 gallons per day/square foot (3 to 40 centimeters/day) of filter surface area. Intermittent filters have the lowest hydraulic loadings, while recirculating filters have the highest. Higher loading rates generally reduce the effluent quality.

## 5.15 Dosing Methods

### *5.150 Gravity Dosing Siphon (Figure 5.4)*

A dosing siphon is a simple mechanical device that delivers fixed amounts of water from one tank or reservoir to another tank or pipeline. Flow rates range from a few gallons per minute to several hundred gallons per minute. Because a siphon operates by gravity, the treatment unit or discharge area being dosed must be located downhill from the dosing tank. Dosing siphons are particularly appropriate for intermittent sand filters where pumping back to a recirculation tank is not necessary.

In on-site wastewater systems, siphons are especially useful in converting small, continuous flows into large, intermittent dosing flows. Today's dosing siphons are made of corrosion-resistant materials, have no moving parts, require no power source, are easy to install, and require very little maintenance. They are a cost-effective alternative to pumps in many situations, especially in remote areas and other sites where electricity is difficult to obtain.

An automatic dosing siphon has two main components, the bell and the trap (see Figure 5.4). The bell includes the bell housing itself, a vertical inlet pipe, an intrusion pipe, and a sniffer pipe. The trap includes a long leg, a short leg, and a discharge fitting with an air vent. Depending on the siphon drawdown, the trap may be fitted with an external trigger trap. The bell and trap are connected with threaded fittings.

Siphons can be installed in almost any type of tank, basin, or reservoir. In small systems (30 GPM or less), a siphon is often

11. *Sedimentation* (SED-uh-men-TAY-shun). The process of settling and depositing of suspended matter carried by water or wastewater. Sedimentation usually occurs by gravity when the velocity of the liquid is reduced below the point at which it can transport the suspended material.

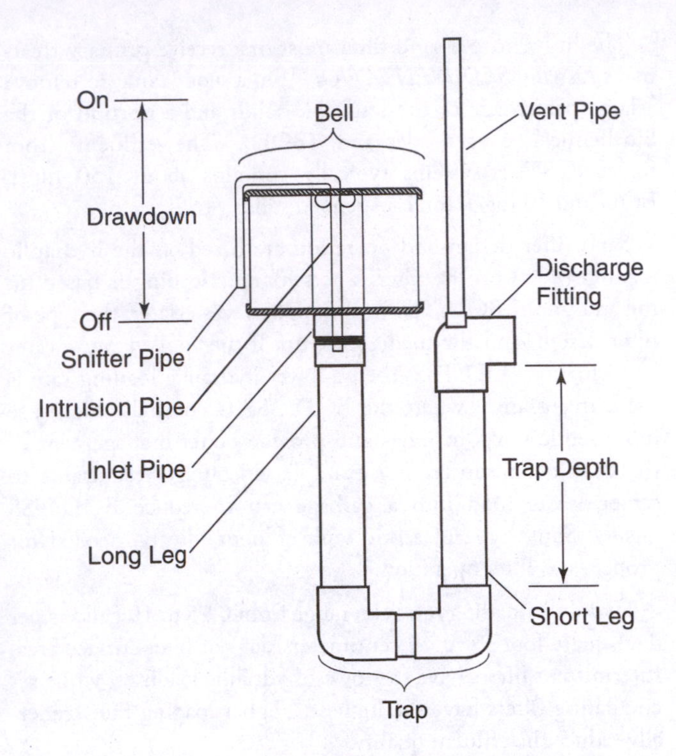


*Fig. 5.4 Dosing siphon*
(Courtesy of Orenco Systems, Inc., www.orenco.com)

mounted in a screened vault, which is placed directly in a single-compartment septic tank so that a second dosing tank or chamber is not required. In larger systems, siphons usually are installed directly in compartmented septic tanks or in separate dosing tanks.

Regardless of the type of siphon or the method of installation, the water must be filtered before it reaches the siphon. A filter helps protect the performance of the siphon, the distribution network, and the discharge area. A key benefit of filtering is keeping the siphon's snifter pipe clear. Even a brief blockage at the end of the discharge cycle may halt the flow of liquid through the siphon.

Following installation in a tank, a siphon must have its trap filled with water. When fluid rises above the open end of the snifter pipe, air is sealed in the bell and long leg of the siphon. As the fluid in the tank rises further, the pressure on the trapped air increases and forces water out of the long leg of the trap. Once the pressure is great enough to force all the water out of the long leg, the trapped air escapes through the short leg to the air release vent pipe. At this point, the siphon has been triggered and water is discharged from the siphon until the liquid level in the tank drops to the bottom of the bell. Air is then drawn under the bell, which breaks the siphoning action and the process begins again. Figure 5.5 shows how a siphon operates.

At the end of a dosing cycle, incoming flow may seal off the bottom of the bell before the bell is fully recharged with air. The snifter pipe, with its open end an inch or more above the bottom of the bell, allows a full recharge of air beneath the bell at the end of each cycle. Because the end of the snifter pipe is the elevation at which air becomes trapped under the bell, shortening or lengthening the snifter pipe is an effective way to increase or decrease the "on" or "trip" level of the siphon. However, you should consult with the siphon manufacturer before changing the length of the snifter pipe.

### 5.151 *Pressure Dosing*

Pumps are another way to deliver fixed amounts of wastewater to a treatment process or to deliver treated effluent to a discharge site. Pumps can dose either uphill or downhill. A programmable timer (PT) installed in a control panel allows precise control of dosing by running a pump's ON and OFF cycles for predetermined lengths of time. Programmable timers allow flows to be discharged more evenly over a period of time. Typically, for example, peak residential flows occur in the morning, evening, or when laundry or dishwashing machines are operated. In a programmable timer-controlled system, effluent is discharged from the septic tank or the recirculation tank in small, uniform doses over the course of the day instead of in the peak flow volumes that would be discharged all at once in a system with a demand float switch.

Automatic monitoring of water usage is a real advantage of programmed dosing. This may be required if the treatment unit or discharge site being dosed has limited hydraulic capacity. If a programmable timer is set to dose the maximum expected volume of effluent each day, any additional wastewater entering the tank beyond this maximum sets off an alarm, alerting the system user to stop or reduce water usage. Since typical washing machines discharge 40 to 60 gallons per load, washing several loads of clothes in a short period of time is a prime cause of hydraulic overloading. Depending on the reserve volume of the tank, a programmable timer will allow a more uniform discharge of effluent or the alarm will sound, alerting the user of the hydraulic overload.

Programmable timer-controlled systems are effective in detecting infiltration from leaky tanks, leaky stormwater connections, and leaky plumbing fixtures. Leaking toilet valves commonly run at a rate of several gallons per minute. Even small infiltration rates are readily detectable; a mere one gallon per minute leaking into a cracked tank during periods of high groundwater, for example, will increase the daily household flow by 1,440 gallons.

Programmed dosing is also commonly retrofitted to existing gravity or pressurized systems that are failing hydraulically or biologically. By limiting the doses to small incremental volumes spread evenly over a 24-hour period, many failing systems can be successfully restored to satisfactory operation. This method has been successful on several different types of systems, including sand filters, gravity and pressurized drain field systems, and mounds.

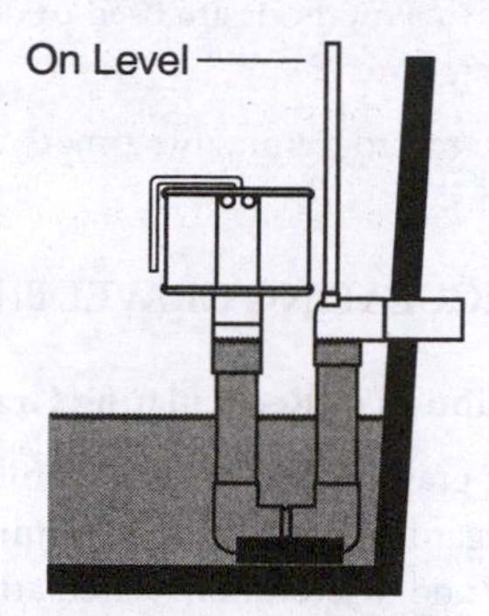


1. Trap must be primed (filled with water) prior to raising liquid level above bottom of snifter pipe.

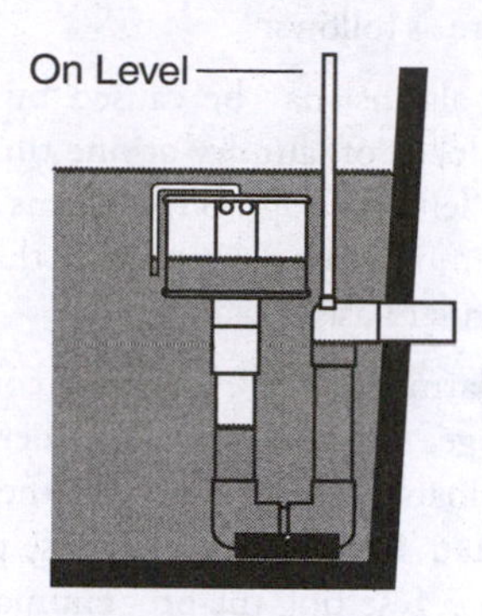


2. Water is discharged from long leg as water level rises above snifter pipe.

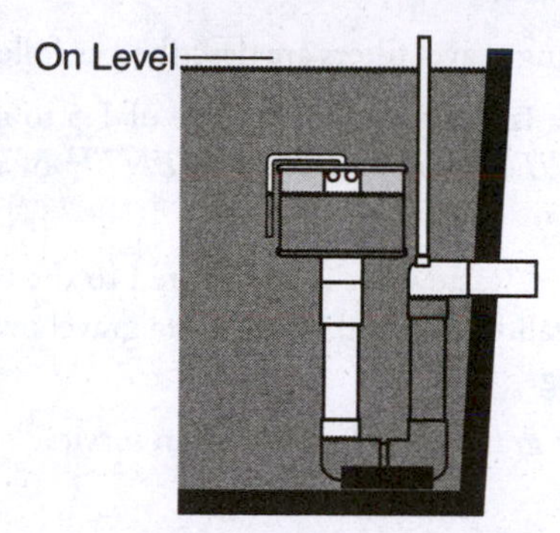


3. Just before triggering, water level in long leg is near bottom of trap.

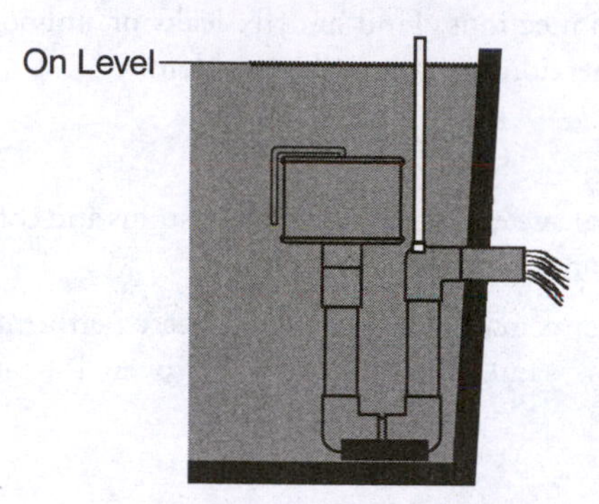


4. Siphon is triggered when air is vented through vent pipe.

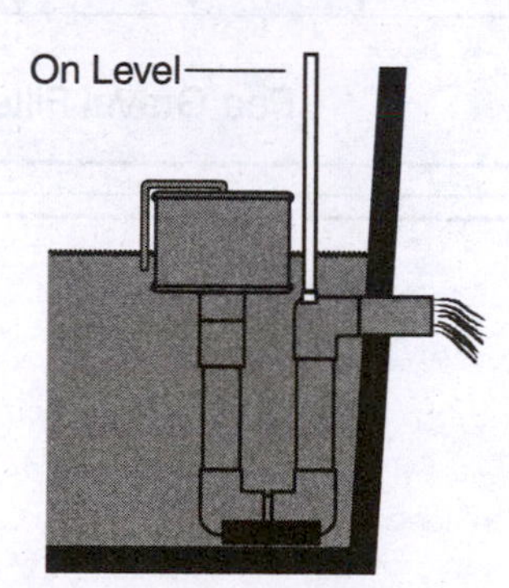


5. Siphon continues to dose until water level drops to bottom of bell.

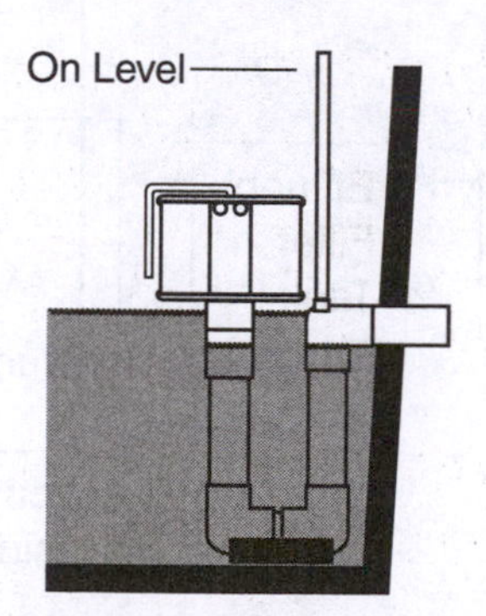


6. Air under bell breaks the siphon. Snifter pipe ensures full recharge of air under bell.

*Fig. 5.5 Operation of a dosing siphon*

(Courtesy of Orenco Systems, Inc., www.orenco.com)

### 5.16 Troubleshooting Intermittent Sand Filters

The most common causes of hydraulic overload alarms on intermittent sand filters are as follows:

- Occasional short alarms may be caused by excessive water usage (too many loads of laundry at one time, large parties, or a water fixture left running). Such alarms simply alert the user that the system is getting more water than it is designed to handle on a regular basis.
- Frequent short alarms (nearly every day) could indicate excessive water usage, a programmable timer out of adjustment, top two floats set too close together, or a clogged screened vault filter. The solutions to these problems are reducing water usage, resetting the programmable timer, repositioning the floats, or cleaning the screened vault filter.
- Short-duration alarms only during storms or very wet periods indicate infiltration from a leaky septic tank, plumbing, or stormwater connections. Find and fix leaks or unhook undesirable connections to correct this problem.

## QUESTIONS

Please write your answers to the following questions and compare them with those on page 242.

5.1D How does recirculating part of the treated effluent back through a sand filter improve the overall treatment results?

5.1E The hydraulic loading of a sand filter is based on what factors?

5.1F What two methods are used to dose treatment units and discharge areas?

5.1G Why are programmable timers used in pressure dosing systems?

## 5.2 RECIRCULATING GRAVEL FILTERS (RGFs)

### 5.20 Description of Recirculating Gravel Filters

Recirculating gravel filters are very similar to recirculating sand filters with regard to pretreatment requirements, physical layout of the plant, and wastewater recirculation method. Figure 5.6 shows a typical layout of a recirculating gravel filter. Similar to the recirculating sand filter, the recirculating gravel filter typically needs an effluent filter after the septic tank and prior to the recirculation tank.

Recirculating gravel filters are described as follows:

1. The media is pea gravel with a size of 1.5 to 4.0 millimeters and a *UNIFORMITY COEFFICIENT*[12] of not more than 2.5.
2. The applied wastewater is distributed to the bed surface just under a shallow (2-inch) layer of pea gravel over the distribution piping.
3. The entire gravel filter bed is kept in service.

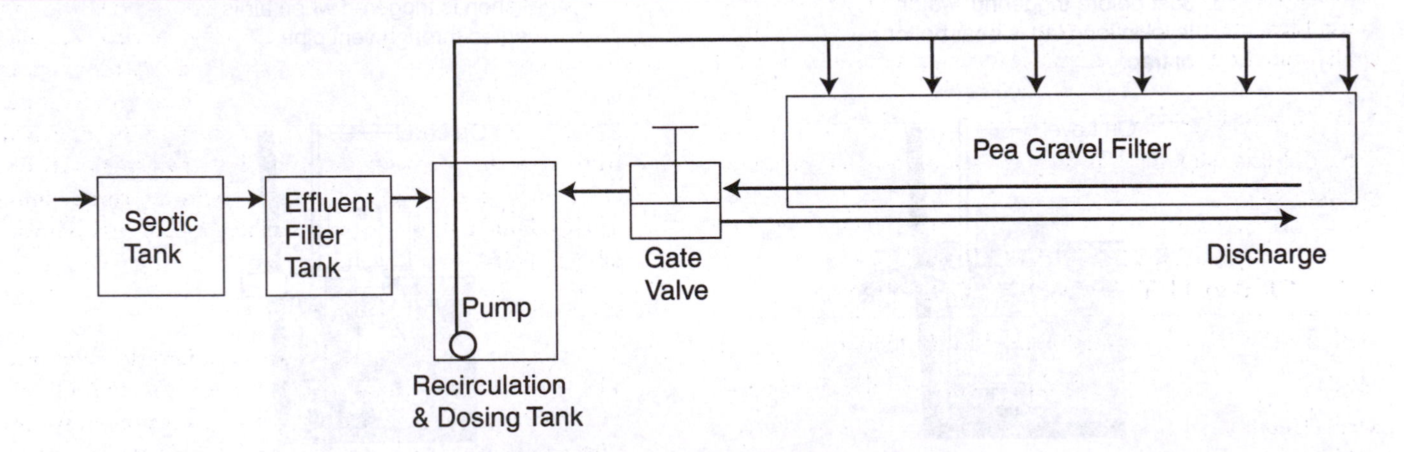


*Fig. 5.6 Typical layout of a recirculating gravel filter system (Portville, New York)*
(Source: *SMALL FLOWS*, January 1991)

12. *Uniformity Coefficient (UC).* The ratio of (1) the diameter of a grain (particle) of a size that is barely too large to pass through a sieve that allows 60 percent of the material (by weight) to pass through, to (2) the diameter of a grain (particle) of a size that is barely too large to pass through a sieve that allows 10 percent of the material (by weight) to pass through. The resulting ratio is a measure of the degree of uniformity in a granular material, such as filter media.

$$\text{Uniformity Coefficient} = \frac{\text{Particle Diameter}_{60\%}}{\text{Particle Diameter}_{10\%}}$$

4. The treatment efficiencies for BOD and SS removal are very good, but not as high as the sand filter.
5. Some gravel filters may develop odor problems when septic tank effluents are applied to the filters and an odor control system may be required.

Recirculating gravel filters are capable of treating and withstanding hydraulic loadings higher than design levels without a significant reduction in treatment efficiency. For this reason, they are used in communities served with conventional or combined wastewater collection systems that have infiltration/inflow (I/I) problems during wet weather conditions. During storm flows, the I/I water acts as dilution water. The filter operation mode is changed from a recirculation of filtrate and influent for multiple passes through the filter to a single pass of the storm-diluted influent and a lesser amount of recycled filtrate. Once the storm flows taper off, the normal filtrate/influent recirculation ratios are reestablished to the filter.

Gravel filters are less prone to pooling or plugging of the media than the finer media sand filters and require less attention to the filter beds. Odors usually are not a problem in properly designed, built, and operated recirculating gravel filters. However, if odor control systems are installed, the additional maintenance they require may mean the gravel filter takes more operator attention than a sand filter.

Gravel filters do not require complex operating procedures. The flow of filtrate from the filter bed can be regulated by adjusting a slide gate valve or proportioning splitter to control recycle and discharge flows. The effluent from a recirculating gravel filter may be discharged in a leach field system; by land application; or discharged to surface waters, after disinfection.

### 5.21 Loading Guidelines

Table 5.2 lists the typical loading guidelines for a recirculating gravel filter.

**TABLE 5.2 RECIRCULATING GRAVEL FILTER LOADING GUIDELINES**

| | |
|---|---|
| Pretreatment | minimum of sedimentation |
| Media | washed round pea gravel, 1.5–4.0 mm |
| Media Depth | |
| (feet) | 1.5–3.0 |
| (meters) | 0.45–1.0 |
| Hydraulic Loading | |
| (gallons/day/sq ft) | 2.5–6.0 |
| (centimeters/day) | 10–24 |
| Dosing Frequency | 5–10 minutes every 30 minutes |
| Recirculation Ratio | 3:1–5:1 |

Odors commonly associated with gravel filters are hydrogen sulfide ($H_2S$) and ammonia ($NH_3$). These two gases give off the most offensive odors. The amount of odor produced depends on the strength of the wastewater (BOD, total suspended solids) being applied to the media, the type of pretreatment provided prior to the filter, the loading rates as determined by the size and number of beds, and the climatic (weather) conditions of the area. Not all gravel filters have odor problems that will require odor control systems. Selection of an odor control system will depend entirely upon local conditions. Several methods for controlling odors are described in the following paragraphs.

1. Masking and Counteraction

   These two methods are suitable for controlling intermittent or seasonal odor problems. Odor masking has been used with limited success for many years. Odor masking is simply applying a more pleasant odor, such as a perfume, to cover the disagreeable odor. The method of applying the masking agent depends on the strength required to do the job. If only a light application is needed, evaporating the material from absorbent wicks or dripping the masking agent in the area of the unpleasant odor may take care of the problem. When higher application rates are needed, atomizers and air blowers can be used. However, this additional equipment increases the operator's maintenance duties and heavy applications of masking agents can become expensive. Some people dislike the masking agents more than the original odor generated by the treatment process.

   Counteraction is the control of odors by adding reactive chemicals to the odor, usually by spraying the chemical into the air over the odor-producing facility. The chemical reacts with the odorous gas and changes its chemical structure, thereby eliminating the odor. Caution must be exercised when considering the application of reactive chemicals because they are usually chlorinated benzene compounds. These compounds may be undesirable from an environmental standpoint.

2. Oxidation

   This method of odor control is designed into the filter bed. The underdrain system is vented through a *MANIFOLD*[13] that discharges the collected air (gases) to a treatment system where an oxidizer such as ozone is used to destroy the odor-producing compounds. These systems must be capable of handling large volumes of air because the volume of air displaced from the filter bed is equal to the volume of water applied to the bed if it is flooded. The *OZONATION*[14] unit is where the gases and ozone are thoroughly mixed in a baffled chamber for 15 to 30 seconds. A limitation of this process is the fact that ozone is very unstable and must be manufactured on site.

13. *Manifold.* A large pipe to which the ends of a series of smaller pipes are connected. Also called a header.
14. *Ozonation* (O-zoe-NAY-shun). The application of ozone to water, wastewater, or air, generally for the purposes of disinfection or odor control.

3. Absorption

Absorption is a process in which the obnoxious compounds are removed from a gas by being taken in or soaked up by a chemical solution. Absorption units are commonly referred to as scrubbers. This method of odor control is one of the more economical processes currently available.

Scrubbers operate in the following manner. Gases are collected from the filter bed and passed through containers that provide for contact between the scrubbing compounds and the air. The container may house just a water spray system, or multiple layers of trays, or it may be filled with media (Figure 5.7) to provide sufficient contact time between the gases and the odor-absorbing compounds. As the gases pass through the absorbent material, the odor compounds become attached to or are oxidized by the chemicals in the scrubber. The volume of gases to be scrubbed determines the number of scrubbers needed.

The chemicals most commonly used to absorb odors include potassium permanganate, sodium hypochlorite, caustic soda, and chlorine dioxide. People using or working near these chemicals must wear appropriate protective clothing, use approved chemical handling procedures, and strictly observe all safety regulations for their own protection and protection of the environment. (Refer to Chapter 3, "Safety," for information about Material Safety Data Sheets (MSDSs) and how to safely work with hazardous chemicals.)

4. Adsorption (Figure 5.8)

Adsorption is a process in which the odorous compounds become attached to a solid surface and are thus removed from the gas. Activated carbon is a solid with an extremely large capability to adsorb obnoxious compounds from gases. The activated carbon is held in containers sized to effectively treat the gas (air) stream. Activated carbon containers used with gravel filters may be cylinders (Figure 5.9) that are three to four feet in diameter or square containers large enough to hold activated carbon to a depth of one to two feet. Gases are collected and passed through the carbon beds where the odors are removed. Figure 5.10 shows the air manifold system used with activated carbon units to remove odors.

After a period of use, the activated carbon media will become plugged, resulting in reduced air flow through the bed. Resting the carbon filter by taking it out of service may partially restore the adsorption capacity of the carbon. Also, spraying water over the carbon bed for approximately an

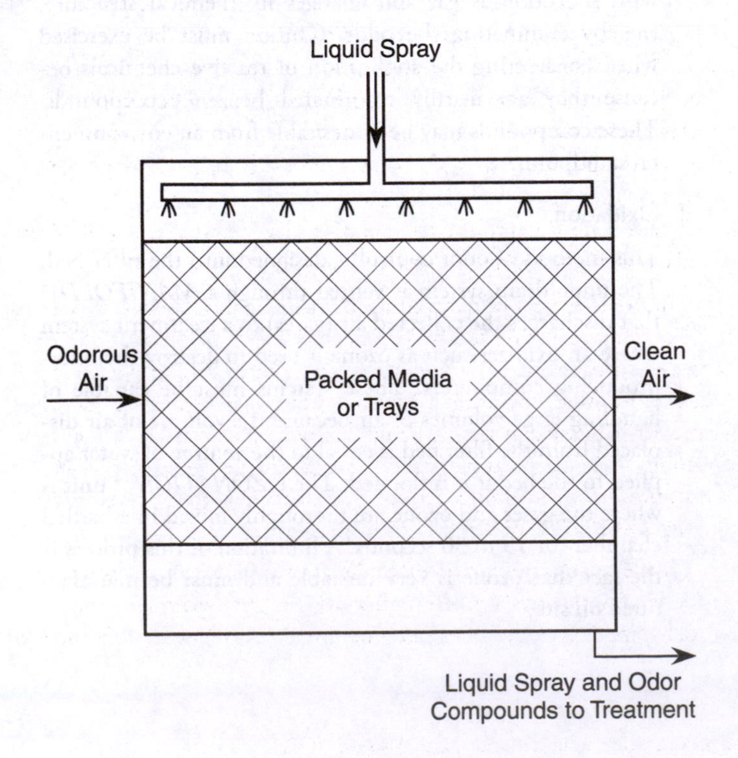


*Fig. 5.7 Packed media tower or tray tower for odor removal by absorption*

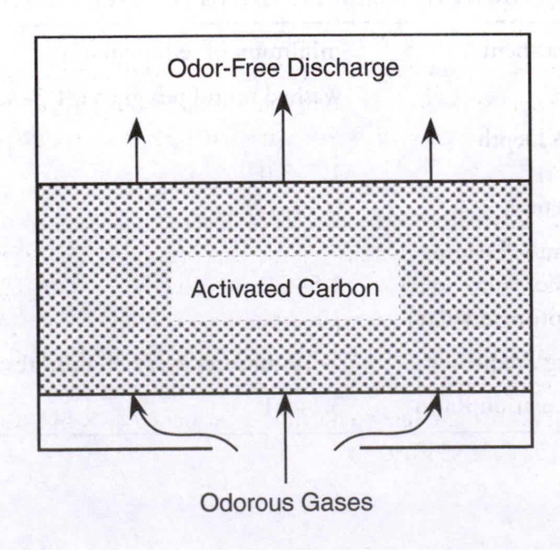


*Fig. 5.8 Activated carbon adsorption process*

*Fig. 5.9 Adsorption columns of activated carbon used to remove odors*

*Fig. 5.10 Air manifold system used with activated carbon units to remove odors*

hour will usually open the *PORES*[15] of the carbon. After activated carbon has been in service for a long time and then is taken out of service, the carbon may take on a gray appearance. The gray appearance is usually caused by salts that form crystals when the activated carbon dries out.

Activated carbon will eventually become so completely saturated with adsorbed compounds that it will lose its adsorption capacity. When this happens, the carbon must be replaced with new material. The spent (used) carbon can be regenerated by heating it in a high-temperature kiln, but often the transportation and handling costs involved in regenerating activated carbon make this option impractical. For small systems it is more economical to discard the spent material and purchase new activated carbon to recharge the filters.

Other adsorption materials such as sand and soil are commonly used on low-volume, low *THRESHOLD ODOR*[16] air streams such as vent gases from air release valves or small pumping stations. Another adsorption material called an iron sponge has occasionally been used. This material is made of wood chips coated with ferric chloride. The disadvantage of iron sponge material is that the ferric chloride is very corrosive and is difficult to handle and store.

## 5.22 Operation and Maintenance

Operational control of a gravel filter is based in part on the desired quality of the effluent that will be discharged from the filter to the environment or further treatment. The factors that determine effluent quality include the appearance, BOD, suspended solids, and nitrogen content of the effluent. Effluent quality can be adjusted by changing the ratio between the amount of filter effluent that is recycled and the amount of effluent that is discharged. Normally, 20 to 40 percent of the effluent is split off and discharged while the remaining 80 to 60 percent is returned to the recirculation chamber where it will be mixed with incoming influent and reapplied to the filter.

Another means of controlling effluent quality from a gravel filter is by adjusting the frequency and length of pump dosing cycles to intermittently apply the wastewater to alternate bed areas. Routine operation with short, frequent dosing cycles is preferred because it keeps the media moist and aerated. For example, the recirculating gravel filter could be dosed every 30 minutes for a 5-minute period. The average hydraulic loading rate could be 3 GPD/sq ft (120 min/day).

If the wastewater applied to a filter contains toxic wastes or if the wastewater has a much higher organic strength than is normally applied to the filter, the microorganisms may die off and

15. *Pore.* A very small open space in a rock or granular material. Also called an interstice, void, or void space. Also see VOID.
16. *Threshold Odor.* The minimum odor of a gas or water sample that can just be detected after successive dilutions with odorless gas or water. Also called odor threshold.

be washed out of the filter. This process is called sloughing (pronounced SLUFF-ing). If the plant is to continue to receive the higher strength wastewater and after several weeks of operation the effluent quality has not improved to an acceptable level, the operator may need to request help to determine if additional pretreatment or expansion of the filter area should be undertaken.

Sloughing is only beneficial when wastewater is pooling in an area of the filter bed and reducing flow through the filter. If the area cannot be opened up by raking or jetting (spraying with a high-velocity stream of water) the media, sloughing may remove enough accumulated material to open up the area.

Maintenance of a gravel filter involves care of the mechanical/electrical equipment. The filter beds themselves should be kept free of debris such as papers, leaves, trash, and weeds.

Distribution piping should be flushed routinely and checked for accumulation of solids. Distribution piping orifices should be uncovered to verify that they are open and delivering equal wastewater applications to the media (Figure 5.11). The orifices usually are ⅛-inch diameter and can be cleared or opened with a piece of wire.

Occasionally, recirculation chambers should be pumped down and checked for accumulation of sludge solids and the top and side walls should be checked for grease accumulation. Although corrosion from hydrogen sulfide is unlikely to be a problem since recirculation chambers are aerobic, check for any signs of surface deterioration. A strategy is needed for discharge of the chamber's contents and what to do with the influent during inspection. Use caution when inspecting a recirculation chamber. An accumulation of wastewater solids can generate toxic or explosive gases. Also, do not enter any chamber or other enclosed space (confined space) without the proper safety equipment. Always observe prescribed safety precautions and be sure additional trained people are readily available to assist you.

Figures 5.12 and 5.13 show a recirculating gravel filter plant in the community of Elkton, Oregon. The plant influent wastewater is screened septic tank effluent from individual homes and businesses. Wastewater is pumped to the plant site by a septic tank effluent pump (STEP) system connected to a small-diameter gravity collection system. In Figure 5.12, the lighter gravel is the filter bed area, which is elevated above the surrounding ground level. The dark gravel around the outer edge (perimeter) of the filter protects the perimeter earth berm from erosion and protects the media from surface water inflow.

The Elkton plant consists of an influent manhole for flow metering, a recirculation chamber equipped with submersible pumps, and a filter effluent flow-proportional splitter to return flow to the recirculation chamber and to divert the remaining portion to the leach field. The effluent distribution system in the leach field (Figure 5.14) permits dosing of separate areas on a rotating basis. Figure 5.15 shows the pump timers that control dosing applications to the areas of the filter.

*Fig. 5.11 Gravel raked away exposing distribution piping orifices discharging a wastestream*

*NOTE:* The building is not a portion of the plant.

*Fig. 5.12 Recirculating gravel filter plant, Elkton, Oregon*

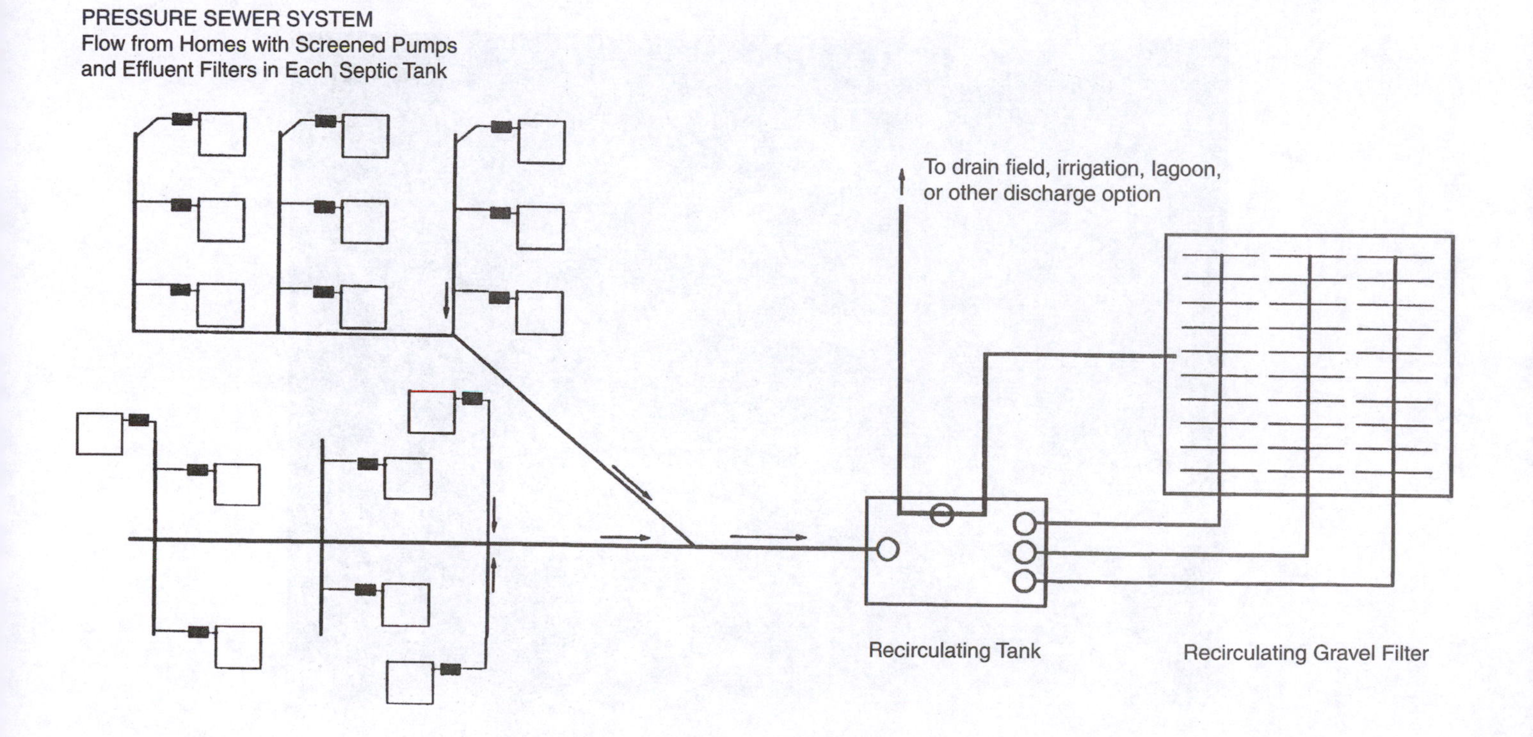


*Fig. 5.13 Layout of Elkton, Oregon, wastewater collection and treatment system*

(Notice the sheep grazing on the leach field surface.)

*Fig. 5.14 Leach field used for effluent discharge*

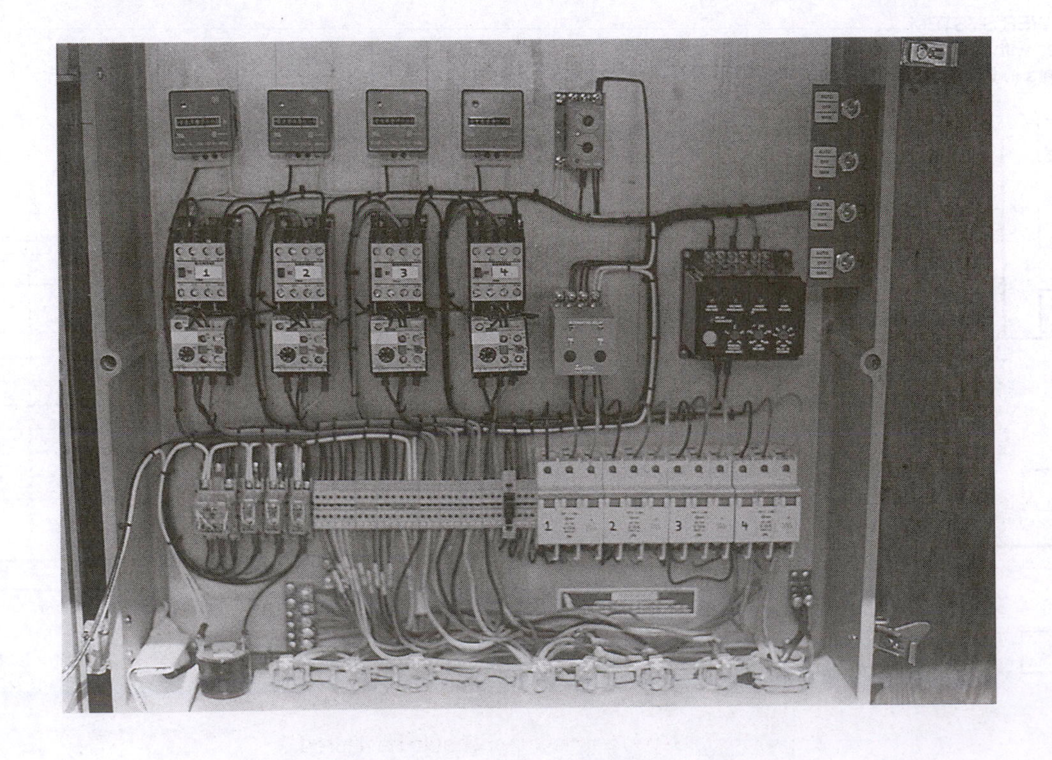

*Fig. 5.15 Pump timers used to control dosing applications*

Another recirculating gravel filter plant in Weott, California, is shown in Figures 5.16 and 5.17. This plant has two large raised media beds contained by concrete perimeter walls (only one bed is shown in Figure 5.16). The Weott plant differs from the Elkton plant in that the community is served by a conventional collection system with a lift station.

The Weott plant site contains a community septic tank where pretreatment is provided to the wastewater before it enters the filter. The large community septic tank, recirculation structure, and chlorine contact chamber occupy the space between the two filter areas. Figure 5.18 shows the access hatches over the septic tank as well as the three activated carbon filters. Figure 5.19 shows the grating covering the chlorine chamber, the other side of the activated carbon filters, and the area that houses the recirculation structure with the plant control building in the background. Effluent from the filters is discharged in a multi-area leach field that covers 24 acres and is very similar to the Elkton leach field shown in Figure 5.14. Plant effluent may be disinfected by chlorination, dechlorinated with sulfur dioxide, and discharged to surface waters during winter months as a permitted seasonal discharge.

## QUESTIONS

Please write your answers to the following questions and compare them with those on page 242.

5.2A List four odor control methods used with recirculating gravel filters.

5.2B How does odor masking work?

5.2C What chemicals are used in absorption scrubbers to control odors?

5.2D Why is routine operation with short, frequent dosing cycles preferred for recirculating gravel filters?

**END OF LESSON 1 OF 2 LESSONS**

**on**

**WASTEWATER TREATMENT AND EFFLUENT DISCHARGE METHODS**

Please answer the discussion and review questions next.

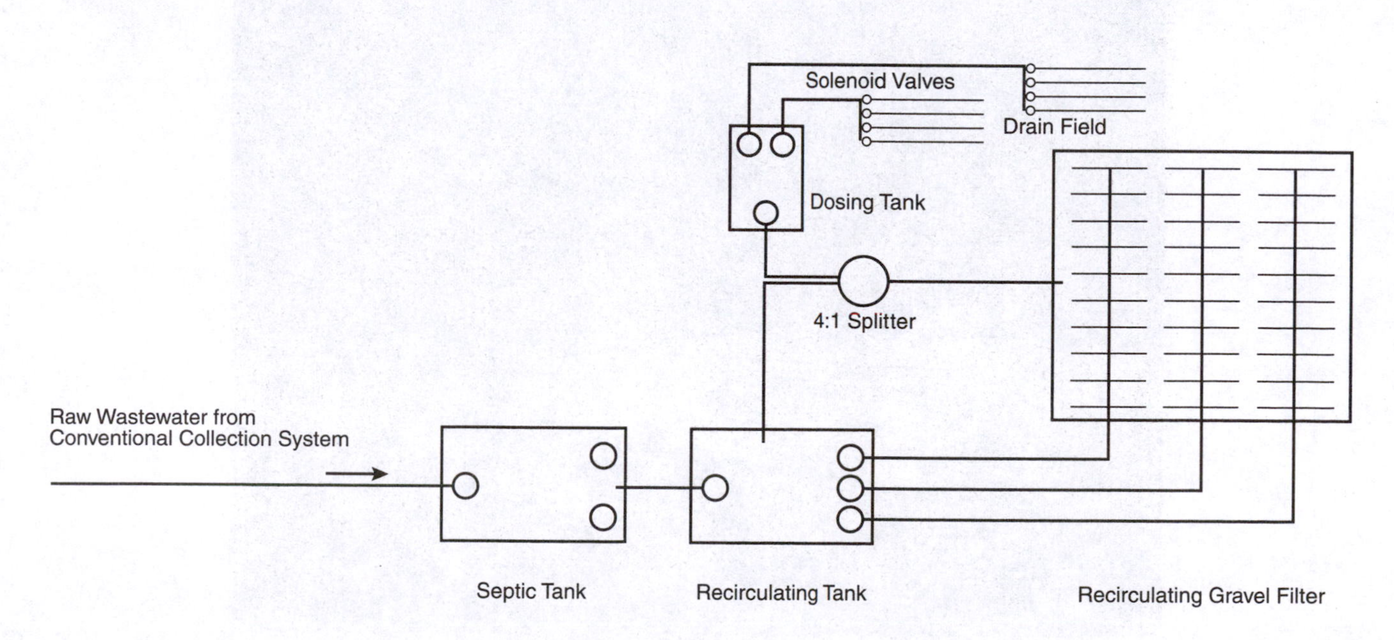


*NOTE:* Septic tank effluent should be filtered.

*Fig. 5.16 Layout of Weott, California, recirculating gravel filter plant*

*Fig. 5.17 Recirculating gravel filter bed (Weott, California)*

*Fig. 5.18 Access hatches over septic tank*

*Fig. 5.19 Grating over chlorine chamber and three activated carbon filters*

## DISCUSSION AND REVIEW QUESTIONS

### Chapter 5. WASTEWATER TREATMENT AND EFFLUENT DISCHARGE METHODS

(Lesson 1 of 2 Lessons)

At the end of each lesson in this chapter, you will find discussion and review questions. Please write your answers to these questions to determine how well you understand the material in the lesson.

1. Why do some recirculating sand filters have a gravel cover?
2. What are the main components of an open, single-pass intermittent sand filter?
3. How are aerobic conditions maintained in the media of a sand filter bed?
4. How do sand filters treat wastewater?
5. What is the purpose of a dosing siphon?
6. Why are recirculating gravel filters used in communities served with conventional or combined wastewater collection systems that have infiltration/inflow (I/I) problems during wet weather conditions?
7. How can an activated carbon bed used to remove odors from air be regenerated?
8. How is effluent quality adjusted or controlled in a recirculating gravel filter?

# CHAPTER 5. WASTEWATER TREATMENT AND EFFLUENT DISCHARGE METHODS

(Lesson 2 of 2 Lessons)

## 5.3 SUBSURFACE WASTEWATER INFILTRATION SYSTEMS (SWIS)

The safe, nuisance-free treatment and discharge of wastewater is of prime importance to protect public health and the environment. Today, many large community treatment systems discharge treated and disinfected effluents to surface waters (for example, rivers, streams, and oceans). In some areas, treated wastewater is applied to land by methods such as irrigation, overland flow, or infiltration beds. Individual on-site wastewater treatment systems typically consist of a septic tank that discharges partially treated effluent to a subsurface system such as a leach field or seepage pit. The widespread public demand for an improved environment and water conservation through recycling and reclamation has led to a better understanding of the environment and the interrelationships between the geology of soils, vegetation, climates, and hydrology as they affect people. Improved methods for treating and discharging wastewater are continually being developed to meet the demand for protection of public health and the environment.

Today, effluent discharge systems based on subgrade or subsurface land application of effluent are commonly referred to as subsurface wastewater infiltration systems (SWIS). The keys to successful use of the SWIS discharge method are reduction of solids and the *PERMEABILITY*[17] of the soil. As the wastewater passes through the soil, it receives additional treatment through filtration, absorption, and biochemical reactions. If the soil is too coarse (large particles of sand, gravel), the effluent will flow down (percolate) too rapidly, not providing the filtration and absorption or the time for adequate biochemical treatment. When this happens, there is a danger that untreated or partially treated wastewater may reach and contaminate the groundwater. If the soil is impermeable (clay or rock), the effluent will pool on the surface creating a nuisance, a public health hazard, and possible contamination of surface waters.[18]

Alternative subsurface wastewater infiltration systems (SWIS) and septic tank effluent treatment and discharge systems have been developed for areas where soil conditions are not suitable for leach field systems. The following types of SWIS are being used for individual homes, groups of homes, and small communities. These systems have been very successful if properly designed, installed, operated, and maintained. (Not all of the following systems are approved by regional public health or environmental protection agencies as acceptable discharge systems.) Types of subsurface wastewater infiltration systems include the following:

1. Leach field or drain field
2. Seepage trench
3. Seepage pit
4. Absorption mound
5. Evapotranspiration (ET) or Evapotranspiration/Infiltration (ETI)

To determine whether a site is suitable for a SWIS, the site must be tested for the following items:

- Soil Type or Classification — best type, sand-silt; poor, coarse sand and clay
- Percolation Rate — faster than 120 min/inch (MPI) or slower than 5 MPI[19]

17. *Permeability* (PURR-me-uh-BILL-uh-tee). The property of a material or soil that permits considerable movement of water through it when it is saturated.
18. Orenco's *Intermittent Sand Filters* video contains animated footage showing how wastewater is treated as it flows through the soil. The video is available for viewing on the Orenco website at http://www.orenco.com/videos/sales/ISF_Video.html.
19. A percolation rate of 60 minutes per inch is a slow percolation rate and 5 minutes per inch is a very fast percolation rate. This technique is being replaced by soil classification in some areas.

- Depth to Groundwater — sufficient to protect groundwater by allowing sufficient soil and microorganism contact to treat the wastewater (typically 5 feet minimum; 10 feet for seepage pits)
- *SLOPE* [20] — SWIS ground slope less than 30 percent
- Depth to Bedrock or Impermeable Layer — sufficient to protect groundwater (typically 5 feet minimum; 10 feet for seepage pits)

Additional items must be considered when evaluating the onsite subsurface wastewater infiltration system (SWIS) for a home. Requirements for these items vary depending on local conditions and local agencies (see Table 5.3.):

- Distance to Surface Water
- Distance to Water Supply Wells
- Distance from SWIS to Property Lines and Dwellings
- Location and Proximity of Existing Water Lines to SWIS
- Relationship of Elevation to SWIS, Home and Septic Tank
- Adequate Replacement Area

**TABLE 5.3 MINIMUM DISTANCES IN FEET BETWEEN COMPONENTS OF THE WASTEWATER DISPOSAL SYSTEM AND OTHER STRUCTURES[a]**

| | US Public Health Service | Uniform Plumbing Code | RWQCB North Coastal Region | RWQCB Lahontan Region | RWQCB Central Valley Region | Contra Costa County | Marin County |
|---|---|---|---|---|---|---|---|
| **SEPTIC TANKS TO:** | | | | | | | |
| Dwelling | 5 | 5 | – | – | – | 10 | 5 |
| Property line | 10 | 5 | – | 25 | 25 | 5 | 5 |
| Surface water | 50 | 50 | 50–100[b] | 25–50[b] | 25–50[b] | 50 | 100 |
| Water well or suction line | 50 | 50 | 100 | 50 | 50–100[b] | 50 | 100 |
| Pressure water supply line | 10 | 5 | – | – | – | 10 | 10 |
| **LEACH FIELD TO:** | | | | | | | |
| Dwelling | 20 | 8 | – | – | – | 10 | 10 |
| Property line | 5 | 5 | 50[c] | 50 | 50 | 5 | 5 |
| Surface water | 50 | 50 | 50–100 | 50–100[d] | 50–100 | 50 | 100 |
| Water well or suction line | 100 | 50 | 100 | 100 | 100 | 50 | 100 |
| Pressure water supply line | 25 | 5 | – | – | – | 10 | 10 |

a From *RURAL WASTEWATER MANAGEMENT TECHNICAL REPORT,* State of California Water Resources Control Board. See local septic tank codes for distances applicable in your area.

b Lower values are for ephemeral streams; higher values are for perennial streams, lakes, and reservoirs.

c Individual wells are used on the same lot.

d 200 ft for reservoirs and lakes.

20. *Slope.* The slope or inclination of a trench bottom or a trench side wall is the ratio of the vertical distance to the horizontal distance or rise over run. Also see GRADE (2).

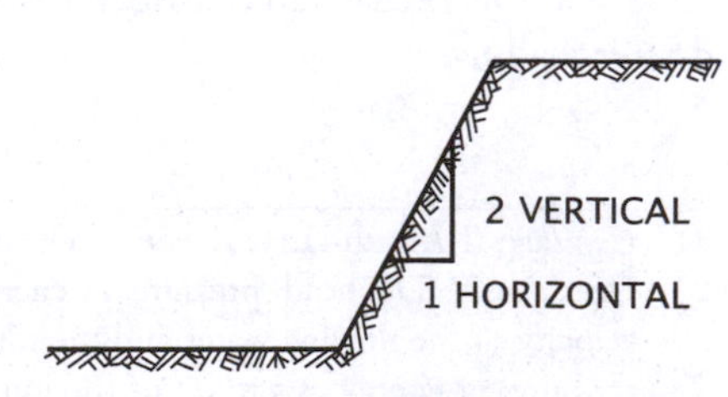

The success or failure of subsurface wastewater infiltration systems depends primarily on soil conditions, the solids content of the effluent, and the quantity of effluent being discharged to the SWIS. SWIS can be operated successfully through responsible operation of pretreatment units (septic tanks or primary treatment units) and by improving the effluent quality with the use of treatment units such as ponds, activated sludge, and rotating biological contactors (described in Volume II, Chapters 9, 10, and 11). Space must be available to install a replacement SWIS if the first one fails.

All new SWIS are constructed with inspection wells at various locations. The purpose of an inspection well is to monitor the system for uniform distribution of effluent and for signs of developing problems. For example, standing water or rising water levels in the monitoring well may indicate a need to reduce the hydraulic loading (effluent flow) or to remove the saturated area from service for a rest period. If the monitoring wells are not routinely inspected, then precautions cannot be taken to avoid failures.

Wastewater effluents applied to subsurface wastewater infiltration systems will eventually enter the groundwater of that area. Therefore, sampling wells are placed around the outside perimeter of a SWIS to determine the groundwater elevation and to obtain samples for laboratory analyses to check for groundwater contamination or pollution from the SWIS.

As with other media beds, organisms will be present in the gravel or rock, but may also inhabit portions of the permeable soil. Ninety-nine percent of all soil organisms are located in the top 16 inches of the soil, which is why drain fields should be very shallow. When the wastewater passes through the gravel and soil, the organisms stabilize the waste materials present in the effluent. Purification is also accomplished by filtering out and absorbing or exchanging some substances as the wastewater travels through the soil. The media area in the SWIS also may serve as a storage reservoir during short-term periods of high hydraulic loadings.

SWIS are effective in removing suspended solids (SS), biochemical oxygen demand (BOD), phosphorus, viruses, and pathogenic (disease-causing) organisms. Nitrate from SWIS systems may be a problem in groundwater contamination if domestic wells are too close or do not have a sanitary seal. High nitrate levels in drinking water are harmful and can be fatal for infants. Nitrate in drinking water above 10 mg/L (as nitrogen) poses an immediate threat to children under three months of age. In some infants, excessive levels of nitrate have been known to react with intestinal bacteria that change nitrate to nitrite, a form that reacts with hemoglobin in the blood. This reaction will reduce the oxygen-carrying ability of the blood and produce an anemic condition commonly known as "blue baby."

## QUESTIONS

Please write your answers to the following questions and compare them with those on page 242.

5.3A How can treated wastewater effluents be discharged?

5.3B What are the keys to successful use of subsurface wastewater infiltration systems?

5.3C List the various types of subsurface wastewater infiltration systems.

5.3D What items should a site be tested for to determine if it is suitable for installation of a subsurface wastewater infiltration system?

### 5.30 How Leach Field Subsurface Wastewater Infiltration Systems Work

The most commonly used and preferred method of effluent discharge with SWIS is one or more trenches. Long narrow trenches provide bottom area as well as a relatively large side wall area. Applied effluent percolates down through the bottom of the trench and also seeps laterally through the side walls of the trench by *CAPILLARY ACTION*.[21]

The shape of the trenches is usually narrow (1 to 3 feet wide) with an excavated depth ranging from 2 to 7 feet. The length of the trenches depends upon how the effluent is transported to the leach field distribution pipes. In gravity-flow systems, the trenches are seldom longer than 100 feet from the point of effluent application.

The bottom of a trench must be level to obtain equal distribution of the effluent across the entire bottom of the trench. In pressure (pumped effluent) leach fields, the trench lengths are determined by the *HEAD LOSS*[22] conditions on the distribution piping created by the effluent flow and discharge orifice size and spacing. A small community with a pressurized leach field may have several thousand feet of trenches divided into two or more leach fields, whereas a single home may have only one or

21. *Capillary* (KAP-uh-larry) *Action.* The movement of water through very small spaces due to molecular forces.

22. *Head Loss.* The head, pressure, or energy (they are the same) lost by water flowing in a pipe or channel as a result of turbulence caused by the velocity of the flowing water and the roughness of the pipe, channel walls, or restrictions caused by fittings. Water flowing in a pipe loses head, pressure, or energy as a result of friction. The head loss through a comminutor is due to friction caused by the cutters or shredders as the water passes through them and by the roughness of the comminutor walls conveying the flow through the comminutor. Also called friction loss.

two trenches in lengths of 50 to 200 feet, depending upon wastewater flow and soil conditions.

The soil infiltration *INTERFACE*[23] surfaces are the exposed areas consisting of the bottom and sides of trenches or, in covered excavations, the areas filled with a porous media. The media may be 3- to 4-inch rock or large (¾- to 2½-inch) gravel. Some agencies prefer a gravel/rock mixture while others accept one or the other. The rock or gravel layer is usually 12 to 36 inches deep. The function of the media is to maintain an open bed for flow of effluent to reach the side walls and bottom soil surfaces after being discharged from the distribution piping.

The effluent distribution pipe in a gravity-dosed SWIS usually consists of 4-inch diameter pipe perforated with ½-inch holes that are spaced 4 inches apart. The pipe is laid level on top of the rock and then covered with 2 to 6 inches of additional rock or gravel (Figure 5.20).

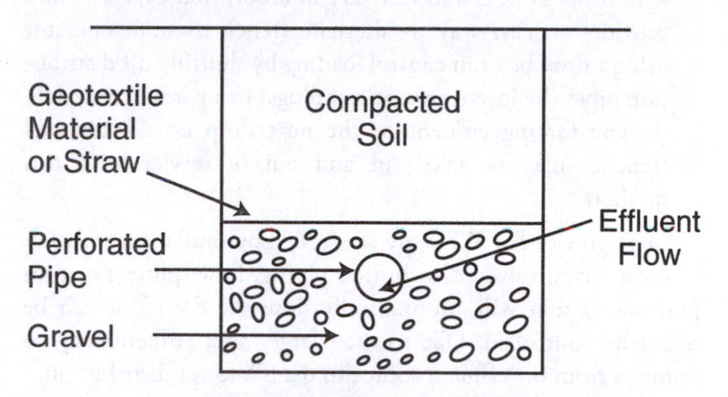


*Fig. 5.20 Leach field trench (end view)*

Before backfilling the trench, a barrier of untreated building paper, geotextile material suitable for drain fields, or a 4- to 6-inch layer of hay or straw is placed on the surface of the media to prevent the cover soil from penetrating into the media, filling the voids, and sealing it off, thus blocking effluent percolation.

Experience with existing systems has shown that narrow, shallow trenches function far better than wide, deep trenches or beds even though deep trenches require less land area. The deeper the trenches, the greater the opportunity for groundwater contamination. Depending on soil conditions, the trenches are constructed several feet apart to prevent oversaturation of the site. The useful life and efficiency of a leach field may be extended by dividing the system into two or more subsystems and controlling (rotating) effluent distribution areas. Effluent distribution to individual trenches or areas may be accomplished by installing valves, flow splitters, a distribution box, or a drop box. This arrangement enables the operator to take a portion of the SWIS out of service to let it rest or dry out while the remaining portions continue to operate normally. The rest period reduces solids and microorganism colonies in the media and on the soil interface, thus partially restoring infiltration rates and reestablishing the ability of the trench to absorb effluent. The time needed to recondition a SWIS in this way may be three months to a year.

Some leach fields use a dosing system (described in Section 5.15, "Dosing Methods") that applies the wastewater to the leach field at regular time intervals during the day.

Another type of leach field trench is called a chamber system. The effluent is distributed across the entire bottom of the trench for very effective percolation or infiltration.

Leach fields are designed to operate until a complete failure occurs. Cleanouts can be installed on the ends of the distribution piping to permit cleaning by rodding or flushing to reestablish distribution flows. If flows cannot be reestablished then the field must be replaced.

Leach field soil can sometimes be rehabilitated by applying chemicals but only experienced contractors should attempt the chemical rehabilitation procedure. The hydrogen peroxide solution that is used is highly corrosive and reactive with organic matter (producing water and oxygen gas). The oxygen gas in a confined space is dangerous and the piping/ground can explode if not properly vented. This procedure is generally successful in renovating a properly designed SWIS that has clogged with age or solids from a poorly maintained septic tank. However, this procedure will not help a leach field that has been installed in poorly draining soils, areas of bedrock, or areas with a high groundwater table.

For your own protection and that of the homeowner, several important steps should be taken before attempting to chemically recondition a leach field. As previously mentioned, working with hydrogen peroxide can be extremely hazardous. If the contractor mistakenly applies too much hydrogen peroxide, the homeowner could be held responsible for a hazardous waste cleanup. Therefore, when selecting a contractor or helping the homeowner select one, verify the contractor's previous experience and reputation for competently performing this procedure. Check with your local health department and, if possible, get written permission to do this work. Ask the contractor for a list of past customers for whom this procedure was performed. Take the time to call these references and ask if they were satisfied with the work of the contractor.

## QUESTIONS

Please write your answers to the following questions and compare them with those on page 242.

5.3E When effluent is applied to a subsurface wastewater infiltration system trench, where does the wastewater percolate out of the trench?

5.3F How is cover soil prevented from penetrating into the media of a subsurface wastewater infiltration system?

5.3G What is the purpose of a rest period for a subsurface wastewater infiltration system leach field?

---

23. *Interface.* The common boundary layer between two substances, such as water and a solid (metal); or between two fluids, such as water and a gas (air); or between a liquid (water) and another liquid (oil).

## 5.31 Monitoring and Control of a SWIS Leach Field

### *5.310 Gravity-Applied Effluent Control*

In a gravity-dosed leach field, distribution of the effluent to the trenches is controlled in one of three ways: (1) by the continuous flow method, (2) by routing flow through a distribution box, or (3) by routing flow through a drop box.

1. When the continuous flow method is used (usually for a single-family residence), all of the required trenches must be constructed on nearly level ground. The bottom elevation of each of the trenches is the same, thereby allowing the applied effluent to flow to any one of the trenches in the SWIS. The problem with this method is that low points are bound to occur. One trench may settle more than the other or the backhoe operator will be off grade and these areas will be overloaded continuously. This method is not preferred.
2. A distribution box is a watertight control structure (box) placed between the septic tank or effluent source and the SWIS. The box has one inlet from the effluent source and a separate outlet to each trench of the SWIS field (Figure 5.21). Note that all the distribution pipes run from one distribution box. The elevation of the box inlet pipe should be at least one inch higher than the outlets for the distribution pipes. The outlet distribution pipes should all be of the same diameter and elevation. The distribution box could settle due to wetting and drying of the soil, freeze and thaw cycles, or the weight of livestock walking on top of the box. If the outlets are at different elevations, one trench may receive more or less flow than the others resulting in overloading or inefficient use of the total SWIS. If the distribution box does not distribute flows properly then stop logs or other devices must be placed in the distribution box to redistribute flows.

   Many installations use a hydrosplitter (Figure 5.22) instead of a distribution box. If a distribution box does not perform properly, replace it with a hydrosplitter. A hydrosplitter (also called a pressure manifold) is an arrangement of pipes from a manifold. Septic tank effluent is pumped or dosed with a siphon into the manifold and the flow is distributed evenly to the other pipes for distribution of flow to each gravity drain field. For pressure dosing systems, a hydrosplitter flow can be split unevenly with the use of flow control orifices in the manifold to accommodate different drain field trench lengths.
3. A drop box is a watertight structure that is installed at the head or middle of each trench (Figure 5.23). Each drop box has a removable cover for access to the pipes inside the box. In a system with multiple trenches, the first box receives effluent from the source. The inlet pipe elevation is approximately two inches above the outlets of the trench dosing pipes. The first box is equipped with an effluent outlet pipe (or multiple pipes at the same elevation) to dose that trench with wastewater.

   The first drop box is also equipped with a larger outlet pipe, set at an elevation one inch below the source influent pipe, to carry wastewater from the first box to the next lower drop box on the next trench, and so on. The trench distribution pipe(s) elevation and dimensions within each box allow for the effluent to drain to the trench at its maximum capacity. Once the first trench has absorbed all that it can, it will refuse additional effluent, thereby raising the effluent level within the drop box to the supply pipeline level of the next drop box. The effluent will then be transferred to the next drop box and SWIS trench. The trenches and drop boxes may be constructed at various elevations to fit the site, but a declining grade line must be maintained on the effluent supply lines between boxes to permit gravity flow; otherwise the effluent must be pumped to the boxes.

   The major advantage of the drop box method is that it uses each trench to its maximum absorptive capacity before another trench is filled. This allows for a natural distribution according to the capacity of the individual trenches (trenches within the same SWIS can vary in absorption capacity) and provides an easy way to alternate trench use. An operator using a drop box can control loading by shutting off distribution pipes (by inserting pipe test plugs) to a particular trench, thereby forcing effluent to the next drop box. Individual trenches may be taken in and out of service with this method.

Large gravity-fed SWIS are controlled by similar means (that is, drop boxes, valved distribution piping, flow splitters or proportioners) that will automatically dose the SWIS or can be manually controlled. Due to their larger size, effluent may be pumped from the effluent source to the SWIS for distribution.

### *5.311 Pressurized Leach Field SWIS*

In pressurized leach fields, the effluent is applied by pumps. Wastewater application rates and loadings may be controlled by automatic valves or pump timers with individual pumps connected directly to a specific portion of a SWIS (Figure 5.24). Also, low-head pressurized distribution can be achieved with a siphon dosing tank.

Pressurized SWIS require a system design with built-in distribution features to apply equal loadings to the system. The main control is hydraulics. Piping is generally smaller diameter ranging from three inches down to one inch. Effluent flows from the main supply manifold distribution laterals for application. At each junction of the supply manifold and the distribution lateral,

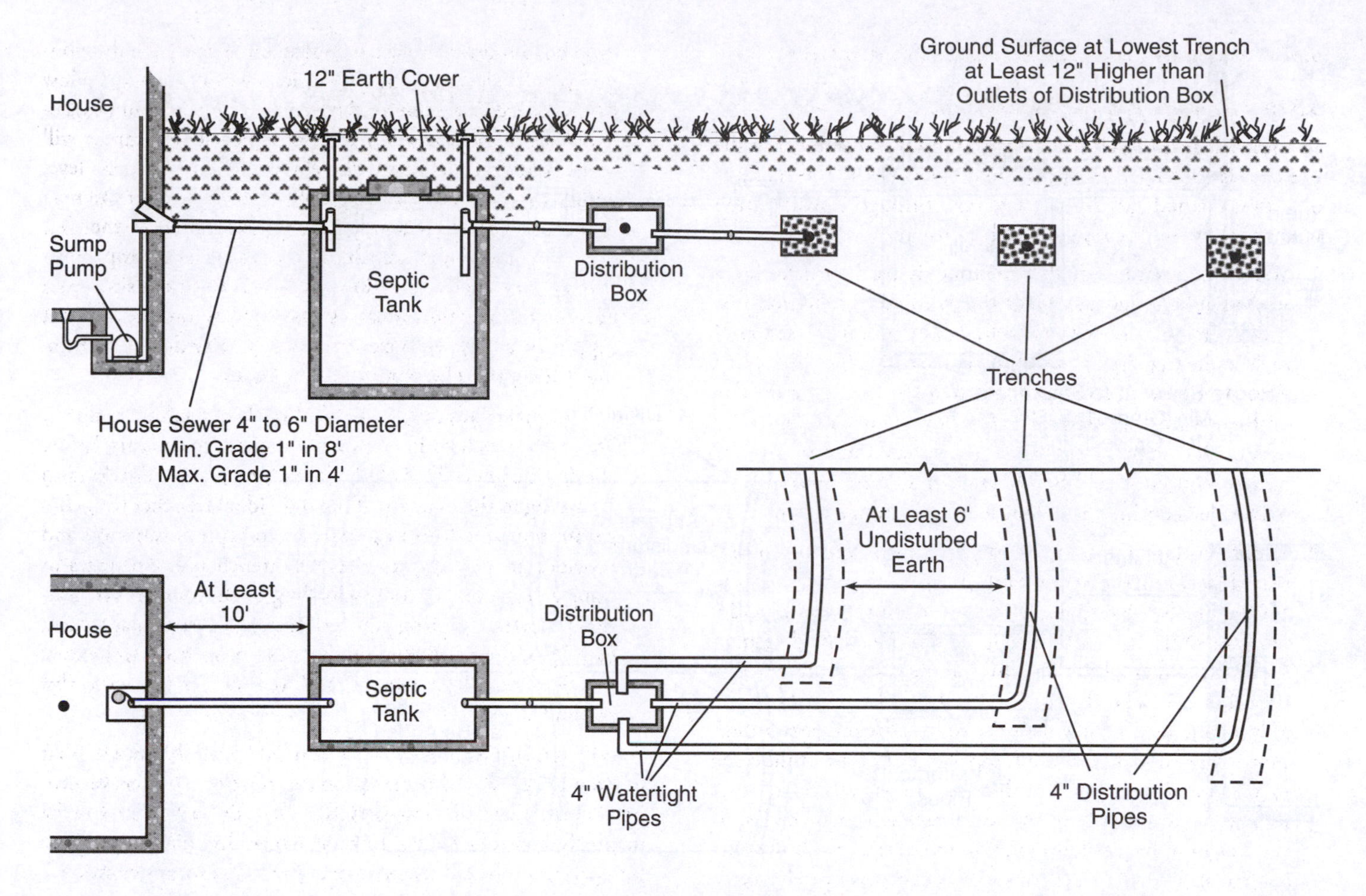


*Fig. 5.21 Typical distribution box/leach field system*

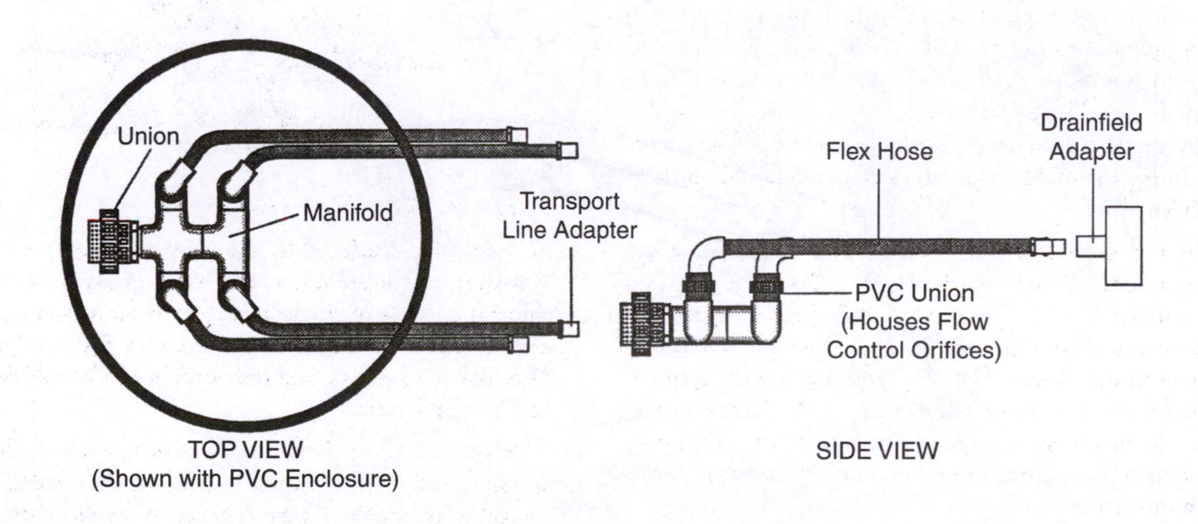


*Fig. 5.22 Hydrosplitter*

(Courtesy of Orenco Systems, Inc., www.orenco.com)

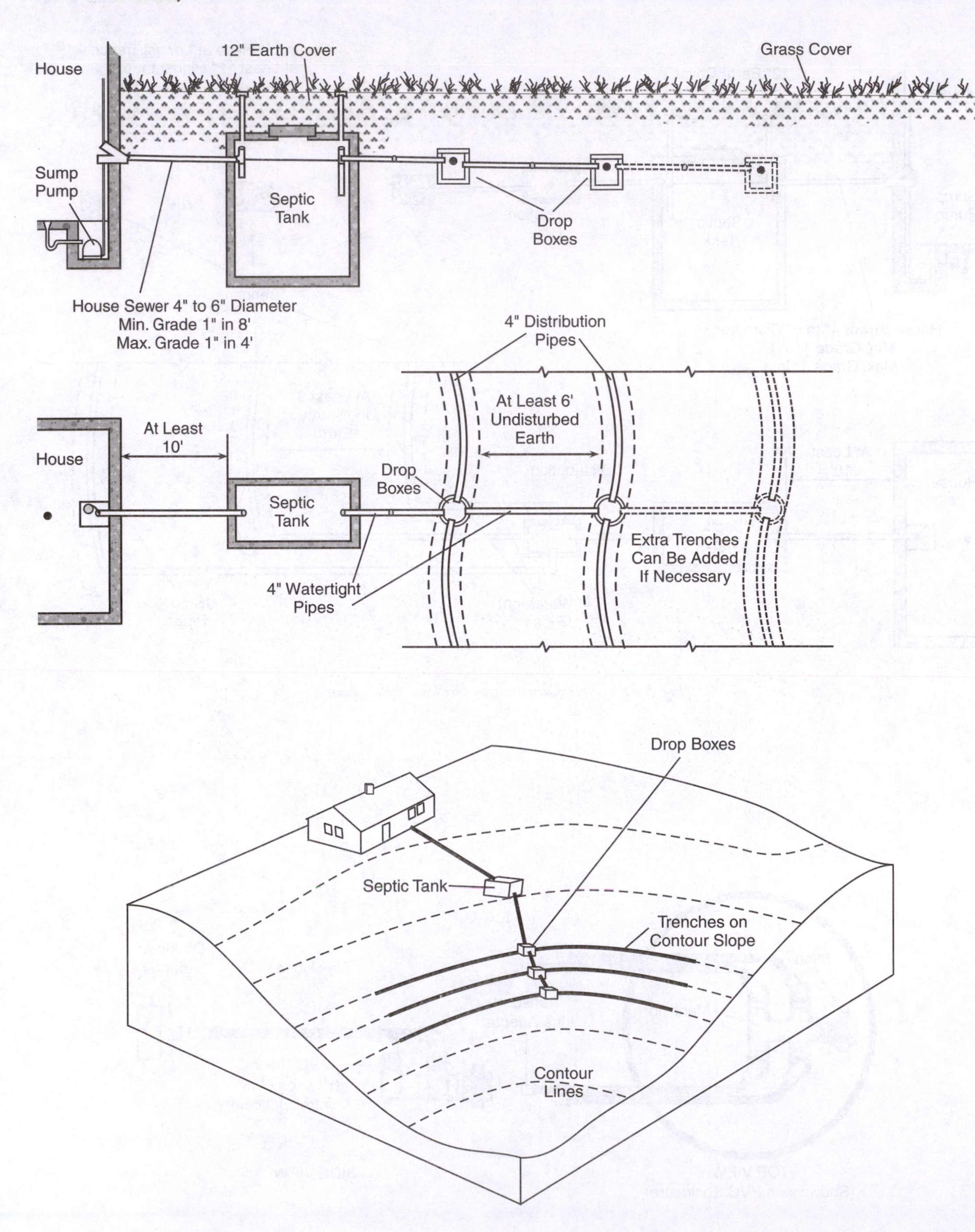


*Fig. 5.23 Typical wastewater treatment system using drop box distribution*

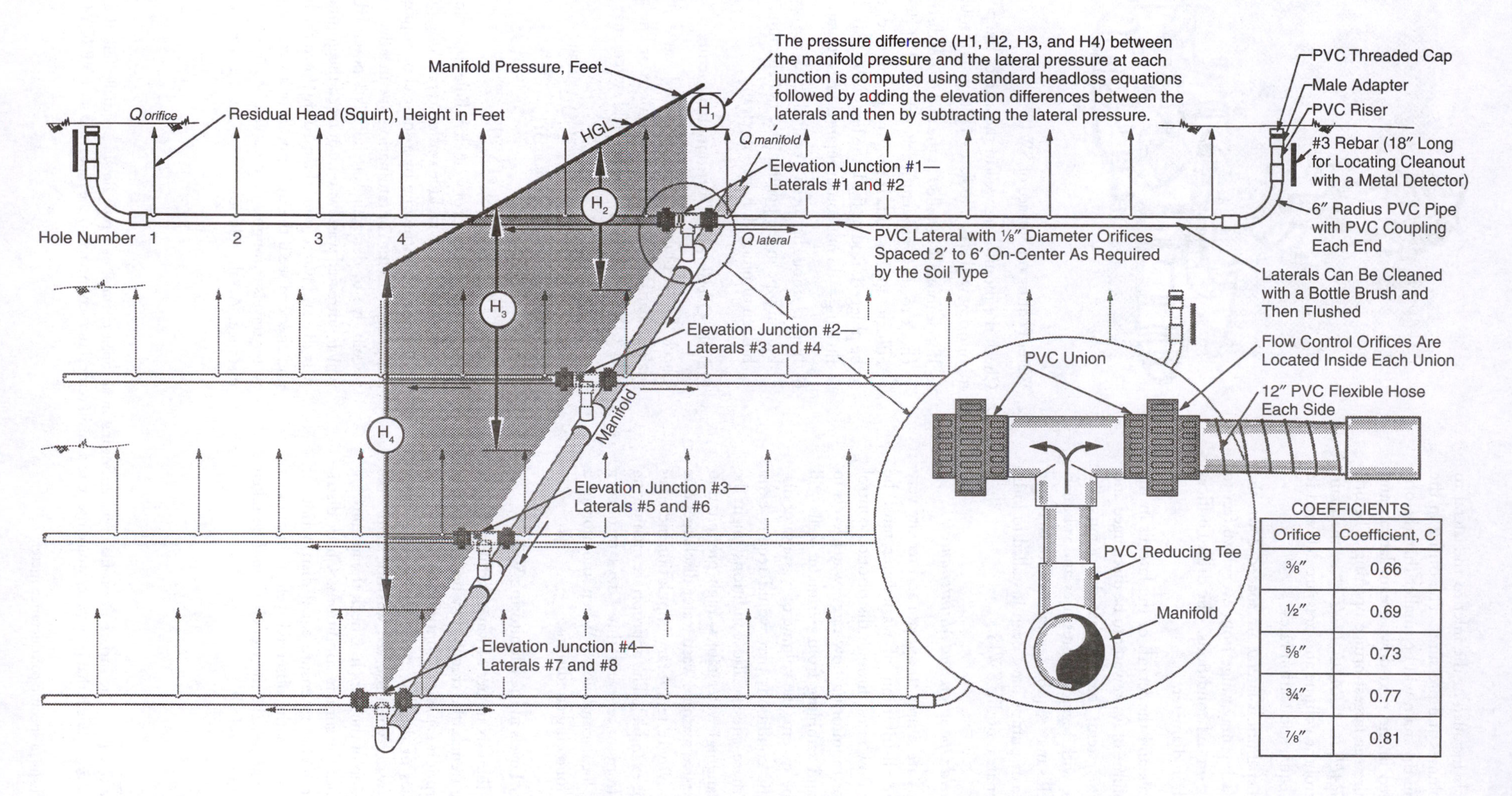


COEFFICIENTS

| Orifice | Coefficient, C |
|---|---|
| 3/8" | 0.66 |
| 1/2" | 0.69 |
| 5/8" | 0.73 |
| 3/4" | 0.77 |
| 7/8" | 0.81 |

*Fig. 5.24 Pressurized drain field laterals*
(Courtesy of Orenco Systems, Inc., www.orenco.com)

an orifice (drilled hole) is installed at the inlet to the lateral to control effluent flow volumes and pressures. The orifice in the first lateral supplied from the main will be smaller than the orifice opening size installed in the next downstream lateral, and the last lateral will require the largest orifice opening. The orifices control the *HYDRAULIC GRADE LINE*[24] in the main manifold, and thus control the volume and pressure delivered to each lateral to maintain equal application rates.

The orifices in the distribution lateral are spaced down the length of the lateral. Spacing may range from every two feet to every six feet, depending on soil conditions. The drilled orifices are almost always ⅛ inch in diameter.

The number of laterals and the length of the laterals in the SWIS depend on the volume of wastewater to be discharged and the size of the pumping system; that is, what the pump's discharge capacities will be with regard to pressures and volumes. Some SWIS pressure fields may require several acres of land (see Figure 5.24). Four effluent pumps have been installed to individually dose one to four areas of this SWIS.

### *5.312 SWIS Leach Field Operation and Maintenance*

SWIS leach field systems are simple to operate and require very little maintenance. The distribution of wastewater must be monitored by checking flow rates through distribution control units. The frequency for monitoring flow rates, water levels in the monitoring wells, and sampling from perimeter wells will depend on the size of the system, the volume of wastewater discharged, and site-specific conditions that would require extra monitoring. Some of these site-specific conditions might be high or fluctuating (changing) water tables, high precipitation rates in the area, or excessive seasonal wastewater discharges due to increased business activity or temporary population increases (an influx of tourists, for example). The appropriate monitoring frequency may range from once every three days to once a month, depending on local conditions. When monitoring a leach field, look for the following conditions:

1. High Standing Water Levels in the Monitoring Wells

   If you observe more than six inches of water (effluent) in the monitoring well of a trench, the trench may be approaching the limit of its absorption capacity. Continue to observe the well at different times of the day over a period of about a week to be sure the water you observed was not simply the result of a recent dosing of the trench. Check the monitoring wells of other trenches for similar conditions. Check distribution flows to all trenches to ensure equal distribution. If the water level increases in the observed well or in another monitoring well in that trench, the trench should be taken out of service for a rest period.

2. Water Pooling on Surface of SWIS

   Check the monitoring well in the area of the pooling. If the monitoring well is dry, look for a plugged distribution line.

   a. If a cleanout is installed at the end of the distribution line, check to see if effluent reaches the end of the line during effluent application. If no flow is observed, the line may be plugged. Rod or flush the line. *NOTE:* Be careful when flushing—do not contaminate domestic water supplies.

   b. If no cleanout has been installed, excavate over the distribution pipe in the pooling area to determine if the pipe is broken or plugged. Repair or clean.

3. Effluent Overflowing Distribution Structure

   Check for a plugged line or closed valve. Occasionally, snakes, toads, or frogs may enter a box or line and block or reduce the flow. Also, the leach field system may be clogged.

4. Water Overflowing at Septic Tank Cleanout

   Plugged line, same response as 3.

5. Crack or Hole in Surface of SWIS

   The most likely causes of a crack or hole appearing on the surface of a leach field include washout, cracking due to the soil drying out, or burrowing by a squirrel or gopher. Check the sampling wells for contamination or run a dye test with fluorescein dye to determine if there is a potential pathway by which untreated wastewater can reach the groundwater table. Mix dye and water, pour the mixture down the hole or crack, and check to see if the dye appears in a sampling well. (If the water travels through soil, dyes will be absorbed after a short distance. A salt solution[25] may be a better test liquid for

24. *Hydraulic Grade Line (HGL).* The surface or profile of water flowing in an open channel or a pipe flowing partially full. If a pipe is under pressure, the hydraulic grade line is that level water would rise to in a small, vertical tube connected to the pipe. Also see ENERGY GRADE LINE (EGL).
25. Salt may be detected using a specific conductance meter.

wastewater traveling through soil.) If the dye (or salt solution) appears in a sampling well within a few hours, this indicates a direct channel exists between the hole and the SWIS. Take that trench out of service, let it dry out, then attempt to seal the hole or crack with soil by jetting or excavating.

6. Creeping Failure of the Leach Field

   A common problem with many conventional gravity distribution leach fields is a creeping failure in the leach field. The front edge of the gravity infiltration overloads segments of the leach field and clogs the pores. The area is kept constantly wet with wastewater and does not get a chance to aerate. This failure area increases as the gravity flow is pushed down the leach field line. Creeping failure can be prevented by proper design and operation.

Other operating precautions you can take to ensure proper operation of a leach field include the following measures:

- Keep heavy equipment, tractors, and vehicles off of the surface of the SWIS. Livestock such as sheep, horses, or cattle can be permitted in dry periods to control the growth of grasses.
- Keep grass or vegetation on the surface to absorb precipitation.
- Monitor the septic tank effluent for solids in the effluent and check sludge and scum levels in the tank. Pump the tank when necessary based on sludge and scum accumulation.

If the soil in a leach field becomes sealed by excessive solids, the leach field may be rehabilitated by pumping the wastewater out of the leach field and providing an extended rest period of at least six months. Try to determine what caused the leach field failure. Refer to the "Do's and Don'ts" list in Appendix B at the end of Chapter 4. Could the cause have been bad household habits, such as dumping too much grease down the sink drain?

## QUESTIONS

Please write your answers to the following questions and compare them with those on page 243.

5.3H How is the effluent distributed to the trenches in a gravity-dosed leach field?

5.3I How are pressurized subsurface wastewater infiltration system leach fields controlled?

5.3J What items should be monitored for a subsurface wastewater infiltration system?

5.3K The frequency of monitoring subsurface wastewater infiltration systems depends on what factors?

## 5.4 SEEPAGE BEDS AND PITS

The seepage system is used where land space for a long narrow trench SWIS is not available. A seepage bed is not recommended if a trench system is feasible. Seepage pits are no longer approved in some areas because of the potential threat of contaminating a drinking water supply. The deeper the seepage pit, the closer the wastewater comes to a groundwater that might be a drinking water supply.

### 5.40 Seepage Beds

A seepage bed is basically a trench that is usually wider than three feet on the bottom and less than one hundred feet long. The wider the bed the greater the chance of disturbing the soil's natural structure, which leads to premature failure. These dimensions contribute to equal distribution of the wastewater throughout the trench. Seepage beds discharge primary treated (septic tank) wastewater effluents through the bottom by gravity forces and through the side walls by capillary action. Long, narrow trenches provide maximum side wall area for a given bottom area.

A limitation of the seepage bed is that it has a much lower ratio of side wall to bottom area than a trench system and side wall capillary action is minimal. If the bed becomes clogged with organic matter, the side walls will offer little added percolation. In addition, use of a seepage bed is limited to relatively flat areas (those with less than a 6-percent slope) to prevent surface weeping or pooling of the effluent from the bed. As with SWIS trench construction, great care should be taken to avoid compacting the soil with heavy machinery.

### 5.41 Seepage Pits

Seepage pits are used only where water tables and bedrock or clay layers are very deep. Some regulatory agencies do not allow the use of seepage pits because of the potential health hazards associated with contamination of the groundwater by the effluent.

Seepage pits are much deeper than trenches. The usual depth is thirty feet, but pits may be constructed to depths of more than fifty feet. Usually, pits are constructed by using a drilling rig to bore holes that are three to four feet in diameter. Seepage pits are filled with leach rock or lined with masonry such as brick in a radial arch without cement mortar so the effluent can escape through the walls of the pit. Due to safety concerns, very deep seepage pits are filled with large cobble rock to within three feet of the top. A brick masonry crown cap with an access cover is constructed over the top of the pit to further reduce the possibility that someone could fall into the pit. One to two feet of backfill soil is then placed over the top of the pit. A home may require from one to four pits to discharge of the wastewater effluent. In general, seepage pits are easy to install with the proper drilling equipment and are often less expensive than trench SWIS.

Some general guidelines for the construction of seepage pits are as follows:

- The pit bottom must be at least ten feet above the groundwater table to provide sufficient treatment by native soil before percolating effluent reaches the groundwater.
- The minimum depth to bedrock must be fifty feet to minimize lateral (sideways) flow or movement of the effluent near the surface.
- Soil percolation must be between 5 and 30 minutes per inch to reduce the chance of clogging and ponding leading to system failure.
- Pit diameter should be greater than three feet and less than twelve feet.

- The distribution box and the piping to each pit must be laid at the same elevation to ensure equal dosing to each pit.
- The top of the pit should be within one foot of the soil surface for ease of access.

### 5.42 Operation and Maintenance

Seepage pits require no operation and maintenance. If the pit becomes saturated due to organic matter sealing, rehabilitation of the pit may be attempted by adding hydrogen peroxide. (See the end of Section 5.30 for suggestions on how to select a contractor to perform this potentially hazardous procedure.) Usually, when a pit fails a new pit is constructed or some alternative wastewater discharge system is used.

### 5.43 Acknowledgment

Information in this section on seepage systems was adapted from *RURAL WASTEWATER MANAGEMENT,* a technical report of the California State Water Resources Control Board.

## QUESTIONS

Please write your answers to the following questions and compare them with those on page 243.

5.4A How do seepage beds discharge septic tank effluent?

5.4B Under what circumstances are seepage pits used?

5.4C Why are very deep seepage pits filled with large cobble rock?

## 5.5 ABSORPTION MOUNDS

### 5.50 Mound Description

Many areas lack native soils suitable for absorption fields. For those areas where a rock layer or the water table is too close to the ground surface, or soil percolation rates are too slow to use conventional absorption field methods, one alternative is the soil absorption mound. This is essentially a leach field elevated by fill dirt of good absorptive quality to provide an adequate separation distance between the mound and the water table or impermeable rock layer (Figures 5.25 and 5.26).

### 5.51 Location

Wherever possible, the mound should be located on a flat area or the crest of a slope. Such a location will have the least impact on surface and groundwater. Mounds should never be constructed on slopes greater than 12 percent. The greater the slope, the greater the percolation rate required to prevent seepage out the side of the mound. Table 5.4 shows suggested percolation rates (in minutes per inch, MPI) in all layers of natural or fill soil to a depth of at least 24 inches below the sand fill.

**TABLE 5.4 PERCOLATION RATES[a]**

| Slope (%) | Percolation Rate (in MPI) |
|---|---|
| Up to 3 | Faster than 120 and slower than 5[b] |
| 3–6 | Faster than 60 and slower than 5 |
| 6–12 | Faster than 30 and slower than 5 |
| Over 12 | Mounds not suitable |

a From *RURAL WASTEWATER MANAGEMENT TECHNICAL REPORT,* State of California Water Resources Control Board.

b This means percolation rates of less than 120 minutes per inch or more than 5 minutes per inch. 120 minutes per inch is considered a very slow percolation rate and 5 minutes per inch is a very fast percolation rate.

### 5.52 Size and Shape

Generally, the dimensions of a wastewater treatment mound depend on the permeability of the soil, the slope of the site, the depth of the fill below the bed, and the distance down to the seasonally high groundwater. A mound more effectively absorbs water if it is relatively long and narrow. This is particularly true when the soil percolation rate is slower than 60 MPI (minutes per inch) and the bed is on a slope. On a flat site, it is acceptable for the bed to be square if the groundwater separation is more than three feet and the percolation rate is faster than 60 MPI. If the shape of the available area will not allow for a long, narrow mound, two rock beds can be built side by side as long as the soil percolation rate is faster than 60 MPI in all layers of natural or fill soil to a depth of at least 24 inches below the sand fill. The two rock beds should be separated by at least four feet of sand.

The width of the rock bed in the mound should not exceed 10 feet. The bottom area of the rock should be calculated by allowing 125 sq ft/bedroom or 0.83 sq ft/gal of wastewater/day (Table 5.5).

**TABLE 5.5 REQUIRED AREA AND LENGTH OF ROCK BED FOR SOIL ABSORPTION MOUND**

| Number of Bedrooms | Rock Bed Area Required (sq ft) | Maximum Width of Rock Bed (ft) | Length of Rock Bed (ft) |
|---|---|---|---|
| 2 | 250 | 10 | 25.0 |
| 3 | 375 | 10 | 37.5 |
| 4 | 500 | 10 | 50.0 |
| 5 | 625 | 10 | 62.5 |
| 6 | 750 | 10 | 75.0 |

*NOTE:* The table is based on 125 sq ft/bedroom and 0.83 sq ft/gal/day. Taken from Minnesota Agricultural Extension Service Bulletin 304.

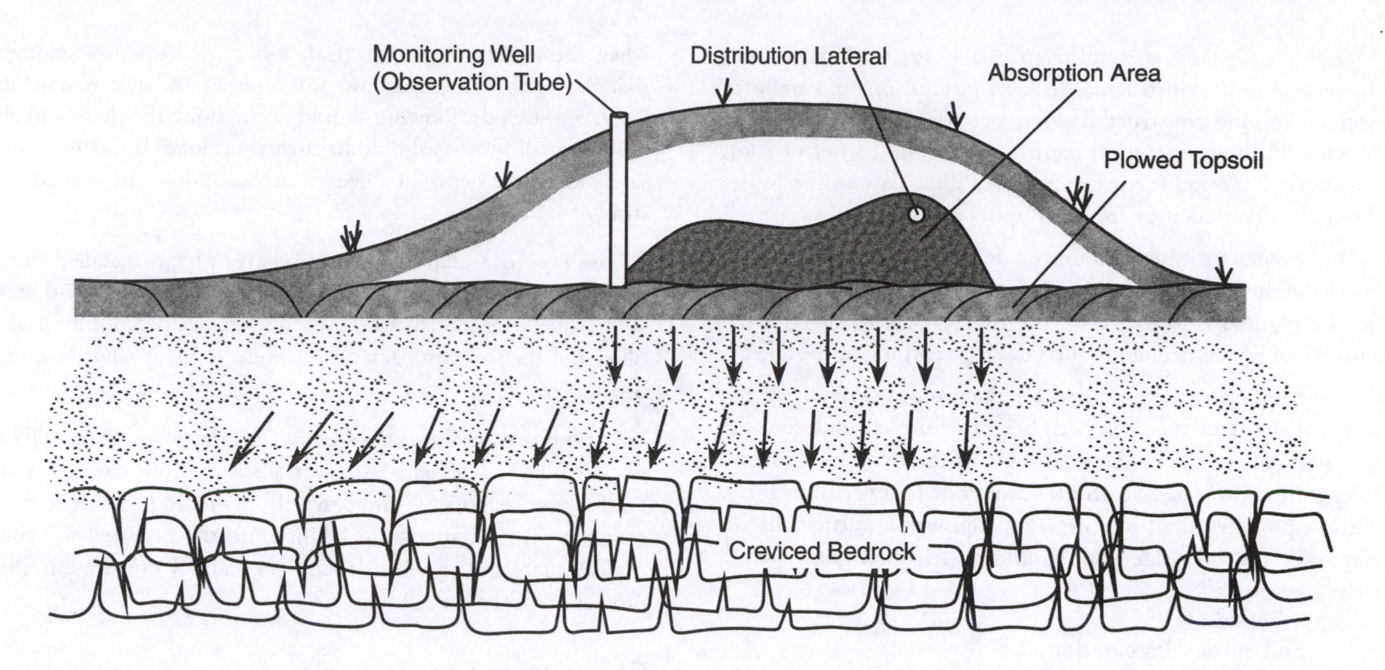


*Fig. 5.25 Mound over water table accessible through fractured rock*
(Adapted from *RURAL WASTEWATER MANAGEMENT TECHNICAL REPORT,* State of California Water Resources Control Board)

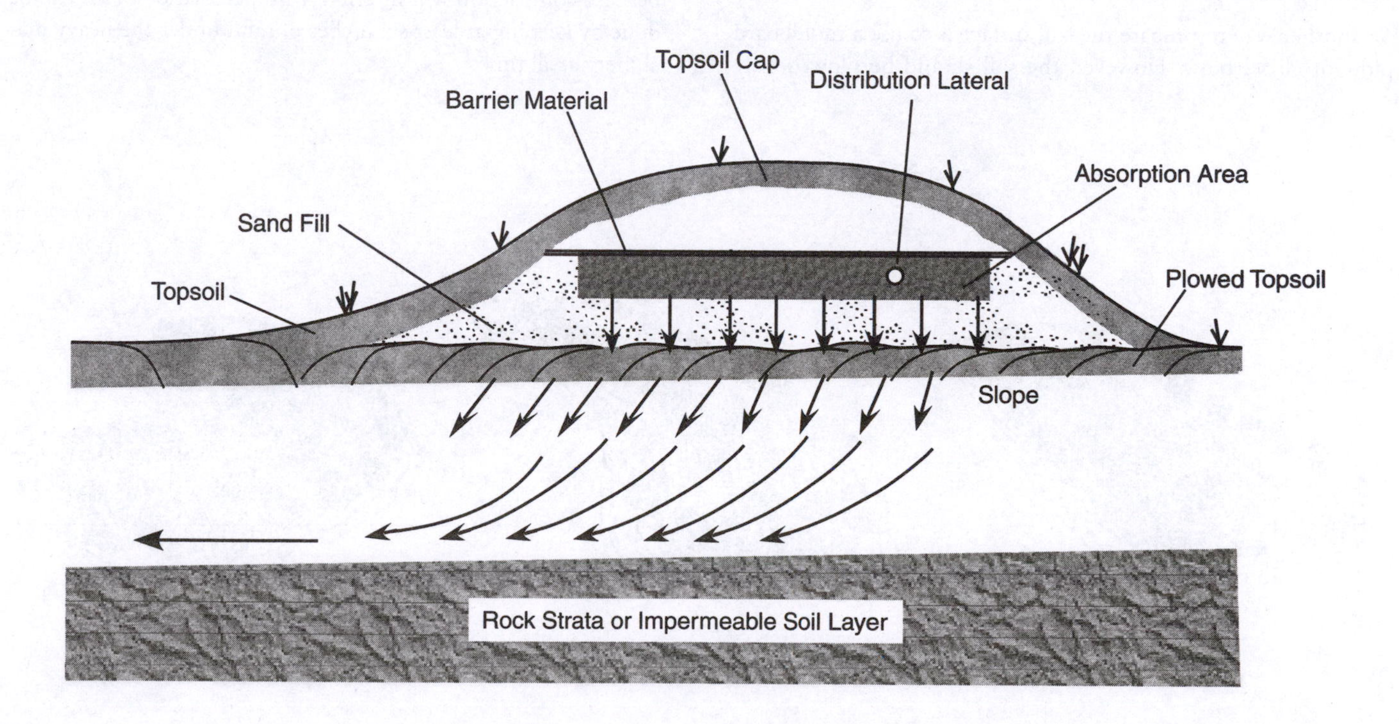


*Fig. 5.26 Mound over impermeable rock layer*
(Adapted from *RURAL WASTEWATER MANAGEMENT TECHNICAL REPORT,* State of California Water Resources Control Board)

Figure 5.27 shows three different shapes for mounds. View A shows a mound constructed on level ground or on a uniform slope. A mound constructed along a contour is shown in View B. View C shows a mound constructed in the corner of a lot. Whenever a mound is constructed on a slope, any surface water should be diverted away from the upper dike.

Wastewater treatment mounds come in many shapes and can be part of the landscape architectural plan. Shrubs or trees should not be planted directly above the rock bed, but low-growing shrubs can be planted along the side slope and toe of the dike.

The total lawn area required for the mound depends on the size and shape of the rock bed, height of the mound, and the side slope of the dike. The side slopes of the dike should be no steeper than three feet horizontally and one foot vertically (3:1). This slope is required to spread out the water, particularly on clay soils. The dike side slope should be 4:1 if the grass cover is to be mowed.

### 5.53 Soil Surface Preparation

Mound systems are used in areas of high groundwater or areas where the water percolation rate is too slow (clays, compacted silts). For clayey soils, it is critical not to smear the soil prior to constructing the mound. (To smear the soil means to smooth the soil surface, making it difficult for water to percolate or infiltrate through the soil.) If there is grass where the mound is to be constructed, the sod must be roughened to provide a good bond with mound sand fill. If the sod is not broken up, lateral seepage will occur along the original grass surface. The grass layer must be broken up without smearing or compacting the soil. One method is to use the teeth on a backhoe bucket to break up the sod and leave the surface rough.

Another way to prepare the soil surface is to use a moldboard plow or chisel plow. However, the soil should be plowed only when the moisture content at an 8-inch depth is low enough that the earth will break into small pieces and not remain in large, moist clods. Plowing should occur along the slope and all the soil in all furrows should be turned upslope. If the topsoil is sandy loam to a depth of at least 8 inches, a disk can be used instead of a plow.

The 1½- to 2-inch diameter discharge pipe is installed from the septic tank effluent pump to the center of the mound area before preparing the surface soil. The trench is carefully backfilled and the soil compacted to prevent effluent seepage along the trench.

Mound construction should start on a day when there will be no rainfall. The prepared soil surface should be covered with sand as soon as possible and every effort should be made to prevent rain (if it occurs) from falling onto the prepared soil surface. A successful mound depends on careful preparation and protection of the soil.

### 5.54 Construction (Figure 5.28)

Suggested guidelines for the construction of the eight layers of a wastewater treatment mound are as follows:

**Layer 1**—The first layer consists of clean sand to distribute the effluent over the soil surface and to provide initial filtration. Use sand that is composed of at least 25 percent particles ranging in size from 0.25 to 2.0 millimeters (mm), less than 50 percent between 0.05 to 0.25 mm, and no more than 10 percent of the particles smaller than 0.05 mm. Care must be taken to avoid a very fine sand that will have a low percolation/infiltration rate. Also, great care should be taken not to compact the ground soil because compaction will interfere with percolation. This can be done by keeping at least six inches of sand under the heavy machinery at all times.

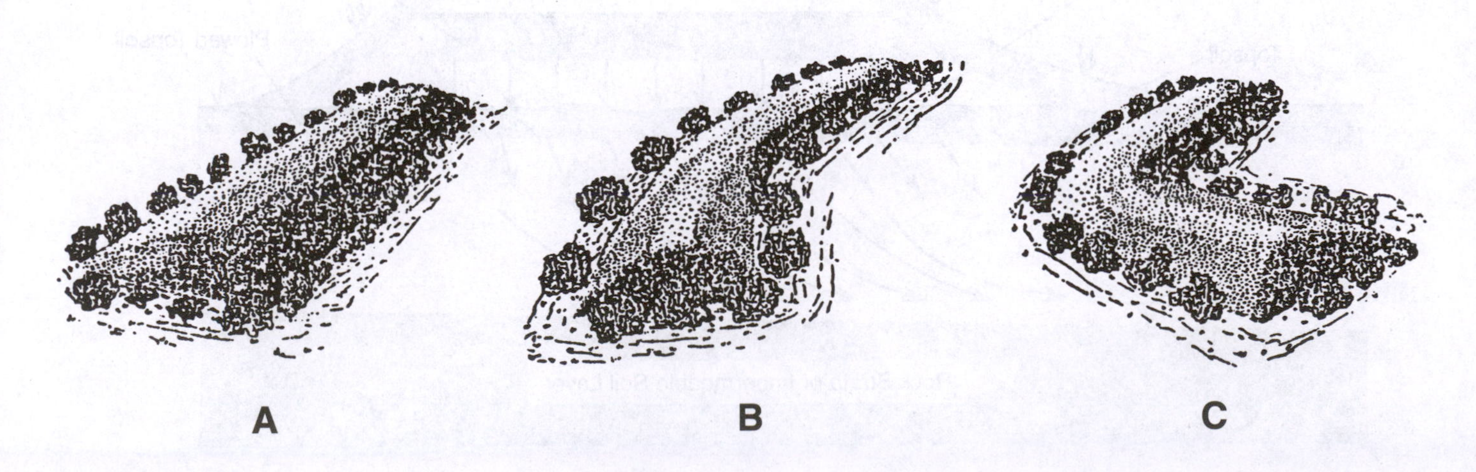


*Fig. 5.27 Three mound shapes*
(Source: *RURAL WASTEWATER MANAGEMENT TECHNICAL REPORT,* State of California Water Resources Control Board)

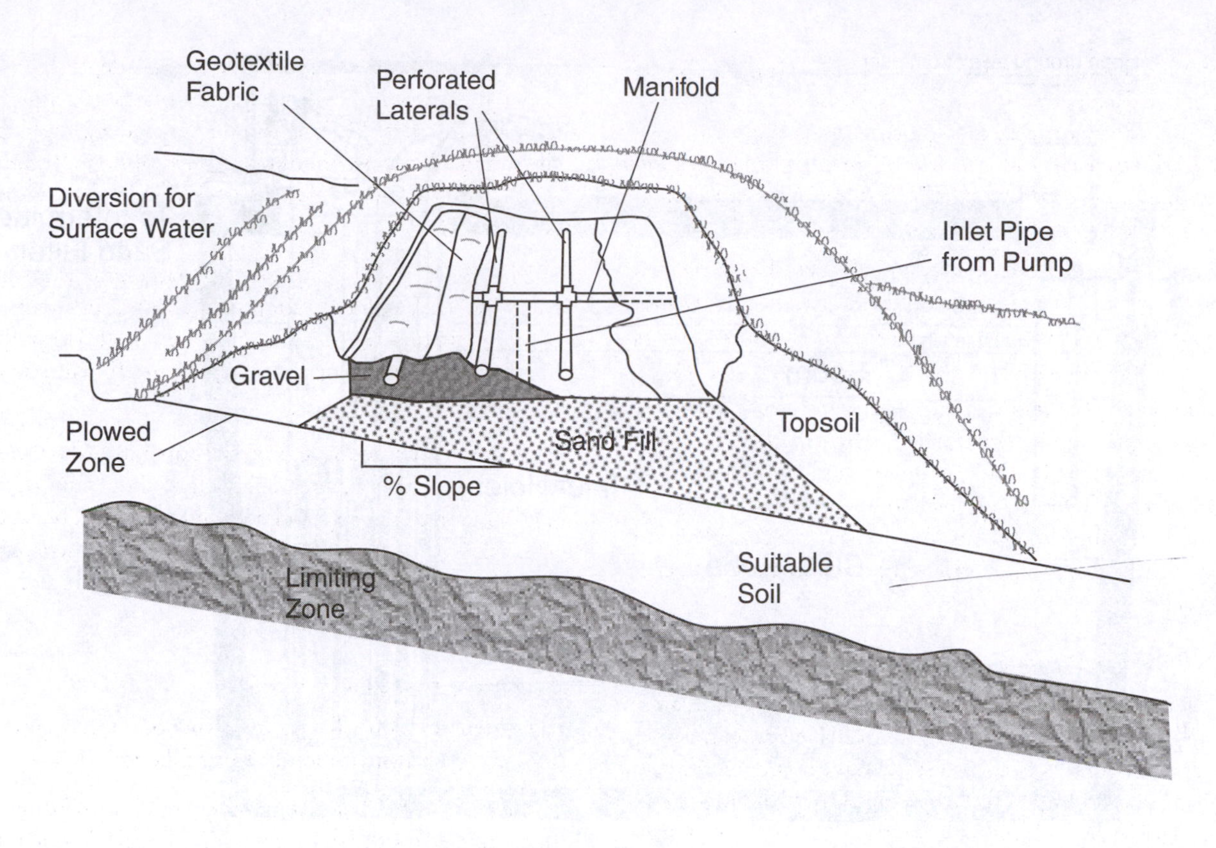


*Fig. 5.28 Layers of a typical wastewater treatment mound*

**Layer 2**—Place nine inches of trench rock or gravel over the sand.

**Layer 3**—Place the perforated plastic distribution pipe on the rock bed. The pipe should have perforations between 3⁄16 and ¼ inch in diameter spaced every three feet in a straight line. When the pipe is in place, the holes should face down on the rock bed. Three laterals are required for good distribution in a 10-foot wide rock bed, one down the length of the bed and the other two 40 inches on each side. Cap the far end of the laterals and connect the head end to a 1½- or 2-inch diameter manifold pipe. All connections must be watertight and able to withstand a pressure head of at least 40 feet.

For proper effluent distribution, no perforated lateral should be longer than 50 feet. The 2-inch manifold pipe in the center will serve a bed up to 100 feet long. Do not perforate the manifold pipe, but slope it slightly toward the supply line from the pump.

**Layer 4**—Carefully place two inches of rock over the perforated pipe.

**Layer 5**—On top of the rock bed, place a barrier of untreated building paper, geotextile material suitable for drain fields, or a 4- to 6-inch layer of hay or straw to prevent soil backfill (Layer 7) from filtering down into the rock layer.

**Layer 6**—Spread sand along the edge of the mound up to the level of the top of the rock bed. As the sand extends out from the top of the rock bed, its slope must not be greater than 3:1 (4:1 if mowing of the grass cover is planned).

**Layer 7**—The next step is to cover the geotextile material or straw (Layer 5) with sandy loam soil. The depth of the soil should be approximately 12 inches at the center of the mound and six inches at the side. If more than one rock bed is installed, the sandy loam soil should be 18 inches deep instead of 12 inches at the center.

**Layer 8**—The final layer is six inches of topsoil over the entire mound. When possible, grass should immediately be planted over the mound as a protective cover against erosion and cold temperatures.

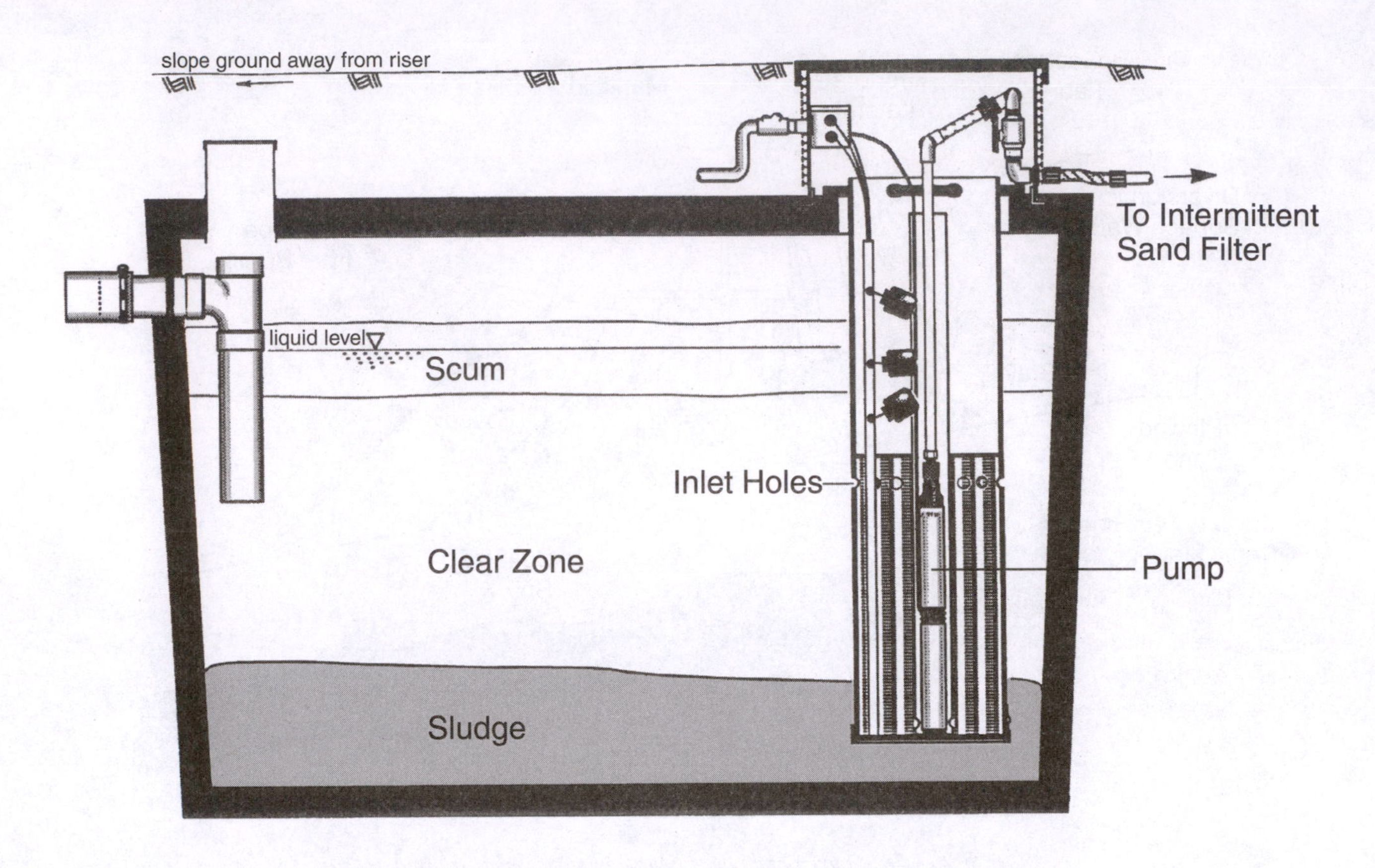


*Fig. 5.29 Typical single-compartment septic tank and pumping system*
(Courtesy of Orenco Systems, Inc., www.orenco.com)

### 5.55 Dosing

It is strongly recommended that a pump be used to lift the septic tank effluent to the discharge field even if the mound is downhill from the septic tanks. There are several reasons for this: pressure provided by the pump gives a more uniform effluent distribution and thus improves the operation of the mound; *SHORT-CIRCUITING*[26] is eliminated; dosing cycles of wetting and drying keep the soil from clogging; and on/off dosing cycles keep the mound from becoming completely saturated. (Also see Section 5.151, "Pressure Dosing," for a discussion of the use of programmable timers to regulate the size and frequency of doses applied to mounds.)

The pumping tank must be gas- and watertight and constructed of materials that will not corrode or decay. Install a manhole level to the ground surface and securely fasten the cover so that unauthorized persons cannot get into the pumping tank. Figure 5.29 shows the details of a typical single-compartment septic tank and pumping system.

No one should descend into the pumping tank without following confined space entry procedures (see Chapter 3, Section 3.5, "Confined Spaces"). Confined space entry procedures include testing the atmosphere for oxygen deficiency/enrichment, explosive gases and toxic gases (hydrogen sulfide, carbon monoxide), ventilating the area, wearing a safety harness, and having retrieval systems or two people on the ground surface strong enough to pull the person up. Gases in a pumping tank can kill you or you could fall into the wastewater and drown.

26. *Short-Circuiting.* A condition that occurs in tanks or basins when some of the flowing water entering a tank or basin flows along a nearly direct pathway from the inlet to the outlet. This is usually undesirable because it may result in shorter contact, reaction, or settling times in comparison with the theoretical (calculated) or presumed detention times.

### 5.56 Operation and Maintenance (From: *TECHNICAL NOTE ON LARGE CLUSTER MOUND SYSTEMS,* by Richard J. Otis)

Many performance problems with large SWIS are due to inadequate maintenance and management. Community-scale SWIS are wastewater treatment facilities and require management, although less than is required for conventional treatment facilities. The facility should be designed to provide maximum flexibility in operation and to permit determination of the following information on a routine basis:

- Accurate flow and dosing event counts for each cell
- Quality of the influent
- Frequency and duration of ponding in each cell
- Groundwater quality in saturated and unsaturated zones (as appropriate) underlying each cell
- Background (upslope) groundwater quality

Facility management and cell rotation decisions (ideally in early summer) should be based on accurate and up-to-date information on the operational status of the system and its impacts on the surrounding environment.

### 5.57 Inspection and Troubleshooting

To analyze problems in a mound system, you must know the location of each portion of your system. Keep a scale drawing (similar to Figure 5.30) of your system handy.

Table 5.6 contains a list of symptoms, probable causes, and possible solutions to common mound problems. Make sure you investigate all possible causes before you attempt a repair. Most of these solutions require an experienced plumber, installer, or electrician.

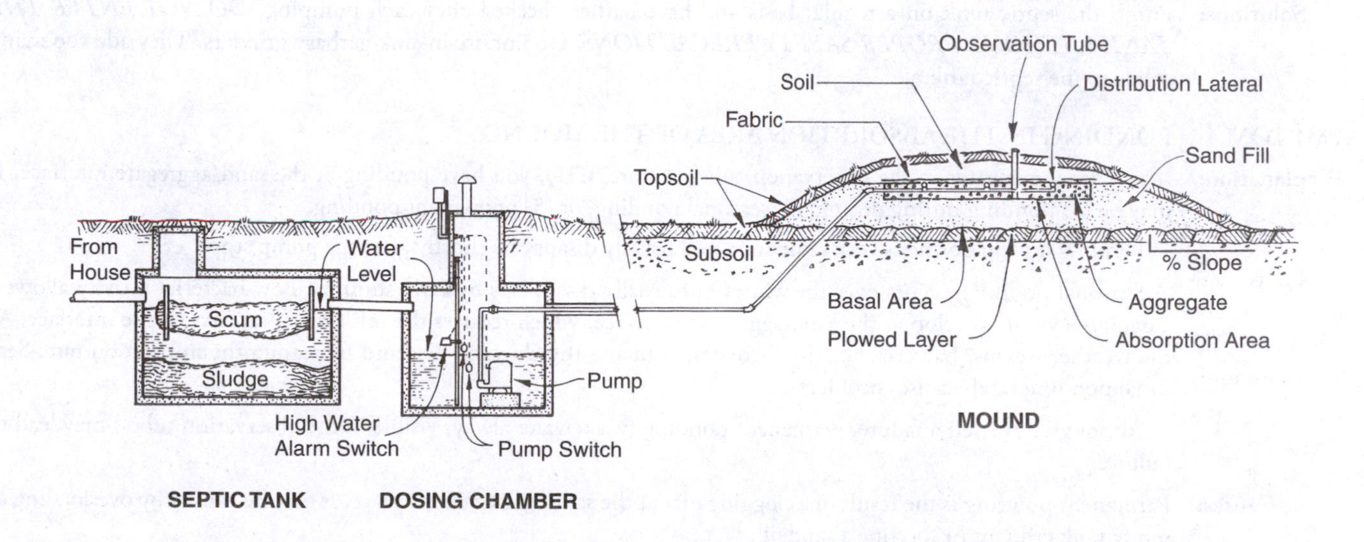


*Fig. 5.30 Typical mound system cross section showing septic tank, dosing chamber, and mound*

(Permission of James C. Converse and E. Jerry Tyler)

TABLE 5.6 TROUBLESHOOTING MOUND SYSTEMS[a]

**SYMPTOM 1: WASTEWATER BACKING UP AT THE HOUSE OR SOURCE**

**Explanation:** Toilets may flush very slowly; wastewater may back up in the floor drain.

**Causes:** If the toilet flushes slowly, the roof vent may be frosted over.

If wastewater backs up in the floor drain and slowly seeps away, tree roots or accumulated solids may be clogging the sewer line to the septic tank. The restriction is often at the inlet to the septic tank. Over time, the blockage prevents wastewater flow from the house. The outlet from the septic tank to the dose chamber may be plugged; or the pump or controls may have failed, causing water to back up into the house.

**Solution:** Check the water level in the septic tank and dose chamber. If the dose chamber is full, the problem is a faulty control unit or pump or a blockage in the force main or mound. The alarm should have sounded. If not, check the alarm system. Inspect the circuit breaker. It may have tripped.

If the liquid level is normal in the dose chamber, but higher than normal in the septic tank, the pipe connecting the septic tank and the dose chamber is plugged. Call a septic tank hauler or plumber to unplug the pipe and check the septic tank baffles.

If the septic tank level is normal, the inlet to the septic tank or the pipe between the house and the septic tank is plugged. Take care when unplugging the inlet or the pipe. ***DO NOT ENTER THE TANK WITHOUT PROPER SAFETY PRECAUTIONS.***

**SYMPTOM 2: ALARM FROM DOSE CHAMBER**

**Explanation:** When the liquid level in the dose chamber reaches a set height above the wastewater level normally needed to activate the pump, it trips an audible alarm or light in the house.

**Causes:** Faulty pump or pump controls, or a malfunctioning alarm.

Blockage in the force main or mound distribution system keeps the pump from moving water to the mound.

**Solutions:** If the problem appears to be a faulty pump or controls, see Symptom 1.

If the pump runs but the water level does not drop, then the force main or distribution laterals are plugged. See Symptom 10.

**SYMPTOM 3: EXCESSIVE SOLIDS ACCUMULATING IN THE DOSE CHAMBER**

**Explanation:** Settled solids should be removed in the septic tank. Solids carried to the dose chamber will be pumped to the mound and may plug the distribution system or the mound infiltrative surface.

**Causes:** Not pumping the septic tank often enough.

Broken baffles in septic tank.

Excessive solids introduced into the system.

**Solutions:** Pump the septic tank on a regular basis and have baffles checked after each pumping. ***DO NOT ENTER THE TANK WITHOUT PROPER SAFETY PRECAUTIONS.*** Do not use in-sink garbage grinders. They add too many solids to the septic tank.

**SYMPTOM 4: PONDING IN THE ABSORPTION AREA OF THE MOUND**

**Explanation:** If you see wastewater in the observation tubes (Figure 5.31), you have ponding at the sand/aggregate interface. It may be (1) ponding during dosing, (2) seasonal ponding, or (3) permanent ponding.

Ponding during dosing is very temporary and usually disappears shortly after the pump stops.

Seasonal ponding occurs over the winter but usually disappears by early summer. Low bacterial activity allows a clogging layer to develop at the sand/aggregate interface, which reduces the infiltration rate across the interface. As the weather warms, bacterial activity increases, reducing the clogging mat and increasing the infiltration rate. Seasonal ponding rarely causes problems.

Although not itself a failure, permanent ponding (wastewater always visible in the observation tubes) may lead to failure.

**Causes:** Permanent ponding is the result of a clogging mat at the sand/aggregate interface. It may be caused by overloading of septic tank effluent or too fine a sand fill.

**Solutions:** Check the observation tubes every 3 months to see if permanent ponding is occurring in the mound's absorption area. If the ponding appears to be permanent, reduce water use. This often reduces permanent ponding.

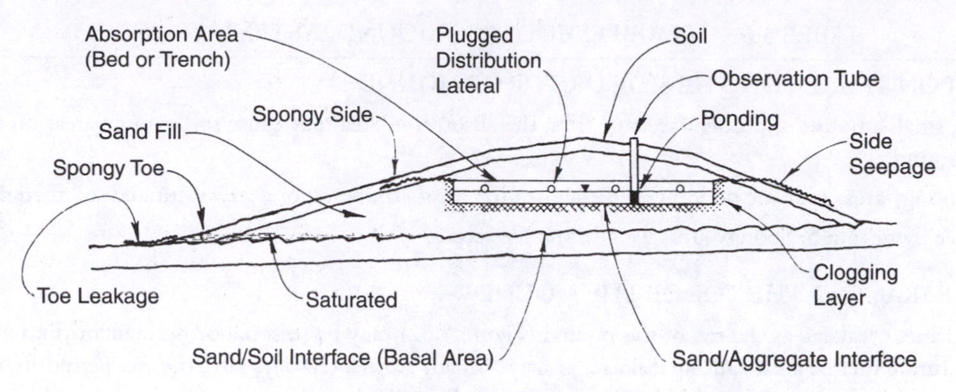


*Fig. 5.31 Cross section of a mound showing potential problems*
(Permission of James C. Converse and E. Jerry Tyler)

TABLE 5.6 TROUBLESHOOTING MOUND SYSTEMS[a] *(continued)*

**SYMPTOM 5:** SEEPAGE OUT THE SIDE OF THE MOUND

**Explanation:** Seepage out the side of the mound is usually black and smelly. It is primarily septic tank effluent that has been pumped into the mound. The breakout normally occurs around an observation tube or at other locations near the top of the mound. The effluent flows down the side of the mound (Figure 5.31).

**Causes:** A clogging mat prevents effluent from infiltrating into the sand as quickly as it is pumped into the mound. Effluent is then forced to the surface of the mound. The clogging mat appears as a black layer at the sand/aggregate interface. The sand several inches below the interface is usually dry and clean.

Temporary or continuous overloading also causes seepage out the side of the mound, even though a clogging mat may not be causing permanent ponding.

**Solutions:** Estimate the effluent entering the system. Look for

(1) Excessive water use in the home
(2) Groundwater entering the dose chamber

Reduce the loading to the mound by conserving water in the home and eliminating infiltration through joints in the riser into the dose chamber. To eliminate infiltration, recaulk all the joints on the outside of the riser including the joint between the riser and the tank cover.

Determine the quality of fill. Sample the sand at several locations and have it analyzed for particle size. (Some experienced people can estimate sand texture in the field.) If the sand beneath the absorption area is fine sand, medium sand with a lot of fines (very small particles) in it, or coarse sand containing a lot of fine and very fine sand plus silt and clay, the mound may have to be partially rebuilt.

To partially rebuild the mound:

(1) Remove the soil above the absorption area
(2) Remove the distribution system and aggregate
(3) Remove the sand beneath the absorption area down to the natural soil
(4) Replace it with an approved sand fill
(5) Replace the distribution system
(6) Cover with a synthetic fabric
(7) Replace, seed, and mulch the topsoil

Another approach may be to lengthen the mound, if you have the space.

(1) Remove the topsoil on the end slope
(2) Till the natural soil
(3) Place the proper quality sand fill
(4) Place the aggregate in the absorption area and extend the laterals
(5) Place fabric on the aggregate
(6) Place topsoil on the mound extension
(7) Seed and mulch

Note that making the absorption area wider may cause leakage, especially on slowly permeable soils. Prior to extending the mound, determine if pump or siphon will provide sufficient head at the end of the distribution laterals.

TABLE 5.6 TROUBLESHOOTING MOUND SYSTEMS[a] *(continued)*

**SYMPTOM 6: SPONGY AREA ON THE SIDE OR TOP OF MOUND**

**Explanation:** A small amount of effluent seepage from the absorption area may cause soft spongy areas on the side or top of the mound.

**Causes:** Spongy areas indicate ponding in the absorption area—the result of nearly saturated soil materials.

**Solutions:** See Symptom 5. Spongy areas usually precede seepage.

**SYMPTOM 7: LEAKAGE AT THE TOE OF THE MOUND**

**Explanation:** Effluent leakage at the toe of the mound (Figure 5.32) may be seasonal or permanent. Extremely wet weather can saturate the toe area, causing leakage. Leakage usually stops a few days after the wet period. In extreme cases, the toe may leak continuously, even during dry weather. Research has shown that the water is of high quality with no odor and few if any fecal bacteria. This leakage is often indistinguishable from natural surface water.

**Causes:** Leakage at the toe may be caused by

(1) Overloading of the mound due to excessive water use or groundwater infiltration
(2) Overestimating the infiltration rate and hydraulic conductivity of the natural soil during design
(3) *HYDROPHOBIC*[27] soils that do not readily accept water
(4) Soil compaction during construction

**Solutions:** Conserve water to add less wastewater to the system.

If the soil accepts the wastewater, but more slowly than anticipated, extending the toe sometimes eliminates the leakage.

To extend the toe:

(1) Remove the existing toe
(2) Allow the soil to dry
(3) Till downslope soil area
(4) Place sand on the tilled area
(5) Place topsoil over the sand
(6) Seed and mulch the topsoil

If the natural soil beneath the mound is dry even though the sand fill above is saturated, the natural soil is hydrophobic, compacted, or accepts the wastewater very slowly. The wastewater is moving horizontally at the sand/soil interface, rather than downward.

Extending the basal area downslope may help. You may also have to increase the length of the mound. This reduces the linear loading rate and reduces the loading at the toe. A combination of both may be required.

In extreme situations, place an interceptor drain at the downslope toe to move leakage away from the toe of the mound to a drainage ditch. Many states prohibit surface discharge of this water, so this approach may not be feasible.

If you know that groundwater is moving laterally downslope on sloping sites, place an interceptor drain on the upslope edge of the mound to intercept the groundwater. This allows the effluent to infiltrate into the soil and replace the intercepted groundwater.

**SYMPTOM 8: SPONGY AREA AT THE TOE OF THE MOUND**

**Explanation:** Saturated sand fill and nearly saturated cover soil at the toe makes it soft and spongy.

**Causes:** Causes are similar to those of Symptom 7, though not as extreme.

**Solutions:** Same as Symptom 7.

**SYMPTOM 9: TOO MUCH EFFLUENT FLOWS BACK INTO THE DOSE CHAMBER AFTER THE PUMP SHUTS OFF**

**Explanation:** The pump pressurizes the absorption area by forcing effluent into the aggregate and soil above the distribution laterals. When the pump shuts off, the effluent flows back into the dose chamber until the effluent level in the absorption area is below the distribution laterals. Side seepage may or may not occur.

**Causes:** Permanent ponding fills the aggregate below the laterals. Verify this by checking for effluent in the observation wells. Rapidly overloading the system may also cause excessive flowback.

**Solutions:** Same as Symptom 5.

---

27. *Hydrophobic* (hi-dro-FOE-bick). Having a strong aversion (dislike) for water. The opposite of HYDROPHILIC.

TABLE 5.6 TROUBLESHOOTING MOUND SYSTEMS[a] *(continued)*

**SYMPTOM 10: THE PUMP RUNS CONTINUOUSLY WITH NO DROP IN THE LIQUID LEVEL IN THE DOSE TANK**

**Explanation:** The observation tubes indicate that the absorption area is not ponded, but the mound does not accept wastewater satisfactorily.

**Causes:** Solids plug the small-diameter holes in the distribution system, and effluent cannot flow into the absorption area. Items such as "flushable" wipes, feminine hygiene products, condoms, or rags will not settle out in the septic tank and are carried over into the dose chamber and forced into the distribution pipes.

**Solutions:** Pump septic tank and dose chamber. (Every 3 years for residential units; more often for heavily used systems.)

Do not flush "flushable" wipes and similar materials down the toilets.

If system is plugged, remove the end caps to the distribution laterals and flush out the solids using a high-volume, high-pressure pump. Recap the laterals and force water or air into the distribution system to unplug the holes. Septic tank pumpers, when pressurized, force water into the laterals to remove the accumulated solids and force water out the holes to unplug them.

Consider installing a ⅛-inch screen around the pump or siphon to keep larger solids out of the system. Other types of filters may also minimize the solids carried over to the dose tank.

**SYMPTOM 11: OCCASIONAL SEPTIC ODORS**

**Explanation:** Biological activity in the septic tank and dose chamber produces ammonia, hydrogen sulfide, and other foul-smelling gases. These gases escape from the dose tank through the vent and possibly the house vent stack.

**Causes:** Odors generated in the septic tank and dose chamber can circulate to occupied areas under certain humidity and wind conditions.

**Solution:** There is no easy solution to this problem, because the odors are usually emitted through the vent of the dose chamber. Extending the dose chamber vent to roof level may minimize these unpleasant odors. If the dose chamber is vented back through the septic tank and house stack, you may be able to plug the dose tank vent during warm weather. Occasionally, the odors may be caused by gases emitted through the house stack. In this case, nothing can be done.

a Adapted from *INSPECTING AND TROUBLE SHOOTING WISCONSIN MOUNDS* by James C. Converse and E. Jerry Tyler, Small Scale Waste Management Project, University of Wisconsin-Madison, 1987.

**WARNING**

Never enter a septic tank or dose chamber without following confined space entry procedures. Septic tanks and dose chambers contain toxic gases and little or no oxygen and people have died in them. Homeowners do not have the necessary equipment or the experience to safely enter these confined spaces.

For additional information about troubleshooting problems with mound systems, refer to Chapter 7, Section 7.31, "Examples—Mound Troubleshooting."

### 5.58 Acknowledgment

Information in this section on mounds was adapted from *RURAL WASTEWATER MANAGEMENT,* a technical report of the California State Water Resources Control Board, materials provided by Orenco Systems, Inc., and *INSPECTING AND TROUBLE SHOOTING WISCONSIN MOUNDS* by James C. Converse and E. Jerry Tyler. Permission to use these materials is greatly appreciated.

## QUESTIONS

Please write your answers to the following questions and compare them with those on page 243.

5.5A Where should a mound system be located?

5.5B Where should low-growing shrubs be planted with a mound system?

5.5C Why should the sod be broken up where a mound is to be constructed?

5.5D Why should a pump be used to apply septic tank effluent to a mound discharge field?

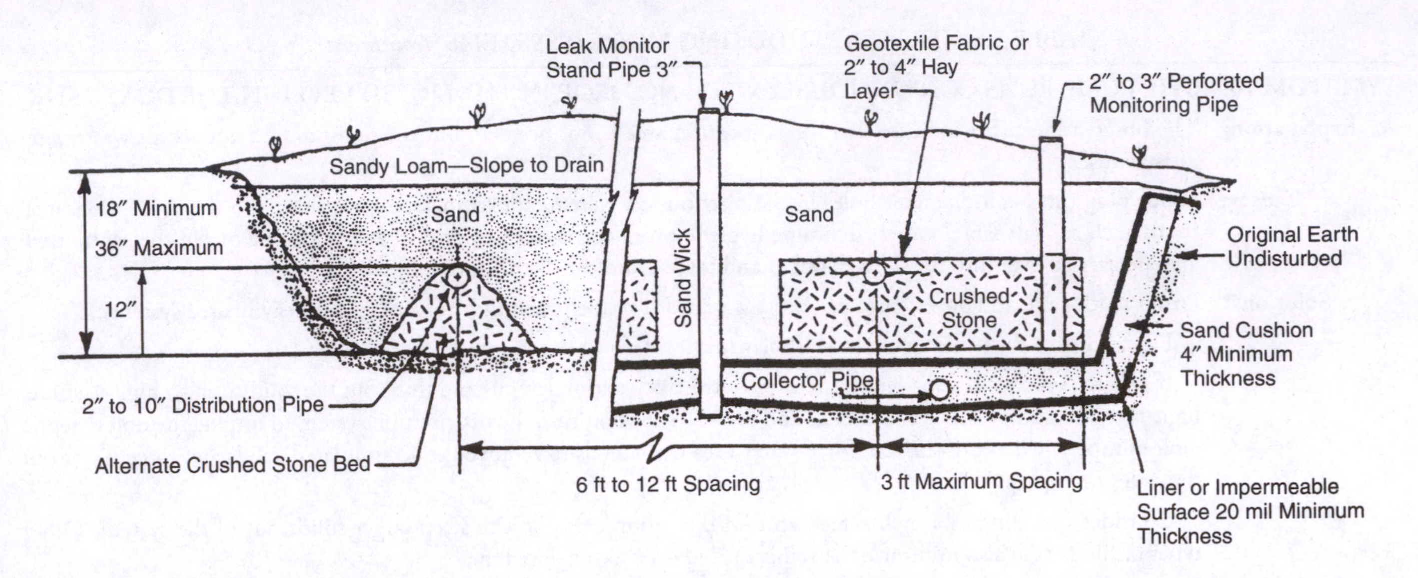


*Fig. 5.32 Typical mounded evapotranspiration system*
(Source: *RURAL WASTEWATER MANAGEMENT TECHNICAL REPORT,* California State Water Resources Control Board)

## 5.6 EVAPOTRANSPIRATION[28] SYSTEMS

A properly designed and operated evapotranspiration (ET) system is another alternative for on-site septic tank effluent treatment and discharge. The units can be operated alone so that there is no discharge to surface or groundwater, or in conjunction with a soil absorption system. Evapotranspiration is the process by which vegetation takes up water or moisture from a soil/water system and evaporates it into the atmosphere. In areas where soil or groundwater characteristics are unsuitable for conventional soil adsorption systems, ET systems offer an alternative method for treated effluent discharge. It is necessary to treat the wastewater in a septic tank or aerobic treatment unit prior to discharge to the ET unit.

Typical ET systems are relatively deep, with gravel pits for storage at the bottom of the system or storage can be provided in a separate lagoon. One type, called a mounded ET system, is shown in Figure 5.32.

Evapotranspiration involves two processes: physical evaporation and biological transpiration by plants. Since ET systems rely on heat from the sun, plant growth, and movement of water up to the soil surface, they work best in hot, dry climates with sandy soils that enhance evaporation and transpiration. They can, however, be designed for more temperate regions if greater storage is provided. Unfortunately, this increases the cost substantially.

Three types of ET systems have been used:

- *Flat-top ET systems with built-in or separate storage facilities.* These consist of three different material layers, the top layer having four to eight inches of topsoil planted with selected plant species. This layer is sloped for proper drainage of surface runoff. The second layer is made up of two to four inches of sand, with the bottom layer having two inches of gravel.
- *Mounded ET systems with built-in or separate storage facilities.* These consist of steep side slopes designed to minimize the percolation of rainwater into the ET bed. Mounded systems are advantageous in areas with very shallow groundwater or bedrock where a below-grade ET system would extend into the rock or groundwater table. Effluent from the septic tank or aerobic unit is usually distributed into the system by pressure effluent pipe.
- *Evaporation lagoons with impermeable liners or coatings into which treated effluent is discharged by gravity or pressure pipe.* If percolation to groundwater is permitted, then the liner can be omitted. Evaporation lagoons are generally only effective in hot, dry climates where evaporation rates are high. Problems with odor and excessive algal growth are common.

The main limitations of ET systems are the large areas required (2,000 to 4,000 sq ft in many areas, depending on evaporation rates), the problem of wastewater storage during seasonal cycles, and the cost of importing large volumes of fill material.

---

28. *Evapotranspiration* (ee-VAP-o-TRANS-purr-A-shun). (1) The process by which water vapor is released to the atmosphere by living plants. This process is similar to people sweating. Also called transpiration. (2) The total water removed from an area by transpiration (plants) and by evaporation from soil, snow, and water surfaces.

Salt buildup can be a problem with these systems and they may need to be pumped out periodically through a well drilled into the storage area.

Information in this section was adapted from *RURAL WASTEWATER MANAGEMENT,* a technical report of the California State Water Resources Control Board.

## QUESTIONS

Please write your answers to the following questions and compare them with those on page 243.

5.6A What is evapotranspiration?

5.6B What type of pretreatment is necessary before effluent is discharged to an evapotranspiration effluent treatment system?

5.6C What types of storage systems are used with evapotranspiration systems?

5.6D Evapotranspiration involves what processes?

## 5.7 ACKNOWLEDGMENTS

We wish to acknowledge the references used to prepare the material in this chapter.

1. *ALTERNATIVE WASTEWATER COLLECTION SYSTEMS,* Technology Transfer Manual, US Environmental Protection Agency.
2. *MANILA COMMUNITY SERVICES DISTRICT, DESIGN-CONSTRUCTION REPORT FOR THE SEPTIC TANK EFFLUENT PUMPING SEWERAGE SYSTEM,* California State Water Resources Control Board.
3. *RURAL WASTEWATER MANAGEMENT,* California State Water Resources Control Board.
4. *HOMEOWNERS' AND USERS' GUIDE FOR ON-SITE WASTEWATER DISPOSAL SYSTEMS,* California State Water Resources Control Board.
5. *ORENCO SYSTEMS, INC., ENGINEERS' CATALOG,* Orenco Systems, Inc., 814 Airway Avenue, Sutherlin, OR 97479, or visit www.orenco.com.
6. "Design, Use and Installation of Dosing Siphons for On-site Wastewater Treatment Systems" and "Pressure Dosing: Attention to Detail" by Eric S. Ball, Orenco Systems, Inc., 814 Airway Avenue, Sutherlin, OR 97479.
7. *INSPECTING AND TROUBLE SHOOTING WISCONSIN MOUNDS* by James C. Converse and E. Jerry Tyler, Small Scale Waste Management Project, University of Wisconsin-Madison, 1987.

## 5.8 ARITHMETIC ASSIGNMENT

Turn to the Appendix, "How to Solve Wastewater System Arithmetic Problems," at the back of this manual and read the following three sections:

1. A.5, "Weight/Volume Relations"
2. A.6, "Force, Pressure, and Head"
3. A.7, "Velocity and Flow Rate"

**END OF LESSON 2 OF 2 LESSONS**

**on**

**WASTEWATER TREATMENT AND EFFLUENT DISCHARGE METHODS**

Please answer the discussion and review questions next.

## DISCUSSION AND REVIEW QUESTIONS

**Chapter 5. WASTEWATER TREATMENT AND EFFLUENT DISCHARGE METHODS**

(Lesson 2 of 2 Lessons)

Please write your answers to the following questions to determine how well you understand the material in the lesson. The question numbering continues from Lesson 1.

9. Why must treated wastewater effluents be properly discharged?
10. What problems are created for a subsurface wastewater infiltration system if the soil is too coarse (large particles of sand, gravel) or too impermeable (clay or rock)?
11. What can happen if a distribution box settles or tilts and changes the elevation of the outlets of the distribution pipes?
12. What health hazards are associated with seepage pits?
13. Under what conditions could absorption mounds be considered to treat septic tank effluents?
14. Why should grass immediately be planted over a mound?
15. Under what conditions do evapotranspiration systems work best?

# SUGGESTED ANSWERS

## Chapter 5. WASTEWATER TREATMENT AND EFFLUENT DISCHARGE METHODS

### ANSWERS TO QUESTIONS IN LESSON 1

Answers to questions on page 206.

5.0A Alternative wastewater treatment and discharge systems include leach fields, seepage beds and pits, absorption mounds, evapotranspiration systems, wetlands, and land treatment systems. Sand and gravel filters are commonly used as a wastewater treatment method.

5.1A Two types of sand filters are intermittent sand filters and recirculating sand filters.

5.1B The four basic sand filter components are (1) filter bed, (2) filter media, (3) distribution piping, and (4) underdrain or collection piping.

5.1C Sand filters treat wastewater by the physical process of filtration and the biological process of stabilization of organic materials by microorganisms living in the media.

Answers to questions on page 210.

5.1D Recirculating part of the effluent back through a sand filter dilutes the wastewater being applied to the filter. This procedure improves overall treatment results and reduces odor problems. In addition, the filtrate contains dissolved oxygen, which helps to maintain the filter bed in an aerobic state.

5.1E The hydraulic loading of a sand filter is based on the amount of BOD/TSS (total suspended solids) in the applied wastewater, the type of filter design, and the media makeup.

5.1F The two methods used to dose treatment units and discharge areas are gravity dosing siphons and pressure dosing. Pumps used for pressure dosing systems may be controlled by programmable timers.

5.1G Programmable timers are used in pressure dosing systems to control the volume and frequency of wastewater discharges from the septic tank. The timer is set to run the pump's ON and OFF cycles for predetermined lengths of time so that effluent is discharged from the tank in small, uniform doses over the course of the day instead of in the peak flow volumes that would be discharged all at once in a system with a demand float switch.

Answers to questions on page 217.

5.2A Four odor control methods used with recirculating gravel filters include (1) masking and counteraction, (2) oxidation, (3) absorption, and (4) adsorption.

5.2B Odor masking is simply applying a more pleasant odor, such as a perfume, to cover a disagreeable odor.

5.2C Chemicals used in absorption scrubbers to control odors include potassium permanganate, sodium hypochlorite, caustic soda, and chlorine dioxide.

5.2D Routine operation of recirculating gravel filters in a sequence of short, frequent dosing cycles is preferred to keep the media moist and aerated.

### ANSWERS TO QUESTIONS IN LESSON 2

Answers to questions on page 222.

5.3A Treated wastewater effluents can be discharged to surface waters, application to land, or use of subsurface wastewater infiltration systems (SWIS).

5.3B The keys to successful use of subsurface wastewater infiltration systems are solids reduction and the permeability of the soil.

5.3C The various types of subsurface wastewater infiltration systems include leach field or drain field, seepage trench, seepage pit, absorption mound, evapotranspiration or evapotranspiration/infiltration systems.

5.3D To determine whether a site is suitable for installation of a subsurface wastewater infiltration system, the site must be tested for soil type, percolation rate, depth to groundwater, slope, and depth to bedrock or impermeable layer.

Answers to questions on page 223.

5.3E Effluent applied to a subsurface wastewater infiltration system trench percolates down through the bottom of the trench and also seeps laterally through the side walls of the trench by capillary action.

5.3F Cover soil is prevented from penetrating into the media of a subsurface wastewater infiltration system by placing a barrier of paper, felt, straw, or fabric on the surface of the media before backfilling.

5.3G The purpose of a rest period for a subsurface wastewater infiltration system leach field is to reduce solids and microorganism colonies in the media and on the soil interface, thus partially restoring infiltration rates and reestablishing the ability of the trench to absorb effluent.

Answers to questions on page 229.

5.3H In a gravity-dosed leach field, distribution of the effluent to the trenches is controlled in one of three ways: (1) by the continuous flow method, (2) by routing flow through a distribution box, or (3) by routing flow through a drop box.

5.3I Pressurized SWIS (subsurface wastewater infiltration system) leach fields, to which the effluent is applied by pump pressure, may be controlled by automatic valves or pump timers.

5.3J Items that should be monitored for a subsurface wastewater infiltration system include flow rates, water levels in monitoring wells, pooling on the surface, and sampling from perimeter wells.

5.3K The frequency of monitoring subsurface wastewater infiltration systems depends on the size of the system, the volume of wastewater discharged, and site-specific conditions that would require extra monitoring.

Answers to questions on page 230.

5.4A Seepage beds discharge septic tank effluents through the bottom by gravity forces and through side walls by capillary action.

5.4B Seepage pits are used only where water tables and bedrock or clay layers are very deep.

5.4C Very deep seepage pits are filled with large cobble rock for safety reasons (someone could fall in).

Answers to questions on page 239.

5.5A A mound system should be located on a flat area or the crest of a slope. Such a location will have the least impact on surface and groundwater.

5.5B Low-growing shrubs can be planted along the side slope and toe of the dike of a mound.

5.5C If the sod in not broken up where a mound is to be constructed, lateral seepage will occur along the original grass surface.

5.5D A pump should be used to apply septic tank effluent to a mound discharge field because the pressure provided by the pump gives a more uniform effluent distribution and thus improves the operation of the mound; short-circuiting is eliminated; dosing cycles of wetting and drying keep the soil from clogging; and on/off dosing cycles keep the mound from becoming completely saturated.

Answers to questions on page 241.

5.6A Evapotranspiration is the process by which vegetation takes up water or moisture from a soil/water system and evaporates it into the atmosphere.

5.6B Necessary pretreatment before an evapotranspiration effluent treatment system should include either a septic tank or an aerobic treatment unit.

5.6C Storage systems used with evapotranspiration systems include either gravel pits at the bottom of the system or a separate storage lagoon.

5.6D Evapotranspiration involves two processes: physical evaporation and biological transpiration by plants.

CHAPTER 6

# COLLECTION SYSTEMS

by

John Brady

# TABLE OF CONTENTS

## Chapter 6. COLLECTION SYSTEMS

# OBJECTIVES

## Chapter 6. COLLECTION SYSTEMS

1. Identify and describe the components of gravity, pressure, and vacuum collection systems.
2. Describe how gravity, pressure, and vacuum collection systems work.
3. Operate and maintain a small-diameter gravity sewer (SDGS) system, a pressure collection system, and a vacuum collection system.
4. Develop a maintenance program for your system.
5. Perform preventive and emergency maintenance on your system.
6. Troubleshoot problems in your system.
7. Accurately record and evaluate operation, maintenance, and cost information.
8. Safely perform your duties.

# WORDS

## Chapter 6. COLLECTION SYSTEMS

AESTHETIC (es-THET-ick) AESTHETIC

Attractive or appealing.

AIR BINDING AIR BINDING

The clogging of a filter, pipe, or pump due to the presence of air released from water. Air entering the filter media is harmful to both the filtration and backwash processes. Air can prevent the passage of water during the filtration process and can cause the loss of filter media during the backwash process.

APPURTENANCE (uh-PURR-ten-nans) APPURTENANCE

Machinery, appliances, structures, and other parts of the main structure necessary to allow it to operate as intended, but not considered part of the main structure.

CAVITATION (kav-uh-TAY-shun) CAVITATION

The formation and collapse of a gas pocket or bubble on the blade of an impeller or the gate of a valve. The collapse of this gas pocket or bubble drives water into the impeller or gate with a terrific force that can knock metal particles off and cause pitting on the impeller or gate surface. Cavitation is accompanied by loud noises that sound like someone is pounding on the impeller or gate with a hammer.

CORPORATION STOP CORPORATION STOP

A water service shutoff valve located at a street water main. This valve cannot be operated from the ground surface because it is buried and there is no valve box. Also called a corporation cock.

CURVILINEAR (KER-vuh-LYNN-e-ur) CURVILINEAR

In the shape of a curved line.

DROP JOINT DROP JOINT

A sewer pipe joint where one part has dropped out of alignment. Also see VERTICAL OFFSET.

DYNAMIC HEAD DYNAMIC HEAD

When a pump is operating, the vertical distance (in feet or meters) from a point to the energy grade line. Also see ENERGY GRADE LINE (EGL), STATIC HEAD, and TOTAL DYNAMIC HEAD (TDH).

DYNAMIC PRESSURE DYNAMIC PRESSURE

When a pump is operating, pressure resulting from the dynamic head.

$$\text{Dynamic Pressure, psi} = (\text{Dynamic Head, ft})(0.433\ \text{psi/ft})$$

or

$$\text{Dynamic Pressure, kPa} = (\text{Dynamic Head, m})(9.8\ \text{kPa/m})$$

EASEMENT EASEMENT

Legal right to use the property of others for a specific purpose. For example, a utility company may have a five-foot (1.5 m) easement along the property line of a home. This gives the utility the legal right to install and maintain a sewer line within the easement.

ENERGY GRADE LINE (EGL) ENERGY GRADE LINE (EGL)

A line that represents the elevation of energy head (in feet or meters) of water flowing in a pipe, conduit, or channel. The line is drawn above the hydraulic grade line (gradient) a distance equal to the velocity head ($V^2/2g$) of the water flowing at each section or point along the pipe or channel. Also see HYDRAULIC GRADE LINE (HGL).

[SEE DRAWING ON PAGE 254]

ENTRAIN ENTRAIN

To trap bubbles in water either mechanically through turbulence or chemically through a reaction.

FLOW LINE FLOW LINE

(1) The top of the wetted line, the water surface, or the hydraulic grade line of water flowing in an open channel or partially full conduit.

(2) The lowest point of the channel inside a pipe, conduit, canal, or manhole. This term is used by some contractors, however, the preferred term for this usage is invert.

FRICTION LOSS FRICTION LOSS

The head, pressure, or energy (they are the same) lost by water flowing in a pipe or channel as a result of turbulence caused by the velocity of the flowing water and the roughness of the pipe, channel walls, or restrictions caused by fittings. Water flowing in a pipe loses head, pressure, or energy as a result of friction. Also called head loss.

HEAD LOSS HEAD LOSS

The head, pressure, or energy (they are the same) lost by water flowing in a pipe or channel as a result of turbulence caused by the velocity of the flowing water and the roughness of the pipe, channel walls, or restrictions caused by fittings. Water flowing in a pipe loses head, pressure, or energy as a result of friction. The head loss through a comminutor is due to friction caused by the cutters or shredders as the water passes through them and by the roughness of the comminutor walls conveying the flow through the comminutor. Also called friction loss.

[SEE DRAWING ON PAGE 255]

HYDRAULIC GRADE LINE (HGL) HYDRAULIC GRADE LINE (HGL)

The surface or profile of water flowing in an open channel or a pipe flowing partially full. If a pipe is under pressure, the hydraulic grade line is that level water would rise to in a small, vertical tube connected to the pipe. Also see ENERGY GRADE LINE (EGL).

[SEE DRAWING ON PAGE 254]

HYDRAULIC JUMP HYDRAULIC JUMP

The sudden and usually turbulent abrupt rise in water surface in an open channel when water flowing at high velocity is suddenly retarded to a slow velocity.

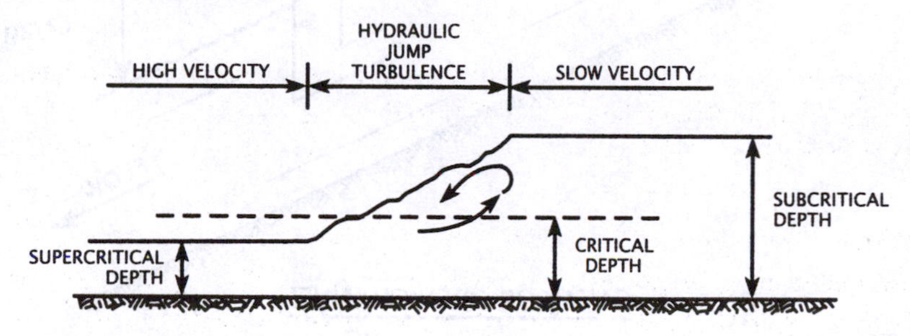


IMPELLER IMPELLER

A rotating set of vanes in a pump or compressor designed to pump or move water or air.

INTERCEPTOR (INTERCEPTING) SEWER INTERCEPTOR (INTERCEPTING) SEWER

A large sewer that receives flow from a number of sewers and conducts the wastewater to a treatment plant. Often called an interceptor. The term interceptor is sometimes used in small communities to describe a septic tank or other holding tank that serves as a temporary wastewater storage reservoir for a septic tank effluent pump (STEP) system.

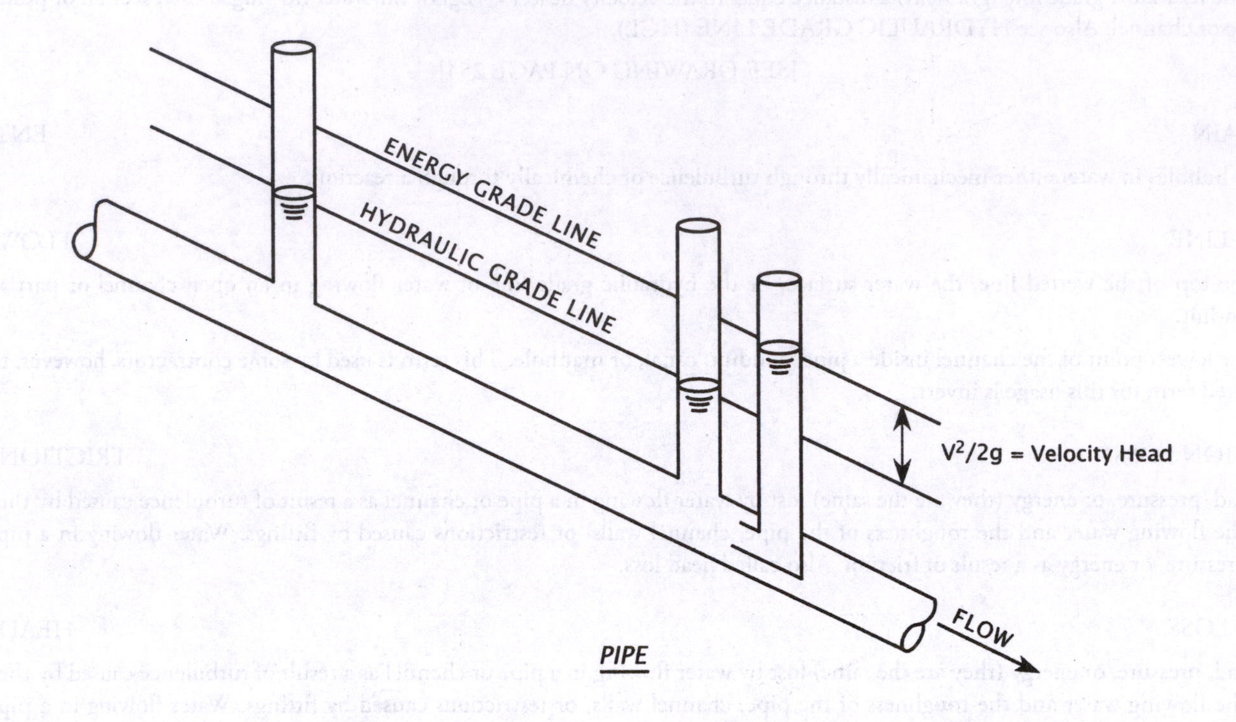


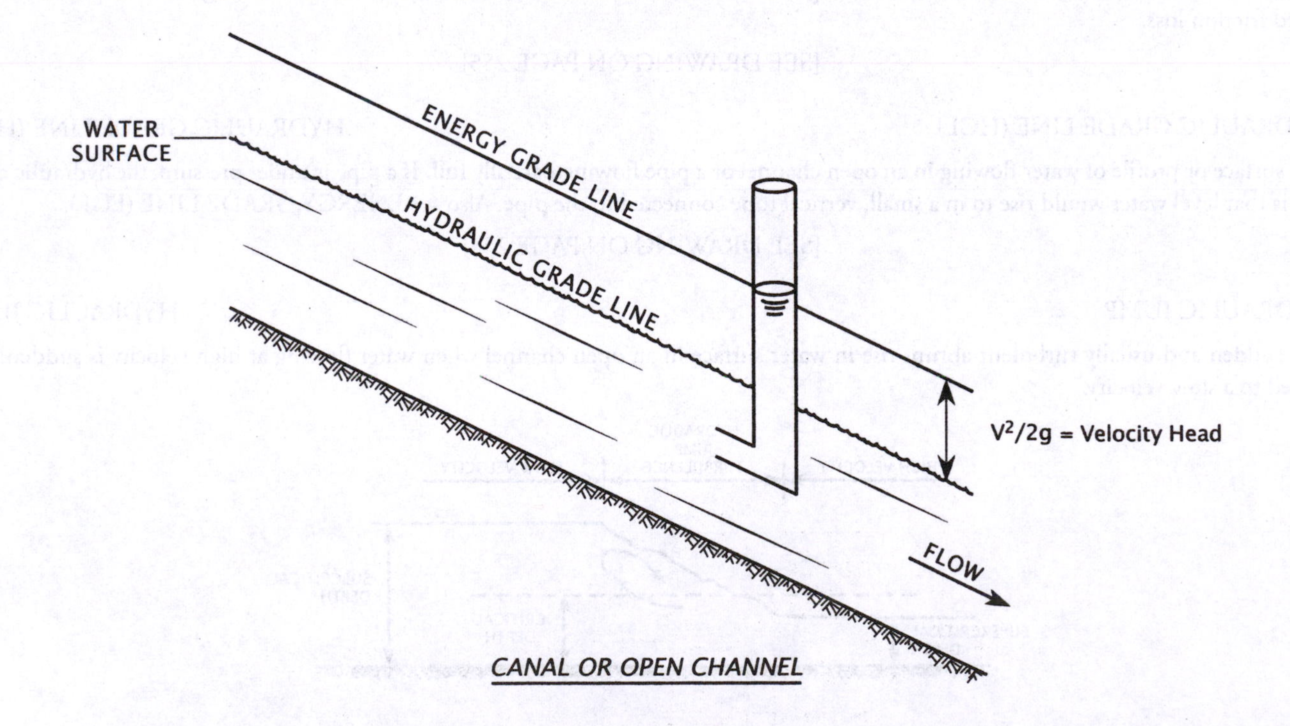


*Energy grade line and hydraulic grade line*

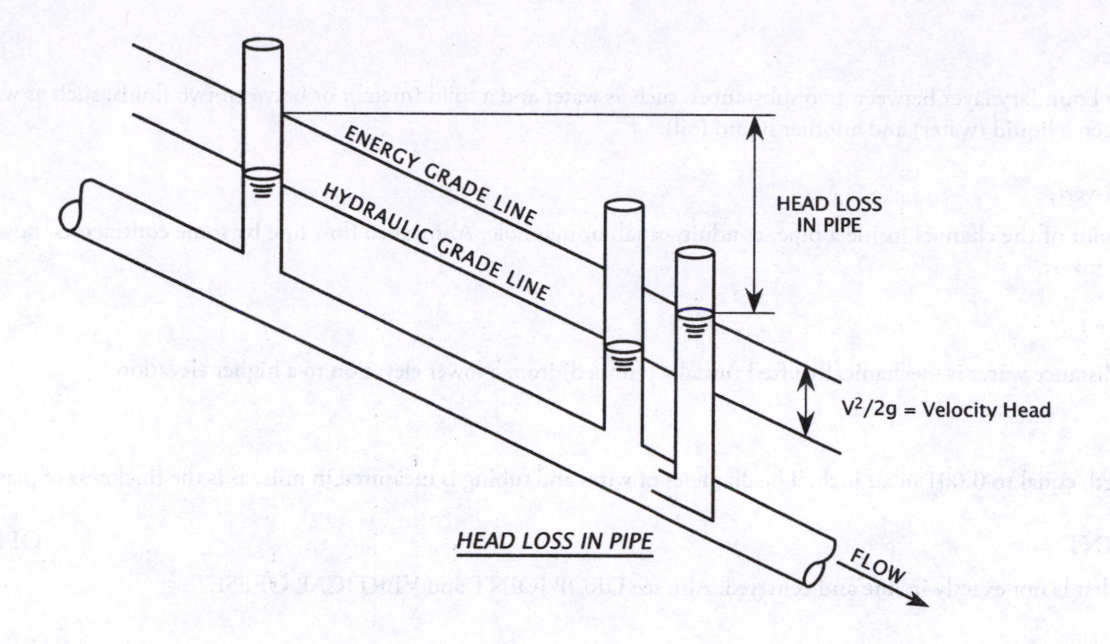
ENERGY GRADE LINE
HYDRAULIC GRADE LINE
HEAD LOSS
IN PIPE
V²/2g = Velocity Head
FLOW
HEAD LOSS IN PIPE

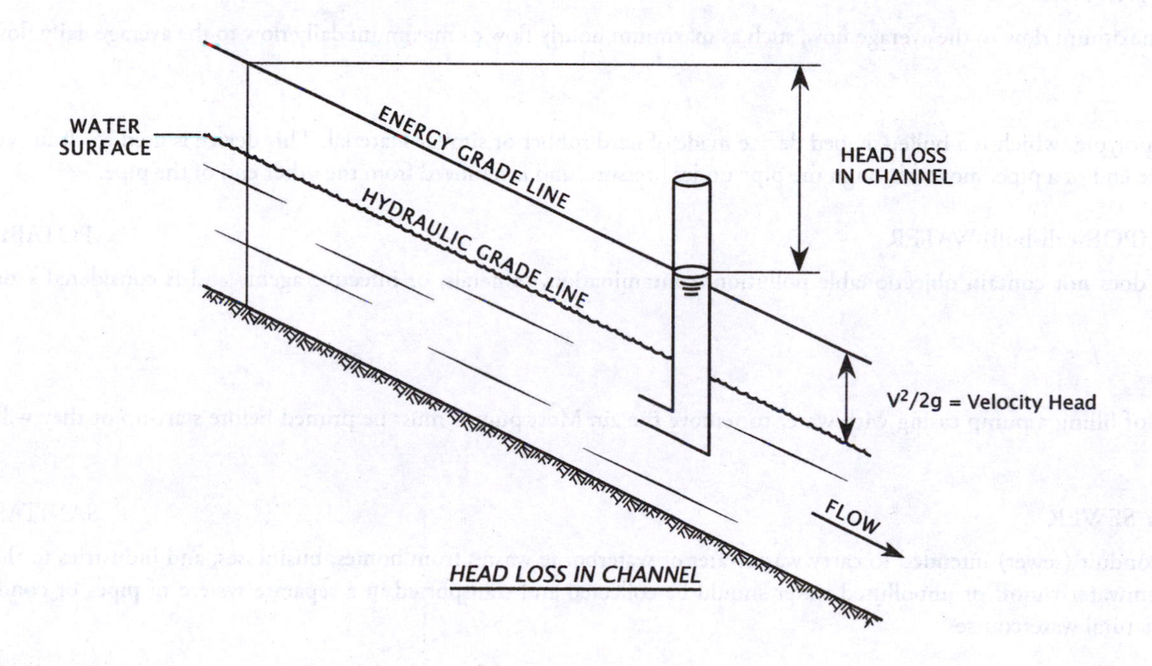
WATER
SURFACE
ENERGY GRADE LINE
HYDRAULIC GRADE LINE
HEAD LOSS
IN CHANNEL
V²/2g = Velocity Head
FLOW
HEAD LOSS IN CHANNEL

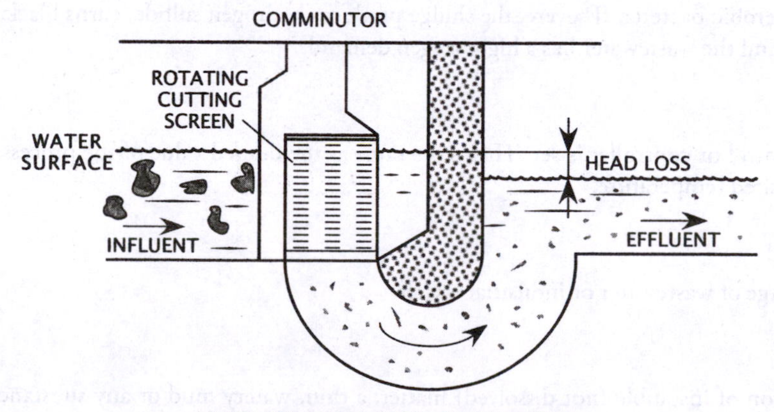
COMMINUTOR
ROTATING
CUTTING
SCREEN
WATER
SURFACE
HEAD LOSS
INFLUENT
EFFLUENT
HEAD LOSS THROUGH COMMINUTOR

*Head loss*

INTERFACE — INTERFACE

The common boundary layer between two substances, such as water and a solid (metal); or between two fluids, such as water and a gas (air); or between a liquid (water) and another liquid (oil).

INVERT (IN-vert) — INVERT

The lowest point of the channel inside a pipe, conduit, canal, or manhole. Also called flow line by some contractors, however, the preferred term is invert.

LIFT — LIFT

The vertical distance water is mechanically lifted (usually pumped) from a lower elevation to a higher elevation.

MIL — MIL

A unit of length equal to 0.001 of an inch. The diameter of wires and tubing is measured in mils, as is the thickness of plastic sheeting.

OFFSET JOINT — OFFSET JOINT

A pipe joint that is not exactly in line and centered. Also see DROP JOINT and VERTICAL OFFSET.

PEAKING FACTOR — PEAKING FACTOR

Ratio of a maximum flow to the average flow, such as maximum hourly flow or maximum daily flow to the average daily flow.

PIG — PIG

Refers to a polypig, which is a bullet-shaped device made of hard rubber or similar material. This device is used to clean pipes. It is inserted in one end of a pipe, moves through the pipe under pressure, and is removed from the other end of the pipe.

POTABLE (POE-tuh-bull) WATER — POTABLE WATER

Water that does not contain objectionable pollution, contamination, minerals, or infective agents and is considered satisfactory for drinking.

PRIME — PRIME

The action of filling a pump casing with water to remove the air. Most pumps must be primed before start-up or they will not pump any water.

SANITARY SEWER — SANITARY SEWER

A pipe or conduit (sewer) intended to carry wastewater or waterborne wastes from homes, businesses, and industries to the treatment works. Stormwater runoff or unpolluted water should be collected and transported in a separate system of pipes or conduits (storm sewers) to natural watercourses.

SEPTIC (SEP-tick) — SEPTIC

A condition produced by anaerobic bacteria. If severe, the sludge produces hydrogen sulfide, turns black, gives off foul odors, contains little or no dissolved oxygen, and the wastewater has a high oxygen demand.

SET POINT — SET POINT

The position at which the control or controller is set. This is the same as the desired value of the process variable. For example, a thermostat is set to maintain a desired temperature.

SLUG — SLUG

Intermittent release or discharge of wastewater or industrial wastes.

SLURRY — SLURRY

A watery mixture or suspension of insoluble (not dissolved) matter; a thin, watery mud or any substance resembling it (such as a grit slurry or a lime slurry).

STATIC HEAD | STATIC HEAD

When water is not moving, the vertical distance (in feet or meters) from a reference point to the water surface is the static head. Also see DYNAMIC HEAD, DYNAMIC PRESSURE, and STATIC PRESSURE.

STATIC LIFT | STATIC LIFT

Vertical distance water is lifted from upstream water surface to downstream water surface (which is at a higher elevation) when no water is being pumped.

STATIC PRESSURE | STATIC PRESSURE

When water is not moving, the vertical distance (in feet or meters) from a specific point to the water surface is the static head. The static pressure in psi (or kPa) is the static head in feet times 0.433 psi/ft (or meters × 9.81 kPa/m). Also see DYNAMIC HEAD, DYNAMIC PRESSURE, and STATIC HEAD.

STATIC WATER HEAD | STATIC WATER HEAD

Elevation or surface of water that is not being pumped.

SURCHARGE | SURCHARGE

Sewers are surcharged when the supply of water to be carried is greater than the capacity of the pipes to carry the flow. The surface of the wastewater in manholes rises above the top of the sewer pipe, and the sewer is under pressure or a head, rather than at atmospheric pressure.

TELEMETERING EQUIPMENT | TELEMETERING EQUIPMENT

Equipment that translates physical measurements into electrical impulses that are transmitted to dials or recorders.

THRUST BLOCK | THRUST BLOCK

A mass of concrete or similar material appropriately placed around a pipe to prevent movement when the pipe is carrying water. Usually placed at bends and valve structures.

TOPOGRAPHY (toe-PAH-gruh-fee) | TOPOGRAPHY

The arrangement of hills and valleys in a geographic area.

TOTAL DYNAMIC HEAD (TDH) | TOTAL DYNAMIC HEAD (TDH)

When a pump is lifting or pumping water, the vertical distance (in feet or meters) from the elevation of the energy grade line on the suction side of the pump to the elevation of the energy grade line on the discharge side of the pump. The total dynamic head is the static head plus pipe friction losses.

VERTICAL OFFSET | VERTICAL OFFSET

A pipe joint in which one section is connected to another at a different elevation, such as a DROP JOINT.

VOLUTE (vol-LOOT) | VOLUTE

The spiral-shaped casing that surrounds a pump, blower, or turbine impeller and collects the liquid or gas discharged by the impeller.

STATIC HEAD

When water is not moving, the vertical distance (in feet or meters) from a reference point to the water surface is the static head. Also see DYNAMIC HEAD, DYNAMIC PRESSURE, and STATIC PRESSURE.

STATIC LIFT

Vertical distance water is lifted from upstream water surface up to downstream water surface (which is at a higher elevation) when no water is being pumped.

STATIC PRESSURE

When water is not moving, the vertical distance (in feet or meters) from a specific point to the water surface is the static head. The static pressure in psi (or kPa) is the static head in feet times 0.433 psi/ft (or meters × 9.81 kPa/m). Also see DYNAMIC HEAD, DYNAMIC PRESSURE, and STATIC HEAD.

STATIC WATER HEAD

Elevation or surface of water that is not being pumped.

SURCHARGE

Sewers are surcharged when the supply of water to be carried is greater than the capacity of the pipes to carry the flow. The surface of the wastewater in manholes rises above the top of the sewer pipe, and the sewer is under pressure or a head, rather than at atmospheric pressure.

TELEMETERING EQUIPMENT

Equipment that translates physical measurements into electrical impulses that are transmitted to dials or recorders.

THRUST BLOCK

A mass of concrete or similar material appropriately placed around a pipe to prevent movement when the pipe is carrying water. Usually placed at bends and valve structures.

TOPOGRAPHY (toe-PAH-gruh-fee)

The arrangement of hills and valleys in a geographic area.

TOTAL DYNAMIC HEAD (TDH)

When a pump is lifting or pumping water, the vertical distance (in feet or meters) from the elevation of the energy grade line on the suction side of the pump to the elevation of the energy grade line on the discharge side of the pump. The total dynamic head is the static head plus pipe friction losses.

VERTICAL OFFSET

A pipe fitting in which one section is connected to another at a different elevation. Also see OFFSET.

VOLUTE (vol-LOOT)

The spiral-shaped casing that surrounds a pump, blower, or turbine impeller and collects the liquid or gas discharged by the impeller.

# Chapter 6. COLLECTION SYSTEMS

(Lesson 1 of 3 Lessons)

## 6.0 NEED TO KNOW

The purpose of a wastewater collection system is to collect the wastewater from a community's homes and convey it to a treatment plant where the pollutants are removed before the treated wastewater is discharged to a body of water or onto land. The wastewater in collection systems is often conveyed by gravity using the natural slope of the land. Wastewater pumps are used when the slope of the land requires lifting the wastewater to a higher elevation for a return to gravity flow. Thus, the location and design of components of a gravity collection system are strongly influenced by the *TOPOGRAPHY*[1] of its service area.

In many cases, construction of a conventional gravity collection system may not be feasible or may not be a cost-effective way to transport wastewater to treatment and disposal facilities. For example, the topography and soil conditions in some regions make excavation for a gravity system extremely expensive. Also, if pumps and other mechanical equipment are needed, they will require extensive maintenance and costly energy to operate. Alternatives to a conventional gravity collection system include a small-diameter gravity sewer system, a pressure collection system, or a vacuum collection system. Each of these alternatives is described in this chapter.

Operators of small wastewater collection systems often are not the ones responsible for the design or construction of the collection system.[2] However, a basic knowledge of the components of collection systems, their purpose, and how engineers design the systems will help you do a better job. At the same time, such knowledge will enable you to discuss design, construction, and inspection with engineers.

Understanding the design and construction of the system you will be responsible for operating and maintaining also will assist you in determining the correct action to take when the system fails to operate properly. In some emergency situations, you may be called upon to design and to construct a small, temporary portion of the wastewater system you are maintaining.

## 6.1 QUANTITY OF WASTEWATER

The amount of wastewater that a collection system will convey is determined by careful analysis of the present and probable future quantities of wastewater from homes and businesses as well as anticipated groundwater infiltration and surface water inflow produced in the service area. The *SANITARY SEWERS*[3] that make up a collection system are usually designed to carry the peak flow from all of these sources.

Wastewater flow from homes, which is referred to as domestic wastewater flow, varies from house to house, community to community, and region to region, sometimes ranging from 50 to 250 gallons per person per day. This flow measurement is also expressed as gallons per capita per day (GPCD). Residential wastewater flow in small rural systems typically is estimated to

1. *Topography* (toe-PAH-gruh-fee). The arrangement of hills and valleys in a geographic area.
2. Collection system operators in small agencies are sometimes required to inspect sewer construction. If you are operating a conventional gravity wastewater collection system or if your duties include inspecting the construction of a collection system, you may find it helpful to read *OPERATION AND MAINTENANCE OF WASTEWATER COLLECTION SYSTEMS,* Volume I, Chapter 5, "Inspecting and Testing Collection Systems." Chapter 5 contains detailed information on inspecting existing sewers for operation and maintenance problems, and inspecting new sewers and replacement sewers for installation as planned by examining line and grade, joint and junction adequacy, and proper installation of manholes and appurtenances. Information on testing for leaks in joints, taps, sewers, and manholes of existing facilities is also included. Detailed descriptions of wastewater collection system components, safe procedures, pipeline cleaning and maintenance methods, and underground repair, including excavation safety requirements, are found in this volume as well. This operator training manual is available from the Office of Water Programs, California State University, Sacramento, 6000 J Street, Sacramento, CA 95819. Price: $49.00.
3. *Sanitary Sewer.* A pipe or conduit (sewer) intended to carry wastewater or waterborne wastes from homes, businesses, and industries to the treatment works. Stormwater runoff or unpolluted water should be collected and transported in a separate system of pipes or conduits (storm sewers) to natural watercourses.

be 40 to 60 gallons per day per person and the typical value used is 50 GPCD. Commercial and industrial wastewater flows are often estimated by multiplying water meter flows by 70 percent.

Some infiltration of groundwater and inflow of surface water occurs in all wastewater collection systems and capacity must be provided for this additional flow. The amount of infiltration entering through cracks in pipes, joints, and manholes will vary with the age and condition of the collection system and the portion of the system that is submerged in groundwater. The amount of inflow will vary with the number of manholes in a collection system that become submerged in surface water and the number of surface water drainage sources, such as roof drains, that are illegally connected to the collection system. Inflow can be reduced by sealing any vent holes in manhole covers that are likely to be flooded.

## QUESTIONS

Please write your answers to the following questions and compare them with those on page 344.

6.0A What is the purpose of a wastewater collection system?

6.0B In areas where the topography or soil conditions are unfavorable for a conventional gravity wastewater collection system, what types of collection systems may be used?

6.0C Why should a collection system operator understand how the system was designed and constructed?

6.1A What are the main sources of wastewater in a collection system?

## 6.2 CONVENTIONAL GRAVITY SEWERS
by Steve Goodman

### 6.20 Components

Wastewater from residences and commercial facilities is collected and conveyed to wastewater treatment plants through the following parts of a gravity wastewater collection system (see Figures 6.1 and 6.2):

BUILDING SEWERS. A building sewer connects a building's internal wastewater drainage system (plumbing) to a septic tank or directly to the larger street sewer. Technically, a building sewer may connect with a lateral sewer, a main sewer, or a trunk sewer, depending on the layout of the system. (All of these sewers are parts of the sewer in the street.) The building sewer may begin immediately at the outside of a building or some distance (such as 2 to 10 feet) from the foundation, depending on local building codes. Where the building sewer officially begins usually marks where the building plumber's responsibility ends and the collection system operator's responsibility starts for maintenance and repair of the system. In many areas, however, collection system maintenance begins at the tap and the building sewer is the property owner's responsibility.

LATERAL AND BRANCH SEWERS are the upper ends of the street sewer components of a collection system. In some instances, lateral and branch sewers may be located in private *EASEMENTS*.[4] This location should be avoided where possible due to maintenance problems created by locations with difficult access and limited work space.

MAIN SEWERS collect the wastewater of several lateral and branch sewers from an area of several hundred acres and convey it (the wastewater) to larger trunk sewers.

TRUNK SEWERS are the main arteries of a wastewater collection system; they convey the wastewater from numerous main sewers either to a treatment plant or an interceptor sewer.

INTERCEPTOR SEWERS receive the wastewater from trunk and other upstream sewers and convey it to the treatment plant.

LIFT STATIONS are used in a gravity collection system to lift (pump) wastewater to a higher elevation when the slope of the route followed by a gravity sewer would cause the sewer to be laid at an insufficient slope or at an impractical depth. Lift stations vary in size and type depending on the quantity of wastewater and the height the wastewater is to be lifted.

### 6.21 Flow Velocity

The wastewater in a conventional gravity sewer should move at a speed that will prevent the solids from settling out in the pipeline; this velocity is called a "scouring velocity." Experience has shown that wastewater velocities of 2.0 feet per second or greater will keep the solids usually contained in wastewater moving in a sewer line. A scouring or self-cleaning velocity may not be reached during periods of minimum daily flows, but a sewer should produce a scouring velocity during the average daily flow or, at the very least, during the maximum daily flow. A velocity of 2.0 feet per second or more will keep normal solids in suspension in the flow. When the velocity is 1.4 to 2.0 feet per second, the heavier materials, such as gravel and sand, begin to accumulate in the pipe. At a velocity of 1.0 to 1.4 feet per second, inorganic grit and organic solids will accumulate. When the velocity drops below 1.0 foot per second, significant amounts of both organic and inorganic solids accumulate in the sewer.

Velocities in sewers should be measured in the field to be sure actual velocities are high enough to prevent the deposition of solids. If roots or other obstructions get into a sewer, design velocities can be reduced and problems could develop. Actual velocities in sewers can be measured by using tracer dyes or floats.

4. *Easement.* Legal right to use the property of others for a specific purpose. For example, a utility company may have a five-foot (1.5 m) easement along the property line of a home. This gives the utility the legal right to install and maintain a sewer line within the easement.

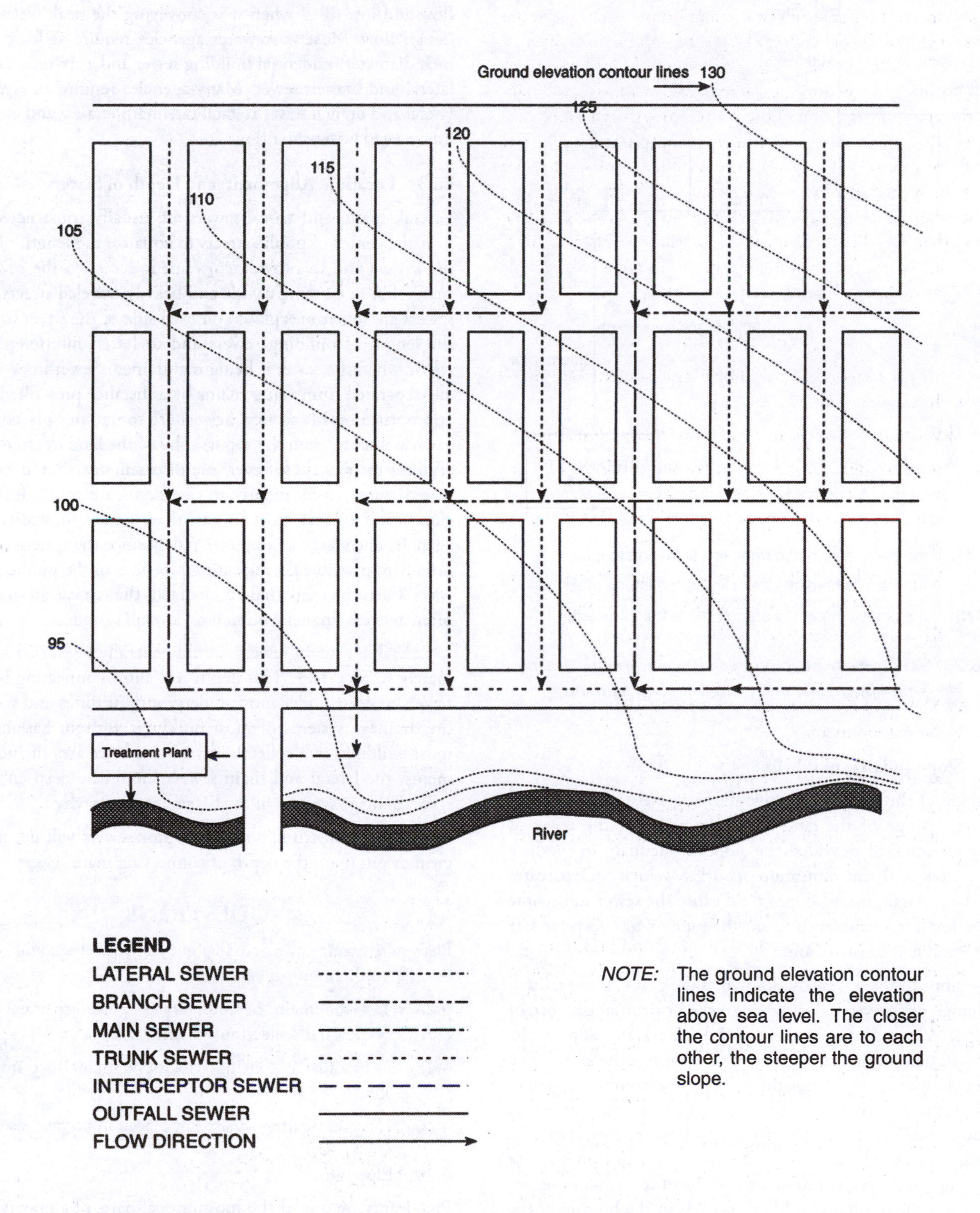


*Fig. 6.1 Schematic of a typical gravity wastewater collection system*

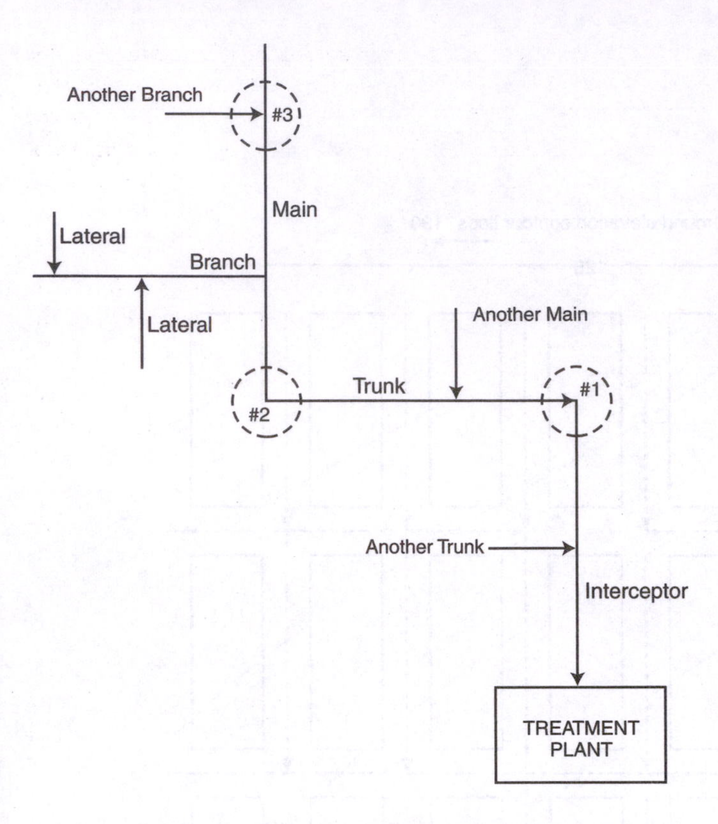
Another Branch
#3
Main
Lateral
Branch
Lateral
Another Main
Trunk
#2
#1
Another Trunk
Interceptor
TREATMENT PLANT
#1: If the plant is here, the trunk line is the interceptor.
#2: If the plant is here, the main line is the interceptor.
#3: If the plant is here, the branch line is the interceptor.

*Fig. 6.2 Gravity wastewater collection system schematic showing alternative interceptor and treatment plant locations*

### 6.22 Slope and Size of Pipeline

The slope of the sewer in a conventional gravity system should follow the slope of the land along the route of the sewer as closely as practical provided the slope is adequate to produce gravity flow and the minimum scouring velocity. Departures from the natural ground slope could cause the sewer to become too shallow for adequate cover on the pipe or too deep for safe and economical construction.

The slope of the sewer, also known as the grade, is the ratio of the change of the vertical distance to the horizontal distance of the pipe. With a given pipe type and diameter, the slope of the sewer will determine the velocity of flow within the sewer. The steeper the slope, the higher the velocity. Conventional gravity sewers are normally designed to maintain minimum velocities to keep the solids suspended. However, in some cases, it may be necessary for certain sections of the line to be laid on a flatter slope. This could create problems with insufficient scouring velocities and the resultant buildup of solids in the bottom of the sewer. Minimum slopes are also related to the diameter of the pipe. For example, the recommended minimum slope that will maintain a scouring velocity in an 8-inch pipe is 0.40 foot for every 100 feet of pipe. Minimum slope for a 16-inch pipe would be 0.14 foot for every 100 feet of pipe, and for a 36-inch pipe it would be 0.046 foot for every 100 feet of pipe.

A sewer line should be large enough for the use of cleaning equipment. A properly sized sewer line will be approximately one-half full when conveying the peak daily dry weather design flow and just filled when it is conveying the peak wet weather design flow. Most wastewater agencies require at least a four-inch diameter residential building sewer and a six-inch diameter lateral and branch sewer. Many agencies require an eight-inch lateral and branch sewer to facilitate maintenance and to convey unexpected wastewater flows.

### 6.23 Location, Alignment, and Depth of Sewers

Lateral, main, and trunk sewers are usually constructed at or near the center of public streets to equalize the length of building sewers and to provide convenient access to the sewer line manholes. In extremely wide and heavily traveled streets, lateral sewers are sometimes placed on each side of the street to reduce the length of building sewers and to lessen interference with traffic when the sewer is being maintained. Sewer lines are usually separated from water mains by a distance prescribed by the appropriate health agency. Sewers are sometimes placed in private easements when the topography of the land or street layout requires their use. However, use of easements should be minimized since sewers in private easements are more difficult to gain access to and to maintain than sewers in streets. Some agencies encourage an opposite policy since the system owner is usually responsible for relocating pipelines in the public right of way. When the road grade is changed, the relocation of pipes is often a severe financial burden on a small system.

Lateral and main sewers should generally be buried approximately six feet deep. This depth will allow connecting building sewers to cross under most underground utilities and will serve the drainage systems of most buildings without basements. If most buildings in a collection system's service area include basements, the lateral and main sewers should be deep enough to serve drainage systems in the basements by gravity.

The depth of trunk and interceptor sewers will depend to a great extent upon the depth of connecting main sewers.

## QUESTIONS

Please write your answers to the following questions and compare them with those on page 344.

6.2A List the main components of a conventional gravity wastewater collection system.

6.2B Why must a scouring velocity be maintained in a sewer line?

6.2C Why are sewers usually not placed in private easements?

### 6.24 Pipe Joints

Pipe joints are one of the most critical parts of a gravity collection system because of the need to control groundwater infiltration, wastewater exfiltration, and root intrusion. Many types of pipe joints are available for the different pipe materials used in sanitary sewer construction. Regardless of the type of sewer pipe used, reliable, tight pipe joints are essential. A good pipe joint must be watertight, root resistant, flexible, and durable. Various forms of gasket (elastomeric seal) pipe joints are now commonly used in sanitary sewer construction. In general, this type of joint can be assembled easily in a broad range of weather conditions and environments with good assurance of a reliable, tight seal that prevents leakage and root intrusion.

### 6.25 Manholes (Figure 6.3)

Manhole bases are made of either precast or cast-in-place unreinforced concrete with wastewater channels formed by hand before the concrete hardens. Manhole barrels are usually made of precast concrete sections with cement mortar or bitumastic joints. Manholes in older collection systems may have been made with clay bricks and cement mortar joints. Some manhole barrels are now being made with fiberglass for use in areas with high groundwater and where the wastewater contains corrosive materials. Manhole frames and covers are made of cast iron with the mating surface between the frame and cover machined to ensure a firm seating. Manhole steps or rungs should be made of wrought iron or reinforced plastic to resist corrosion. In some areas, steps are prohibited and ladders are used to eliminate the hazard of corroded steps. Steps could possibly fail and maintenance personnel could be injured by falling to the bottom of the manhole.

### 6.26 Sewer Design Considerations

A civil engineer is usually the person who prepares plans and specifications for the construction of a wastewater collection system. Typically, the engineer will prepare both a plan and a profile of the entire system. The plan view is the top view of a sewer; it shows the pipe, manholes, and street location. The profile is the side view of a sewer; it shows the pipe, manholes, *INVERT*[5] elevations, street surface, and underground pipes and other underground utilities such as storm sewers, gas mains, water mains, and electrical cables.

Operators of conventional gravity wastewater collection systems should request the opportunity to review the design of collection systems that they will be responsible for maintaining after construction. In order to make an effective review, an operator must be familiar with the elements of design that can create problems during the life of a collection system. The rest of this section gives some suggestions for the review of plans and profiles for construction of a collection system.

When you receive a set of plans and profiles for review, look them over to get an overview of how the system is to work. Then, make a detailed review by asking yourself how the design may be improved for the elimination of potential maintenance problems. Make a list of items that are not clear to you and ask the designer for clarification.

After you have reviewed the collection system plans and profiles, walk the proposed route and think about how the sewer lines and equipment will be maintained. Ask yourself how the actual field conditions will influence maintenance.

During both the office and field review of plans for a wastewater collection system, ask yourself the following questions concerning the design:

1. Is the route or alignment satisfactory?
2. Could there be a better route?
   a. Consider the effect of pipe diameters and depth of excavation on maintenance.
   b. Think about maintenance equipment access.
3. Is there sufficient overhead clearance for maintenance and repair equipment?
4. How could repairs be made in case of failures?
   a. Availability of large-diameter pipe?
   b. Depth of excavation?
   c. Shoring requirements?
5. Can the system be easily and properly operated and maintained?
6. Will there be any commercial or industrial waste dischargers? If so, what will they discharge and will the discharge cause any problems? How can these problems be controlled?
7. Where can vehicles be parked when operators and equipment require access to a manhole?
8. Is traffic a problem around or near a manhole and, if so, can it be relocated to minimize the problem? Will night work be necessary?
9. Can the atmosphere in the manhole be safely tested for oxygen deficiency/enrichment, explosive gases, and toxic gases? Can large ventilation blowers be used?
10. Can an operator reach the bottom of the manhole with a ladder if steps are not provided?
11. If there are steps in the manhole wall for access, will they be damaged by corrosion and become unsafe?
12. Is the shelf in the bottom of the manhole properly sloped? If the shelf is too flat, it will get dirty; if the shelf is too steep, it can be a slipping hazard to operators.
13. Are the channels properly shaped to provide smooth flow through a manhole to minimize splashing and turbulence that can cause the release of toxic, odorous, and corrosive hydrogen sulfide?

5. *Invert* (IN-vert). The lowest point of the channel inside a pipe, conduit, canal, or manhole. Also called flow line by some contractors, however, the preferred term is invert.

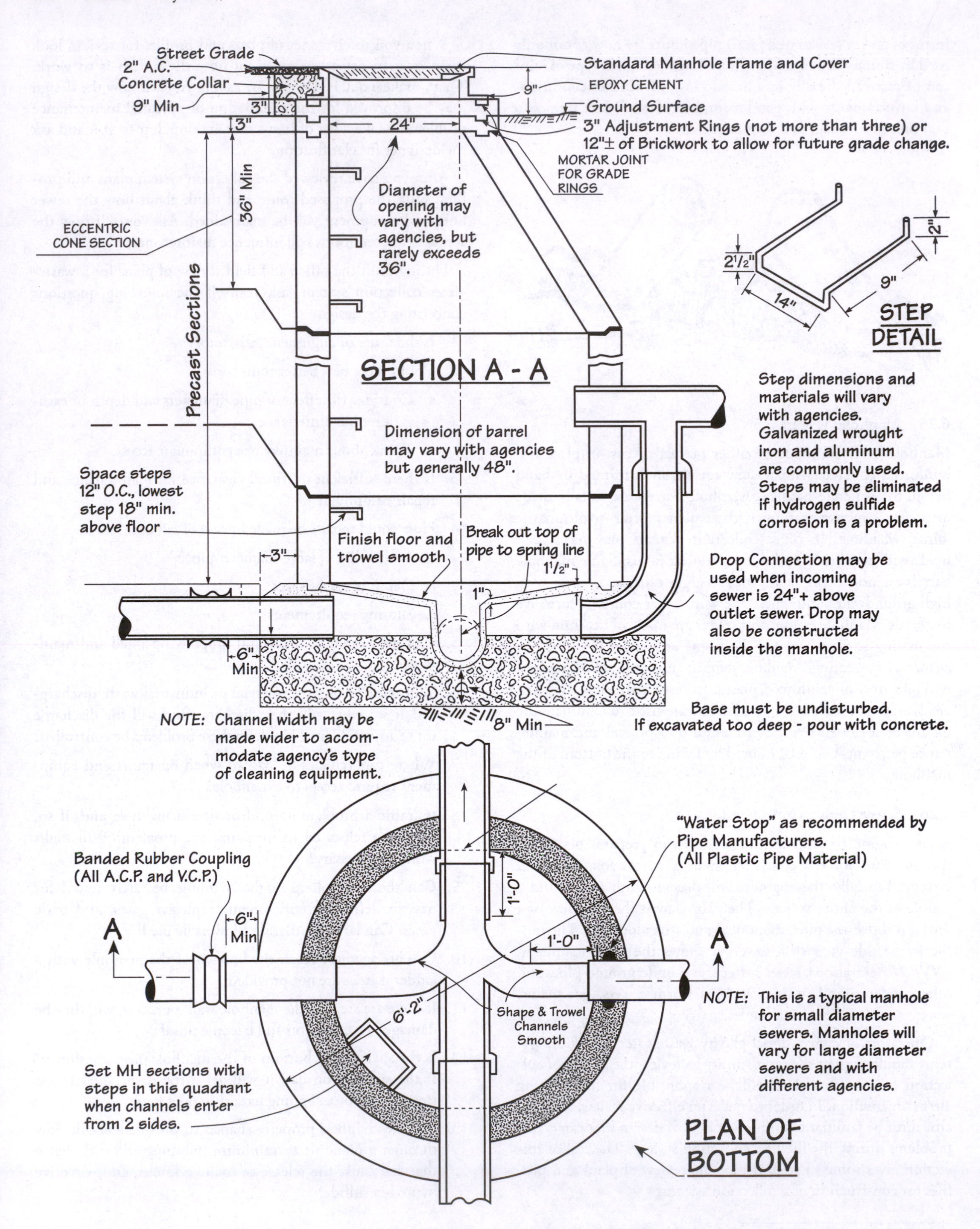


*Fig. 6.3 Precast concrete manhole*

14. Are the channels at least as large as the pipe so that maintenance equipment and plugs can be fitted or inserted into the pipes?
15. Can the available sewer maintenance equipment operate effectively between the manholes or is the distance too great?
16. Is the manhole opening large enough to allow access for tools and equipment used to maintain a large sewer line? Is the opening sufficient to handle equipment and personnel during routine maintenance or inspection?
17. Does the area flood? How often? Can flows be diverted? Stored? How? For how long? What tools, equipment, and procedures are needed?

In addition to the above questions, use your experience to raise other questions on how the design of a collection system can affect your operation and maintenance activities and emergency repairs.

Your questions and ideas concerning improvements in the design of a collection system should be given to the designer in a written and understandable form with an offer to meet to discuss them. The designer should be capable of explaining and justifying a design in terms of economics and installation. You should be able to justify your concerns on the basis of operation and maintenance requirements.

### 6.27 Construction, Testing, and Inspection

Wastewater collection facilities sometimes have not functioned as intended due to poor or improper construction materials and techniques. Good construction, inspection, and testing procedures are very important in order to eliminate operation and maintenance problems. Whenever problems (stoppages, failures, or odors) develop in your collection system, try to identify the cause so it can be corrected now and in the design of new systems. Did the problem start with the design, construction, or inspection of the facility, or was it caused by a waste discharger, by aging of the system, or by your maintenance and repair program? Whatever the cause, identify it and try to correct it.

For more information on the construction, inspection, and testing of conventional gravity collection systems, see *OPERATION AND MAINTENANCE OF WASTEWATER COLLECTION SYSTEMS,* Volume I, Chapter 5, "Inspecting and Testing Collection Systems." Ordering information can be found in footnote 2 on page 259 of this chapter.

## QUESTIONS

Please write your answers to the following questions and compare them with those on page 344.

6.2D List the problems that can be caused by poor sewer pipe joints.

6.2E Why are some manhole barrels made with fiberglass?

6.2F What is meant by the plan view and profile view of a sewer?

6.2G Why is the slope of the shelf in a manhole important?

## 6.3 SMALL-DIAMETER GRAVITY SEWER (SDGS) SYSTEM

### 6.30 Description of System

A small-diameter gravity sewer (SDGS) system consists of a series of septic tanks discharging effluent by gravity, pump, or siphon to a small-diameter wastewater collection main. The wastewater flows by gravity to a lift station, a manhole in a conventional gravity collection system, or directly to a wastewater treatment plant.

Septic tanks perform two important functions that make it possible to use small-diameter sewer lines. First, the septic tank removes floating and settleable solids from the raw wastewater. The flow velocity may be much lower in SDGS systems than in conventional gravity collection systems because the wastewater contains much smaller quantities of grit, large solids, greases, and other debris. These materials settle out in conventional gravity collection system pipelines unless the flow velocity is high enough to keep them suspended in the liquid. SDGS systems are commonly designed for a minimum flow velocity of 0.5 foot per second. At this velocity, the solids that enter the collection system and the slime growths that develop within the pipelines are easily conveyed without settling out and blocking the lines. Grinder pumps are never permitted to discharge into a small-diameter gravity sewer system because the increased solids load would quickly plug the relatively small sewer pipes due to the low flow velocity.

The second important function of SDGS septic tanks is attenuation (leveling out) of flows. In normal operation, a septic tank retains wastewater long enough for the solids to separate from the liquid portion of the wastewater. Even when the tank is two-thirds full of solids, it usually will still provide 24 hours of detention time. Because the tank is designed to provide this temporary storage capacity, wide variations in the amount of wastewater entering the septic tank have little impact on the rate of flow out of the tank. Occasional heavy loadings are leveled out and the rate of flow discharged from an individual septic tank rarely exceeds one gallon per minute. In contrast, conventional gravity sewers must be designed to accommodate not only more suspended solids but also a much wider range of flow volumes. Typically, a conventional gravity sewer is designed with a *PEAKING FACTOR*[6] of four times the average flow. For this reason, much larger sewer pipes are needed for conventional gravity systems than for SDGS systems.

In most SDGS systems, only one type of septic tank is used throughout the system. However, some hybrid SDGS systems have been built to handle flows from septic tank effluent pump (STEP) systems as well as from septic tank effluent filter (STEF) systems. In hybrid systems such as these, raw wastewater solids or solids discharged through a grinder pump are not tolerated or permitted.

## QUESTIONS

Please write your answers to the following questions and compare them with those on page 344.

6.30A What is a small-diameter gravity sewer (SDGS) system?

6.30B Where may a small-diameter gravity sewer terminate?

6.30C What is the minimum flow velocity in a small-diameter gravity sewer system?

6.30D Why are grinder pumps never permitted to discharge into a small-diameter gravity sewer system?

### 6.31 Components

A small-diameter gravity sewer (SDGS) system (Figure 6.4) consists of building sewers, septic tanks, service laterals, collection mains, cleanouts, manholes and vents, and, if required, lift stations (not shown in Figure 6.4).

#### *6.310 Building Sewer*

The building sewer carries the raw wastewater from the building to the septic tank. Typically, it is a four- to six-inch diameter pipe laid on a prescribed slope according to local code. Usually, the slope is at least a one percent slope (one foot drop per 100 feet of length). Pipe material may be cast iron, ABS, PVC, or vitrified clay.

#### *6.311 Septic Tank*

The construction, installation, operation, and maintenance of standard residential septic tanks are described in detail in Chapter 4 of this manual. The septic tanks described there are suitable for use in small-diameter gravity sewer systems with one important exception: grinder pumps may not be used in SDGS systems because the quantity of solids discharged by grinder pumps will quickly plug the small-diameter sewer pipes.

Inlet and outlet baffles (Figure 6.5) are provided in conventional septic tanks to retain solids within the tank. These baffles are adequate for small-diameter gravity sewer applications. The inlet baffle must be open at the top to allow venting of the septic tank gases through the building plumbing stack. On the outlet, outlet screens should be used to capture suspended solids that might otherwise pass through the tank. Solids that pass through an unscreened septic tank outlet can and do cause problems. The type of air release valves used on STEP systems will plug if the tank outlets are not screened or do not have filters. Also, pressure-sustaining valves will not function properly with unscreened effluent. Septic tank effluent solids can increase the amount of maintenance required downstream of the septic tanks and interfere with downstream treatment processes such as sand or gravel filters.

Another type of device used to reduce the amount of effluent solids is the septic tank Biotube™ effluent filter system (Figure 6.6). In a septic tank equipped with this device, the effluent passes through a filter screen by gravity flow or by a dosing siphon before it enters the small-diameter gravity collection system.

Flow control orifices are sometimes needed in septic tanks to limit peak effluent flow rates to a predetermined maximum. One such method of controlling flow is the use of modulating discharge plate as illustrated in Figure 6.6.

Watertightness is a critical element in the proper operation of small-diameter gravity sewer systems. Unless all components are watertight, water leaking into the system will contribute extra clear water inflow that the system was not designed to carry. For this reason, existing in-service septic tanks are rarely used in new SDGS systems. Earlier systems attempted to use the existing septic tank at each home to reduce construction costs. It was found that old septic tanks are difficult to inspect and repair properly. Also, SDGS systems that had significant numbers of old tanks were all found to have high ratios of wet weather to dry weather flows, indicating that excessive infiltration was occurring.

Common practice now is to replace all tanks. Currently, there is no standard procedure for inspecting existing tanks for leakage. The practice of installing new septic tanks has the added advantage of requiring the property owner to replace the building sewer, which helps to ensure watertightness of the system. Some projects require replacement of the building sewer so that the building plumbing can be inspected to eliminate roof leaders, foundation drains, and other unwanted connections that contribute clear water inflow.

6. *Peaking Factor.* Ratio of a maximum flow to the average flow, such as maximum hourly flow or maximum daily flow to the average daily flow.

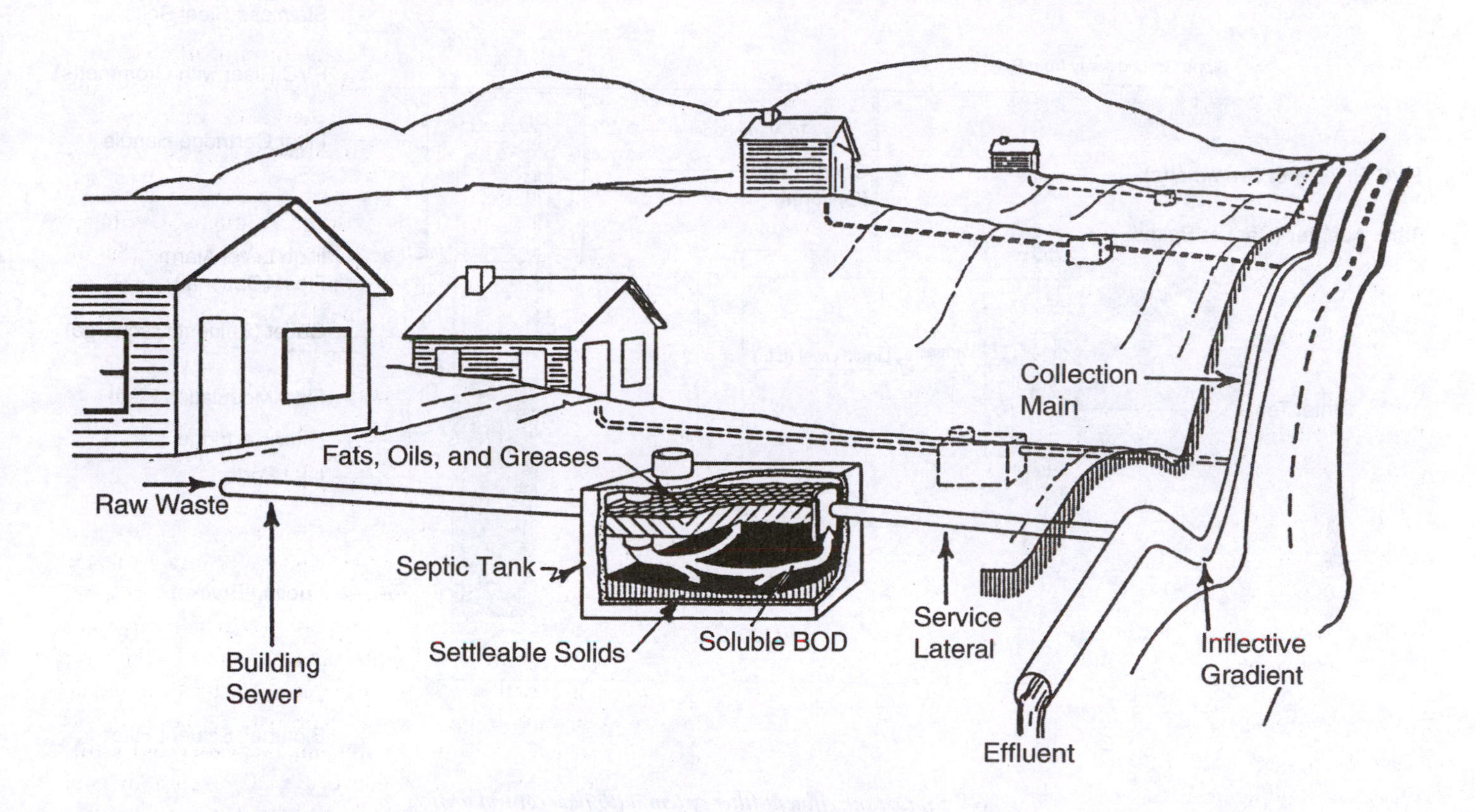


*Fig. 6.4 Components of a small-diameter gravity sewer (SDGS) system*
(Source: *ALTERNATIVE WASTEWATER COLLECTION SYSTEMS*, Technology Transfer Manual, US Environmental Protection Agency)

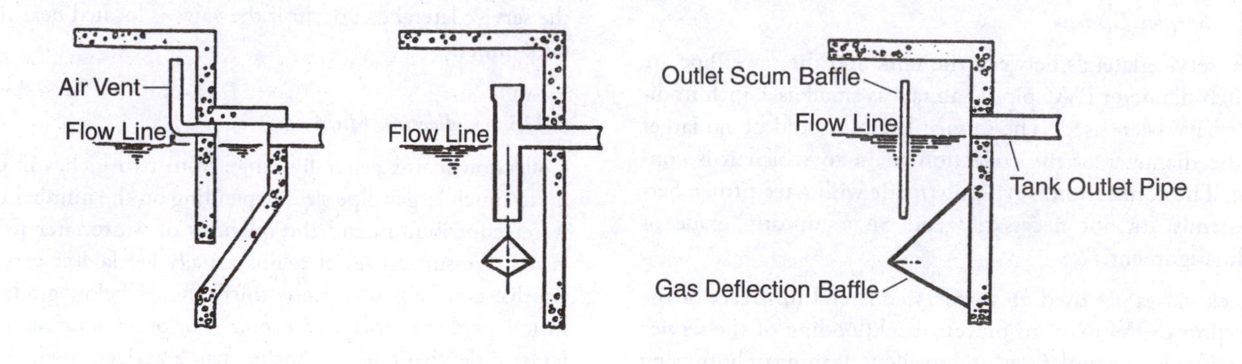


*Fig. 6.5 Typical septic tank outlet baffles*
(Source: *ALTERNATIVE WASTEWATER COLLECTION SYSTEMS*, Technology Transfer Manual, US Environmental Protection Agency)

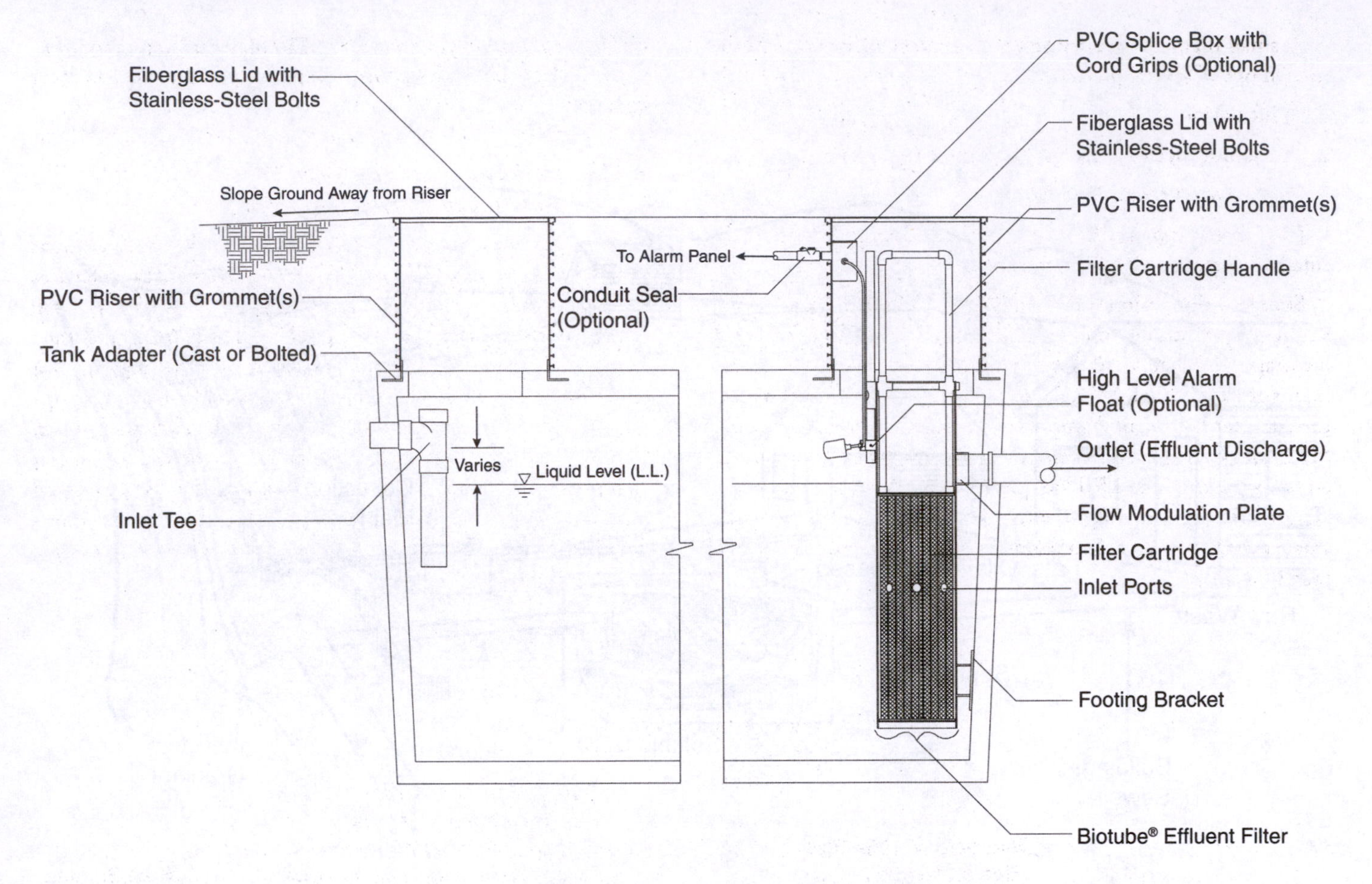


*Fig. 6.6 Septic tank effluent filter system with flow control orifices*
(Courtesy of Orenco Systems, Inc., www.orenco.com)

### 6.312 *Service Lateral*

Typical service laterals between the tank and the sewer line are four-inch diameter PVC pipe. Laterals as small as 1 inch in diameter have been used. The service lateral should be no larger than the diameter of the collection main to which it is connected. The connection is typically made with a tee fitting. Service laterals are not necessarily laid on a uniform grade or straight alignment.

Check valves are used on the service lateral upstream of the connection to the main to prevent backflooding of the service connection during peak flows and to allow cleaning of lines using a device called a *PIG*.[7] It is important that the valve be located very close to the collection main connection. *AIR BINDING*[8] of the service lateral can occur if the valve is located near the septic tank outlet.

### 6.313 *Collection Main*

Collection mains generally range from two inches in diameter up to much larger pipe sizes depending on the number of wastewater contributors and the quantity of wastewater to be conveyed. Pressurized sewer mains usually are laid at very shallow depths; common practice is thirty inches below grade. In rare instances, due to rock outcroppings or other construction problems, a depth of twelve inches has been accepted. The main factor controlling depth is prevailing soil freezing or the frost line depth.

7. *Pig.* Refers to a polypig, which is a bullet-shaped device made of hard rubber or similar material. This device is used to clean pipes. It is inserted in one end of a pipe, moves through the pipe under pressure, and is removed from the other end of the pipe.
8. *Air Binding.* The clogging of a filter, pipe, or pump due to the presence of air released from water. Air entering the filter media is harmful to both the filtration and backwash processes. Air can prevent the passage of water during the filtration process and can cause the loss of filter media during the backwash process.

A small-diameter gravity sewer collection main may terminate in one of several places:

1. A lift station
2. A manhole in a conventional gravity sewer system
3. A wastewater treatment plant

PVC plastic pipe is the most commonly used pipe material in small-diameter gravity sewer systems.

Standard dimension ratio (SDR) 21[9] Class 200 pipe is used in most applications, but SDR 26 may be specified for road crossings or where water lines are within 10 feet of the sewer line. For deep burial, SDR 21 may be necessary. Where the use of septic tank effluent pump (STEP) units is anticipated, only SDR 21 or 26 should be used for collector mains because the added strength is needed for pressurized segments. Typically, elastomeric (rubber ring) joints are used; however, for pipe smaller than three inches in diameter, only solvent-welded joints may be available.

High density polyurethane (HDPE) pipe has been used infrequently, but successfully. Pipe joining is accomplished by heat fusion.

### *6.314 Cleanouts (Figures 6.7 and 6.8) and Pig Launch Ports (Figure 6.8)*

One of the advantages STEP and STEG (septic tank effluent gravity) systems have over conventional gravity sewer collection systems is that manholes are not necessary. Cleanouts and pig launch ports are less costly to install than manholes and they provide adequate access to the collection main for pigging and other maintenance tasks. Cleanouts are typically located at upstream ends of mains, junctions of mains, and changes in main diameter. Where depressed sections occur on the collection main, the main must be vented. Cleanouts may be combined with air release valves (Figure 6.9) at high points on the collection main. Cleanouts are usually extended to the ground surface within valve boxes.

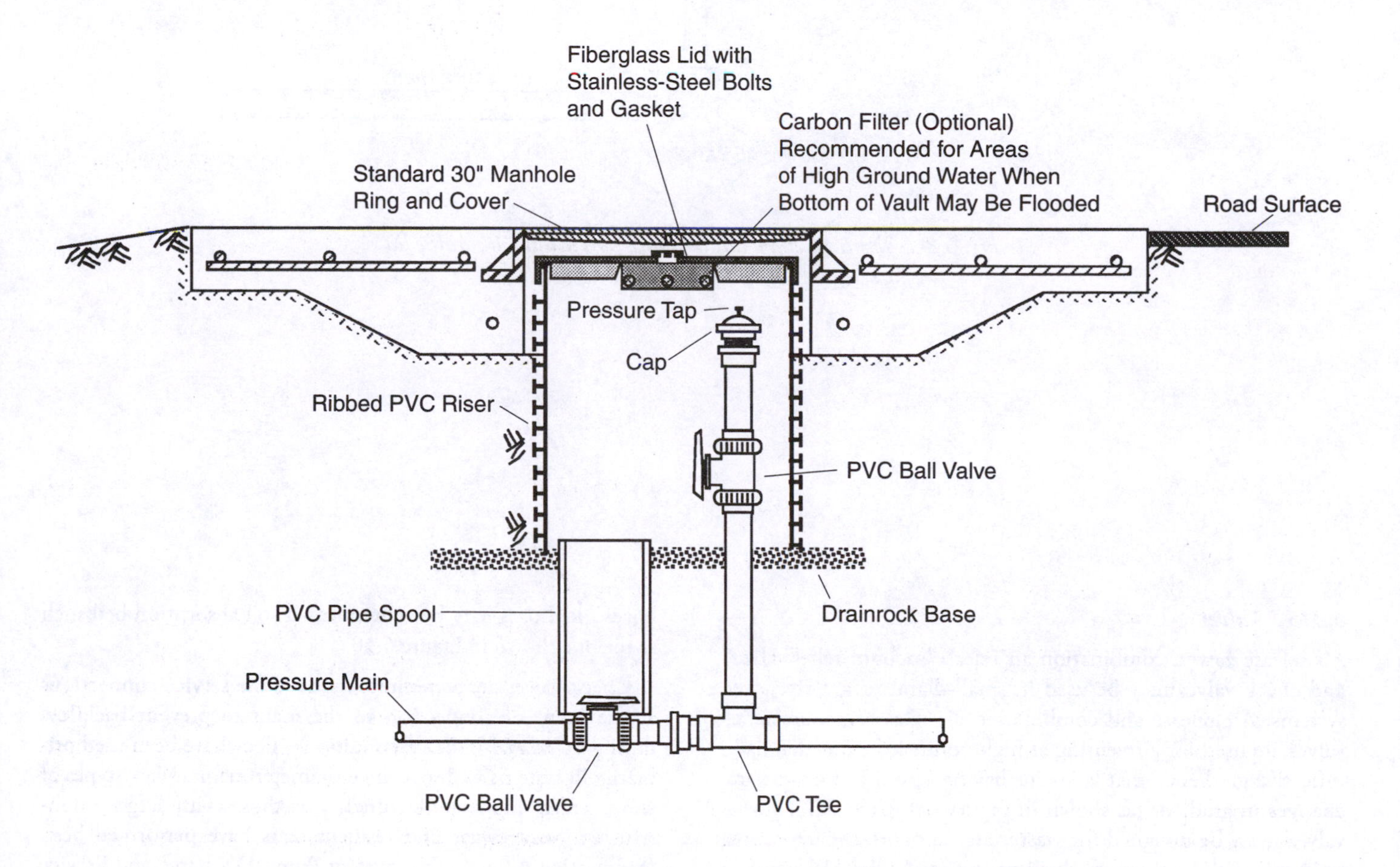


*Fig. 6.7 Typical in-line cleanout detail*

(Courtesy of Orenco Systems, Inc., www.orenco.com)

9. SDR 21 pipe commonly refers to ASTM (American Society for Testing and Materials) D2241 pipe.

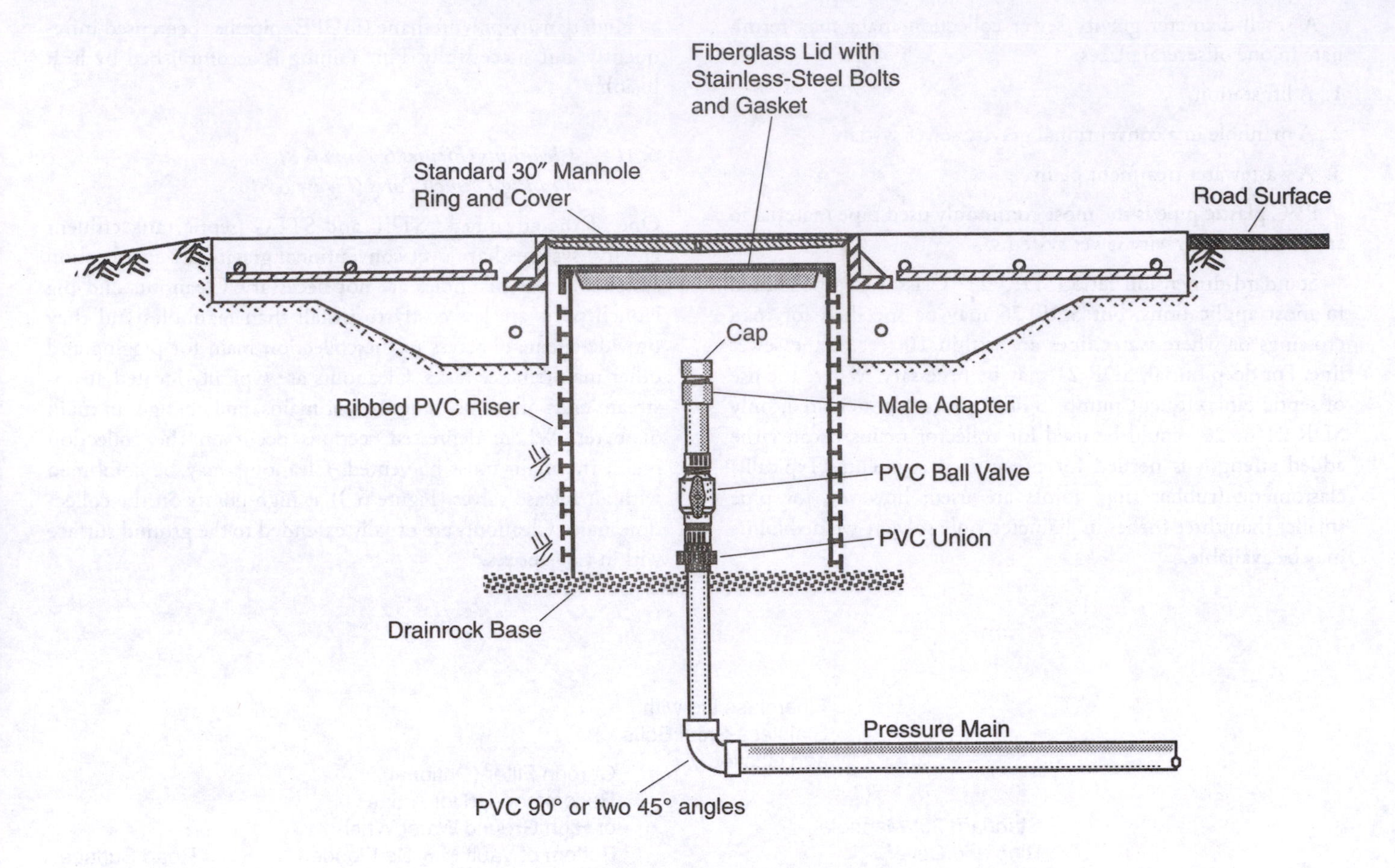


*Fig. 6.8 Terminal cleanout/pig launch port detail with traffic lid*
(Courtesy of Orenco Systems, Inc., www.orenco.com)

### *6.315 Valves*

Air release valves, combination air release/vacuum relief valves, and check valves may be used in small-diameter gravity sewer systems. Air release and combination air release/vacuum relief valves are used for air venting at high points in mains that have inflective gradients, that is, mains having upward or downward changes in gradient (as shown in Figure 6.4, page 267). These valves must be designed for wastewater applications (corrosion resistant) with working mechanisms made of type 316 stainless steel or of a plastic proven to be suitable. The valves are installed within meter or valve boxes set flush to grade and covered with a watertight lid (Figure 6.9). If odors are detected from the valve boxes, the boxes may be vented into a soil absorption bed such as the one shown in Figure 6.10.

Check valves are sometimes used on the service connections at the point of connection to the main to prevent backflow during *SURCHARGED*[10] conditions. They have been used primarily in systems with two-inch diameter mains. Many types of check valves are manufactured, but those with large, unobstructed passageways and resilient seats have performed best. (See Section 4.25, "STEP System Pump Discharge and Service Line," for a discussion of several types of check valves.) Wye pattern swing check valves are preferred over tee pattern valves when installed horizontally.

---

10. *Surcharge.* Sewers are surcharged when the supply of water to be carried is greater than the capacity of the pipes to carry the flow. The surface of the wastewater in manholes rises above the top of the sewer pipe, and the sewer is under pressure or a head, rather than at atmospheric pressure.

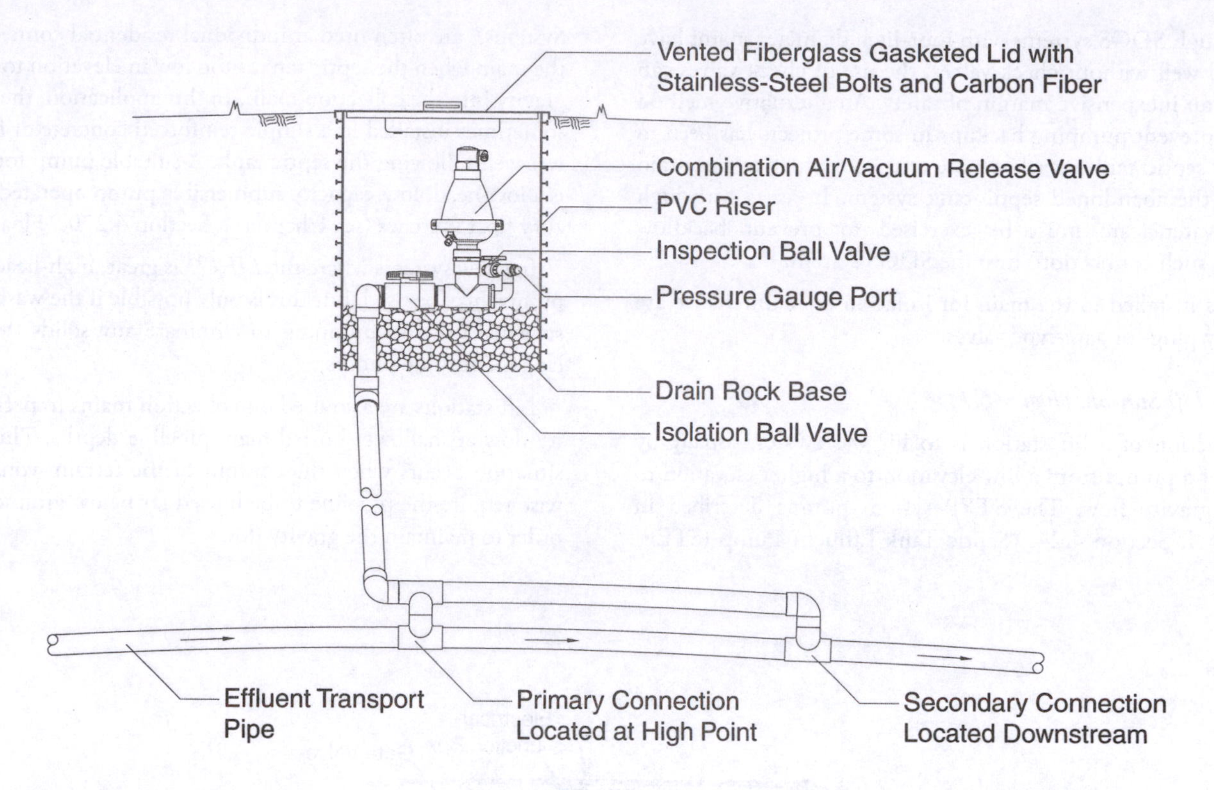


*Fig. 6.9 Automatic air release assembly*

(Courtesy of Orenco Systems, Inc., www.orenco.com)

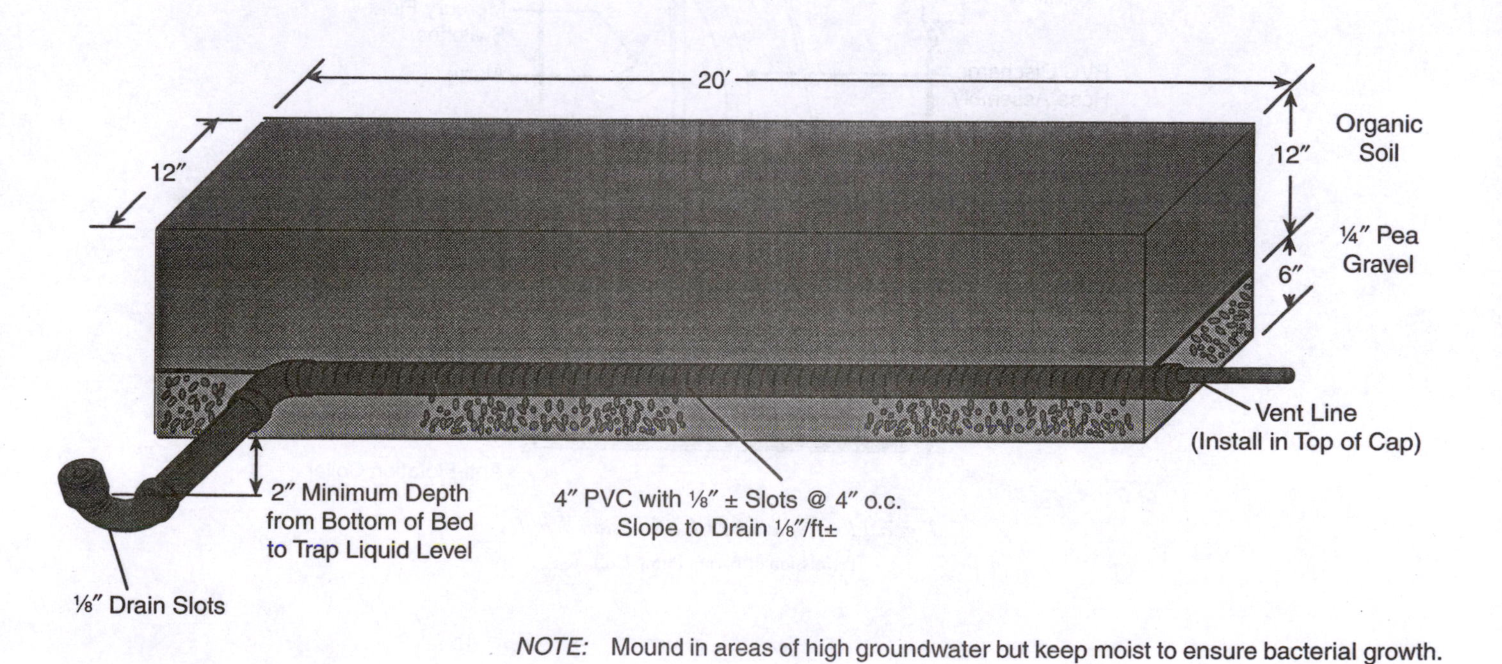


*Fig. 6.10 Soil absorption bed for odor control*

(Courtesy of Orenco Systems, Inc., www.orenco.com)

Although SDGS systems with four-inch diameter mains have operated well without check valves, the use of check valves can provide an inexpensive margin of safety. An alternative method used to prevent pumping backups in some projects has been to install a septic tank overflow pipe and connect it to the drain field of the abandoned septic tank system. In areas with high groundwater, care must be exercised to prevent backflow through such connections into the SDGS system.

Valves installed in the main for isolation or maintenance are generally plug- or gate-type valves.

### *6.316 Lift Stations (Figure 6.11)*

The function of a lift station is to lift wastewater, usually by means of a pump, from a low elevation to a higher elevation to permit gravity flow. The STEP system pumps described in Chapter 4, Section 4.24, "Septic Tank Effluent Pump (STEP) Systems," are often used at individual residential connections to the main when the septic tank is too low in elevation to drain by gravity into the collection main. In this application, the pump is sometimes installed in a simple reinforced concrete or fiberglass wet well following the septic tank. A suitable pump for this use is a low-head, low-capacity submersible pump operated by mercury float switches (see Chapter 4, Section 4.270, "Floats").

In a few systems where the *LIFT*[11] is great, high-head turbine pumps may be used, but this is only possible if the wastewater is screened prior to pumping to eliminate any solids that might clog the turbine pumps.

Lift stations are also used on collection mains to permit gravity flow at shallower buried main pipeline depths. This type of situation occurs when the contour of the terrain would otherwise require the pipeline to be buried far below ground level in order to maintain the gravity flow.

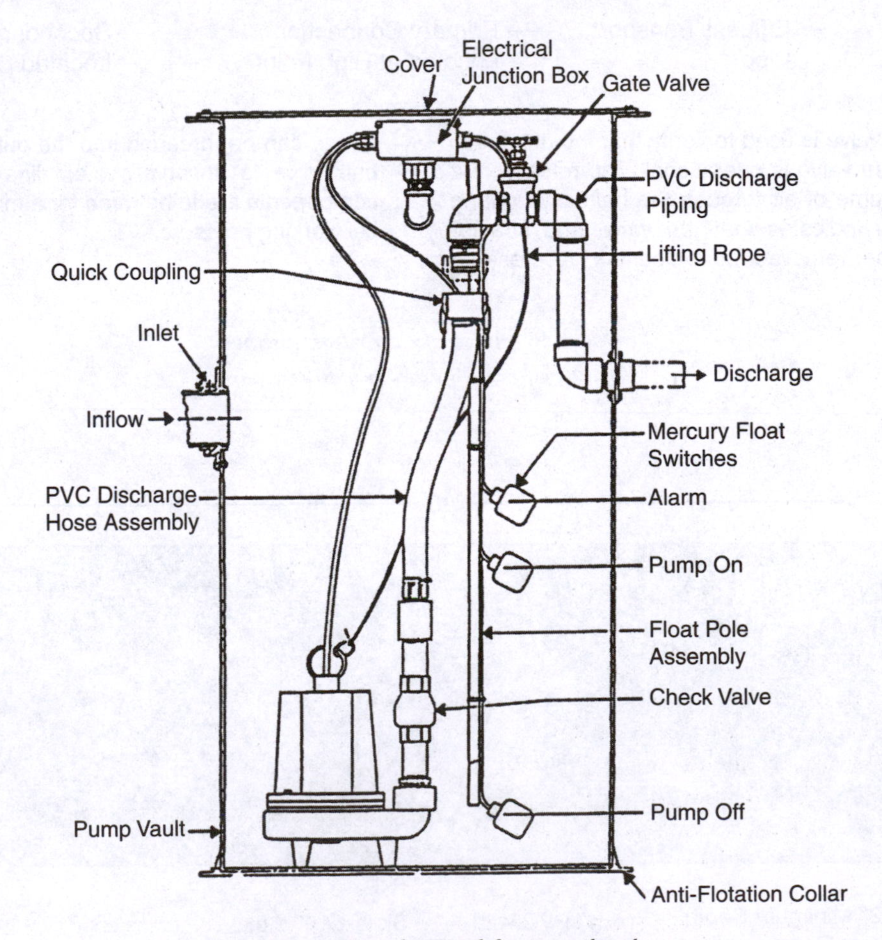


*Fig. 6.11 Typical STEP lift station detail*

(Permission of Barnes Pump Co.)

11. *Lift.* The vertical distance water is mechanically lifted (usually pumped) from a lower elevation to a higher elevation.

## QUESTIONS

Please write your answers to the following questions and compare them with those on page 344.

6.31A List the components of a small-diameter gravity sewer system.

6.31B Why are existing in-service septic tanks usually replaced when a new SDGS system is installed?

6.31C What is the purpose of cleanouts?

6.31D What is the purpose of check valves on service connections?

### 6.32 Operation and Maintenance

The sewer utility agency usually is responsible for maintenance of the entire collection system. This includes all septic tanks and any appurtenances such as STEP units located on private property. The utility district assumes responsibility for all SDGS components downstream from the septic tank inlet. In some instances, the responsibility for maintenance of the components located on private property has been left to the individual property owner. This avoids the need for utility operators to enter private property. However, since the septic tank is critical to the performance of the SDGS system, responsibility for maintenance should be retained by the district. If your duties as an operator require you to maintain or repair equipment located on private property, review Appendix A at the end of Chapter 4 to refresh your memory about your legal responsibilities with regard to entering and protecting private property.

#### *6.320 Operation and Maintenance (O&M) Manual*

Although most maintenance tasks in small-diameter gravity sewer systems are relatively simple and usually do not involve mechanical equipment, an O&M manual is still a valuable reference for the operator. When kept up to date, an O&M manual provides a complete inventory of system components, drawings showing the exact location of pipelines and components, maintenance requirements and procedures, important information about safety, and many other types of information about the system.

One portion of the O&M manual is particularly important for small-diameter gravity sewer system maintenance workers: the record drawings (also called as-built drawings). SDGS pipelines often are laid following the contours of the terrain in a *CURVILINEAR*[12] fashion rather than in grids of straight lines. SDGS systems also have relatively few manholes or cleanouts. These two characteristics make it difficult for the operator to quickly pinpoint the location of collection mains unless accurate drawings of the system identify pipeline locations in relation to permanent structures nearby. Record drawings of each individual service connection and all on-lot facilities should also be made. They will be needed when making repairs and to identify the location of system components when other construction activity occurs in the area. Record drawings filed in an up-to-date O&M manual can provide needed information quickly and conveniently.

If your SDGS system does not have an O&M manual, or if it has not been kept up to date, use the information presented in Chapter 4, Appendix C, "The Operation and Maintenance (O&M) manual," as a guide to prepare or update an O&M manual for your system. Any operator who has wasted hours or days searching for an equipment repair manual or trying to locate a buried collection pipe knows firsthand how valuable this type of resource can be. (In a properly designed and constructed system, however, toning wire would have been buried with the pipe making the buried pipe much easier to locate.)

#### *6.321 Staff and Equipment Requirements*

Operation and maintenance requirements of SDGS systems usually are not complicated. Specialized training for SDGS maintenance personnel is not necessary. However, basic plumbing skills and familiarity with the system operation are desirable.

If a significant number of service connections include STEP units, an understanding of pumps and electrical controls also is helpful. If the system uses only a small number of STEP units, it is common for the utility district to hire local plumbing and electrical contractors who have been trained to work on these system to make any necessary repairs.

The SDGS system operator's responsibilities are limited largely to service calls, new service connection inspections, and administrative duties. In most systems, septic tank pumping is usually performed by an outside contractor under the direction of the utility district. Maintenance equipment is seldom required and usually can be provided by outside contractors as needed.

#### *6.322 Spare Parts*

Because SDGS systems have few mechanical parts, the need to maintain a supply of spare parts is limited. If individual STEP units are included in the system, spare pumps and controls must be available for emergency repairs. A minimum of two spare pumps and the associated float switches and controls should be available for small systems. Pipe and pipe fittings should be kept on hand to repair any pipeline breaks that may occur. Spare septic tank lids and riser rings should also be readily available.

12. *Curvilinear* (KER-vuh-LYNN-e-ur). In the shape of a curved line.

### 6.323 *Safety*

Operators of small-diameter gravity sewer systems are exposed to the usual hazards associated with raw wastewater, such as risk of infections and the possible risk of contracting infectious diseases. Operators may also encounter hazardous gases released by *SEPTIC*[13] wastewater, namely, hydrogen sulfide gas and methane gas. SDGS system construction activities may involve excavations, which can present hazards from cave-ins, street traffic, and confined spaces. These and other potential hazards are described in detail in Chapter 3, "Safety." All operators must be trained in and use safe procedures at all times. For your own health and well-being and that of your co-workers, review Chapter 3 from time to time and make safety a frequent topic of discussion.

### 6.324 *Recordkeeping*

It is difficult to overemphasize the need for complete, well-organized records. Good recordkeeping of all operation and maintenance duties performed is essential for effective preventive maintenance and troubleshooting when problems occur. A daily log should be kept and maintenance reports on all equipment should be filed in a timely manner. Flows at the main line lift stations should be estimated daily or weekly by recording the pump running times. This is helpful in evaluating whether infiltration or inflow problems are developing. A record of each service call and corrective action taken should be filed by service connection identification number. This record should include tank inspection and pumping reports. These records are particularly useful if they are reviewed just prior to responding to a service call.

### 6.325 *Hybrid SDGS System O&M*

Many SDGS collection systems serve a combination of septic tank effluent pump units and septic tank effluent filter (screened) units. Usually, effluent from these septic tank units flows into the collection system by gravity or as a controlled discharge by a siphon. The operation and maintenance of hybrid SDGS systems such as these is very similar to the O&M procedures described above for standard SDGS systems.

If a community has combined all of the above components along with a conventional gravity wastewater collection system, more operator attention to O&M will be required. These hybrid systems are not common today and in all cases the small-diameter gravity sewers (SDGS) will terminate into the conventional gravity wastewater collection system. Larger solids and greases should never be discharged into the SDGS system because the system could not function under those circumstances. If you are the operator responsible for a hybrid system, additional instructional material is available in a similar operator training course titled, *OPERATION AND MAINTENANCE OF WASTEWATER COLLECTION SYSTEMS* (two volumes), which is available from the Office of Water Programs at California State University, Sacramento.

## QUESTIONS

Please write your answers to the following questions and compare them with those on pages 344 and 345.

6.32A Why should the maintenance of the septic tanks in a SDGS system be the responsibility of the utility district?

6.32B Why are record (as-built) drawings important for the operator of a SDGS system?

6.32C What skills/training are necessary for SDGS maintenance personnel?

### 6.33 Preventive Maintenance

Preventive maintenance includes inspection and pumping of the septic tanks, inspection and cleaning of the collection mains, inspection and servicing of any STEP units or lift stations, testing and servicing of air release valves, and exercising of isolation valves and other *APPURTENANCES*.[14]

### 6.330 *Septic Tanks*

Septic tanks must be pumped periodically to prevent solids from entering the collector mains. Prescribed pumping frequencies are typically three to ten years. Restaurants and other high-use facilities such as taverns require more frequent pumping. Common practice is to require an additional septic tank to serve as a grease trap for grease removal and to pump tanks serving these high-use facilities every three to six months. Before the pumping operations, utility district personnel measure and record the depth of sludge and thickness of any scum in each tank so that the pumping can be scheduled when needed and the schedule can be altered in the future according to actual sludge and scum accumulation rates.

Tank inspection is usually performed by agency personnel immediately after a septic tank has been pumped out to check for cracks, leaks, baffle integrity, and general condition of the tank. Effluent screens used on the tank outlet must be pulled

---

13. *Septic* (SEP-tick). A condition produced by anaerobic bacteria. If severe, the sludge produces hydrogen sulfide, turns black, gives off foul odors, contains little or no dissolved oxygen, and the wastewater has a high oxygen demand.
14. *Appurtenance* (uh-PURR-ten-nans). Machinery, appliances, structures, and other parts of the main structure necessary to allow it to operate as intended, but not considered part of the main structure.

and cleaned regularly by flushing with water if they are to be effective. Follow the screen manufacturer's recommendations or directions in your O&M manual regarding the frequency of cleaning effluent screens.

### *6.331 Collection Mains*

Periodic inspection of the collection mains is an important maintenance function even though reported performance of SDGS systems has been good. In systems where the mains have been inspected, no noticeable solids accumulations have been noted and cleaning of the collection mains has seldom been needed. Many large systems have been operating for more than 30 years without main cleaning. However, regular inspection and cleaning still is recommended for long flat sections in which daily peak flow velocities are less than 0.5 FPS (foot per second).

The most common method of cleaning collection mains in SDGS systems is with the use of pigs (Figure 6.12). (This method is not recommended if the collection mains are constructed of SDR (standard dimension ratio) 35 pipe because it is not considered strong enough to withstand the force that develops during the cleaning operation.)

*Fig. 6.12 Types of swabs and pigs*
(Courtesy of Girard Industries)

Pigging devices are slightly larger than or the same size as the inside diameter of the main to be cleaned. Normally, the pig is placed in a blind flange or valved access port (Figure 6.13) (also see Figure 6.8 on page 270). The access or entry station is sealed on the insertion side and a valve is opened to allow the pig to enter the collector main. High water pressure applied behind the pig forces the pig into and down the collection main or pipeline, pushing accumulated grit or debris ahead of it leaving a clean pipe full of water behind it. The pig will continue to move down the pipe until the inside pipe diameter changes (to either a larger or smaller diameter), until sufficient debris is collected ahead of the pig to stall it, or until an obstacle such as a closed valve, *OFFSET JOINT*,[15] or root intrusion stops the pig.

When purchasing a pigging device, be sure that some means of tracing the device has been incorporated into its design. If the pig is accidentally lost in the pipe during the cleaning operation, many hours can be spent searching for its location. During the search time, the segment of pipeline above the pig will be out of use.

Schedule pigging operations during periods of low flow so that the water used in the cleaning operation does not overload the collection system. Also, coordinate your maintenance activities with the operator of the wastewater treatment facility to prevent overloading the treatment plant with the cleaning water and solids discharged by the cleaning operation.

### *6.332 Lift Stations*

Main line lift stations should be inspected on a daily or weekly basis. Pump operation, alarms, and switching functions should be checked and running times of the pumps should be recorded. The flowmeter of each pump, if provided, should be calibrated annually. If a pump is found to be 20 percent lower in flow than when it was originally installed, install a new pump or rebuild the existing one to restore it to its design capacity. Clean wet wells by pumping down the wastewater and hosing out the structure. Check for debris or grit deposits and, if present, attempt to determine the source and eliminate the entry point.

Corrosion is a problem that is usually most visible at lift stations. (*NOTE:* Corrosion may also occur at air release stations if the valves have metal parts.) Concrete wet wells may need to be coated with corrosion-resistant materials and nonmetallic hardware should be used. Corrosion problems can be reduced in lift stations by using wet well/dry well construction with an adequately vented wet well.

## 6.34 Emergency Calls

Emergency call-outs are usually due to plumbing backups into homes or odors. In nearly every case, the plumbing backups are due to obstructions in the building sewer. Although the building sewer usually is the property owner's responsibility, most utilities have assisted the owner in clearing the obstruction. Main line or service lateral obstructions and lift station failures require that emergency actions be taken to limit the time the system is out of service to prevent environmental or property damage that might occur. Each utility must develop its own emergency operating procedures.

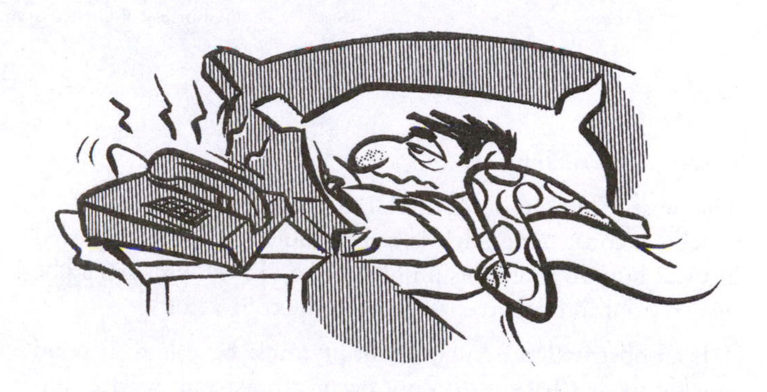

15. *Offset Joint.* A pipe joint that is not exactly in line and centered. Also see DROP JOINT and VERTICAL OFFSET.

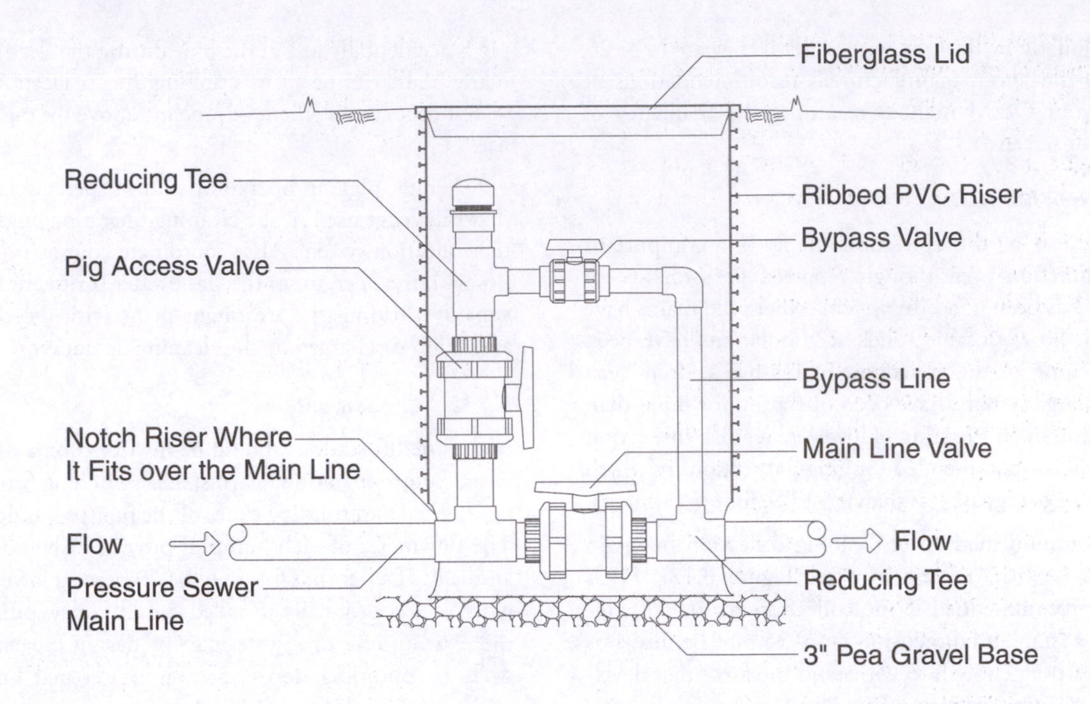


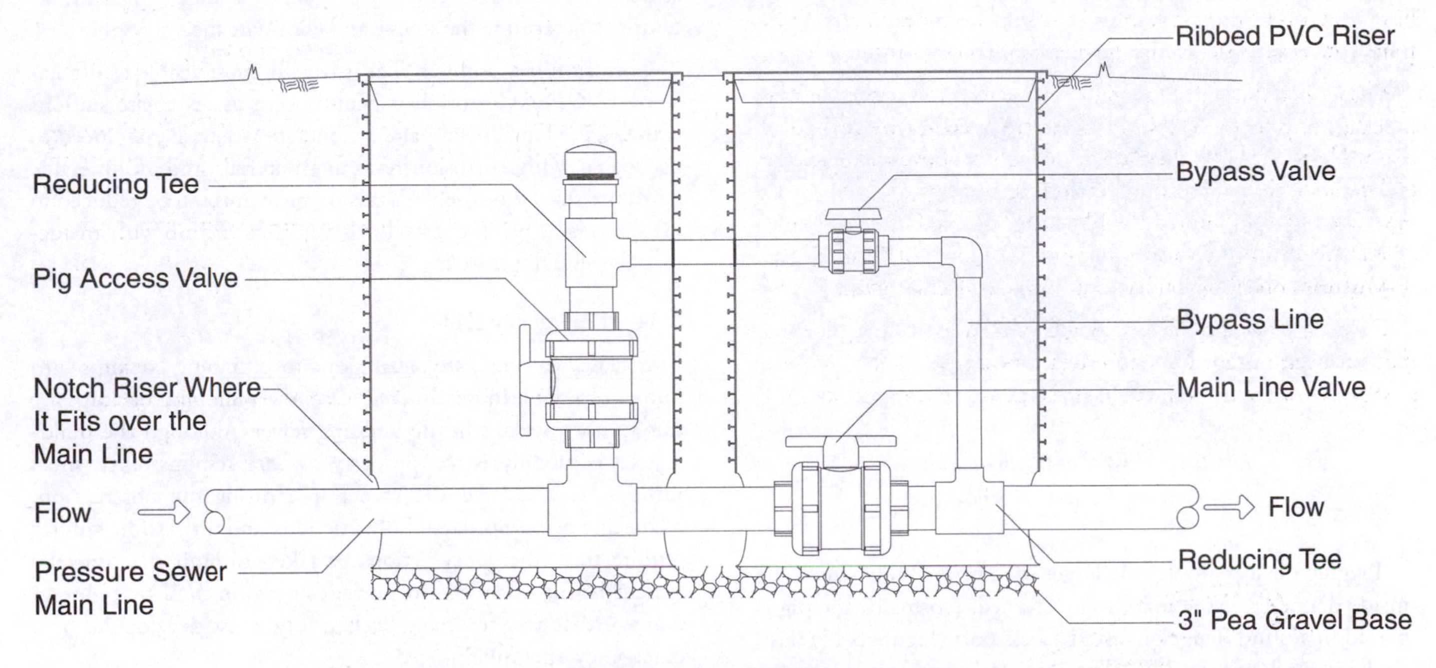


*Fig. 6.13 Typical pigging ports*
(Courtesy of Orenco Systems, Inc., www.orenco.com)

### *6.340 Obstructions*

The most common source of blockages in newly installed STEP systems is construction debris, which should have been removed but was not. To avoid this problem, all STEP systems should be pigged prior to final acceptance from the contractor.

If an obstruction occurs, the utility must be able to respond quickly so that backups do not occur at upstream service connections. Experience has shown that most obstructions caused by construction debris cannot be removed by simple flushing. It is often necessary to excavate the main and break it open to remove the obstruction. The most difficult part of this task is identifying exactly where the obstruction is located in the collection main. Begin by looking for a cleanout that has standing or rising wastewater in it. Then, check the next cleanout downstream. If it is dry or has reduced flow through it, the blockage is between the dry cleanout and the flooded upstream cleanout.

The exact location of the obstruction may be identified by the use of hand rods. The rods must be inserted at the upstream (full or filling) cleanout. Cleanouts are equipped with a 45-degree or 90-degree ell entering with the direction of flow, which makes it almost impossible to push rods upstream. To locate the obstruction, measure how far the rods were pushed into the main to hit the obstruction. Next, lay the rods on the surface above the route of the main and subtract the distance equal to the height of the cleanout to the main. This procedure should provide a fairly accurate location to excavate the main.

Once you have determined the probable location of the obstruction but before you break into the main, insert the hand rods into the main again until they are stopped by the obstruction. Remain at the site of the blockage and listen while another person pulls the rods back a foot or so and rapidly pushes them in. Often, it is possible to hear the rod cutter or tip hitting the obstruction or pipe, thus confirming the blockage location.

When you are confident you know where the blockage is located, prepare to break open the main several feet below the indicated blockage. First, clear the excavation of all debris, rocks, and as much loose soil or bedding material as possible. You may even wish to line the excavation with 1 or 2 *MIL*[16] plastic sheeting to provide a lined reservoir for the wastewater that will flow out when you break the pipe. It would also be appropriate to lay a hose from the excavation to the next downstream cleanout and have a pump at the excavation in case it becomes necessary to pump the wastewater to the downstream access point.

Break open the collection main several feet downstream of the blockage to provide access to a clear and open gravity collection main. Next, cut open the pipe at or slightly above the blockage and remove the cut section (including the obstruction if you have correctly pinpointed the blockage). Backed up wastewater should flow into the excavation and will flow out through the downstream opening you have cut in the collection main.

Clearing blockages or repairing a broken main in a pipeline that has been installed under a stream or creek bed may require a slightly different approach. Emergency pumping stations (Figure 6.14) are sometimes installed above and below such locations (Figure 6.15) so that the section of pipe under the creek can be bypassed temporarily while repairs are made or the blockage is being removed.

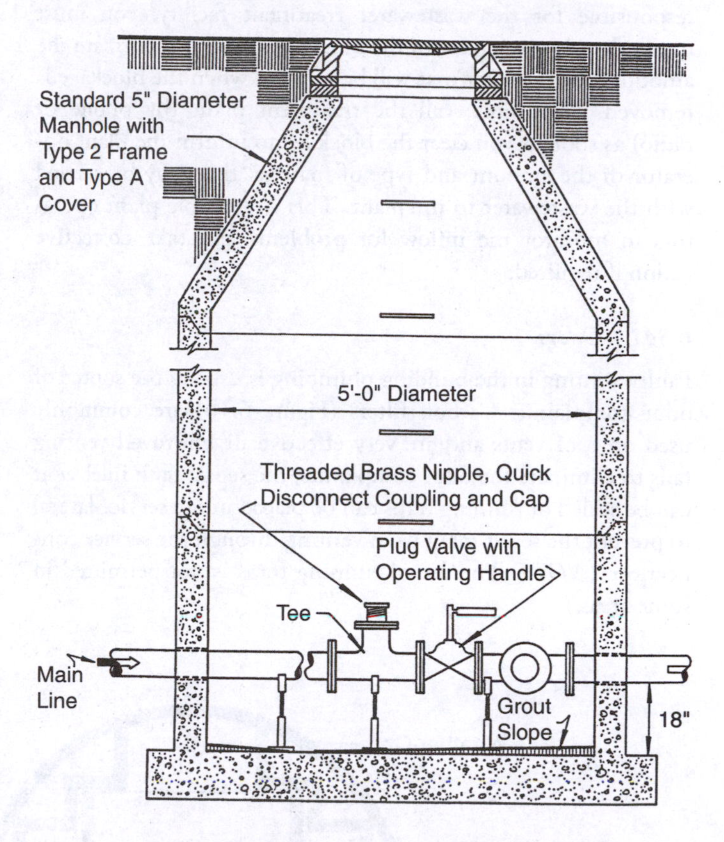


*Fig. 6.14 Emergency pumping station*
(Source: *ALTERNATIVE WASTEWATER COLLECTION SYSTEMS*, Technology Transfer Manual, US Environmental Protection Agency)

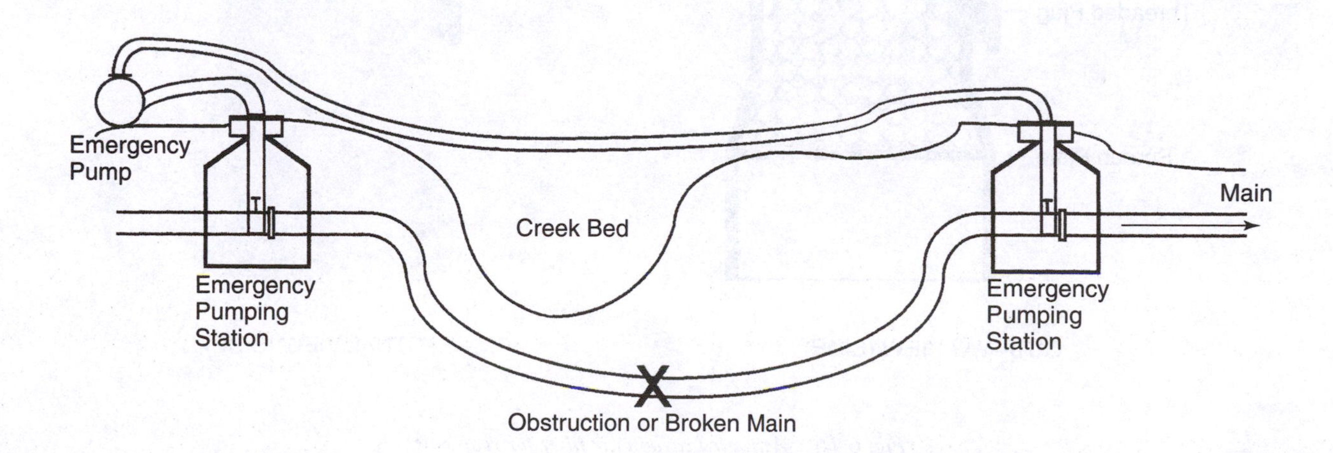


*Fig. 6.15 Setup for temporary pumping around a blockage or broken main*

16. *Mil.* A unit of length equal to 0.001 of an inch. The diameter of wires and tubing is measured in mils, as is the thickness of plastic sheeting.

After removing a blockage and repairing the pipeline, it is good practice to check the downstream treatment facility for normal operation. *SLUGS*[17] of debris, grit, or other undesirable objects discharged during the cleaning operation could interfere with the treatment processes or equipment. If you are not responsible for the wastewater treatment facility, you must notify the plant operator of the blockage and try to estimate the amount of wastewater that will be released when the blockage is removed. If possible, call the treatment plant (by phone or radio) as soon as you clear the blockage to inform the plant operator of the amount and type of material that may be carried with the wastewater to the plant. This will enable plant operators to monitor the inflow for problems and take corrective action if required.

### *6.341 Odors*

Faulty venting in the building plumbing is usually the source of odor complaints. Carbon filters (Figure 6.16) are commonly used on roof vents and are very effective. If improved venting fails to eliminate the odor complaints, the septic tank inlet vent can be sealed or running traps can be placed in the service lateral to prevent the sewer main from venting through the service connection. (*NOTE:* The use of running traps is not permitted in some areas.)

### *6.342 Lift Station Failures*

Lift stations may fail due to loss of power or a mechanical failure. Portable emergency generators can be used to provide power during long outages. Many small communities have provided added storage at the lift station (Figure 6.17) or truck-mounted pumps that can pump from the wet well to a downstream hose connection in an emergency pumping manhole (Figure 6.14) on the force main. This latter method also works well for mechanical failures.

## QUESTIONS

Please write your answers to the following questions and compare them with those on page 345.

6.33A Preventive maintenance of a small-diameter gravity sewer system usually includes what activities?

6.33B Why should utility district personnel be present at septic tank pumping operations?

6.33C What is the most common method used to clean SDGS collection mains?

6.33D Where does corrosion usually occur in a SDGS system?

6.34A What are the most common types of SDGS emergency call-outs?

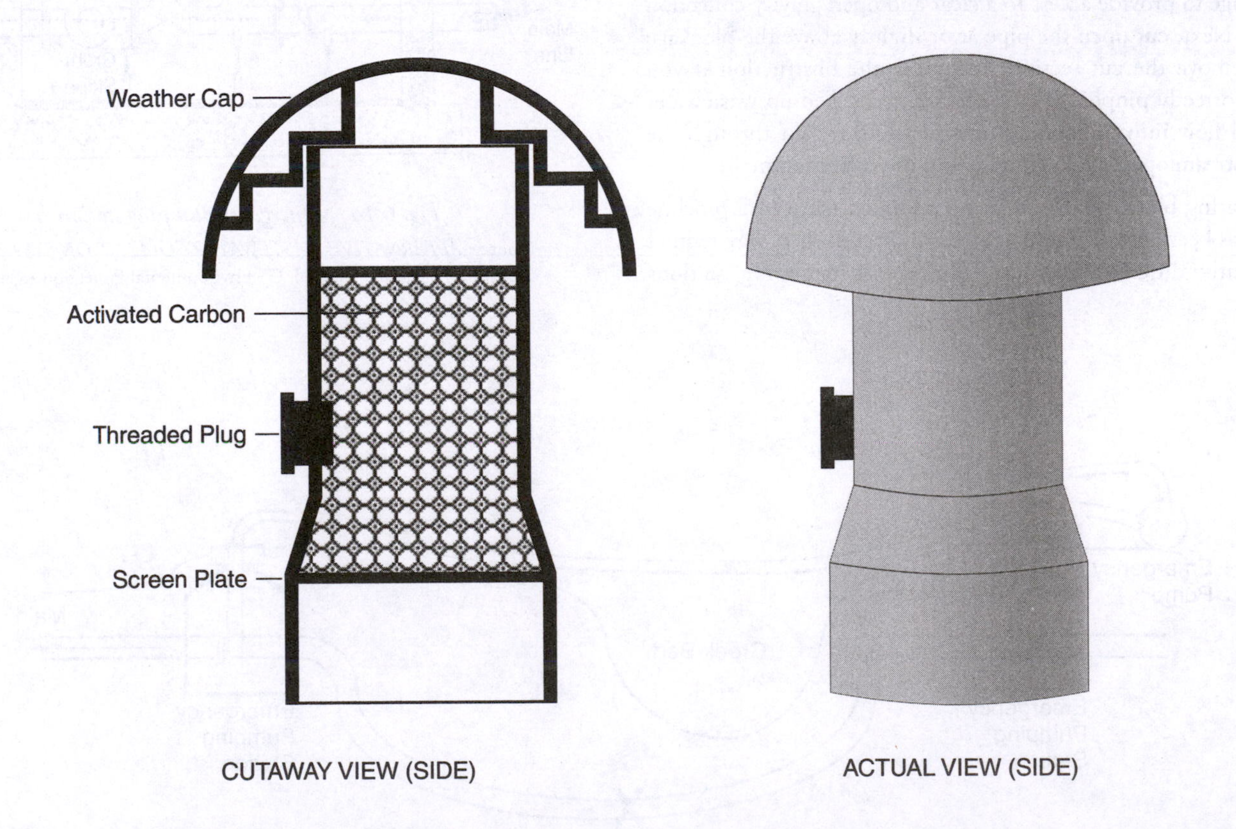


*Fig. 6.16 Activated carbon air filter for roof vent*

(Courtesy of Orenco Systems, Inc., www.orenco.com)

17. *Slug.* Intermittent release or discharge of wastewater or industrial wastes.

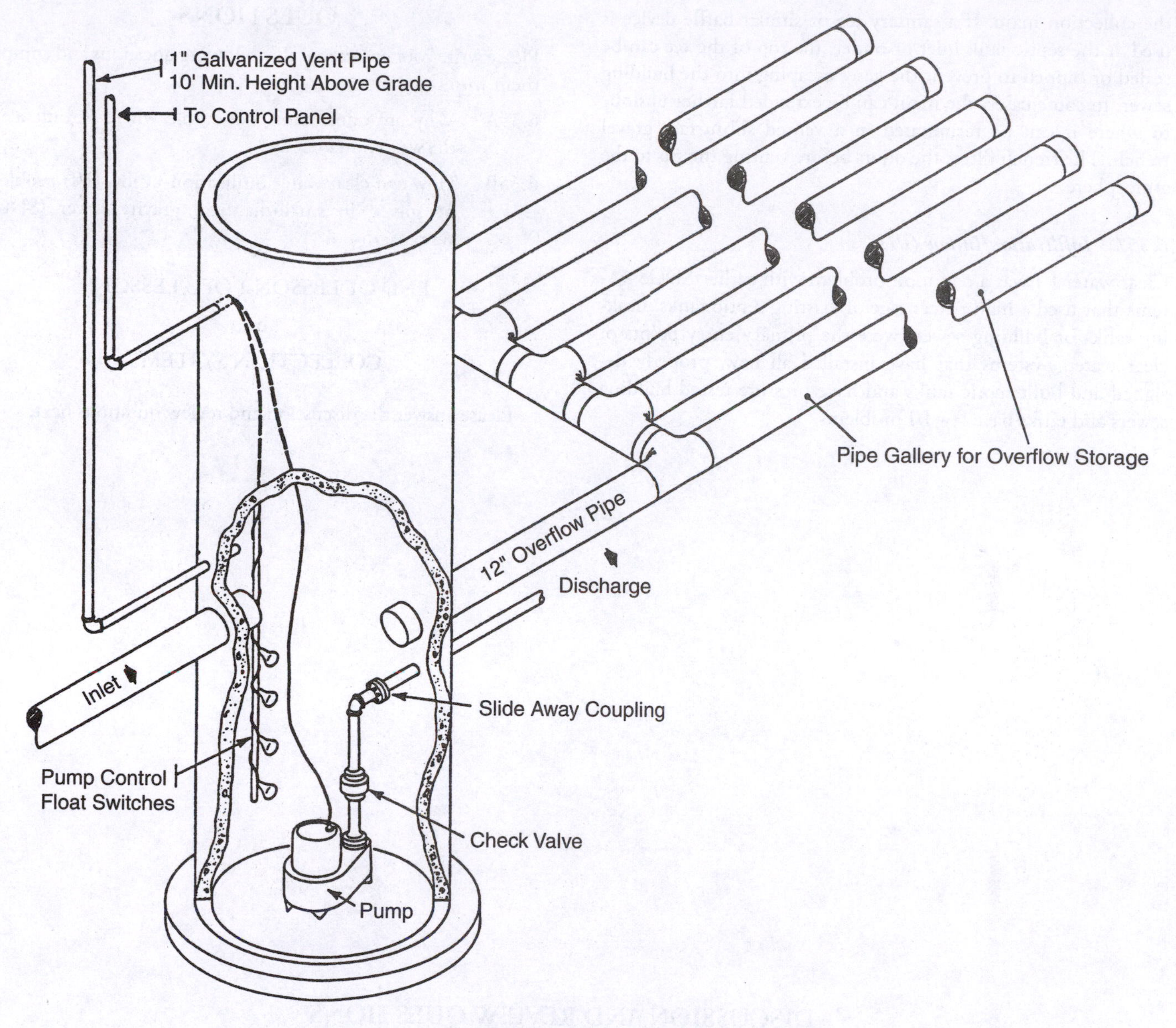


*Fig. 6.17 Main line lift station with emergency storage*
(Source: *ALTERNATIVE WASTEWATER COLLECTION SYSTEMS*, Technology Transfer Manual, US Environmental Protection Agency)

## 6.35 Troubleshooting

### *6.350 Odors*

Well-designed small-diameter gravity sewer (SDGS) systems do not have odor problems although odors are a commonly reported problem with poorly designed SDGS systems. The settled wastewater collected by a SDGS is septic and therefore contains dissolved hydrogen sulfide and other foul-smelling gases. These gases tend to be released to the atmosphere in larger amounts where turbulent conditions occur such as lift stations, drop cleanouts, and *HYDRAULIC JUMPS*.[18] The gases tend to escape through house plumbing stack vents, particularly at homes located at higher elevations or ends of lines. The turbulence releases the obnoxious gases dissolved in the wastewater.

Measures that can be taken to correct odor problems include installing soil odor filters (see Figure 6.10 on page 271), air-tight wet well covers, and carbon filter vents (see Figure 6.13 on page 276). Odors at individual connections often originate in

18. *Hydraulic Jump.* The sudden and usually turbulent abrupt rise in water surface in an open channel when water flowing at high velocity is suddenly retarded to a slow velocity. (See drawing on page 253.)

the collection main. If a sanitary tee or similar baffle device is used at the septic tank inlet or outlet, the top of the tee can be sealed or capped to prevent the gases escaping into the building sewer. In some cases, the main can be extended farther upslope to where it can be terminated in a vented subsurface gravel trench. The trench filters the odors before venting the gas to the atmosphere.

*6.351 Infiltration/Inflow (I/I)*

Clear water I/I was a common problem with earlier SDGS systems that used a high percentage of existing septic tanks. Leaking tanks or building sewers were the primary entry points of clear water. Systems that have installed all new, properly designed and built septic tanks and have pressure tested building sewers and tanks have few I/I problems.

## QUESTIONS

Please write your answers to the following questions and compare them with those on page 345.

6.35A Why are odors most noticeable where turbulence in the wastewater occurs?

6.35B How can clear water infiltration/inflow (I/I) problems be solved in small-diameter gravity sewer (SDGS) systems?

**END OF LESSON 1 OF 3 LESSONS**

**on**

**COLLECTION SYSTEMS**

Please answer the discussion and review questions next.

## DISCUSSION AND REVIEW QUESTIONS

**Chapter 6. COLLECTION SYSTEMS**

(Lesson 1 of 3 Lessons)

At the end of each lesson in this chapter, you will find discussion and review questions. Please write your answers to these questions to determine how well you understand the material in the lesson.

1. Why should collection system operators know how engineers design wastewater collection systems?
2. What are the sources of infiltration of groundwater and inflow of surface water into wastewater collection systems?
3. Why does the slope of conventional gravity sewers generally follow the slope of the land?
4. Why are lateral, main, and trunk sewers in a conventional gravity collection system usually constructed at or near the center of a street?
5. Why are good construction, inspection, and testing procedures very important for wastewater collection facilities?
6. Successful use of small-diameter sewer lines depends on what two important functions of the septic tank?
7. Why is it important to ensure that all components of a small-diameter gravity sewer system are watertight?
8. Describe the basic function of a lift station.
9. Why should SDGS system record drawings (as-builts) identify collection mains in relation to permanent structures nearby?
10. What are the general types of preventive maintenance activities associated with small-diameter gravity collection systems?

# CHAPTER 6. COLLECTION SYSTEMS

(Lesson 2 of 3 Lessons)

## 6.4 PRESSURE SEWERS

### 6.40 Description of System

Where the topography and ground conditions of an area are not suitable for a conventional gravity collection system due to flat terrain, rocky conditions, or extremely high groundwater, low-pressure collection systems are now becoming a practical alternative. These systems may serve septic tanks equipped with effluent pumps (STEP systems) or with grinder pumps (GP systems). Smaller pipe sizes can be used in a pressure collection system because the wastewater flows under pressure created by pumps and because infiltration is not a problem in a pressurized system.

In some localities, a pressure system is combined with a gravity collection system. The pressure sewer serves portions of the community where topography limits the use of gravity sewers. In most cases, the pressure system discharges into the gravity system. In other instances, the two systems are kept totally separate, with both discharging at the inlet to the wastewater treatment facility.

#### *6.400 STEP Systems (Figure 6.18)*

At each service connection in a STEP system, a small (normally ⅓ to ½ horsepower) submersible pump discharges septic tank effluent to a completely pressurized collection main. The pumps may be installed inside the septic tank using an internal vault or in a pump vault outside the tank. Electrical service providing 110 to 120 volts is needed at each connection to power the pump. In most installations, three mercury float liquid level sensors are installed in the pump vaults: pump ON, pump OFF, and a high water alarm. The discharge line from the pump is equipped with at least one check valve and one gate valve. Cleanouts are installed to provide access for maintenance. Automatic air release valves are required at, or slightly downstream of, high points along the main pipeline.

#### *6.401 Grinder Pump Systems (Figure 6.19)*

Grinder pump (GP) systems are cost-effective for small communities and can be used in areas with hilly or flat terrain. No deep excavation is necessary. Septic tanks are not used with GP systems. Instead, the building sewer is connected to a fiberglass pump vault that is approximately 3 feet in diameter with a liquid capacity of about 40 gallons. A grinder pump is suspended inside this pump vault. (Typically, the pump for a single residence is a 2-horsepower pump that requires 220 volts. One-horsepower units are also used.) The vault also contains liquid level sensors that operate the pump. The grinder pump shreds wastewater solids and pumps the wastewater to a pressurized collection system. Pump discharge lines to the sewer main are plastic and may be as small as 1.25 inches in diameter. A small check valve and a gate valve on the discharge line prevent backflow. Cleanouts are installed to provide access for flushing, and automatic air release valves are installed at, or slightly downstream of, high points in the system.

### 6.41 Pipe Material

Pipe material and pressure rating should be determined by the design requirements, construction techniques, ground conditions, and other factors. The information in this section should be considered as informational only.

The most widely used pipe material is PVC (polyvinyl chloride) with a working pressure rated at 200 psi. PVC pipe with a working pressure rated at 160 psi has been used successfully in pipe sizes 2 inches and larger. The operating pressure of pipes is usually two-thirds or less of the rated working pressure of the pipe. From a pressure standpoint, the lower rated pipe is acceptable, but thin-wall pipe is more likely to be damaged during installation. Since there is little cost difference between the two pipe sizes when the excavation, backfill, and surface restoration are considered, 200 psi pipe is recommended.

Polyethylene pipe has also been used, especially when it is important to minimize the number of joints, such as in areas with a high water table or unstable soil conditions, and for lake or river crossings. The reduced number of pipe joints enhances pipe integrity, minimizes infiltration/inflow, and minimizes the possibility of leaks (exfiltration) from the system.

As with all wastewater collection systems, the pipe diameter in the head or upper reaches of the pressure system starts out relatively small and increases in size to handle additional flows collected along the way to the wastewater treatment plant or discharge site. Pressure sewer pipe sizes are very small compared to

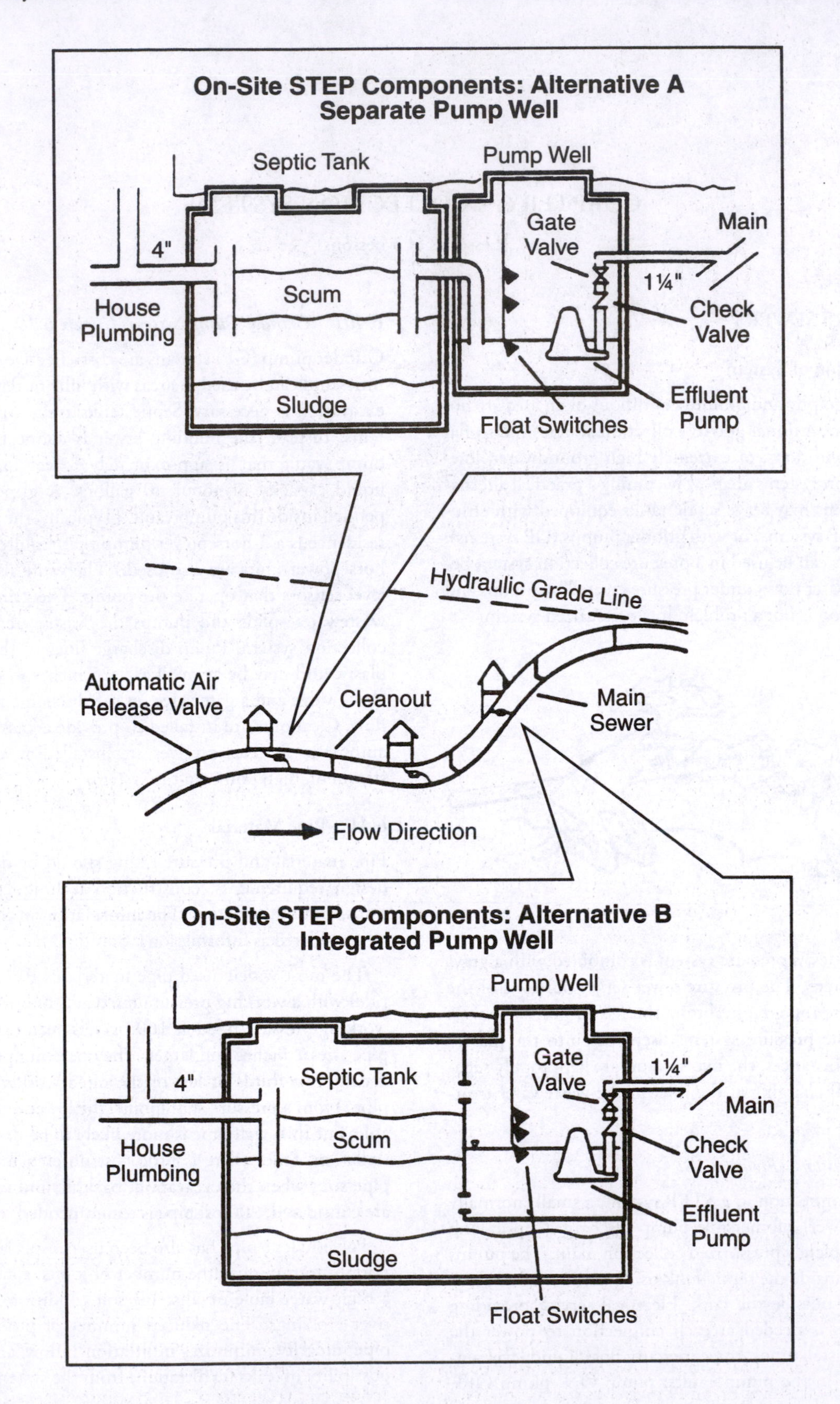


*Fig. 6.18 On-site STEP system components*

(Source: *SMALL COMMUNITY WASTEWATER COLLECTION, TREATMENT AND DISPOSAL OPTIONS*, Rural Community Assistance Corporation)

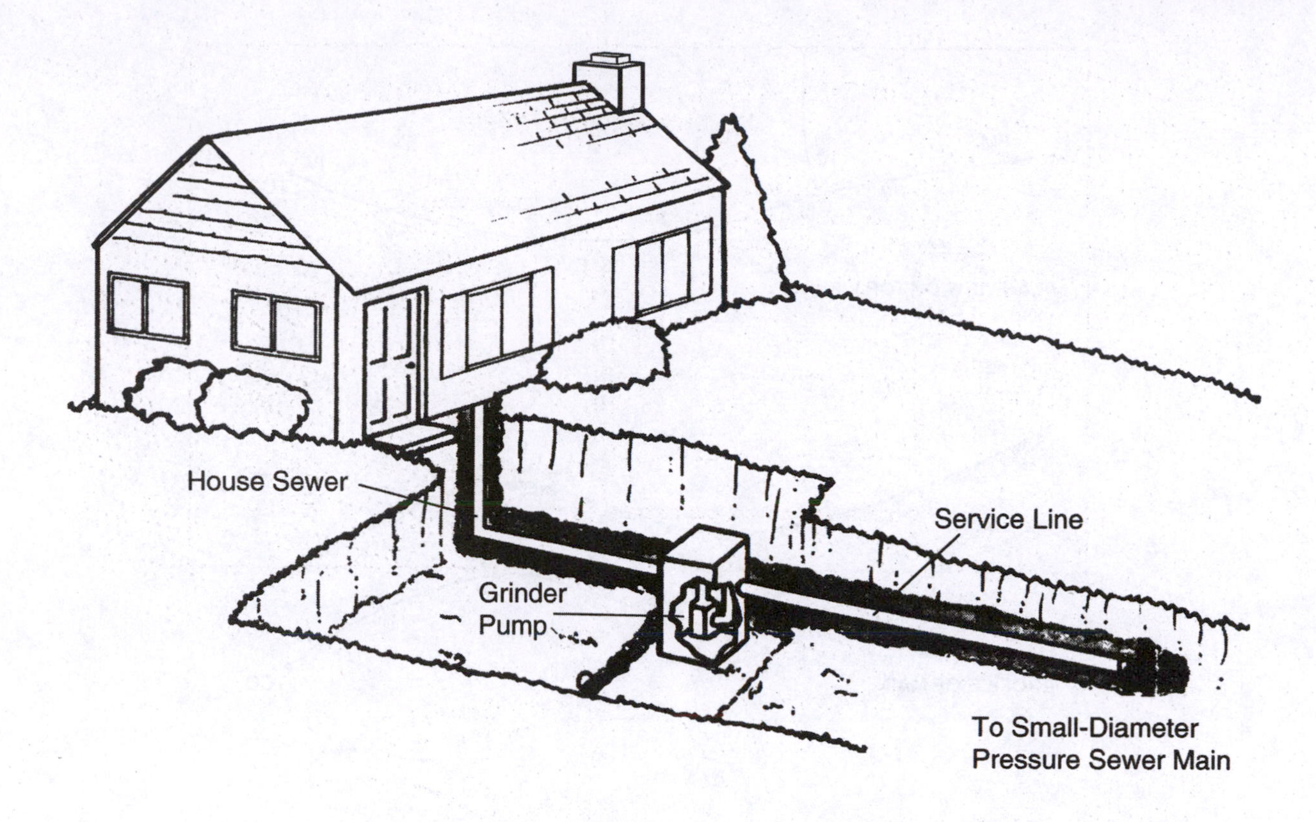


*Fig. 6.19 On-site grinder pump (GP) pressure system components*
(Source: *SMALL COMMUNITY WASTEWATER COLLECTION, TREATMENT AND DISPOSAL OPTIONS*, Rural Community Assistance Corporation)

gravity collection systems. Two-inch diameter pipe is common at the upper end (beginning) of the pressure main and the pipe size increases to whatever is necessary to convey the community's wastewater at a reasonable system pressure to the treatment plant or discharge site.

The critical parts of any wastewater collection system piping are the joints. The design engineer selected the most suitable pipe material and jointing method for each particular locality and geographic conditions. Whether the originally installed system used rubber ring joints (for thermal expansion or contraction) or solvent-welded joints, the operators should use the same methods and materials for additions or repairs.

## 6.42 Collection Main

Figure 6.20 illustrates a simplified pressure main layout and the location of components. The plan view shows how the system layout would look from a bird's-eye view, that is, looking at it from above the surface. The profile shows the elevation of the main, that is, how it moves up and down over the terrain. Note that the automatic air release valve is at an upper elevation on the main between two depressions. The air release valve permits venting of air or gases trapped in the main and prevents an air lock or blockage in the main.

Many STEP systems are located in areas with a high groundwater problem where it would be difficult and expensive to install a single pressure main that gradually increases in diameter. A high groundwater table makes trenching very difficult and dangerous. The pipes tend to float out of alignment and then are likely to be damaged when the excavation is backfilled. In high groundwater areas, several smaller diameter mains are sometimes installed to service particular segments of the community in place of the standard approach of laying a single pressure main.

### *6.420 Service Line or Lateral Connections to the Main*

Service saddles, tees, or tapped couplings are used to join the service line to the wastewater collection main. From an operation and maintenance point of view, the service saddle provides the best method of connecting new services or repairing existing connections (Figure 6.21). In some instances, nipples, *CORPORATION STOPS,*[19] and check valves are installed on the lateral

19. *Corporation Stop.* A water service shutoff valve located at a street water main. This valve cannot be operated from the ground surface because it is buried and there is no valve box. Also called a corporation cock.

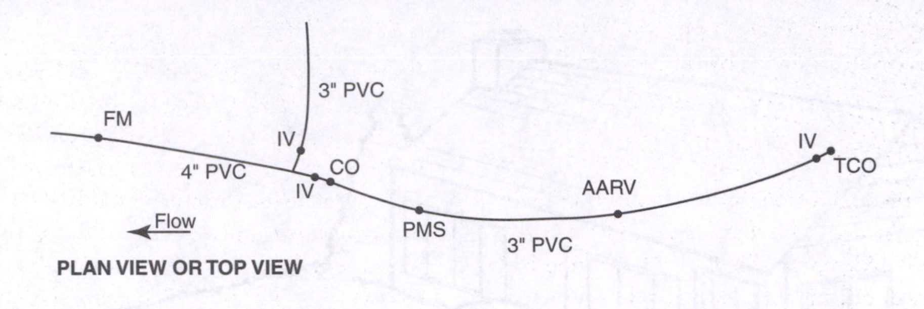


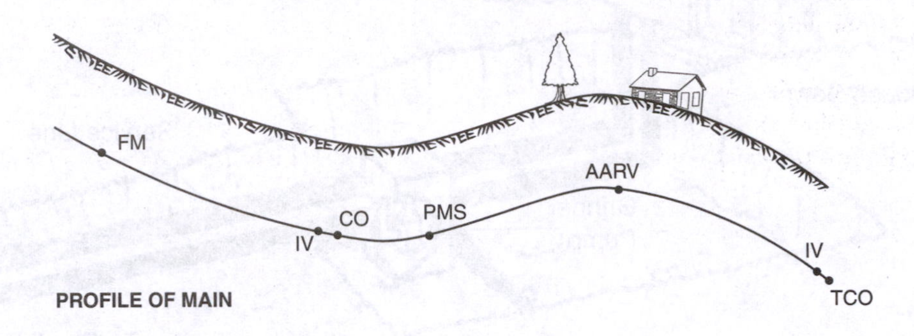


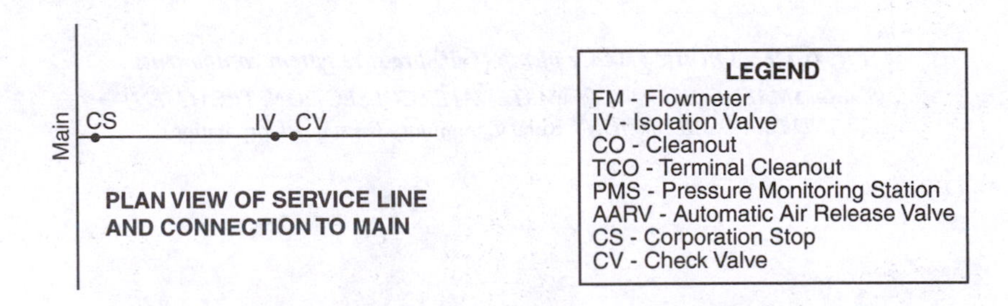


*Fig. 6.20 Plan and profile of STEP piping system and appurtenances*
(Source: *ALTERNATIVE WASTEWATER COLLECTION SYSTEMS,* Technology Transfer Manual, US Environmental Protection Agency)

Lateral connection at main line showing saddle, ball valve, and check valve

Lateral connection at main line

*Fig. 6.21 Lateral connections*
(Source: "Manila Community Service District, Design-Construction Report for the Septic Tank Effluent Pumping Sewerage System," California State Water Resources Control Board)

next to the saddle. However, many agencies prefer to install only a service line isolation valve and check valve on the lateral at the property line.

### 6.421 *Main Line Isolation Valves*

Isolation valves are installed in the wastewater collection main to isolate other branch mains at their intersections or junctions in case of breaks or required repairs. Small-diameter pipe ball valves are sometimes used, but in larger mains, gate valves that can withstand the system working pressure are usually installed (Figure 6.22). Isolation valves should be installed wherever problems might develop from unstable ground conditions, at river crossings, at bridges, or where future system additions would require flow isolation to make connections. Main isolation valves are sometimes installed on long pipe runs to break the system into shorter segments to facilitate repairs.

### 6.422 *Cleanouts*

In-line cleanouts (Figure 6.23; also see Chapter 2, Figure 2.17 on page 44) are installed in the main line to facilitate cleaning the line with a device called a pig (shown previously in Figure 6.12 on page 275). Each time the main line pipe size increases, a cleanout is required to handle the larger size pig or other cleaning equipment. However, one size pig can clean more than one pipe size as long as the pipe diameters increase in size in the direction the pig is moving. Pigging stations or cleanout locations vary depending on the preference of the designer or district. Multiple valve inlets separated by a pipe spool and blind flange accesses are commonly used. Another way to provide access for cleaning, as illustrated in Figure 6.24, is to install a vault over a pipe spool installed in the main line. Use bypass piping and valving to convey wastewater around the vault while the spool is being removed to install a pig for cleaning the main. In general,

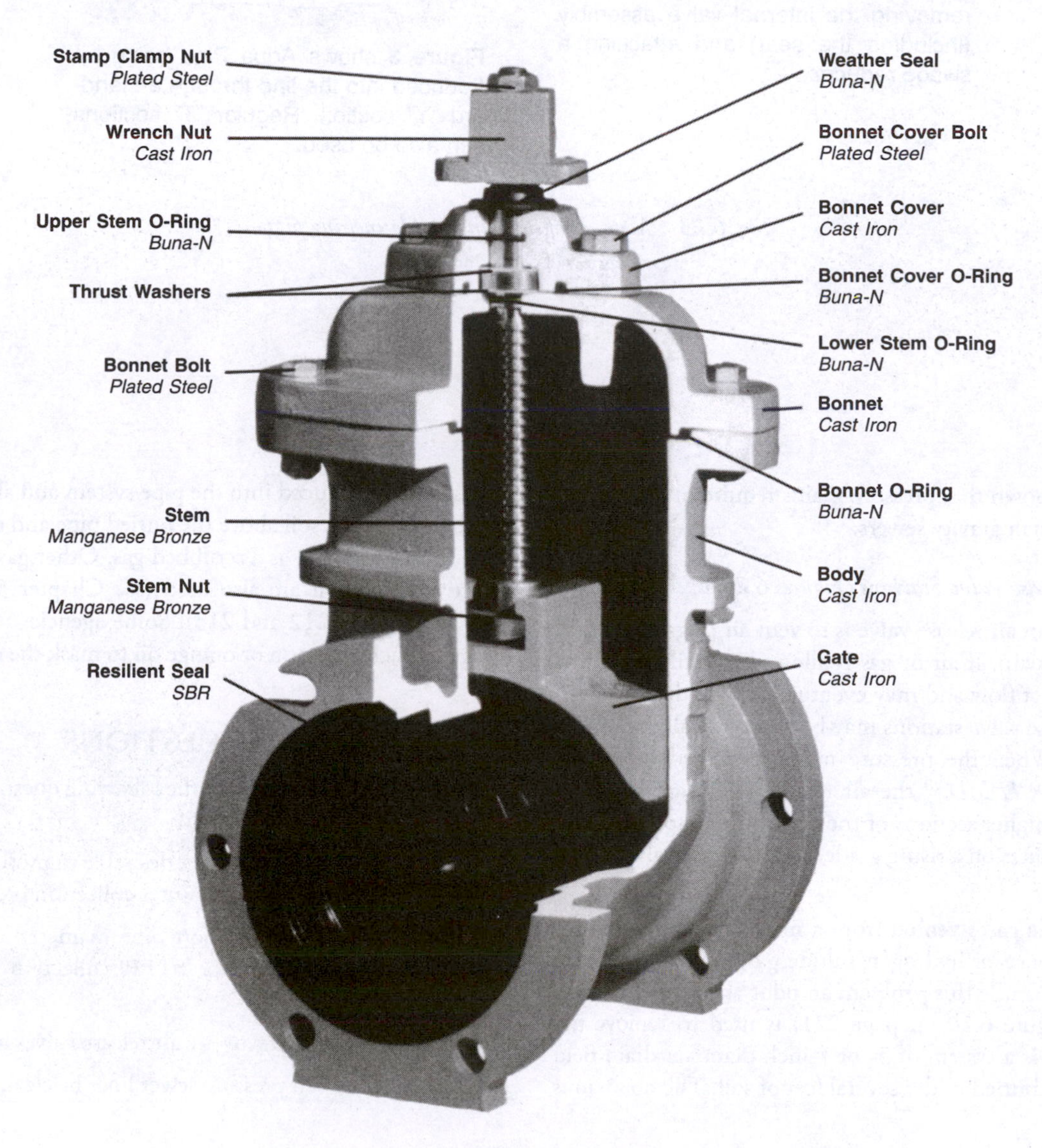


*Fig. 6.22 Gate valve parts*

(Permission of American Valve and Hydrant, A Division of American Cast Iron Pipe Company)

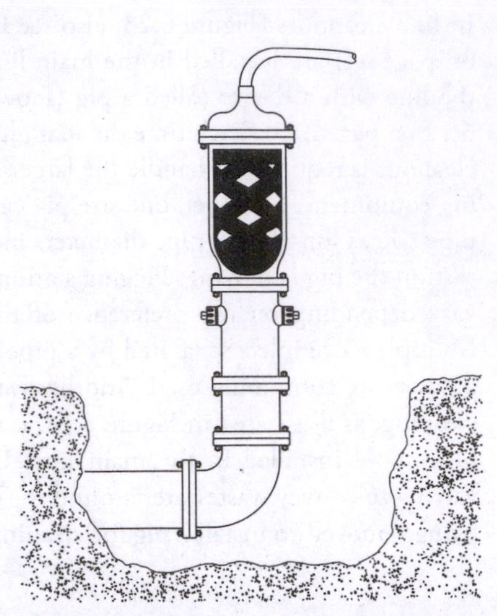

Figure 1 shows how a fire hydrant can be used to launch the Aqua Pig, by removing the internal valve assembly (including the seat) and attaching a swage reducer.

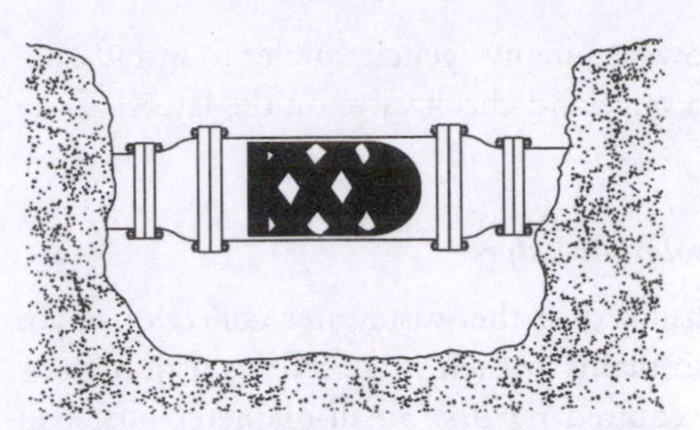

Figure 2 shows how an oversized spool can be coupled into the line for launching.

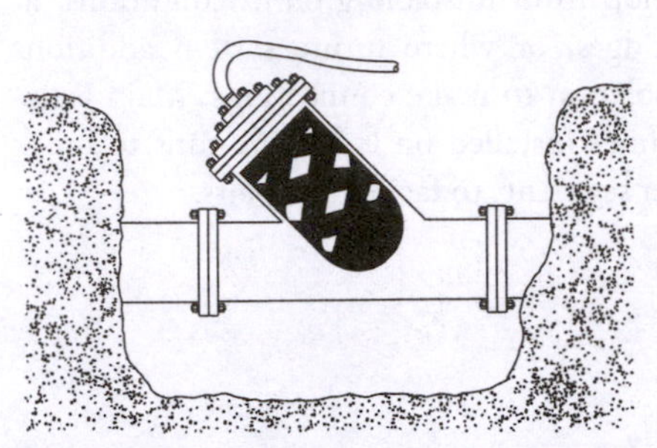

Figure 3 shows Aqua Pigs being introduced into the line through a standard "Y" section. Regular "T" sections can also be used.

*Fig. 6.23 Methods of inserting and launching pigs*
(Courtesy of Girard Industries)

experience has shown that pressure mains require much less frequent cleaning than gravity sewers.

### *6.423 Air Release Valve Stations (Figures 6.25, 6.26, and 6.27)*

The purpose of an air release valve is to vent air or gases trapped in the pressure main. If air or gas is allowed to build up in the line, it will restrict flow and may eventually block the line completely. Air release valve stations may be automatically or manually operated. When the pressure main is located below the *STATIC WATER HEAD*,[20] the air release valves are placed at summits or the higher sections of the main. If the upstream end of a main terminates on a rising grade, an air release valve is used at this location.

Frequently, the gases vented from a main are quite offensive due to the presence of hydrogen sulfide gas, which smells like rotten eggs. To handle this problem an odor absorption field or soil bed (see Figure 6.10 on page 271) is used to remove the odor. A soil bed is a system of 3- or 4-inch diameter drain field perforated pipes buried under several feet of soil. The obnoxious gases are introduced into the pipe system and allowed to permeate through the soil above the buried pipe and eventually escape to the atmosphere as a scrubbed gas. Other gas filter systems of activated carbon are also used (see Chapter 5, Figures 5.7 to 5.10 on pages 212 and 213). Some agencies use odor masking agents such as lemon or orange oil to mask the odors.

## QUESTIONS

Please write your answers to the following questions and compare them with those on page 345.

6.41A Why is it sometimes desirable to minimize the number of joints in a wastewater collection pressure pipeline?

6.41B What is a common pipe diameter at the upper end (beginning) of a STEP collection system pressure main?

6.42A Where and why are air release valves installed?

6.42B How can pressure sewer lines be cleaned?

20. *Static Water Head.* Elevation or surface of water that is not being pumped.

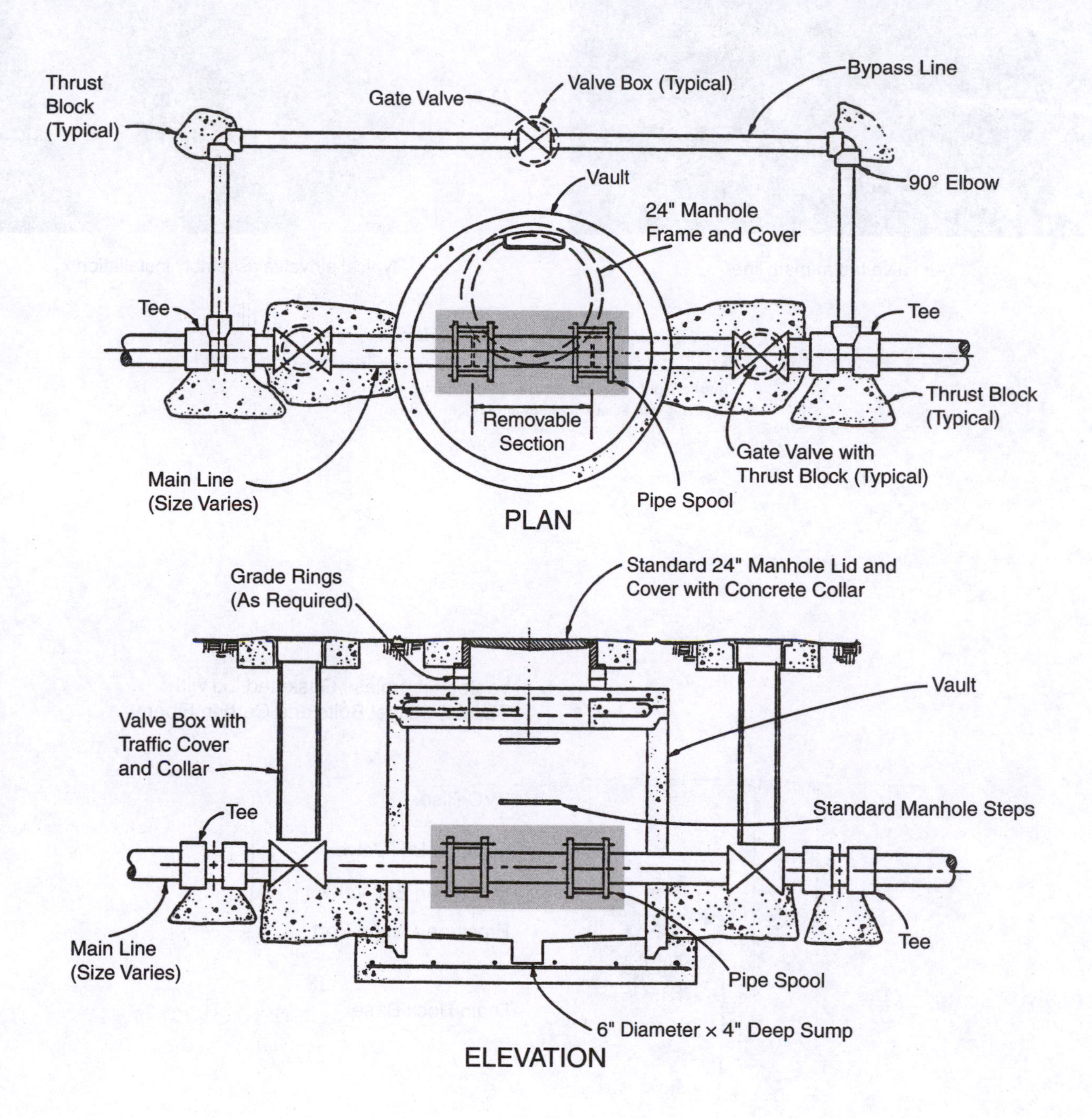


*Fig. 6.24 Removable pipe section detail*

(Source: "Manila Community Service District, Design-Construction Report for the Septic Tank Effluent Pumping Sewerage System," California State Water Resources Control Board)

Air valve tap at main line

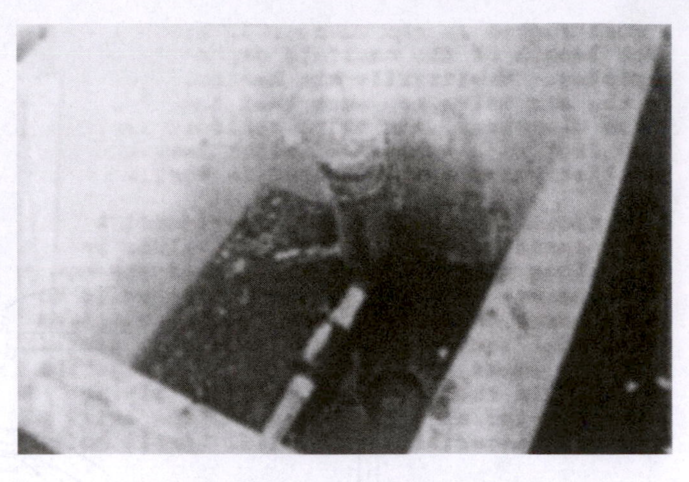

Typical air valve assembly installation

*Fig. 6.25 Air release valve installation*

(Source: "Manila Community Service District, Design-Construction Report for the Septic Tank Effluent Pumping Sewerage System," California State Water Resources Control Board)

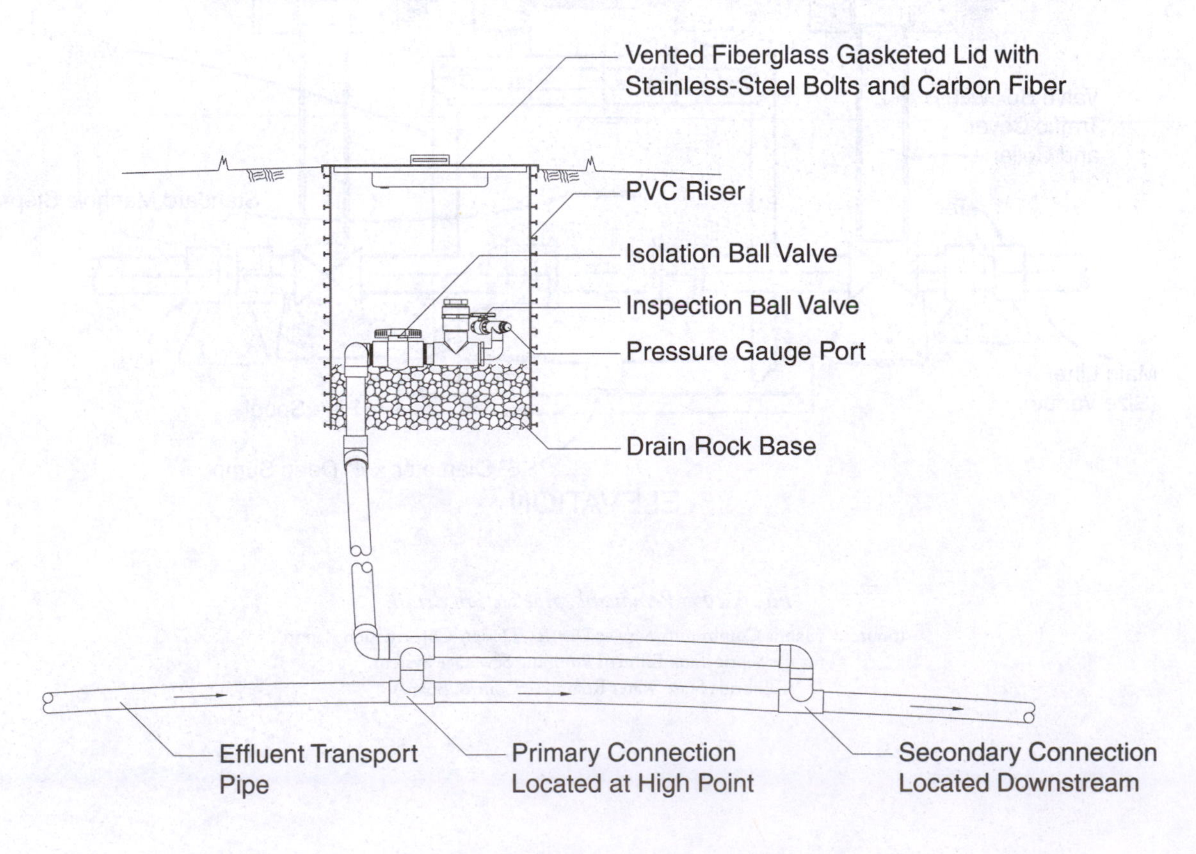


*Fig. 6.26 Manual air release assembly*

(Courtesy of Orenco Systems, Inc., www.orenco.com)

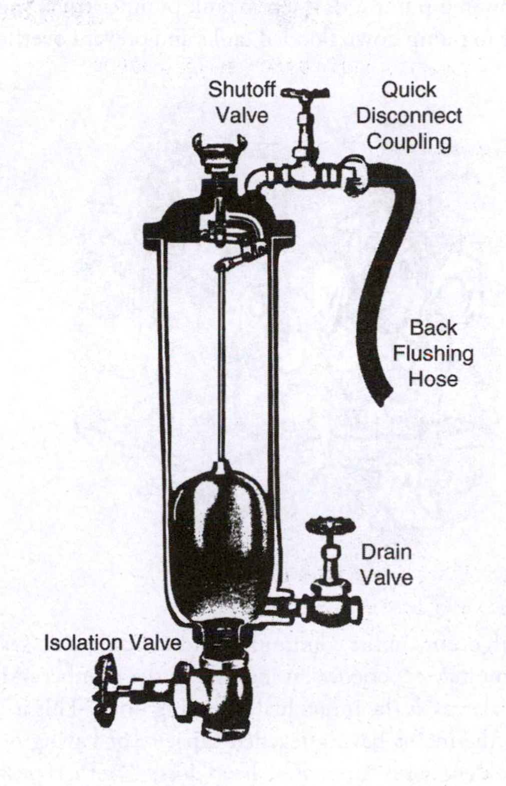


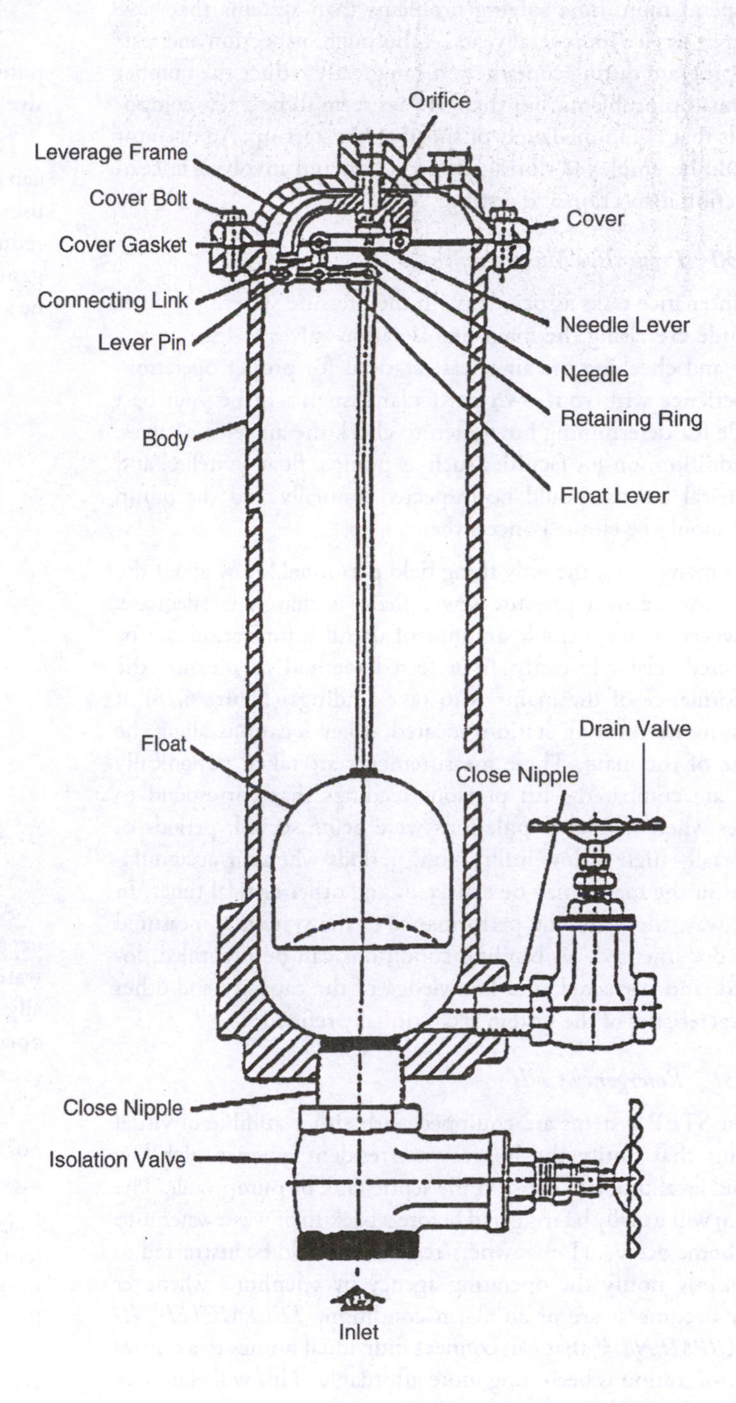


*Fig. 6.27 Air release valve details*
(Permission of APCO/Valve and Primer Corporation)

### 6.43 Pressure Sewer Maintenance

Pressure collection system mains operate reliably with only a few problems, most of which are connected with air release valves in the system. Newly constructed systems tend to require operators to spend more time solving problems than systems that have been in service for several years. A thorough inspection and testing program during construction can greatly reduce the number of start-up problems, but there always seem to be a few components that fail immediately or shortly after start-up. An operator should be employed during construction and involved in construction inspection and testing.

#### *6.430 Preventive Maintenance*

Maintenance tasks associated with the pressure system pipelines include exercising the main line isolation valves at least once a year and checking the air release stations for proper operation. Experience with your own particular system will be your best guide for determining how often to check the air release valves. In addition, on-lot facilities such as pumps, float switches, and electrical panels should be inspected annually and the pump well should be cleaned once a year.

In many cases, the only thing field personnel know about the performance of a pressure sewer main is that it is adequate. However, a considerable amount of useful information can be gathered relatively easily. A preferred method to measure the performance of the mains is to take readings (Figure 6.28) at pressure monitoring stations located at key locations along the route of the main. These measurements are taken periodically and are compared with previous readings that correspond to times when smaller populations were being served, periods of especially high or low infiltration, periods when air accumulations in the mains may be expected, and other critical times. In that way, the hydraulic performance of the system is measured and documented; air binding conditions can be identified, located, and corrected; and knowledge of the capacity and other characteristics of the system is continually refined.

#### *6.431 Emergency Calls*

Most STEP systems are equipped with either audible or visual alarms that notify the homeowner/resident when a high/low liquid level problem exists in the septic tank or pump vault. The alarm will usually be triggered before a backup of wastewater into the home occurs. Homeowners/residents should be instructed to promptly notify the operating agency by telephone whenever they become aware of an alarm condition. *TELEMETERING EQUIPMENT* [21] that can connect individual homes to a central control station is becoming more affordable. This will eliminate having the resident respond to the alarm.

An operator should attend to the emergency as quickly as possible to minimize damage from an overflow or backup and inconvenience to the customer. Prompt operator response (within 4 to 6 hours) is crucial when the pump vaults used in the system have small reserve volumes between the high level alarm *SET POINT* [22] and the top of the basin. In many cases, maintenance personnel must respond to such calls whenever the call is received, even during late night or early morning hours.

Grinder pump systems may experience problems due to pump blockages or float switch malfunctions caused by excessive grease accumulation in the pump vault.

During extended power outages, pump basins may fill to capacity and overflow or back up into the home. However, water use and corresponding wastewater flows usually are greatly reduced during power outages. Portable standby generators, gasoline-powered pumps, or a septic tank pumper truck should be available to pump down flooded tanks and prevent overflows.

As growth occurs in the community and the collection system expands, you may experience an increase in the number of high water level alarms at the individual pumping units. This is usually due to the mains having reached capacity or having developed a problem with air-caused head loss. Another possible cause of this problem may be that the pumps are too small.

Even though they may be infrequent, main line ruptures are possible. Repair materials and equipment should be reasonably accessible for such needs.

In some localities, a wastewater treatment operating agency is required to notify the local or state water quality control or environmental control agencies whenever a wastewater spill or overflow occurs. Often, such notification must be made within prescribed time limits. Check with your local regulatory agencies to verify the spill reporting requirements in your area.

#### *6.432 Maintenance Safety*

Septic sewer gases are commonly encountered at STEP septic tanks, at relatively inactive grinder pump (GP) vaults, and

21. *Telemetering Equipment.* Equipment that translates physical measurements into electrical impulses that are transmitted to dials or recorders.
22. *Set Point.* The position at which the control or controller is set. This is the same as the desired value of the process variable. For example, a thermostat is set to maintain a desired temperature.

# DOUGLAS COUNTY DEPARTMENT OF PUBLIC WORKS
# GLIDE PRESSURE SEWER
# PRESSURE MONITORING WORKSHEET

DATE: ______________________

INSPECTOR: ______________________

| NO. | STATION | DESCRIPTION | STATIC ELEV. | GAUGE NO. | GAUGE ELEV. | GAUGE READING | EGL ELEV. | TIME |
|---|---|---|---|---|---|---|---|---|
| 1 | 65+00I | AARV W.R. | 703.30 | | 675.20 | | | |
| 2 | 15+00ID-2 | CO Sch.L. | 740.00 | | 716.00 | | | |
| 3 | 55+00IE | AARV N.U.H. | 785.12 | | 767.86 | | | |
| 4 | 44+00IG | MARV Lo. | 740.00 | | 717.00 | | | |
| 5 | 261+58I | MARV N.U.H. | 740.00 | | 730.00 | | | |
| 6 | 350+04I | AARV N.U.H. | 790.04 | | 763.33 | | | |
| 7 | 387+04I | MARV N.U.H. | 790.04 | | 785.00 | | | |
| 8 | 19+14 IE-2D | AARV Gv. | 855.02 | | 843.95 | | | |

NOTE: EGL ELEV. = GAUGE ELEV. + GAUGE READING

COMMENTS: ______________________

*Fig. 6.28 Pressure sewer pressure monitoring worksheet*

(Permission of Glide-Idleyld Park Sewer District)

points where pressure sewers discharge. The proper practices of pressure sewer maintenance are usually such that the gases are not hazardous due to dilution with surrounding air, but service personnel need to be aware of the potential hazards of these gases. For example, hydrogen sulfide gas (which smells like rotten eggs) is known to deaden a person's ability to smell it after a brief exposure. This may cause the person to wrongly think the gas has drifted away. Many deaths have been attributed to exposure to hydrogen sulfide gas. All service personnel should be advised of procedures and precautions regarding exposure to hazardous gases. In addition, the entire system should be designed so that no exposure is required.

Extensive electrical connections and equipment are used at each pumping unit, typically including an electrical control panel, mercury float switches, and electrical junction boxes. Service personnel are exposed to the possibility of electrical burn or shock in their maintenance duties. They may be standing on wet ground while servicing electrical components and working under adverse conditions such as darkness, rain, or snow. Electrical safety should be emphasized in the O&M manual and system designs should minimize the exposure of O&M personnel to live electrical systems.

As with any project involving contact with wastewater, maintenance of septic tanks exposes operators to the possibility of infections or infectious diseases. For additional information about safety hazards associated with septic tank systems and recommended safety precautions, see Chapter 3, "Safety."

#### 6.433 *Maintenance Staff Requirements*

A minimum of two people should be trained and available. On some systems, one person can attend to the service calls. Some pumping units, especially grinder pumps, may require two persons to remove the pump from the tank or vault due to the weight of the pump. Similarly, two or more operators may be needed if it is necessary to remove pumping units surrounded by vault strainers. If the strainer drain valve fails to open when the unit is lifted, the weight of the retained water adds considerably to the weight that must be lifted. However, newer designs require only one operator.

A fully trained and experienced backup person is needed for times when the lead person is unavailable. The two people do not necessarily have to be full-time employees, but at least one person has to be available on call.

### 6.44 Troubleshooting

See Table 6.1, "Troubleshooting STEP Systems: Laterals and Main Pressure Sewers."

Systematic maintenance of properly located air release valves will reduce the incidence of problems with air- or gas-bound service lines and pressure mains. When a service line does become air-bound, the solution is to remove or release the air. This is accomplished by first shutting off electrical power to the pump, then disconnecting the discharge hose in the riser where it enters the service line, thus releasing the trapped gas or air. After the air is bled out of the line, reconnect the discharge hose to the pump, restore electrical power, and check to determine that the pump is functioning properly by pumping the tank level down to the normal shutoff position.

*NOTE:* A ⅛-inch orifice is drilled in the pump discharge pipe below the check valve. This allows air to escape from the pump volute, which prevents air binding problems. Since 1-inch pipe is used for pump discharge plumbing and service lines, air entrapment is not a problem there. However, an antisiphon valve (Figure 6.29, below the center of the lid) is required when pumping downhill.

### 6.45 Information Resources

If you, the operator, are fortunate enough to be on hand during construction of a new system, learn as much as you can about all of the components. The design engineer will most likely be more than willing to help you understand how the system is intended to work and the function of each component.

The operator's most valuable resource for maintenance information is a complete, up-to-date operation and maintenance manual. If yours is out of date or has been lost, use the guidelines presented in Appendix C at the end of Chapter 4 to update an old manual or prepare a new one.

Equipment suppliers are an excellent source of information for operators. They want their products to be properly installed and maintained and will often provide literature on the products they manufacture and sell. A supplier may also be aware of methods other operators have used to solve a particular problem or improve the performance of a piece of equipment. Operators who take the time to establish good working relationships with their suppliers can benefit from their expertise and may receive better support service when problems arise. Good working relationships also benefit suppliers and they usually appreciate comments from operators who use their products and have discovered a better way of maintaining them or have suggestions for improving a product.

The increasing use of pressure sewers increases the number of people who, like yourself, are responsible for the proper operation and maintenance of these systems. Visiting or communicating with other system operators often provides ideas or methods

TABLE 6.1 TROUBLESHOOTING STEP SYSTEMS: LATERALS AND MAIN PRESSURE SEWERS

| Indicator/Observation | Probable Cause | Check | Solution |
|---|---|---|---|
| Alarm on—high liquid level. Septic tank pump vault or pumping unit runs for long cycle on | Every item previously covered in septic tank and pump vaults OK (see Chapter 4, Tables 4.1, 4.3, and 4.4). Broken lateral restricting discharge from unit | Look for pooling or damp spots along lateral route | 1. Septic tank or pump vault setting can cause shearing of lateral pipe, usually within first year of construction or installation of system, particularly in spot where line is subject to vehicle surface load. Excavate and repair broken line.<br>2. Some locations use band clamps on small-diameter septic tank to lateral service line. Corrosion or electrolysis destroys clamp adjusting screw, releasing pipe connection and causing leak. Replace clamp or solvent weld new connection to service line. |
| Complaint from citizen, odor or pooling of wastewater | Air release valve inoperative | Operation of valve | Isolate valve by closing inlet valve. Drain valve. Replace valve or remove cap. Clean float, orifice seat, and needle assembly. Replace gasket and cover, close drain valve, open inlet isolation valve. |
| Wastewater flow through main is stopped or sluggish | Air- or gas-bound main | Head and flow through main (pressure gauge or flowmeter) | Go through operation of manual air release valves, check automatic valves for proper operation. Close inlet—open drain valve—all air no liquid, valve needs service—little air, more water, put valve back on line by opening inlet valve. Find faulty valve, repair and be sure bound air is released. |
| No flow through main | Main line valve closed or broken main | Check valve and main | Be sure isolation valves are open. Look for wet spots that indicate a damaged or broken main. |

that you or they can use to make the job easier and the system perform better and more economically. Be aware, however, that while there may be other wastewater systems similar to yours, the problems and solutions are never identical. Different geographical locations, terrain, products, equipment, and community makeup can and often do dictate different operational duties and problems. Just as there is no single document that can tell you what must be done and when, there is no other system just like yours. You will have to evaluate other operators' suggestions in the context of your own system and try to find your own best operating strategy.

## QUESTIONS

Please write your answers to the following questions and compare them with those on page 345.

6.43A What preventive maintenance tasks are associated with pressure system pipelines?

6.43B What types of emergency calls should maintenance personnel be prepared to respond to?

6.43C What are the safety hazards associated with pressure collection systems?

6.45A Where could an operator find information about the operation and maintenance of STEP systems and pressure sewers?

**END OF LESSON 2 OF 3 LESSONS**

**on**

**COLLECTION SYSTEMS**

Please answer the discussion and review questions next.

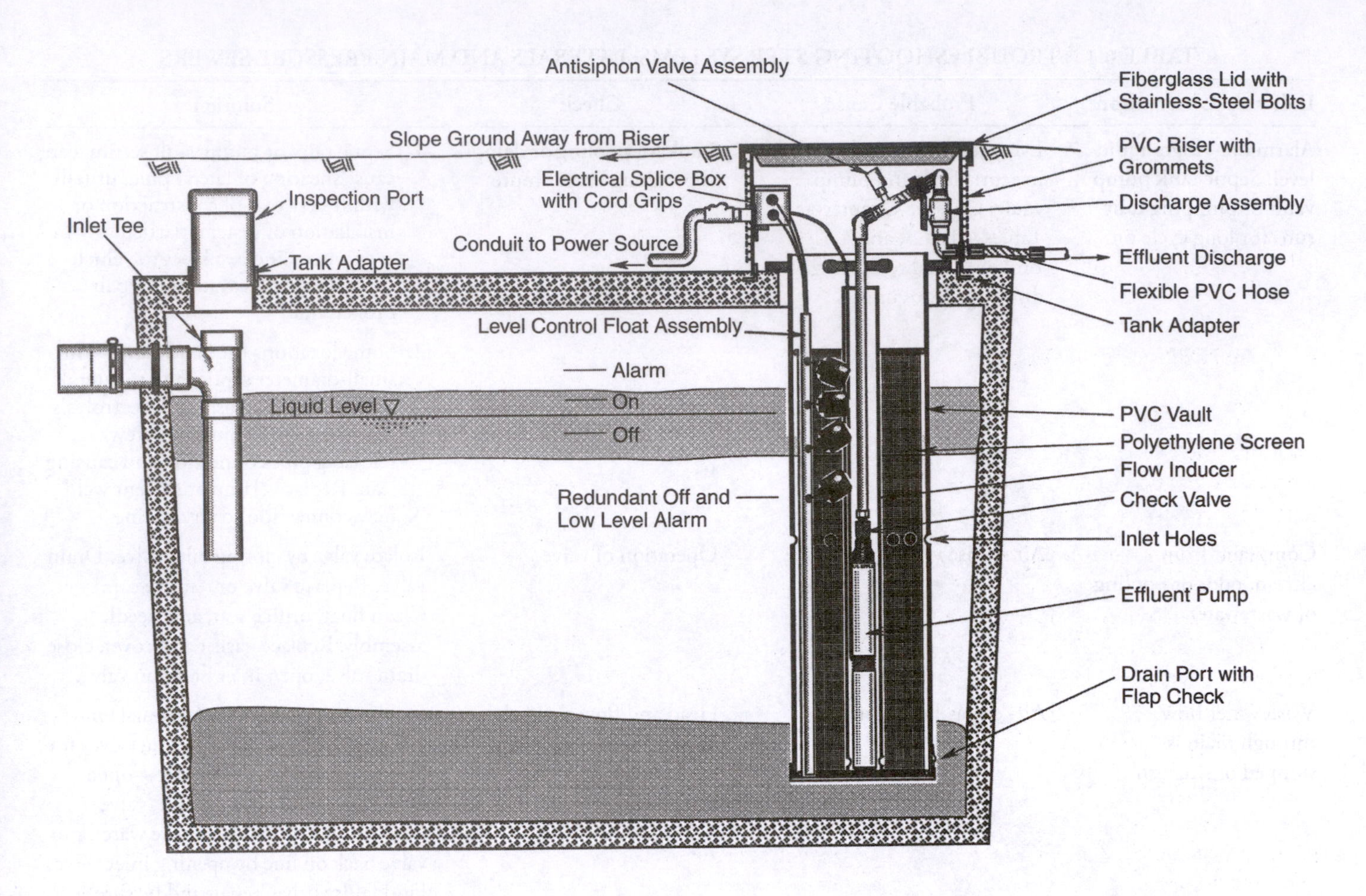


*Fig. 6.29 STEP system septic tank showing installation of antisiphon valve*

(Permission of Orenco Systems, Inc.)

# DISCUSSION AND REVIEW QUESTIONS

## Chapter 6. COLLECTION SYSTEMS

(Lesson 2 of 3 Lessons)

Please write your answers to the following questions to determine how well you understand the material in the lesson. The question numbering continues from Lesson 1.

11. Why are pressure collection systems sometimes used as an alternative to gravity collection systems?
12. In what types of areas would it be important to minimize the number of pipe joints in a pressure collection system?
13. Why are isolation valves installed in wastewater collection mains?
14. List three methods for handling the offensive gases (such as hydrogen sulfide) that are vented from a pressure main.
15. What are the general types of preventive maintenance activities associated with pressure system pipelines?
16. What procedure should be used to correct the problem of an air-bound service line?

# CHAPTER 6. COLLECTION SYSTEMS

(Lesson 3 of 3 Lessons)

## 6.5 VACUUM SEWER SYSTEMS[23]

### 6.50 Description

Vacuum sewers have been used in Europe for over 100 years. However, it has been only since about 1975 that vacuum transport of wastewater has been used in the United States.

A vacuum sewer system (Figure 6.30) has three major parts: the central vacuum station, the collection pipeline network, and the on-site facilities. A vacuum is created at the central collection station and is transmitted by the collection network throughout the area to be served. Raw wastewater from a residence flows by gravity to an on-site holding tank. When about 10 gallons of wastewater has been collected, the vacuum *INTERFACE*[24] valve opens for a few seconds allowing the wastewater and a volume of air to be pulled through the service pipe and into the main. It is the difference between the atmospheric pressure behind the wastewater and the vacuum ahead that provides the drawing force. Since both air and wastewater flow at the same time, a high pressure is produced, which prevents blockages. When the interface valve closes, the system returns to a motionless state and the wastewater comes to rest at the low points of the collection network. After several valve cycles, the wastewater reaches the central collection tank, which is under vacuum. When the wastewater reaches a certain level, a conventional nonclog wastewater pump pushes it through a force main to a treatment facility or gravity interceptor sewer.

A vacuum system is very similar to a water distribution system, only the flow is in reverse direction (Figure 6.31). This relationship would be complete if the vacuum valve was manually opened, like a water faucet.

Vacuum sewers are considered where one or more of the following conditions exist:

- Unstable soil
- Flat terrain
- Rolling land with many small elevation changes
- High water table
- Restricted construction conditions
- Rocky terrain
- Urban development in rural areas

The advantages of vacuum sewer systems compared to other types of collection systems may include substantial reductions in water used to transport wastes, material costs, excavation costs, and treatment expenses. Specifically, vacuum systems offer the following advantages:

- Small pipe sizes are used.
- No manholes are necessary.
- Field changes can easily be made so that underground obstacles can be avoided by going over, under, or around them.
- Installation at shallow depths (2 to 4 feet) eliminates the need for wide, deep trenches, thus reducing excavation costs and environmental impacts.
- High scouring velocities are reached, reducing the risk of blockages and keeping wastewater aerated (containing dissolved air) and mixed.
- Unique features of the system eliminate exposing maintenance personnel to the risk of toxic hydrogen sulfide gas.
- The system will not allow major leaks to go unnoticed, resulting in a very environmentally sound situation.
- Only one source of power, at the central vacuum station, is required (in cold climates, 10°F, heaters or pipe heating tape around the valves keeps them from freezing).
- The elimination of infiltration permits a reduction in the size and cost of the treatment plant.

## QUESTIONS

Please write your answers to the following questions and compare them with those on page 345.

6.50A Vacuum sewer systems are considered where what types of conditions exist?

6.50B List the potential advantages of vacuum sewer systems.

23. Information in this lesson on vacuum sewer systems was adapted by John Brady from *ALTERNATIVE WASTEWATER COLLECTION SYSTEMS,* Technology Transfer Manual, US Environmental Protection Agency.

24. *Interface.* The common boundary layer between two substances, such as water and a solid (metal); or between two fluids, such as water and a gas (air); or between a liquid (water) and another liquid (oil).

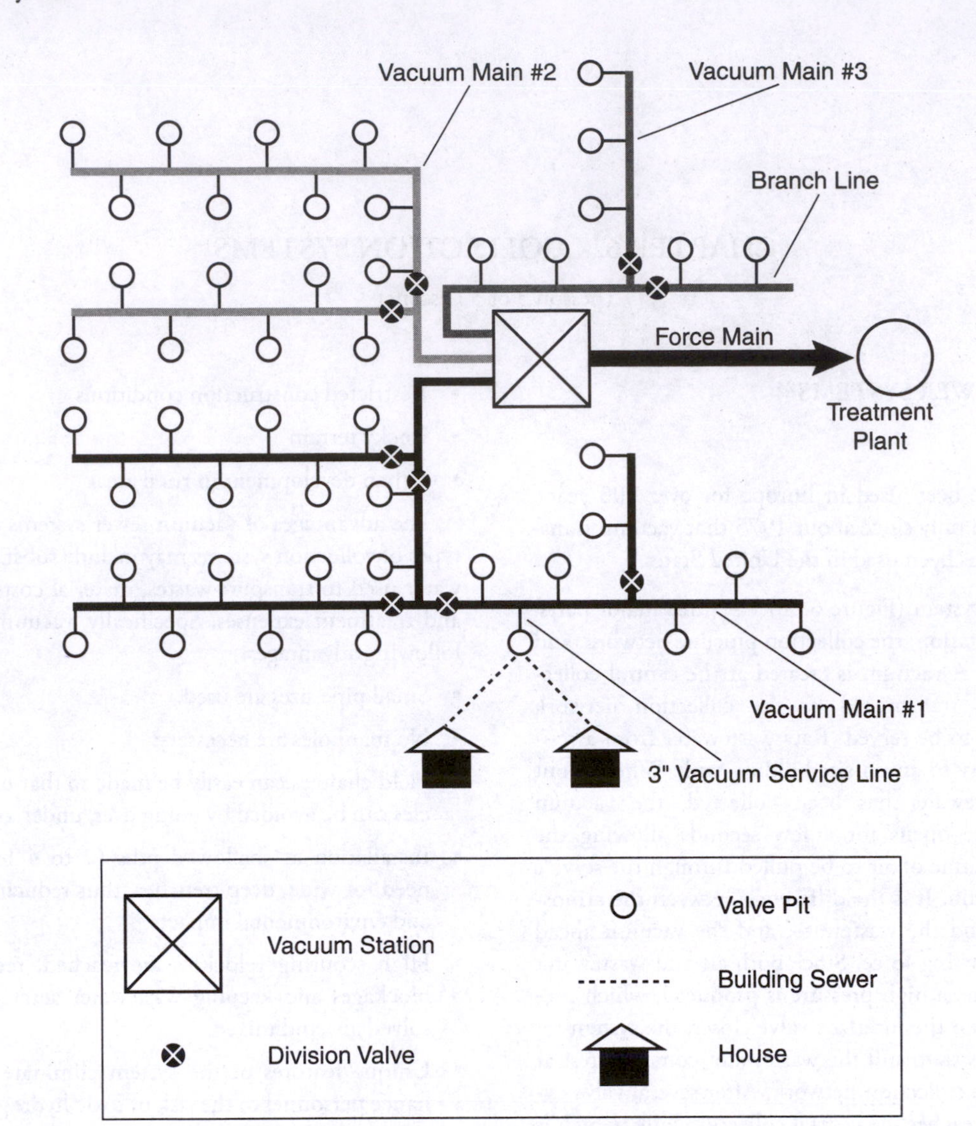


*Fig. 6.30 Major components of a vacuum sewer system*
(Source: *ALTERNATIVE WASTEWATER COLLECTION SYSTEMS,* Technology Transfer Manual, US Environmental Protection Agency)

### 6.51 Components

### *6.510 On-Site Services (Figure 6.32)*

The on-site services in a vacuum system consist of the following components:

- Building sewer (usually this is the homeowner's responsibility)
- Valve pit/sump or buffer tank
- Vacuum interface valve
- Auxiliary vent
- Service lateral

### *6.5100 BUILDING SEWER*

The term "building sewer" refers to the gravity flow pipe extending from the home to the valve pit/sump. For residential service, the building sewer should be 4 inches in diameter and slope continuously downward at a rate of not less than 0.25 inch/foot (two-percent grade). Line size for commercial users will depend on the amount of flow and local code requirements. Bends should be avoided in building sewers but if the line changes direction, a cleanout should be installed each time the direction changes by 57 degrees or more. Figure 6.33 shows typical layouts for making two-house and single house building sewer connections to the valve pit.

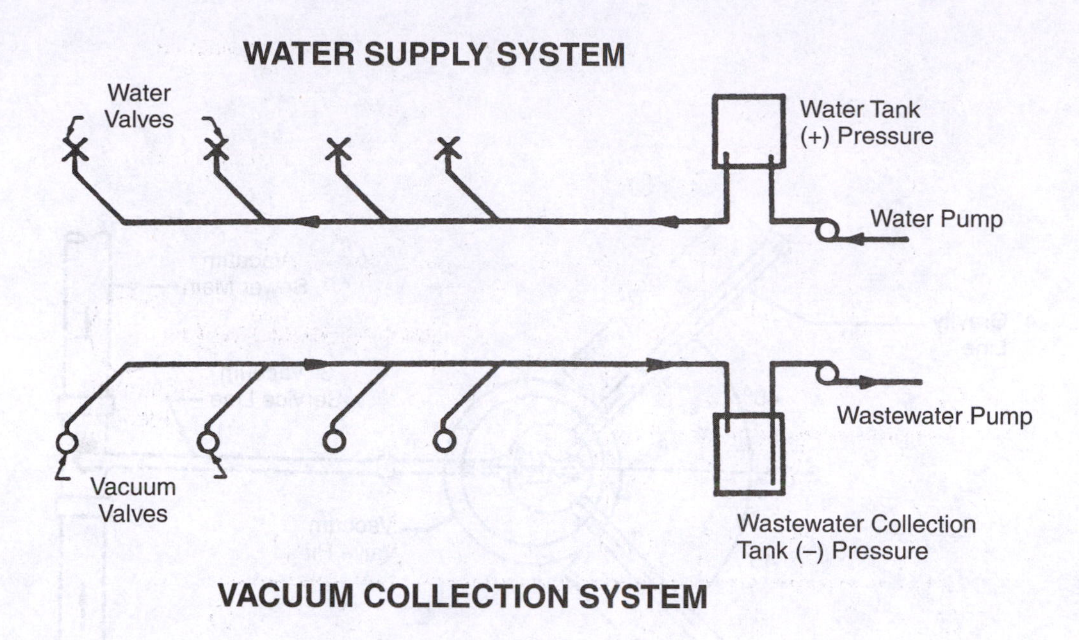


*Fig. 6.31 Water sewer/vacuum system similarities*
(Drawing provided courtesy of AIRVAC)

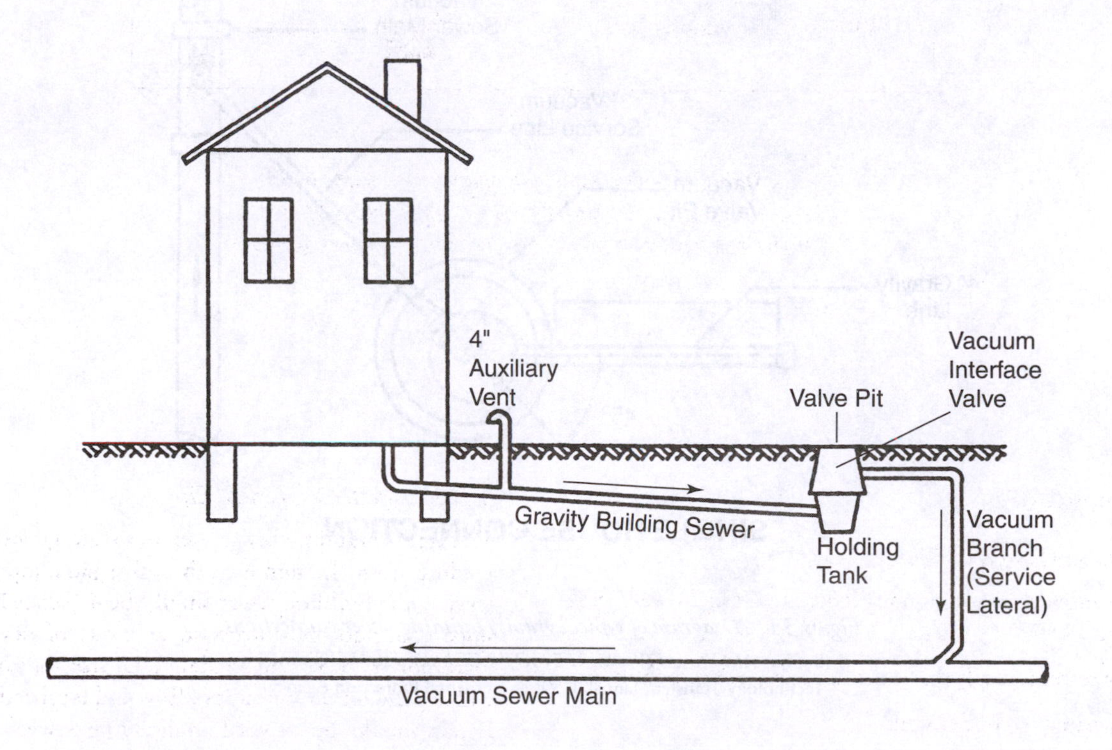


*Fig. 6.32 On-site portion of a vacuum collection system*
(Source: *ALTERNATIVE WASTEWATER COLLECTION SYSTEMS,* Technology Transfer Manual, US Environmental Protection Agency)

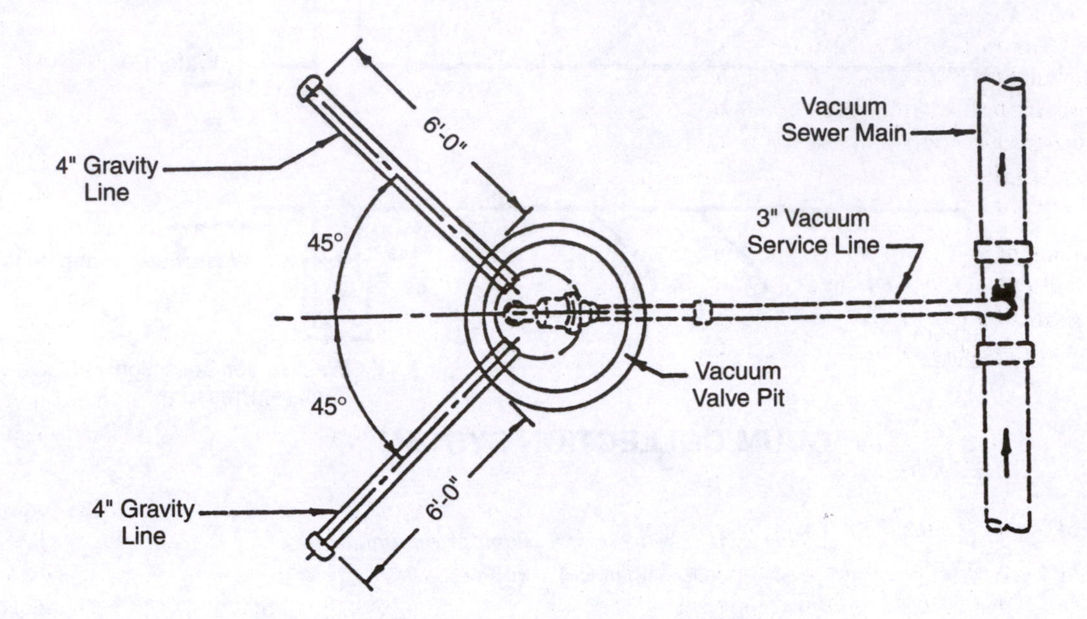


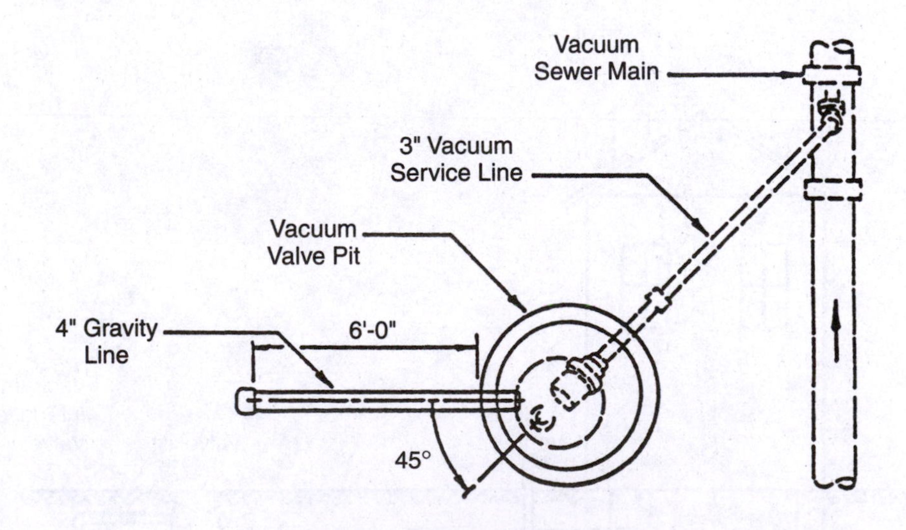


*Fig. 6.33 Diagram of typical gravity building sewer connections*

(Source: *ALTERNATIVE WASTEWATER COLLECTION SYSTEMS,* Technology Transfer Manual, US Environmental Protection Agency)

The valve pit should be located near the home, if possible, because a short building sewer will require less maintenance and there will be less opportunity for infiltration and inflow (I/I) or unwanted water entering the collection system. Infiltration through leaking building sewers is a common occurrence and should be avoided. Inflow from illegal connections of roof and yard drains to the building sewer is also common and should be prevented. An inspection should be conducted before final connection of a residence to the system to determine if these situations exist. If so, steps should be taken to remove the connections to roof and yard drains.

If the home piping network does not have a cleanout within it, one should be placed outside and close to the home. Some agencies prefer to install a cleanout at the dividing line where agency maintenance begins (for example, the end of the six-foot stub-out pipe).

### *6.5101 VALVE PITS AND SUMPS*

**Fiberglass Pits and Sumps** (Figure 6.34)

The package fiberglass type of valve pit and sump is by far the most common. This type of pit and sump is composed of four main parts: the bottom chamber (sump), the top chamber (valve pit), the plate that separates the two chambers (pit bottom), and the lid.

Wastes from the home flow through the building sewer and enter the sump through an inlet, which is located 18 inches above the bottom of the sump. Up to four separate building sewers can be connected to one sump, each at an angle of 90 degrees to the next one.

The sump has a wall thickness of 3⁄16 inch and is designed for appropriate traffic loading with 2 feet of cover. Elastomer (a rubber-like material) connections are used for the entry of the building sewer. Holes for the building sewers are field cut at the position directed by the engineer.

Package valve pits/sumps are typically available in two different heights, 30 and 54 inches. Both are 18 inches in diameter at the bottom and 36 inches in diameter at the top. The smaller size has a capacity of 55 gallons and the larger one has a capacity of 100 gallons.

The valve pit houses the vacuum valve and controller. The pit itself usually is made of filament-wound fiberglass and, like the sump, it has a wall thickness of 3⁄16 inch in order to withstand the stress of traffic loading. The cone-shaped valve pit is normally 36

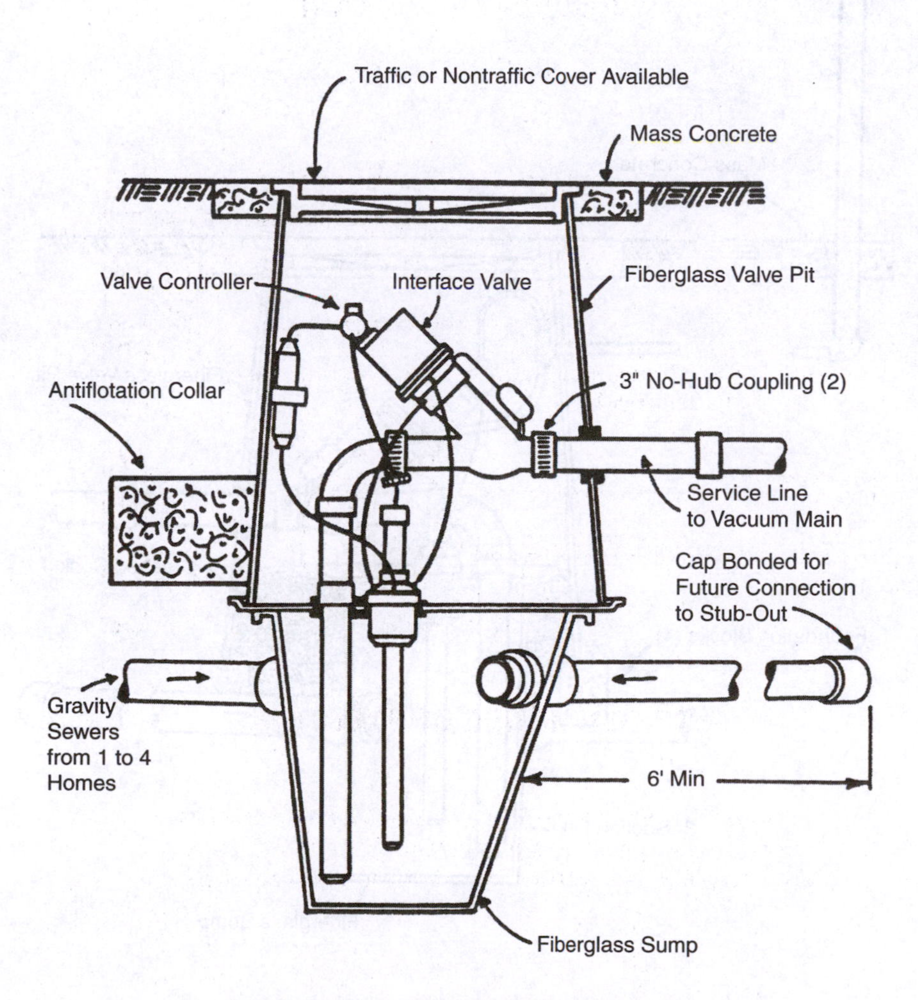


*Fig. 6.34 Typical fiberglass valve pit and sump*

(Drawing provided courtesy of AIRVAC)

inches in diameter at the bottom and narrows to a 24-inch diameter at the top where it is fitted with a clear opening cast-iron frame and cover. A tapered shape helps facilitate the backfilling procedure. Depth of burial is normally 42 inches. One 3-inch diameter opening, with an elastomer seal, is pre-cut to accept the 3-inch vacuum service line.

The pit bottom is made of reinforced fiberglass that is ¼-inch thick at the edges and 5⁄16-inch thick in the center. These bottoms have been molded by the resin injection process and have pre-cut holes for the 3-inch suction line, the 4-inch cleanout/sensor line, and the sump securing bolts. The seal between the bottom of the valve pit and the top of the sump is made in the field using silicone, a butyl tape rubber sealant, or a neoprene O-ring. The pit bottom should have a lip that allows the valve pit to rest on top of it.

Cast-iron covers and frames, designed for heavy traffic loading, are typically used. The frame weight is generally 90 pounds and the lid weight about 100 pounds. When a lighter lid is desired, such as in nontraffic situations, a lightweight aluminum or cast-iron lid may be used. These two types of lids do not have frames, but rather are attached to the valve pit with two J-bolts. These lids should clearly be marked "Nontraffic."

A shallower valve pit/sump arrangement is possible, if so desired (Figure 6.35). This arrangement would be used in areas where the depth of the building sewers is very shallow due to a high groundwater table or poor soils.

Antiflotation collars (see Figure 6.34) are sometimes installed on fiberglass valve pits and sumps. Experience has shown that these collars are usually not needed but buoyancy calculations should be performed by the designer to be sure. If antiflotation collars are needed, care must be taken during the valve pit installation because poor bedding and backfill may lead to soil settlement problems. Settlement of the concrete ring may damage the building sewer or the pit itself.

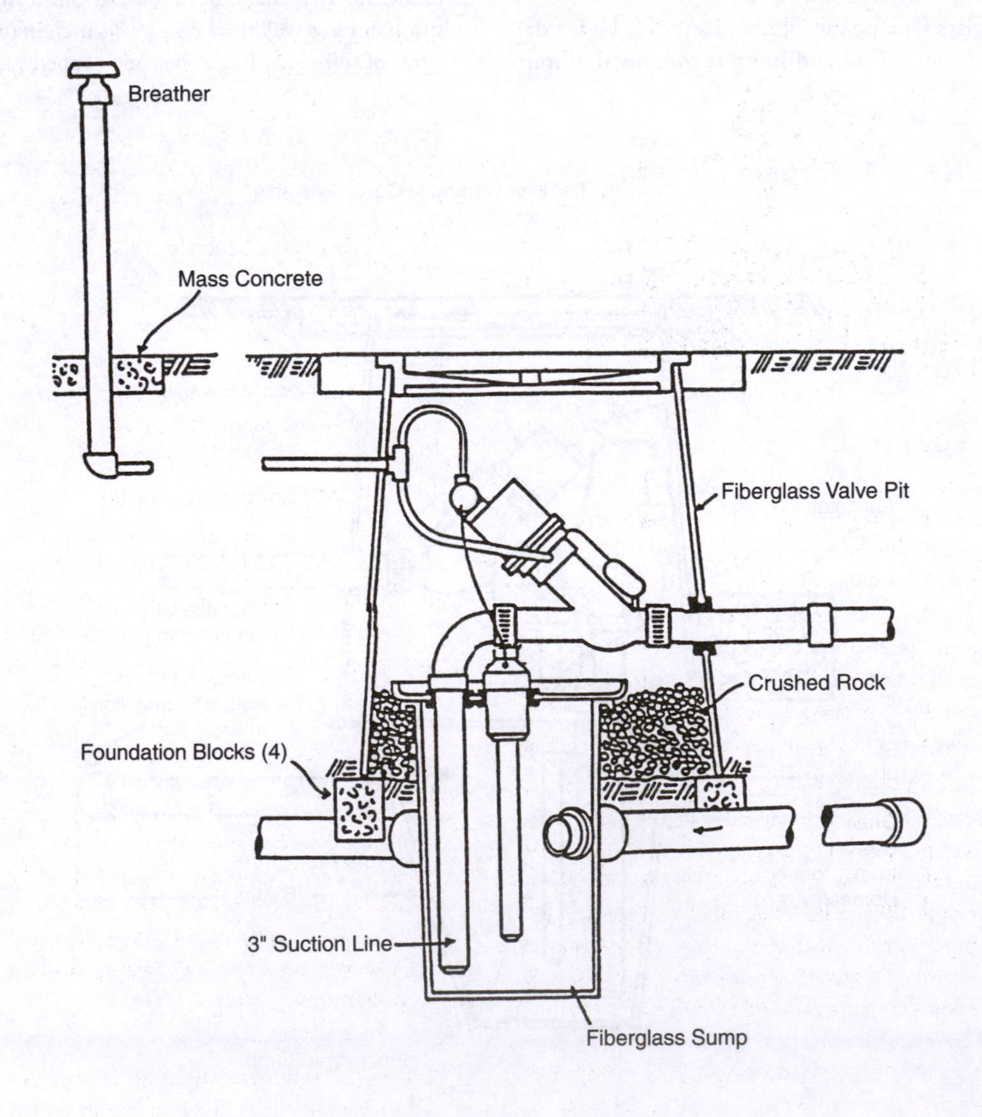


*Fig. 6.35 Shallow fiberglass valve pit and sump*
(Drawing provided courtesy of AIRVAC)

### Concrete Pits and Sumps

Certain situations call for the use of a concrete valve pit:

- When the deepest fiberglass pit and sump is not sufficient to accept the building sewer
- When a large flow is anticipated requiring flow leveling
- When an interface between two system types (for example, pressure and vacuum) is needed

The deepest possible fiberglass pit and sump is 8 feet. The building sewer depth at the sump is therefore limited to 6.5 feet since the building sewer enters 18 inches above the bottom. If a deeper pit and sump is necessary, a concrete valve pit and sump may be used. The maximum recommended depth for a concrete sump is 10 feet.

Concrete pits and sumps are typically constructed of 4-foot diameter manhole sections, with the bottom section having a preshaped 18-inch diameter sump. It is very important that all joints and connections be watertight to eliminate groundwater infiltration. Equally important is the need for a well-designed pipe support system, since these tanks are open from top to bottom. The support hardware should be of stainless steel or plastic.

**Buffer Tanks** (Figure 6.36)

A buffer tank should be used for large flows that require leveling out, such as flows from schools, apartments, nursing homes, and other large dischargers to the vacuum sewer system. Buffer tanks are designed with a small operating sump in the lower portion and additional emergency storage space in the tank. Like the deep concrete valve pit and sump, the buffer tank is typically constructed of 4-foot diameter manhole sections, with the bottom section having a pre-poured 18-inch diameter sump. As with concrete pits and sumps, connections must be watertight and pipes must be well supported in buffer tanks.

A dual buffer tank (Figure 6.37) is similar to a single buffer tank except that it is large enough to accommodate two vacuum valves. These tanks typically use 5-foot diameter manhole sections. A dual buffer tank should be used when an interface between two system types is needed, for example, when vacuum sewers serve the majority of the service area but pressure sewers are required for the low-lying fringes. At some point, a transition will be needed between the pressure flow and vacuum flow. Dual buffer tanks also may be used if a single buffer tank does not have the capacity for the large flows. A single buffer tank has a 30 GPM (gallons per minute) capacity while a dual buffer tank has a 60 GPM capacity.

## QUESTIONS

Please write your answers to the following questions and compare them with those on pages 345 and 346.

6.51A List the on-site components of a vacuum collection system.

6.51B When might a concrete valve pit/sump be installed rather than the more common fiberglass type?

6.51C What types of facilities might require the use of a buffer tank?

### *6.5102 VACUUM VALVES*

Most vacuum valves are entirely pneumatic; that is, they are vacuum operated on opening and spring assisted on closing. No electricity is required. System vacuum ensures positive valve seating. Typically, vacuum valves have a 3-inch diameter opening. Some states have made this a minimum size requirement because 3 inches is the throat diameter of the standard toilet. Figure 6.38 shows the parts of a vacuum valve.

The controller/sensor is the key component of the vacuum valve. The device relies on three forces for its operation: pressure, vacuum, and atmosphere. As the wastewater level rises in the sump, it compresses air in the sensor tube. This pressure begins to open the valve by overcoming spring tension in the controller and activating a three-way valve. Once opened, the three-way valve allows the controller/sensor to take vacuum from the downstream side of the valve and apply it to the actuator chamber to fully open the valve. The controller/sensor is capable of maintaining the valve fully open for a fixed period of time, which is adjustable over a range of 3 to 10 seconds. After the time period has elapsed, atmospheric air is admitted to the actuator chamber permitting spring-assisted closing of the valve. All materials of the controller/sensor are made of plastic or elastomer that is chemically resistant to the usual solids and gases in domestic wastewater.

Many types of vacuum valve arrangements are available; as an example, the Model D and the Model S from one manufacturer are described in the following paragraphs. The valves in both arrangements are identical but each relies on a different piping/plumbing arrangement for their source of atmospheric air.

1. Model D (Figure 6.39)

   The Model D arrangement gets atmospheric air through a tube, which is connected to an external breather pipe. The Model D arrangement is the most reliable since there is little chance of water entering the controller. However, some operators dislike this type of arrangement because of the appearance of the external breather tube above ground and because the exposed tube could be an inviting target for vandalism. However, experience has shown that these problems are relatively minor.

   One type of external breather (Figure 6.40) consists of 1¼-inch polyurethane pipe that is anchored in the ground by concrete. This material is very flexible, making it virtually vandal proof. No matter what arrangement is used, two items require attention. First, the entire breather piping

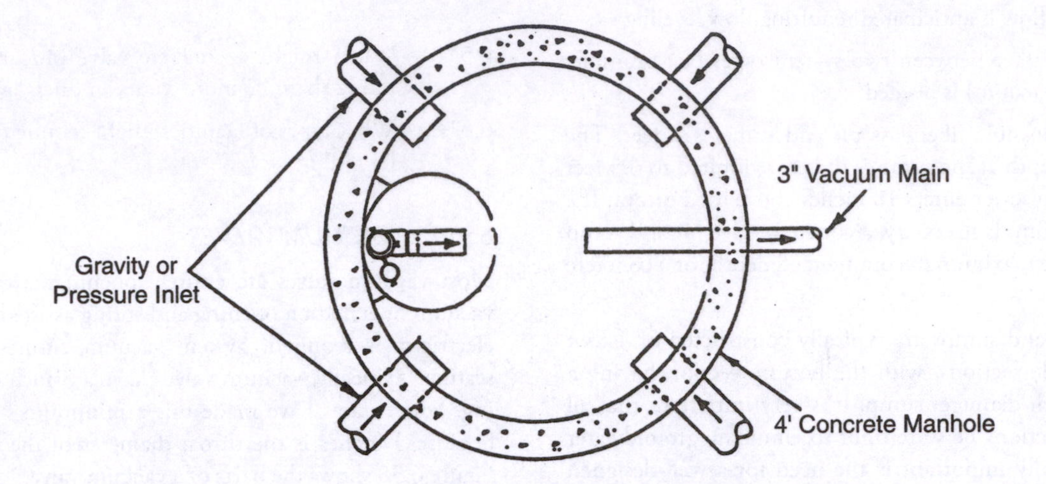


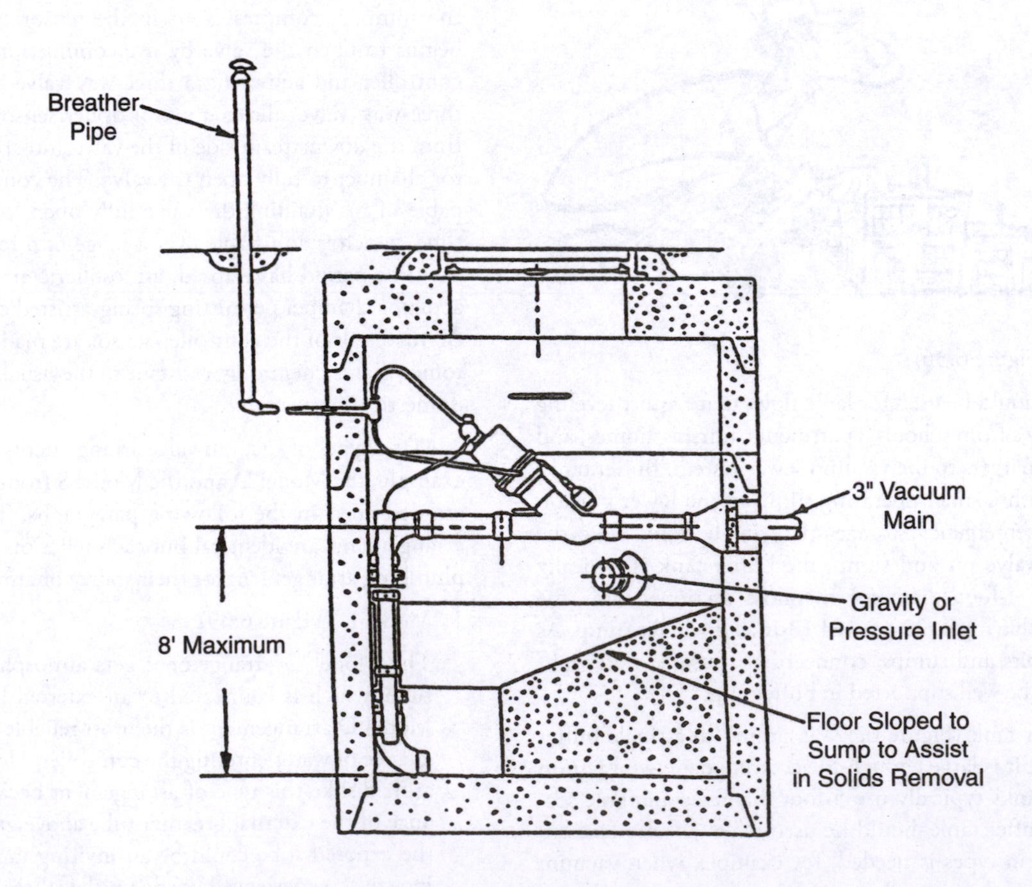


*Fig. 6.36 Plan and elevation views of typical concrete buffer tank*

(Drawing provided courtesy of AIRVAC)

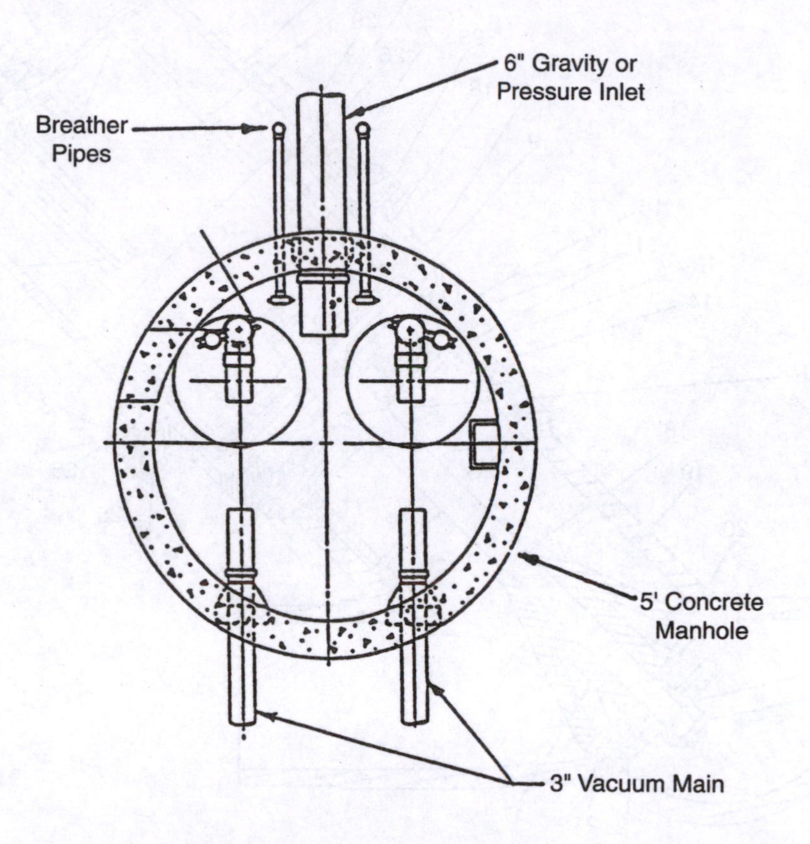


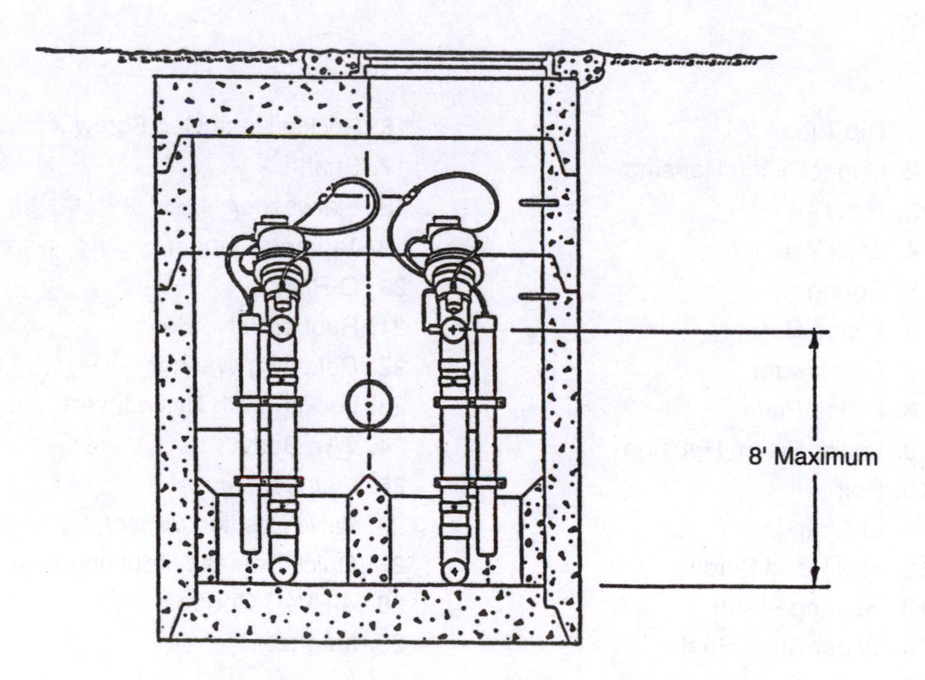


*Fig. 6.37 Typical concrete dual buffer tank*

(Drawing provided courtesy of AIRVAC)

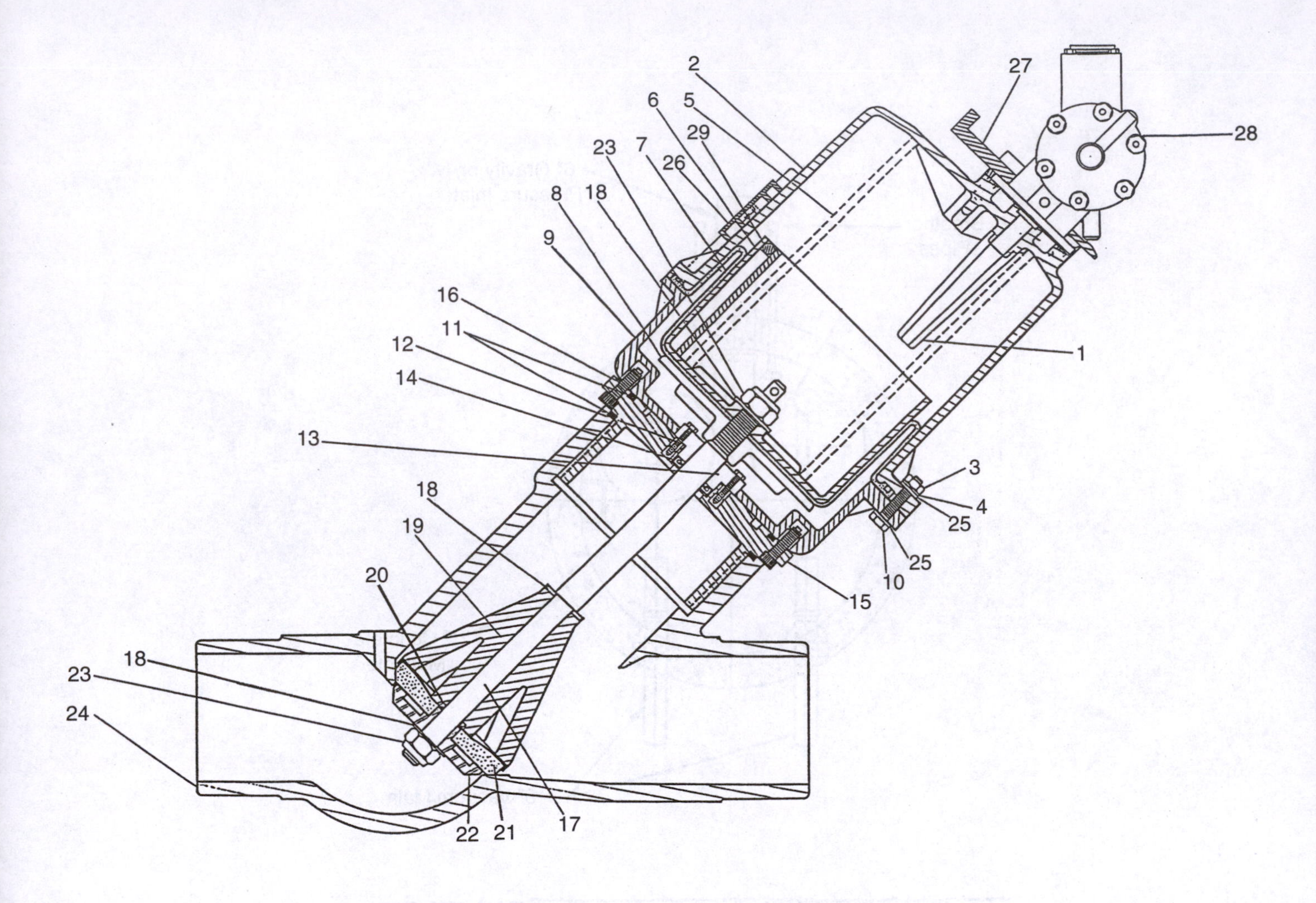


1. Dip Tube
2. Upper Piston Housing
3. Hex Nut
4. Lock Washer
5. Spring
6. Piston Cup
7. Diaphragm
8. Piston Plate
9. Lower Piston Housing
10. Bolt
11. O-Ring
12. Hex Head Screw
13. Bearing - Blue
14. Wiper Shaft Seal
15. Screw Plug
16. Socket Head Cap Screw
17. Shaft
18. Flatwasher
19. Tapered Plunger
20. O-Ring
21. Rubber Valve Seat
22. Retaining Washer
23. Locknut with Nylon Insert
24. Wye Body
25. Flat Washer
26. Valve Position Sensor
27. Quick Release Mounting Key
28. AIRVAC Controller
29. Magnet

*Fig. 6.38 Parts of a vacuum valve*
(Drawing provided courtesy of AIRVAC)

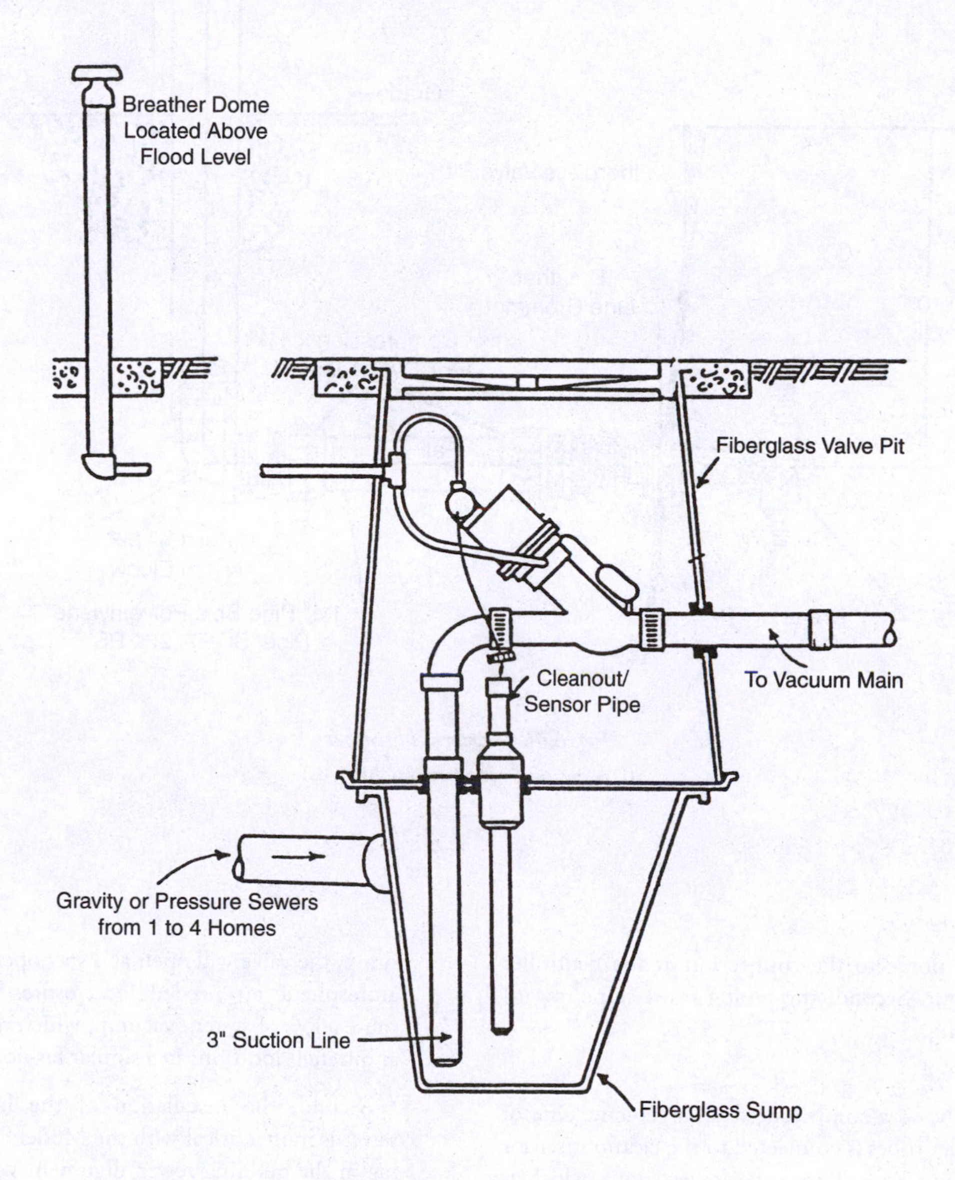


*Fig. 6.39 Model D valve arrangement with external breather*

(Drawing provided courtesy of AIRVAC)

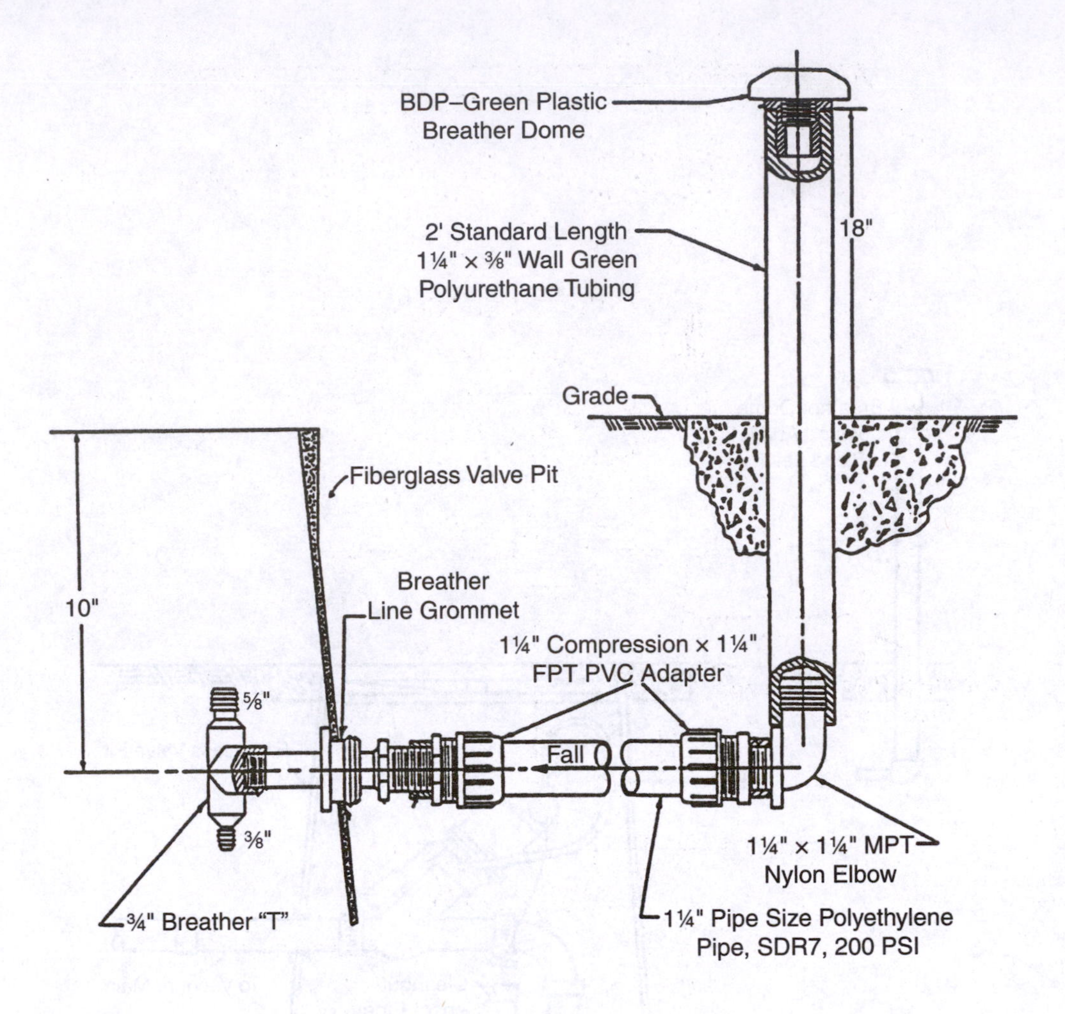


*Fig. 6.40 External breather*
(Drawing provided courtesy of AIRVAC)

system from the dome to the connection at the controller must be watertight. Second, the piping must slope toward the valve pit setting.

2. Model S (Figure 6.41)

The Model S valve is a sump-vented arrangement. One of the three controller tubes is connected to the cleanout/sensor piping. This piping extends into the lower sump, which is connected to the building sewer, and the building sewer is open to atmospheric air through the 4-inch auxiliary vent.

While eliminating some of the concerns associated with the external breather, the Model S valve has some potential problems. First, the sump must be airtight and watertight. If the system vacuum at the valve drops to less than 5 inches of mercury vacuum, the valve will not operate and wastewater will continue to fill the sump. If the sump is watertight, it will become pressurized with a bubble of air trapped at the top of the sump. When system vacuum is restored, the bubble of air will be used by the controller in the valve closing process. However, if water completely fills the bottom sump, the valve will open and stay open since it will lack the atmospheric air needed for closure. This open valve will cause a loss of system vacuum, which may affect other valves at different locations in a similar fashion.

Second, the installation of the homeowner's building sewer is more critical with the Model S valve arrangement. A sag in the building sewer alignment will trap water and not allow the free flow of atmospheric air.

Some engineers have experimented with a blend of the two concepts, by using a breather that gets its air from the top chamber. This has been successful in areas where there is no chance for surface water to enter the top chamber. The danger is that water may enter the top chamber and fill it to a level above the breather. This water will directly enter the controller and cause problems with valve closure. In either the Model D or the Model S valve, water in the top chamber would be of no concern if the controller itself were watertight. However, this is not always the case if a problem has occurred. Since preventing water from entering the top

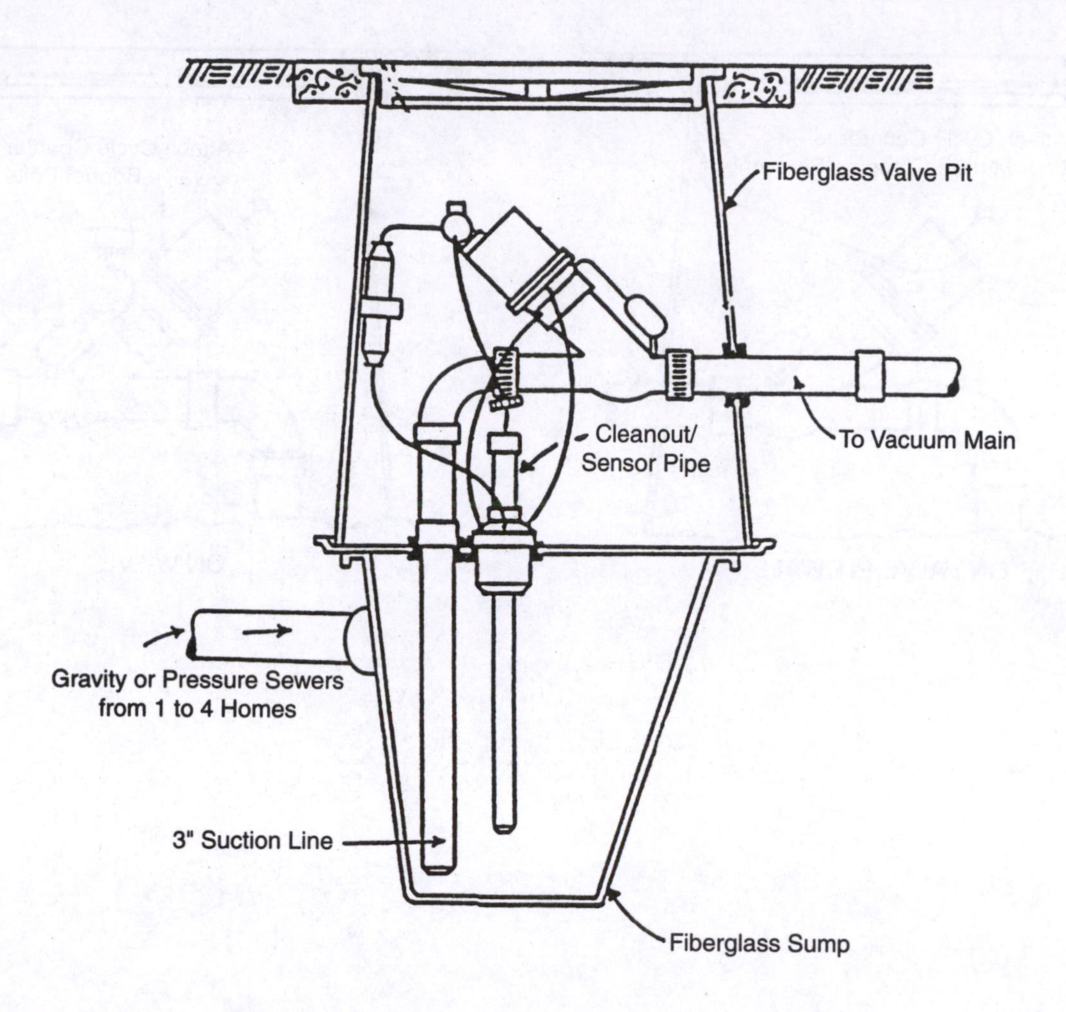


*Fig. 6.41 Model S valve with sump-vented arrangement*
(Drawing provided courtesy of AIRVAC)

**chamber has proven to be a difficult task, venting from the top chamber is discouraged.**

**Because the Model D arrangement is less likely to experience problems than the Model S, it is the recommended type. Since the two valves are physically identical, it is possible to convert from a Model S to a Model D or from a Model D to a Model S, if needed.**

A device called a cycle counter (Figure 6.42) can be mounted directly on the vacuum valve or the valve pit wall to monitor the number of cycles of a particular valve. The unit is enclosed in a watertight housing with a clear nylon top. Cycle counters typically are used where a large water use is expected in order to determine if the valve is reasonably capable of keeping up with the flow.

Some agencies use the cycle counter as a metering device. Knowing the number of cycles and the approximate volume per cycle, one can estimate the amount of wastewater passing through the vacuum valve over a given period. Other agencies use the device as a method of determining illegal storm connections to the vacuum sewer. The flow through the valve can be estimated and compared to metered water use. From this, it is possible to determine if extra water is entering the vacuum sewer and generally in what amounts.

It is not necessary to have a cycle counter for each valve (unless they are being used as a metering device for billing purposes) because they are small and can easily be moved from location to location.

### *6.5103 AUXILIARY VENT (Figure 6.43)*

A 4-inch PVC vent is required on the building sewer. This auxiliary vent prevents the water in home plumbing fixture traps from being drawn into the vacuum system. The auxiliary vent also permits air to enter the building sewer where it acts as the driving force behind the wastewater that is evacuated from the lower sump. With a Model S valve arrangement, the vent provides the necessary atmospheric air for proper controller operation. Most agencies require the auxiliary vent to be located against a permanent structure, such as the house, a wall, or a tree

*Fig. 6.42 AIRVAC cycle counter—two methods of connection*
(Drawing provided courtesy of AIRVAC)

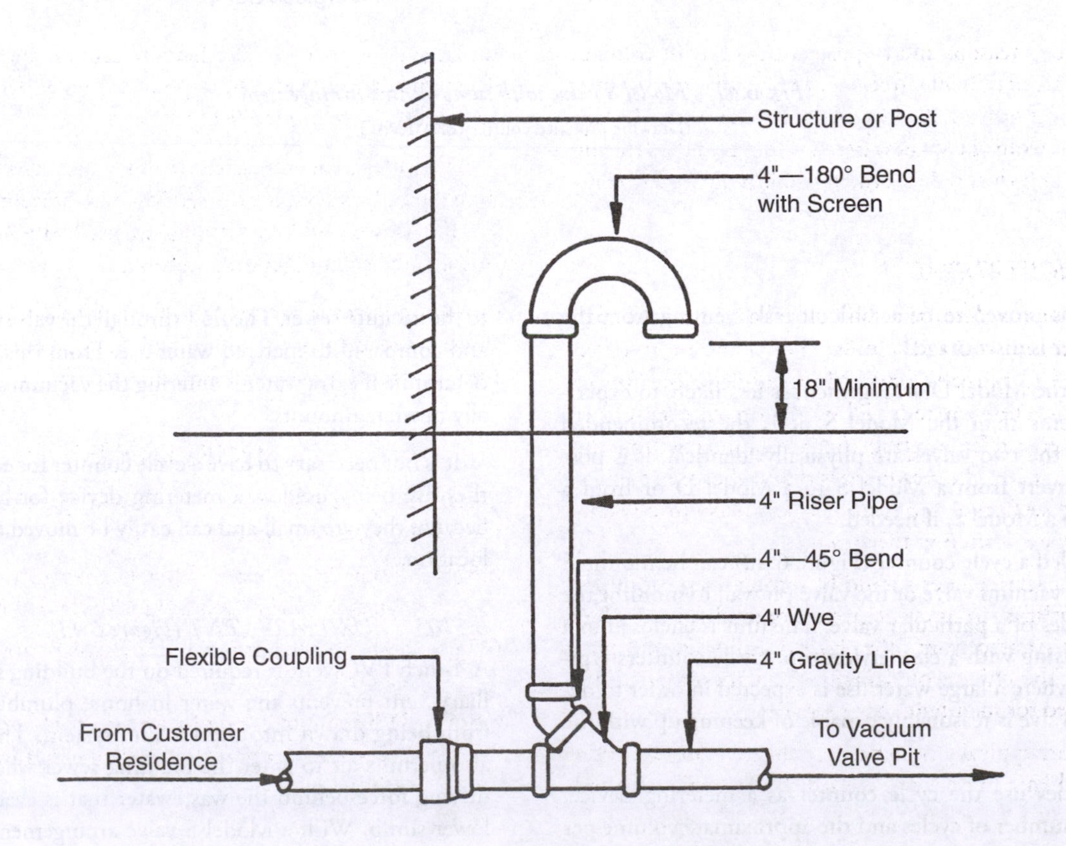


*Fig. 6.43 Auxiliary air vent detail*
(Drawing provided courtesy of AIRVAC)

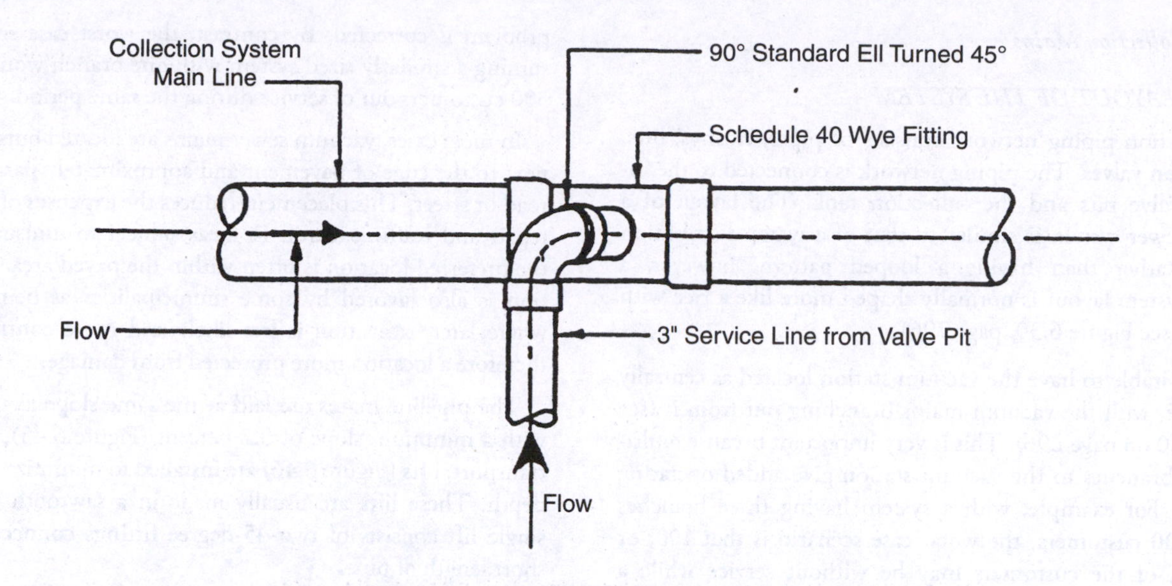


*Fig. 6.44 Top view of crossover connection*
(Source: *ALTERNATIVE WASTEWATER COLLECTION SYSTEMS,* Technology Transfer Manual, US Environmental Protection Agency)

for *AESTHETIC*[25] reasons and to protect the vent. In climates where temperatures fall below freezing, the auxiliary vent must be located a maximum of 20 feet from the valve pit. In this manner, the heat from the wastewater acts to warm the freezing atmospheric air and thus reduces the possibility that valve components will freeze.

### *6.5104 SERVICE LATERAL*

Vacuum sewer service lines (laterals) run from the valve pit/sump to the vacuum main. Typically, these lines are installed near and parallel to property lines. Most municipalities prefer locating the service line where it will not be driven over. Other agencies, however, prefer locating the service line within the paved area of the road. The reasoning is that if the service line is under the paved section of the road, it is less likely to be damaged by any later excavation in the area.

Vacuum sewer service lines should be located away from potable water pipelines to prevent cross contamination. They should also be separated from other buried utility lines, if possible, to reduce the chance of damage when the other utility lines are excavated for maintenance or repair.

Service lines are typically 3 inches in diameter. An exception to this occurs when a buffer tank is used. Buffer tanks are used for large flows that would otherwise cause frequent valve cycles. To maintain good vacuum response at the buffer tank, a 6-inch service line is recommended.

As with the vacuum sewer mains, Class 200, SDR 21 PVC pipe typically is used for the service lines. Solvent-welded DWV (Drain, Waste, and Vent) fittings are used, although rubber ring fittings are becoming more common.

All connections to the main, for example, crossover connections, are made over the top (Figure 6.44). This is accomplished using a vertical wye and a long-radius elbow.

## QUESTIONS

Please write your answers to the following questions and compare them with those on page 346.

6.51D What is the main difference between the two main types (Model D and Model S) of vacuum valves?

6.51E What is the purpose of the auxiliary vent on a building sewer?

6.51F Why should vacuum sewer service lines be located at a distance from potable water pipelines and other buried utility lines?

---

25. *Aesthetic* (es-THET-ick). Attractive or appealing.

### *6.511 Collection Mains*

#### *6.5110 LAYOUT OF THE SYSTEM*

The collection piping network consists of pipes, fittings, lifts, and division valves. The piping network is connected to the individual valve pits and the collection tank. The layout of a vacuum sewer system is similar to that of a water distribution system. Rather than having a looped pattern, however, a vacuum system layout is normally shaped more like a tree with branches (see Figure 6.30, page 296).

It is desirable to have the vacuum station located as centrally as possible, with the vacuum mains branching out from it (see Figure 6.30 on page 296). This is very important because multiple main branches to the vacuum station give added operating flexibility. For example, with a system having three branches serving 300 customers, the worst case scenario is that 100, or one-third, of the customers may be without service while a problem is corrected. By contrast, the worst case scenario assuming a similarly sized system with one branch would have all 300 customers out of service during the same period.

In most cases, vacuum sewer mains are located outside of and next to the edge of pavement and approximately parallel to the road or street. This placement reduces the expenses of pavement repair and traffic control. In areas subject to unusual erosion, the preferred location is often within the paved area. This location is also favored by some municipalities as being an area where later excavation is less likely and more controlled, and therefore a location more protected from damage.

The pipeline mains are laid at the same slope as the ground with a minimum slope of 0.2 percent (Figure 6.45). For uphill transport, lifts (Figure 6.46) are installed to minimize excavation depth. These lifts are usually made in a sawtooth fashion. A single lift consists of two 45-degree fittings connected with a short length of pipe.

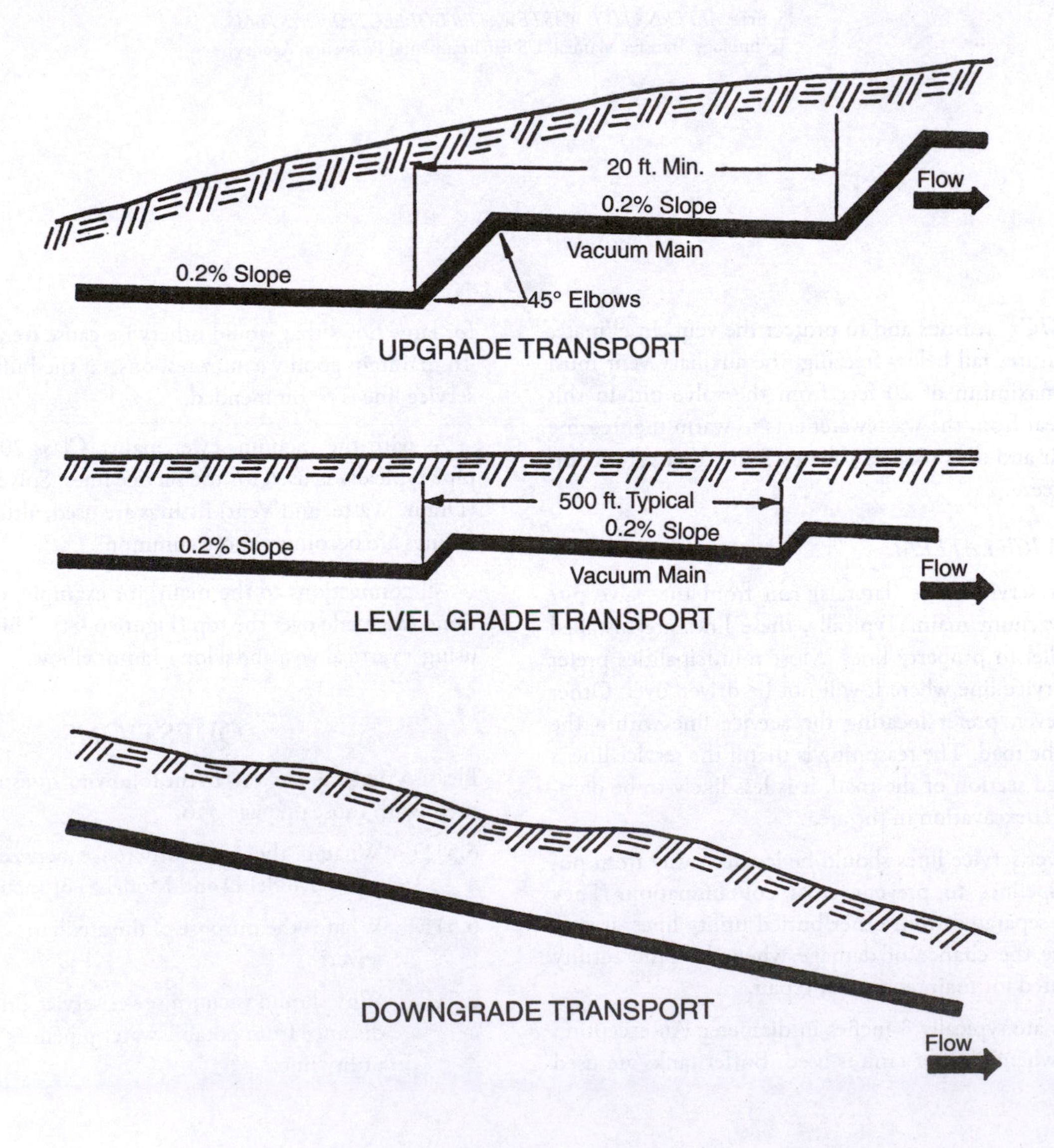


*Fig. 6.45 Detailed upgrade/level/downgrade profile of vacuum sewer*
(Drawing provided courtesy of AIRVAC)

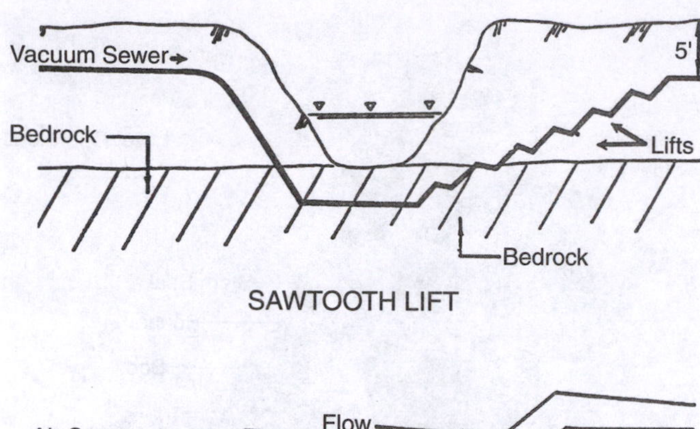


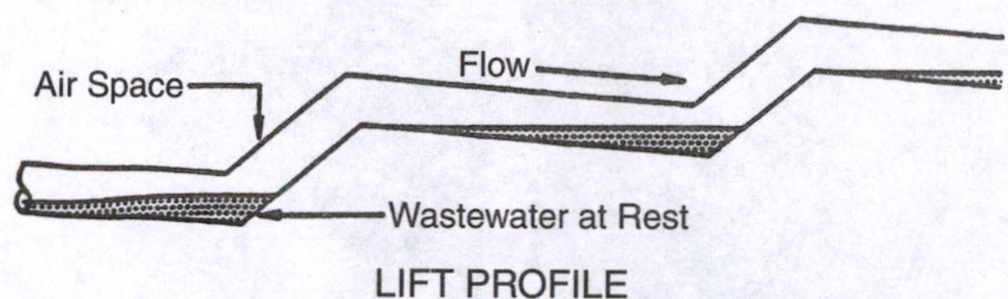


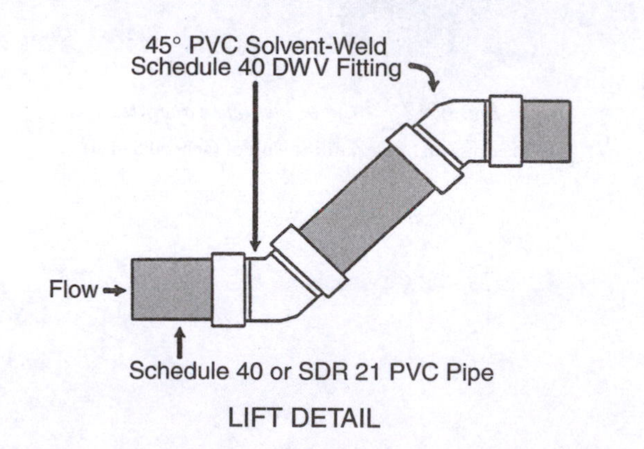


*Fig. 6.46 Lift detail*

(Source: *ALTERNATIVE WASTEWATER COLLECTION SYSTEMS,* Technology Transfer Manual, US Environmental Protection Agency)

One of the major expenses of a vacuum collection system is the valve pit. When two or more homes share one valve, overall system construction costs can be significantly reduced, resulting in a major cost advantage. To do this, however, may require the main line to be located in private property, typically in the back yard of a home. There are two disadvantages to this type of placement. First, it requires a permanent easement from one of the property owners, which may be difficult to obtain. Second, experience has shown that multiple-house hookups are a source of neighborhood friction unless the valve pit is located on public property. Routing is not an operator decision, but the designer should carefully weigh the tradeoff of reduced costs with the social issues and possible maintenance and operation problems before making the final routing decision.

There are no manholes in a vacuum collection system but access for maintenance can be gained at each valve pit or at the end of a line where an access pit may be installed.

When reviewing the layout of a vacuum system, the operator should be sure that the pipe runs selected minimize lift, minimize length, and equalize flows in each main branch. The length of collection lines is governed by two factors: *STATIC LIFT*[26] and *FRICTION LOSS*.[27]

## 6.5111 PIPE AND FITTINGS

PVC thermoplastic pipe is normally used for vacuum sewers in the United States. The most common PVC mains are iron pipe size (IPS) 200 psi working pressure rated, standard dimension ratio 21 (Class 200 SDR 21 PVC). Class 160, SDR 26, and Schedule 40 PVC have also been used. From a pressure standpoint, the lower class pipe is acceptable. However, thin-wall pipe is more likely to be damaged during installation. Since there is little cost difference between SDR 21 and SDR 26 when the excavation, backfill, and surface restoration are considered, Class 200, SDR 21 pipe is recommended.

PVC pressure fittings are needed for directional changes as well as for the crossover connections (previously shown in Figure 6.44) from the service lines to the main line. These fittings may be solvent-welded or gasketed. Experience has shown that there are fewer problems with gasketed pipe fittings than with solvent-welded fittings. Where gasketed pipe is used, the gaskets must be certified for use under vacuum conditions.

In the past, solvent-welded fittings were most often used with DWV pipe. Presently, however, there is a move toward eliminating all solvent welding. These fittings are still more widely available than gasketed joint fittings. Until recently, one major line component, the wye required for each service connection to the main, was not available in gasketed PVC. At least one of the major fitting manufacturers is now making gasketed wye fittings. Spigot adapters (an adapter fitting that can be solvent welded in a controlled environment into each of the three legs of the wye resulting in three-gasketed joints) can also be used.

Expansion and contraction may be a concern when the entire system is solvent welded, but allowances are made for these forces in gasketed main line pipe joints.

---

26. *Static Lift.* Vertical distance water is lifted from upstream water surface to downstream water surface (which is at a higher elevation) when no water is being pumped.

27. *Friction Loss.* The head, pressure, or energy (they are the same) lost by water flowing in a pipe or channel as a result of turbulence caused by the velocity of the flowing water and the roughness of the pipe, channel walls, or restrictions caused by fittings. Water flowing in a pipe loses head, pressure, or energy as a result of friction. Also called head loss.

### 6.5112 DIVISION (ISOLATION) VALVES

Division valves are used on vacuum sewer mains to isolate portions of the system for repairs and maintenance. Plug valves (Figure 6.47) and resilient-seated gate valves (Figure 6.22, page 285) have both been used successfully. Typical locations for division valves are at branch/main intersections, at both sides of a bridge crossing, at both sides of areas of unstable soil, and at periodic intervals on long routes. The intervals vary with the judgment of the engineers, but a typical interval is 1,500 to 2,000 feet.

Some systems have installed a gauge tap (Figure 6.48) just downstream of the division valve. The purpose for installing this device is to allow vacuum monitoring by one person in the field, rather than requiring two people (one to operate the valve in the field and one to read the vacuum gauge at the vacuum station). This greatly reduces the staffing costs for emergency maintenance activities. Therefore, gauge taps are recommended. Gauges should be positioned so that they are easily viewed when the isolation (division) valves are operated. The operator must know the location of the gauge taps. Refer to the system O&M manual for frequency of use.

Division valves 4 inches and smaller may be directly actuated while all 6-inch and larger valves should be provided with gear

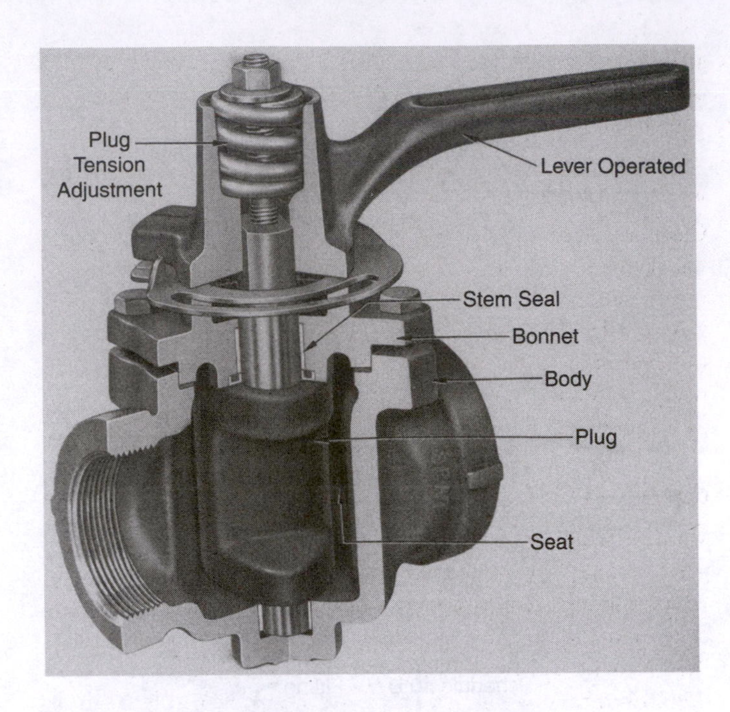


*Fig. 6.47 Plug valve, lever operated*
(Permission of DeZurik, a Unit of General Signal)

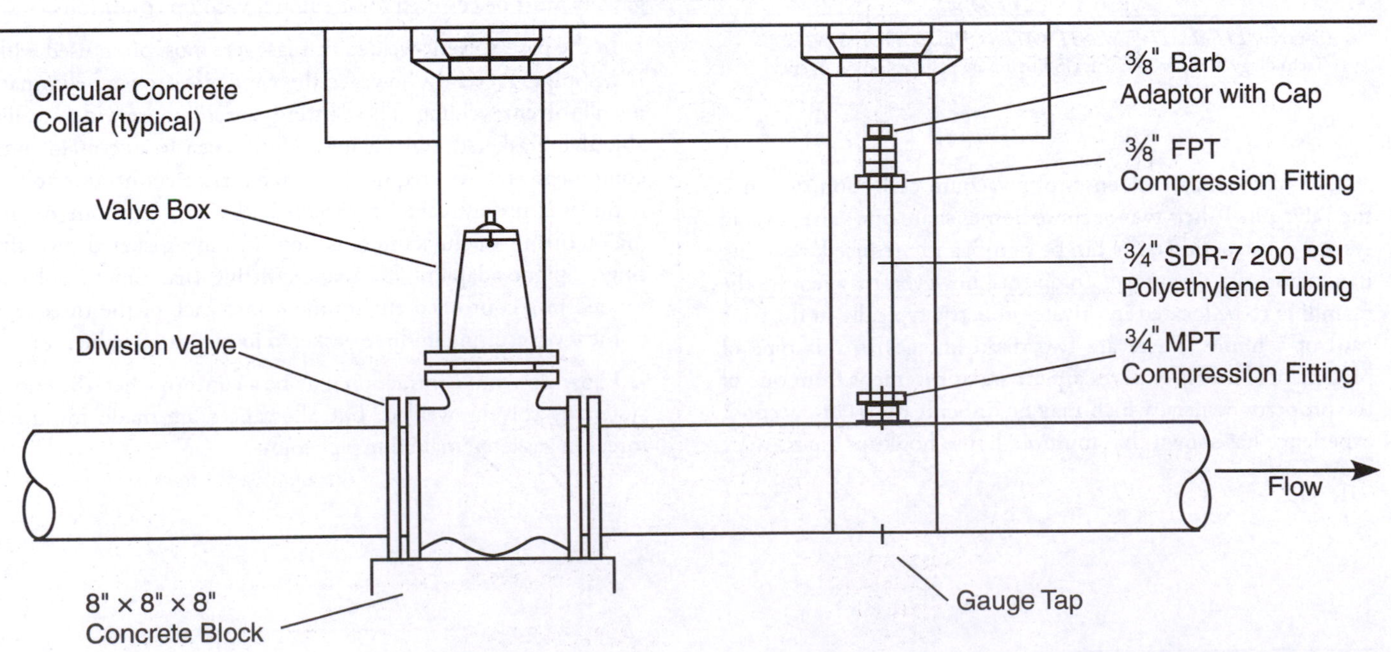


*Fig. 6.48 Division (isolation) valve with gauge tap detail*
(Adapted from *ALTERNATIVE WASTEWATER COLLECTION SYSTEMS*, Technology Transfer Manual, US Environmental Protection Agency)

actuators. The valves should be installed in a valve box with the operating nut extended to a position where it can be reached with a standard valve wrench.

### *6.5113 CLEANOUTS*

Cleanouts (also called access points) have been used in the past but their use is no longer recommended in systems with many valves since access to the vacuum main can be gained at any valve pit. However, some state codes still require cleanouts to be installed at specified intervals. In these cases, and in stretches where valves are nonexistent, access points similar to the one shown in Figure 6.49 should be constructed.

### *6.5114 INSTALLATION*

Vacuum sewers are normally buried 3 feet deep. In a few cases, where economy is important and the risk of damage is small, vacuum sewers are buried with less cover. The depth of burial in colder climates usually depends on frost penetration depths. In the northern United States, sewers are often placed 4 to 5 feet deep. Even though line freezing is a concern, it usually is not a problem with vacuum mains because the retention time in small-diameter lines is relatively short and turbulence is high. Long periods of disuse could still lead to freezing conditions, however, unless the system is designed to ensure relatively complete emptying of the mains (a technique used for winterizing some resort systems).

The separation of vacuum sewers from water supply mains and laterals often requires the vacuum sewer to be buried deeper than would be required for other reasons. Horizontal and vertical separation requirements are set by regulatory agencies. These requirements vary from state to state.

Trenching may be accomplished by using a backhoe, wheel trencher, or chain-type trencher. The contractor's staff usually selects the type of equipment based on the material to be excavated, depth requirements, topography, and available working space.

An advantage to the use of vacuum sewers is that the small-diameter PVC pipe used is flexible and can be easily routed around obstacles. This feature allows vacuum sewers to follow a winding path if necessary. The pipe should be bent in as long a radius as possible, never less than the angle recommended by the pipe manufacturer. Good record drawings (as-builts) are essential to be able to locate pipe bent around obstacles.

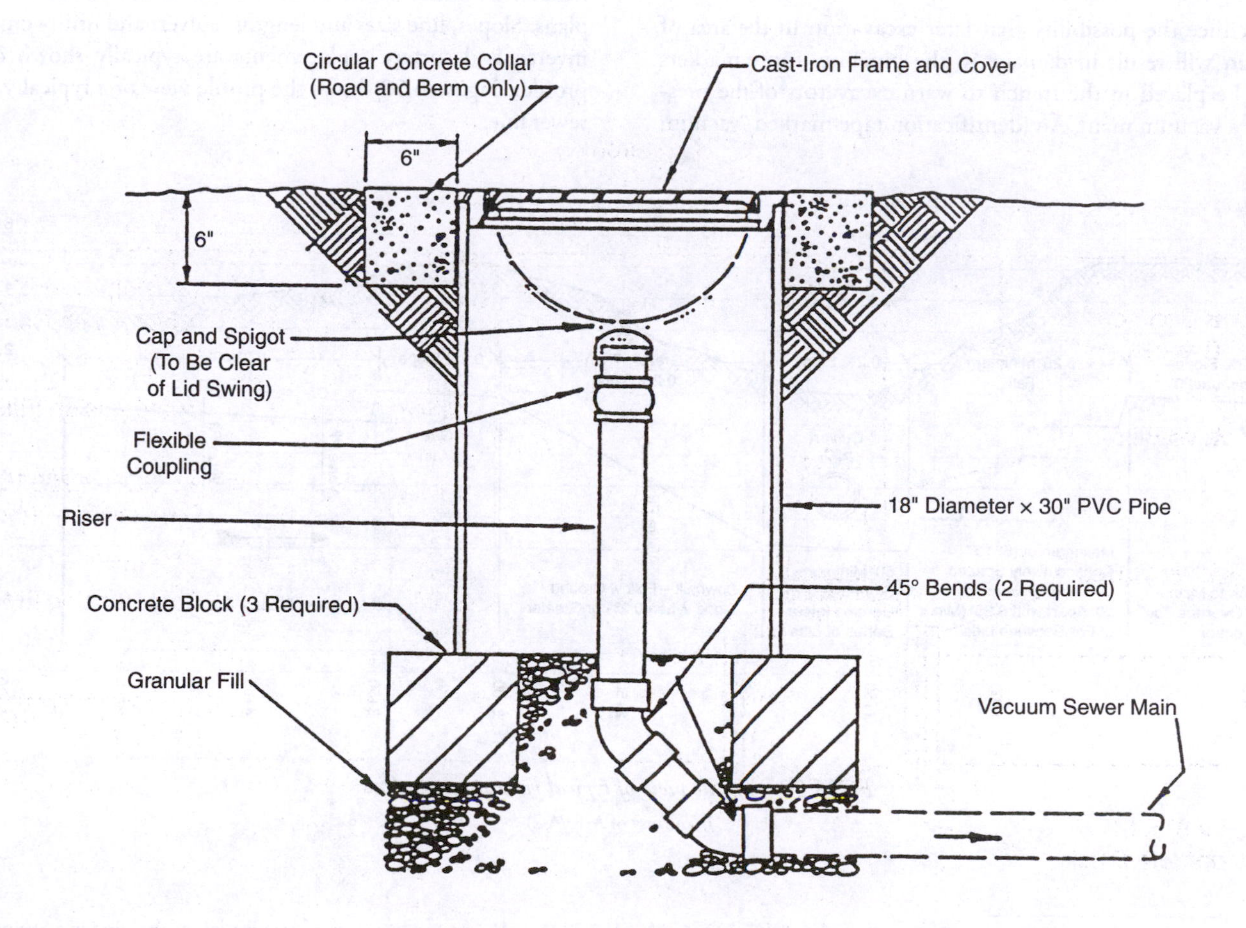


*Fig. 6.49 Terminal access point detail*

(Source: *ALTERNATIVE WASTEWATER COLLECTION SYSTEMS*, Technology Transfer Manual, US Environmental Protection Agency)

Culvert and utility crossings often make it necessary to vary the depth at which vacuum sewer mains are buried. As a result, there may be many sags and high points (summits) in the pipeline profile. Unlike pressure mains, where air accumulates at a summit and must be vented by an air release valve, vacuum sewers are not affected by high points in profile. The sags, however, may present a problem since they typically will add lift to the system. The greater the lift, the more the vacuum has to work to lift the wastewater. In addition, a sag may trap wastewater during low-flow periods and thus block off the low part of the sewer.

Because vacuum sewers are exposed to repeated surges of wastewater, pipe movement is possible if proper installation practices are not followed. Each fitting is a point of possible joint failure. In early systems, concrete *THRUST BLOCKS*[28] were installed at each fitting. More recent systems have been installed without thrust blocking. The idea behind this is that the pressure is inward rather than outward as would be the case in a positive-pressure situation. If thrust blocking is used, the pipe and fitting should be covered with a thin sheet of plastic before the concrete thrust blocks are poured. When thrust blocks are not used, failure of the fitting due to trench settlement is an important concern. Care must be exercised in the backfill and compaction operations. Fittings can be protected by covering them with granular backfill material and then mechanically compacting the backfill.

To reduce the possibility that later excavation in the area of the main will result in damage to the pipeline, route markers should be placed in the trench to warn excavators of the presence of a vacuum main. An identification tape marked "vacuum sewer" is sometimes placed shallowly in the pipeline trench. The tape can be metalized so that it can be detected with common utility locating equipment. Most tapes cannot be induced with a tone at significant depth, so metalized tape should be placed just below the ground surface for reliable detection. Other protective measures include installing a toning wire along the top of the pipeline or color coding the pipes themselves.

Imported material called "pipe zone backfill" is often placed in the area right around the main if material excavated from the trench is regarded as unsuitable for this purpose. Pipe zone backfill is usually granular material such as pea gravel or coarse sand. Fine sand or soil is generally not as desirable because it clumps or piles up rather than flowing smoothly into place around and under the curved pipe. The agency controlling the road or street usually specifies what type of material can be used for the remaining backfill, especially if the mains are located within the pavement. In some cases, a cement-sand *SLURRY*[29] is used for backfill. This option is particularly attractive when a trencher is used, the mains are located within the pavement, and prompt restoration for traffic control is important.

### *6.5115 RECORD (AS-BUILT) PLANS*

Precise record (as-built) drawings must be prepared to record the actual location of all collection system pipelines and appurtenances. Profiles of the mains should always be shown on the plans. Slopes, line sizes and lengths, culvert and utility crossings, inverts, and surface replacements are typically shown on the profiles. Figure 6.50 shows the profile view of a typical vacuum sewer line.

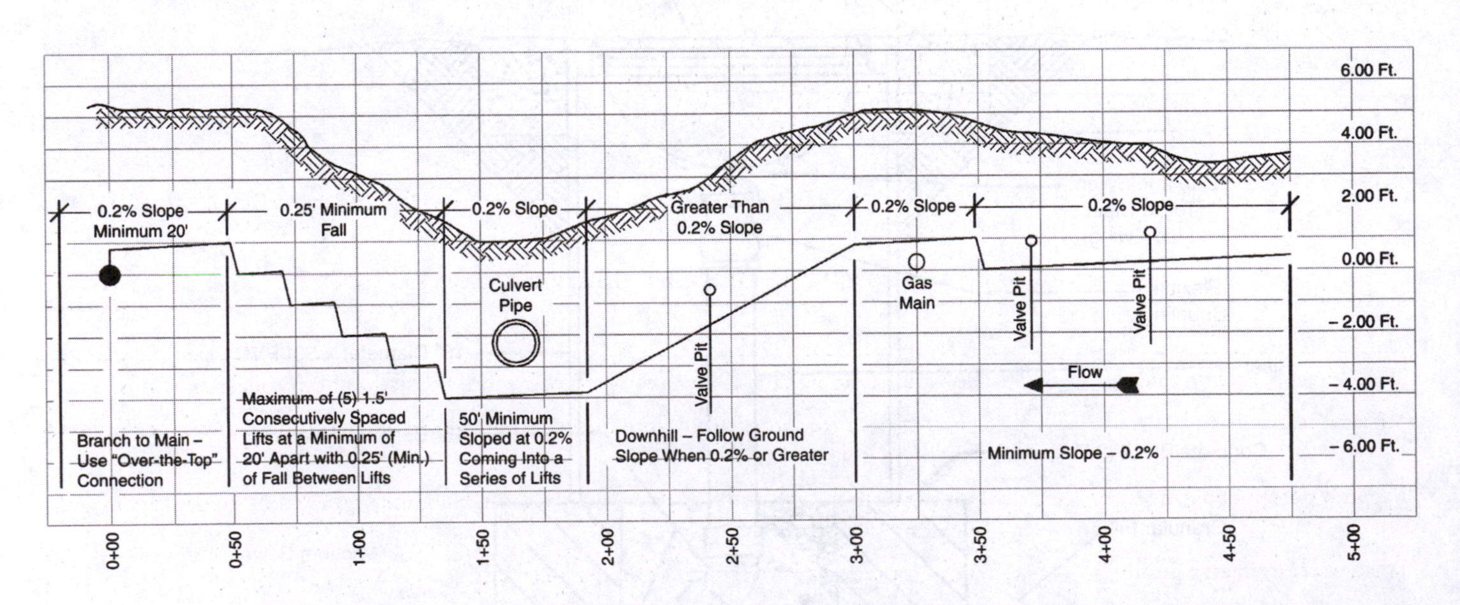


*Fig. 6.50 Profile view of typical vacuum sewer line*

(Courtesy of AIRVAC)

28. *Thrust Block.* A mass of concrete or similar material appropriately placed around a pipe to prevent movement when the pipe is carrying water. Usually placed at bends and valve structures.

29. *Slurry.* A watery mixture or suspension of insoluble (not dissolved) matter; a thin, watery mud or any substance resembling it (such as a grit slurry or a lime slurry).

## QUESTIONS

Please write your answers to the following questions and compare them with those on page 346.

6.51G A major cost savings can be made when two or more homes share one valve pit. What are the two major disadvantages to this type of routing?

6.51H What is the purpose of installing division valves?

6.51I What methods are used to identify the location of collection main pipes?

### *6.512 Central Vacuum Station*

Vacuum stations function as transfer facilities between a central collection point for all vacuum sewer lines and a pressurized line leading directly or indirectly to a wastewater treatment facility. Table 6.2 lists the equipment in a central vacuum station and describes the purpose of each component (refer to Figures 6.51 and 6.52).

The central collection vacuum station is similar to a conventional wastewater pumping station. These stations are typically two-story concrete and block buildings. The basement of the vacuum station should be provided with a sump to collect washdown water. This sump is usually emptied by a vacuum valve that is connected by piping to the wastewater collection tank. A check valve and eccentric plug valve should be fitted between the sump valve and the wastewater collection tank.

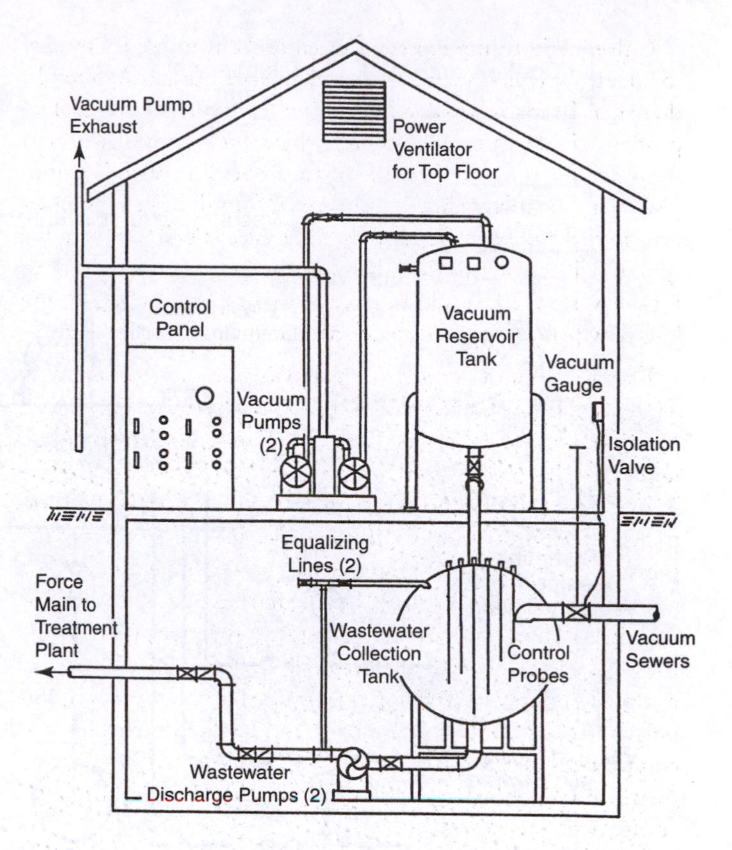


*Fig. 6.51 Diagram of a typical vacuum station*
(Drawing provided courtesy of AIRVAC)

**TABLE 6.2 COMPONENTS OF A CENTRAL VACUUM STATION**

| Part | Purpose |
|---|---|
| Wastewater collection tank | Serves as a holding tank for wastewater flowing into the station from collection mains |
| Vacuum reservoir tank | Prevents carryover of moisture from the wastewater collection tank and reduces the frequency of vacuum pump starts |
| Vacuum pumps and gauges | Pumps create the vacuum that draws wastewater through the system and the gauges display operating vacuum levels |
| Wastewater discharge pumps | Pump wastewater to treatment or discharge facilities |
| Wastewater level control probes | Regulate operation of the wastewater discharge pumps |
| Motor control center (control panel) | Houses the motor starters, overloads, control circuitry, the hours run meter for each vacuum and wastewater pump, vacuum chart recorders, collection tank level control relays, and an alarm system |
| Chart recorder | Records pump operating data |
| Standby generator | Provides power for continuous vacuum station operation in the event of a power failure |
| Fault monitoring system | Warns of equipment malfunctions |

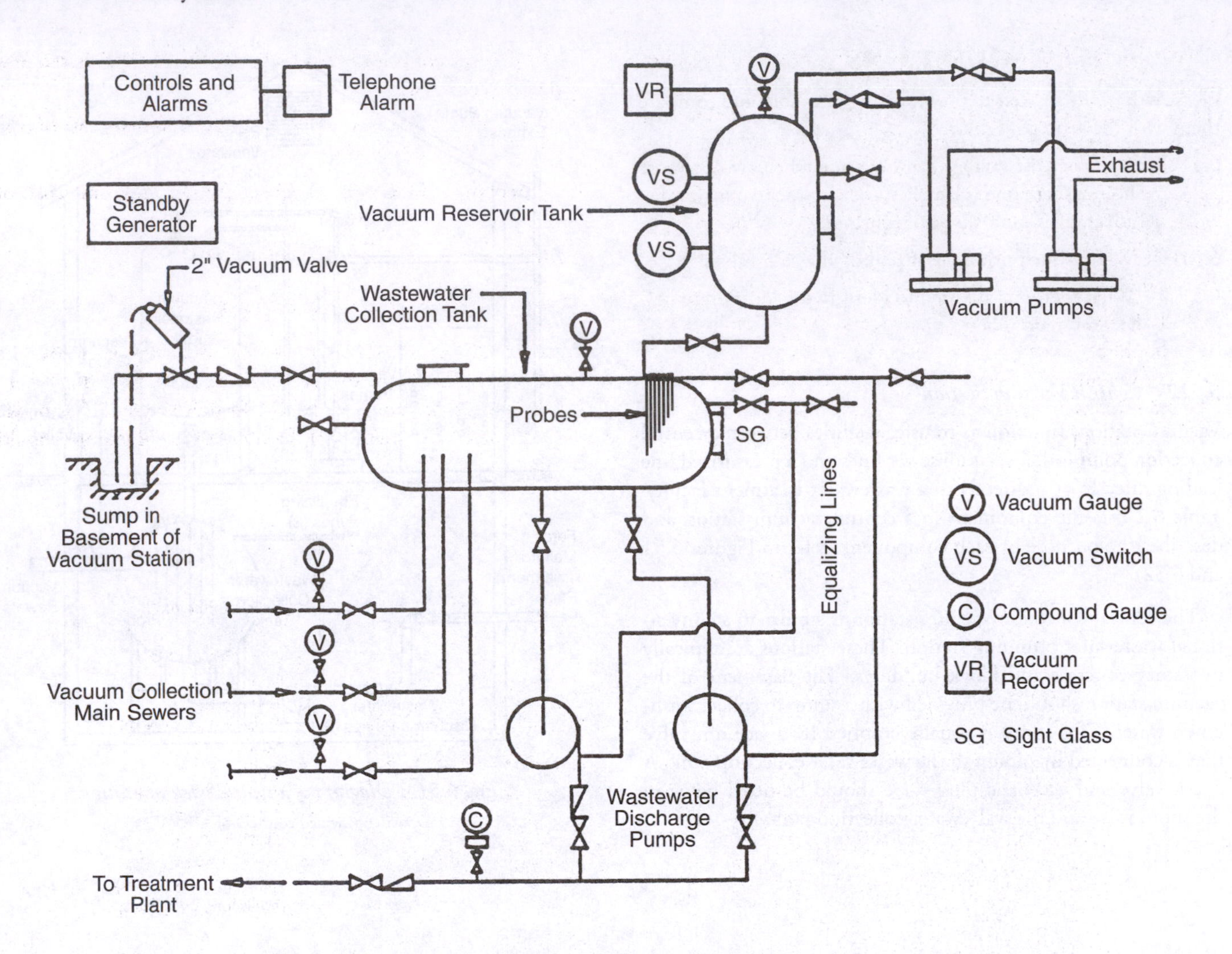


*Fig. 6.52 Line diagram of a typical vacuum station layout*

(Drawing provided courtesy of AIRVAC)

### *6.5120 VACUUM AND WASTEWATER COLLECTION TANKS*

Two vacuum tanks are usually required for each vacuum station: the wastewater collection tank and the vacuum reservoir tank. Some stations may not require the vacuum reservoir tank.

Wastewater is stored in the collection tank until a sufficient volume accumulates, at which point the tank is emptied. The wastewater collection tank is a sealed, vacuum-tight vessel made either of fiberglass or steel. Fiberglass tanks are generally more expensive but do not require the periodic maintenance (painting) of a steel tank. Vacuum, produced by the vacuum pumps, is transferred to the collection system through the top part of this tank. The part of the tank below the invert of the incoming collection lines acts as the wet well. A bolted hatch provides access to the tank for maintenance. The operating volume of the collection tank is the amount of wastewater required to restart the discharge pump. The tank usually is sized so that at minimum design flow the pump will operate every 15 minutes.

Most wastewater collection tanks are located at a low elevation in relation to most of the components of the vacuum station. This minimizes the lift required for the wastewater to enter the collection tank, since wastewater must enter at or near the top of the tank to ensure that vacuum can be restored upstream. The wastewater pump suction lines should be placed at the lowest point on the tank and as far away as possible from the main line inlets. The main line inlet elbows inside the tank should be turned at an angle away from the pump suction openings to remove as much as possible of the air *ENTRAINED*[30] in the wastewater entering the wastewater pump suction line.

30. *Entrain.* To trap bubbles in water either mechanically through turbulence or chemically through a reaction.

The vacuum reservoir tank is located between the vacuum pumps and the collection tank. This tank has three functions: (1) to reduce carryover of moisture and small particles into the vacuum pumps, (2) to act as an emergency reservoir, and (3) to reduce the frequency of vacuum pump starts. Like the wastewater collection tank, the vacuum reservoir tank can be made of either fiberglass or steel. The recommended size of the vacuum reservoir tank for most applications is 400 gallons, but in some cases it may be larger.

Both the wastewater collection tank and the vacuum reservoir tank should be designed for a working pressure of 20 inches of mercury vacuum and tested to 28 inches of mercury vacuum. Fiberglass tanks must have 150 psi rated flanges. Each tank should be equipped with a sight glass and its associated valves.

To prevent corrosion, steel tanks should be sandblasted and painted as follows:

- Internally: One coat of epoxy primer and two coats of coal tar epoxy
- Externally: One coat of epoxy primer and one coat of epoxy finish

Painting may be required every 5 to 6 years.

### 6.5121 *VACUUM PUMPS AND GAUGES*

Vacuum pumps produce the vacuum necessary to move the wastewater through the collection system to the vacuum station. These pumps may be either the sliding-vane (Figure 6.53) or the liquid-ring type (Figure 6.54). In either case, the pumps should

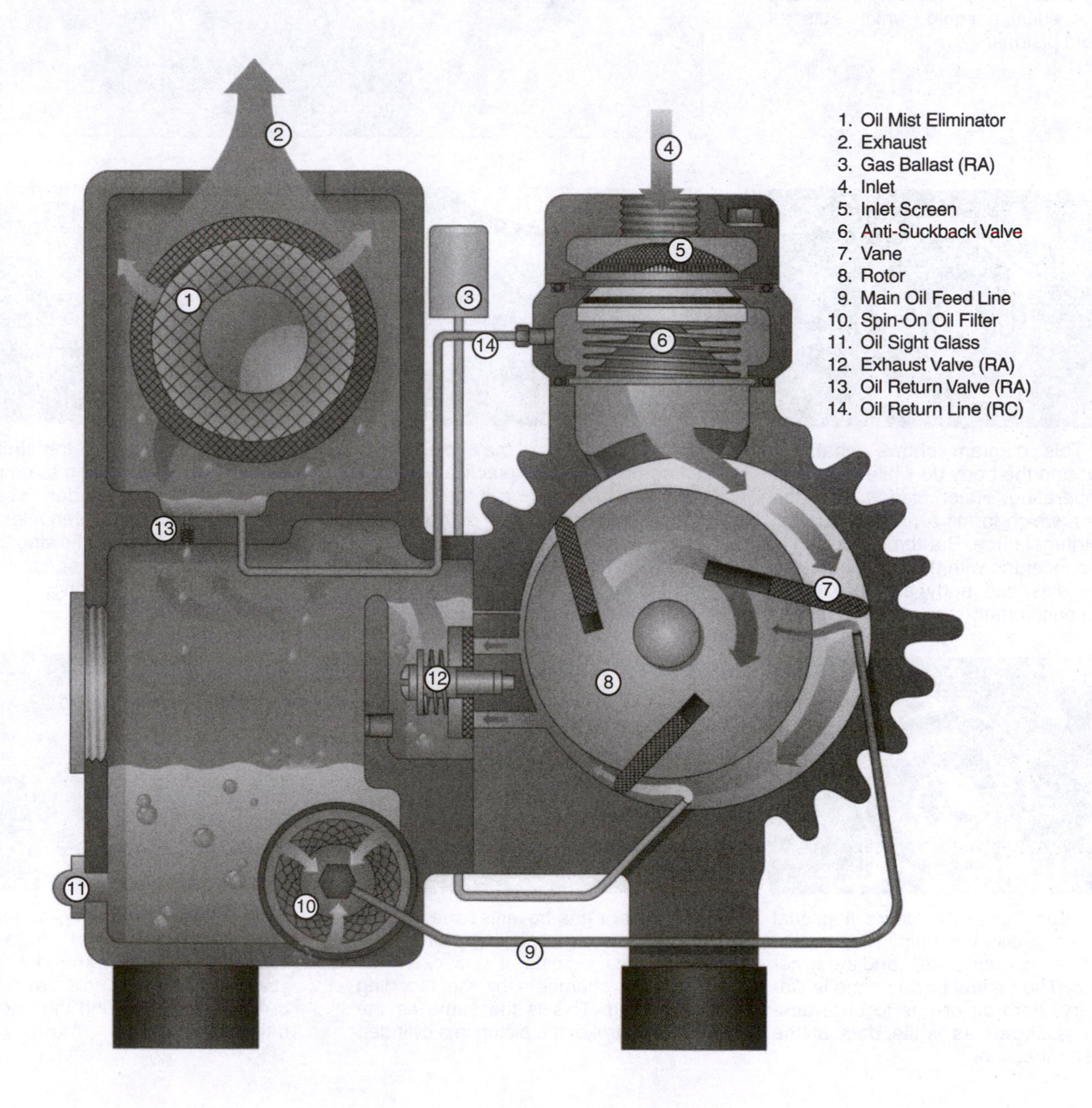


*Fig. 6.53 Sliding-vane vacuum pump*
(Permission of Busch, Inc.)

**1** From the outside, a Nash vacuum pump has the look of rugged industrial equipment, but it gives few hints about how the internal parts function. Shaft bearings on heavy cast brackets are obviously accessible for servicing. The two connections at the top are pump inlets, and two flanges on the far side of the pump are discharge connections. Internally, this machine is actually a dual vacuum pump. The two inlets can, within limits, serve two systems at different vacuum levels. The Nash pump uses water or any other suitable liquid, which acts as "liquid pistons."

**5** This diagram shows what the rotor and the body do while the pump is operating. Rotor blades whirl the water, which forms a ring because of centrifugal force. But the water ring is not concentric with the rotor, because rotor axis and body axis are offset from each other.

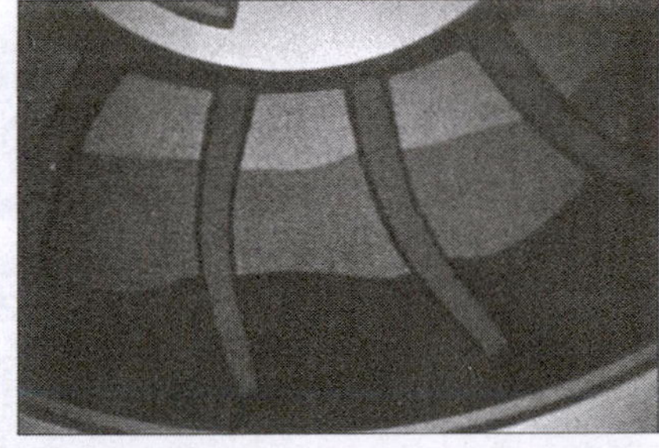

**6** At the bottom of the circle, the rotating ring of water practically touches the stationary cone in the center of rotation. There is little or no empty space in the chambers between the rotor blades. We can start to trace the pumping cycle at this point.

**7** Moving up around the circle, you can see the water begin to empty the rotor chambers. The dark sector at the left indicates the open inlet port in the cone. Air or gas is flowing through the inlet port into the space vacated by the receding water ring.

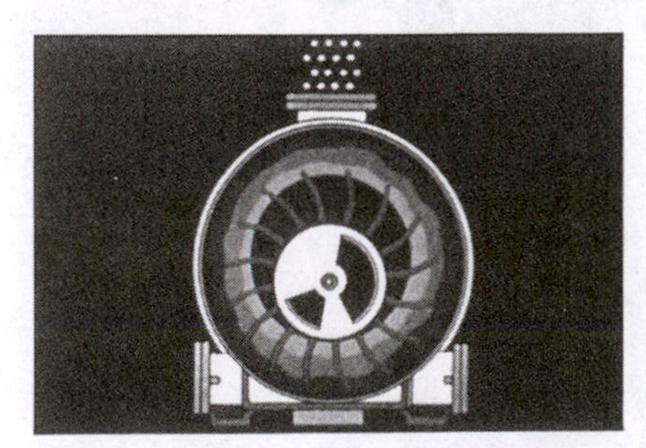

**11** Looking at the entire sectional diagram again, it is helpful to remember that the rotor blades and the water rotate. The central ported cone is stationary. Here air or gas to be evacuated is shown as white dots at the inlet connection.

**12** Air or gas traverses the internal passage to the cone inlet port. As the white dots indicate, it is sucked into the rotor chambers by the receding water ring. This is the same as the suction stroke of a piston in a cylinder.

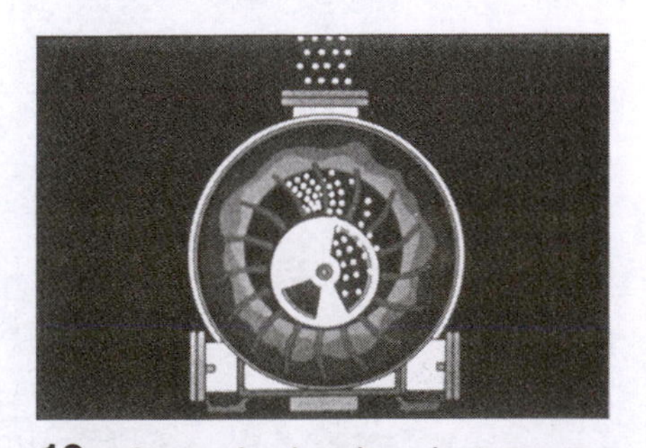

**13** As each chamber, in turn, rotates past the inlet port, it carries a volume of air or gas around with it. As you see, the white dots are confined between the cone and the ring of rotating water.

*Fig. 6.54 Liquid-ring vacuum pump*

(Permission of The Nash Engineering Company)

**2** The one moving part in a Nash vacuum pump is its cast rotor. Here, the pump body has been taken away to reveal the rotor. The two heads have been pulled apart so that you can see the cones. When the pump is assembled, those two cones project inside the ring of rotor blades.

**3** It is apparent in this picture that the chambers between rotor blades are open around the periphery. It is not so easy to see, but the inner ends of the chambers are open, too. Inner edges of the rotor blades are machined to run around the cone surfaces with a close, noncontact fit.

**4** An internal passage joins the openings indicated by two hands here. That passage goes from the external pump inlet to an inlet port on the cone. Correspondingly, there is a passage from the cone discharge to the discharge connections on the head.

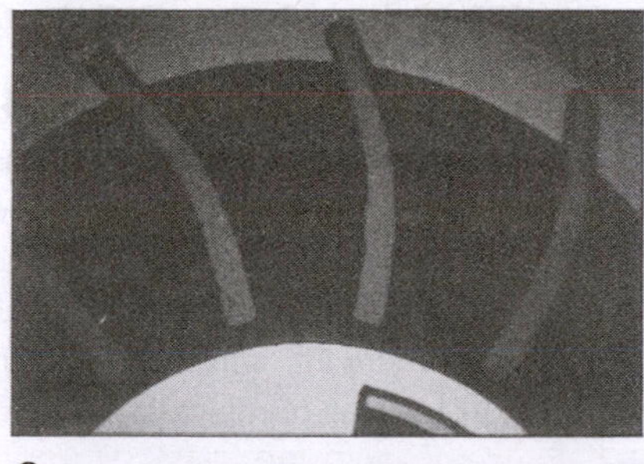

**8** At the top of the circle, the water has almost completely emptied the rotor chambers. You can see that the chambers at the left have rotated past the inlet port opening. They have finished drawing in air or gas, and they are ready to compress it.

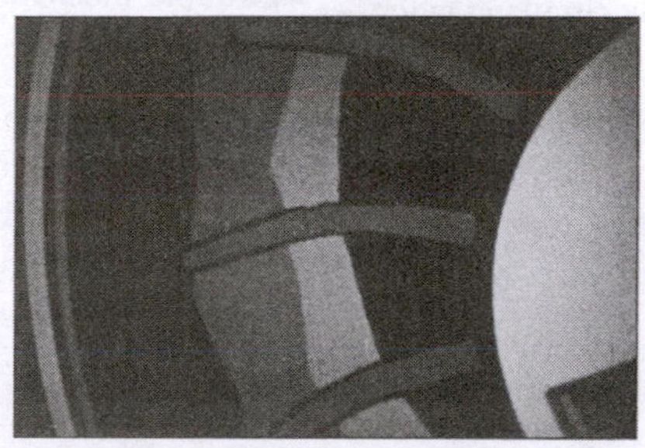

**9** As the water closes in again, it begins to fill the rotor chambers. There is less and less empty space between the water ring and the cone. Trapped air or gas is compressed as the available volume in each chamber is progressively reduced.

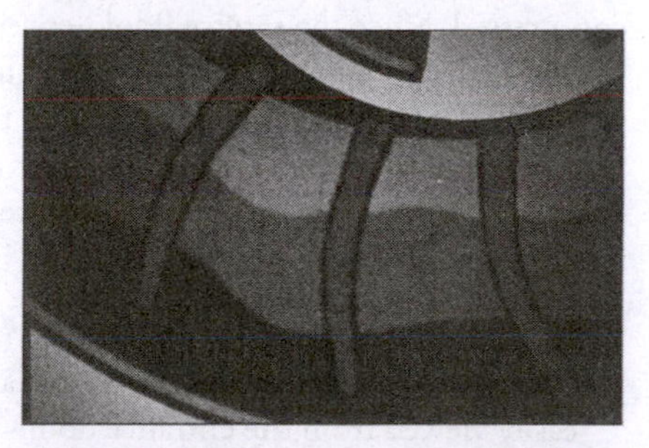

**10** At the bottom of the circle, the inner end of each chamber goes past the discharge port in the cone. It is indicated by the black sector at the top of this picture. Here the compressed air or gas escapes to discharge. Then the cycle repeats.

**14** The dots get squeezed together as the water ring closes in. This represents, schematically, the compression of air or gas from vacuum up to atmospheric pressure. The water ring does the job of pistons, and the rotor chambers play the part of cylinders.

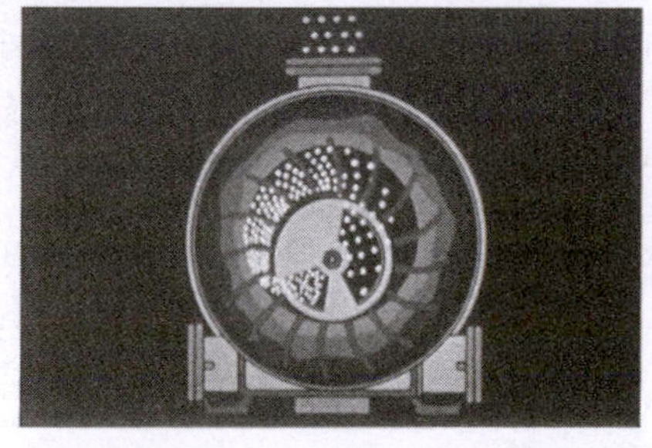

**15** When each chamber, in turn, rotates to the discharge port opening, the compressed air or gas escapes from that chamber through the discharge port to the internal discharge passage.

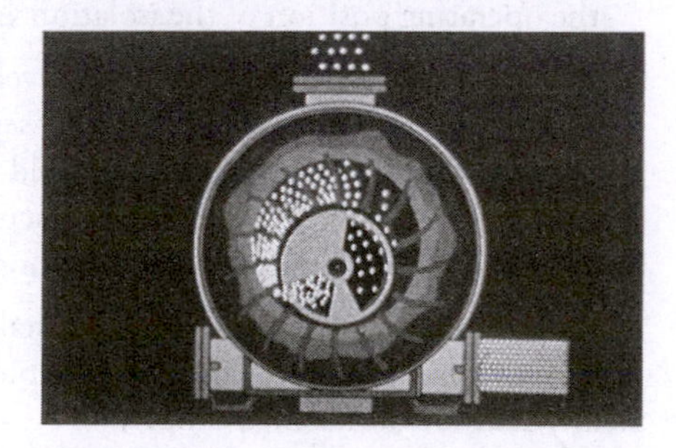

**16** Air or gas, with the dots closely packed to indicate higher pressure, is shown here flowing out of the discharge connection at right. In this completed schematic, you can see the entire vacuum pumping cycle.

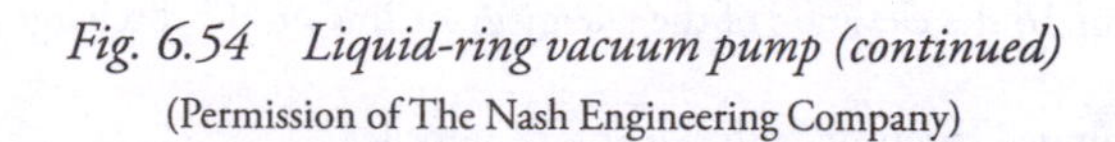

*Fig. 6.54 Liquid-ring vacuum pump (continued)*

(Permission of The Nash Engineering Company)

be air-cooled and capable of continuous operation. Duplicate pumps, each capable of delivering 100 percent of the required air flow, should be provided.

Sliding-vane vacuum pumps require an air filter installed on the inlet line between the reservoir tank and the vacuum pump to remove any small solid particles that could cause excessive impeller wear if they were to enter the pump *VOLUTE*.[31] Sliding-vane vacuum pumps use less power for a given pump capacity than liquid-ring vacuum pumps. On the other hand, sliding-vane pumps are easily damaged if liquid reaches the inside of the pump. If liquid gets into a sliding-vane pump, it can shorten the life of the pump and the pump must be taken out of service to remove the foreign liquid. In contrast, the liquid-ring pump usually can withstand an accident of this type with very little damage. Design precautions, such as an electrically controlled plug valve between the collection tank and the reservoir tank, can be added to the piping system in order to protect sliding-vane pumps.

A liquid-ring vacuum pump uses a service liquid (usually oil) as a sealing medium between an offset *IMPELLER*[32] and the pump casing. As the impeller spins, the service liquid is forced against the pump's outer casing by centrifugal force and air is compressed. Since the service liquid continually circulates when the pump is in operation, a service liquid tank must be provided. The tank is vented with an outlet to the outside. The service liquid carries a significant quantity of heat away from the pump and a heat exchanger also is required.

To enable the operator to monitor the system, vacuum gauges should be installed at the following locations:

- On the side of the vacuum reservoir tank in a position that is easily viewed from the entrance door.
- On the wastewater collection tank in a position that is easily viewed.
- On each incoming collection system main line to the collection tank, immediately upstream of the isolation valve on the line. These gauges should be in a position above the incoming collection system main lines that is easily viewed from the operating position of the isolation valves.

The usual vacuum pump operating level is 16 to 20 inches of mercury vacuum, with a low-level alarm set at 14 inches of mercury vacuum. However, the pumps should have the capability of providing up to 25 inches of mercury vacuum because this level is sometimes needed in the troubleshooting process.

Noise abatement devices may be installed to control noise. They may be installed on the vacuum pump or as part of the facility that surrounds the pump.

## QUESTIONS

Please write your answers to the following questions and compare them with those on page 346.

6.51J What is the function of vacuum stations?

6.51K What is the operating volume of the collection tank?

6.51L What is the purpose of a vacuum pump?

### *6.5122 WASTEWATER DISCHARGE PUMPS*

Wastewater pumps transfer the liquid that is pulled into the collection tank by the vacuum pumps to its point of treatment or disposal. Dry pit pumps have been used extensively, although submersible wastewater pumps mounted on guide rails inside the collection tank may be used as an alternative. The most frequently used pump has been the nonclog type. Both vertical (Figure 6.55) and horizontal pumps (Figure 6.56) can be used.

Wastewater discharge pumps should be made of cast iron with stainless-steel shafts; aluminum, bronze, and brass pumps should be avoided. Fiber packing is not recommended; instead, double mechanical seals that are adaptable to vacuum service should be used.

Where possible, horizontal wastewater pumps should be used because they have smaller suction losses compared to vertical pumps. Suction losses are reduced by installing suction pipes two inches larger in diameter than the discharge line. Wastewater pump shafts should be fitted with double mechanical shaft seals with the seal chamber pressurized with light oil to prevent air leakage around the pump shaft into the pump casing.

Wastewater pumps are typically installed at an elevation significantly below the collection tank to minimize the net positive suction head (NPSH) requirement.

To ensure continuous operation of the vacuum collection system, each pump must be capable of pumping 100 percent of the system's design capacity at the specified *TOTAL DYNAMIC HEAD (TDH)*.[33]

Each pump should be equipped with an enclosed nonclog, two-port impeller,[34] statically and dynamically balanced and capable of passing a 3-inch sphere. Pumps should have an inspection opening in the pump volute to facilitate cleaning or clearing a plugged impeller.

The level controls are set for a minimum of two minutes' pump running time to protect the pumps from the increased wear that results from too frequent pump starts.

---

31. *Volute* (vol-LOOT). The spiral-shaped casing that surrounds a pump, blower, or turbine impeller and collects the liquid or gas discharged by the impeller.
32. *Impeller.* A rotating set of vanes in a pump or compressor designed to pump or move water or air.
33. *Total Dynamic Head (TDH).* When a pump is lifting or pumping water, the vertical distance (in feet or meters) from the elevation of the energy grade line on the suction side of the pump to the elevation of the energy grade line on the discharge side of the pump. The total dynamic head is the static head plus pipe friction losses.
34. Two-port impeller means water being pumped enters on both sides of the impeller.

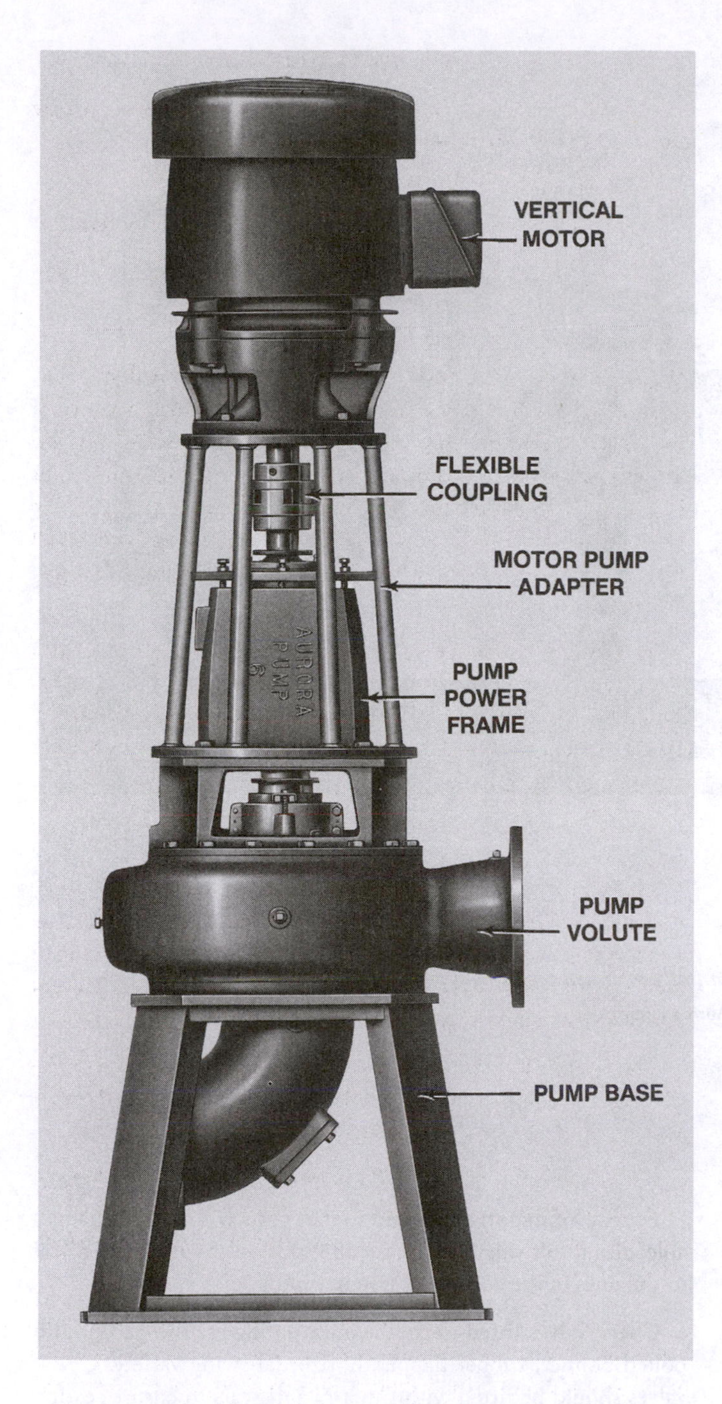


*Fig. 6.55 Flexibly coupled, vertically mounted centrifugal pump*
(Courtesy of Aurora Pump)

The pumps should have shutoff valves on both the suction and discharge piping so the pumps can be removed for maintenance without affecting the vacuum level.

Check valves are used on each pump discharge line and on a common manifold after the discharge lines are joined to it.

Equalizing bleeder lines should be installed on each pump. Their purpose is to remove air from the pump and equalize the vacuum across the impeller. In addition, they will prevent the loss of *PRIME*[35] should a check valve leak. Since this setup will result in a small part of the discharge flow being recirculated to the collection tank, a decreased net pump capacity results. Clear PVC pipe is recommended for the bleeder lines because small air leaks and blockages will be clearly visible to the system operator. On small discharge pumps (generally less than 100 GPM), the equalizing lines should be fitted with motorized full-port valves that close when the pumps are in operation to reduce recycle flow.

### 6.5123 *WASTEWATER LEVEL CONTROL PROBES*

The wastewater discharge pumps and alarms are often controlled by seven probes inside the collection tank. These probes are ¼-inch stainless steel with a PVC coating. They perform the following functions:

1. Ground probe
2. Both discharge pumps stop
3. Lead discharge pump start
4. Lag discharge pump start
5. High level alarm
6. Reset for probe #7
7. High level cutoff—stops all discharge pumps (auto position only) and vacuum pumps (auto and manual positions)

Figure 6.57 presents approximate elevations of these probes in the collection tank in relation to the discharge pumps and incoming vacuum mains.

An acceptable alternative to the seven probes is a single capacitance-inductive type probe capable of monitoring all seven set points. This type of probe requires a transmitter/transducer to send a signal to the motor control center (MCC) and is a more complex system to maintain.

### 6.5124 *MOTOR CONTROL CENTER*

The motor control center (MCC) houses all of the motor starters, overloads, control circuitry, and the hours-run meter for each vacuum and wastewater pump. The vacuum chart recorders, collection tank level control relays, and an alarm system are also normally located within the motor control center.

Chart recorders for both the vacuum and wastewater pumps are needed so that system characteristics can be established and monitored. Like gauges, these recorders are vital in the troubleshooting process. The vacuum pump recorder should be a seven-day circular chart recorder with a minimum chart diameter of 12 inches. The recording range should be 0 to 30 inches of mercury vacuum, with the 0 position at the center of the chart. The chart recorder should have stainless-steel bellows.

35. *Prime.* The action of filling a pump casing with water to remove the air. Most pumps must be primed before start-up or they will not pump any water.

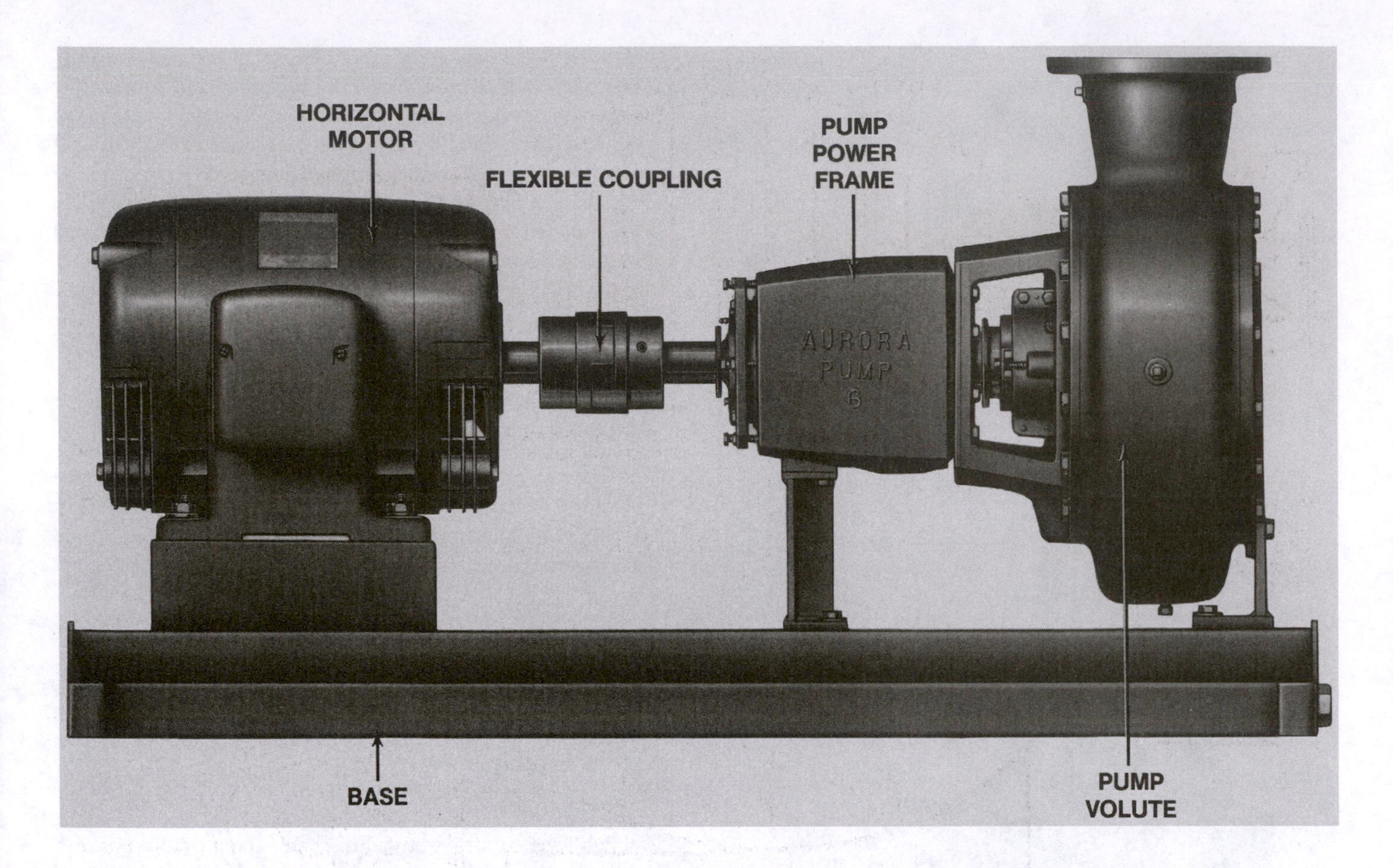


*Fig. 6.56 Flexibly coupled, horizontally mounted centrifugal pump*
(Courtesy of Aurora Pump)

### 6.5125 *STANDBY GENERATOR*

A standby generator is a must. It ensures the continuing operation of the system in the event of a power outage. The generator should be capable of providing 100 percent of the standby power required for the station operation. It typically is located inside the station, although generators housed in an enclosure outside the station and portable generators are common.

### 6.5126 *VACUUM STATION PIPING*

Vacuum station piping includes piping, valves, fittings, pipe supports, fixtures, and drains involved in providing a complete installation. The station piping should be adequately supported to prevent unnecessary strains or loads on mechanical equipment or sagging and vibration. The piping also should be installed in a manner that permits expansion, venting, drainage, and accessibility of lifting eyes for removal of equipment, tank valves, and pipe segments for maintenance. For fiberglass tanks, all piping must be supported so that no weight is supported by the tank flanges. Flange bolts should only be tightened to the manufacturer's recommendations to prevent damage to the fiberglass tanks.

For ease of maintenance and so that parts will be interchangeable, all shutoff valves in the vacuum station should be identical to the ones in the collection system piping.

Check valves fitted to the vacuum piping are the 125-pound bolted bonnet, rubber flapper, horizontal swing variety. Check valves should be fitted with Buna-N soft seats to ensure seating and prevent leakage.

Check valves fitted to the wastewater discharge piping should be supplied with an external lever and weight to ensure positive closing. They also should be fitted with soft rubber seats.

### 6.5127 *FAULT MONITORING SYSTEM*

A fault monitoring system is needed to alert the operator to any changes from the normal operating guidelines, such as a low vacuum level. The fault monitoring system activates an automatic alarm system. A voice communication-type automatic telephone dialing alarm system is usually provided and mounted on a wall next to the motor control center. The system should be self-contained and capable of automatically monitoring up to four independent alarm conditions.

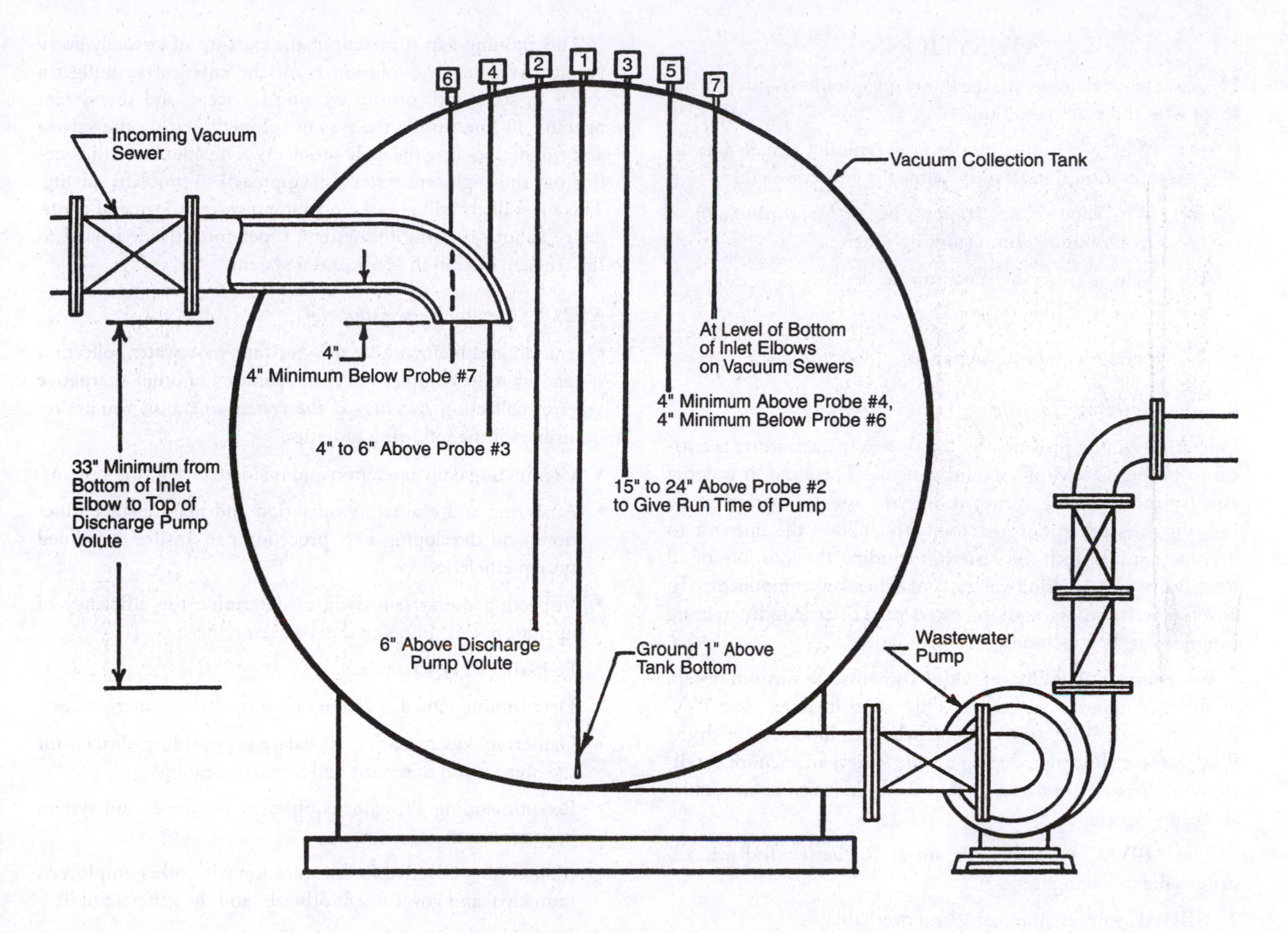


*Fig. 6.57 Typical elevations of level control probes*

(Drawing provided courtesy of AIRVAC)

The monitoring system should, upon the opening of any one alarm point, access the telephone lines, wait for the dial tone, and begin to dial the first of four field-programmed telephone numbers. The system will then deliver a voice message indicating a two-digit station number and the fault status at that station. The message will be repeated a preset number of times with sufficient space between messages to allow the called individual to acknowledge receipt of the call. Acknowledgment of the message is accomplished by pressing a touch tone star (*) key on the telephone between messages. Following the acknowledgment, the system will sign off and hang up. The system then enters a 30-minute delay to allow adequate time for follow-up measures to be taken. If another fault occurs during the 30-minute delay period, the system will begin calling again. Additionally, the system can be called at any time from a standard telephone, whereupon it will answer the call and deliver a vocalized (spoken) message indicating the station number and fault status at the location.

If the delay period passes and faults still exist, the system will begin dialing in one-minute intervals attempting to deliver the fault message. If no acknowledgment is received, the system will hang up, wait 60 seconds, and call the next priority number. After dialing the last priority number, the system will, if necessary, return to the first priority number and repeat the sequence indefinitely.

If the monitoring system is to be housed in the motor control center, provisions must be made to isolate the system from interference. Also, the monitoring system should be provided with continuously charged batteries for 24 hours of standby operation in the event of a power outage.

## QUESTIONS

Please write your answers to the following questions and compare them with those on page 346.

6.51M Why are wastewater pumps typically installed at an elevation significantly below the collection tank?

6.51N Why should level controls be set for a minimum of two minutes' pump running time?

6.51O Why is a standby generator needed?

### 6.52 Operation and Maintenance

#### *6.520 Operator Training*

Operator training provided by the system manufacturer is critical to proper O&M of vacuum systems. Therefore, it is desirable for the operating agency to hire the system operator while the system is under construction. This allows the operator to become familiar with the system, including the locations of all lines, valve pits, division valves, and other key components. To add further training, manufacturers may offer lengthy training programs at their facilities.

For example, AIRVAC provides a small-scale vacuum system at their manufacturing center. This setup includes clear PVC pipe with various lift arrangements where trainees can watch the flow inside a clear pipe during a wide variety of vacuum conditions. Faults are simulated so that the trainee can gain troubleshooting experience.

The AIRVAC training program at Rochester, Indiana, addresses the following topics:

1. AIRVAC valve maintenance and overhaul
2. AIRVAC controller/sensor unit overhaul and adjustment
3. Use of the AIRVAC controller test box in troubleshooting the controller, testing the breather line, and adjusting the valve and sensor for proper performance (starting and stopping)
4. Collection station maintenance
5. Installation of AIRVAC valves, valve pit, holding tanks, crossovers, and sewers
6. Troubleshooting procedures—faults are set up in the demonstration rig for trainees to locate and correct
7. Recordkeeping

The best training is gained by actual operating experience. Many times, however, the knowledge is gained at the expense of costly mistakes. This is especially true at start-up time. During this time, the engineer, who provided day-to-day inspection services during construction, is gradually spending less time on the system. The operator is busy setting (adjusting the operation of) vacuum valves and inspecting customer hookups. During this start-up period, the operating characteristics of the system continually change until all of the customers are connected and all of the valves are fine-tuned. However, with the operator(s) being preoccupied with other tasks, this fine-tuning sometimes is not done, problems develop, and both customers and operators can lose faith in the system.

This training gap is present at the start-up of virtually every vacuum system. One solution is for the engineer to budget a three- to six-month on-site training service to aid the system operator in fine-tuning the system (properly setting the valves) and troubleshooting the early problems. The operator will benefit from the engineer's systematic approach to problem solving. This most likely will give the operator a certain degree of confidence about operating the system. Operator attitude is vital to the efficient operation of a vacuum system.

#### *6.521 Operator Responsibilities*

Operation and maintenance of a vacuum wastewater collection system is a more complex job than operation of other alternative types of collection systems. As the system operator, you are responsible for the following activities:

- Conducting daily operation and maintenance of the system
- Analyzing and evaluating operation and maintenance functions and developing new procedures to ensure continued system efficiency
- Inspecting the system daily to determine the efficiency of operation, cleanliness, and maintenance needs
- Preparing work schedules
- Determining remedial action necessary during emergencies
- Gathering and reviewing all data and recording all data for the preparation of reports and purchase requests
- Recommending all major equipment purchases and system improvements
- Maintaining effective communication with other employees, municipal and government officials, and the general public
- Preparing operational reports and maintenance reports

#### *6.522 Understanding How the System Works*

To operate and maintain a vacuum collection system, an operator must have a good understanding of how the vacuum transport process works. Let us briefly review this process. We will assume the collection mains have been installed using the sawtooth type of lift construction shown previously in Figure 6.46 on page 311.

With the sawtooth profile, and as long as no vacuum valves are operating, no wastewater transport (flow) takes place. All wastewater remaining in the collection system lines will lie in the low spots. Since the wastewater will not seal the bore of the pipe in these static conditions, little vacuum loss is experienced throughout the system when low or no flow is occurring.

When a sufficient volume of wastewater accumulates in the sump, the vacuum valve cycles. The difference in pressure that exists between the vacuum sewer main and the atmosphere propels (forces) the sump contents and a quantity of air into the main. As the moving wastewater picks up speed, it quickly becomes a frothy foam and soon occupies only part of the pipe cross section. At this point, the transfer of momentum from air to water takes place. The strength of the propulsive force starts to decline noticeably when the vacuum valve closes, but remains important as the air continues to expand within the pipe. Eventually, friction and gravity bring the wastewater to rest at a low spot. Another valve cycle (which forces sump wastewater into the main), at any location upstream of the low spot, will cause this wastewater to continue its movement toward the vacuum station.

Vacuum systems are designed to operate on two-phase (air/liquid) flows with the air being admitted for a time period twice that of the liquid. Air to liquid ratios can be changed by adjusting the open time of the vacuum valve using the adjustment screw on the valve. A typical open time is six seconds.

Normally, the vacuum pumps are set to operate at 16 to 20 inches of mercury (Hg) vacuum. At 16 inches of mercury vacuum, the total available head loss is 18 feet; 5 feet of this head loss is required to operate the vacuum valve, leaving 13 feet available for wastewater transport (Figure 6.58). The combined total of friction losses and static lift (Figure 6.59) from any one point on the sewer network to the vacuum station must not exceed 13 feet. Table 6.3 shows the recommended lift heights for various pipe sizes.

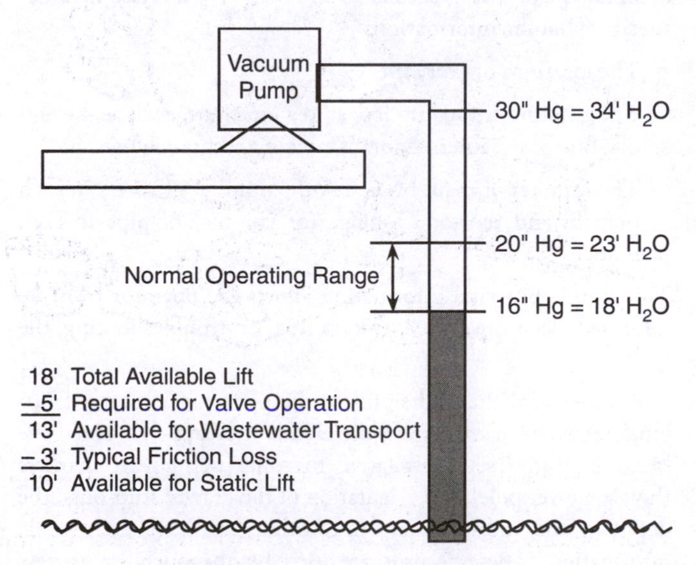


*Fig. 6.58 Vacuum lift capacity*

(Source: *ALTERNATIVE WASTEWATER COLLECTION SYSTEMS*, Technology Transfer Manual, US Environmental Protection Agency)

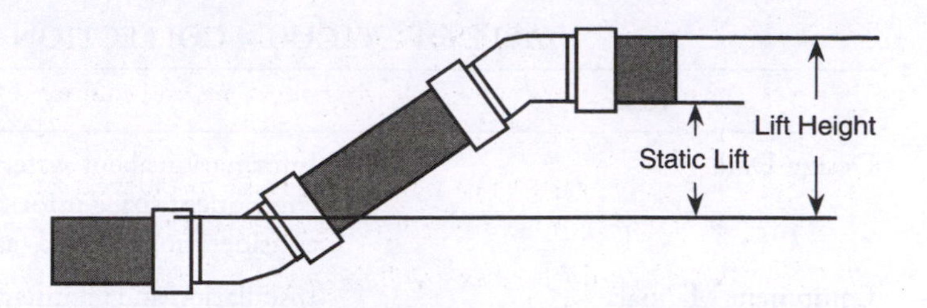


*Fig. 6.59 Static lift determination*

(Source: *ALTERNATIVE WASTEWATER COLLECTION SYSTEMS*, Technology Transfer Manual, US Environmental Protection Agency)

**TABLE 6.3 RECOMMENDED LIFT HEIGHT[a]**

| Pipe Diameter (in) | Lift Height (ft) |
|---|---|
| 3 | 1.0 |
| 4 | 1.0 |
| 6 | 1.5 |
| 8 | 1.5 |
| 10 | 2.3 |

a From *ALTERNATIVE WASTEWATER COLLECTION SYSTEMS*, Technology Transfer Manual, US Environmental Protection Agency.

### 6.523 *Operation and Maintenance (O&M) Manual*

An O&M manual should be assembled specifically for each vacuum collection system. A well-written O&M manual should contain the information necessary to achieve the following goals:

- To provide an accessible reference for the wastewater collection system operators in developing standard operating and maintenance procedures and schedules
- To provide a readily available source of data, including permits, design data, and equipment shop drawings of the particular system
- To provide the system operators assistance and guidance in analyzing and predicting the system efficiency
- To provide the system operators assistance and guidance in troubleshooting the system

While an O&M manual is a valuable tool, it should not be viewed as the source of a solution to every problem. Efficient operation of the system also depends on the initiative, ingenuity, and sense of responsibility of the system operator. To be most useful, however, the manual should be constantly updated to reflect actual operational experience, equipment data, problems, and solutions.

Table 6.4 lists the minimum information that should be included in the O&M manual for a vacuum collection system. (Also see Chapter 4, Appendix C, for information that can be

TABLE 6.4 VACUUM COLLECTION SYSTEM O&M MANUAL CONTENTS

| Topic | What Should Be Included |
|---|---|
| Design Data | Information about system make-up, such as the number of valves, line footage, and line sizes; component sizing information; anticipated operating ranges; any other important design considerations; record (as-built) drawings showing all system components |
| Equipment Manuals | Installation and maintenance manuals from the manufacturers of the major equipment; a list showing the manufacturer and supplier as well as contact persons, addresses, and phone numbers |
| Warranty Information | All warranties, including effective dates |
| Shop Drawings | A list of all approval drawings and a copy of each drawing identifying the manufacturer, model number, and a general description of the equipment |
| Permits and Standards | All permits, such as the National Pollutant Discharge Elimination System (NPDES) permit; applicable water quality standards |
| Operation and Control Information | Description of the overall system and identification of the major components; a list of the following information for each major component: relationship to adjacent units, operation, controls, problems and troubleshooting guides, maintenance, preventive maintenance schedule, and equipment data sheet |
| Personnel Information | Staffing requirements, including qualifications and responsibilities |
| Records | A list of the type of records as well as a list of important reference materials |
| Preventive Maintenance | Established maintenance schedules; list all equipment and cross-reference each item to equipment catalogs |
| Emergency Operating and Response Plan | A description of actions and responses to be followed during emergency situations; a list of contact persons, including addresses and phone numbers, of persons responsible for various community services |
| Safety Information | A safety plan that includes practices, precautions, and reference materials |
| Utility Listing | A list of all utilities in the system area, including contact persons, addresses, and phone numbers |

used to help you develop an O&M manual for your collection system.)

Much of the information contained in the O&M manual can easily be recorded on a computer database program. Computer software is becoming more affordable and can provide powerful aids to the O&M staff in small communities.

It is common in the industry for changes to be made during construction. Any such changes should be recorded on the record (as-built) drawings. As the name implies, these drawings show exactly how the system was built. This is a vital tool to the operating agency for maintenance, troubleshooting, and future improvements or extensions to the system.

An index map showing the entire system should be included in the record drawings. Shown on this map will be all key components, line sizes, line identifications, valve pit numbering and locations, and division valve locations. Detailed plan sheets of each line of the collection system should be included, with dimensions necessary to allow the operator to locate the line as well as all related appurtenances.

Unique to a vacuum system is the need for an as-built hydraulic map. This is similar to an index map but also includes special hydraulic information:

- The locations of every lift
- The amount of vacuum loss at key locations, such as the end of a line or the intersection of a main and branch line
- The number of main branches, the number of valves in each branch, and the total footage (or volume) of pipe in each branch

This simple but vital information allows the operator to make informed decisions when fine-tuning or troubleshooting the system.

Another tool that is helpful to the operator is a record (as-built) drawing of each valve pit. These drawings will show the location of the setting relative to some permanent markers (house, power pole), the orientation of the gravity stub-outs, the depth of the stub-outs, and any other pertinent site-specific information. These records are used by the operator as new customers connect to the system.

The vacuum station drawings also should be altered to reflect changes made during construction, especially any changes in dimensions, since any future modification will depend on available space.

## QUESTIONS

Please write your answers to the following questions and compare them with those on pages 346 and 347.

6.52A Why should the operator of a vacuum collection system be present during the construction of the system?

6.52B How does the vacuum transport process work?

6.52C A well-written O&M manual should contain the information necessary to achieve what goals?

6.52D Why should the O&M manual be constantly updated?

6.52E What information should be included on an as-built hydraulic map?

### *6.524 Spare Parts*

For optimum operating efficiency, a supply of spare parts for the collection system should be kept on hand. Some of the spare parts, such as fittings and pipe, can be purchased through local builder's supply companies. However, some parts are unique to vacuum systems and cannot be purchased locally. Table 6.5 is a list of recommended spare parts.

TABLE 6.5 SPARE PARTS LIST PER EVERY 50 VALVES[a]

| Quantity | Part |
|---|---|
| 2 each | Vacuum valves |
| 4 each | Controllers/sensors |
| 2 each | Sensor/cleanout tubes |
| 2 each | Controller/sensor rebuilding kits |
| 5 each | Valve cycle counters |
| 4 each | 3-inch no-hub couplings |
| 2 each | Valve pits |
| 2 each | Valve pit bottom plates |
| 1 each | Standard collection sump |
| 1 each | Deep collection sump |
| 2 each | Valve pit covers |
| 5 feet | Clear valve tubing: ⅜-inch |
| 10 feet | Breather tubing: ⅝-inch |

a From *ALTERNATIVE WASTEWATER COLLECTION SYSTEMS*, Technology Transfer Manual, US Environmental Protection Agency.

In addition to spare parts, certain specialty maintenance tools and equipment are needed; these are listed in Table 6.6.

TABLE 6.6 SPECIALTY MAINTENANCE TOOLS AND EQUIPMENT FOR COLLECTION SYSTEMS[a]

| Quantity | Part |
|---|---|
| 1 each | Portable vacuum pump |
| 2 each | Portable vacuum chart recorders |
| 100 each | Vacuum charts |
| 3 each | Chart pens |
| 2 each | 0–20 in W.G. magnahelic gauges |
| 2 each | 0–50 in W.G. magnahelic gauges |
| 1 each | 12-volt DC submersible pump |
| 15 feet | Pump discharge hose |
| 2 each | No-hub torque wrenches |
| 2 each | Calibrated vacuum gauges |
| 1 each | Controller test box |
| 1 each | Pipe locator |

a From *ALTERNATIVE WASTEWATER COLLECTION SYSTEMS*, Technology Transfer Manual, US Environmental Protection Agency.

The vacuum station also requires spare parts. These range from spare pump seals to fuses. Specialty items that should be considered are listed in Table 6.7.

TABLE 6.7 SPECIALTY EQUIPMENT FOR VACUUM STATION[a]

| Quantity | Part |
|---|---|
| 1 each | Inductance probe |
| 1 each | Probe transmitter |
| 1 each | Probe microprocessor card |
| 2 each | Vacuum switches |
| 2 each | Vacuum gauges |
| 1 each | Auto dialer microprocessor card |

a From *ALTERNATIVE WASTEWATER COLLECTION SYSTEMS*, Technology Transfer Manual, US Environmental Protection Agency.

Especially vital for the vacuum station are spare microprocessor-based electronic components. This type of equipment is used for the level controls and the fault monitoring systems and is very sensitive to power spikes. These system components are critical to the station, as they essentially operate and monitor the system. Losing level controls due to some type of failure will quickly cause severe operational problems such as loss of vacuum or discharge pump malfunctions.

Not having spare equipment only makes a problem worse because the system will need to be operated manually. This would require an operator on a continual basis (until the spare part arrived) to manually cycle both the vacuum pumps and the wastewater pumps. Loss of the fault monitoring system also increases operator O&M time and response time to system malfunctions.

### *6.525 Routine Operation and Preventive Maintenance*

Routine operation includes visiting the central vacuum station daily. Some daily maintenance procedures include recording pump running hours and checking oil and block temperatures. Once you are familiar with the operating characteristics of your system, a simple visual check of the gauges and the charts in the station will alert you to any problems. This visual check, along with recording operating data, generally takes about 30 minutes. Weekly procedures include checking battery terminals, battery conditions, and operational testing of the standby generator. Monthly procedures include cleaning the collection tank sight glass, checking the mechanical seal pressurizers on the wastewater discharge pumps, and testing the telephone alarm system.

A properly designed vacuum station will be equipped with a fault monitoring or alarm system, such as a telephone dialer (alarm system). This system monitors the operation of both the vacuum station and the collection system and automatically notifies the operator of low vacuum, high levels of wastewater in the collection tank, and power outages.

On an average day, you will not be required to visit the onsite system components or the collection system mains. Normal station gauge and chart readings are an indication that the collection system is fine. Depending on a system's history of breakdown maintenance, some periodic inspection may be required. This would include inspecting and manually operating each valve at some regular interval. The breather lines should be inspected for the accumulation of moisture. With a little experience, you will quickly learn the sounds of a properly functioning valve. Table 6.8 summarizes the daily, weekly, and monthly tasks you will be expected to perform.

Wastewater collection systems operate and must be maintained 365 days a year. Variations in operation and maintenance workloads occur, making it imperative that preventive maintenance be planned and scheduled. This will ensure that there is no idle time during off-peak workload periods. Inspection and maintenance planning and scheduling involve time, personnel, equipment, costs, work orders, and priorities.

A preventive maintenance schedule for all major equipment should be developed and a work order system should be established. This system identifies the required work, indicates the priority of each task, and provides any other information such as special tools or the parts required for the job. Work orders provide a means to initiate needed maintenance as well as a record of work completed.

Scheduled maintenance on the collection piping should be minimal. Areas where difficult or unusual conditions were encountered during construction should be visited periodically. Other pipeline areas to be visited include steep slopes and potential slippage areas.

At least twice a year, the division (isolation) valves should be exercised. This is done by moving the valve through the entire opening and closing cycle at least once. This procedure will keep valves in operating condition. In addition, it will familiarize the operating personnel with the location of all valves.

All vacuum valves should be inspected once a year. They should be manually cycled to see that they are operating properly. Record the controller timing cycle and compare it to the original setting. If necessary, reset the timing. This entire procedure can be done by one person; it requires about 10 to 15 minutes per valve.

About every five to six years, each vacuum valve should be removed and a spare put in its place. Take the old valve to the workshop where it can be taken apart and inspected for wear. If needed, replace the seat. When the valve is reassembled, install a new shaft seal and bearing. The seals and diaphragms of the controller/sensor should be replaced every five years. One person can examine and repair a valve in about 45 minutes and a controller in about an hour.

Always follow the manufacturers' recommendations for preventive maintenance on the major equipment at the vacuum station. Yearly maintenance might include removal from service and a thorough inspection of check valves, plug valves, vacuum pumps, wastewater pumps, standby generator, and the telephone dialer alarm system.

**TABLE 6.8 TYPICAL VACUUM STATION OPERATING AND MAINTENANCE TASKS**

## DAILY

**Motor Control Center**

Visually check charts (flow/vacuum) and record readings for previous 24 hours

Record vacuum pumps' run time

Record wastewater pumps' run time

Check equipment switch positions for proper settings, power available/tripped breakers

Check panel ground fault

Test alarm annunciator

Check building environment, lighting, heating, air conditioning

**Vacuum Pumps**

Operation

Vibration, oil and block temperatures, motor temperatures

Check and record vacuum gauge readings

Check heat exchanger: temperature and water flow

Check oil reservoirs: level and temperature

If belt driven, look for broken belts, worn belts (fine dust or particulate matter under belts, sheaves, and guard)

**Vacuum Reservoir Tanks**

Vacuum gauge readings

Check sight glass or inspection cock for accumulated moisture

**Wastewater Collection Tank**

Vacuum gauge readings

Wastewater collection mains vacuum readings

Sight glass water level of tank

Valve positions

**Wastewater Pumps**

Operation

Gas bound (bleeder lines flowing with pump running)

Vibration when pumping check valve unseats

Discharge pressure of each pump and force main

Pump/motor temperature

Pump shutoff, check valve seats

Position of pump suction and discharge valves (full open)

Pump shaft mechanical seal, leaking seal, fluid reservoir at proper level, check seal piping and valves for leaks and proper positions

**Building Sump**

Water level in sump

Sump dewaters, discharge valve seats

## WEEKLY

Change vacuum/flow recorder charts

Alternate lead/lag vacuum pumps

Alternate lead/lag wastewater pumps

Check all fluid reservoirs for proper levels

Check battery conditions on all emergency systems, generator, lighting, alarm

Check generator oil, fuel, coolant tanks

Operate generator under load for one-half hour

House cleaning: sweep out motor control center (MCC) room, hose down vacuum pump and wastewater pump rooms

Dust equipment

Clean sliding-vane vacuum pumps' inlet filters

Mow lawns

## MONTHLY

Complete preventive maintenance schedule for month

Prepare monthly reports

Check building security, door locks, fences, gates, fence grounding connections

Review data on reports for normal/abnormal readings or trends. Investigate unusual conditions such as irregular equipment operating times, high flows occurring on certain weekdays or times, unusual maintenance problems

Clean grounds, prune shrubs

Table 6.9 summarizes the preventive maintenance tasks and their recommended frequencies.

**TABLE 6.9 PREVENTIVE MAINTENANCE TASKS AND FREQUENCIES**[a]

| |
|---|
| **Every 6 Months** |
| Exercise division valves |
| **Once Each Year** |
| Inspect vacuum valves |
| Check controller timing |
| Check plug and check valves at station |
| Inspect vacuum and wastewater pumps for wear |
| Inspect standby generator |
| Inspect dialer alarm system |
| **Every 5 to 6 Years** |
| Remove valve and replace worn parts |
| Rebuild controller |

a Adapted from *ALTERNATIVE WASTEWATER COLLECTION SYSTEMS,* Technology Transfer Manual, US Environmental Protection Agency.

### *6.526 Emergency Maintenance*

Although a vacuum collection system requires very little operator maintenance on a day-to-day basis, there will be times when emergency maintenance is necessary. Responding to emergencies usually requires more than one person, particularly when it involves searching for a malfunctioning valve. Emergency or breakdown maintenance may involve the piping system, the vacuum station, or the vacuum valves.

Assuming proper design and construction, there is very little that can go wrong in the piping system. Occasionally, a line break caused by landslides or excavation for other utilities will cause a loss of system vacuum. Using the division valves, you can easily isolate the defective section and make the necessary repairs.

Malfunctions at the vacuum station are generally caused by pump, motor, or electrical control breakdowns. Extra equipment (for example, having two identical pumps, vacuum valves, or motors) allows for the continued operation of the system when this occurs.

Most emergency maintenance is related to malfunctioning vacuum valves caused by either low system vacuum or water in the controller. Failure of the valve is possible in either the closed or open position. A valve failing in the closed position will give the same symptoms as a blocked gravity line, that is, the customer will experience problems with toilet flushing or a backup of wastewater on the property. A phone call from the affected party makes diagnosing this problem relatively easy. However, nearly all valve failures occur in the open position, usually due to leakage of water into the controller. When this happens, a loss of system vacuum occurs because the system is temporarily open to the atmosphere. The fault monitoring system will recognize this low-vacuum condition and alert the operator to the problem.

## QUESTIONS

Please write your answers to the following questions and compare them with those on page 347.

6.52F Why is it important to have spare parts on hand for the microprocessor-based electronic components of the system?

6.52G List the routine daily operating tasks for a vacuum sewer system.

6.52H Planning and scheduling inspection and maintenance activities involve what items?

6.52I Where can emergency or breakdown maintenance occur?

## 6.53 Troubleshooting Guides

The following material on troubleshooting a vacuum collection system was adapted from procedures developed by AIRVAC. The authors wish to thank AIRVAC for permitting reproduction of this material.

Malfunctions of the AIRVAC system can be divided into three groups: (1) vacuum collection lines, (2) vacuum valve, and (3) central vacuum station.

### *6.530 Vacuum Collection Lines*

Malfunctions of the collection lines may be further divided into three categories: (1) vacuum valve malfunction, (2) break in the vacuum lines, and (3) closed isolation valves.

1. Vacuum Valve Malfunction
   a. Valve failed to close. If the valve failed to close, it will show up as a low vacuum alarm.
   b. Valve failed to open. If the valve failed to open, it will show up the same as a blocked gravity lateral, that is, as the 30-gallon sump fills, the homeowner will experience problems in flushing toilets. When this happens, the homeowner should contact the system operator.
2. Break in the Vacuum Lines

   Usually due to excavation work in the area. Check utility companies for areas where work is in progress. Use isolation valves if necessary to locate leaks and repair.
3. Closed Isolation Valves

   An isolation valve may accidentally be left shut, in which case a section of vacuum line will not have vacuum. This will give the same symptoms as a valve(s) failed to open.

*NOTE:* A low vacuum does not indicate that wastewater flow has stopped. It simply indicates that the system vacuum has fallen below the start point of the vacuum pumps and that the pumps are in continuous operation.

When you reach the collection station that has signaled the low vacuum alarm, use the following procedure to locate and repair the leak. At the collection tank, isolate each incoming collection sewer in turn to determine which one is leaking. Close off the leaking collection line. Build up the vacuum in the other collection system lines to clear out as much wastewater as possible. Close off the nonleaking lines. Open the leaking line. Go to the division valve located halfway on the leaking collection system sewer. Close it off. Check on which side of the division valve the leak is located. Isolate the leak by continuing to open and close division valves. When the leak is isolated to a small area, drive to each valve pit, listen for the malfunctioning valve, and correct the problem.

If no vacuum valves are found to be malfunctioning, a break in the vacuum piping exists. Check for underground construction in the area by utility companies. It is possible that a collection system line has been cut.

The procedure described above appears lengthy. At a demonstration supervised by an EPA consultant on a vacuum sewer line 6,200 feet long fitted with only one isolating valve, the operator located and repaired the leak in 15 minutes.

As you gain experience with your system, you will be able to watch the central station vacuum gauge and know the approximate location of the malfunction.

### 6.531 *Vacuum Valve (see Figure 6.38, page 304)*

When a valve or controller is found to be defective, it should be replaced with another one and taken to the workshop for an overhaul. Use the following procedure to identify and correct problems with vacuum valves.

1. Valve Fails to Close
   a. Disconnect vacuum from the controller. If the valve closes, then (1) the controller is faulty; (2) pressure is present on the sensor due to a blocked sensor line or a blocked suction line; or (3) the breather line is broken, blocked, or restricted by dirt on the vent dome or a water trap in the breather line because of improper slope.
   b. If the valve still fails to close after checking item a above, the valve is faulty. Remove the valve and fit a spare. At the workshop, strip the valve and check for a blocked controller port, damaged shaft or bearing, rags or rocks jammed in the body, nuts off shaft, or other damage. Repair and put valve into spare inventory.
2. Valve Fails to Open
   a. Remove vacuum hose from the controller. Remove ⅝-inch breather hose from the controller air port and insert vacuum hose. If the valve opens, the problem is not in the valve.
   b. If an S model valve cycles when the ⅝-inch hose is removed, then pressure is present in the holding sump possibly due to a blocked gravity line. (See Figure 6.41 on page 307 for an illustration of a valve vented to a holding sump.)
   c. If the valve does not open, check to ensure 5-inch mercury (Hg) vacuum is available by fitting a vacuum gauge to the surge tank. If 5-inch Hg vacuum is not available, remove the surge tank to check if vacuum is available in the sewer. If vacuum is available to the sewer, the problem is in the surge tank. With no vacuum available at the sewer, the problem could be (1) station failure, (2) closed isolation valve, (3) damaged vacuum sewer, or (4) valve open at a different location.
   d. If the problem is not in the valve, reconnect the vacuum hose to the controller. Remove the hose from the 2-inch sensor cap and apply pressure at this hose to cycle the controller. If the valve opens, this indicates the problem is in the 2-inch sensor line. This line may be blocked or leaking. With the separate holding tank and valve pit installation, excavate to the 30-gallon tank and correct the line. On the combined holding tank and valve pit installation, pull out the 2-inch sensor line and inspect for blockage or leakage.
   e. If the valve fails to operate when pressure is applied to the sensor hose, the controller is faulty and should be replaced.

Valve failures, if not located and corrected within a couple of hours, may cause failures in other parts of the system. A valve that is stuck open or that continuously cycles will cause system vacuum to drop. If the vacuum pumps cannot keep up with this vacuum loss, there will not be enough vacuum to open other valves. This may lead to backups. When vacuum is finally restored, a large amount of wastewater (in relation to the amount of air) will be introduced into the system. This surge could cause a condition called "waterlogging" and the resulting loss of vacuum.

Loss of vacuum due to waterlogging is one of the most severe problems that can occur with the collection system. Waterlogging causes ever-decreasing vacuum levels beyond the waterlogged section. This may lead to insufficient vacuum to open the vacuum valves and ultimately to wastewater backup in the home. In most cases, waterlogging is caused by too much liquid (extra water) and not enough air entering the system at one time. This is a very difficult problem to correct since it typically is related to illegal, and difficult-to-find, stormwater connections by the customer. An aggressive infiltration/inflow (I/I) program during the hookup stage will keep problems such as this to a minimum. If I/I continues to be a problem, smoke or dye testing of the suspected service area will locate the source of the problem, but you will have to test each individual building sewer in that area.

When a system becomes waterlogged it must be manually operated, allowing the vacuum pumps to run longer than usual. The manual repetitive cycling is continued until the system catches up. At that point, the system can be returned to its automatic mode.

The following procedure is a systematic approach to locating the source of vacuum loss. The time for the first four steps is generally about two to three minutes, while the entire procedure can usually be completed within 30 minutes. Sometimes, a shortcut can be taken. Experience has shown that it is often the same valve(s) that fails repeatedly. This is usually due to some particular hydraulic condition at the specific location. In these situations, the operators check these valves before any other isolation is done.

Another shortcut sometimes used by a skilled operator is estimating how far the problem site is from the vacuum station by analyzing the vacuum charts at the station. The degree of vacuum loss is inversely proportional to distance from the station. That is, the greater the vacuum loss, the closer the problem is to the station. Knowing the approximate location of the problem greatly simplifies the job of isolating that part of the system.

To help you diagnose system problems, Figure 6.60 lists three common symptoms of vacuum system problems; the numbers following each symptom refer to the numbered problems described in Table 6.10.

### 6.532 *Central Vacuum Station*

Table 6.11 (page 335) identifies problems that may occur in the central vacuum station, describes what symptoms to look for, and identifies the probable cause of the problem.

## QUESTIONS

Please write your answers to the following questions and compare them with those on page 347.

6.53A What are the three categories of collection line malfunctions?

6.53B What is the meaning of "waterlogging"?

6.53C List two common ways a vacuum sewer system may become waterlogged.

| Symptom/Problem | Conduct Tests in Sequence from Left to Right. | | | | | | | | | |
|---|---|---|---|---|---|---|---|---|---|---|
| AIRVAC VALVE WILL NOT OPEN | 3.1 | 6.1 | 1.1 | 1.3 | 1.2 | 1.4 | 2.1 | 4.1 | 5.1 | 1.8 |
| AIRVAC VALVE CYCLES | 5.2 | 7.1 | 7.2 | 4.3 | 2.5 | | | | | |
| AIRVAC VALVE WILL NOT CLOSE | 1.4 | 1.5 | 1.6 | 1.7 | 1.8 | 2.2 | 4.2 | 5.2 | 1.2 | |

*Fig. 6.60 Vacuum collection system troubleshooting chart*

(Reprinted courtesy of AIRVAC)

TABLE 6.10 VACUUM COLLECTION SYSTEM TROUBLESHOOTING GUIDE

| Component | Problem | Symptom or Effect |
|---|---|---|
| 1. AIRVAC Valve. The purpose of the AIRVAC valve is to isolate vacuum from the gravity lateral. | 1.1 Nuts or bolts off shaft | 1.1 Valve will not open |
| | 1.2 Shaft out of round, nicked, or dirt buildup | 1.2a Valve may not open<br>1.2b Valve may not close |
| | 1.3 Torn rubber diaphragm | 1.3 Valve will not open |
| | 1.4 Foreign material in wye body | 1.4a Valve will not open<br>1.4b Valve will not close |
| | 1.5 Defective bearing | 1.5 Hanging valve on downward travel preventing closing |
| | 1.6 Broken seat preventing valve from closing | |
| | 1.7 Not adequate vacuum to seat valve causing vacuum leak | |
| | 1.8 Blocked dip tube or lower housing ports | 1.8a Valve will not open<br>1.8b Valve will not close |
| 2. AIRVAC Controller/ Sensor Unit. The purpose of the AIRVAC controller/ sensor unit is to activate the valve. | 2.1 Valve will not open | 2.1a Low vacuum (sump vent valve closed if present)<br>2.1b Sensor air pressure blocked<br>2.1c Sensor diaphragm damaged<br>2.1d 0.016-inch orifice blocked or tube kinked<br>2.1e Water in sensor chamber<br>2.1f 0.093-inch air passage orifice blocked<br>2.1g Leaking valve diaphragm |
| | 2.2 Valve will not close | 2.2a Water in timing volume and needle valve<br>2.2b 0.093-inch air passage orifice blocked<br>2.2c Needle valve closed or blocked<br>2.2d Shaft in controller sticking<br>2.2e Sensor seat leaking<br>2.2f 0.016-inch orifice blocked or tube kinked |
| | 2.3 Water in sensor line and controller fails | 2.3a Leaking sensor tubing connection or sensor diaphragm. Clamps on all connections<br>2.3b Leaking surge suppressor |
| | 2.4 Unable to adjust controller for long timing (12 seconds) | 2.4a Leak in 3-way valve diaphragm loose on shaft (closing needle valve does not prevent controller from timing out)<br>2.4b Cracked diaphragm plate<br>2.4c Leaking Chamber B seal #4 (in field use when the valve is cycled, Chamber B vacuum will drop faster than usual because of dropping line vacuum if seal is leaking)<br>2.4d Bad controller check valve—leaking<br>2.4e Check valve in vacuum line to controller is leaking |
| | 2.5 After valve is cycled and closes, it triggers again for a second short cycle | 2.5a Third seal is leaking<br>2.5b Bad surge suppressor or excessive sensor line back pressure (consult AIRVAC) |
| | 2.6 Covering valve connection and the atmosphere port sets off the controller | 2.6a Bad #2 shaft seal—leaking<br>2.6b Sensor seat leaking |
| | 2.7 Controller works but vacuum flow is low and vacuum is leaking through atmosphere port | 2.7 Bad shaft seal on air port #1 |

TABLE 6.10 VACUUM COLLECTION SYSTEM TROUBLESHOOTING GUIDE *(continued)*

| Component | Problem | Symptom or Effect |
|---|---|---|
| 3. Surge Tank (AIRVAC 3-inch Valve Only). The purpose of the surge tank is to prevent backwash due to differential pressure when the valve opens and the wastewater passes. | 3.1 Valve will not open | 3.1 Wastewater in the surge tank inlet is blocking vacuum flow or check valves are reversed |
| 4. 2-inch Sensor Line. The purpose of the 2-inch sensor line is to trap air within to operate the sensor. | 4.1 Valve will not open | 4.1a No-hub clamp installed incorrectly if present<br>4.1b Solvent-bonded joints leak<br>4.1c 2-inch line too close to bottom of tank. If sensor line is closer to the bottom of the tank than the 3-inch suction line, a vacuum may be created in the sensor line<br>4.1d ⅜-inch hose from 2-inch sensor cap to controller installed incorrectly or not clamped |
| | 4.2 Valve will not close | 4.2a Line incorrectly graded, creating blockage holding pressure on sensor line<br>4.2b 3-inch suction line blocked. Wastewater not being removed from tank |
| | 4.3 Valve cycles frequently | 4.3a Length of 2-inch line too long<br>4.3b Ground infiltration |
| 5. Breather Line. The purpose of the breather line is to supply atmospheric air for the controller and valve operation. | 5.1 Valve will not open | 5.1 Breather piping blocked |
| | 5.2 Valve will not close | 5.2 Breather piping blocked |
| 6. Vacuum. The purpose of vacuum is to operate the valve and to aid in the transport of wastewater. | 6.1 Valve will not open | 6.1a No vacuum at collection station<br>6.1b ⅜-inch vacuum hose blocked<br>6.1c Surge tank blocked<br>6.1d Isolation valve closed<br>6.1e Broken vacuum line<br>6.1f Less than 5 inches mercury vacuum available<br>6.1g Sump vent on S model valve closed |
| 7. Gravity Line. The purpose of the gravity line is to allow the wastewater from the home to flow by gravity to the AIRVAC valve holding tank. | 7.1 Valve cycles a second cycle after sensor is triggered | 7.1 Excessive back pressure on gravity line due to extremely long gravity line. Check or replace surge suppressor. Remove surge suppressor if problem persists; retry. If problem continues, call AIRVAC. |
| | 7.2 Valve cycles several cycles after sensor is triggered and when vacuum gauge is hooked into side. Check valve hose. Vacuum does not drop on air cycle | 7.2a Gravity line is not properly laid. Pockets in gravity line are collecting wastewater. When valve cycles, it empties holding tank, applies vacuum to gravity line and closes. Then, wastewater is pulled from gravity line pocket to tank, which triggers valve.<br>7.2b If no valves are malfunctioning, check for underground construction that could have caused a break in the transport piping. Walk the route of the problem sewer and look for evidence of a break, such as a sunken area. |

**TABLE 6.11 CENTRAL VACUUM STATION TROUBLESHOOTING GUIDE**

| Component | | Problem/Symptom | | Probable Cause |
|---|---|---|---|---|
| 1. Vacuum System | 1.1 | Loss of vacuum, pumps running | 1.1a | If the vacuum is low and the vacuum pumps are running, the leak is in one of the sewers. |
| | | | 1.1b | If no leaks are found in the sewers, the problem could be insufficient liquid inside the vacuum pumps (liquid-ring vacuum pumps only) or leaking check valves. |
| | 1.2 | Low vacuum reading, pumps not running | 1.2 | The vacuum pumps will be locked off by the high wastewater level probe, by overloads in the pump starters, by a faulty vacuum switch, or no electric power present. |
| 2. Discharge Pumps | 2.1 | Loss of prime | 2.1 | Loss of prime can be caused by:<br>a. Incorrect or faulty seats fitted to the force main check valves<br>b. Faulty mechanical shaft seals<br>c. Seal pressurizing system malfunction<br>d. Blocked equalizing line<br>e. Leaking gaskets between the check valve flanges and the pump discharge connection<br>f. Sand in collection tank |
| | 2.2 | Pumps locked off by the motor overloads | 2.2 | This situation may be caused by a blocked pump or a pump in which a mechanical failure has occurred. |
| 3. Other Problems That Could Occur | 3.1 | High wastewater level alarm operating. One wastewater discharge pump running, problem with other discharge pump | 3.1a | Failure of mechanical alternator |
| | | | 3.1b | Failure of second duty (or first duty) start electrode |
| | | | 3.1c | Blockage of pump and failure on overload |
| | 3.2 | Both wastewater discharge pumps running, vacuum pumps off, high wastewater level alarm operating, both pumps discharging a small volume of wastewater | 3.2 | Problem caused by a partial blockage of pump suction lines in collection tank. This problem can occur during system start-up if the contractor allows excessive sand, rocks, or debris to enter the system through a valve connection or a damaged sewer.<br>Remedy: shut down the system, open and clean the collection tank. |
| | 3.3 | Wastewater level probe malfunction | 3.3 | Grease insulating probe or pinholes in PVC probe insulation, which may cause probe to short circuit |

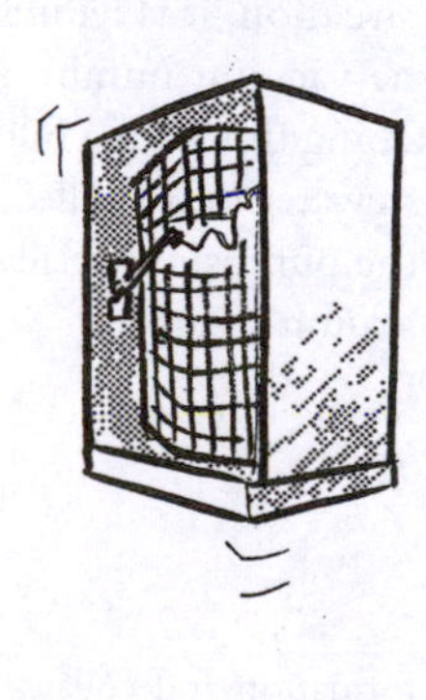

### *6.533 Troubleshooting Other Vacuum System Problems*

- **Discharge Pump Cavitation**[36]

  Some systems have experienced problems with wastewater discharge pump cavitation. This is due to the characteristics of the pump itself. A vortex plate installed in the bottom of the wastewater collection tank can correct this problem. The real solution lies in proper design. Net positive suction head (NPSH) calculations should be performed and a pump having sufficient NPSH available should be selected.

- **Leaking Check Valves**

  If check valves are installed in a vertical pipe run between the vacuum pumps and the reservoir tank, vacuum loss problems could occur in the station piping. Installing the check valves in a horizontal run of station piping will correct this problem.

- **Oversized Vacuum Pumps Keeping Up with Open Valve**

  If the vacuum pumps have sufficient capacity to keep up with an open valve, low vacuum will not be recorded at the station even though it is occurring at the location of the open valve. In addition, the automatic alarm system will not be activated. This results in increased run time of the vacuum pump and increased power costs. The operator must be very observant when analyzing the daily vacuum charts to notice that the pumps are cycling more frequently than normal. Otherwise, a valve that is hung open may go unnoticed for days. A simple solution to this problem is to install another relay in the control wiring, which will activate the alarm and call the operator if both pumps are operating at the same time for a predetermined length of time (for example, 10 minutes) regardless of the vacuum level.

- **Wastewater Being Pulled into Vacuum Pumps**

  A high wastewater level in the collection tank can be caused by pump failure (usually control related rather than a pump malfunction) or by more flow coming in than the wastewater pumps are capable of pumping out. Most system designs call for an automatic shutdown of both vacuum and discharge pumps if this happens. Some systems, however, do not have provisions to prevent the automatic mode from being overridden by manual operation.

  When faced with a zero-vacuum situation, it is natural for the operator to manually operate the vacuum pumps in an attempt to restore system vacuum. Doing this with a full collection tank, however, can cause wastewater to be pulled into the vacuum pumps. Wastewater in the pumps, especially the sliding-vane type, will damage the vacuum pumps.

  The proper procedure, assuming the discharge pumps work, is to valve off the incoming lines and turn the vacuum pumps off. Then, use the discharge pumps to pump the collection tank down to the normal operating level. Once the normal operating level is reached, open the line valves and return the vacuum pumps to automatic mode. Good design will include electrical provisions to prevent overriding the automatic mode without these steps first being taken.

- **Component Defects**

  Possible component defects include (1) broken spring valve failure in open position, (2) unreliable valve controllers, (3) controller shaft/seal problems in multifiring valves, (4) plug valve problems, and (5) defective tubing valve failure in open position. Any defective parts should be replaced. If a part continues to fail, try to locate another manufacturer or try to use a different device or approach.

- **Faulty Level Control Probes and Transmitter**

  A problem with level control probes sometimes occurs. The probes fail to send the proper signal to the motor control center to turn the vacuum pumps off during periods when high wastewater levels exist in the collection tank. This results in wastewater being pulled into the vacuum pumps and damaging them. Also, a faulty transmitter may fail to send the proper signal to the motor control center to turn off a discharge pump. This results in the pump continuing to run until the tank is dry. The faulty transmitter must be replaced.

  Since failure of the probe can lead to severe damage to the vacuum or wastewater pumps, the operator must develop a simple, inexpensive backup system to regulate the pumps. One such system involves strapping magnets to the sight glass above and below the highest and lowest set points of the probe. A float-mounted magnet (inside a plastic tube) located inside the sight glass moves with the level of the tank. If the probes fail, the floating magnets and one of the fixed magnets will meet, causing a circuit to close and starting or stopping the appropriate pumps.

- **Telephone Dialer (Alarm System) Malfunction—Interference**

  Electrical spikes (power surges) may cause interference in the motor control center, which causes the automatic telephone dialer to call the operator only to report that the system is all clear. Shielded cable added to the dialer wiring will not help. To correct the problem, the dialer must be taken out of the MCC and mounted on a nearby wall. Microprocessor-based

---

36. *Cavitation (kav-uh-TAY-shun).* The formation and collapse of a gas pocket or bubble on the blade of an impeller or the gate of a valve. The collapse of this gas pocket or bubble drives water into the impeller or gate with a terrific force that can knock metal particles off and cause pitting on the impeller or gate surface. Cavitation is accompanied by loud noises that sound like someone is pounding on the impeller or gate with a hammer.

equipment is very sensitive to power spikes. Provisions must be made to filter the power supply or to provide a separate smooth power supply to this kind of equipment.

- **Telephone Dialer Malfunction—Undetermined Cause**

  Telephone dialers (alarms) are considered a critical component in a vacuum station. Without a dialer, which provides 24-hour per day coverage, problems can go unnoticed until the next workday. While this significantly reduces the amount of operator overtime, it also causes increased pump run times. The operating personnel may feel that the system is reliable enough (that is, oversized vacuum pumps can keep up with a valve failure) and valve failures are so infrequent that replacing the dialers cannot be economically justified. This reasoning is a case of false economy. Allowing a low-vacuum situation to go undetected for hours will increase power costs and put unnecessary wear on the vacuum pumps, thereby shortening their useful life.

  A high wastewater level in the collection tank typically results in total system shutdown until the operator corrects whatever is wrong. With no dialer (alarm) to notify the operator of this situation, the system may be down for hours at a time. This will lead to waterlogging, which will require a significant effort to correct.

  Even more serious is the potential liability for damage to the customers' homes if wastewater backs up.

- **Water in Controller—Valve Failure in Open Position**

  Component defects were the main cause of valve failures in the past but improvements in the manufacture of valves has drastically reduced the failure rate. One of the remaining causes of valve failure is water in the controller. When water gets into the controller, it will prevent the controller from completing its cycle and the valve will remain open. This problem is more likely to occur with the Model S valve than with the Model D valve.

  In either the Model S or Model D valve, water can enter the controller in two general ways. First, water that is present in the upper valve chamber enters directly into the controller as a result of the controller not being air- or vacuum-tight. A tightness test is normally performed at installation, during the annual preventive maintenance visit, and any time emergency maintenance is performed on the valve. The other way water enters the controller is through condensation in the breather tubing. A properly installed breather line, with a drain leading to the valve, will prevent this second case from occurring.

  With the Model S valve, there is a third possibility. Water comes directly from the lower sump to the controller. The Model S valve requires a watertight seal between the upper and lower sections of the valve pit so that a bubble of air can be trapped for use by the valve controller during its cycle. If the seal is broken and water takes the place of air, the water will be drawn into the controller by vacuum when the valve cycles.

## QUESTIONS

Please write your answers to the following questions and compare them with those on page 347.

6.53D What problem can be caused by a leaking check valve?

6.53E What problems result when a low-vacuum situation is allowed to go undetected?

6.53F How can wastewater be pulled into vacuum pumps?

### *6.534 O&M Problems Related to Construction*

- **Solvent Welding in Cold Weather**

  Solvent welding PVC pipe in temperatures approaching or below freezing will lead to vacuum leaks. Some contractors make the mistake of keeping the glue warm, only to apply it to a pipe that is much colder. This procedure leads to leaking joints and difficulty in passing leakage tests. This problem can be avoided by not solvent welding joints when the temperature is 40°F (4°C) or colder. However, since this may not be practical for some jobs, a better solution would be to minimize the number of solvent-welded joints in the system. There has been a significant move in the vacuum industry toward the use of gasketed pipe and fittings. In all cases, installation of piping should be in accordance with the proper specification of the Plastic Pipe Institute, ASTM F645 or ASTM D2774. *NOTE:* ASTM International, formerly known as the American Society for Testing and Materials (ASTM), is a globally recognized leader in the development and delivery of international voluntary consensus standards (http://www.astm.org/).

- **Line Break**

  Line breaks caused by a trench settlement are most likely the result of inadequate bedding materials or poor compaction during construction.

- **Broken Fittings**

  There have been cases of loss of system vacuum due to broken fittings. In all but a few cases, the failures have occurred at fittings at or near the valve pit. This is most likely due to insufficient soil compaction around the valve pit coupled with system rigidity due to solvent-welded fittings.

- **Construction Debris—Valve Failure in Open Position**

  Vacuum collection systems experience more problems than others on start-up. Construction debris (stones, small pieces of pipe) in the homeowner's building sewer causes the vacuum valve to hang open until it is manually cleared. These problems usually disappear within a few weeks of start-up. Problems of this nature are easily discovered by the operator simply by listening to the auxiliary vent. Hearing the constant rushing of air is a good indication that the valve is hung open. Opening the valve pit, unscrewing the valve body, and clearing the obstruction is a 10-minute procedure.

- **Valve Pit Settlement**

  Poor workmanship by the contractor can lead to valve pit settlement. Settlement causes alignment problems for the owner at the time of the valve installation because the vacuum line entering the pit has moved from its level position. Improper valve alignment can lead to future valve malfunctions.

  Valve pit settlement problems can be avoided by quality control on both the contractor's and inspector's part during construction. Taking time to ensure proper alignment and proper compaction around the pit will greatly reduce the likelihood of this problem occurring.

- **Broken Cleanouts**

  Breaks in the fittings next to the cleanouts can occur if the pit settles due to poor compaction of the backfill around the pit.

- **Excessive Use of Fittings**

  Two fixed points, at varying inverts and locations, but requiring rigid connection piping, can result in the contractor using an excessive amount of fittings. These fittings can be located within the pit excavation. If there is a lack of compaction in the pit area, any future soil settlement could easily lead to fitting failures.

  The use of fittings in the service lines can be minimized by proper planning and coordination between line and pit crews. To minimize the difficulties, some contractors install the valve pits first. The use of gasketed fittings, which adds a certain degree of flexibility, will also help to minimize the number of fittings.

### *6.535 Odors, Corrosion, and Grease Accumulation*

Odors usually are not a problem with vacuum collection sewers for three reasons: (1) the system is sealed, (2) a large volume of air is brought into the system at each flow input, and (3) wastewater travel and detention times are short.

The entire system, from the valve pit setting to the vacuum station, is sealed. The valve pit sump containing the wastewater should be tested for tightness both at the factory and in the field after installation. The piping system contains no air releases, and the collection tank at the vacuum station, into which all of the sewer mains empty, is a vacuum-tight vessel.

A large amount of air is drawn into the system at each valve pit and sump. The air to liquid ratio, by volume, typically is 2:1 or higher. This amount of air throughout the collection system aids in the prevention of septic wastewater.

In a typical valve pit/sump, only a small amount of wastewater (about ten gallons) accumulates before the valve cycles. Once these small quantities of wastewater are drawn into the collection system main, the wastewater travels at a speed of more than 15 ft/sec. Compared to a gravity collection system, the wastewater remains in the collection system pipelines a relatively short time. The combination of air mixed with the wastewater and a short detention time usually prevents the wastewater from becoming septic and producing foul odors.

Odors may sometimes be a problem when concrete buffer tanks are used. Unlike fiberglass pits and sumps, concrete buffer tanks are open from the sump to the top of the pit. Although the volume of wastewater in the tank usually is not large enough to produce dangerous levels of hydrogen sulfide, gases can build up in the sewer lines to these tanks. Operating personnel, therefore, must take appropriate precautions to test the atmosphere and ventilate the area when performing maintenance. Also, since buffer tanks typically are used to level out large flows, some wastewater may well remain in the tank long enough to turn septic.

Corrosion has not been a problem in vacuum collection sewers because all of the system parts in contact with wastewater are made of PVC, ABS, FRP, rubber, or stainless steel, which are corrosion resistant.

The accumulation of grease is a cause for concern in conventional lift stations as well as in vacuum stations. Grease builds up on level controls and on the sides of the vacuum collection tank. Grease traps are typically required in applications such as restaurants to minimize these problems.

Grease has not presented problems in vacuum sewer mains. When the wastewater is evacuated from the sump, the suction pulls floatable grease into the vacuum mains. Since the wastewater moves through the mains at high velocities, there is little opportunity for grease in the sewer to build up in the system to any level that could cause a blockage. However, grease buildups in the wastewater pumps and force mains to the wastewater treatment plant can be a problem. Monitoring of the force main pressure may give you an early indication that a restriction in the wastewater flow is developing and that the force main needs to be cleaned by pigging.

## QUESTIONS

Please write your answers to the following questions and compare them with those on page 347.

6.53G Why should solvent welding not be done in cold weather?

6.53H What is an alternative to solvent welding?

6.53I What problem can result from improper valve alignment?

6.53J Why are odors not usually a problem in vacuum collection sewers?

6.53K Why is corrosion not usually a problem in vacuum collection sewers?

### 6.54 Recordkeeping

#### *6.540 Importance of Recordkeeping*

Good records are important for the efficient, orderly operation of the system. Pertinent and complete records provide a necessary aid to control procedures as they are used as a basis of the system operation. The very first step of any troubleshooting procedure is an analysis of the records. This is especially true of the collection system. A wealth of information is contained in the basic records kept on a daily basis. The following types of records should be kept:

- Routine maintenance records
- Preventive maintenance records
- Emergency maintenance records
- Operating cost records

These records should be carefully preserved and filed where they are readily available to operating personnel. All records should be neat and accurate and made at the time the data are obtained. It is good practice to summarize this data in a brief monthly report and a more complete annual report. These reports should be submitted to local authorities such as the city council and the mayor to keep them informed of progress, problems, and long-term needs of the facility.

#### *6.541 Routine Maintenance Records*

The following information should be recorded on a daily basis using a form such as the one shown in Figure 6.61:

- Date
- Personnel on duty
- Weather conditions
- Routine duties performed
- Complaints received and the remedies
- Facility visitors
- Accidents or injuries
- Operating range of vacuum pumps
- Run times of vacuum pumps
- Run times of wastewater discharge pumps
- Run time of standby generator
- Flow data
- Unusual conditions
- Alterations to the system

#### *6.542 Preventive Maintenance Records (Figure 6.62, top)*

Adequate records provide information about when service was last performed on each system component and approaching service or preventive maintenance requirements. Knowing this information makes it possible to schedule maintenance tasks efficiently so they do not interfere with other important aspects of system operation.

Results of periodic inspections should be recorded and filed. This would include a list of all potential problems, the likely causes of these problems, the repairs necessary to solve the problems, and recommendations for improvements that would minimize these problems in the future.

#### *6.543 Emergency Maintenance Records (Figure 6.62, bottom)*

Records containing the following information should be kept concerning all emergency maintenance:

- Date and time of occurrence
- Person(s) responding to problem
- Description of problem
- Remedy of problem
- Parts and equipment used
- Total time to correct problem
- Recommendations for future improvements

#### *6.544 Operating Cost Records*

To ensure budget adequacy, it is very important to keep accurate information concerning the costs of all of the following operation and maintenance items:

- Wages and fringe benefits
- Power and fuel consumption
- Utility charges
- Equipment purchases
- Repair and replacement expenses
- Chemicals
- Miscellaneous costs

**Daily Operating Log**

| Date and Time | Hours Run by Vacuum Pumps | | | Hours Run by Discharge Pumps | | Hours Run by Generator | Flowmeter | Power Meter Reading | Weather | Operator's Initials |
|---|---|---|---|---|---|---|---|---|---|---|
| | #1 | #2 | #3 | #1 | #2 | | | | | |
| | | | | | | | | | | |

**Plant Failures and Complaints Register**

| Date and Time of Report | Failure or Complaint | Causes | Spares Used | Time Repairs Completed | Operator's Signature |
|---|---|---|---|---|---|
| | | | | | |

*Fig. 6.61 Routine maintenance report form*

**Valve Number:** **Line:**

**Location:**

| Maintenance Due | Maintenance Completed | Spare Parts Used | Initials |
|---|---|---|---|
| | | | |
| | | | |
| | | | |
| | | | |
| | | | |
| | | | |
| | | | |
| | | | |
| | | | |

FRONT

**Failures and Complaints Log**

| Date | Problem | Reason | Remedy | Initials |
|---|---|---|---|---|
| | | | | |
| | | | | |
| | | | | |
| | | | | |
| | | | | |
| | | | | |
| | | | | |
| | | | | |
| | | | | |

REAR

*Fig. 6.62 Typical valve maintenance record card*

## QUESTIONS

Please write your answers to the following questions and compare them with those on page 347.

6.54A What is the very first step of any troubleshooting procedure?

6.54B List the four different types of records that should be kept by vacuum sewer system operators.

6.54C How often should collected data be summarized in reports?

### 6.55 System Management Considerations

#### *6.550 Collection System Agency Responsibilities*

A sewer-use ordinance is an essential tool that will enable a wastewater collection agency to operate efficiently. This document sets consistent rules for all users to follow. Included are material specifications for the building sewer, minimum slope requirements, and vent locations. A provision of the sewer-use ordinance that is especially important to the collection agency is a limitation on use of the vacuum sewer to sanitary wastes only. Connections from roof and yard drains should not be permitted because the extra water will cause operational problems.

Once all customers are connected, the collection system agency's focus is on providing reliable, efficient service to their customers. To achieve this, the operating personnel must be capable, dependable, and knowledgeable. The operator's attitude is very important. An operator who does not believe in the system will ultimately cause the system to operate below its potential in terms of reliability and costs. On the other hand, an operator with a good attitude finds creative ways to get more out of the system than was originally planned.

An active program for the identification of extra water sources should be developed. This may include smoke testing and dye testing. To identify and quantify sources of extra water, the agency can take advantage of the cycle counter (see Figure 6.42 on page 308) built into the collection system at the valve pit. The cycle counter records the number of times the valve opens in a given period. Each time the valve opens, approximately ten gallons of wastewater enters the collection system. To estimate the total amount of wastewater discharged in a certain period, multiply the number of cycles times 10 gallons. Then, examine the customer's water consumption records during the same time period. Divide the total amount of water consumed by 10 gallons to estimate the minimum number of cycles you would expect to occur over that period. If the number of discharge cycles is much greater than the number of cycles you would expect based on water consumption, this may be a sign that extra water is being added to the collection system. Listening to the auxiliary vent for sounds of running water when no wastewater-generating activity is taking place may also provide clues for sources of extra water.

The collection system agency also is responsible for future extensions of the system. This includes proper planning, design, and construction of such extensions. Finally, future connections to the existing system are made by the collection agency in accordance with the provisions of the sewer-use ordinance.

#### *6.551 Homeowner Responsibilities*

Use of the system should comply with requirements of the sewer-use ordinance. Typically, the homeowner is required to protect the on-site facilities from damage and not to drive or build over the tank. Discharge of flammables, acids, excessive amounts of grease, feminine hygiene products, condoms, rags, or other nonwastewater items is discouraged. User ordinance requirements for vacuum collection systems are very similar to user ordinances for conventional sewers. Proper use of the system results in lower user charges and improved reliability.

#### *6.552 Other Agencies*

Many different agencies and individuals, including regulatory agencies and the engineer, are involved during the planning, design, and construction of sewer systems. It is during these times that many decisions are made and details finalized. Often, these agencies view the start-up of a system as their final involvement. While this attitude is understandable, it is not acceptable. Continuing involvement is needed to help develop an experience base with newer systems, which permits informed applications in the future.

The engineer should spend a significant amount of time with the collection system agency during the start-up of the system. Tests should be run and problems simulated to see if the system is operating as designed and problems and their solutions should be recorded. The operating records should be analyzed periodically to make sure operating costs are within budget estimates. In short, the engineer should use experience with operating the vacuum sewer system to help develop improvements in future designs.

## QUESTIONS

Please write your answers to the following questions and compare them with those on page 347.

6.55A What is the purpose of a sewer-use ordinance?

6.55B What does a sewer-use ordinance require of homeowners?

### 6.6 ACKNOWLEDGMENTS

Major portions of this chapter were adapted from Chapter 3, "Vacuum Sewer Systems," of the United States Environmental Protection Agency's Technology Transfer Manual, *ALTERNATIVE WASTEWATER COLLECTION SYSTEMS.* Much of the material in *ALTERNATIVE WASTEWATER COLLECTION SYSTEMS* came from AIRVAC's copyrighted *DESIGN MANUAL.* AIRVAC has kindly permitted the reproduction and use of this material. The author gratefully recognizes these two valuable resources used in preparation of this chapter.

1. *ALTERNATIVE WASTEWATER COLLECTION SYSTEMS,* United States Environmental Protection Agency, Office of Research and Development, Center for Environmental Research Information, Risk Reduction Engineering Laboratory, Cincinnati, OH 45628. EPA/625/1-91/024. October 1991.

2. *AIRVAC DESIGN MANUAL,* AIRVAC Vacuum Sewer Systems, PO Box 528, Old US 31 N, Rochester, IN 46975.

Other references used in preparing this chapter included:

1. *MANILA COMMUNITY SERVICES DISTRICT, DESIGN-CONSTRUCTION REPORT FOR THE SEPTIC TANK EFFLUENT PUMPING SEWERAGE SYSTEM,* California State Water Resources Control Board.
2. *RURAL WASTEWATER MANAGEMENT,* California State Water Resources Control Board.
3. *ORENCO SYSTEMS, INC., ENGINEERS' CATALOG,* Orenco Systems, Inc., 814 Airway Avenue, Sutherlin, OR 97479.
4. Glide-Idleyld Park Sewer District Forms and Publications, Douglas County Department of Public Works, Roseburg, OR 97479.

## 6.7 ADDITIONAL READING

Useful information on check valve and air valve selection is available on the APCO Willamette Valve and Primer Corporation website at www.apcovalves.com.

## 6.8 ARITHMETIC ASSIGNMENT

Turn to the Appendix, "How to Solve Wastewater System Arithmetic Problems," at the back of this manual and read Section A.8, "Pumps."

**END OF LESSON 3 OF 3 LESSONS**

**on**

**COLLECTION SYSTEMS**

Please answer the discussion and review questions next.

# DISCUSSION AND REVIEW QUESTIONS

## Chapter 6. COLLECTION SYSTEMS

(Lesson 3 of 3 Lessons)

Please write your answers to the following questions to determine how well you understand the material in the lesson. The question numbering continues from Lesson 2.

17. List the three major components of a vacuum sewer system.
18. Why should the valve pit be located near the home?
19. What are the four main parts of a package fiberglass type of valve pit and sump?
20. Describe how the controller/sensor in a vacuum valve operates.
21. Why is it desirable for the vacuum station to be located near the center of the system?
22. List the main components of a central vacuum station.
23. What types of pumps are used to produce the vacuum needed to move wastewater through a vacuum collection system?
24. Why is it important to have a standby generator at the central vacuum station?
25. Describe the vacuum transport process, that is, how wastewater is moved through a vacuum collection system.
26. What types of daily, weekly, and monthly routine preventive maintenance activities are associated with vacuum collection systems?
27. What problem is the most common cause of emergency maintenance?
28. What problems could be caused by the failure to quickly locate and fix a malfunctioning vacuum valve?
29. What are some types of construction-related O&M problems?
30. What procedures can be used to identify sources of extra water being discharged to the collection system?

# SUGGESTED ANSWERS

## Chapter 6. COLLECTION SYSTEMS

### ANSWERS TO QUESTIONS IN LESSON 1

Answers to questions on page 260.

6.0A The purpose of a wastewater collection system is to collect the wastewater from a community's homes and convey it to a treatment plant where the pollutants are removed before the treated wastewater is discharged to a body of water or onto land.

6.0B In areas where the topography or soil conditions are unfavorable for a conventional gravity wastewater collection system, a small-diameter gravity sewer system, a pressure collection system, or a vacuum collection system may be used.

6.0C Collection system operators should understand how the systems they are responsible for operating and maintaining were designed and constructed so they can determine the correct action to take when the system fails to operate properly.

6.1A The main sources of wastewater in a collection system are homes or businesses, infiltration of groundwater, and inflow of surface water.

Answers to questions on page 262.

6.2A The parts or components of a conventional gravity wastewater collection system include building sewers, lateral and branch sewers, main sewers, trunk sewers, interceptor sewers, and lift stations.

6.2B A scouring velocity must be achieved at some time during the day to prevent solids from settling out of the wastewater and accumulating in the sewer line.

6.2C Sewers usually are not placed in private easements because they are more difficult to gain access to and to maintain than sewers in streets.

Answers to questions on page 265.

6.2D Poor sewer pipe joints can lead to problems caused by groundwater infiltration, wastewater exfiltration, and root intrusion.

6.2E Some manhole barrels are made with fiberglass for use in areas with high groundwater and where the wastewater contains corrosive materials.

6.2F The plan view is the top view of a sewer; it shows the pipe, manholes, and street. The profile is the side view of a sewer; it shows the pipe, manholes, invert elevations, street surface, and underground pipes and other underground utilities such as storm sewers, gas mains, water mains, and electrical cables.

6.2G The slope of the shelf in a manhole is important because if the shelf is too flat, it will get dirty; if the shelf is too steep, it can be a slipping hazard to operators.

Answers to questions on page 266.

6.30A A small-diameter gravity sewer (SDGS) system consists of a series of septic tanks discharging effluent by gravity, pump, or siphon to a small-diameter wastewater collection main. The wastewater flows by gravity to a lift station, a manhole in a conventional gravity collection system, or directly to a wastewater treatment plant.

6.30B A small-diameter gravity sewer may terminate at a lift station, at a manhole in a conventional gravity collection system, or at a wastewater treatment plant.

6.30C The minimum flow velocity in a small-diameter gravity sewer system is 0.5 foot per second.

6.30D Grinder pumps are never permitted to discharge into a small-diameter gravity sewer system because grinder pumps discharge small solids and grease that could cause blockages due to the low flow velocities.

Answers to questions on page 273.

6.31A The components of a small-diameter gravity sewer system include building sewers, septic tanks, service laterals, collection mains, cleanouts, manholes and vents, and lift stations if required.

6.31B Existing in-service septic tanks are usually replaced when a new SDGS system is installed because watertightness is critical to proper operation of the system. Unless all components are watertight, water leaking into the system will contribute extra clear water inflow that the system was not designed to carry. In addition, old septic tanks are difficult to inspect and repair properly.

6.31C Cleanouts provide access to the collection main for various maintenance tasks.

6.31D Check valves are used on service connections at the point of connection to the main to prevent backflow during surcharged conditions.

Answers to questions on page 274.

6.32A Since the septic tanks are critical to the proper performance of the SDGS system, responsibility for maintenance should be retained by the utility district.

6.32B SDGS system record (as-built) drawings are important because the system has relatively few manholes or cleanouts and pipelines are laid following the contours of the terrain. These characteristics make it difficult to pinpoint the location of collection mains unless accurate drawings of the system relate pipeline locations to permanent structures nearby.

6.32C Skills/training necessary for SDGS maintenance personnel include basic plumbing skills and familiarity with the system operation. If the system includes STEP units, an understanding of pumps and electrical controls is also helpful.

Answers to questions on page 278.

6.33A Preventive maintenance of small-diameter gravity sewer (SDGS) systems usually includes inspection and pumping of the septic tanks, inspection and cleaning of the collection mains, inspection and servicing of any STEP units or lift stations, testing and servicing of air release valves, and exercising of isolation valves and other appurtenances.

6.33B Utility district personnel should be present during septic tank pumping operations to measure and record the depth of sludge and thickness of any scum in each tank so that the pumping schedule can be altered in the future according to actual sludge and scum accumulation rates.

6.33C The most common method of cleaning collection mains in SDGS systems is with the use of pigs.

6.33D In SDGS systems, corrosion usually occurs at lift stations and at air release stations if the valves have metal parts.

6.34A Emergency SDGS call-outs are usually due to plumbing backups into homes or odors.

Answers to questions on page 280.

6.35A Odors are most pronounced where turbulence occurs due to the release of obnoxious gases (such as hydrogen sulfide) dissolved in the wastewater.

6.35B Leaking tanks or building sewers are the primary entry points of clear water infiltration/inflow (I/I) problems in small-diameter gravity sewer (SDGS) systems. Therefore, by installing all new, properly designed and built septic tanks and pressure testing building sewers and tanks, clear water infiltration/inflow (I/I) problems can be solved.

## ANSWERS TO QUESTIONS IN LESSON 2

Answers to questions on page 286.

6.41A Minimizing the number of joints in a wastewater collection pressure pipeline enhances pipe integrity, minimizes infiltration/inflow, and minimizes the possibility of leaks (exfiltration) from the system.

6.41B Two-inch diameter pipe is common at the upper end (beginning) of a STEP collection system pressure main.

6.42A Air release valves are installed at high points in a main between depressions. They are installed to permit the venting of air or gases trapped in the main and to prevent an air lock or blockage in the main.

6.42B Pressure sewer lines can be cleaned by inserting a pig in the line and allowing water pressure to push the pig through the line.

Answers to questions on page 293.

6.43A Preventive maintenance tasks associated with pressure system pipelines include exercising the main line isolation valves at least once a year and checking the air release stations for proper operation.

6.43B Pressure collection system maintenance personnel should be prepared to respond to the following types of emergency calls: (1) high/low liquid level alarms, (2) extended power outages, and (3) main line ruptures.

6.43C Operators of pressure collection systems could be exposed to hazardous gases such as hydrogen sulfide gas, electrical burn or shock from controls, switches, and junction boxes, and infections and infectious diseases from exposure to wastewater.

6.45A An up-to-date operation and maintenance manual is the operator's best source of information about the operation and maintenance of STEP systems and pressure sewers. Other sources of information include the design engineer, equipment suppliers, and the operators of similar systems.

## ANSWERS TO QUESTIONS IN LESSON 3

Answers to questions on page 295.

6.50A Vacuum sewer systems are considered where one or more of the following conditions exist:

1. Unstable soil
2. Flat terrain
3. Rolling land with many small elevation changes
4. High water table
5. Restricted construction conditions
6. Rocky terrain
7. Urban development in rural areas

6.50B The potential advantages of vacuum sewer systems may include substantial reductions in water used to transport wastes, material costs, excavation costs, and treatment expenses.

Answers to questions on page 301.

6.51A On-site components of a vacuum collection system include the valve pit/sump or buffer tank, a vacuum interface valve, an auxiliary vent, and a service lateral. The building sewer connecting the residence to the collection system is usually considered the homeowner's responsibility.

6.51B A concrete valve pit may be needed when (1) the deepest fiberglass pit and sump is not sufficient to accept the building sewer, (2) a large flow is anticipated requiring flow leveling, or (3) an interface between two system types (for example, pressure and vacuum) is needed.

6.51C A buffer tank should be used for large flows that require leveling out, such as flows from schools, apartments, nursing homes, and other large dischargers to the vacuum sewer system.

Answers to questions on page 309.

6.51D The valves in both the Model D arrangement and Model S arrangement are identical but each relies on a different piping/plumbing arrangement for their source of atmospheric air. The Model D gets atmospheric air through a tube, which is connected to an external breather pipe. The Model S is a sump-vented arrangement. One of the three controller tubes is connected to the cleanout/sensor piping. This piping extends into the lower sump, which is connected to the building sewer, and the building sewer is open to atmospheric air through the 4-inch auxiliary vent.

6.51E The auxiliary vent on a building sewer prevents the water in home plumbing fixture traps from being drawn into the vacuum system. The auxiliary vent also permits air to enter the building sewer where it acts as the driving force behind the wastewater that is evacuated from the lower sump.

6.51F Vacuum sewer service lines should be located away from potable water pipelines to prevent cross contamination. They should also be separated from other buried utility lines, if possible, to reduce the chance of damage when the other utility lines are excavated for maintenance or repair.

Answers to questions on page 315.

6.51G Routing two or more homes to share one valve pit has two major disadvantages: (1) it requires a permanent easement from one of the property owners (which may be difficult to obtain), and (2) experience has shown that multiple-house hookups are a source of neighborhood friction unless the valve pit is located on public property.

6.51H Division valves are installed to permit portions of the system to be isolated for repairs and maintenance.

6.51I To identify the location of collection main pipes, an identification tape marked "vacuum sewer" is sometimes placed shallowly in the pipeline trench. The tape can be metalized so that it can be detected with common utility locating equipment. Other methods include installing a toning wire along the top of the pipeline or color coding the pipes themselves.

Answers to questions on page 320.

6.51J Vacuum stations function as transfer facilities between a central collection point for all vacuum sewer lines and a pressurized line leading directly or indirectly to a wastewater treatment facility.

6.51K The operating volume of the collection tank is the amount of wastewater required to restart the discharge pump at a minimum design flow of every 15 minutes.

6.51L The purpose of a vacuum pump is to produce the vacuum necessary to move the wastewater through the collection system to the vacuum station.

Answers to questions on page 324.

6.51M Wastewater pumps are typically installed at an elevation significantly below the collection tank to minimize the net positive suction head (NPSH) requirement.

6.51N Level controls are set for a minimum of two minutes' pump running time to protect the pumps from the increased wear that results from too frequent pump starts.

6.51O A standby generator is needed to ensure continuing operation of the system in the event of a power outage.

Answers to questions on page 327.

6.52A The operator of a vacuum collection system should be present during the construction of the system to become familiar with the system, including the locations of all lines, valve pits, division valves, and other key components.

6.52B The vacuum transport process works as follows: when a sufficient volume of wastewater accumulates in the sump, the vacuum valve cycles. The difference in pressure between the vacuum sewer main and the atmosphere forces the sump contents into the main. While accelerating, the wastewater is rapidly transformed into foam and soon occupies only part of the pipe cross section. Eventually, friction and gravity bring the wastewater to rest at a low spot. Another valve cycle, at any location upstream of the low spot, will cause this wastewater to continue its movement toward the vacuum station.

6.52C A well-written O&M manual should contain the information necessary to achieve the following goals:

1. To provide an accessible reference for the wastewater collection system operators in developing standard operating and maintenance procedures and schedules
2. To provide a readily available source of data, including permits, design data, and equipment shop drawings of the particular system
3. To provide the system operators assistance and guidance in analyzing and predicting the system efficiency
4. To provide the system operators assistance and guidance in troubleshooting the system

6.52D The O&M manual should be constantly updated to reflect actual operational experience, equipment data, problems, and solutions.

6.52E As-built hydraulic maps include the following information:

1. The locations of every lift
2. The amount of vacuum loss at key locations
3. The number of main branches, the number of valves in each branch, and the total footage (or volume) of pipe in each branch

Answers to questions on page 330.

6.52F It is important to have on hand a supply of spare parts for the microprocessor-based electronic components of the system because these components (level controls and the fault monitoring systems) essentially operate and monitor the system. Losing level controls due to some type of failure will quickly cause severe operational problems such as loss of vacuum or discharge pump malfunctions. Without spare parts to make repairs, an operator must manually cycle the vacuum and wastewater pumps. Loss of the fault monitoring system increases operator O&M time and response time to system malfunctions.

6.52G The following routine operating tasks should be performed daily:

1. Visually check gauges/charts
2. Record pump run times
3. Check oil and block temperatures

6.52H Inspection and maintenance planning and scheduling involve time, personnel, equipment, costs, work orders, and priorities.

6.52I Emergency or breakdown maintenance can occur in the piping system, at the vacuum station, or at the vacuum valves.

Answers to questions on page 332.

6.53A Three categories of collection line malfunctions are (1) vacuum valve malfunction, (2) break in the vacuum lines, and (3) closed isolation valves.

6.53B In most cases, waterlogging means there is too much liquid (extra water) in the collection system pipelines and not enough air entering the system at one time.

6.53C Two common causes of waterlogging are valve failure and illegal stormwater connections to the vacuum collection system.

Answers to questions on page 337.

6.53D A leaking check valve can result in vacuum loss problems in the station piping.

6.53E If a low-vacuum situation is allowed to go undetected, this will result in unnecessary additional run time of the vacuum pump. This will result in increased power costs as well as increased wear on major equipment.

6.53F Wastewater can be pulled into vacuum pumps when a zero-vacuum situation occurs and the operator tries to manually operate the vacuum pumps when the collection tank is full.

Answers to questions on page 339.

6.53G Solvent welding should not be done in cold weather because this leads to leaking joints and difficulty in passing leakage tests.

6.53H An alternative to solvent welding is the use of gasketed pipe and fittings.

6.53I Improper valve alignment can lead to future valve malfunctions.

6.53J Odors usually are not a problem in vacuum collection sewers because (1) the system is sealed, (2) a large volume of air is brought into the system at each flow input, and (3) wastewater travel and detention times are short.

6.53K Corrosion is not a problem in vacuum collection sewers because all system parts in contact with wastewater are made of corrosion-resistant material.

Answers to questions on page 342.

6.54A The very first step of any troubleshooting procedure is an analysis of the records.

6.54B The following types of records should be kept:

1. Routine maintenance records
2. Preventive maintenance records
3. Emergency maintenance records
4. Operating cost records

6.54C Collected data should be summarized in a brief monthly report and a more complete annual report.

Answers to questions on page 342.

6.55A The purpose of a sewer-use ordinance is to set consistent rules for all users of the system to follow.

6.55B The sewer-use ordinance typically requires homeowners to protect the on-site collection system facilities from damage and not to drive or build over the tank. The ordinance also discourages homeowners from discharging flammables, acids, excessive amounts of grease, feminine hygiene products, condoms, rags, or other nonwastewater items.

CHAPTER 7

# MAINTENANCE AND TROUBLESHOOTING

by

Peg Hannah

and

Ken Kerri

# TABLE OF CONTENTS

## Chapter 7. MAINTENANCE AND TROUBLESHOOTING

# OBJECTIVES

## Chapter 7. MAINTENANCE AND TROUBLESHOOTING

1. Develop a preventive maintenance program for your on-site system, wastewater collection system, or treatment system.
2. Set up a maintenance recordkeeping system.
3. Use a systematic approach when troubleshooting equipment or system problems.
4. Recognize the serious consequences that could occur when inexperienced, unqualified, or unauthorized persons attempt to troubleshoot or repair electrical panels, controls, circuits, wiring, or equipment.
5. Properly select and safely use the following pieces of equipment (if qualified and authorized):
   a. Multimeter
   b. Ammeter
   c. Megger
   d. Ohmmeter
6. Communicate effectively with electricians.

*NOTE:* Special maintenance information was presented in the previous chapters for specific types of collection systems and treatment processes, where appropriate.

# WORDS

## Chapter 7. MAINTENANCE AND TROUBLESHOOTING

AMPLITUDE | AMPLITUDE

The maximum strength of an alternating current during its cycle, as distinguished from the mean or effective strength.

AUGER (AW-grr) | AUGER

A sharp tool used to go through and break up or remove various materials that become lodged in sewers.

CYCLE | CYCLE

A complete alternation of voltage or current in an alternating current (AC) circuit.

HERTZ (Hz) | HERTZ (Hz)

The number of complete electromagnetic cycles or waves in one second of an electric or electronic circuit. Also called the frequency of the current.

MEGOHM (MEG-ome) | MEGOHM

Millions of ohms. Mega- is a prefix meaning one million, so 5 megohms means 5 million ohms.

NAMEPLATE | NAMEPLATE

A durable, metal plate found on equipment that lists critical installation and operating conditions for the equipment.

ROD, SEWER | ROD, SEWER

A light metal rod, three to five feet long, with a coupling at each end. Rods are joined and pushed into a sewer to dislodge obstructions.

SURCHARGE | SURCHARGE

Sewers are surcharged when the supply of water to be carried is greater than the capacity of the pipes to carry the flow. The surface of the wastewater in manholes rises above the top of the sewer pipe, and the sewer is under pressure or a head, rather than at atmospheric pressure.

# CHAPTER 7. MAINTENANCE AND TROUBLESHOOTING

## 7.0 NEED FOR A MAINTENANCE PROGRAM

The operators of small on-site wastewater treatment systems cannot afford to ignore equipment and system maintenance. The wastewater collection and treatment system must operate every day and equipment breakdowns cannot be tolerated if the system is to be effective. Therefore, the operator should strive to eliminate equipment breakdowns or at least minimize the time they cause the system to be out of operation. Most equipment problems can be avoided if the operator will regularly inspect, clean, and lubricate the equipment. Maintenance inspections should be performed according to a regular schedule, which should be written down so inspections are not overlooked. Unusual noises, vibrations, leaks, and malfunctions should receive prompt attention. Spare parts should be kept on hand so breakdowns can be repaired quickly and so worn parts can be replaced when a pump or other piece of equipment is returned to the shop for overhaul.

The operator of a small wastewater collection and treatment system is also responsible for maintaining clean and tidy conditions in and around the areas where utility equipment is installed. Pipes, plumbing fittings, tools, and other materials must be stored in a safe and orderly manner. Junk and clutter should be eliminated.

A good maintenance program will reduce breakdowns, extend equipment life, and provide for more efficient use and performance of operators. The basic features of a maintenance program include an equipment record system, planning and scheduling, a storeroom and inventory system, maintenance operators, and costs and budgets for the maintenance program. The maintenance program provides information to aid in troubleshooting and solving maintenance problems and developing cost and budget recommendations.

## 7.1 TYPES OF MAINTENANCE

Although there are many different views about what is a good maintenance program, try to think in terms of what needs to be done. First, a maintenance program needs to ensure that regular preventive maintenance is performed to prevent problems and equipment failures. A good maintenance program tries to anticipate what is going to go wrong and prevent it from happening. If something goes wrong, the problem is identified and corrected.

### 7.10 Preventive Maintenance

Preventive maintenance is the most effective and efficient type of maintenance program. It is a systematic approach for conducting maintenance activities before any equipment failures occur. The purposes of preventive maintenance are to extend equipment life, reduce maintenance costs, and increase reliability of the system.

Every wastewater agency, no matter how small, should devote an appropriate level of resources to developing a preventive maintenance program. Unfortunately, many small agencies find this difficult to do because their resources are already committed to emergency maintenance. However, experiences in both the public and private wastewater agencies show that as more resources are dedicated to preventive maintenance, less and less time and money will have to be spent on emergency maintenance.

### 7.11 Emergency Maintenance

Even with all our modern equipment and an excellent preventive maintenance program, emergencies will sometimes occur. They may be caused by natural catastrophes such as fires, floods, hurricanes, or earthquakes, or they may be caused by equipment breakdowns that are unpredictable and difficult to detect during routine inspections and maintenance activities (for example, a pump becomes plugged with rags or a force main pipeline unexpectedly breaks).

The cost of responding to emergencies can be enormous. For example, the total cost of a lift station failure could include not only the cost of restoring the station to operation, but also the costs of cleaning up an overflow or spill of raw wastewater into lakes or streams, cleaning up backups of wastewater into homes or businesses, and paying for the loss or destruction of personal property. These costs can easily amount to thousands of dollars. In addition, regulatory agencies frequently view an overflow or spill as a violation of the NPDES permit and significant fines and accompanying bad publicity for these violations can follow.

Naturally, we cannot schedule emergency maintenance. We can only plan to handle it quickly and effectively. As your preventive maintenance scheduling techniques develop and improve, the need for emergency maintenance will be constantly decreasing. In addition, planning ahead and preparing for natural disasters by establishing a comprehensive emergency response plan is an effective way to minimize the effects of the disaster on the community's wastewater collection and treatment system performance.

## QUESTIONS

Please write your answers to the following questions and compare them with those on page 376.

7.0A What should an operator's maintenance program strive to achieve?

7.1A What is preventive maintenance?

7.1B What is emergency maintenance?

## 7.2 ELEMENTS OF A PREVENTIVE MAINTENANCE PROGRAM

### 7.20 Operation and Maintenance (O&M) Manual

The operator's most valuable resource for maintenance information is a complete, up-to-date operation and maintenance (O&M) manual. If yours is out of date or has been lost, use the guidelines presented in Appendix C at the end of Chapter 4 to update an old manual or prepare a new one.

### 7.21 Manufacturers' Recommendations

Suppliers of equipment and materials are often experts in their field and frequently give generously of their time to assist an operator in using their products successfully. They want their products to be properly installed and maintained and will usually provide literature on the products they manufacture and sell. A supplier may be able to indicate how other operators have solved particular problems or suggest ways to improve the performance of a piece of equipment. Operators who take the time to establish good working relationships with their suppliers can benefit from their expertise and may receive better support service when problems arise. Good working relationships also benefit suppliers and they usually appreciate comments from operators who use their products and have discovered a better way of maintaining them or have suggestions for improving a product.

### 7.22 Records

Preventive maintenance programs help operating personnel keep equipment in satisfactory operating condition and aid in detecting and correcting malfunctions before they develop into major problems.

A frequent occurrence in a preventive maintenance program, particularly in very small wastewater systems, is the failure of the operator to record the work after it is completed. When this happens, the operator must rely on memory to know when to perform each preventive maintenance function. As days pass into weeks and months, the preventive maintenance program is lost in the turmoil of everyday operation.

The only way an operator can keep track of a preventive maintenance program is by good recordkeeping. Whatever record system is used, it should be kept up to date on a daily basis and not left to memory for some other time.

Equipment service records such as the ones shown in Figure 7.1 are easy to set up and require little time to keep up to date. The *EQUIPMENT SERVICE CARD* tells what should be done and when, while the *SERVICE RECORD CARD* is a record of what you did and when you did it. Many agencies keep this information on a computer.

An *EQUIPMENT SERVICE CARD* (master card) should be filled out for each piece of equipment in the system. Each card should have the equipment name on it, such as Raw Wastewater Lift Pump No. 1.

1. List each required maintenance service with an item number.
2. List maintenance services in order of frequency of performance. For instance, show daily service as items 1, 2, and 3 on the card; weekly items as 4 and 5; monthly items as 6, 7, 8, and 9; and so on.
3. Describe each type of service under work to be done.

Make sure all necessary inspections and services are shown and that the frequency of service is noted. For reference data, list page or section numbers as shown in the manufacturer's literature. Under time, enter the day or month service is due. Service card information may be changed to fit the needs of your system or particular equipment as recommended by the equipment manufacturer. Be sure the information on the cards is complete and correct.

The *SERVICE RECORD CARD* should have the date and work done, listed by item number and signed by the operator who performed the service. Some operators prefer to keep both cards clipped together, while others place the service record card near the equipment. When the service record card is completed, it should be filed for future reference and a new card attached to the master card.

Another important type of record is an equipment identification (ID) card. The ID card should contain the following vital information for each piece of equipment:

- Equipment function
- Name
- Location
- Size (GPM, HP)
- Voltage, amperage, RPMs
- Vendor's name
- Service representative's name, telephone number, FAX number, and email address
- Spare parts on hand
- Delivery time for parts orders

In addition to setting up a formal recordkeeping system, many operators keep a daily diary to explain information recorded in the other records and any unusual events or activities

| EQUIPMENT SERVICE CARD | | | | |
|---|---|---|---|---|
| EQUIPMENT: #1 Raw Wastewater Lift Pump | | | | |
| **Item No.** | **Work to Be Done** | **Reference** | **Frequency** | **Time** |
| 1 | Check water seal and packing gland | | Daily | |
| 2 | Operate pump alternately | | Weekly | Monday |
| 3 | Inspect pump assembly | | Weekly | Wednesday |
| 4 | Inspect and lube bearings | | Quarterly | 1-4-7-10* |
| 5 | Check operating temperature of bearings | | Quarterly | 1-4-7-10 |
| 6 | Check alignment of pump and motor | | Semiannually | 4 & 10 |
| 7 | Inspect and service pump | | Semiannually | 4 & 10 |
| 8 | Drain pump before shutdown | | | |
| | | | | |
| | | | | |

*1-4-7-10 represent the months of the year when the equipment should be serviced—1–January, 4–April, 7–July, and 10–October.

| SERVICE RECORD CARD | | | | | |
|---|---|---|---|---|---|
| EQUIPMENT: #1 Raw Wastewater Lift Pump | | | | | |
| **Date** | **Work Done (Item No.)** | **Signed** | **Date** | **Work Done (Item No.)** | **Signed** |
| 1-5-11 | 1 & 2 | J.B. | | | |
| 1-6-11 | 1 | J.B. | | | |
| 1-7-11 | 1-3-4-5 | R.W. | | | |
| | | | | | |
| | | | | | |
| | | | | | |
| | | | | | |
| | | | | | |
| | | | | | |

*Fig. 7.1 Equipment service card and service record card*

not recorded elsewhere. The operator's diary usually consists of informal notes and the operator is free to record any important comments, observations, reminders, or explanations. The types of information recorded may include notes about equipment breakdowns; changes in treatment; severe weather; changes in effluent quality; or visits by repair personnel, health inspectors, customers, and community officials. This information is very helpful to the relief operator who must take over operation of the system suddenly or on occasion.

### 7.23 Tools

Nothing makes a maintenance or repair job tougher than to have the wrong tools or no tools at all. The list of tools in Table 7.1 is provided as a guide to help you determine the tools you will need to do your job. This list is not complete. You may need additional tools to do your job and some of the tools you will not need.

Proper care of tools is essential. You must maintain your tools and hand equipment as necessary and keep everything in proper working order. After a job is completed, clean your tools and store them properly so they can be found and used when needed again.

### 7.24 Safety

When operating any type of wastewater collection and treatment system, *safety must always come first.*

In even the smallest wastewater collection and treatment systems, it is important for the operator to be prepared to respond quickly and effectively in an emergency. If you are the only operator responsible for a small wastewater collection and treatment system, you must take care of yourself. There frequently is no one else present to make sure that you follow safe procedures and that there are no safety hazards that can injure you. There may be no one else available to help you or to rescue you if you become injured.

Studies have shown that the types of injuries operators suffer most frequently are muscle and back strains caused by lifting and injuries resulting from slips and falls. These types of injuries can easily be prevented by alert and safety-conscious operators.

If you have not thought about what you would do in various types of emergencies, do it now. Consider what actions you would take in the event of a serious injury to yourself or another operator. Prepare a list of emergency phone numbers and keep a copy with you at all times if you frequently work alone. Post a copy of the emergency phone numbers in your service vehicle in plain sight and in other locations where an accident might occur. Emergency phone numbers that should be on this list include the following:

1. Police Department
2. Fire Department
3. Hospital or Physician
4. Responsible Utility Agency Officials
5. Local Emergency Disaster Office
6. *CHEMTREC,* (800) 424-9300
7. Emergency Team (if your system has one)

The *CHEMTREC* toll-free number may be called at any time. Personnel at this number will give information on how to handle emergencies created by hazardous materials and will notify appropriate emergency personnel.

Review Chapter 3, "Safety," for additional information about possible safety hazards around wastewater collection and treatment equipment and safe procedures for doing your job.

### 7.25 Information Resources

One of the best ways to learn about maintenance of a wastewater collection and treatment system is to be present during construction of the system. If you are fortunate enough to participate in the design and construction of your system, learn as much as you can about all of the components and use this opportunity to discuss operation and maintenance intentions and requirements with the manufacturers' representatives and the design engineer. In cases where it is not possible for the operator to be present during construction, the operator should know the names of those who were there—the inspector, engineer, contractor, and superintendent. Inspection reports and daily logs may also be useful.

It is important to remember that it is the owner and you, the operator, who are responsible for the ongoing successful operation and maintenance of the system. No matter how good the equipment may be, or how well a system is designed, poor O&M will result in system failures. Learn the acceptable limits of your system as designed and constructed and routinely monitor critical elements. As you learn more about your system, ongoing adjustments and improvements in O&M should be considered routine.

Visiting or communicating with the operators of wastewater systems similar to yours often provides ideas or methods that you or they can use to make the job easier and the system perform better and more economically. Be aware, however, that every system is different. The problems that arise in each system and the best solutions are never identical. Different geographical locations, terrain, products, equipment, and community makeup mean each system is unique. Just as there is no single document

TABLE 7.1 LIST OF OPERATORS' TOOLS

1. *WRENCHES*
   a. Deep-well socket set, ⅜″ to 1¼″ with ½″ drive ratchet and breaker bar
   b. Pipe wrenches, assorted sizes 6″ to 36″
   c. Adjustable end wrenches (known as crescent wrenches), assorted sizes 6″ to 18″
   d. Combination open and box end wrench set ¼″ to ¾″
   e. Allen wrench set
2. *SCREWDRIVERS*—Set, assorted sizes
3. *HAMMERS*
   a. Claw—16 oz
   b. Ball peen—16 oz and 8 oz
   c. Brass, rawhide, or soft plastic head
4. *SLEDGEHAMMERS*
   a. 2 pound
   b. 8 pound
5. *PLIERS*
   a. Slip joint common pliers
   b. Groove joint pliers
   c. Electric lineman's pliers
   d. Diagonal cutter pliers
   e. Needle nose pliers
   f. Vise grip pliers
   g. Electric terminal end pliers with wire stripper
   h. Fuse puller pliers
   i. Internal snap-ring pliers
   j. External snap-ring pliers
6. *HACKSAW*
7. *METAL FILES*—set of round and flat
8. *LIGHTS*
   a. 2-cell flashlight
   b. 6-volt battery lantern
9. *MEASURING TAPE*
   a. 10′ retractable steel
   b. 100′ cloth
10. *SHOVEL*
    a. Round point, heavy duty
    b. Square end
11. *PICK*
12. *PIPE VISE*—Truck mounted
13. *FLARING TOOLS AND CUTTER*—For copper and plastic pipe
14. *TUBING BENDER*
15. *PROPANE TORCH AND CYLINDERS*
16. *SOLDER AND FLUX*
17. *PVC GLUE CLEANER AND GLUE*
18. *TEFLON TAPE*—For pipe threads
19. *CRIMPING TOOL*—For emergency shutoff of service pipes
20. *POCKET KNIFE*—Folding with blade lock feature
21. *GASKET SCRAPER TOOL*
22. *COLD CHISELS AND PUNCHES*—Set, assorted sizes
23. *PUTTY KNIFE*
24. *MANHOLE HOOK*
25. *ELECTRIC DRILL*—½″ reversible, variable speed
26. *DRILL BITS* — Set, assorted sizes
27. *TAP AND DIE SET*
28. *EASY OUTS*—Set
29. *STUD BOLT PULLER*
30. *PACKING REMOVAL HOOKS*—Flexible, set of sizes
31. *LANTERN RING REMOVAL TOOLS*—Set
32. *PIPE CUTTING TOOLS*—Manual or electric power
33. *WIRE BRUSHES*
34. *STEEL PRY BARS*—8′ long
35. *EXTENSION LADDER*—Aluminum
36. *STEP LADDER*—Nonconductive
37. *COME-A-LONG*—2-ton capacity
38. *CHAIN HOIST*—2-ton capacity
39. *NYLON ROPE*—½″ diameter
40. *WATER PRESSURE GAUGE*—With connection to hose faucet
41. *FIRE HOSE*—50′
42. *MUD PUMP*—3″ positive displacement, gasoline powered
43. *CLOTHING*
    a. Rain suit—pants and jacket with hood or hat
    b. Rubber boots—calf length, hip length
    c. Rubber gloves
    d. Hard hat
    e. Face mask, clear plastic
    f. Eye goggles
    g. Leather gloves
44. *PORTABLE 5,000-WATT, 110/220-VOLT GRINDING SET*
45. *MULTIMETER*—110 to 600 volts
46. *MULTIMETER/AMMETER TESTER*
47. *ELECTRICAL HEAT SHRINK SPLICERS*
48. *SELF-CONTAINED BREATHING APPARATUS (SCBA)*
49. *RESPIRATOR*—With replaceable filters
50. *ATMOSPHERIC MONITOR*—$O_2$, $H_2S$, explosive limits
51. *GREASE GUN*—With graphite cartridges
52. *ELECTRONIC CALCULATOR*
53. *OPERATION MANUALS*—For plant system and plant equipment
54. *TOOL BOXES*—Lockable steel
55. *WIPING RAGS*
56. *PLASTIC ELECTRICAL TAPE*
57. *FIRST-AID KIT*
58. *PUSH BROOM*—Heavy duty
59. *GASKET ADHESIVE/PIPE SEALANT*

that can tell you what must be done and when, there is no other system just like yours. You will have to evaluate other operators' suggestions in the context of your own system and try to find your own best operating strategy.

Regulatory agencies, such as the health department or pollution control agency, also can assist you in solving problems that arise. A lot of good advice is available without charge and you should not hesitate to ask for assistance from these public agencies when the need arises.

## QUESTIONS

Please write your answers to the following questions and compare them with those on page 376.

7.2A How can an operator keep track of a preventive maintenance program?

7.2B What is the difference between an *EQUIPMENT SERVICE CARD* and a *SERVICE RECORD CARD*?

7.2C What is the purpose of an operator's daily diary?

7.2D What should an operator do with tools after a job is completed?

7.2E What types of injuries do operators suffer most frequently?

7.2F Prepare a list of emergency phone numbers for your wastewater collection and treatment system.

## 7.3 EFFECTIVE TROUBLESHOOTING PROCEDURES

### 7.30 General Approach

Practical step-by-step procedures combined with a commonsense approach is the key to effective troubleshooting.

**Never take anything for granted.**

If you observe changes in the operation of your equipment during a routine maintenance inspection or when a piece of equipment breaks down, use a systematic approach to investigate the problem and identify the best solution.

1. *GATHER PRELIMINARY INFORMATION:* What is supposed to happen, operate, and so forth when this piece of equipment is working properly? What is it doing now?
2. *EXAMINE ALL OTHER FACTORS:* Make a visual inspection of the area. What other unusual things are happening in the system now that this equipment does not work properly? Other pumps stopped, lift station flooded? Is action required, such as isolation, valve changes, or chemical feeding shutdown to protect equipment, system, and safety?
3. *ANALYZE WHAT YOU KNOW:* What part of the system is working correctly? Is this a mechanical failure or an electrical problem caused by a mechanical failure?
4. *SELECT SIMPLE PROCEDURES:* To localize the problem, select logical ways to investigate the problem that can be simply and quickly accomplished.
5. *CLOSE IN ON THE SOURCE OF TROUBLE:* Mechanical or electrical, motor or control system, whatever it might be. Electrical problems usually result from some type of mechanical failure.
6. *PINPOINT THE PROBLEM:* Exactly where is the problem and what do you need for repair?
7. *FIND THE CAUSE:* What caused the problem? Poor design, wear, voltage, overloading?
8. *SOLVE THE PROBLEM AND ELIMINATE THE CAUSE:* Repair the failed piece of equipment to get the system operating again and make whatever changes are needed to prevent this problem from occurring in the future.

### 7.31 Examples—Mound Troubleshooting[1]

This section presents two examples of how to go about troubleshooting a problem in a mound system. As you read each example, notice how the problem is investigated in a logical, step-by-step manner. Possible sources of the problem are examined systematically until the actual problem has been identified.

**EXAMPLE 1: Alarm Activation**

You are called out for service work on a Sunday afternoon. Every minute that passes is another precious moment of the ball game missed and you need to diagnose the problem quickly and solve it so that your customers can rest assured that the expert they have called is competent to perform the maintenance required. The alarm's shrieking confirms that there is something wrong. Most alarms have a Push to Silence button to stop the shrieking. But what is the problem? Is it in the mound itself, the distribution piping, the pump tank, or the electrical supply?

The alarm signals a high wastewater level in the pump tank or a low water level due to a leaking or siphoning tank. This is caused by many things:

- Electrical—power or switch failure, connections, wiring, pump failure
- Distribution system—plugged or broken force main or laterals
- Pump chamber—float anchors, clogged pump filter
- Soil infiltration—severe ponding or *SURCHARGE*[2] flowback
- Mechanical—impeller damaged or plugged
- Hydraulic—high wastewater flows, groundwater infiltration

As the service professional, you need to begin with the most obvious, most easily solved problems and advance through the possibilities. Figure 7.2 is a flow chart to assist you through alarm activation troubleshooting.

1. This material on mound troubleshooting was provided by Lynita M. Docken, Master Plumber and Plumbing Designer, Wisconsin Department of Commerce, LaCrosse, WI. Her contribution is greatly appreciated.
2. *Surcharge.* Sewers are surcharged when the supply of water to be carried is greater than the capacity of the pipes to carry the flow. The surface of the wastewater in manholes rises above the top of the sewer pipe, and the sewer is under pressure or a head, rather than at atmospheric pressure.

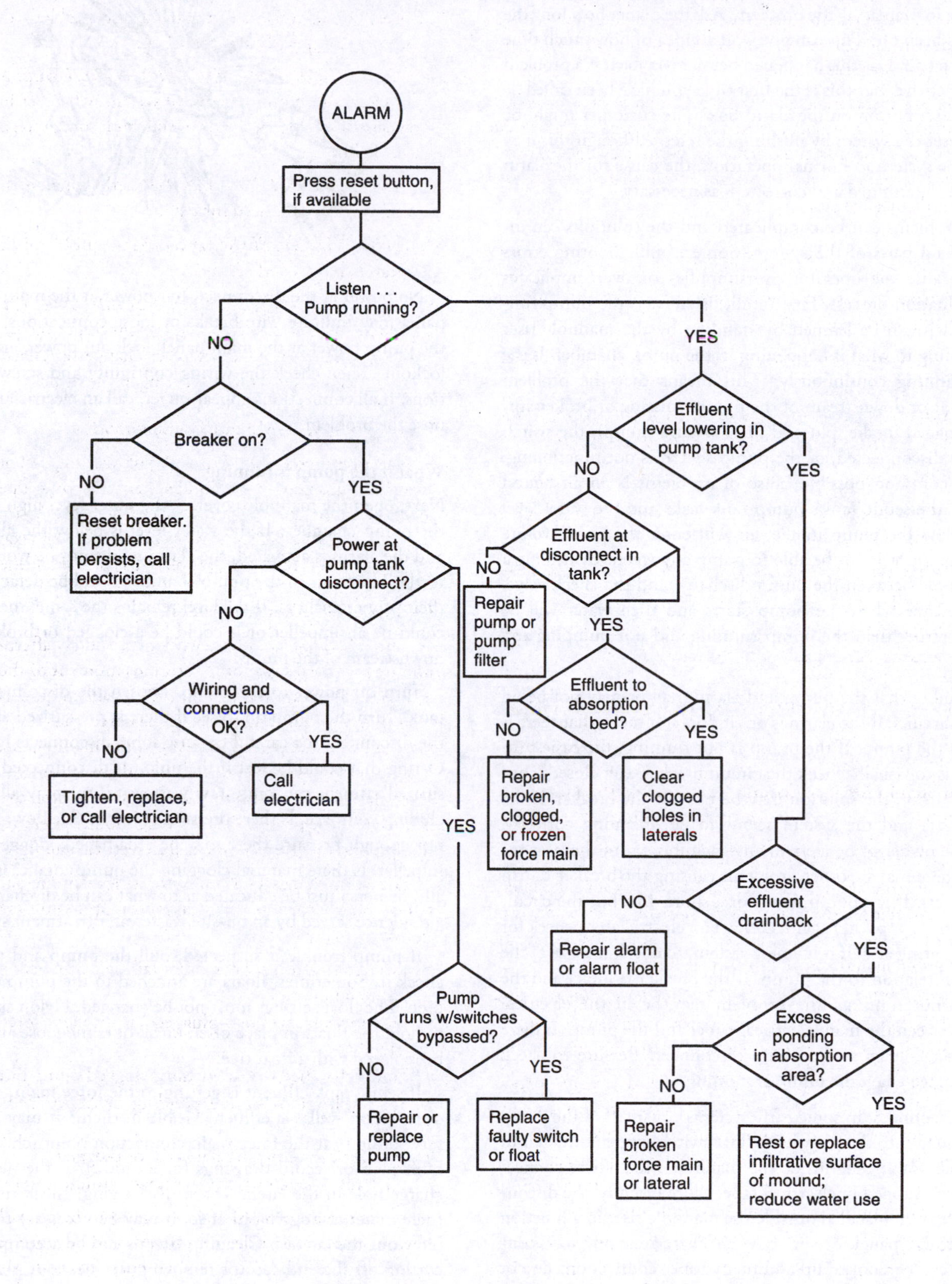


*Fig. 7.2 Alarm troubleshooting*

(Reprinted with permission of Lynita M. Docken)

When a customer calls, some information can be collected that may assist in evaluating the problem. Ask the owner how long the alarm has been on? This can give you an idea of how much time is left for repair. Has this happened before? Maybe it is a problem that is repetitive, but this is the first time you have been called. Is there a Reset button on the alarm box? The customer might be able to reset the system by pushing the reset. Although this may restore the system to normal operation, the cause for the alarm should be determined and corrected, as necessary.

The problems can be complicated and the solutions can involve several parts of the system. Some troubleshooting errors occur because the operator oversimplifies or overcomplicates their evaluation process. First of all, listen … is the pump running? Much can be learned by standing by the manhole riser and listening to what is happening in the pump chamber. Is the pump running continuously? This means that the problem could be at or downstream of the pump (missing or broken impeller, clogged intake, plugged filter around pump), the switch floats are disconnected, or the pump switch is not functioning. Another common possible cause of problems is an air-bound pump. If the septic tank's pump tank leaks and the water level drops below the pump intake, air will enter the pump volute and the pump will not be able to pump any water. By drilling a ⅛-inch hole between the pump discharge and the check valve, air can escape when the pump starts and then water will be pumped rather than the pump running and not pumping any water.

Remember that the pump and alarm may or may not be on separate circuits. If the alarm is on, it does not mean that there is power to the pump. If the pump is not running, the easiest inspection is to visually check the circuit breaker. Is it closed (connected)? If not, the solution may be to reset the breaker. If the pump starts and the system seems to be operating correctly, advise the owner to be alert to any malfunction in the near future, as there may be other problems causing the breaker to trip (disconnect). If the pump starts for a short time but the circuit breaker trips (pops off) again, contact an electrician to check for short circuiting and, if it is a new system, to check the size of the breaker in relation to the pump. If the breaker is intact and the pump is not running, the problem may be in the electrical wiring between the main electrical panel and the pump. Is there the correct voltage at the pump disconnect? Be sure to check pump voltage when the pump is running.

The disconnects in some early systems consisted of the pump wire plugged into the piggyback float switch plug and then into the female power plug inside the manhole riser. These connections were wrapped in electrical tape. More recently, the disconnects have been installed outside the manhole riser in a junction box or breaker panel. Be very careful. There are many ways connections can be hooked up and many times connections do not meet electrical code requirements. Such connections are very dangerous.

If there is power at the disconnect, bypass the switches to see if the pump runs without automatic control. If it does, the pump ON switch is the problem and may need to be replaced or reconnected to the force main or float tree.

No power at the disconnect, but power at the main electrical panel may indicate wire breaks or loose connections. Turn off the pump power at the main panel, lock out power, and tag the lockout. Then check the wiring continuity and screw connections. If all connections appear intact, call an electrician to diagnose the problem.

**What if the pump is running?**

Next, open the manhole cover. Is the water level high or low in the pump chamber? If the water level is below the alarm level and the alarm is activated, the alarm might not be working correctly. The floats in the pump chamber could be detached from their proper anchors. If the level remains the same, the problem could be an impeller or it could be a clogged or broken point downstream of the pump.

Turn off power to the pump. Loosen the disconnect in the tank. Turn the pump on to see if there is pressurized effluent to the disconnect (be careful because some disconnects include an O-ring that could be lost if the union is disconnected). A pressurized effluent can cause raw wastewater to spray all over, including your face. Is there pressure or just trickle flow? If there is not enough pressure there may be a damaged, clogged, or lost impeller. Is there material clogging the pump intake? Occasionally, owners must be educated as to what can be discharged from a residence served by an on-site wastewater treatment system.

If pump trouble is suspected, pull the pump and physically check it. Sometimes, floats are attached to the pump discharge pipe. The float settings must not be changed. Nylon straps used to hold the floats in place often break or come loose and should be replaced with a float tree.

If septic tank effluent is getting to the force main, check the observation wells. Is effluent visible in them? It may be necessary to dig near the force main connection point (check the as-built plan or record drawings for location). If the nearest discharge hole in the lateral is not discharging, check the hole. Is there material clogging it? If so, it may be necessary to clean or blow out the laterals. Cleaning laterals can be accomplished by cutting off the end cap (or removing the cleanout plugs in the turn-ups) and pushing (or pulling) brushes through the laterals.

Pulling is accomplished by first pushing a sewer tape or electrical fish tape through the lateral and then pulling the brush back. The brush is attached to the end of a snake or tape and pushed and pulled using a scrubbing motion. The use of a high-pressure water nozzle on the end of a ⅜-inch diameter hose is

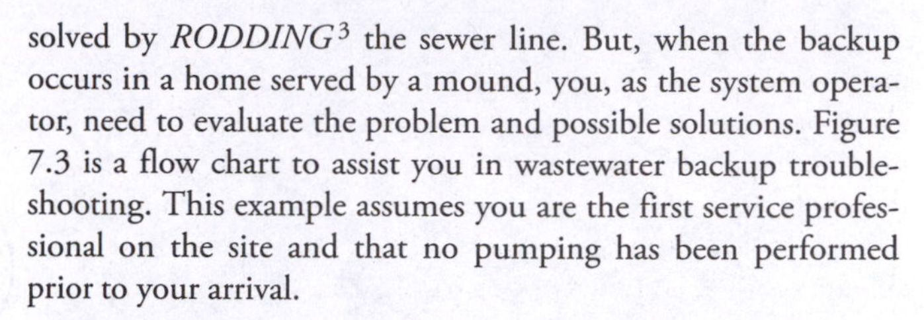

very effective. Blowing out the laterals can be accomplished by applying positive pressure at the force main connection. A vacuum can be applied to the laterals by a pump truck, also.

If the effluent level drops when the pump is running and there is no excessive drainback into the tank, then the problem could be a faulty alarm. If possible, repair or replace the alarm box or check the alarm float position.

Excessive drainback can be caused by ponding in the soil absorption system that exceeds the height of the laterals. This could be caused by a clogged infiltration surface (thick biomat), frozen soil absorption system, groundwater or surface water infiltration, or hydraulic overload. Evaluate the biomat at the infiltrative surface, check for leaking plumbing fixtures, and take corrective action. Corrective action may include a reduction in water use, replacement of the infiltrative surface and clogging mat, or replacement of the entire mound with another system.

**Problem solved?**

After the system appears to be working, it is the operator's responsibility to evaluate the cause of the system failure and the possibility that a combination of problems exists.

- Did the pump quit because of short cycling caused by movement of the float settings?
- Or did the pump run constantly because the laterals were nearly or completely plugged?
- Did the system pond excessively because fine sand was used for mound construction?
- Has the system been hydraulically overloaded?
- Is the effluent a high-strength waste that led to premature soil absorption system failure?

All service work requires evaluation of the entire system and the final determination that the correct course of action was or will be taken.

Alarm activation is only one aspect of mound troubleshooting. Operators should evaluate any other indications of problems in a similar manner. Troubleshooting in an efficient manner can save time and money and make your system operate more smoothly. Try to attend any training center sessions in your area to learn more about advanced troubleshooting methods.

**EXAMPLE 2: Wastewater Backup**

There is a stench-laden mixture that infiltrates a family's basement. Carpet, treasured relics, and squirreled-away "I'll use it sometime" junk become dumpster material. The call you receive as a system manager is a plea for help. No one looks forward to the inevitable time when the sewer backs up. If the home is served by a municipal sewer, the problem stoppage is usually solved by *RODDING*[3] the sewer line. But, when the backup occurs in a home served by a mound, you, as the system operator, need to evaluate the problem and possible solutions. Figure 7.3 is a flow chart to assist you in wastewater backup troubleshooting. This example assumes you are the first service professional on the site and that no pumping has been performed prior to your arrival.

The following is a series of questions that will help in the evaluation process.

- **Is the alarm on?** If the alarm is on, refer to Example 1, "Alarm Activation." If the alarm is not sounding, the next step is to find the dose or pump tank, take off the cover, and examine the contents.
- **Is the water level in the dose or pump tank low?** Normal levels in a pump tank can range from the level of the pump OFF float to the pump activation elevation. A siphon dose tank's normal level can vary from the bottom edge of the bell to a short distance below the overflow pipe. Pump tanks very often leak, which can cause low water levels.
- **Is it a pump tank or a siphon?** If the level is excessive in a pump tank, return to Example 1, "Alarm Activation." Excessive level in a siphon can be a complicated problem. If the siphon wastewater level is well above the overflow, the problem is downstream of the siphon tank or in the siphon trap. This could be in the force main or laterals (see Example 1), or in the absorption area.

Although trickle flow from a siphon will not cause wastewater to back up into the home, trickle flow can initiate clogging of distribution laterals and perforations. Also, trickle flow in cold climates can build up as ice and eventually plug the force main. Pressurized flow can flush laterals, but the trickle flow from a malfunctioning siphon can promote bacterial slime accumulation. Compounding the problem is the fact that siphons designed in some areas are not required to include any alarm or monitoring device. Some siphons can be reactivated by blowing air under the bell. At other times, more radical procedures are needed to reset the siphon, such as replacing the bolts, resetting the tank, or replacing the siphon vent pipe. Contact the manufacturer of the siphon for assistance.

If a siphon has been trickling, the first discharge should be observed to check for a slow discharge rate. This would indicate some plugging in the discharge line or laterals or air entrapment in downstream transport pipes.

What if the water level is low in the dose tank? If everything seems to be functioning properly at this point, it is time to investigate upstream possibilities. Check the septic tank. Ask the owner when the septic tank was last pumped. The last time the septic tank was pumped should be recorded in the system records. The operator can obtain some of the data from the local septic tank pump contractors. Homeowners

3. *Rod, Sewer.* A light metal rod, three to five feet long, with a coupling at each end. Rods are joined and pushed into a sewer to dislodge obstructions.

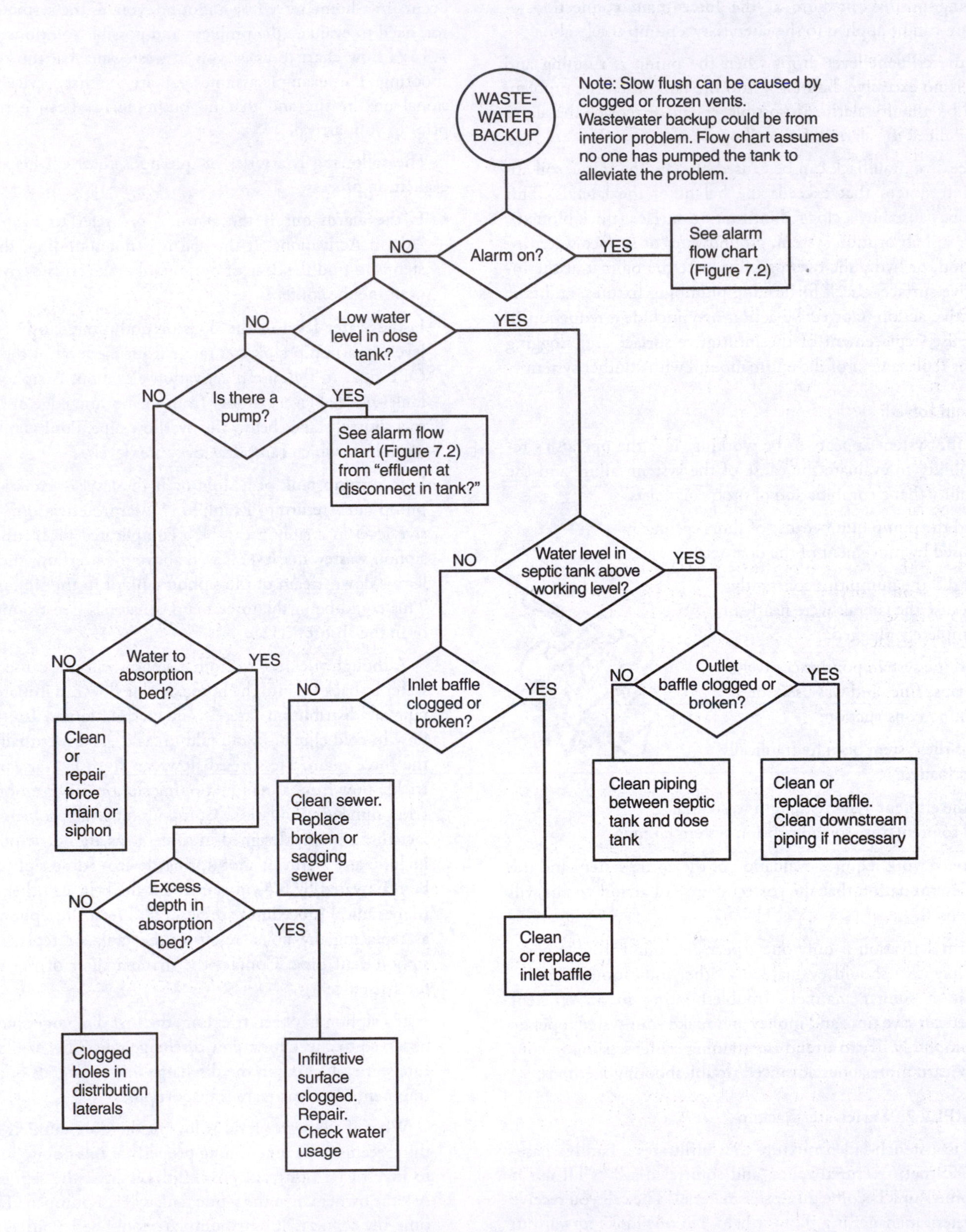


*Fig. 7.3 Troubleshooting a wastewater backup*

(Reprinted with permission of Lynita M. Docken)

sometimes take pride in the fact that their "septic tank has not been pumped in 25 years," but when sludge and scum occupy a large volume of a tank, the system cannot function properly. The tank should be pumped before any scum or sludge can escape; generally this is when the sludge and scum occupy about one-third of the tank volume. If this is the problem, call a pumper to pump and check the tank. In cold climates, the septic tank may freeze, especially during the first year of operation when the biological activity may not be adequately established.

- **What is the water level in the septic tank?** Is the water level above the normal working level in the septic tank? This could be caused by a plugged septic tank outlet or filter. There are many household articles that can cause a filter to plug. If foreign material is found, it is necessary to educate the owners about the use of the private wastewater system. The rule of thumb is, "only wastewater and toilet paper should go down the drain." The garbage disposal has become a common appliance in residences, but its use with private wastewater systems should be discouraged. Many food particles are buoyant (tend to float) and can become lodged in filters or lateral perforations, and vegetable matter may speed up the rate of sludge accumulation in the tank.
- **What is the condition of the baffles or any filtration device in the septic tank?** If the plugged discharge from the septic tank is caused by a broken or missing outlet baffle, the baffle must be replaced and the downstream piping cleaned, if necessary.

**CAUTION**

Never enter a septic tank or dose tank without following confined space entry procedures. Tanks may contain hazardous atmospheres.

If the septic tank level appears normal (about 1 to 4 inches below the inlet piping), check the inlet baffle for clogs. It is possible that objects like paper towels, diapers, or other materials such as a buildup of excessive grease and cooking oils could plug the inlet baffle.

- **Is the sewer plugged?** If everything is normal in the septic tank, the problem is identical to a sewer blockage in the city. The sewer must be cleaned and if it is impassible with an *AUGER* [4] or rodder or if the blockage returns, the sewer may need to be excavated and a sag in the piping repaired or broken pipe replaced. Frozen sewers can be thawed by use of warm water or steam. Care must be taken when thawing plastic piping because excessive temperatures can damage the piping. Once a sewer is known to be prone to freezing it may require insulation. Although PVC sewers have not been plagued with tree root intrusion, older sewer materials like cast iron and clay tile are more likely to experience such problems. If root intrusion becomes excessive it may be necessary to replace the sewer. In some cases it has been found that occasional wastewater backup into a floor drain or slow water closet flush can be caused by a frozen or plugged stack vent. If all other causes are ruled out, this may be the culprit.

With careful examination of the causes of blockages, repeat performances can be avoided in most cases. The success of your repair and service activities relies on your skill in diagnosing the problem and resolving it.

## QUESTIONS

Please write your answers to the following questions and compare them with those on page 376.

7.3A What is the key to effective troubleshooting?

7.3B When a customer calls with an alarm problem, what questions should be asked?

7.3C What is the first item to check if a pump is not running?

7.3D What should be your major concern when required to enter a septic tank?

## 7.4 TROUBLESHOOTING ELECTRICAL PROBLEMS

Operators of small wastewater collection, treatment, and discharge systems often work alone and do not have a staff of experts to assist them with maintenance and repair activities. Many have a good working knowledge of electrical equipment and have learned by experience how electric motors and electrical circuits work. They may be tempted to try to fix just about any piece of equipment even though they may not fully understand the equipment or the risks associated with it. When electricity is involved, the risks are particularly serious because the consequences of an accident can be death by electrocution or severe, disabling burns.

The purpose of this section is to give you a basic understanding of electricity, which will help you provide electricians with the information they will need when you request their assistance. Commonly used electrical terms are explained and suggestions are provided for investigating the source of electrical problems. Also, the appendix at the end of this chapter describes step-by-step procedures for troubleshooting an electrical control panel.

4. *Auger* (AW-grr). A sharp tool used to go through and break up or remove various materials that become lodged in sewers.

### 7.40 Recognize Your Limitations

In the water and wastewater departments of all small systems, there is a need for maintenance operators to know something about electricity. Duties could range from repairing a taillight on a trailer or vehicle to repairing complex pump controls and motors. Very few maintenance operators do the actual electrical repairs or troubleshooting. This is a highly specialized field and unqualified people can seriously injure themselves and damage costly equipment. For these reasons, you must be familiar with electricity, know the hazards, and recognize your own limitations when you must work with electrical equipment.

Most municipalities employ electricians or contract with a commercial electrical company that they call when major problems occur. However, the maintenance operator should be able to explain how the equipment is supposed to work and what it is doing or is not doing when it fails. After studying this section, you should be able to tell an electrician what appears to be the problem with electrical panels, controls, circuits, and equipment.

The need for safety should be apparent. If proper safe procedures are not followed in operating and maintaining electrical equipment, accidents can happen that cause injuries, permanent disability, or loss of life. Serious accidents that could have been avoided (see examples below) have happened because machinery was not shut off, locked out, and tagged (Figure 7.4) properly.

1. A maintenance operator could be cleaning a pump and have it start, thus losing an arm, hand, or finger.
2. Electric motors or controls not properly grounded could lead to possible severe shock, paralysis, or death.
3. Improper circuits such as a wrong connection, safety devices jumped, wrong fuses, or improper wiring can cause fires or injuries due to incorrect operation of machinery.

It may seem like a waste of time to tag a piece of equipment (such as an alarm panel) that you have shut off while you inspect some other component of the system. After all, you are the only person working on the system and you can easily remember what you have just done. However, suppose you are injured before you can reset the alarm or your work is interrupted by another emergency and it becomes necessary for an electrician or another operator to finish the job. Will you still remember in a day or

**DANGER**

**OPERATOR WORKING ON LINE**

**DO NOT CLOSE THIS SWITCH WHILE THIS TAG IS DISPLAYED**

TIME OF: ____________________

DATE: ____________________

SIGNATURE: ____________________

This is the ONLY person authorized to remove this tag.

INDUSTRIAL INDEMNITY/INDUSTRIAL UNDERWRITERS/
INSURANCE COMPANIES

4E210—R66

*Fig. 7.4 Typical warning tag*

(Source: Industrial Indemnity/Industrial Underwriters/Insurance Companies)

two that you shut off the alarm? How will the electrician know what you have done? Make it a habit to properly turn off, lock out, and tag any piece of equipment, particularly electrical equipment, that has the potential to injure someone. (See Chapter 3, Section 3.22, "Basic Lockout/Tagout Procedure," for additional information about how to properly lock out equipment. Also see Section 3.7, "Electricity," for a description of the effects of electricity on a human body.)

Another reason the operator of a wastewater collection or treatment system needs a basic working knowledge of electricity is to prevent financial losses resulting from motors burning out and from damage to equipment, machinery, and control circuits. Additional costs result when damages have to be repaired, including payments for outside labor.

**WARNING**

Never work in electrical panels or on electrical controls, circuits, wiring, or equipment unless you are qualified and authorized. By the time you find out what you do not know about electricity, you could find yourself too dead to use the knowledge.

## QUESTIONS

Please write your answers to the following questions and compare them with those on page 376.

7.4A Why must unqualified or inexperienced people be extremely careful when attempting to troubleshoot or repair electrical equipment?

7.4B What could happen when machinery is not shut off, locked out, and tagged properly?

### 7.41 Understanding Electricity

Most electrical equipment used in wastewater collection and treatment systems is labeled by the manufacturer with a *NAMEPLATE*.[5] All of the information on the nameplate should be recorded and placed in a file for future reference. Many times, the nameplate is painted, corroded, or missing from the unit when the information is needed to repair the equipment or replace parts. Record the serial or model number, the date of installation, and the date the equipment was placed in service on the equipment service card or service record card or file this information in a safe and convenient location.

The nameplate on a piece of electrical equipment will always indicate the proper voltage and allowable current in amps for that particular piece of equipment. The following sections describe the meaning of the terms "voltage" and "amps" and the basic differences between direct current and alternating current.

#### *7.410 Volts*

Voltage (E) is also known as electromotive force (EMF). It is the electrical pressure available to cause a flow of current (amperage) when an electric circuit is closed.[6] This pressure can be compared with the pressure or force that causes water to flow in a pipe. Some pressure in a water pipe is required to make the water move. The same is true of electricity. A force is necessary to push electricity or electric current through a wire. This force is called voltage. There are two types of current: direct current (DC) and alternating current (AC).

#### *7.411 Direct Current (DC)*

Direct current (DC) flows in one direction only and is essentially free from pulsation, that is, the strength of the current remains steady. Direct current is seldom used in wastewater collection and treatment systems except possibly in electronic equipment, some control components of pump drives, and standby lighting. Direct current is used exclusively in automotive equipment, certain types of welding equipment, and a variety of portable equipment. Direct current is found in various voltages such as 6 volts, 12 volts, 24 volts, 48 volts, and 110 volts. All batteries are direct current. DC voltage can be measured by holding the positive and negative leads of a DC multimeter on the corresponding terminals of the DC device such as a battery. Direct current usually is not found in higher voltages (over 24 volts) around treatment plants except in motor generator sets. Care must be taken when installing battery cables and wiring that positive (+) and negative (–) poles are connected properly to wires marked (+) and (–). If not properly connected, you could get an arc of electricity across the unit that could cause an explosion.

#### *7.412 Alternating Current (AC)*

An alternating current circuit is one in which the voltage and current periodically change direction and *AMPLITUDE*.[7] In other words, the current goes from zero to maximum strength, back to zero, and to the same strength in the opposite direction. Most AC circuits have a frequency of 60 *CYCLES*[8] per second. "Hertz" is the term we use to describe the frequency of cycles completed per second so our AC voltage would be 60 Hertz (Hz).

There are three classifications of alternating current:

1. Single phase
2. Two phase
3. Three phase or polyphase

The most common of these are single phase and three phase. The various voltages you probably will find on your job are 110 volts, 120 volts, 208 volts, 220 volts, 240 volts, 277 volts, 440 volts, 480 volts, 2,400 volts, and 4,160 volts.

5. *Nameplate.* A durable, metal plate found on equipment that lists critical installation and operating conditions for the equipment.
6. Electricians often talk about closing an electric circuit. This means they are closing a switch that actually connects circuits together so electricity can flow through the circuit. Closing an electric circuit is like opening a valve on a water pipe.
7. *Amplitude.* The maximum strength of an alternating current during its cycle, as distinguished from the mean or effective strength.
8. *Cycle.* A complete alternation of voltage or current in an alternating current (AC) circuit.

Single-phase power is found in lighting systems, small pump motors, various portable tools, and throughout our homes. This power is usually 120 volts and sometimes 240 volts. Single phase means that only one phase of power is supplied to the main electrical panel at 240 volts and it has three wires or leads. Two of these leads have 120 volts each; the other lead is neutral and usually is coded white. The neutral lead is grounded. Many appliances and power tools have an extra ground (commonly a green wire) on the case for additional protection.

Three-phase power is generally used with motors and transformers found in wastewater treatment plants, but is generally not available in rural areas. Phase converters allow three-phase electric motors to be operated on single-phase power. A converter is wired in between the motor and the power line. The advantages of three-phase motors are lower current draw, simple reversing, and lighter weight. Heavy-duty converters are suitable for long starting, frequent starting, and reversing.

Circuit breakers (Figure 7.5) are used to protect electric circuits from overloads. Most circuit breakers are metal conductors that de-energize the main circuit when excess current passes through a metal strip causing it to overheat and open the main circuit.

Two-phase systems will not be discussed because they are seldom found in wastewater collection and treatment facilities.

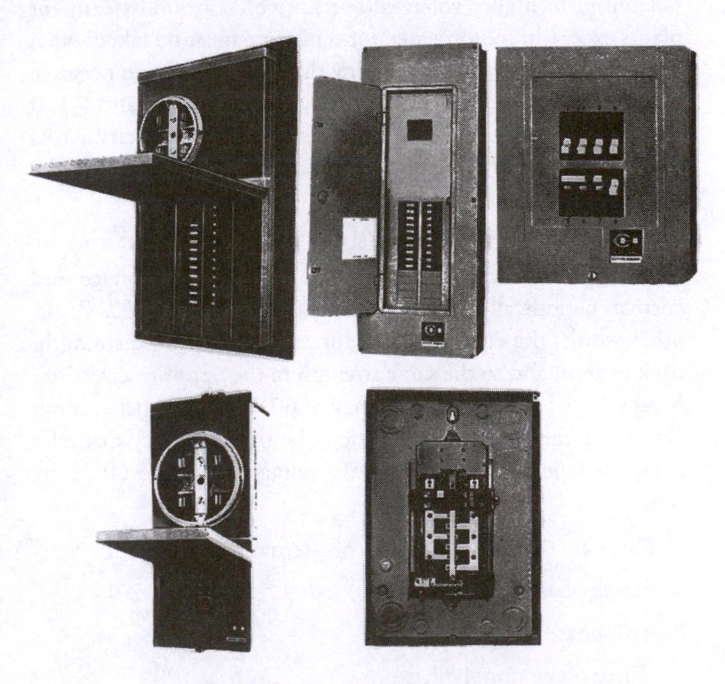

*Fig. 7.5 Circuit breakers*
(Courtesy of Consolidated Electrical Distributors, Inc.)

### *7.413 Amps*

An ampere (A) is the SI (*Le Système International d'Unités* or International System of Units)[9] unit of electric current (I). This is the current produced by a pressure of one volt in a circuit having a resistance of one ohm. Amperage is the measurement of current or electron flow and is an indication of work being done or how hard the electricity is working.

In order to understand amperage, one more term must be explained. The *OHM* is the practical unit of electrical resistance (R). "Ohm's Law" states that in a given electric circuit, the amount of current (I) in amperes is equal to the pressure in volts (E) divided by the resistance (R) in ohms. The following three formulas are given to provide you with an indication of the relationships among current, resistance, and EMF (electromotive force).

$$\text{Current, amps} = \frac{\text{EMF, volts}}{\text{Resistance, ohms}} \qquad I = \frac{E}{R}$$

$$\text{EMF, volts} = (\text{Current, amps})(\text{Resistance, ohms}) \qquad E = IR$$

$$\text{Resistance, ohms} = \frac{\text{EMF, volts}}{\text{Current, amps}} \qquad R = \frac{E}{I}$$

These equations are used by electrical engineers for calculating circuit characteristics. If you memorize the following relationship, you can always figure out the correct formula.

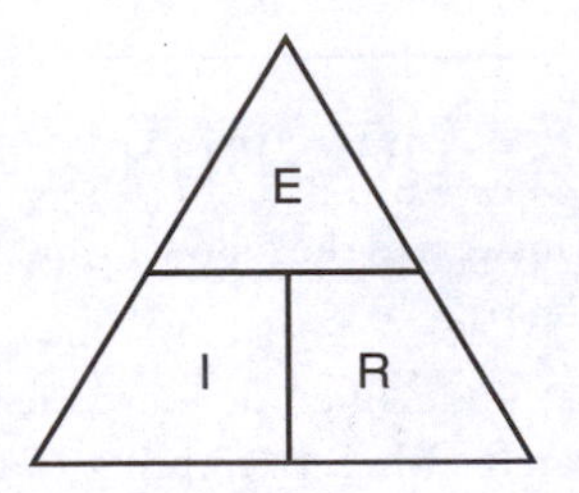


To use the above triangle, you cover up with your finger the term you do not know or are trying to find out. The relationship between the other two known terms will indicate how to calculate the unknown. For example, if you are trying to calculate the current, cover up I. The two knowns (E and R) are shown in the triangle as E/R. Therefore, I = E/R. The same procedure can be used to find E when I and R are known or to find R when E and I are known.

### *7.414 Watts*

Watts (W) and kilowatts (kW) are the units of measurement of the rate at which power is being used or generated. In DC circuits, watts (W) equal the voltage (E) multiplied by the current (I).

$$\text{Power, watts} = (\text{Electromotive Force, volts})(\text{Current, amps})$$

$$\text{or P, watts} = (\text{E, volts})(\text{I, amps})$$

In AC polyphase circuits, the formula becomes more complicated because of the inclusion of two additional factors. First, there is the square root of 3, for three-phase circuits, which is equal to 1.73. Second, there is the power factor, which is the ratio of the true or actual power passing through an electric circuit to the product of the voltage times the amperage in the circuit. For

9. SI (*Le Système International d'Unités* or International System of Units) is a system of physical units (SI Units) based on the meter, kilogram, second, ampere, kelvin, candela, and mole, together with a set of prefixes to indicate multiplication or division by a power of ten.

standard three-phase induction motors, the power factor will be somewhere near 0.9. The formula for power input to a three-phase motor is:

$$\text{Power, kilowatts} = \frac{(\text{E, volts})(\text{I, amps})(\text{Power Factor})(1.73)}{1{,}000 \text{ watts/kilowatt}}$$

Since 0.746 kilowatt equals 1.0 horsepower, then the power output of a motor is:

$$\text{Power Output, horsepower} = \frac{(\text{Power Input, kilowatts})(\text{Efficiency, \%})}{(0.746 \text{ kilowatt/horsepower})(100\%)}$$

### *7.415 Power Requirements*

Power requirements (PR) are expressed in kilowatt hours. One watt for 1,000 hours or 500 watts for two hours equals one kilowatt hour. The power company charges so many cents per kilowatt hour.

$$\text{Power Req, kW-hr} = (\text{Power, kilowatts})(\text{Time, hours})$$

$$\text{PR, kW-hr} = (\text{P, kW})(\text{T, hr})$$

### *7.416 Conductors and Insulators*

A material, like copper, that permits the flow of electric current is called a conductor. Material that will not permit the flow of electricity, like rubber, is called an insulator. Such material when wrapped or cast around a wire is called insulation. Insulation is commonly used to prevent the loss of electrical flow by two conductors coming into contact with each other.

## QUESTIONS

Please write your answers to the following questions and compare them with those on page 376.

7.4C How can you determine the proper voltage and allowable current in amps for a piece of equipment?

7.4D What are two types of current?

7.4E Amperage is a measurement of what?

### 7.42 Tools, Meters, and Testers

**WARNING**

Never enter any electrical panel or attempt to troubleshoot or repair any piece of electrical equipment or any electric circuit unless you are qualified and authorized.

### *7.420 Voltage Testing*

In order to maintain, repair, and troubleshoot electrical equipment and circuits, the proper tools are required. You will need a *MULTIMETER* to check for voltage (Figures 7.6 and 7.7). There are several types on the market and all of them work. They are designed to be used on energized circuits and care must be exercised when testing. By holding one lead on ground and the other on a power lead, you can determine if the circuit is energized.

Be sure the multimeter that you are using has sufficient range to measure the voltage you would expect to find. In other words, do not use a multimeter with a limit of 600 volts on a circuit that normally is energized at 2,400 volts. With the multimeter, you can tell if the current is AC or DC and the intensity or voltage, which will probably be one of the following: 120, 208, 240, 480, 2,400, or 4,160.

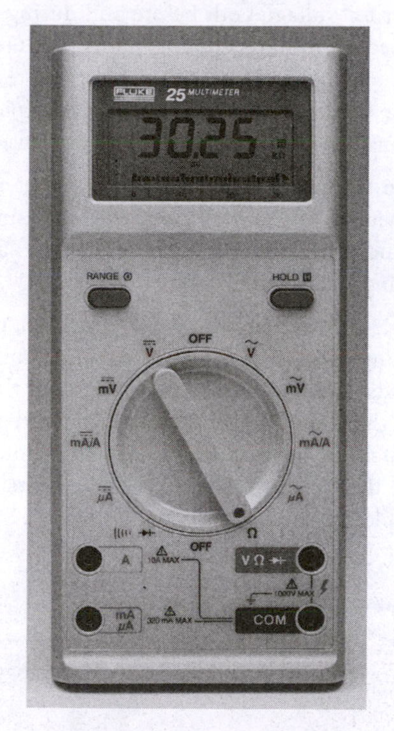


*Fig. 7.6 Digital multimeter*
(Reproduced with permission of Fluke Corporation)

*Fig. 7.7 Analog clamp-on multimeter*
(Permission of Simpson Electric)

Do not work on any electric circuits unless you are qualified and authorized. Use a multimeter and other circuit testers to determine if a circuit is energized, or if all voltage is off. This should be done after the main switch is turned off to make sure it is safe to work inside the electrical panel. Always be aware of the possibility that even if the disconnect to the unit you are working on is off, the control circuit may still be energized if the circuit originates at a different distribution panel. Also, a capacitor in the unit such as a power factor correction capacitor on a motor may have sufficient energy stored to cause considerable harm to an operator. Test for voltage both before and during the time the switch is pulled off to have a double-check. This procedure ensures that the multimeter is working. Use circuit testers to measure voltage or current characteristics to a given piece of equipment and to make sure that you have or do not have a live circuit.

In addition to checking for power, a multimeter can be used to test for open circuits, blown fuses, single phasing of motors, grounding, and many other uses. Some examples are illustrated in the following paragraphs.

In the single-phase circuit shown in Figure 7.8, test for power by first opening the switch as shown. Using a multimeter with clamp-on leads, clamp one wire (lead) between the switch and the light and the other lead between the light and ground. Now, close the switch. In general, the multimeter should register at 110 volts. If there is no reading, then the switch is probably faulty. Try replacing the switch.

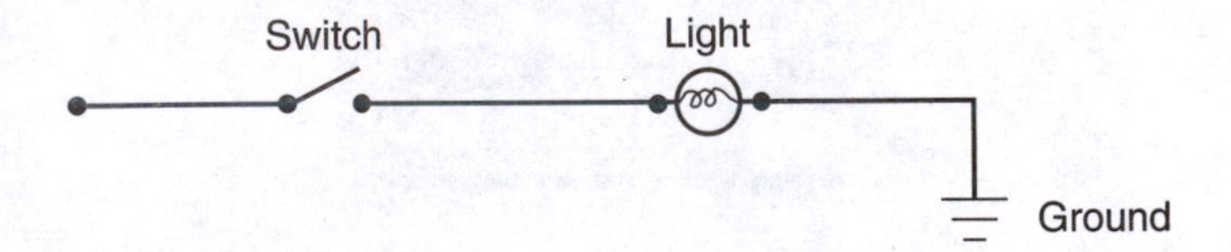


*Fig. 7.8 Single-phase circuit (switch in open position)*

In the single-phase, three-lead circuit shown in Figure 7.9, test for power by first opening the switches as shown.

1. Using a multimeter with clamp-on leads, clamp one lead on L1 and the other lead on L2 between the fuses and the load (equipment). With the connections made, bring the multimeter and attached leads out of the panel and close the panel door as far as possible without cutting or damaging the meter leads. Now, close the switches. (*NOTE:* Some switches cannot be closed if the panel door is open. The panel door is closed when testing because hot copper sparks could seriously injure you when the circuit is energized and the voltage is high.)
2. The multimeter should register at 220 volts. If there is no reading, either a switch or both switches could be faulty, or a fuse or both fuses could be blown. Let us first determine if the problem is with the fuses.
3. To test the fuses, first open the switches. Clamp a lead on L1 and a lead on L2 between the switches and the line (panel). Close the panel door, then close the switches.

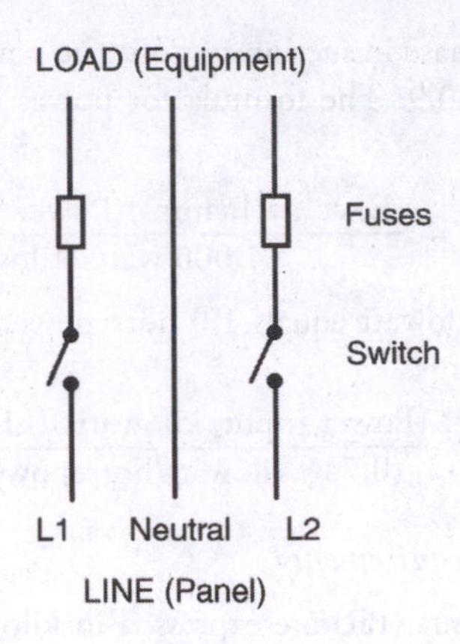


*Fig. 7.9 Single-phase, three-lead circuit*

4. If no reading is registered, test the fuse on L1 by first opening the switches. Move the test lead on L1 to a position between the fuse and the load; move the test lead on L2 to a position between line and load on the neutral lead. Close the panel door, then close the switches. If a reading of 110 volts is observed, then the fuse on L1 is OK; we also know that the switch on L1 is OK. If there is no reading, the switch on L1 could be faulty or the fuse on L1 could be blown. Open the switches and try replacing the fuse on L1. Retest for power as described in Step 1.
5. The multimeter should register at 220 volts. If there is still no reading, a switch or both switches could be faulty or the fuse on L2 could be blown.
6. Test the fuse on L2 by first opening the switches. Move the test lead on L1 to a position between the fuse and the load on L2. (The other test lead should remain on the neutral lead.) Close the panel door, then close the switches. If a reading of 110 volts is observed, then the fuse on L2 is OK; we also know that the switch on L2 is OK. If there is no voltage reading, the switch on L2 could be faulty or the fuse on L2 could be blown. Open the switches and try replacing the fuse on L2. Retest for power as described in Step 1.
7. The multimeter should register at 220 volts. If there is still no reading, the problem could be faulty switches, a broken neutral line, an open switch in another location, or a power outage.

**WARNING**

Turn off power and be sure that there is no voltage in either power line before changing fuses. Use a fuse puller. Test circuit again in the same manner to make sure fuses or circuit breakers are OK. 220 volts power or voltage should be present between L1 and L2. If fuse or circuit breaker trips again, shut off and determine the source of the problem.

Test for power at points A, B, and C in a three-phase circuit (Figure 7.10). Place the multimeter leads on lines A and B. Close all switches. 220 volts should register on the multimeter. Check between lines A and C, and between lines B and C. 220 volts should be recorded between all of these points. If voltage is not present, one or all of the fuses are blown or the circuit breaker has been tripped. First, check for voltage above the fuses at all of these points, A to B, A to C, and B to C, to make sure power is available (see 220 readings in Figure 7.10). If voltage is recorded, move leads back down to the bottom of the fuses. If voltage is present from A to B, but not at A to C and B to C, the fuse on line C is blown. If there were no voltage readings at any of the test points, all the fuses could be blown.

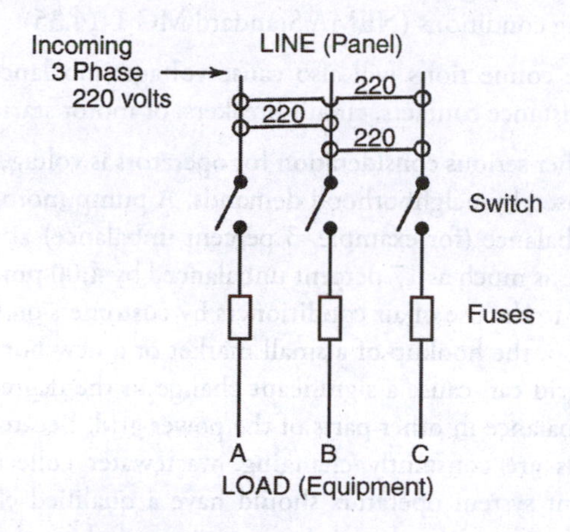


*Fig. 7.10 Three-phase circuit, 220 volts*

Another way of checking the fuses with the load connected on this three-phase circuit would be to take your multimeter and place one lead on the bottom, and one lead on the top of each fuse. You should not get a voltage reading on the multimeter. This is because electricity takes the path of least resistance. If you get a reading across any of the fuses (top to bottom), that fuse is bad.

Always make sure that when you use a multimeter it is set for the proper voltage. If voltage is unknown and the meter has different scales that are manually set, always start with the highest voltage range and work down. Otherwise, the multimeter could be damaged. Look at the equipment instruction manual or nameplate for the expected voltage. Actual voltage should not be much higher than given unless someone made a mistake when the equipment was wired and inspected.

### *7.421 Ammeter*

Another meter used in electrical maintenance and testing is the *AMMETER.* The ammeter records the current or amps flowing in the circuit. There are several types of ammeters, but only two will be discussed in this section. The ammeter generally used for testing is called a clamp-on type. The term "clamp-on" means that it can be clamped around a wire supplying a motor, and no direct electrical connection needs to be made. Each "leg" or lead on a three-phase motor must be individually checked.

The first step should be to read the motor nameplate data and find what the amperage reading should be for the particular motor or device you are testing. After you have this information, set the ammeter to the proper scale. Set it on a higher scale than necessary if the expected reading is close to the top of the meter scale. Place the clamp around one lead at a time. Record each reading and compare with the nameplate rating. If the readings are not similar to the nameplate rating, find the cause, such as low voltage, bad bearings, poor connections, or excessive load. If the ammeter readings are higher than expected, the high current could produce overheating and damage to the equipment. Try to find the problem and correct it.

Current imbalance is undesirable because it causes uneven heating in a motor that can shorten the life expectancy of the insulation. However, a small amount of current imbalance is to be expected in the leads to a three-phase motor. This imbalance can be caused by either peculiarities in the motor or by a power company imbalance. To isolate the cause, make the following test. Note that this test should be done by a qualified electrician. Refer to Figure 7.11.

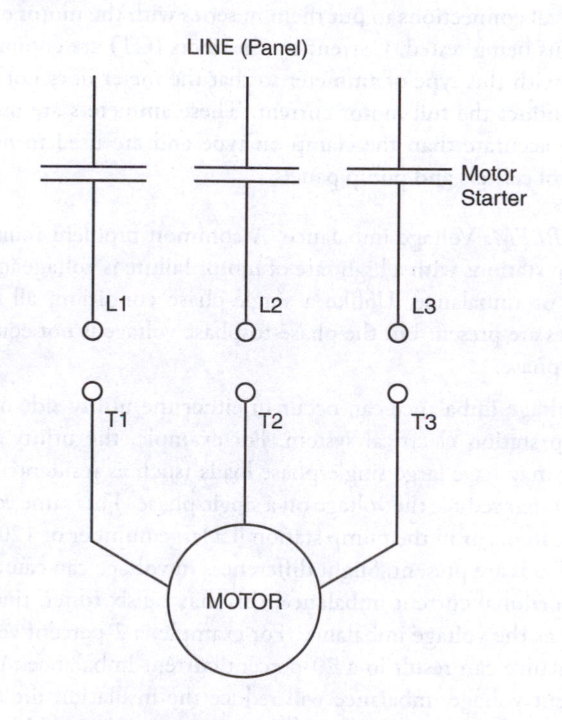


1. With the motor wired to its starter, L1 to T1, L2 to T2, and L3 to T3, measure and record the amperage on L1, L2, and L3.
2. De-energize the circuit and reconnect the motor as follows: L1–T3, L2–T1, L3–T2. This wiring change will not change the direction of the rotation of the motor.
3. Start up the motor and again measure and record the amperage on L1, L2, and L3.

*Fig. 7.11 Determination of current imbalance*

If the current readings on lines L1, L2, and L3 are about the same both before and after the wiring change, this is an indication that the imbalance is being caused by the power company and they should be asked to make adjustments to correct the condition. However, if the current readings followed the motor terminal (T) numbers rather than the power line (L) numbers, the problem is within the motor and there is not much that can be done except contact the motor manufacturer for a possible exchange.

When using a clamp-on ammeter, be sure to set the meter on a high enough range or scale for the starting current if you are testing during start-up. Starting currents range from 500 to 700 percent higher than running currents and using too low a range can ruin an expensive and delicate instrument. Newer clamp-on ammeters automatically adjust to the proper range and can measure both starting or peak current and normal running current.

Another type of ammeter is one that is connected in line with the power lead or leads. Generally, they are not portable and are usually installed in a panel or piece of equipment. They require physical connections to put them in series with the motor or apparatus being tested. Current transformers (CT) are commonly used with this type of ammeter so that the meter does not have to conduct the full motor current. These ammeters are usually more accurate than the clamp-on type and are used in motor control centers and pump panels.

*PROBLEM:* Voltage imbalance. A common problem found in pump stations with a high rate of motor failure is voltage imbalance or unbalance. Unlike a single-phase condition, all three phases are present but the phase-to-phase voltage is not equal in each phase.

Voltage imbalance can occur in either the utility side or the pump station electrical system. For example, the utility company may have large single-phase loads (such as residential services) that reduce the voltage on a single phase. This same condition can occur in the pump station if a large number of 120/220 volt loads are present. Slight differences in voltage can cause disproportional current imbalance; this may be six to ten times as large as the voltage imbalance. For example, a 2-percent voltage imbalance can result in a 20-percent current imbalance. A 4.5-percent voltage imbalance will reduce the insulation life to 50 percent of the normal life. This is the reason a dependable voltage supply at the motor terminals is critical. Even relatively slight variations can greatly increase the motor operating temperatures and burn out the insulation.

It is common practice for electrical utility companies to furnish power to three-phase customers in open delta or wye configurations. An open delta or wye system is a two-transformer bank that is a suitable configuration where lighting loads are large and three-phase loads are light. This is the exact opposite of the configuration needed by most pumping facilities where three-phase loads are large. (Examples of three-transformer banks include Y-delta, delta-Y, and Y-Y.) In most cases, three-phase motors should be fed from three-transformer banks for proper balance. The capacity of a two-transformer bank is only 57 percent of the capacity of a three-transformer bank. The two-transformer configuration can cause one leg of the three-phase current to furnish higher amperage to one leg of the motor, which will greatly shorten its life.

Operators should acquaint themselves with the configuration of their electric power supply. When an open delta or wye configuration is used, operators should calculate the degree of current imbalance existing between legs of their polyphase motors. If you are unsure about how to determine the configuration of your system or how to calculate the percentage of current imbalance, always consult a qualified electrician. Current imbalance between legs should never exceed 5 percent under normal operating conditions (NEMA Standard MG 1-14.35).

Loose connections will also cause voltage imbalance as will high-resistance contacts, circuit breakers, or motor starters.

Another serious consideration for operators is voltage fluctuation caused by neighborhood demands. A pump motor in near perfect balance (for example, 3 percent unbalance) at 9:00 am could be as much as 17 percent unbalanced by 4:00 pm on a hot day due to the use of air conditioners by customers on the same grid. Also, the hookup of a small market or a new home to the power grid can cause a significant change in the degree of current unbalance in other parts of the power grid. Because energy demands are constantly changing, wastewater collection and treatment system operators should have a qualified electrician check the current balances between legs of their three-phase motors at least once a year.

*SOLUTION:* Motor connections at the circuit box should be checked frequently (semiannually or annually) to ensure that the connections are tight and that vibration has not caused the insulation on the conductors to wear away. Measure the voltage at the motor terminals and calculate the percentage unbalance (if any) using the procedures below.

Do not rely entirely on the power company to detect unbalanced current. Complaints of suspected power problems are frequently met with the explanation that all voltages are within the percentages allowed by law and no mention is made of the percentage of current unbalance, which can be a major source of problems with three-phase motors. A little research of your own can pay big dividends. For example, a small water company in Central California configured with an open delta system (and running three-phase unbalances as high as 17 percent as a result) was routinely spending $14,000 a year for energy and burning out a 10-HP motor on the average of every 1.5 years (six 10-HP motors in 9 years). After consultation, the local power utility agreed to add a third transformer to each power board to bring the system into better balance. Pump drop leads were then rotated, bringing overall current unbalances down to an average of 3 percent; heavy-duty three-phase capacitors were added to absorb the voltage surges common in the area; and computerized controls were added to the pumps to shut them off when pumping volumes got too low. These modifications resulted in a saving in energy costs the first year alone of $5,500.

## FORMULAS

Percentage of current unbalance can be calculated by using the following formulas and procedures:

$$\text{Average Current} = \frac{\text{Total of Current Value Measured on Each Leg}}{3}$$

$$\text{\% Current Unbalance} = \frac{\text{Greatest Amp Difference from the Average}}{\text{Average Current}} \times 100\%$$

## PROCEDURES

1. Measure and record current readings in amps for each leg. (Hookup 1.) Disconnect power.
2. Shift or roll the motor leads from left to right so the drop cable lead that was on terminal 1 is now on 2, lead on 2 is now on 3, and lead on 3 is now on 1. (Hookup 2.) Rolling the motor leads in this manner will not reverse the motor rotation. Start the motor, measure and record current reading on each leg. Disconnect power.
3. Again, shift drop cable leads from left to right so the lead on terminal 1 goes to 2, 2 goes to 3, and 3 to 1. (Hookup 3.) Start pump, measure and record current reading on each leg. Disconnect power.
4. Add the values for each hookup.
5. Divide the total by 3 to obtain the average.
6. Compare each single leg reading to the average current amount to obtain the greatest amp difference from the average.
7. Divide this difference by the average to obtain the percentage of unbalance.
8. Use the wiring hookup that provides the lowest percentage of unbalance.

## CORRECTING THE THREE-PHASE POWER UNBALANCE

Example: Check for current unbalance for a 230-volt, 3-phase, 60-Hz submersible pump motor, 18.6 full load amps.

Solution: Steps 1 to 3 measure and record amps on each motor drop lead for Hookups 1, 2, and 3 (Figure 7.12).

| | Step 1 (Hookup 1) | Step 2 (Hookup 2) | Step 3 (Hookup 3) |
|---|---|---|---|
| ($T_1$) | $DL_1$ = 25.5 amps | $DL_3$ = 25 amps | $DL_2$ = 25.0 amps |
| ($T_2$) | $DL_2$ = 23.0 amps | $DL_1$ = 24 amps | $DL_3$ = 24.5 amps |
| ($T_3$) | $DL_3$ = 26.5 amps | $DL_2$ = 26 amps | $DL_1$ = 25.5 amps |
| Step 4 | Total = 75 amps | Total = 75 amps | Total = 75 amps |

Step 5 Average Current = $\frac{\text{Total Current}}{\text{3 readings}} = \frac{75}{3}$ = 25 amps

| | | | |
|---|---|---|---|
| Step 6 | Greatest amp difference from the average: | (Hookup 1) = 25 – 23 | = 2 |
| | | (Hookup 2) = 26 – 25 | = 1 |
| | | (Hookup 3) = 25.5 – 25 | = .5 |
| Step 7 | % Unbalance | (Hookup 1) = 2/25 × 100 | = 8 |
| | | (Hookup 2) = 1/25 × 100 | = 4 |
| | | (Hookup 3) = 0.5/25 × 100 | = 2 |

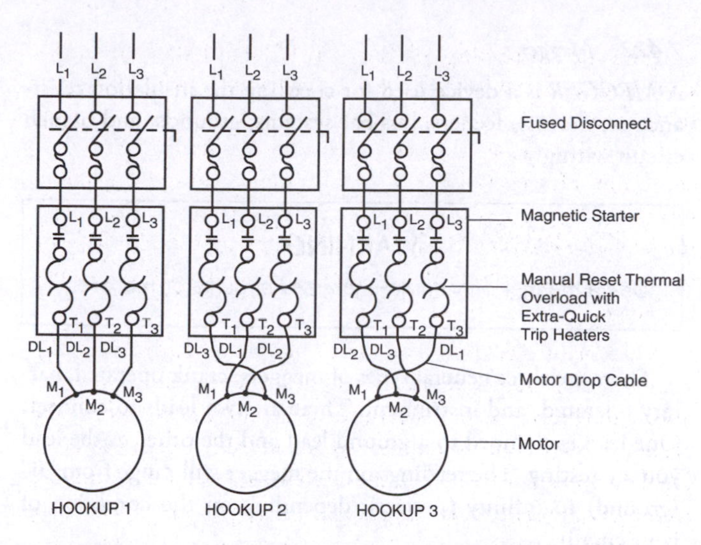


*Fig. 7.12 Three hookups used to check for current unbalance*

As can be seen, Hookup 3 should be used since it shows the least amount of current unbalance. Therefore, the motor will operate at maximum efficiency and reliability on Hookup 3.

By comparing the current values recorded on each leg, you will note the highest value was always on the same leg, $L_3$. This indicates the unbalance is in the power source. If the high current values were on a different leg each time the leads were changed, the unbalance would be caused by the motor or a poor connection.

If the current unbalance is greater than 5 percent, contact your power company for help.

**Acknowledgment**

Material on unbalanced current was provided by James W. Cannell, President, Canyon Meadows Mutual Water Company, Inc., Bodfish, California. His contribution is greatly appreciated.

## QUESTIONS

Please write your answers to the following questions and compare them with those on page 376.

7.4F How can you determine if there is voltage in a circuit?

7.4G What are some of the uses of a multimeter?

7.4H What precautions should be taken before attempting to change fuses?

7.4I How do you test for voltage with a multimeter when the voltage is unknown?

7.4J What could be the cause of amp readings different from the nameplate rating?

### *7.422 Megger*

A *MEGGER* is a device used for checking the insulation resistance on motors, feeders, bus bar systems, grounds, and branch circuit wiring.

**WARNING**

Use a megger only on de-energized circuits or motors.

There are three general types of meggers: crank operated, battery operated, and instrument. There are two leads to connect. One lead is clamped to a ground lead and the other to the lead you are testing. The readings on the megger will range from "0" (ground) to infinity (perfect), depending on the condition of your circuit.

The megger is usually connected to a motor terminal at the starter, and the other lead to the ground lead. Results of this test indicate if the insulation is deteriorating or cut.

Insulation resistance of electrical equipment is affected by many variables such as the equipment design, the type of insulating material used, including binders and impregnating compounds, the thickness of the insulation and its area, cleanliness (or uncleanliness), moisture, and temperature. For insulation resistance measurements to be conclusive in analyzing the condition of equipment being tested, these variables must be taken into consideration.

Such factors as the design of the equipment, the kind of insulating material used, and its thickness and area cease to be variables after the equipment has been put into service, and minimum insulation resistance values can be established within reasonable tolerances. The variables that must be considered after the equipment has been put into service, and at the time that the insulation resistance measurements are being made, are uncleanliness, moisture, temperature, and damage such as fractures.

The most important steps an operator can take to ensure reliable operation of electrical equipment are keeping the equipment clean, preventing moisture from getting into equipment insulation, and keeping control panels and cabinets closed and well sealed with weatherstripping. This is merely good housekeeping but it is essential in the maintenance of all types of electrical equipment. The very fact that insulation resistance is affected by moisture and dirt, with due allowances for temperature, makes the megger insulation test the valuable tool that it is in electrical maintenance. The test is an indication of cleanliness and good housekeeping as well as a detector of deterioration and impending trouble.

Suggested safe levels or minimum values of insulation resistance have been developed. These values should be provided by the equipment manufacturer and should serve as a guide for equipment in service. However, periodic tests on equipment in service will usually reveal readings considerably higher than the suggested minimum safe values. Records of periodic tests must be kept because persistent downward trends in insulation resistance usually give fair warning of impending trouble, even though the actual values may be higher than the suggested minimum safe values.

Also, allowances must be made for equipment in service showing periodic test values lower than the suggested minimum safe values, so long as the values remain stable or consistent. In such cases, after due consideration has been given to temperature and humidity conditions at the time of the test, there may be no need for concern. This condition may be caused by uniformly distributed leakages of a harmless nature, and may not be the result of a dangerous localized weakness. Here again, records of insulation resistance tests over a period of time reveal changes that may justify investigation. The trend of the curve may be more significant than the numerical values themselves.

For many years, one *MEGOHM*[10] has been widely used as a fair allowable lower limit for insulation resistance of ordinary industrial electrical equipment rated up to 1,000 volts. This value is still recommended for those who may not be too familiar with insulation resistance testing practices, or who may not wish to approach the problem from a more technical point of view.

For equipment rated above 1,000 volts, the "one megohm" rule is usually stated: "A minimum of one megohm per thousand volts." Although this rule is somewhat arbitrary, and may be criticized as lacking an engineering foundation, it has stood the test of a good many years of practical experience. This rule gives some assurance that equipment is not too wet or not too dry and has saved many an unnecessary breakdown.

More recent studies of the problem, however, have resulted in formulas for minimum values of insulation resistance that are based on the kind of insulating material used and the electrical and physical dimensions of the types of equipment under consideration.[11]

---

10. *Megohm* (MEG-ome). Millions of ohms. Mega- is a prefix meaning one million, so 5 megohms means 5 million ohms.
11. Portions of the preceding paragraphs were taken from *INSTRUCTION MANUAL FOR MEGGER INSULATION TESTERS,* No. 21-J, pages 42 and 43, published by Biddle Instruments, no longer in print. For additional information, see *A STITCH IN TIME: THE COMPLETE GUIDE TO ELECTRICAL INSULATION TESTING* (Revised 2006). A PDF of the guide is available from the Megger© division of MeterCenter© at http://www.biddlemegger.com/biddle/Stitch-new.pdf.

Motors and wiring should be megged at least once a year, and twice a year, if possible. The readings taken should be recorded and plotted in some manner so that you can determine when insulation is breaking down. Meg motors and wirings after a pump station has been flooded. If insulation is wet, excessive current could be drawn and cause pump motors to kick out.

#### *7.423 Ohmmeter*

*OHMMETERS,* sometimes called circuit testers, are valuable tools used for checking electric circuits. An ohmmeter is used only when the electric circuit is OFF, or de-energized. The ohmmeter supplies its own power by using batteries. An ohmmeter is used to measure the resistance (ohms) in a circuit. These are most often used in testing the control circuit components such as coils, fuses, relays, resistors, and switches. They are used also to check for continuity.

An ohmmeter has several scales that can be used. Typical scales are: R × 1, R × 10, R × 1,000, and R × 10,000. Each scale has a level of sensitivity for measuring different resistances. To use an ohmmeter, set the scale (start at the low point—R × 1), put the two leads across the part of the circuit to be tested, such as a coil or resistor, and read the resistance in ohms. A reading of infinity would indicate an open circuit, and a zero would read no resistance. Ohmmeters usually would be used only by skilled technicians because they are very delicate instruments.

All meters should be kept in good working order and calibrated periodically. They are very delicate, susceptible to damage, and should be well protected during transportation. When readings are taken, they should always be recorded on a machinery history card for future reference. Meters are a good way to determine pump and equipment performance. *CAUTION: Never use a meter for testing electrical equipment unless you are qualified and authorized.*

## QUESTIONS

Please write your answers to the following questions and compare them with those on page 376.

7.4K How often should motors and wirings be megged?

7.4L An ohmmeter is used to measure the resistance (ohms) in what control circuit components?

### 7.5 ADDITIONAL INFORMATION

For additional information on switch gear, electric motors, auxiliary power, electric meter troubleshooting, high voltage, and electrical safety, see *WATER TREATMENT PLANT OPERATION,* Volume II, Chapter 18, "Maintenance," in this series of operator training manuals.

Another book that describes the principles of electricity is *BASIC ELECTRICITY* by Van Valkenburgh, Nooger & Neville, Inc. Published by Delmar, an imprint of Cengage Learning. Obtain from Cengage Learning, Attn.: Order Fulfillment, PO Box 6904, Florence, KY 41022-6904. ISBN 978-0-7906-1041-2. Price, $78.95, plus shipping and handling.

### 7.6 ARITHMETIC ASSIGNMENT

Turn to the Appendix, "How to Solve Wastewater System Arithmetic Problems," at the back of this manual and read Section A.9, "Steps in Solving Problems."

Please answer the discussion and review questions next.

## DISCUSSION AND REVIEW QUESTIONS

### Chapter 7. MAINTENANCE AND TROUBLESHOOTING

Please write your answers to the following discussion and review questions to determine how well you understand the material in the chapter.

1. What does a good maintenance program achieve?
2. Why should operators take the time to establish good working relationships with their suppliers?
3. Why should an operator be present during the construction of a system?
4. What could be the cause of an alarm for a mound system?
5. What should an operator do after a failed system starts working again?
6. Why should inexperienced, unqualified, or unauthorized persons and even qualified and authorized persons be extremely careful around electrical panels, circuits, wiring, and equipment?
7. Why must motor nameplate data be recorded and filed?
8. What is the purpose of a circuit breaker and how does it work?
9. What is amperage?
10. List three uses of a multimeter.
11. What type of device is used for checking insulation resistance?

# SUGGESTED ANSWERS

## Chapter 7. MAINTENANCE AND TROUBLESHOOTING

Answers to questions on page 356.

7.0A The operator's maintenance program should strive to achieve the elimination of equipment breakdowns or at least minimize the time breakdowns cause the system to be out of operation.

7.1A Preventive maintenance is a systematic approach for conducting maintenance activities before any equipment failures occur.

7.1B Emergency maintenance is the response to system emergency problems caused by natural catastrophes and equipment breakdowns that are unpredictable and difficult to detect during routine inspections and maintenance activities.

Answers to questions on page 360.

7.2A The only way an operator can keep track of a preventive maintenance program is by good recordkeeping.

7.2B The *EQUIPMENT SERVICE CARD* tells what should be done and when, while the *SERVICE RECORD CARD* is a record of what you did and when you did it.

7.2C The purpose of an operator's daily diary is to explain information recorded in other records and any unusual events or activities not recorded elsewhere.

7.2D After a job is completed, the operator should clean the tools and store them properly so they can be found and used when needed again.

7.2E Studies have shown that the types of injuries operators suffer most frequently are muscle and back strains caused by lifting and injuries resulting from slips and falls.

7.2F Emergency phone numbers for a wastewater collection and treatment system include the following:

1. Police Department
2. Fire Department
3. Hospital or Physician
4. Responsible Utility Agency Officials
5. Local Emergency Disaster Office
6. *CHEMTREC,* (800) 424-9300
7. Emergency Team (if your system has one)

Answers to questions on page 365.

7.3A The key to effective troubleshooting is following practical step-by-step procedures combined with a commonsense approach.

7.3B When a customer calls to report an alarm problem, ask these questions:

1. How long has the alarm been on?
2. Has this happened before?
3. Is there a Reset button on the alarm box?

7.3C The first item to check if a pump is not running is the circuit breaker.

7.3D Never enter a septic tank without following confined space entry procedures. Tanks may contain hazardous atmospheres.

Answers to questions on page 367.

7.4A Unqualified or inexperienced people must be extremely careful when attempting to troubleshoot or repair electrical equipment because they can be seriously injured and damage costly equipment if a mistake is made.

7.4B When machinery is not shut off, locked out, and tagged properly, the following accidents could occur:

1. A maintenance operator could be cleaning a pump and have it start, thus losing an arm, hand, or finger.
2. Electrical motors or controls not properly grounded could lead to possible severe shock, paralysis, or death.
3. Improper circuits such as a wrong connection, safety devices jumped, wrong fuses, or improper wiring can cause fires or injuries due to incorrect operation of machinery.

Answers to questions on page 369.

7.4C The proper voltage and allowable current in amps for a piece of equipment can be determined by reading the nameplate information or the instruction manual for the equipment.

7.4D The two types of current are direct current (DC) and alternating current (AC).

7.4E Amperage is a measurement of current or electron flow and is an indication of work being done or how hard the electricity is working.

Answers to questions on page 373.

7.4F You test for voltage by using a multimeter.

7.4G A multimeter can be used to test for voltage, open circuits, blown fuses, single phasing of motors, and grounding.

7.4H Before attempting to change fuses, turn off power and check both power lines for voltage. Use a fuse puller.

7.4I If the voltage is unknown and the multimeter has different scales that are manually set, always start with the highest voltage range and work down. Otherwise, the multimeter could be damaged.

7.4J Amp readings different from the nameplate rating could be caused by low voltage, bad bearings, poor connections, or excessive load.

Answers to questions on page 375.

7.4K Motors and wiring should be megged at least once a year, and twice a year, if possible.

7.4L An ohmmeter is used to measure the resistance (ohms) in control circuit components such as coils, fuses, relays, resistors, and switches.

# APPENDIX

## Control Panel Troubleshooting Guide

(Orenco Systems, Inc., Control Panels A–1 and A–2)

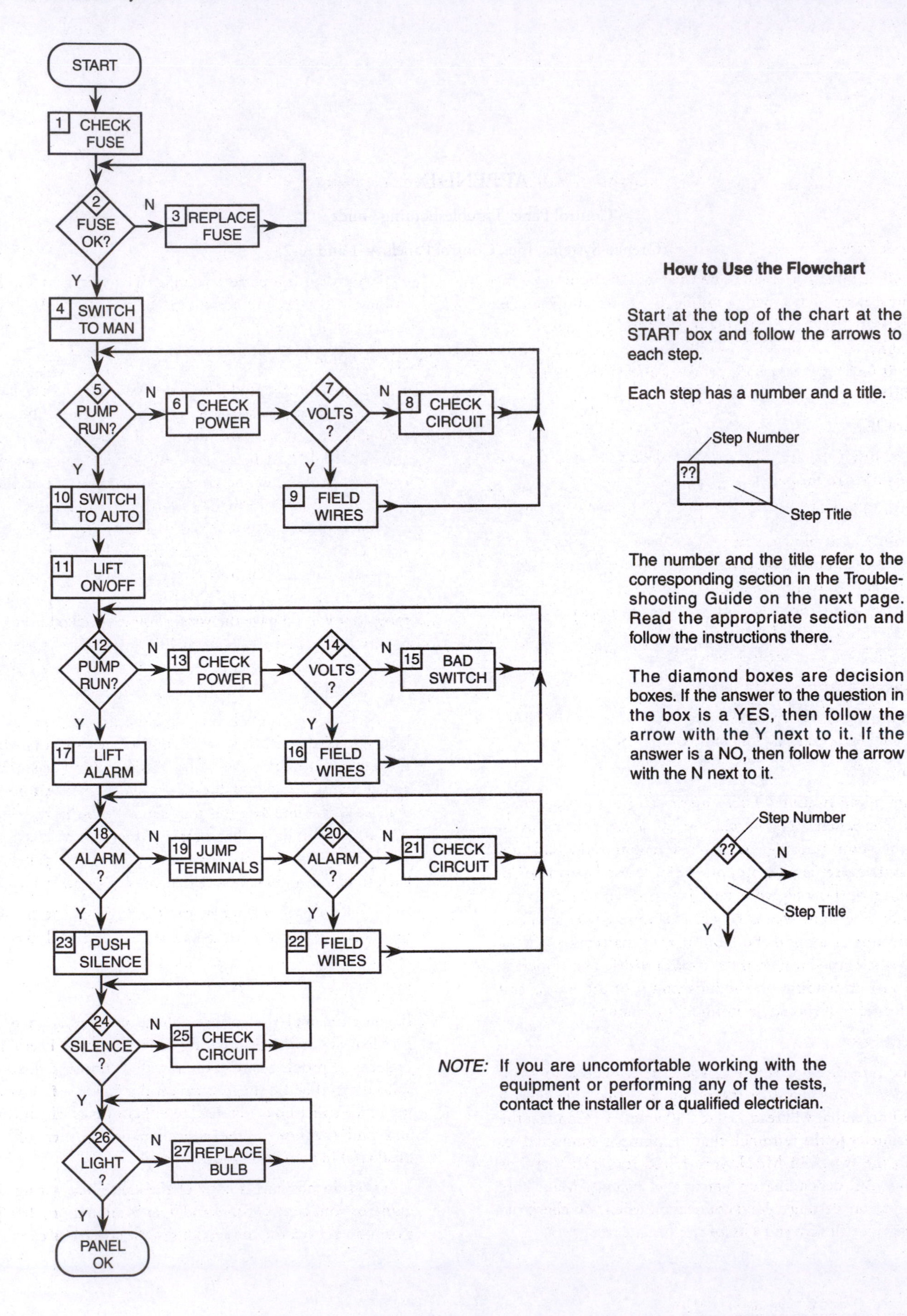


*Control panel troubleshooting flowchart (Orenco Systems, Inc., Models A-1 and A-2)*

(Permission of Orenco Systems, Inc.)

## Control Panel Troubleshooting Guide[12]

1. Check Fuse

   The first thing is to make sure that you have power to the control panel. This includes both the pump and the controls. First, make sure that the fuse is good and that the circuit breaker is in the ON position. Also, make sure that the circuit breaker in the main breaker panel is also in the ON position.

2. Fuse OK?

   Is the fuse OK, are the circuit breakers ON, and do you have power to the panel?

3. Replace Fuse

   If your fuse is blown, you will not have any power to the controls. In the A–1, this only means that your high water alarm will not work; the pump is not affected by the fuse. If your breaker is tripped or off, you need to reset it. You also need to keep in mind that a tripped breaker often indicates that you have a problem somewhere else.

4. Switch to MAN

   Flip the HOA switch from the OFF position to the MAN (MANUAL) position. This should turn the pump on.

5. Pump Run?

   Is the pump running? There are several ways to determine this. The easiest way is to take the lid off the tank and listen for the pump running. You can also look into the tank and see if the water level is dropping. This can take a while depending on how big your tank is and the flow rate of the pump. The third way is to get a clamp-on type ammeter and clamp it around the loop of wire that comes off the load side of the breaker in the control panel. The amperage will vary depending on the horsepower of the motor and the load, but it should be around 9 to 13 amps.

6. Check Power

   If the pump is not running, you need to check to see if there is power going to the pump. For this, you need a multimeter that will read 120 to 240 volts AC. Connect the multimeter to the terminals that the pump is connected to. With the switch in MAN, you should read 120 V AC or 230 V AC depending on your panel voltage. Make sure that you are getting a good connection when you check the voltage; it will save you a lot of trouble later on.

7. Volts?

   Did you read the correct voltage? The voltage should be around 120 V AC for an A–1 panel and 240 V AC for an A–2 panel.

8. Check Circuit

   If your multimeter read zero volts, you need to check the voltage in a couple of other spots to determine the exact problem. If you read the wrong voltage, like 120 V AC when it should have been 230 V AC, or vice versa, then you probably have the wrong voltage hooked up to your panel. Go back to Step 6 to make sure that you were testing for the right panel. There is a gold sticker on the inside of the front cover of the control panel that tells you the voltage that the panel is built for. Remember, an A–1 is built for 120 volts, and an A–2 is built for 240 volts. If you determine that you do have the wrong voltage hooked up to the panel, you need to have a qualified electrician make the necessary changes. Attempting to run a 120 volt pump on 230 volts, or vice versa, will damage your pump.

   To check your panel, first measure the voltage on the load side of the circuit breaker (that is the top side, where the loop of wire comes out of it). If you do not measure any voltage, and the circuit breaker is on, then measure the voltage at the line side (bottom side). If you have voltage there, then you have a bad breaker. If you do not have any voltage, then you do not have any power to the panel. Go back to Step 1 and make sure that you have power.

   If you do have voltage at the load side of the breaker, then you probably have a bad switch and will have to replace it.

9. Field Wires

   If you measured the correct voltage in Step 6, then the problem lies in the field wiring or in the pump itself. The majority of problems are in the way the pump and floats are spliced together. Be sure to turn off the power before working in the splice box. You need to either turn off the breaker and pull the fuse in the control panel or turn off the breaker(s) in your main breaker panel.

   If you do not have a copy of the splice box wiring diagram for your control panel and float combination, then get a copy and check the connections in the splice box carefully.

12. The procedures described in this appendix were adapted from a troubleshooting guide provided by Orenco Systems, Inc. Permission to use these materials is greatly appreciated.

The splices should match the diagram; be good, strong connections; and be watertight. If the splices are correct and good, then you either have a broken wire somewhere between the control panel and the splice box, the wires are miswired at the control panel, or the pump is bad.

First, check to make sure that the wires are connected to the correct terminals in the control panel. If they are not, then fix them and go back and re-check the panel to see if the motor works now.

If the wires are OK, then you will have to check to see if you have power at the motor leads. For this you will have to disconnect the motor and connect a multimeter to the wires from the control panel. If you read the correct voltage when you turn the power back on, then your pump motor is the problem. You will need to remove the pump for testing by an authorized service center.

If you did not read the correct voltage, then you have a wiring problem somewhere between the splice box and the control panel. You will need to have a qualified electrician check it out.

Be sure of your tests. You do not want to have costly and unnecessary work done on your wiring or pump.

10. Switch to AUTO

Flip the HOA switch to the AUTO position. The next tests will check to see if the pump works with the floats.

11. Lift ON/OFF

This test requires you to lift the ON/OFF float in the septic tank. Usually, the float is out of reach down in the tank, so you will need to get some sort of hook on the end of a pole or, with persistence, you can use just a long stick. Using your pole or stick, reach down and lift the float up, and hold it there. Your pump should turn on.

12. Pump Run?

Is your pump running? See Step 5 for instructions on how to tell if your pump is running.

13. Check Power

If the pump is not running, you need to check to see if there is power going to the pump. For this, you need a multimeter that will read 120 to 240 volts AC. Connect the multimeter to terminals #3 and #5. With the switch in AUTO, you should read 120 V AC or 230 V AC depending on your panel voltage. Make sure that you are getting a good connection when you check the voltage; it will save you a lot of trouble later on.

14. Volts?

Did you read the correct voltage? The voltage should be around 120 V AC for an A–1 panel and 240 V AC for an A–2 panel.

15. Bad Switch

If your multimeter reads zero volts, then you probably have a bad switch. First, make sure that you actually had the

power on, and the HOA switch in the AUTO position. If you did not, then go back to Step 12 and see if the pump runs. If you need a new switch, you can order one from the manufacturer.

The above check assumes that you have gone through the previous steps, and that the pump was running on MANUAL. If you have not gone through the above steps, do so now.

16. Field Wires

If you measured the correct voltage in Step 14, then the problem lies in the field wiring or the ON/OFF float. The majority of problems are in the way the pump and floats are spliced together. Be sure to turn off the power before working in the splice box. You need to either turn off the breaker and pull the fuse in the control panel or turn off the breaker(s) in your main breaker panel.

If you do not have a copy of the splice box wiring diagram for your control panel and float combination, then get a copy and check the connections in the splice box carefully. The splices should match the diagram; be good, strong connections; and be watertight. If the splices are correct and good, then you either have a broken wire somewhere between the control panel and the splice box, the wires are miswired at the control panel, or the ON/OFF float is bad.

First, check to make sure that the wires are connected to the correct terminals in the control panel. If they are not, then fix them and go back and re-check the panel to see if the pumps work now.

If the wires are OK, then you will have to check to see if your float is working correctly. For this, you will have to disconnect the float and connect a continuity tester or ohmmeter to the float. The meter should show an open circuit (no continuity) when the float is down, and a closed circuit (continuity) when the float is lifted up. If the float does not do that, then it is probably bad and needs to be replaced.

If the float checks out OK, then you have a wiring problem somewhere between the splice box and the control panel. You will need to have a qualified electrician check it out.

Be sure of your tests. You do not want to have costly and unnecessary work done on your wiring or pump. If you have any questions, contact the equipment manufacturer for assistance.

17. Lift Alarm

Using a pole or the stick mentioned in Step 11, lift the High Water Alarm float. It is the black float at the top of the float stem. Lifting the float will make the alarm sound. If the control panel is a long distance away, or you are in a really noisy area, you might need someone to help you.

18. Alarm?

Did the alarm sound when you lifted the float?

19. Jump Terminals

If the alarm did not sound, you will need to check to see if the alarm works in the panel. The easiest way to do this is to use a wire to jumper the alarm terminals.

First, take a piece of insulated wire, and strip some of the insulation off each end, and then jumper terminals #1 and #2.

Be careful. There are 120 volts on the terminals. Do not touch any metal part of the jumper wire or the metal link rails on the center of the terminal blocks.

The best way to perform this test is to turn off the power first (by pulling out the fuse), make the connection, and then turn the power back on.

20. Alarm?

Did the alarm sound when you jumpered the terminals?

21. Check Circuit

If the alarm did not sound when you jumpered the terminals, then the problem lies somewhere in the panel.

The first thing to check is to see if the red light on the front cover of the control panel turns on when you perform the test in Step 19. If it does, but the buzzer does not sound, then the problem lies with the silence relay or the buzzer itself. The first thing to do is to check to see if the silence relay (the little yellow ice-cube relay) is firmly plugged into its socket. If it is not, then fix it and perform the test in Step 19. If the buzzer sounds, then go back and perform Steps 17 and 18 again to see if the float works the buzzer.

If the relay is OK, then the problem is probably in the buzzer. With the alarm circuit operating, use the multimeter to check the voltage across the buzzer's terminals. There should be 120 volts there. If there is, then you have a bad buzzer. If you do not read 120 volts, then the problem lies in the silence relay, or some other part of the circuit.

If the red light does not turn on, then you want to check to make sure that there is power to the alarm circuit. Take your multimeter and check the voltage between terminals #2 and 'N'. You should read 120 volts. If you do not, then check the fuse to see if it is good and has power. If you do read 120 volts, then the problem lies elsewhere in the panel.

22. Field Wires

If the alarm sounded when you jumpered the terminals, then the problem lies somewhere in the connections between the control panel and the splice box, or in the High Water Alarm float itself. The majority of problems are in the way the floats are spliced. Be sure to turn off the power before working in the splice box. You need to either turn off the breaker and pull the fuse in the control panel or turn off the breaker(s) in your main breaker panel.

If you do not have a copy of the splice box wiring diagram for your control panel and float combination, then get a copy and check the connections in the splice box carefully. The splices should match the diagram; be good, strong connections; and be watertight. If the splices are correct and good, then you either have a broken wire somewhere between the control panel and the splice box, the wires are miswired at the control panel, or the High Water Alarm float is bad.

First, check to make sure that the wires are connected to the correct terminals in the control panel. If they are not, then fix them and go back and re-check the panel to see if the pumps work now.

If the wires are OK, then you will have to check to see if your float is working correctly. For this, you will have to disconnect the float and connect a continuity tester or ohmmeter to the float. The meter should show an open circuit (no continuity) when the float is down, and a closed circuit (continuity) when the float is lifted up. If the float does not do that, then it is probably bad and needs to be replaced.

If the float checks out OK, then you have a wiring problem somewhere between the splice box and the control panel. You will need to have a qualified electrician check it out.

Be sure of your tests. You do not want to have costly and unnecessary work done on your wiring or floats.

23. Push Silence

With the alarm still sounding from your test in Step 17, push the illuminated red button on the front of the control panel. This should silence the buzzer, but the light should stay on. Lowering the High Water Alarm float will turn off the light and reset the alarm. Lifting the float again will illuminate the lamp and sound the buzzer again.

24. Silence?

    When you pushed the button, did the alarm silence?

25. Check Circuit

    If the buzzer did not silence when you pushed the button, then you need to check the circuit to find the problem. The Silence button is a switch that makes a connection between the two black wires on the back side of the button. Take a piece of insulated wire and strip some of the insulation off each end. Jumper the two screw terminals that the black wires are connected to. You only have to make a brief contact, and the buzzer should silence.

    Be careful. There are 120 volts on the terminals. Do not touch any metal part of the jumper wire or the metal screw terminals on the push button.

    If the buzzer does silence when you do the test, then you have a bad contact block in the push button. The contact block will have to be replaced.

    If the buzzer did not silence, then the problem lies in the silence relay. Make sure that the relay (the little yellow ice-cube relay) is firmly plugged into its socket. If the buzzer still fails to be silenced when you perform the above test, then the silence relay is probably bad and should be replaced.

26. Light?

    Does the red push-button light on the front cover of the control panel light up when the alarm sounds?

27. Replace Bulb

    If the red light does not turn on with the buzzer, then the bulb is probably burned out and will have to be replaced.

    The bulb is a 3-watt, 120-volt T1¾ bayonet-mounted bulb. You can get to it by unscrewing the red lens on the push button. The bulb is very hard to get at using your fingers, but you can use a small piece of rubber hose, or an eraser cap to remove it.

If you need to contact the manufacturer for assistance or to order replacement parts, first take a few minutes to collect the following information:

- Write down a clear description of the problem. Be specific. Without specific information, the manufacturer will have a hard time helping you. The following are some hints to make figuring out your problem easier.
- Write down exactly which control panel you have and what voltage the pumps are. The easiest way to figure out which model you have is by looking at the title on the wiring diagram. If you can find the date of the diagram, that will be helpful too. If you do not have the diagram or cannot figure out what model of control panel you have, then write down the number on the Underwriters' Laboratory sticker. The sticker usually is located on the inside of the front cover of the control panel. The sticker will say "Underwriters' Laboratory Listed Enclosed Industrial Control Panel No. ________ ."
- Write down the symptoms: what it is doing and what it is not doing. Write down everything it is doing. Even little things that may not seem important to you might help the manufacturer figure out the problem.
- Record which troubleshooting steps you have performed and what the results of the tests were.
- If you need to order replacement parts, know the name of the manufacturer and the part number of the component.

CHAPTER 8

# SETTING RATES FOR SMALL WASTEWATER UTILITIES

by

Ken Kerri

# TABLE OF CONTENTS

## Chapter 8. SETTING RATES FOR SMALL WASTEWATER UTILITIES

# OBJECTIVES

## Chapter 8. SETTING RATES FOR SMALL WASTEWATER UTILITIES

1. Determine revenue needs for a small utility.
2. Itemize various system expenses.
3. Relate costs to level of service.
4. Distribute costs to customers.
5. Keep accurate records of costs and sources of revenues.
6. Develop rates for a small wastewater utility.
7. Provide stable financial management for a small utility.
8. Determine where and how to look for sources of financial assistance.

# WORDS

## Chapter 8. SETTING RATES FOR SMALL WASTEWATER UTILITIES

BASE-EXTRA CAPACITY METHOD BASE-EXTRA CAPACITY METHOD

A cost allocation method used by utilities to determine rates for various groups. This method considers base costs (O&M expenses and capital costs), extra capacity costs (additional costs for maximum day and maximum hour demands), customer costs (meter maintenance and reading, billing, collection, accounting), and fire protection costs.

BOND BOND

(1) A written promise to pay a specified sum of money (called the face value) at a fixed time in the future (called the date of maturity). A bond also carries interest at a fixed rate, payable periodically. The difference between a note and a bond is that a bond usually runs for a longer period of time and requires greater formality. Utility agencies use bonds as a means of obtaining large amounts of money for capital improvements.

(2) A warranty by an underwriting organization, such as an insurance company, guaranteeing honesty, performance, or payment by a contractor.

COMMODITY-DEMAND METHOD COMMODITY-DEMAND METHOD

A cost allocation method used by water utilities to determine water rates for the various water user groups. This method considers the commodity costs (water, chemicals, power, amount of water use), demand costs (treatment, storage, distribution), customer costs (meter maintenance and reading, billing, collection, accounting), and fire protection costs.

COVERAGE RATIO COVERAGE RATIO

The coverage ratio is a measure of the ability of the utility to pay the principal and interest on loans and bonds (this is known as debt service) in addition to any unexpected expenses.

DEBT SERVICE DEBT SERVICE

The amount of money required annually to pay the (1) interest on outstanding debts, or (2) funds due on a maturing bonded debt or the redemption of bonds.

DEPRECIATION DEPRECIATION

The gradual loss in service value of a facility or piece of equipment due to all the factors causing the ultimate retirement of the facility or equipment. This loss can be caused by sudden physical damage, wearing out due to age, obsolescence, inadequacy, or availability of a newer, more efficient facility or equipment. The value cannot be restored by maintenance.

EQUITY EQUITY

The value of an investment in a facility.

FIXED COSTS FIXED COSTS

Costs that a utility must cover or pay even if there is no service provided. Also see VARIABLE COSTS.

OPERATING RATIO OPERATING RATIO

The operating ratio is a measure of the total revenues divided by the total operating expenses.

OVERHEAD OVERHEAD

Indirect costs necessary for a utility to function properly. These costs are not related to the actual collection, treatment, and discharge of wastewater, but include the costs of rent, lights, office supplies, management, and administration.

RATE OF RETURN RATE OF RETURN

A value that indicates the return of funds received on the basis of the total equity capital used to finance physical facilities. Similar to the interest rate on savings accounts or loans.

VARIABLE COSTS VARIABLE COSTS

Costs that a utility must cover or pay that are associated with the actual collection, treatment, and discharge of wastewater. These costs vary or fluctuate on the basis of the volume of wastewater collected, treated, and discharged or reused. Also see FIXED COSTS.

# CHAPTER 8. SETTING RATES FOR SMALL WASTEWATER UTILITIES

## 8.0 DEVELOPING RATES

### 8.00 General Rate Setting Philosophy

When setting utility service rates, the goal is to develop a rate setting method that fairly distributes costs to various classes of users and then recovers these costs through rates that will sustain the utility and not discriminate against any particular class of users. To put this goal in simpler terms—rates should be cost based and nondiscriminatory.

As wastewater utility managers, we must develop rate setting practices and procedures that fairly allocate costs to all of the users who benefit from our services. Again, the burden of service costs should not be unfairly thrust upon any individual class of customers whether they be small or large, new or old, domestic or commercial/industrial users.

A second goal in rate setting is stability of rates. Careful analysis of past, projected, and actual costs will help you avoid frequent, unexpected rate increases. No matter how fair and equitable the rate structure is, sudden large increases upset consumers and raise questions about the competence of the utility manager. A rate structure that is fair to all customers reflects well on your integrity, and a rate structure that is stable and predictable over a period of years reflects well on your professional ability.

In this chapter, we will describe rate setting practices and procedures that can be readily used by small wastewater utilities as guidelines to establish rates and charges. Each wastewater utility is unique and will have to deal with its own special considerations. By using these procedures, the small utility should be able to develop a cost-based, fair, and reasonable set of rates and charges that accurately reflect the cost of service for each class of customer service provided.

### 8.01 General Action Strategy

The first step in setting utility service rates is to develop a general action strategy. That is to say, we need to outline the specific steps we will take to reach our goal. Fortunately, many of the resources that we will need to use are readily available in the form of budgets, budget reports showing conformance to the spending plan, and customer-related records. Hopefully, the customer records will include such essential information as customer-related operation and maintenance expenses.

Armed with a good working knowledge of our past financial and operating history, we can begin to develop a schedule of rates and charges by using the following procedures:

1. Determine total annual revenue requirements.
2. Determine reserve fund requirements.
3. Allocate revenue requirements to cost components.
4. Distribute component costs to customer classes.
5. Design rates that will recover these costs from each class of customer.

The following sections describe each of these procedures in enough detail that you can apply them to your own wastewater utility. Figure 8.1 outlines the action strategy or procedures developed in this chapter for a manager to determine rates for a small utility agency. As you begin this procedure, keep in mind one basic difference between budget planning for a utility and budgeting your own personal expenses. You probably begin with a fixed annual income and try to adjust your expenses to fit within that amount. In budgeting for a utility, you will begin by figuring out all present and anticipated expenses and then set fair and equitable rates that will produce the amount of annual income needed.

1. DETERMINE REVENUE NEEDS (Table 8.2, page 394). IDENTIFY EXPECTED COSTS FOR
   a. O&M expenses, including administration
   b. Debt service on borrowed funds
   c. Rehabilitation, replacement, and expansion costs
   d. Cash reserves (10 percent of O&M)

2. DETERMINE RESERVE FUND REQUIREMENTS
   a. Equipment replacement
   b. Emergency repairs

3. SELECT A METHOD FOR A REVENUE BASE
   a. Cash basis
   b. Utility basis

4. ALLOCATE COSTS

5. DETERMINE RATES PER CUSTOMER ON BASIS OF LEVEL OF SERVICE
   a. Fixed costs
   b. Variable costs

*Fig. 8.1 Procedure to determine rates*

### 8.02 Developing a Hypothetical Utility District

To better illustrate the application of the rate setting procedures in this chapter, a hypothetical utility district will be used as a model utility.

Assume that the XYZ Utility District has 800 dwellings (services) and serves a predominantly residential community with a population of approximately 2,250 people. It has no industrial users or small commercial users.

All of the dwellings served by this utility are approximately the same size. Each dwelling is served by a septic tank and either a subsurface leaching system, a small area subsurface leaching system serving a number of dwellings, or a mound system. All dwellings are assumed to receive the same level of service. There is no metering of either drinking water consumption or wastewater flows. Therefore, there are no *VARIABLE COST*[1] categories.

If the utility district served commercial establishments such as a restaurant, service station, campground, or school, a multiplication factor could be developed if these establishments received a higher level of service and the utility district believed higher rates were appropriate. Utility rates could be adjusted depending on sizes of lots or sizes of dwellings if more than one level of service was provided.

## QUESTIONS

Please write your answers to the following questions and compare them with those on page 403.

8.0A What are the goals of a fair and equitable rate setting program?

8.0B List the five steps that should be followed to determine utility service rates.

## 8.1 REVENUE REQUIREMENTS

### 8.10 Determining Revenue Needs

Developing a statement of revenue requirements that accurately reflects the cash needs of our utility is the first step in our action strategy for setting rates. If we have developed a good budget and spending plan, this should merely be an extension of existing budgeting and accounting practices.

Our objective is to determine the cash needs and timing of expenditures (cash flow) for the utility. Typically, revenue is needed in these categories:

- O&M expenses
- *DEBT SERVICE*[2] on borrowed funds
- Rehabilitation and replacement costs
- Administrative expenses related to the operation of our system
- Cash reserves

To avoid cash flow shortages during the rate period, we need to be sure to project our revenue needs far enough into the future and allow some cash reserve for emergencies and contingencies (unplanned events). Sometimes, it is best to project several years into the future even though the rate period is typically only one year.

We also need to allow for inflation and system growth because these factors play a significant role in setting stable rates. A good historical database from which to project future needs will greatly assist in setting long-term rates. This database will provide indications of historical trends.

Revenue requirements may be broadly divided into two major categories: operation and maintenance (O&M) expenses and capital costs. A description of these costs is given in Table 8.1.

**TABLE 8.1 MAJOR REVENUE REQUIREMENTS**

| |
|---|
| OPERATION AND MAINTENANCE (O&M) EXPENSES |
| Costs of operating the collection, treatment, and discharge systems |
| Costs of maintaining facilities and equipment |
| Administration |
| Personnel |
| CAPITAL COSTS |
| Annual costs associated with plant construction investment |
| Debt service expenses (principal and interest on bonds or other loans) |
| Facilities replacement costs (based on expected service life of equipment and facilities) |
| Facilities extension (expansion) costs |
| Facilities improvement costs, including new facilities |
| Cash reserves |

1. *Variable Costs.* Costs that a utility must cover or pay that are associated with the actual collection, treatment, and discharge of wastewater. These costs vary or fluctuate on the basis of the volume of wastewater collected, treated, and discharged or reused. Also see FIXED COSTS.
2. *Debt Service.* The amount of money required annually to pay the (1) interest on outstanding debts, or (2) funds due on a maturing bonded debt or the redemption of bonds.

## 8.11 Forecasting Expenditures

As previously mentioned, it is necessary to forecast expenses carefully in establishing a stable system of rates and charges. A stable system is one in which rates are adequate to cover all anticipated expenses and still maintain a positive cash position at the end of the rate period. This avoids bumping rates up in the middle of the rate period, and allows the utility to build a modest cash reserve, which can be called upon in times of serious financial need. A good rule of thumb for a small utility is to establish a general fund reserve that is a minimum of 10 percent of one year's O&M expenses. Special reserve funds may also need to be established for capital improvement program needs.

In forecasting future expenditures, the following items should be considered:

- System growth rate
- Anticipated inflation
- Long-range capital improvement program needs
- Long-range financing requirements

A major concern when projecting future revenue needs is the sensitivity of population growth estimates and the ability of revenues from services provided to meet projected target values. A drop in planned revenue resulting from people moving out of the area could have a significant impact on projected revenue for a small utility. Provisions may have to be made for dwelling owners to pay utility service charges regardless of whether or not the dwelling is occupied.

The growth rate of your community may have a significant impact on service demands, capital facility requirements, and staffing needs. If your community is growing rapidly, you will need both a long-range and a short-range capital improvement program designed to meet system demands as they occur over time. A short-range program is generally forecast for one to five years into the future, whereas a long-range program may estimate future system demands and facility needs for 25 years or more. Obviously, short-range planning is much better focused than long-range programs. A long-range plan should be reviewed periodically and revised as appropriate.

Just as important as long-range capital facility planning is the need to finance these programs. Utility rates alone usually are not an adequate method of financing most major capital improvements. Additional financing methods such as *BONDS*[3] and development-related charges and fees are often required to cover growth-related capital expenditures. These will be discussed later in Section 8.62, "Capital Improvements and Funding in the Future."

Inflation is another important consideration in setting stable utility rates. There are two general components of inflation that will affect rates and charges: the consumer price index (CPI) and the construction cost index (CCI). The CPI is established by the federal government and is published for most areas of the United States. The CPI should be applied to general O&M expense items such as materials and supplies as well as labor. The CCI, published by *ENGINEERING NEWS RECORD* magazine, is generally used for estimating construction-related capital improvement costs. It is advisable to use both of these indices because inflation rates may affect these two major sectors of the economy differently.

## 8.12 Itemizing System Expenses

In order to develop a complete statement of our revenue needs, we need to look more closely at our collection, treatment, and discharge expenses. As you read through this section, look at Table 8.2 to see how each expense item is listed in an overall statement of projected revenue requirements for our example utility serving a community of approximately 800 dwellings.

An examination of Table 8.2 reveals that the major revenue requirement for many small utilities is to meet labor costs. The total labor costs are $124,200 and non-labor costs are $47,800. Included in the non-labor costs is $8,500 for capital replacement. This utility does not have any debt service expenses because all of the initial construction costs were paid for by the developers and homeowners.

## 8.13 Establishing the Revenue Base

There are two basic accounting methods for establishing the revenue base: (1) the cash basis and (2) the utility basis. Using the cash basis, the utility must raise enough money to cover all cash needs including debt service. The cash basis is generally used by publicly owned utilities and is, therefore, the focus of this chapter. It is essentially an extension of the cash-oriented budgeting and accounting system used by many governmental agencies.

3. *Bond.* (1) A written promise to pay a specified sum of money (called the face value) at a fixed time in the future (called the date of maturity). A bond also carries interest at a fixed rate, payable periodically. The difference between a note and a bond is that a bond usually runs for a longer period of time and requires greater formality. Utility agencies use bonds as a means of obtaining large amounts of money for capital improvements. (2) A warranty by an underwriting organization, such as an insurance company, guaranteeing honesty, performance, or payment by a contractor.

TABLE 8.2 PROJECTED REVENUE REQUIREMENTS FOR XYZ UTILITY DISTRICT

| Expense Item | Actual | Adjusted to Actual Expenses | Revised Budget |
|---|---|---|---|
| Supervision and Labor | $59,900 | $67,150 | $69,000 |
| Overtime | 3,000 | 1,900 | 2,500 |
| Office Salaries | 7,300 | 7,800 | 8,000 |
| Construction Inspection | 15,000 | 1,800 | 3,500 |
| Retirement Fund Contribution | 16,000 | 14,200 | 16,500 |
| Payroll Taxes | 7,200 | 6,500 | 6,500 |
| Insurance (Health and Life) | 8,300 | 7,300 | 7,500 |
| Insurance (Workers' Compensation) | 2,000 | 1,900 | 2,000 |
| Retiree Benefits | | | |
| Employee Benefit—Deferred Compensation | 1,200 | 1,100 | 2,500 |
| Employee Dental Benefits | 4,300 | 4,300 | 4,500 |
| Accounting and Audit Fees | 2,400 | 2,400 | 2,500 |
| Engineering Fees | 0 | 0 | 0 |
| O&M | 3,000 | 5,000 | 4,000 |
| Insurance (General) | 5,300 | 5,300 | 5,300 |
| Legal | 700 | | |
| Materials and Research (Mounds) | 1,000 | 500 | 1,000 |
| Materials and Supplies | 1,500 | 1,800 | 2,000 |
| Office Supplies | 3,000 | 2,300 | 3,000 |
| Staff Development | | | 1,000 |
| Travel | 1,500 | 600 | 1,500 |
| Utilities | 9,200 | 7,800 | 8,000 |
| Vehicle Maintenance | 1,400 | 900 | 1,500 |
| Vehicle Operations | 1,200 | 900 | 1,000 |
| Bank Payroll Charges | 300 | 150 | 500 |
| Computer Billing Service | 1,800 | 1,400 | 1,500 |
| Capital Replacement | 8,500 | 6,200 | 8,500 |
| Outside Services | 6,500 | | 4,000 |
| Government Regulatory Fees | 0 | 2,000 | 2,000 |
| Other | 500 | 400 | 500 |
| Total | $172,000 | $151,600 | $170,300 |
| Labor | $124,200 | $113,950 | $122,500 |
| Non-Labor | $47,800 | $37,650 | $47,800 |

*CASH BASIS*

Revenue base requirements include the following:

- O&M expenses

  Operation and maintenance expenses are usually based on actual expenditures and are adjusted to reflect conditions expected to occur during the rate period. Historical spending patterns are normally used as a starting point for developing cost estimates. System O&M expense items include salaries and fringe benefits for operation and maintenance personnel; power; outside services (contracts); equipment rental; and materials and supplies.

- Administration and general expenses

  Administration and general expenses are usually listed under operation and maintenance expenses. They include salaries and fringe benefits for administrative, secretarial, and janitorial personnel; building rental; and billing and collection services.

- Debt service expenses

  Debt service expenses on bonds or other loans include principal and interest payments.

- Capital expenses

  Capital expense items include equipment replacements; main extensions and improvements; and major rehabilitation of facilities.

- Cash reserves

Routine repair and replacement of capital facilities of a minor nature are generally financed from current revenues. Most major capital projects are financed by bonds or developer-related fees and charges in order to extend the repayment period over the useful life of the facilities. This helps to avoid putting the burden of cost for these facilities on current users when facilities are designed to meet current as well as future system demands.

*UTILITY BASIS*

The utility basis for determining revenue requirements is used primarily by investor-owned utilities, although some publicly owned utilities may also use this method to establish rates for customers outside the utility's service area. The major difference in this accounting method is that it entitles the utility owner to a reasonable rate of return (profit) on the *EQUITY*[4] in the system. To accommodate this additional cost item, capital costs are additionally broken down into two other components: (1) depreciation expenses and (2) return on rate base.

*DEPRECIATION*[5] expense is related to the gradual loss in service value of facilities and equipment over time due to age (wear and tear, obsolescence). Depreciation expense essentially provides for recovery of the capital investment over the useful life of the facility and equipment.

The *RATE OF RETURN*[6] component is designed to pay the cost of debt service and to provide a fair rate of return for total equity capital used to finance physical facilities.

The utility method requires development of a "rate base" (value of assets on which the utility is entitled to earn a return). The rate base is usually based on the value of the utility's plant and property actually used to serve the utility's customers.

In contrast to the cash basis accounting method, capital improvement costs in the utility basis accounting method are recovered through depreciation (wear and tear) and rate of return (debt service), and no provision is made for reserves. Because these cost items tend to be offsetting, the revenue requirement could actually be less under the utility basis for capital items. This possibility should be thoroughly investigated if you are considering using the utility basis for a public or municipal utility.

The total revenue requirement for investor-owned utilities includes the following:

- O&M expenses
- Administration and general expenses
- Federal and state income taxes
- Other taxes (such as property taxes)
- Depreciation expenses
- Return on rate base

4. *Equity.* The value of an investment in a facility.
5. *Depreciation.* The gradual loss in service value of a facility or piece of equipment due to all the factors causing the ultimate retirement of the facility or equipment. This loss can be caused by sudden physical damage, wearing out due to age, obsolescence, inadequacy, or availability of a newer, more efficient facility or equipment. The value cannot be restored by maintenance.
6. *Rate of Return.* A value that indicates the return of funds received on the basis of the total equity capital used to finance physical facilities. Similar to the interest rate on savings accounts or loans.

## QUESTIONS

Please write your answers to the following questions and compare them with those on page 403.

8.1A What is the meaning of "debt service"?

8.1B What types of costs are considered operation and maintenance expenses?

8.1C List the types of capital costs that must be covered by revenue.

8.1D What are the two general components of inflation that affect setting of rates and charges?

## 8.2 COST ALLOCATION METHODS

If all dwellings made similar demands on the utility system, as they do in our hypothetical XYZ Utility District, we could simply divide the total revenue requirements of the utility by the total number of dwellings and thus allocate costs evenly across the board. This is usually not the case, however, so costs need to be allocated to customers on the basis of their particular service requirements.

Water utilities may use more complicated methods than wastewater utilities to determine rates because of consumer demands. The concepts used by water utilities are described in this section for the purpose of providing wastewater utilities with an awareness of the potential applicability of more complicated methods.

Water service to customers is influenced by the relationship between the peak and average demand characteristics of each customer class. These demand characteristics differ among customer classes and require the utility to provide different levels of service to meet the demands. Two accepted methods of analyzing the cost to serve various customer classes and allocating costs to varying levels of service are the base-extra capacity method and the commodity-demand method.

- Base-extra capacity
  1. Base costs—tend to vary with the total quantity of water used, reflecting average load conditions
  2. Extra-capacity costs—associated with meeting rate-of-use requirements in excess of average
     a. Maximum day
     b. Maximum hour
  3. Customer-related costs
  4. Direct fire-protection costs
- Commodity demand
  1. Commodity costs—tend to vary with the quantity of water produced
     a. Chemicals
     b. Power
     c. Reservoir
  2. Demand costs—associated with facilities to meet peak rates of use
  3. Customer-related costs
  4. Direct fire-protection costs

In the base-extra capacity method, costs of service are separated into four primary cost components: (1) base costs, (2) extra-capacity costs, (3) customer-related costs, and (4) direct fire-protection costs. Base costs are those that vary with the total quantity of water used plus operation and maintenance expenses and capital costs associated with average load conditions. Extra-capacity costs are the operation and maintenance expenses and capital costs for system capacity in excess of the average rate of use. Extra-capacity costs are further divided between maximum-day demand and maximum-hour demand. Customer-related costs stem from services to customers, regardless of the amount of water used or the demand on the system. They include meter reading, billing, and customer service activities. Finally, direct fire-protection costs include the maintenance and replacement costs of public fire hydrants and the mains and valves that serve them.

In the commodity-demand method, costs of service are separated into four primary cost components as well: (1) commodity costs, (2) demand costs, (3) customer-related costs, and (4) direct fire-protection costs. Commodity costs include only those items that vary directly with the quantity of water produced such as water treatment chemicals, power, and reservoir costs. Demand costs are those associated with the rate of water consumed. They include expenses related to providing water production and transmission facilities, operating and maintaining them, and replacing them to meet peak rates of use. The definitions of customer-related costs and direct fire-protection costs for this method are identical to the base-extra capacity method.

Of the two, the commodity-demand method is more sensitive to the relationship between the peak and average demand characteristics of each customer class. For example, higher costs may be assigned to residential users by the commodity-demand method compared with the base-extra capacity method due to greater demand placed on the utility by residential irrigation of lawns and gardens during the summer months. The commodity-demand cost allocation method may be more appropriate for use by small utilities because it is less complicated to administer, but it should be noted that it may not be as accurate as the base-extra capacity method in some instances.

## 8.3 DISTRIBUTION OF COSTS TO CUSTOMERS

### 8.30 Establishing Customer Classes

The next step in our planning effort is to establish different classes of service by customer groups. In setting up customer classes, you must be sure to create sufficient classes and broad enough category definitions to cover all user groups. Always keep in mind that the overall intent of establishing customer classes is to group customers into categories of similar use or service demand patterns. Thus, classes should separate users by service demands and requirements. If a facility services large crowds on weekends (such as a sporting event or concert), special peak charges may be appropriate.

While the process of establishing customer classes will vary depending on a system's unique mix of customers, the following general customer classes are frequently used by many utilities:

1. Residential
2. Commercial
3. Industrial
4. Agricultural (irrigation)
5. Other (public authority, government, schools, etc.)

Residential customers are typically single-family residences. In some instances, this class includes multiple-family dwelling units. Commercial customers are nonindustrial, business, and service-type enterprises. Industrial customers are usually manufacturing and processing enterprises. Agricultural customers may be heavy irrigation users or livestock or poultry feeding operations. The "Other" category may be used for schools, governmental uses, parks, and recreational facilities.

### 8.31 Special Considerations

In distributing costs to customers, you will need to give special consideration to unusual service characteristics and demand patterns.

Unusual service characteristics are unique factors that require special consideration for a particular customer or class of users. For example, demand patterns could influence rates for seasonal users such as canneries.

Special rate schedules may also need to be established for certain customers such as those who reside outside district boundaries. As mentioned in a previous section, you may want to assess a surcharge to these customers to recover a portion of the system capital costs above the actual investment in the form of a rate of return.

### 8.32 Capital Contributions

Traditionally, bonds have been used to finance major capital improvements to the wastewater system. In recent years, however, there has been increasing pressure to include capital costs that are general in nature when calculating the general wastewater utility rate. Other capital costs incurred for such uses as system expansion required by new development might better be met by new-user capital contributions in the form of connection fees or other related charges.

Connection fees are generally up-front fees paid by new users connecting to the system to cover new capital costs incurred to provide service to their property. Or, new users are sometimes required to pay a buy-in fee at the time of connection to the system to pay back an appropriate portion of capital already invested in the wastewater collection, treatment, and discharge systems. User fee-based financing systems are often referred to as "pay-as-you-go" financing. The advantages of this method of assessment are that it avoids payment of interest expense on long-term debt and it does not put the entire burden of system expansion on current ratepayers.

## QUESTIONS

Please write your answers to the following questions and compare them with those on page 403.

8.2A List two accepted methods of analyzing the cost to serve various customer classes and allocating costs to varying levels of service.

8.3A Who would be included in a commercial customer class?

8.3B What is a connection fee?

## 8.4 RATE DESIGN

### 8.40 General

The final step in our action strategy is to design utility rates that will recover appropriate costs from each class of customer. In this section, we will develop a typical bill for our hypothetical district.

Simplified methods of rate design can be used in small utilities when the cost of preparing a comprehensive rate study is prohibitive. However, when a utility, regardless of size, has special circumstances such as a large industrial base, a comprehensive or partial rate study may be in order.

While rates should always be based on the cost of service, good judgment must also be used in rate setting. Remember that the goal is to develop a rate structure that has the maximum degree of equity among all customer classes.

Miscellaneous charges might include a fee for establishing a new account or other one-time costs that should be segregated from the other rate components.

### 8.41 Information and Data Requirements

In order to set accurate rates that differentiate between customer classes, we need a good historical database from which information on the various user classes can be taken. If the majority of the system users are residential customers, you may not find it necessary or possible to set up different customer classes. In many cases, a uniform rate for all dwellings and customers may prove to be quite satisfactory.

### 8.42 Rate Design Example for a Very Small System

To illustrate the application of the concepts presented in this chapter, a rate structure will be developed for a very small wastewater utility.

1. Determine the service area statistics.

| | |
|---|---|
| Customers (services) | 800 |
| Population | 2,250 people |

Sometimes, the customers (services) number used for calculating a rate may be slightly lower than the actual number. This allows for people moving out of the service area or people not paying their utility bill. This is a conservative approach and could generate some additional revenue.

2. Estimate the annual revenue needs based on current and expected expenses (Table 8.2).

| | |
|---|---|
| Labor, including fringe benefits | $124,200 |
| Professional fees | 3,100 |
| O&M costs | 3,000 |
| Materials and supplies | 5,500 |
| Insurance | 5,300 |
| Travel | 1,500 |
| Utilities | 9,200 |
| Vehicles | 2,600 |
| Services | 8,600 |
| Regulatory fees | 0 |
| Other | 500 |
| Debt service | 0 |
| Capital replacement | 8,500 |
| Total Revenue Requirement | $172,000 |

3. Determine the monthly utility service charge.

$$\text{Service Charge, \$/mo} = \frac{\text{Total Revenue Requirement, \$/yr}}{(\text{Number of Dwellings})(12\text{ mo/yr})}$$

$$= \frac{\$172{,}000}{(800\text{ dwellings})(12\text{ mo/yr})}$$

$$= \$17.92\text{/mo per dwelling}$$

Recommend establishing a utility service charge of $18.00 per month per dwelling, plus any additional charges to build up a reserve fund.

## 8.5 ADMINISTRATION OF RATES AND CHARGES

### 8.50 Recordkeeping

As described in Section 8.41, good recordkeeping is the basis for setting accurate wastewater utility rates. Until fairly recently, wastewater utility recordkeeping has been done manually. The current availability of low-cost personal computer systems puts automation of many manual bookkeeping functions within the means of small utilities.

To automate your recordkeeping functions as they relate to customer billing, you will need to develop a simple database management system that will create tables similar to those illustrated in this chapter. This can be readily accomplished by using standard spreadsheet software programs, which are available in the marketplace at a cost of $300 to $400. Hardware, including a personal computer (PC), maintenance contract, printer, and data backup solution, can be purchased for under $3,000, if purchased as separate components. A bundled package, one for which a company offers limited choices on components, can cost much less. Computer systems should be upgraded regularly or replaced every 3 to 5 years.

### 8.51 Computer Uses

Excellent computer software packages are being developed and offered to assist utility managers. SURF (Small Utility Rates and Finances) has been developed by the American Water Works Association (AWWA). SURF is a self-guided, interactive spreadsheet application designed to assist small drinking water systems in developing budgets, setting user rates, and tracking expenses. SURF requires very little computer or software knowledge and can be used by system operators, bookkeepers, and managers to improve the financial management practices of their utilities. Although SURF was prepared for small water utilities, small wastewater utilities can use this program also. Wastewater utilities will not have water usage (from meters) information, so all costs will be *FIXED COSTS*[7] (as with our example wastewater utility) and there will be no variable costs. SURF can print out three separate modules: (1) system budget, (2) user rate(s), and (3) system expenses.

SURF hardware and software requirements are modest.

| | |
|---|---|
| Hardware Requirements: | Windows compatible PC. Ink jet or laser jet printer. |
| Software Requirements: | Windows operating system and Microsoft Excel 97 or higher. |

The SURF spreadsheet and user's guide are available free from the American Water Works Association. They can be downloaded from the AWWA website (www.AWWA.org). For more information, contact the AWWA Small Systems Program at (303) 347-6191.

### 8.52 Presenting Rates to the Ratepayer

If you have followed the procedures properly up to this point, then presenting your rate needs to the ratepayer should not be a problem, but rather a routine matter. An important aspect in presenting revenue needs to the ratepayer is having the ability to justify your system's revenue needs. This is best accomplished by practicing sound financial management and being able to back up your needs with factual records.

7. *Fixed Costs.* Costs that a utility must cover or pay even if there is no service provided. Also see VARIABLE COSTS.

In making rate increase presentations to an elected board of directors or other elected officials at a public hearing, it is helpful to prepare illustrations of the utility's needs in the form of charts and graphs that show the following information clearly and simply:

- Recent cost performance of the utility (budget comparisons)
- Current revenue requirements versus projected needs
- Effect of proposed rate increases on typical user groups
- Comparison of proposed rates with other similar utilities in your area
- Trends in rate increases

It is good practice to consider potential rate increases on a regular basis, usually annually. Current revenues will not be enough to account for future inflation. Therefore, it is recommended to increase rates gradually, which will promote good public relations. This approach will promote a stable rate adjustment program and will avoid increasing rates by 15 to 20 percent or more during periods of high inflation or greater demand on utility funds. The latter sometimes occurs when rates are not adjusted for multiple-year periods. The consumer expects utility costs to rise moderately over time, but significant one-time bumps in rates every three or four years can be judged by the consumer to reflect poor fiscal management.

## QUESTIONS

Please write your answers to the following questions and compare them with those on page 404.

8.4A When might a utility need to conduct a comprehensive rate study?

8.5A What factors should be presented to officials when making a rate increase presentation?

## 8.6 PLANNING FOR FINANCIAL STABILITY

Financial management for a utility should include providing financial stability for the utility, careful budgeting, and providing capital improvement funds for future utility expansion. These three areas must be examined on a routine basis to ensure the continued operation of the utility. They may be formally reviewed on an annual basis or more frequently when the utility is changing rapidly. The utility manager should understand what is required for each of the three areas and be able to develop record systems that keep the utility on track and financially prepared for the future.

### 8.60 Measuring Stability

How do you measure financial stability for a utility? Two very simple calculations can be used to help you determine how healthy and stable the finances are for the utility. These two calculations are the *OPERATING RATIO* and the *COVERAGE RATIO.* The operating ratio is a measure of the total revenues divided by the total operating expenses. The coverage ratio is a measure of the ability of the utility to pay the principal and interest on loans and bonds (this is known as debt service) in addition to any unexpected expenses. A utility that is in good financial shape will have an operating ratio and coverage ratio above 1.0. In fact, most bonds and loans require the utility to have a coverage ratio of at least 1.25. As state and federal funds for utility improvements have become much more difficult to obtain, these financial indicators have become more important for utilities. Being able to show and document the financial stability of the utility is an important part of getting funding for more capital improvements.

The operating ratio is perhaps the simplest measure of a utility's financial stability. In essence, the utility must be generating enough revenue to pay its operating expenses. The actual ratio is usually computed on a yearly basis because many utilities may have monthly variations that do not reflect the overall performance. The total revenue is calculated by adding up all revenue generated by user fees, hook-up charges, taxes or assessments, interest income, and special income. Next, determine the total operating expenses by adding up the expenses of the utility, including administrative costs, salaries, benefits, energy costs, chemicals, supplies, fuel, equipment costs, equipment replacement fund, principal and interest payments, and other miscellaneous expenses.

*EXAMPLE 1*

The total revenues for a utility are $1,686,000 and the operating expenses for the utility are $1,278,899. The debt service expenses are $560,000. What is the operating ratio? What is the coverage ratio?

| Known | | Unknown |
|---|---|---|
| Total Revenue, $ | = $1,686,000 | Operating Ratio |
| Operating Expenses, $ | = $1,278,899 | Coverage Ratio |
| Debt Service Expenses, $ | = $560,000 | |

1. Calculate the operating ratio.

$$\text{Operating Ratio} = \frac{\text{Total Revenue, \$}}{\text{Operating Expenses, \$}}$$

$$= \frac{\$1,686,000}{\$1,278,899}$$

$$= 1.32$$

2. Calculate nondebt expenses.

$$\text{Nondebt Expenses, \$} = \text{Operating Exp, \$} - \text{Debt Service Exp, \$}$$

$$= \$1,278,899 - \$560,000$$

$$= \$718,899$$

3. Calculate coverage ratio.

$$\text{Coverage Ratio} = \frac{\text{Total Revenue, \$} - \text{Nondebt Expenses, \$}}{\text{Debt Service Expenses, \$}}$$

$$= \frac{\$1,686,000 - \$718,899}{\$560,000}$$

$$= 1.73$$

These calculations provide a good starting point for looking at the financial strength of the utility. Both of these calculations use the total revenue for the utility, which is an important component for any utility budgeting. As managers, we often focus on the expense side and forget to look carefully at the revenue side of utility management. The fees collected by the utility, including hook-up fees and user fees, must accurately reflect the cost of providing service. These fees must be reviewed annually and they must be increased as expenses rise to maintain financial stability. Some other areas to examine on the revenue side include how often and how well user fees are collected, the number of delinquent accounts, and the accuracy of meters (for drinking water utilities). Some small communities have found they can cut their administrative costs significantly by switching to a quarterly billing cycle. The utility must have the support of the community to determine and collect user fees, and the utility must keep track of revenue generation as carefully as resource spending.

## 8.61 Budgeting

Budgeting for the utility is perhaps the most challenging task of the year for many managers. The list of needs usually is much larger than the possible revenue for the utility. The only way for the manager to prepare a good budget is to have good records from the year before. A system of recording or filing purchase orders or a requisition records system must be in place to keep track of expenses and prevent spending money that is not in the budget.

To budget effectively, a manager needs to understand how the money has been spent over the last year, the needs of the utility, and how the needs should be prioritized. The manager also must take into account cost increases that cannot be controlled while trying to minimize the expenses as much as possible. The following problem is an example of the types of decisions a manager must make to keep the budget in line while also improving service from the utility.

*EXAMPLE 2*

A wastewater pump in operation for 25 years pumps a constant 600 GPM through 47 feet of dynamic head. The pump uses 6,071 kilowatt-hours of electricity per month, at a cost of $0.085 per kilowatt-hr. The old pump efficiency has dropped to 63 percent. Assuming a new pump that operates at 86 percent efficiency is available for $9,730.00, how long would it take to pay for replacing the old pump?

| Known | | Unknown |
|---|---|---|
| Electricity, kW-hr/mo | = 6,071 kW-hr/mo | New Pump Payback Time, yr |
| Electricity Cost, $/kW-hr | = $0.085/kW-hr | |
| Old Pump Efficiency, % | = 63% | |
| New Pump Efficiency, % | = 86% | |
| New Pump Cost, $ | = $9,730 | |

1. Calculate old pump operating costs in dollars per month.

$$\text{Old Pump Operating Costs, \$/mo} = \left(\text{Electricity, kW-hr/mo}\right)\left(\text{Electricity Cost, \$/kW-hr}\right)$$

$$= (6{,}071 \text{ kW-hr/mo})(\$0.085/\text{kW-hr})$$

$$= \$516.04/\text{mo}$$

2. Calculate new pump operating electricity requirements.

$$\text{New Pump Electricity, kW-hr/mo} = (\text{Old Pump Elect, kW-hr/mo})\frac{(\text{Old Pump Eff, \%})}{(\text{New Pump Eff, \%})}$$

$$= (6{,}071 \text{ kW-hr/mo})\frac{(63\%)}{(86\%)}$$

$$= 4{,}447 \text{ kW-hr/mo}$$

3. Calculate new pump operating costs in dollars per month.

$$\text{New Pump Operating Costs, \$/mo} = \left(\text{Electricity, kW-hr/mo}\right)\left(\text{Electricity Cost, \$/kW-hr}\right)$$

$$= (4{,}447 \text{ kW-hr/mo})(\$0.085/\text{kW-hr})$$

$$= \$378.03/\text{mo}$$

4. Calculate annual cost savings of new pump.

$$\text{Cost Savings, \$/yr} = (\text{Old Costs, \$/mo} - \text{New Costs, \$/mo})(12 \text{ mo/yr})$$

$$= (\$516.04/\text{mo} - \$378.03/\text{mo})(12 \text{ mo/yr})$$

$$= \$1{,}656.12/\text{yr}$$

5. Calculate the new pump payback time in years.

$$\text{Payback Time, yr} = \frac{\text{Initial Cost, \$}}{\text{Savings, \$/yr}}$$

$$= \frac{\$9{,}730.00}{\$1{,}656.12/\text{yr}}$$

$$= 5.9 \text{ years}$$

In this example, a payback time of 5.9 years is acceptable and would probably justify the expense for a new pump. This calculation was a simple payback calculation that did not take into account the maintenance on each pump, depreciation, and inflation. Many excellent references are available from EPA to help utility managers make more complex decisions about purchasing new equipment.

The annual report should be used to help develop the budget so that long-term planning will have its place in the budgeting process. The utility manager must track revenue generation and

expenses with adequate records to budget effectively. The manager must also get input from other personnel in the utility as well as community leaders as the budgeting process proceeds. This input from others is invaluable to gain support for the budget and to keep the budget on track once adopted.

## 8.62 Capital Improvements and Funding in the Future

The need for capital improvements and planning was discussed earlier in this chapter. The discussion here will focus on how to determine the needs of the utility and the steps needed to provide a financial base for future improvements. The first step in the process is to make a list of the expected capital improvements that will need to be made over the next year, two years, five years, and ten years. You can use the information in your annual reports and operational logs to help compile and prioritize the list. The calculations you have made previously, including corrective to preventive maintenance ratios, operating ratio, coverage ratio, and payback time, will all be useful in prioritizing and streamlining your list of needs. Once you have made your list of capital improvements for a five-year period, estimated the cost of these improvements, and prioritized the list, then you can determine the total cost for these improvements.

The capital improvement fund must be a part of the utility budget and included in the operating ratio. Be sure that everyone—your governing body and the public—understands the capital improvement fund is not a profit for the utility, but a replacement fund to keep the utility operating in the future.

Long-term capital improvements such as a new plant or a new treatment process are usually anticipated in your 10-year or 20-year projection. These long-term capital improvements usually require some additional financing. The basic ways for a utility to finance capital improvements are through general obligation bonds, revenue bonds, or loan funding programs and grants.

General obligation bonds or *ad valorem* (based on value) taxes are assessed based on property taxes. These bonds usually have a lower interest rate and longer payback time, but the total bond limit is determined for the entire community. This means that the water or wastewater utility will have available only a portion of the total bond capacity of the community. These bonds are not often used for funding water and wastewater utility improvements today.

The second type of bond, the revenue bond, is commonly used to fund utility improvements. This bond has no limit on the amount of funds available and the user charges provide repayment on the bond. To qualify for these bonds, the utility must show sound financial management and the ability to repay the bond. As the utility manager, you should be aware of the provisions of the bond. Be sure the bond has a call date, which is the first date when you can pay off the bond. The common practice is for a 20-year bond to have a 10-year call date and for a 15-year bond to have an 8-year call date. The bond will also have a call premium, which is the amount of extra funds needed to pay off the debt on the call date. You should try to get your bonds a call premium of no more than 102 percent par. This means that for a debt of $200,000 on the call date, the total payoff would be $204,000, which includes the extra two percent for the call premium. You will need to get help from a financial advisor to prepare for and issue the bonds. These advisors will help you negotiate the best bond structure for your community.

Special assessment bonds may be used to extend services into specific areas. The direct users pay the capital costs and the assessment is usually based on frontage or area of real estate. These special assessments carry a greater risk to investors but may be the best way to extend service to some areas.

The most common way to finance water and wastewater improvements in the past has been federal and state grant programs. The block grants from HUD are still available for some projects and Rural Utilities Service (RUS) loans may also be used as a funding source. In addition, state revolving fund (SRF) programs provide loans (but not direct grants) for improvements. The SRF program has already been implemented with wastewater improvements and the Safe Drinking Water Regulations include an SRF program for funding water treatment improvements. These SRF programs are very competitive and utilities must provide evidence of sound financial management to qualify for these loans. You should contact your state regulatory agency to find out more about the SRF program in your state.

## 8.63 Financial Assistance

Many small wastewater utility systems need additional funds to repair and upgrade their systems. Potential funding sources include loans and grants from federal and state agencies, banks, foundations, and other sources. Some federal funding programs for small public utility systems are listed below:

- Appalachian Regional Commission (ARC)
- Department of Housing and Urban Development (HUD) (provides community development block grants)
- Economic Development Administration (EDA)
- Indian Health Service (IHS)
- Rural Utilities Service (RUS) (formerly Farmer's Home Administration (FmHA)) and Rural Development Administration (RDA))

Another valuable contact is the Environmental Financing Information Network (EFIN), which provides information on financing alternatives for state and local environmental programs and projects in the form of abstracts of publications, case studies, and contacts. EFIN is a component of the Center for Environmental Finance. For further information, contact US Environmental Protection Agency, Office of the Chief Financial Officer, Center for Environmental Finance, 1200 Pennsylvania Avenue, NW, Mail Code 2731A, Washington, DC 20460. Phone (202) 564-4994 and Fax (202) 565-2587.

Also, many states have one or more special financing mechanisms for small public utility systems. These funds may be in the form of grants, loans, bonds, or revolving loan funds. Contact your state for more information.

## 8.7 ARITHMETIC ASSIGNMENT

Turn to the Appendix, "How to Solve Wastewater System Arithmetic Problems," at the back of this manual and read the following four sections:

1. A.10, "Basic Conversion Factors (English System)"
2. A.11, "Basic Formulas"
3. A.12, "How to Use the Basic Formulas"
4. A.14, "Basic Conversion Factors (Metric System)"

Solve all of the problems in Section A.13, "Typical Wastewater System Problems (English System)," and Section A.15, "Typical Wastewater System Problems (Metric System)" using a calculator. You should be able to get the same answers.

## 8.8 ACKNOWLEDGMENTS

Much of this material was written by Jim Beard for *SMALL WATER SYSTEM OPERATION AND MAINTENANCE,* Chapter 8, "Setting Water Rates and System Security for Small Water Utilities," in this series of operator training manuals. The original chapter was reviewed by Leonard Ainsworth, Mick Cherniak, and Dave Davidson.

Material for this revised chapter was provided by Joe Borgerding, Georgetown Divide Public Utility District, and Tom Cooke, Humboldt Community Services District. All of these contributions are greatly appreciated.

## 8.9 ADDITIONAL READING

The following publications of the American Water Works Association may be obtained by contacting Customer Services, AWWA, 6666 West Quincy Avenue, Denver, CO 80235. Prices include the cost of shipping and handling.

1. *PRINCIPLES OF WATER RATES, FEES, AND CHARGES* (M1). Order No. 30001. ISBN 978-1-58321-069-7. Price to members, $95.50; nonmembers, $146.50
2. *FUNDAMENTALS OF WATER UTILITY CAPITAL FINANCING* (M29). Order No. 30029. ISBN 978-1-58321-545-6. Price to members, $73.50; nonmembers, $108.50.

For a more detailed rate setting procedure for a similar small water utility, see Chapter 8, "Setting Water Rates and System Security for Small Water Utilities," in *SMALL WATER SYSTEM OPERATION AND MAINTENANCE,* available from Office of Water Programs, California State University, Sacramento, 6000 J Street, Sacramento, CA 95819-6025. Price, $49.00. Since water utilities typically use water meters to measure water usage or consumption, variable costs based on water use and fixed costs are used to set water rates.

Two other useful publications available from the Office of Water Programs are *UTILITY MANAGEMENT* ($29.00) and *MANAGE FOR SUCCESS* ($49.00).

*FORMULATE GREAT RATES: THE GUIDE TO CONDUCTING A RATE STUDY FOR A WATER SYSTEM* is an excellent resource produced by the Rural Community Assistance Partnership (RCAP). The guide is available as a PDF on the RCAP website at www.rcap.org.

## QUESTIONS

Please write your answers to the following questions and compare them with those on page 404.

8.6A List the three main areas of financial management for a utility.

8.6B How is a utility's operating ratio calculated?

8.6C Why is it important for a manager to consult with other utility personnel and with community leaders during the budgeting process?

8.6D How can long-term capital improvements be financed?

Please answer the discussion and review questions next.

# DISCUSSION AND REVIEW QUESTIONS

## Chapter 8. SETTING RATES FOR SMALL WASTEWATER UTILITIES

Please write your answers to the following discussion and review questions to determine how well you understand the material in the chapter.

1. How can utility rates be set that accurately reflect the cost of service for each customer?
2. How can future needs be estimated when setting utility service rates?
3. What is the difference between establishing a revenue base on a cash basis and a utility basis?
4. What are the main differences between the base-extra capacity method and the commodity-demand method of allocating costs?
5. What factors would you consider in determining the customer classes for utility rates?
6. Why is it thought to be a good practice to consider rate increases on a regular basis, usually annually?
7. How can a utility's financial stability be measured?
8. What steps should be taken to determine the long-term capital improvement needs of a utility?

# SUGGESTED ANSWERS

## Chapter 8. SETTING RATES FOR SMALL WASTEWATER UTILITIES

Answers to questions on page 392.

8.0A The main goal of a fair and equitable rate setting program is to determine rates on a cost basis and be nondiscriminatory to all users. A second goal in rate setting is stability of rates.

8.0B Below are the five steps that should be followed to determine utility service rates:

1. Determine total annual revenue requirements.
2. Determine reserve fund requirements.
3. Allocate revenue requirements to cost components.
4. Distribute component costs to customer classes.
5. Design rates that will recover these costs from each class of customer.

Answers to questions on page 396.

8.1A Debt service is the amount of money required annually to pay the (1) interest on outstanding debts or (2) funds due on a maturing bonded debt or the redemption of bonds.

8.1B Operation and maintenance expenses are the costs of actually operating and maintaining the wastewater collection, treatment, and discharge facilities, including personnel and administrative costs.

8.1C Capital costs that must be covered by revenue include costs associated with plant construction investment, debt service, replacement, facility expansion and improvement, and cash reserves.

8.1D The two general components of inflation that affect setting of rates and charges are consumer price index (CPI) and construction cost index (CCI). The CPI should be applied to general O&M expense items such as labor, materials, and supplies. The CCI is used for estimating construction-related capital improvement costs.

Answers to questions on page 397.

8.2A Two accepted methods of analyzing the cost to serve various customer classes and allocating costs to varying levels of service are the base-extra capacity method and the commodity-demand method.

8.3A A commercial customer class includes nonindustrial, business, and service-type enterprises.

8.3B A connection fee is an up-front fee paid by new users connecting to the system to cover capital costs to provide service to their property.

Answers to questions on page 399.

8.4A When a utility has special circumstances such as a large industrial base, a comprehensive or partial rate study may be in order.

8.5A When making a rate increase presentation, illustrate the utility's needs in the form of charts and graphs that show the following information clearly and simply:

1. Recent cost performance of the utility (budget comparisons)
2. Current revenue requirements versus projected needs
3. Effect of proposed rate increases on typical user groups
4. Comparison of proposed rates with other similar utilities in your area
5. Trends in rate increases

Answers to questions on page 402.

8.6A The three main areas of financial management for a utility include providing financial stability for the utility, careful budgeting, and providing capital improvement funds for future utility expansion.

8.6B The operating ratio for a utility is calculated by dividing total revenues by total operating expenses.

8.6C It is important for a manager to get input from other personnel in the utility as well as community leaders as the budgeting process proceeds in order to gain support for the budget and to keep the budget on track once adopted.

8.6D The basic ways for a utility to finance long-term capital improvements are through general obligation bonds, revenue bonds, or loan funding programs and grants.

# APPENDIX

# SMALL WASTEWATER SYSTEM OPERATION AND MAINTENANCE

(VOLUME I)

Comprehensive Review Questions and Suggested Answers

How to Solve Wastewater System Arithmetic Problems

Wastewater Abbreviations

Wastewater Words

Subject Index

# COMPREHENSIVE REVIEW QUESTIONS

This section was prepared to help you review the material in this manual. The questions are divided into four types:

1. True-False
2. Best Answer
3. Multiple Choice
4. Short Answer

To work this section, complete these steps:

1. Write the answer to each question in your notebook.
2. After you have worked a group of questions (you decide how many), check your answers with the suggested answers at the end of this section.
3. If you missed a question and do not understand why, reread the material in the manual.

You may wish to use this section for review purposes when preparing for civil service and certification examinations.

Since you have already completed this course, please *DO NOT SEND* your answers to California State University, Sacramento.

**True-False**

1. The goal of a small wastewater system operator is to keep the community healthy.
   1. True
   2. False
2. Small wastewater system operators are in a responsible position to protect the public health and the environment.
   1. True
   2. False
3. The amount of settleable solids in raw wastewater will determine how fast solids will build up in a septic tank.
   1. True
   2. False
4. The bacteria that grow in the intestinal tract of diseased humans are likely to find the environment in the wastewater treatment plant or receiving waters favorable for their growth and reproduction.
   1. True
   2. False
5. Pressure sewer mains are dependent on a slope to create a gravity flow.
   1. True
   2. False
6. The wastewater treatment settling process and biological processes remove very few organisms from the wastewater flow.
   1. True
   2. False
7. If you must work alone as a small wastewater collection and treatment system operator, then you and only you are responsible for your safety.
   1. True
   2. False
8. Tetanus is a waterborne disease.
   1. True
   2. False
9. Only explosion-proof electrical equipment and fixtures should be used in areas potentially capable of developing explosive mixtures.
   1. True
   2. False
10. A variable valve such as a gate valve on a water pipe is like a variable resistor on an electrical wire.
    1. True
    2. False
11. Wastewater treatment facilities contain some equipment that produces high noise levels, intermittently or continuously.
    1. True
    2. False
12. Persons using septic tank systems must be encouraged not to use the septic tank system for anything that can be disposed of safely as trash.
    1. True
    2. False
13. Pump vaults and covers are manufactured of fiberglass or thin-wall fiberglass reinforced polyester (FRP) and are commonly supplied as a complete package containing the pump and associated controls and alarms.
    1. True
    2. False
14. The operator has certain legal responsibilities with regard to entering and protecting private property.
    1. True
    2. False

15. Maintenance of septic tank systems exposes operators to the possibility of infections or disease.
    1. True
    2. False
16. Most of the organisms in sand filters are anaerobic and therefore do not require oxygen to live and function.
    1. True
    2. False
17. Because a siphon operates by gravity, the treatment unit or discharge area being dosed must be located downhill from the dosing tank.
    1. True
    2. False
18. A programmable timer allows a more uniform discharge of effluent to a discharge site.
    1. True
    2. False
19. If a seepage bed becomes clogged with organic matter, the side walls will offer little added percolation.
    1. True
    2. False
20. A sewer line should be large enough for the use of cleaning equipment.
    1. True
    2. False
21. After you have reviewed a set of collection system plans and profiles, walk the proposed route and think about how the sewer lines and equipment will be maintained.
    1. True
    2. False
22. Air release valves, combination air release/vacuum relief valves, and check valves may be used in small-diameter gravity sewer systems.
    1. True
    2. False
23. Air release valve stations may be automatically or manually operated.
    1. True
    2. False
24. In vacuum sewer systems, it is the difference between the atmospheric pressure behind the wastewater and the vacuum ahead that provides the drawing force.
    1. True
    2. False
25. Expansion and contraction may be a concern when the entire vacuum system is solvent welded, but allowances are made for these forces in gasketed main line pipe joints.
    1. True
    2. False
26. Operator attitude is vital to the efficient operation of a vacuum system.
    1. True
    2. False
27. The large amount of air throughout the vacuum collection system aids in the creation of septic wastewater.
    1. True
    2. False
28. Preventive maintenance is the most effective and efficient type of maintenance program.
    1. True
    2. False
29. After a job is completed, clean your tools and store them properly so they can be found and used when needed again.
    1. True
    2. False
30. The success of your mound repair and service activities relies on your skill in diagnosing the problem and resolving it.
    1. True
    2. False
31. Voltage imbalance can occur in either the utility side or the pump station electrical system.
    1. True
    2. False
32. A small wastewater utility rate structure that is stable and unpredictable over a period of years reflects well on your professional ability.
    1. True
    2. False
33. When using the cash basis of accounting, the utility must raise enough money to cover all cash needs including debt service.
    1. True
    2. False
34. In order to set accurate utility rates that differentiate between customer classes, a good historical database is needed from which information on the various user classes can be taken.
    1. True
    2. False
35. A utility that is in good financial shape will have an operating ratio and a coverage ratio below 1.0.
    1. True
    2. False

**Best Answer** (Select only the closest or best answer.)

1. How does a small wastewater system operator protect the environment and the public health of the community?
   1. By an extensive public health education program
   2. By correct treatment and proper disposal or discharge of wastewater from the community
   3. By providing school children with interesting tours of the wastewater treatment facilities
   4. By working with environmental organizations to protect the environment
2. Why is an operator able to help a consulting engineer when an old wastewater treatment plant is being remodeled?
   1. The operator has the knowledge, skills, and abilities to treat the wastewater
   2. The operator is familiar with the expectations of the regulatory agencies
   3. The operator knows the sizes of the facilities needed to treat the wastewater
   4. The operator's experience provides valuable technical knowledge concerning the characteristics of the wastewater, its sources, and the limitations of the present facilities
3. After heating a sample of raw wastewater enough to evaporate all the water, what does the weight of the remaining residue include?
   1. Both dissolved and suspended solids
   2. Both suspended solids and inactive pathogens
   3. Only dissolved solids
   4. Only floatable solids
4. What is one of the principal objectives of wastewater treatment?
   1. To allow FOG to enter the receiving water
   2. To keep the temperature of the receiving water as cool as possible
   3. To minimize the presence of salinity in the plant effluent
   4. To prevent as much oxygen-demanding organic material as possible from entering the receiving water
5. What is the public health concern regarding a high nitrate concentration in drinking water?
   1. Creates intestinal distress in susceptible humans
   2. Fertilizes pathogenic organisms
   3. Interferes with the oxygen uptake in the blood of infants
   4. Provides a critical nutrient in the evolution of algal blooms in water
6. Which item best defines "inflow"?
   1. The arrangement of hills and valleys in a geographic area
   2. The lowest point of the channel inside a pipe, conduit, canal, or manhole
   3. The seepage of groundwater into a sewer system, including service connections
   4. The water discharged into a sewer system and service connections from sources other than regular connections
7. Why do some wastewater treatment plants neutralize the chlorine in the effluent?
   1. To allow recycling of the effluent
   2. To detoxify the effluent
   3. To eliminate chlorine tastes and odors
   4. To prevent the regrowth of disease organisms
8. How can gas explosions be avoided?
   1. By eliminating all sources of ignition in areas potentially capable of developing explosive mixtures
   2. By eliminating all sources of oxygen in areas potentially capable of developing explosive mixtures
   3. By eliminating most of the potential ignition sources
   4. By providing operator safety training on how to eliminate explosive conditions
9. What does the federal worker Right-to-Know Law state?
   1. That a person using a chemical has the right to know the hazards associated with that chemical
   2. That a person using a chemical has the right to know the manufacturer of the chemical
   3. That a person working in a plant has the right to know the sources of hazards discharged into the collection system
   4. That a person working with other operators has the right to know the safety training received by the other operators
10. What is shoring?
    1. A complete framework of wood or metal that is designed to support the walls of a trench
    2. A line of material used to hold up the walls of a trench
    3. A procedure using prefabricated containers to minimize soil movement
    4. An excavation procedure used to allow minimum flow of groundwater and soils
11. What is the purpose of a hearing conservation program?
    1. To educate operators of the potential damage from excessive noise levels
    2. To prevent hearing loss that might affect an operator's ability to hear and understand normal speech
    3. To protect the utility from costly litigation
    4. To provide operators with information needed to pass certification examinations
12. Why do steel septic tanks require a coating or other corrosion-resistant treatment and cathodic protection in corrosive (low pH) soils?
    1. To prevent rusting and possible leakage
    2. To prevent stray current electrons from corroding household plumbing
    3. To prevent toxic effluent from killing aquatic life
    4. To prevent toxic heavy metals from entering the septic tank effluent

13. What could cause a septic tank to become buoyant (float)?
    1. Excessive leakage of groundwater into the septic tank
    2. Improper backfilling procedures
    3. Use of extremely lightweight construction materials
    4. Use of improper septage pumping procedures in areas with a high groundwater table

14. What is the only sure way to determine when a septic tank needs to be pumped?
    1. Actual inspection of sludge and scum levels in the tank
    2. Evaluation of the microorganism population
    3. Measurement of the supernatant volume in the tank
    4. Observation of wastewater backing up into the home

15. What is the benefit of installing a new building sewer?
    1. Ensures the system is always maintained
    2. Ensures the system is never maintained
    3. Ensures the system is watertight to prevent infiltration/inflow
    4. Ensures the system is watertight to promote infiltration/inflow

16. Under what circumstances is a pipe saddle or hot tap connection used?
    1. When a new tap is inserted using special devices and materials for hot work
    2. When a service line is connected to a system main during the heat of the day
    3. When heat is used to facilitate the connection of a new service line to a main under pressure
    4. When new service lines must be added to a system while the main remains under pressure

17. What problems can be created by pumping a septic tank too often?
    1. May fill up solids handling and storage areas too soon
    2. May inconvenience the resident by the frequent presence of septage hauling trucks
    3. May interfere with the development of a healthy population of microorganisms and thus interfere with proper treatment of the wastewater solids
    4. May prevent landscape vegetation from fully developing

18. When applying heat to shrink the insulated heat shrink tubing used to seal electrical wire splices, why should the propane torch or heat gun be kept moving?
    1. To allow for uniform flow of the sealant oozing out the end of the tube
    2. To avoid breaking the wiring connection
    3. To ensure the splice is watertight
    4. To prevent too much concentrated heat from damaging the tubing

19. What is a dosing siphon?
    1. A mechanical device that delivers fixed amounts of water from one tank or reservoir to another tank or pipeline
    2. A mechanical device that mechanically transfers water from a lower elevation to a higher elevation
    3. A mechanical device that moves water from a septic tank to a leach field
    4. A mechanical device that uses pressure differential to lift water from a lower location to a higher one

20. How can the effluent quality from a gravel filter be adjusted?
    1. By changing the magnitude of the allowable BOD discharged into the sewer
    2. By changing the ratio between the amount of filter effluent that is recycled and the amount of effluent that is discharged
    3. By implementing a public education program for the homeowner
    4. By increasing the size of the gravel filter

21. What happens to the effluent from subsurface wastewater infiltration systems (SWIS) when it is discharged to long narrow trenches?
    1. It evaporates during warm weather
    2. It flows out the ends of the trenches to surface waters
    3. It percolates down through the bottom of the trench and also seeps laterally through the side walls
    4. It transpires through ground surface vegetation

22. Why are rest periods recommended for subsurface wastewater infiltration systems (SWIS)?
    1. To allow microorganisms to rest so they will be more active when the next load is applied
    2. To increase the wastewater storage space or capacity of the trench
    3. To provide time for vegetation to increase the evapotranspiration rate
    4. To reduce solids and microorganism colonies in the media and on the soil interface, thus partially restoring infiltration rates

23. Why should care be taken to avoid a very fine sand in the first layer of a mound system?
    1. Because it will allow rapid percolation
    2. Because it will encourage biological growth and slimes
    3. Because it will have a low percolation/infiltration rate
    4. Because it will support vegetative growth

24. What is the purpose of a wastewater collection system?
    1. To collect sewer overflows and convey them to a treatment facility
    2. To collect solid wastes from a community's homes and convey them to a treatment plant
    3. To collect stormwater and also inflow and infiltration and convey these waters to a discharge site
    4. To collect the wastewater from a community's homes and convey it to a treatment plant
25. Why are much larger sewer pipes needed for conventional gravity sewer systems than for small-diameter gravity sewer (SDGS) systems?
    1. Because of the flow velocities associated with the slope of conventional gravity systems
    2. Because of the requirement to allow ventilation air to flow above the water surface
    3. Because the conventional gravity sewer system is designed to convey odorous air to a treatment system
    4. Because the conventional gravity sewer system is designed with a peaking factor greater than the average flow
26. What is a hybrid small-diameter gravity sewer (SDGS) system? A system that serves
    1. A combination of force mains and pressure sewers
    2. A combination of lift stations and gravity flow sewer systems
    3. A combination of pressure and vacuum sewer systems
    4. A combination of septic tank effluent pump units and septic tank effluent filter (screened) units
27. Why are automatic air release valves installed at upper elevations on the main between depressions in a septic tank effluent pump (STEP) piping system?
    1. To facilitate maintenance of the pressure sewer
    2. To facilitate the draining of the pressure sewer
    3. To permit venting of air or gases trapped in the main and prevent an air lock or blockage in the main
    4. To prevent formation of high concentrations of odorous or toxic gases
28. Where are cycle counters typically used in vacuum sewer systems?
    1. Where a large water use is expected in order to determine if the valve is reasonably capable of keeping up with the flow
    2. Where maintenance is based on the hours of valve operation
    3. Where sewer service charges are based on the flow
    4. Where water conservation programs require the detection of leaky systems
29. Why should wastewater discharge pumps in vacuum systems have an inspection opening in the pump volute?
    1. To facilitate cleaning or clearing a plugged impeller
    2. To facilitate examining the volute for corrosion
    3. To facilitate inspecting the suction and discharge lines
    4. To facilitate viewing the impeller for cavitation damage
30. Which item could cause waterlogging in a vacuum collection system?
    1. A pipe that is too small to transmit the hydraulic load
    2. A system that becomes hydraulically overloaded and floods
    3. A valve that closes and causes the wastewater to build up behind the valve
    4. A valve that is stuck open or that continuously cycles and causes the system vacuum to drop
31. What information does an equipment service card provide? The equipment service card tells
    1. What equipment service procedures are used to prioritize work
    2. What equipment service should be done and when
    3. What equipment service you did and when you did it
    4. What procedure to use when equipment service work must be rescheduled
32. When operators are attempting to fix equipment and electricity is involved, the risks are particularly serious for what reason?
    1. Because nearby fields could be set on fire
    2. Because of the potential for causing a power outage to nearby homes
    3. Because the consequences of an accident can be death by electrocution or severe, disabling burns
    4. Because the power company may need to assign a crew to correct any damages
33. How can you determine if an electric circuit is energized?
    1. By checking the readings on the watt meter
    2. By flipping a switch and observing any changes
    3. By using a multimeter and holding one lead on ground and the other on a power lead
    4. By using an ammeter and checking the phases
34. What is the first step in setting utility service rates?
    1. Determine the rates water users' would be willing to pay
    2. Establish procedures to allocate costs fairly
    3. Estimate the funds available to water users
    4. Outline the specific steps needed to reach the goal

35. What is rate of return?
    1. A measure of the total life cycle costs of the facility or equipment
    2. A value that indicates the return of funds received on the basis of the total equity capital used to finance physical facilities
    3. The gradual loss in service value of a facility or piece of equipment due to all the factors causing the ultimate retirement of the facility or equipment
    4. The value of an investment in a facility

36. How can a utility manager justify the system's revenue needs to the ratepayers?
    1. By documenting that total expenditures have always been less that the budgeted amounts
    2. By practicing sound financial management and being able to back up needs with factual records
    3. By retaining a professional to present and justify the needs
    4. By showing that the utility has always spent less than the total revenue

**Multiple Choice** (Select all correct answers.)

1. What are the main components of a conventional on-site wastewater system?
    1. Control panels with alarms
    2. Domestic well
    3. Pit privy
    4. Septic tank
    5. Subsurface disposal field

2. What information or benefits should a manual on small wastewater system operation and maintenance provide?
    1. A better tomorrow, both for the operator and for the public whose health and environment will benefit
    2. How to better communicate with local and state environmental agencies
    3. How various wastewater collection and treatment processes and systems function
    4. The opportunity to achieve state certification for the level of competency required by your system
    5. The terminology needed to communicate effectively with operators of similar systems or larger municipal treatment systems

3. For normal municipal wastewater, which contains domestic wastewater as well as some industrial and commercial wastes, what are the goals of the treatment plant designer and operator?
    1. To kill or inactivate the pathogenic organisms
    2. To remove dissolved inorganic solids
    3. To remove dissolved organic solids
    4. To remove inorganic suspended solids
    5. To remove organic suspended solids

4. NPDES permits regulate discharges from what types of point sources of pollution?
    1. Industries
    2. Large agricultural feedlots
    3. Return irrigation flows
    4. Sanitary landfills
    5. Small businesses discharging into sewers

5. Compared with a gravity collection system, pressure sewers offer which advantages?
    1. Deep trenching to maintain flow is unnecessary
    2. Inverted siphons are not needed to cross roads and rivers
    3. Pressurized systems experience fewer stoppages
    4. Root intrusion problems are eliminated
    5. Smaller diameter pipe may be used

6. How can excess activated sludge from an oxidation ditch be treated or disposed of?
    1. Dried directly on sludge drying beds
    2. Recycled as biosolids for disposal on land
    3. Returned to the primary clarifier for settling with primary sludge
    4. Stored in a holding tank for later transfer to larger treatment plants
    5. Stored in sludge lagoons for later disposal in approved sanitary landfills

7. Which items are common forms of stored energy that could be encountered by a small wastewater treatment system operator?
    1. Compressed gases that are stored under pressure
    2. Electric energy
    3. Horse power
    4. Hydraulic pressure
    5. Spring-loaded equipment

8. Toxic or suffocating gases may come from which sources?
    1. From industrial waste discharges
    2. From process chemicals
    3. From the decomposition of domestic wastewater
    4. From winds blowing over sanitary landfills
    5. From winds caused by shifts in atmospheric pressure gradients

9. When inspecting the mechanical condition of a work vehicle, which items should be inspected?
    1. Brakes on the vehicle and trailer, if so equipped
    2. Mirrors and windows
    3. Towed vehicle lighting system
    4. Ventilation blowers
    5. Windshield wipers

10. What information should be included on a Material Safety Data Sheet (MSDS)?
    1. Fire and explosion hazard data
    2. Hazardous ingredients/identity information
    3. Health hazard data
    4. Physical/chemical characteristics
    5. Precautions for safe handling and use

11. What are the main elements of a comprehensive hearing conservation program?
    1. Access to noise level information
    2. Audiometric testing
    3. Hearing protection
    4. Monitoring and recordkeeping
    5. Training in the use of protective equipment and procedures

12. Why must there be a reduction of the solids content in the wastewater entering the second compartment of a septic tank?
    1. To maximize the production of methane gas biofuel
    2. To prevent a wastewater backup of the system into the home
    3. To prevent solids from plugging the leach field's distribution piping
    4. To prevent solids from sealing the soil in the leach field
    5. To reduce the quantity of solids entering the leach field

13. What happens when solids clog the discharge leach field system?
    1. Liquid may break through to the ground surface
    2. Organisms treating wastes in the leach field become more robust
    3. Rate of solids removal from the leach field must increase
    4. Treatment detention time increases in the leach field
    5. Wastewater may back up into the home's plumbing fixtures

14. What are the most common causes of pipe blockages in septic tank systems?
    1. Grease or detergent buildup in the pipe
    2. Leach field system cave-ins
    3. Settling of septic tank effluent pipes
    4. Tree roots entering the pipe
    5. Washout of leach field berms

15. The selection of septic tank effluent pumps is greatly dependent upon which conditions?
    1. The flow rate to be pumped
    2. The number of operators available
    3. The qualifications of the available operators
    4. The total dynamic head (TDH) of the collection system
    5. The type of wastewater to be pumped

16. During extended power outages when STEP system pump basins may fill to capacity, which items should be available to pump down flooded tanks and prevent overflows?
    1. Activated sludge treatment process
    2. Gasoline-powered pumps
    3. Portable standby generators
    4. Septic tank pumper truck
    5. Venturi pumps

17. Which items are symptoms of poor electrical connections?
    1. Alarm light on, high level
    2. Alarm sounds intermittently
    3. Control box breaker trips
    4. Pump does not run
    5. Short pump cycles

18. Which wastewater treatment and discharge options are appropriate for use either as on-site treatment systems or as small community treatment systems?
    1. Anaerobic digestion systems
    2. Evapotranspiration systems
    3. Leach fields
    4. Mounds
    5. Seepage beds and pits

19. Why must the water be filtered before it reaches the dosing siphon?
    1. To collect and recycle valuable solids
    2. To help protect the performance of the discharge area
    3. To help protect the performance of the distribution network
    4. To help protect the performance of the siphon
    5. To keep the siphon's snifter pipe clear

20. In an effluent discharge system based on subgrade or subsurface land application of effluent, what problems will or may occur if the soil is too coarse (large particles of sand, gravel)?
    1. Absorption for adequate biochemical treatment will not be provided
    2. Effluent will flow down (percolate) too rapidly
    3. Filtration for adequate biochemical treatment will not be provided
    4. Time for adequate biochemical treatment will not be provided
    5. Untreated or partially treated wastewater may reach and contaminate the groundwater

21. In a gravity-dosed leach field, distribution of the effluent to the trenches may be controlled by which methods?
    1. By routing and distributing flow through a Palmer-Bowlus meter
    2. By routing flow through a distribution box
    3. By routing flow through a drop box
    4. By the continuous flow method
    5. By the V-notch weir method

22. When monitoring a subsurface wastewater infiltration system (SWIS) leach field, operators should look for which conditions?
    1. Creeping failure of the leach field
    2. Effluent overflowing the distribution structure
    3. High standing water levels in the monitoring wells
    4. Water overflowing at the septic tank cleanout
    5. Water pooling on the surface of the SWIS

23. Which systems are possible alternatives to a conventional gravity collection system?
    1. Force main systems
    2. Pressure collection systems
    3. Small-diameter gravity sewer systems
    4. Suction lift systems
    5. Vacuum collection systems

24. Which questions should be asked concerning the design of a wastewater collection system?
    1. Can the atmosphere in the manhole be safely tested for oxygen deficiency/enrichment, explosive gases, and toxic gases?
    2. Can the available sewer maintenance equipment operate effectively between the manholes or is the distance too great?
    3. Does the area flood and, if so, how often?
    4. If there are steps in the manhole wall for access, will they be damaged by corrosion and become unsafe?
    5. Is the shelf in the bottom of the manhole properly sloped?

25. What does preventive maintenance of a small-diameter gravity sewer (SDGS) system include?
    1. Exercising of isolation valves
    2. Inspection and cleaning of the collection mains
    3. Inspection and pumping of the septic tanks
    4. Inspection and servicing of any STEP units or lift stations
    5. Testing and servicing of air release valves

26. What are the major causes of lift station failures?
    1. Failure of the preventive maintenance program
    2. Loss of power
    3. Mechanical failure
    4. Settling due to earthquakes
    5. Uncontrolled station fires

27. What are the advantages of vacuum sewer systems compared to other types of collection systems? Substantial reductions in
    1. Excavation costs
    2. Material costs
    3. Number of operators required
    4. Treatment expenses
    5. Water used to transport wastes

28. What are the operation and maintenance responsibilities of a vacuum wastewater collection system operator?
    1. Analyzing and evaluating operation and maintenance functions and developing new procedures to ensure continued system efficiency
    2. Determining remedial action necessary during emergencies
    3. Gathering and reviewing all data and recording all data for the preparation of reports and purchase requests
    4. Inspecting the system daily to determine the efficiency of operation, cleanliness, and maintenance needs
    5. Preparing work schedules

29. Emergency or breakdown maintenance may involve which aspects of a vacuum collection system?
    1. Anxious regulator inspectors
    2. Piping system
    3. Upset homeowners
    4. Vacuum station
    5. Vacuum valves

30. Which types of records should be kept for a vacuum collection system?
    1. Emergency maintenance records
    2. Operating cost records
    3. Preventive maintenance records
    4. Routine maintenance records
    5. Social activities records

31. What does a good maintenance program need to accomplish? A good maintenance program
    1. Ensures that regular preventive maintenance is performed to prevent problems and equipment failures
    2. Generates sufficient revenue to fund the maintenance program
    3. Identifies and corrects the problem if something goes wrong
    4. Schedules emergency maintenance well ahead of time
    5. Tries to anticipate what is going to go wrong and prevent it from happening

32. Which persons should be present during the construction of a wastewater collection and treatment system?
    1. Contractor
    2. Engineer
    3. Inspector
    4. Operator
    5. Superintendent

33. Why does the operator of a wastewater collection or treatment system need a basic working knowledge of electricity?
    1. To prevent financial losses resulting from damage to control circuits
    2. To prevent financial losses resulting from damage to equipment
    3. To prevent financial losses resulting from damage to machinery
    4. To prevent financial losses resulting from motors burning out
    5. To prevent operators from seriously injuring themselves

34. Which procedures should be used to develop a small wastewater utility's schedule of rates and charges?
    1. Allocate revenue requirements to cost components
    2. Design rates that will recover required costs from each class of customer
    3. Determine reserve fund requirements
    4. Determine total annual revenue requirements
    5. Distribute component costs to customer classes

35. What are the general components of inflation that affect utility rates and charges?
    1. Construction cost index (CCI)
    2. Consumer price index (CPI)
    3. Gross national product (GNP)
    4. Growth of commodities (GOC)
    5. Industrial cost extension (ICE)
36. In the commodity-demand method of cost allocation, costs of service include which primary cost components?
    1. Base costs
    2. Commodity costs
    3. Customer-related costs
    4. Demand costs
    5. Direct fire-protection costs
37. What are the advantages of a user fee-based financing system?
    1. Allows current ratepayers to finance a portion of the debt
    2. Avoids payment of interest expense on long-term debt
    3. Does not put the entire burden of system expansion on current ratepayers
    4. Encourages current ratepayers to contribute toward attracting new industries (jobs)
    5. Simplifies accounting and cost allocation methods
38. To determine a utility's total operating expenses, add up the expenses of the utility, including which items?
    1. Administrative costs
    2. Chemicals
    3. Equipment replacement fund
    4. Salaries
    5. Supplies

**Short Answer**

1. Why should an operator have an understanding of natural cycles?
2. What problems could be caused in receiving waters by discharging wastewater containing too much nitrogen?
3. What are the advantages and limitations of a pressure collection system as compared with a gravity collection system?
4. What type of stabilization pond is normally used in most areas?
5. What is the purpose of the rotor in an oxidation ditch?
6. What is the operator's responsibility with regard to safety?
7. What precautions should you take to avoid transmitting disease to your family?
8. Any time traffic may be affected by road work activities, who must you notify before work begins or before traffic is diverted or disrupted?
9. Why should wastewater system operators be familiar with Material Safety Data Sheets?
10. How can wastewater system operators protect their hearing from loud noises?
11. How is the volume of wastewater solids reduced in a septic tank?
12. What steps should homeowners be advised to take to protect their septic tank systems?
13. Why are new building sewers and new septic tanks usually installed when a septic tank effluent pump (STEP) system is installed?
14. What types of STEP system maintenance situations require that more than one operator be present?
15. When would it be necessary to install two septic tank effluent pumps at the same location?
16. Routine preventive maintenance of septic tanks and grinder pumps in a STEP system includes what tasks?
17. What are the main components of an open, single-pass intermittent sand filter?
18. What is the purpose of a dosing siphon?
19. Why must treated wastewater effluents be properly discharged?
20. What can happen if a distribution box settles or tilts and changes the elevation of the outlets of the distribution pipes?
21. Why should grass immediately be planted over a mound?
22. Why should collection system operators know how engineers design wastewater collection systems?
23. Why does the slope of conventional gravity sewers generally follow the slope of the land?
24. Why should SDGS system record drawings (as-builts) identify collection mains in relation to permanent structures nearby?
25. Why are pressure collection systems sometimes used as an alternative to gravity collection systems?
26. What are the general types of preventive maintenance activities associated with pressure system pipelines?
27. Describe how the controller/sensor in a vacuum valve operates.
28. Why is it important to have a standby generator at the central vacuum station?
29. What problem is the most common cause of emergency maintenance?
30. What are some types of construction-related O&M problems?
31. What does a good maintenance program achieve?
32. What could be the cause of an alarm for a mound system?
33. Why must motor nameplate data be recorded and filed?
34. List three uses of a multimeter.
35. How can utility rates be set that accurately reflect the cost of service for each customer?
36. What factors would you consider in determining the customer classes for utility rates?
37. How can a utility's financial stability be measured?

# SUGGESTED ANSWERS TO COMPREHENSIVE REVIEW QUESTIONS

**True-False**

1. True — The goal of a small wastewater system operator is to keep the community healthy.
2. True — Small wastewater system operators are in a responsible position to protect the public health and the environment.
3. True — The amount of settleable solids in raw wastewater will determine how fast solids will build up in a septic tank.
4. False — The bacteria that grow in the intestinal tract of diseased humans are NOT likely to find the environment in the wastewater treatment plant favorable for their growth and reproduction.
5. False — Pressure sewer mains are NOT dependent on a slope to create a gravity flow.
6. False — The wastewater treatment settling process and biological processes remove a great number of organisms (NOT very few organisms) from the wastewater flow.
7. True — If you must work alone as a small wastewater collection and treatment system operator, then you and only you are responsible for your safety.
8. False — Tetanus is NOT a waterborne disease.
9. True — Only explosion-proof electrical equipment and fixtures should be used in areas potentially capable of developing explosive mixtures.
10. True — A variable valve such as a gate valve on a water pipe is like a variable resistor on an electrical wire.
11. True — Wastewater treatment facilities contain some equipment that produces high noise levels, intermittently or continuously.
12. True — Persons using septic tank systems must be encouraged not to use the septic tank system for anything that can be disposed of safely as trash.
13. True — Pump vaults and covers are manufactured of fiberglass or thin-wall fiberglass reinforced polyester (FRP) and are commonly supplied as a complete package containing the pump and associated controls and alarms.
14. True — The operator has certain legal responsibilities with regard to entering and protecting private property.
15. True — Maintenance of septic tank systems exposes operators to the possibility of infections or disease.
16. False — Most of the organisms in sand filters are aerobic (NOT anaerobic) and therefore require oxygen to live and function.
17. True — Because a siphon operates by gravity, the treatment unit or discharge area being dosed must be located downhill from the dosing tank.
18. True — A programmable timer allows a more uniform discharge of effluent to a discharge site.
19. True — If a seepage bed becomes clogged with organic matter, the side walls will offer little added percolation.
20. True — A sewer line should be large enough for the use of cleaning equipment.
21. True — After you have reviewed a set of collection system plans and profiles, walk the proposed route and think about how the sewer lines and equipment will be maintained.
22. True — Air release valves, combination air release/vacuum relief valves, and check valves may be used in small-diameter gravity sewer systems.
23. True — Air release valve stations may be automatically or manually operated.
24. True — In vacuum sewer systems, it is the difference between the atmospheric pressure behind the wastewater and the vacuum ahead that provides the drawing force.
25. True — Expansion and contraction may be a concern when the entire vacuum system is solvent welded, but allowances are made for these forces in gasketed main line pipe joints.
26. True — Operator attitude is vital to the efficient operation of a vacuum system.
27. False — The large amount of air throughout the vacuum collection system aids in the prevention (NOT creation) of septic wastewater.
28. True — Preventive maintenance is the most effective and efficient type of maintenance program.
29. True — After a job is completed, clean your tools and store them properly so they can be found and used when needed again.

30. True The success of your mound repair and service activities relies on your skill in diagnosing the problem and resolving it.

31. True Voltage imbalance can occur in either the utility side or the pump station electrical system.

32. False A small wastewater utility rate structure that is stable and predictable (NOT unpredictable) over a period of years reflects well on your professional ability.

33. True When using the cash basis of accounting, the utility must raise enough money to cover all cash needs including debt service.

34. True In order to set accurate utility rates that differentiate between customer classes, a good historical database is needed from which information on the various user classes can be taken.

35. False A utility that is in good financial shape will have an operating ratio and a coverage ratio above (NOT below) 1.0.

**Best Answer**

1. 2 A small wastewater system operator protects the environment and the public health of the community by correct treatment and proper disposal or discharge of wastewater from the community.

2. 4 The operator can help a consulting engineer when an old wastewater treatment plant is being remodeled because the operator's experience provides valuable technical knowledge concerning the characteristics of the wastewater, its sources, and the limitations of the present facilities.

3. 1 After heating a sample of raw wastewater enough to evaporate all the water, the weight of the remaining residue includes both dissolved and suspended solids.

4. 4 One of the principal objectives of wastewater treatment is to prevent as much oxygen-demanding organic material as possible from entering the receiving water.

5. 3 A high nitrate concentration in drinking water interferes with the oxygen uptake in the blood of infants.

6. 4 Inflow is defined as the water discharged into a sewer system and service connections from sources other than regular connections.

7. 2 Some wastewater treatment plants neutralize the chlorine in the effluent to detoxify the effluent.

8. 1 Gas explosions can be avoided by eliminating all sources of ignition in areas potentially capable of developing explosive mixtures.

9. 1 The federal worker Right-to-Know Law states that a person using a chemical has the right to know the hazards associated with that chemical.

10. 1 Shoring is a complete framework of wood or metal that is designed to support the walls of a trench.

11. 2 The purpose of a hearing conservation program is to prevent hearing loss that might affect an operator's ability to hear and understand normal speech.

12. 1 Steel septic tanks require a coating or other corrosion-resistant treatment and cathodic protection in corrosive (low pH) soils to prevent rusting and possible leakage.

13. 4 In areas with a high groundwater table, use of improper septage pumping procedures can cause a septic tank to become buoyant (float).

14. 1 The only sure way to determine when a septic tank needs to be pumped is actual inspection of sludge and scum levels in the tank.

15. 3 Installing a new building sewer ensures the system is watertight to prevent infiltration/inflow.

16. 4 A pipe saddle or hot tap is used when new service lines must be added to a system while the main remains under pressure.

17. 3 Pumping a septic tank too often may interfere with the development of a healthy population of microorganisms and thus interfere with proper treatment of the wastewater solids.

18. 4 When applying heat to shrink the insulated heat shrink tubing used to seal electrical wire splices, the propane torch or heat gun should be kept moving to prevent too much concentrated heat from damaging the tubing.

19. 1 A dosing siphon is a mechanical device that delivers fixed amounts of water from one tank or reservoir to another tank or pipeline.

20. 2 The effluent quality from a gravel filter can be adjusted by changing the ratio between the amount of filter effluent that is recycled and the amount of effluent that is discharged.

21. 3 When effluent from subsurface wastewater infiltration systems (SWIS) is discharged to long narrow trenches, it percolates down through the bottom of the trench and also seeps laterally through the side walls.

22. 4 Rest periods are recommended for subsurface wastewater infiltration systems (SWIS) to reduce solids and microorganism colonies in the media and on the soil interface, thus partially restoring infiltration rates.

23. 3 Care should be taken to avoid a very fine sand in the first layer of a mound system because it will have a low percolation/infiltration rate.

24. 4 The purpose of a wastewater collection system is to collect the wastewater from a community's homes and convey it to a treatment plant.

25. 4 Conventional gravity sewer systems require larger pipes than small-diameter gravity sewer (SDGS) systems because the conventional gravity sewer system is designed with a peaking factor greater than the average flow.

26. 4 A hybrid small-diameter gravity sewer (SDGS) system is a system that serves a combination of septic tank effluent pump units and septic tank effluent filter (screened) units.

27. 3 Automatic air release valves are installed at upper elevations on the main between depressions in a septic tank effluent pump (STEP) piping system to permit venting of air or gases trapped in the main and prevent an air lock or blockage in the main.

28. 1 Cycle counters are typically used in vacuum sewer systems where a large water use is expected in order to determine if the valve is reasonably capable of keeping up with the flow.

29. 1 Wastewater discharge pumps in vacuum systems should have an inspection opening in the pump volute to facilitate cleaning or clearing a plugged impeller.

30. 4 A valve that is stuck open or that continuously cycles and causes the system vacuum to drop could cause waterlogging in a vacuum collection system.

31. 2 The equipment service card tells what equipment service should be done and when.

32. 3 When operators are attempting to fix equipment and electricity is involved, the risks are particularly serious because the consequences of an accident can be death by electrocution or severe, disabling burns.

33. 3 You can determine if an electric circuit is energized by using a multimeter and holding one lead on ground and the other on a power lead.

34. 4 The first step in setting utility service rates is to outline the specific steps needed to reach the goal.

35. 2 Rate of return is a value that indicates the return of funds received on the basis of the total equity capital used to finance physical facilities.

36. 2 A utility manager can justify the system's revenue needs to the ratepayers by practicing sound financial management and being able to back up needs with factual records.

**Multiple Choice**

1. 4, 5 The main components of a conventional onsite wastewater system are the septic tank and the subsurface disposal field.

2. 1, 2, 3, 4, 5 A manual on small wastewater system operation and maintenance should provide all of these items: a better tomorrow, both for the operator and for the public whose health and environment will benefit; information on how to better communicate with local and state environmental agencies; information on how various wastewater collection and treatment processes and systems function; the opportunity to achieve state certification for the level of competency required by your system; and the terminology needed to communicate effectively with operators of similar systems or larger municipal treatment systems.

3. 1, 3, 4, 5 For normal municipal wastewater, which contains domestic wastewater as well as some industrial and commercial wastes, the goals of the treatment plant designer and operator are these: to kill or inactivate pathogenic organisms, to remove dissolved organic solids, and to remove inorganic and organic suspended solids.

4. 1, 2, 3, 4 NPDES permits regulate discharges from point sources of pollution such as industries, large agricultural feedlots, return irrigation flows, and sanitary landfills.

5. 1, 2, 3, 4, 5 Compared with a gravity collection system, pressure sewers offer these advantages: deep trenching to maintain flow is unnecessary, inverted siphons are not needed to cross roads and rivers, pressurized systems experience fewer stoppages, root intrusion problems are eliminated, and smaller diameter pipe may be used.

6. 1, 4, 5 Excess activated sludge from an oxidation ditch may be dried directly on sludge drying beds, stored in a holding tank for later transfer to larger treatment plants, or stored in sludge lagoons for later disposal in approved sanitary landfills.

7. 1, 2, 4, 5 Common forms of stored energy that could be encountered by a small wastewater treatment system operator include compressed gases that are stored under pressure, electric energy, hydraulic pressure, and spring-loaded equipment.

8. 1, 2, 3 Toxic or suffocating gases may come from industrial waste discharges, process chemicals, or the decomposition of domestic wastewater.

9. 1, 2, 3, 5 When inspecting the mechanical condition of a work vehicle, items that should be inspected include the brakes on the vehicle and trailer, if so equipped; the mirrors and windows; the towed vehicle lighting system; and the windshield wipers.

10. 1, 2, 3, 4, 5 A Material Safety Data Sheet (MSDS) should include fire and explosion hazard data; hazardous ingredients/identity information; health hazard data; physical/chemical characteristics; and precautions for safe handling and use.

11. 1, 2, 3, 4, 5 The main elements of a comprehensive hearing conservation program include access to noise level information, audiometric testing, hearing protection, monitoring and record-keeping, and training in the use of protective equipment and procedures.

12. 2, 3, 4, 5 There must be a reduction of the solids content in the wastewater entering the second compartment of a septic tank for these reasons: to prevent a wastewater backup of the system into the home, to prevent solids from plugging the leach field's distribution piping, to prevent solids from sealing the soil in the leach field, and to reduce the quantity of solids entering the leach field.

13. 1, 5 When solids clog the discharge leach field system, liquid may break through to the ground surface and wastewater may back up into the home's plumbing fixtures.

14. 1, 4 The most common causes of pipe blockages in septic tank systems are grease or detergent buildup in the pipe and tree roots entering the pipe.

15. 1, 4, 5 The selection of septic tank effluent pumps is greatly dependent upon the flow rate to be pumped, the total dynamic head (TDH) of the collection system, and the type of wastewater to be pumped.

16. 2, 3, 4 During extended power outages when STEP system pump basins may fill to capacity, gasoline-powered pumps, portable standby generators, and a septic tank pumper truck should be available to pump down flooded tanks and prevent overflows.

17. 1, 2, 3, 4, 5 Symptoms of poor electrical connections include the following: "alarm light on, high level," "alarm sounds intermittently," "control box breaker trips," "pump does not run," and "short pump cycles."

18. 2, 3, 4, 5 Wastewater treatment and discharge options appropriate for use either as on-site treatment systems or as small community treatment systems include evapotranspiration systems, leach fields, mounds, and seepage beds and pits.

19. 2, 3, 4, 5 The water must be filtered before it reaches the dosing siphon to help protect the performance of the discharge area, the distribution network, and the siphon, and to keep the siphon's snifter pipe clear.

20. 1, 2, 3, 4, 5 In an effluent discharge system based on subgrade or subsurface land application of effluent, if the soil is too coarse (large particles of sand, gravel), the following problems may occur: absorption for adequate biochemical treatment will not be provided, the effluent will flow down (percolate) too rapidly, filtration for adequate biochemical treatment will not be provided, time for adequate biochemical treatment will not be provided, and untreated or partially treated wastewater may reach and contaminate the groundwater.

21. 2, 3, 4 In a gravity-dosed leach field, distribution of the effluent to the trenches may be controlled by the following methods: by routing flow through a distribution box, by routing flow through a drop box, or by the continuous flow method.

22. 1, 2, 3, 4, 5 When monitoring a subsurface wastewater infiltration system (SWIS) leach field, operators should look for the following conditions: creeping failure of the leach field, effluent overflowing the distribution structure, high standing water levels in the monitoring wells, water overflowing at the septic tank cleanout, and water pooling on the surface of the SWIS.

23. 2, 3, 5 Alternatives to a conventional gravity collection system include pressure collection systems, small-diameter gravity sewer systems, and vacuum collection systems.

24. 1, 2, 3, 4, 5 Ask the following questions concerning the design of a wastewater collection system: Can the atmosphere in the manhole be safely tested for oxygen deficiency/enrichment, explosive gases, and toxic gases? Can the available sewer maintenance equipment operate effectively between the manholes or is the distance too great? Does the area flood and, if so, how often? If there are steps in the manhole wall for access, will they be damaged by corrosion and become unsafe? and Is the shelf in the bottom of the manhole properly sloped?

25. 1, 2, 3, 4, 5 Preventive maintenance of a small-diameter gravity sewer (SDGS) system includes exercising of isolation valves, inspection and cleaning of the collection mains, inspection and pumping of the septic tanks, inspection and servicing of any STEP units or lift stations, and testing and servicing of air release valves.

26. 2, 3 The major causes of lift station failures are loss of power and mechanical failure.

27. 1, 2, 4, 5 The advantages of vacuum sewer systems compared to other types of collection systems include substantial reductions in excavation costs, material costs, treatment expenses, and water used to transport wastes.

28. 1, 2, 3, 4, 5 The operation and maintenance responsibilities of a vacuum wastewater collection system operator include the following: analyzing and evaluating operation and maintenance functions and developing new procedures to ensure continued system efficiency; determining remedial action necessary during emergencies; gathering and reviewing all data and recording all data for the preparation of reports and purchase requests; inspecting the system daily to determine the efficiency of operation, cleanliness, and maintenance needs; and preparing work schedules.

29. 2, 4, 5 Emergency or breakdown maintenance may involve the piping system, the vacuum station, or the vacuum valves.

30. 1, 2, 3, 4 Records that should be kept for a vacuum collection system include emergency maintenance records, operating cost records, preventive maintenance records, and routine maintenance records.

31. 1, 3, 5 A good maintenance program accomplishes the following: ensures that regular preventive maintenance is performed to prevent problems and equipment failures, identifies and corrects the problem if something goes wrong, and tries to anticipate what is going to go wrong and prevent it from happening.

32. 1, 2, 3, 4, 5 Persons who should be present during the construction of a wastewater collection and treatment system include the contractor, the engineer, the inspector, the operator, and the superintendent.

33. 1, 2, 3, 4, 5 The operator of a wastewater collection or treatment system needs a basic working knowledge of electricity for these reasons: to prevent financial losses resulting from damage to control circuits, equipment, and machinery; to prevent financial losses resulting from motors burning out; and to prevent operators from seriously injuring themselves.

34. 1, 2, 3, 4, 5 Procedures to use when developing a small wastewater utility's schedule of rates and charges include the following: allocate revenue requirements to cost components, design rates that will recover required costs from each class of customer, determine reserve fund and annual revenue requirements, and distribute component costs to customer classes.

35. 1, 2 There are two general components of inflation that affect utility rates and charges: construction cost index (CCI) and consumer price index (CPI).

36. 2, 3, 4, 5 In the commodity-demand method of cost allocation, costs of service include four primary cost components: commodity costs, customer-related costs, demand costs, and direct fire-protection costs.

37. 2, 3 The advantages of a user fee-based financing system are that it avoids payment of interest expense on long-term debt and it does not put the entire burden of system expansion on current ratepayers.

38. 1, 2, 3, 4, 5 To determine a utility's total operating expenses, add up the expenses of the utility, including administrative costs, chemicals, equipment replacement fund, salaries, and supplies.

**Short Answer**

1. Operators need an understanding of natural cycles in order to control wastewater treatment processes and odors and also to protect receiving waters.
2. If too much nitrogen is discharged to receiving waters, too many algae could be produced. Water with excessive algae can be unsightly. Bacteria decomposing dead algae from occasional die-offs can use up all the dissolved oxygen and cause a fish kill and possibly create septic waters that stink and look terrible.
3. Compared with a gravity collection system, pressure systems offer the following advantages:

   1. Deep trenching to maintain flow is unnecessary.
   2. Inverted siphons are not needed to cross roads and rivers.
   3. Smaller diameter pipes may be used.
   4. Pressurized systems experience fewer stoppages.
   5. Root intrusion problems are eliminated.

   The limitations of pressure collection systems compared with gravity systems depend on the topography of the area. Energy requirements and capital cost comparisons depend on whether flow is by gravity or if lift stations are required. Technical expertise and maintenance for each system depends on the design and installed equipment.
4. The type of stabilization pond found in most areas is a facultative pond in which the top layer is aerobic and the bottom layer is anaerobic.
5. The rotor in an oxidation ditch is a brush-like device that has two functions: (1) it entrains oxygen into the liquid and (2) it keeps the contents of the ditch mixed and moving to prevent settling of the solids.
6. Operators must know how to recognize dangerous situations and take the precautions necessary to ensure their safety.
7. To avoid transmitting disease to your family, do not wear your work clothes home. If you must take your work clothes home, launder them separately from your regular family wash.

8. Any time traffic may be affected by road work activities, you must notify appropriate authorities in your area before work begins. These could be state, county, or local authorities depending on whether it is a state, county, or local street. In some cases, when traffic diversion or disruption may obstruct access by emergency response agencies such as fire and police departments, these agencies must be notified before traffic is diverted or disrupted as well.
9. Wastewater system operators should be familiar with Material Safety Data Sheets because they provide important safety information about hazardous chemicals to which the operator may be exposed.
10. Wastewater system operators can protect their hearing from loud noises by using hearing protectors that eliminate or reduce noise to acceptable levels.
11. Bacteria in the first compartment of a septic tank break down wastewater solids and thereby reduce the volume of solids in the tank.
12. To protect their septic tank systems, homeowners should be advised to practice water conservation measures, not to use the septic tank system for anything that can be disposed of safely some other way, and not to dump toxic materials down the drain to dispose of them.
13. When a septic tank effluent pump (STEP) system is installed, installation of new building sewers prevents infiltration/inflow and reduces problems such as broken pipe, offset joints, root intrusion, and improper grade. Similarly, installation of new septic tanks ensures watertight tanks that reduce the possibility of infiltration. Installation of all new tanks and building sewers also means that components throughout the new system will all be similar. This makes for more dependable operation and easier maintenance of the STEP system.
14. Two operators may be required in the following STEP system maintenance situations: (1) to remove grinder pumps, (2) to remove the strainer assembly if a large buildup of rags or stringy material is attached to the filter screen unit, or (3) if the drain port at the bottom of the pump assembly is blocked and the liquid cannot drain out as the unit is lifted from the tank.
15. It may be necessary to install two effluent pumps at any location where a commercial user produces a high volume of wastewater or multiple septic tanks discharge to a single effluent vault. Two pumps are needed for systems with peaks over the average daily flow.
16. Routine preventive maintenance of STEP septic tanks and grinder pumps includes visual inspections, pumping, and cleaning of tanks and vaults; cleaning effluent screens; running pumps and controllers through their cycles; checking pump voltage and amperage readings; and checking the motor starter contacts.
17. The main components of an open, single-pass intermittent sand filter are the bed, media, distribution piping, and underdrain or collection piping.
18. A dosing siphon is a mechanical device that converts small, continuous flows from a septic tank into large, intermittent dosing flows to a treatment process or discharge area.
19. Treated wastewater effluents must be properly discharged to protect the public health and the environment.
20. If a distribution box settles or tilts and changes the elevation of the outlets of the distribution pipes, one trench may receive more or less flow than the other trenches, thus resulting in overloading or inefficient use of the total subsurface wastewater infiltration system.
21. Grass should immediately be planted over a mound as a protective cover against erosion and cold temperatures.
22. A knowledge of how engineers design wastewater collection systems will assist the operator of the system in determining the correct action to take when the system fails to operate properly. A basic knowledge of the components of the collection system will help the operator do a better job and will enable the operator to discuss design, construction, and inspection with engineers.
23. The slope of the sewer in a conventional gravity system generally follows the slope of the land along the route of the sewer as closely as practical provided the slope is adequate to produce gravity flow and the minimum scouring velocity. Departures from the natural ground slope could cause the sewer to become too shallow for adequate cover on the pipe or too deep for safe and economical construction.
24. SDGS pipelines are often laid following the contours of the terrain in a curvilinear fashion rather than in grids of straight lines. SDGS systems also have relatively few manholes or cleanouts. These two characteristics make it difficult for the operator to quickly pinpoint the location of collection mains unless accurate drawings of the system identify pipeline locations in relation to permanent structures nearby.
25. Pressure collection systems are sometimes used as an alternative to gravity collection systems where the topography and ground conditions of an area are not suitable for a conventional gravity collection system due to flat terrain, rocky conditions, or extremely high groundwater.
26. Maintenance activities associated with pressure system pipelines include exercising the main line isolation valves at least once a year and checking the air release stations for proper operation. In addition, on-lot facilities such as pumps, float switches, and electrical panels should be inspected annually and the pump well should be cleaned once a year.

27. The controller/sensor in a vacuum valve relies on three forces for its operation: pressure, vacuum, and atmosphere. As the wastewater level rises in the sump, it compresses air in the sensor tube. This pressure begins to open the valve by overcoming spring tension in the controller and activating a three-way valve. Once opened, the three-way valve allows the controller/sensor to take vacuum from the downstream side of the valve and apply it to the actuator chamber to fully open the valve. The controller/sensor is capable of maintaining the valve fully open for a fixed period of time, which is adjustable over a range of 3 to 10 seconds. After the time period has elapsed, atmospheric air is admitted to the actuator chamber permitting spring-assisted closing of the valve.
28. A standby generator capable of providing 100 percent of the standby power needed for station operation should be available at the central vacuum station to ensure the continuing operation of the system in the event of a power outage.
29. Most emergency maintenance is related to malfunctioning vacuum valves caused by either low system vacuum or water in the controller.
30. Types of construction-related O&M problems could include leaking joints due to improper solvent welding in cold weather, line breaks due to trench settlement, broken fittings due to insufficient soil compaction, valve failure due to construction debris, valve pit settlement, broken cleanouts, and excessive use of fittings.
31. A good maintenance program tries to anticipate what is going to go wrong and prevent it from happening. If something goes wrong, the problem is identified and corrected.
32. Alarms for a mound system could be caused by a problem in the mound itself, the distribution piping, the pump tank, or the electrical supply.
33. Many times, the nameplate on a piece of equipment is painted, corroded, or missing from the unit. It is therefore important to record and file the motor nameplate data so that the information will be available when needed to repair the motor or to obtain replacement parts.
34. A multimeter is used to check for power in a circuit and to test for open circuits, blown fuses, single phasing of motors, and grounding.
35. Utility rates that accurately reflect the cost of service for each customer can be set by identifying costs and properly allocating the costs to each customer. Begin by using the following procedures:
    1. Determine total annual revenue requirements.
    2. Determine reserve fund requirements.
    3. Allocate revenue requirements to cost components.
    4. Distribute component costs to customer classes.
    5. Develop rates that will recover these costs from each class of customer.
36. When determining customer classes, be sure to use sufficient classes and broad enough category definitions to cover all user groups. Customer classes should be separated into categories of similar use or service demand patterns.
37. A utility's financial stability can be measured by calculating its operating ratio and its coverage ratio. The operating ratio is a measure of the utility's total revenues divided by the total operating expenses. The coverage ratio is a measure of the ability of the utility to pay the principal and interest on loans and bonds plus any unexpected expenses.

APPENDIX

# HOW TO SOLVE SMALL WASTEWATER SYSTEM ARITHMETIC PROBLEMS

(VOLUME I)

by

Ken Kerri

# TABLE OF CONTENTS

## HOW TO SOLVE SMALL WASTEWATER SYSTEM ARITHMETIC PROBLEMS

## OBJECTIVES

### HOW TO SOLVE SMALL WASTEWATER SYSTEM ARITHMETIC PROBLEMS

After completion of this Appendix, you should be able to:

1. Add, subtract, multiply, and divide.
2. List from memory basic conversion factors and formulas.
3. Solve wastewater system arithmetic problems.

# HOW TO SOLVE SMALL WASTEWATER SYSTEM ARITHMETIC PROBLEMS (VOLUME I)

## A.0 HOW TO STUDY THIS APPENDIX

This appendix may be worked early in your training program to help you gain the greatest benefit from your efforts. Whether to start this appendix early or wait until later is your decision. The chapters in this manual were written in a manner requiring very little background in arithmetic. You may wish to concentrate your efforts on the chapters and refer to this appendix when you need help. Some operators prefer to complete this appendix early so they will not have to worry about how to do the arithmetic when they are studying the chapters. You may try to work this appendix early or refer to it while studying the other chapters.

The intent of this appendix is to provide you with a quick review of the addition, subtraction, multiplication, and division needed to work the arithmetic problems in this manual. This appendix is not intended to be a math textbook. There are no fractions because you do not need fractions to work the problems in this manual. Some operators will be able to skip over the review of addition, subtraction, multiplication, and division. Others may need more help in these areas. If you need help in solving problems, read Section A.9, "Steps in Solving Problems." Basic arithmetic textbooks are available at every local library or bookstore and should be referred to if needed. Most instructional or operating manuals for pocket electronic calculators contain sufficient information on how to add, subtract, multiply, and divide.

After you have worked a problem involving your job, you should check your calculations, examine your answer to see if it appears reasonable, and, if possible, have another operator check your work before making any decisions or changes.

## A.1 BASIC ARITHMETIC

In this section, we provide you with basic arithmetic problems involving addition, subtraction, multiplication, and division. You may work the problems by hand if you wish, but we recommend you use a calculator. The operating or instructional manual for your calculator should outline the step-by-step procedures to follow. All calculators use similar procedures, but most of them are slightly different from others.

We will start with very basic, simple problems. Try working the problems and then comparing your answers with the given answers. If you can work these problems, you should be able to work the more difficult problems in the text of this training manual by using the same procedures.

## A.10 Addition

|  |  |  |  |  |
|---|---|---|---|---|
| 2 | 6.2 | 16.7 | 6.12 | 43 |
| 3 | 8.5 | 38.9 | 38.39 | 39 |
| 5 | 14.7 | 55.6 | 44.51 | 34 |
|  |  |  |  | 38 |
|  |  |  |  | 39 |
| 2.12 | 0.12 | 63 | 120 | 37 |
| 9.80 | 2.0 | 32 | 60 | 29 |
| 11.92 | 2.12 | 95 | 180 | 259 |
|  |  |  |  | 70 |
| 4 | 23 | 16.2 | 45.98 | 50 |
| 7 | 79 | 43.5 | 28.09 | 40 |
| 2 | 31 | 67.8 | 114.00 | 80 |
| 13 | 133 | 127.5 | 188.07 | 240 |

## A.11 Subtraction

|  |  |  |  |  |
|---|---|---|---|---|
| 7 | 12 | 25 | 78 | 83 |
| −5 | 3 | 5 | −30 | −69 |
| 2 | 9 | 20 | 48 | 14 |
| 61 | 485 | 4.3 | 3.5 | 123 |
| −37 | −296 | −0.8 | −0.7 | −109 |
| 24 | 189 | 3.5 | 2.8 | 14 |
| 8.6 | 11.92 | 27.32 | 3.574 | 75.132 |
| −8.22 | −3.70 | −12.96 | −0.042 | −49.876 |
| 0.38 | 8.22 | 14.36 | 3.532 | 25.256 |

## A.12 Multiplication

(3)(2)* = 6
(4)(7) = 28
(10)(5) = 50
(10)(1.3) = 13
(2)(22.99) = 45.98
(6)(19.5) = 117
(16)(17.1) = 273.6
(50)(20,000) = 1,000,000
(40)(2.31) = 92.4
(80)(0.433) = 34.64

(40)(20)(6) = 4,800
(4,800)(7.48) = 35,904
(1.6)(2.3)(8.34) = 30.6912
(0.001)(200)(8.34) = 1.668
(0.785)(7.48)(60) = 352.308
(12,000)(500)(60)(24) = 8,640,000,000 or $8.64 \times 10^9$
(4)(1,000)(1,000)(454) = 1,816,000,000 or $1.816 \times 10^9$

*NOTE:* The term, $\times 10^9$, means that the number is multiplied by $10^9$ or 1,000,000,000. Therefore, $8.64 \times 10^9 = 8.64 \times 1{,}000{,}000{,}000 = 8{,}640{,}000{,}000$.

*(3)(2) is the same as 3 × 2.

### A.13 Division

$$\frac{6}{3} = 2 \qquad \frac{48}{12} = 4$$

$$\frac{50}{25} = 2 \qquad \frac{300}{20} = 15$$

$$\frac{20}{7.1} = 2.8 \qquad \frac{11,400}{188} = 60.6$$

$$\frac{1,000,000}{17.5} = 57,143 \qquad \frac{861,429}{30,000} = 28.7$$

$$\frac{4,000,000}{74,880} = 53.4 \qquad \frac{1.67}{8.34} = 0.20$$

$$\frac{80}{2.31} = 34.6 \qquad \frac{62}{454} = 0.137$$

$$\frac{250}{17.1} = 14.6 \qquad \frac{4,000,000}{14.6} = 273,973$$

*NOTE:* When we divide 1 by 3 (⅓), we get a long row of 3s (0.3333). Instead of the row of 3s, we round off our answer so ⅓ = 0.33. For a discussion of rounding off numbers, see Section A.95, "Significant Figures."

### A.14 Order of Arithmetic Calculations

Most of the arithmetic problems we work in the wastewater collection and treatment field require us to plug numbers into formulas and calculate the answer. To simplify arithmetic problems, it is important to calculate arithmetic problems in a specific order. The correct order of calculations is as follows:

1. Work all calculations that lie within parentheses ( ). If parentheses are enclosed within other enclosure symbols such as brackets [ ], work the calculations from the inside out.
2. Simplify any exponents or square roots, working from left to right. Remember that the exponent is attached to the number or variable it is closest to. Treat any square roots as exponents.
3. Perform all multiplications and divisions, working from left to right.
4. Perform all additions and subtractions, working from left to right.
5. If the problem includes a fraction or division bar, divide the number above the bar (the numerator) by the number below the bar (the denominator). Be sure to perform all calculations above and below the fraction bar before dividing.

Let us look at an example problem to see how the order of calculations applies. This year, one of the responsibilities of the operators at our plant is to paint both sides of the wooden fence across the front of the facility. The fence is 145 feet long and 9 feet high. The steel access gate, which does not need painting, measures 14 feet wide by 9 feet high. Each gallon of paint will cover 150 square feet of surface area. How many gallons of paint should be purchased?

STEP 1: Identify the correct formula.

$$\text{Paint Req, gal} = \frac{\text{Total Area, sq ft}}{\text{Coverage, sq ft/gal}}$$

or

$$\text{Paint Req, gal} = \frac{(\text{Fence L, ft} \times \text{H, ft} \times \text{No. Sides}) - (\text{Gate L, ft} \times \text{H, ft} \times \text{No. Sides})}{\text{Coverage, sq ft/gal}}$$

STEP 2: Plug numbers into the formula.

$$\text{Paint Req, gal} = \frac{(145 \text{ ft} \times 9 \text{ ft} \times 2) - (14 \text{ ft} \times 9 \text{ ft} \times 2)}{150 \text{ sq ft/gal}}$$

STEP 3: Work the multiplication within the parentheses.

$$\text{Paint Req, gal} = \frac{(2,610 \text{ sq ft}) - (252 \text{ sq ft})}{150 \text{ sq ft/gal}}$$

STEP 4: Work the subtraction above the line.

$$\text{Paint Req, gal} = \frac{2,358 \text{ sq ft}}{150 \text{ sq ft/gal}}$$

STEP 5: Divide the numerator by the denominator.

Paint Req, gal = 15.72 gal
or 16 gallons of paint will be needed.

Instructions for your calculator can provide you with the detailed procedures for working the practice problems below.

$$\frac{(3)(4)}{2} = 6 \qquad \frac{64}{(8)(4)} = 2$$

$$\frac{(2+3)(4)}{5} = 4 \qquad \frac{54}{(4+2)(3)} = 3$$

$$\frac{(7-2)(8)}{4} = 10 \qquad \frac{48}{(8-3)(4)} = 2.4$$

$$\frac{(90)(54)(10.25)}{1,728} = 28.83$$

$$(28.83)(7.48) = 215.6$$

$$\left(\frac{216}{150}\right)(24) = 34.6$$

$$\frac{(225-25)(100)}{225} = 88.9$$

### A.15 Actual Problems

*NOTE:* Very little arithmetic is involved in the operation and maintenance of small wastewater systems described in this training manual, *SMALL WASTEWATER SYSTEM OPERATION AND MAINTENANCE,* Volume I. Therefore, some of the example formulas and problems in this Arithmetic Appendix relate to processes described in this volume but others relate to the wastewater treatment processes described in Volume II.

Let us look at some arithmetic problems as they might be encountered by an operator.

The wastewater entering a septic tank must be retained in the tank at least long enough for the solid portion to separate from the liquids. A minimum detention time of 6 to 24 hours is recommended. Detention time in a single-compartment septic tank can be calculated using the following formulas:

$$\text{Clear Zone Vol, cu ft} = \frac{\left(\text{Tank Length, in}\right)\left(\text{Tank Width, in}\right)\left(\text{Clear Zone Depth, in}\right)}{1{,}728 \text{ cu in/cu ft}}$$

$$\text{Clear Zone Vol, gal} = (\text{Clear Zone Volume, cu ft})(7.48 \text{ gal/cu ft})$$

$$\text{Detention Time, hr} = \left(\frac{\text{Clear Zone Volume, gal}}{\text{Influent Flow, gal/day}}\right)(24 \text{ hr/day})$$

*EXAMPLE:* In a small residential community served by individual septic tanks and subsurface wastewater infiltration systems (SWIS), the average wastewater flow per household is 150 gallons per day. The average width of a septic tank is 54 inches and the average length is 90 inches. If the clear zone depth is 10.25 inches, is the wastewater detention time adequate to allow for liquid/solids separation to occur?

| Known | | Unknown |
|---|---|---|
| Influent Flow, gal/day | = 150 gal/day | 1. Detention Time, hr |
| Tank Length, in | = 90 in | |
| Tank Width, in | = 54 in | 2. Is detention time adequate? |
| Clear Zone Depth, in | = 10.25 in | |

1. Calculate the clear zone volume in cubic feet. (*NOTE:* There are 1,728 cubic inches per cubic foot.)

$$\text{Clear Zone Vol, cu ft} = \frac{\left(\text{Tank Length, in}\right)\left(\text{Tank Width, in}\right)\left(\text{Clear Zone Depth, in}\right)}{1{,}728 \text{ cu in/cu ft}}$$

$$= \frac{(90 \text{ in})(54 \text{ in})(10.25 \text{ in})}{1{,}728 \text{ cu in/cu ft}}$$

$$= \frac{49{,}815 \text{ cu in}}{1{,}728 \text{ cu in/cu ft}}$$

$$= 28.83 \text{ cu ft}$$

2. Calculate the clear zone volume in gallons.

$$\text{Clear Zone Vol, gal} = (\text{Clear Zone Volume, cu ft})(7.48 \text{ gal/cu ft})$$

$$= (28.83 \text{ cu ft})(7.48 \text{ gal/cu ft})$$

$$= 215.6 \text{ gal or approximately 216 gallons}$$

3. Calculate the detention time in hours.

$$\text{Detention Time, hr} = \left(\frac{\text{Clear Zone Volume, gal}}{\text{Influent Flow, gal/day}}\right)(24 \text{ hr/day})$$

$$= \left(\frac{216 \text{ gal}}{150 \text{ gal/day}}\right)(24 \text{ hr/day})$$

$$= 34.6 \text{ hr or approximately 35 hour}$$

Is the detention time of approximately 35 hours adequate? Yes, this detention time exceeds the recommended minimum detention time range of 6 to 24 hours.

*EXAMPLE:* The influent BOD to an activated sludge plant is 225 mg/L and the effluent BOD is 25 mg/L. What is the BOD removal efficiency of the plant?

| Known | | Unknown |
|---|---|---|
| Influent BOD, mg/L | = 225 mg/L | Plant BOD Efficiency, % |
| Effluent BOD, mg/L | = 25 mg/L | |

Calculate the efficiency of the plant in removing BOD.

$$\text{Plant BOD Efficiency, \%} = \left(\frac{\text{In} - \text{Out}}{\text{In}}\right)(100\%)$$

$$= \left(\frac{225 \text{ mg/L} - 25 \text{ mg/L}}{225 \text{ mg/L}}\right)(100\%)$$

$$= \left(\frac{225 - 25}{225}\right)(100\%)$$

$$= 88.9\%$$

*EXAMPLE:* The total revenues for a utility are $1,467,000 and the operating expenses for the utility are $1,176,000. The debt service expenses are $590,000. What is the operating ratio? What is the coverage ratio?

| Known | | Unknown |
|---|---|---|
| Total Revenue, $ | = $1,467,000 | 1. Operating Ratio |
| Operating Expenses, $ | = $1,176,000 | |
| Debt Service Expenses, $ | = $590,000 | 2. Coverage Ratio |

1. Calculate the operating ratio.

$$\text{Operating Ratio} = \frac{\text{Total Revenue, \$}}{\text{Operating Expenses, \$}}$$

$$= \frac{\$1{,}467{,}000}{\$1{,}176{,}000}$$

$$= 1.25$$

2. Calculate nondebt expenses.

$$\text{Nondebt Expenses, \$} = \text{Operating Exp, \$} - \text{Debt Service Exp, \$}$$

$$= \$1{,}176{,}000 - \$590{,}000$$

$$= \$586{,}000$$

3. Calculate the coverage ratio.

$$\text{Coverage Ratio} = \frac{\text{Total Revenue, \$} - \text{Nondebt Expenses, \$}}{\text{Debt Service Expenses, \$}}$$

$$= \frac{\$1{,}467{,}000 - \$586{,}000}{\$590{,}000}$$

$$= 1.49$$

### A.16 Percentage

Expressing a number in percentage is just another, and sometimes simpler, way of writing a fraction or a decimal. It can be thought of as parts per 100 parts, since the percentage is the numerator of a fraction whose denominator is always 100. Twenty-five parts per 100 parts is more easily recognized as 25/100 or 0.25. However, it is also 25%. In this case, the symbol % takes the place of the 100 in the fraction and the decimal point in the decimal fraction.

For the above example, it can be seen that changing from a fraction or a decimal to percent is not a difficult procedure.

**To change a fraction to percent, multiply by 100%.**

*EXAMPLE:* $\frac{2}{5} \times 100\% = \frac{200\%}{5} = 40\%$

*EXAMPLE:* $\frac{5}{4} \times 100\% = \frac{500\%}{4} = 125\%$

**To change percent to a fraction, divide by 100%.**

*EXAMPLE:* $15\% \div 100\% = 15\% \times \frac{1}{100\%} = \frac{15}{100} = \frac{3}{20}$

*EXAMPLE:* $0.4\% \div 100\% = 0.4\% \times \frac{1}{100\%} = \frac{0.4}{100} = \frac{4}{1{,}000} = \frac{1}{250}$

In these examples, note that the two percent signs cancel each other.

Following is a table comparing common fractions, decimal fractions, and percent to indicate their relationship to each other:

| Common Fraction | Decimal Fraction | Percent |
|---|---|---|
| $\frac{285}{100}$ | 2.85 | 285% |
| $\frac{100}{100}$ | 1.0 | 100% |
| $\frac{20}{100}$ | 0.20 | 20% |
| $\frac{1}{100}$ | 0.01 | 1% |
| $\frac{1}{1{,}000}$ | 0.001 | 0.1% |
| $\frac{1}{1{,}000{,}000}$ | 0.000001 | 0.0001% |

### A.17 Sample Problems Involving Percent

Problems involving percent are usually not complicated since their solution consists of only one or two steps. The principal error made is usually a misplaced decimal point. The most common types of percentage problem include the following:

1. Finding what percent one number is of another

   In this case, the problem is simply one of reading carefully to determine the correct fraction and then converting to a percentage:

*EXAMPLE:* What percent is 20 of 25?

$\frac{20}{25} = \frac{4}{5} = 0.8$

$0.8 \times 100\% = 80\%$

*EXAMPLE:* Four is what percent of 14?

$\frac{4}{14} = 0.2857$

$0.2857 \times 100\% = 28.57\%$

*EXAMPLE:* Influent BOD to a clarifier is 200 mg/L. Effluent BOD is 140 mg/L. What is the percent removal in the clarifier.

$$\text{Removal, \%} = \left(\frac{\text{In} - \text{Out}}{\text{In}}\right) \times 100\%$$

$$= \left(\frac{200\text{ mg/L} - 140\text{ mg/L}}{200\text{ mg/L}}\right) \times 100\%$$

$$= 30\%$$

2. Finding a percent of a given number

   In this case, the percent is expressed as a decimal, and the two numbers are multiplied together.

*EXAMPLE:* Find 7% of 32.

$\frac{7\%}{100\%} \times 32 = 2.24$

*EXAMPLE:* Find 90% of 5.

$\frac{90\%}{100\%} \times 5 = 4.5$

*EXAMPLE:* What is the weight of dry solids in a ton (2,000 pounds) of wastewater sludge containing 5% solids and 95% water? *NOTE:* 5% solids means there are 5 pounds of dry solids for every 100 pounds of wet sludge. Therefore,

$2{,}000\text{ lbs} \times \frac{5\%}{100\%} = 100\text{ lbs of solids}$

A variation of the preceding problem is:

3. Finding a number when a given percent of it is known

   Since this problem is similar to the previous problem, the solution is to convert to a decimal and divide by the decimal.

*EXAMPLE:* If 5% of a number is 52, what is the number?

$\left(\frac{100\%}{5\%}\right)(52) = 1{,}040$

A check calculation may now be performed—what is 5% of 1,040?

$\left(\frac{5\%}{100\%}\right)(1{,}040) = 52\text{ (Check)}$

*EXAMPLE:* 16 is 80% of what amount?

$$\left(\frac{100\%}{80\%}\right)(16) = 20$$

*EXAMPLE:* Percent removal of BOD in a clarifier is 35%. If 70 mg/L are removed, what is the influent BOD?

$$\text{Influent BOD, mg/L} = \left(\frac{70\text{ mg/L}}{35\%}\right)(100\%)$$

$$= 200\text{ mg/L}$$

Check:

$$\text{Original Load} \times \%\text{ Removal} = \text{Load Removed}$$

$$200\text{ mg/L} \times \frac{35\%}{100\%} = 70\text{ mg/L}$$

## A.2 AREAS

### A.20 Units

Areas are measured in two dimensions or in square units. In the English system of measurement, the most common units are square inches, square feet, square yards, and square miles. In the metric system, the units are square millimeters, square centimeters, square meters, and square kilometers.

### A.21 Rectangle

The area of a rectangle is equal to its length (L) multiplied by its width (W).

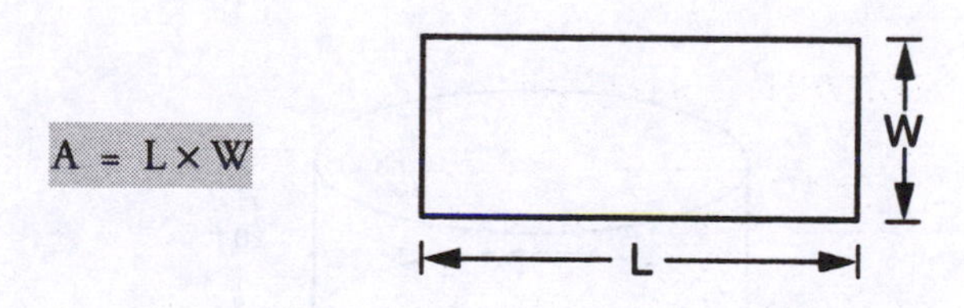


*EXAMPLE:* Find the area of a rectangle if the length is 5 feet and the width is 3.5 feet.

$$\text{Area, sq ft} = \text{Length, ft} \times \text{Width, ft}$$

$$= 5\text{ ft} \times 3.5\text{ ft}$$

$$= 17.5\text{ ft}^2$$

or $$= 17.5\text{ sq ft}$$

*EXAMPLE:* The surface area of a settling basin is 330 square feet. One side measures 15 feet. How long is the other side?

$$A = L \times W$$

$$330\text{ sq ft} = \text{L, ft} \times 15\text{ ft}$$

$$\frac{\text{L, ft} \times 15\text{ ft}}{15\text{ ft}} = \frac{330\text{ sq ft}}{15\text{ ft}}$$ Divide both sides of the equation by 15 ft.

$$\text{L, ft} = \frac{330\text{ sq ft}}{15\text{ ft}}$$

$$= 22\text{ ft}$$

### A.22 Triangle

The area of a triangle is equal to ½ the base multiplied by the height. This is true for any triangle.

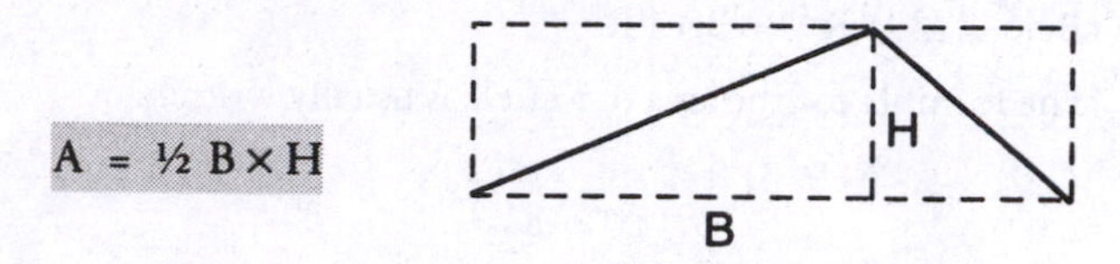


*NOTE:* The area of any triangle is equal to ½ the area of the rectangle that can be drawn around it. The area of the rectangle is B × H. The area of the triangle is ½ B × H.

*EXAMPLE:* Find the area of triangle ABC.

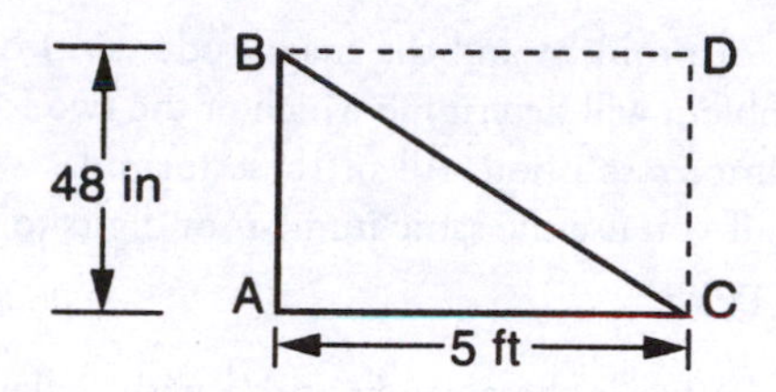


The first step in the solution is to make all the units the same. In this case, it is easier to change inches to feet.

$$48\text{ in} = 48\text{ }\cancel{\text{in}} \times \frac{1\text{ ft}}{12\text{ }\cancel{\text{in}}} = \frac{48}{12}\text{ ft} = 4\text{ ft}$$

*NOTE:* All conversions should be calculated in the above manner. Since 1 ft/12 in is equal to unity, or 1, multiplying by this factor changes the form of the answer but not its value.

$$\text{Area, sq ft} = \tfrac{1}{2}(\text{Base, ft})(\text{Height, ft})$$

$$= \tfrac{1}{2} \times 5\text{ ft} \times 4\text{ ft}$$

$$= \frac{20}{2}\text{ sq ft}$$

$$= 10\text{ sq ft}$$

*NOTE:* Triangle ABC is ½ the area of rectangle ABCD. The triangle is a special form called a "right triangle" since it contains a 90° angle at point A.

### A.23 Circle

A square with sides of 2R can be drawn around a circle with a radius of R.

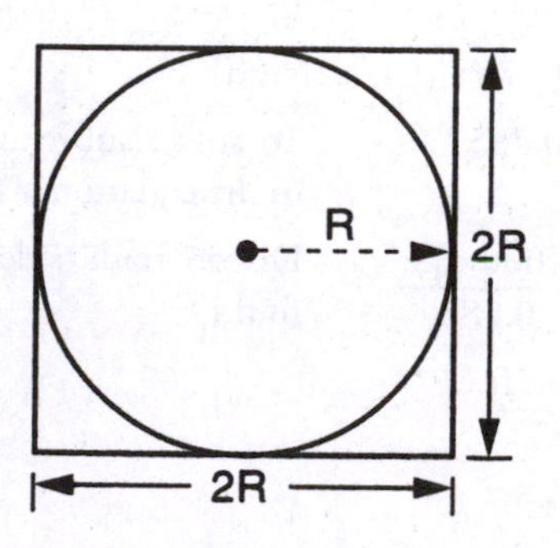

The area of the square is: $A = 2R \times 2R = 4R^2$.

It has been found that the area of any circle inscribed within a square is slightly more than ¾ of the area of the square. More precisely, the area of the preceding circle is:

$$\text{A circle} = 3\tfrac{1}{7} R^2 = 3.14 R^2$$

The formula for the area of a circle is usually written:

$$A = \pi R^2$$

The Greek letter π (pronounced pie and written as "pi") merely substitutes for its standard or constant value of 3.1416.

Because the diameter of any circle is equal to twice the radius, the formula for the area of a circle can be rewritten as follows:

$$A = \pi R^2 = \pi \times R \times R = \pi \times \frac{D}{2} \times \frac{D}{2} = \frac{\pi D^2}{4} = \frac{3.14}{4} D^2 = 0.785\ D^2$$

The type of problem and the magnitude (size) of the numbers in a problem will determine which of the two formulas will provide a simpler solution. All of these formulas will give the same results if you use the same number of digits to the right of the decimal point.

*EXAMPLE:* What is the area of a circle with a diameter of 20 centimeters?

In this case, the formula using a radius is more convenient because it takes advantage of multiplying by 10.

$$\text{Area, sq cm} = \pi(\text{R, cm})^2$$
$$= 3.14 \times 10 \text{ cm} \times 10 \text{ cm}$$
$$= 314 \text{ sq cm}$$

*EXAMPLE:* What is the area of a clarifier with a 50-foot radius?

In this case, the formula using diameter is more convenient.

$$\text{Area, sq ft} = (0.785)(\text{Diameter, ft})^2$$
$$= 0.785 \times 100 \text{ ft} \times 100 \text{ ft}$$
$$= 7{,}850 \text{ sq ft}$$

Occasionally, the operator may be confronted with a problem giving the area and requesting the radius or diameter. This presents the special problem of finding the square root of the number.

*EXAMPLE:* The surface area of a circular clarifier is approximately 5,000 square feet. What is the diameter?

$A = 0.785\ D^2$, or

$$\text{Area, sq ft} = (0.785)(\text{Diameter, ft})^2$$

$$5{,}000 \text{ sq ft} = 0.785\ D^2$$

To solve, substitute given values in the equation.

$$\frac{0.785\ D^2}{0.785} = \frac{5{,}000 \text{ sq ft}}{0.785}$$

Divide both sides by 0.785 to find $D^2$.

$$D^2 = \frac{5{,}000 \text{ sq ft}}{0.785}$$
$$= 6{,}369 \text{ sq ft}$$

Therefore,

D = square root of 6,369 sq ft, or

$$\text{Diameter, ft} = \sqrt{6{,}369 \text{ sq ft}}$$

Press the √ sign on your calculator and get

$$\text{D, ft} = 79.8 \text{ ft}$$

Sometimes, a trial-and-error method can be used to find square roots. Since $80 \times 80 = 6{,}400$, we know the answer is close to 80 feet.

Try $79 \times 79 = 6{,}241$

Try $79.5 \times 79.5 = 6{,}320.25$

Try $79.8 \times 79.8 = 6{,}368.04$

The diameter is 79.8 feet, or approximately 80 feet.

## A.24 Cylinder

With the formulas presented thus far, it would be a simple matter to find the number of square feet in a room that was to be painted. The length of each wall would be added together and then multiplied by the height of the wall. This would give the surface area of the walls (minus any area for doors and windows). The ceiling area would be found by multiplying length times width and the result added to the wall area gives the total area.

The surface area of a circular cylinder, however, has not been discussed. If we wanted to know how many square feet of surface area are in a tank with a diameter of 60 feet and a height of 20 feet, we could start with the top and bottom.

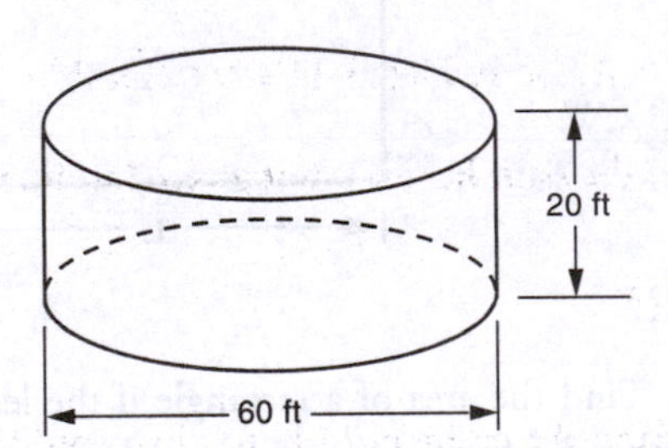


The area of the top and bottom ends are both $\pi \times R^2$.

$$\text{Area, sq ft} = (2 \text{ ends})(\pi)(\text{Radius, ft})^2$$
$$= 2 \times \pi \times (30 \text{ ft})^2$$
$$= 5{,}652 \text{ sq ft}$$

The surface area of the wall must now be calculated. If we made a vertical cut in the wall and unrolled it, the straightened wall would be the same length as the circumference of the floor and ceiling.

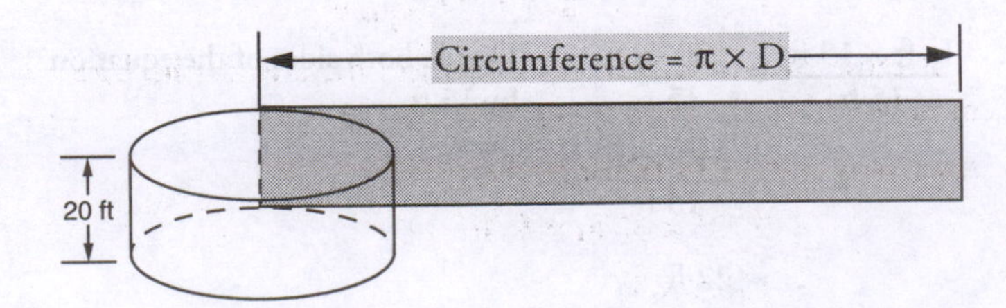

This length has been found to always be $\pi \times D$. In the case of the tank, the length of the wall would be:

Length, ft = $\pi$(Diameter, ft)

= $3.14 \times 60$ ft

= 188.4 ft

Area would be:

$A_w$, sq ft = Length, ft × Height, ft

= 188.4 ft × 20 ft

= 3,768 sq ft

Outside Surface Area to Paint, sq ft = Area of Top and Bottom, sq ft + Area of Wall, sq ft

= 5,652 sq ft + 3,768 sq ft

= 9,420 sq ft

A container has inside and outside surfaces and you may need to paint both of them.

### A.25 Cone

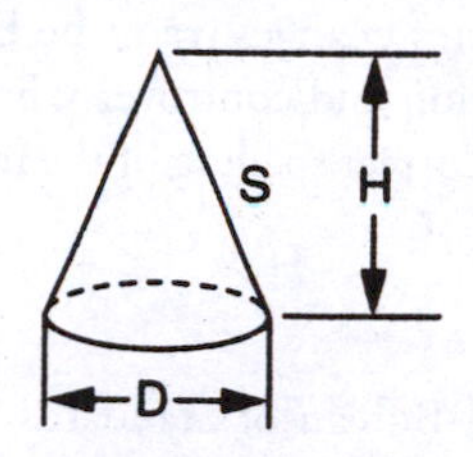


The lateral area of a cone is equal to ½ of the slant height (S) multiplied by the circumference of the base.

$$A_L = \tfrac{1}{2}S \times \pi \times D = \pi \times S \times R$$

In this case, the slant height is not given, but it may be calculated by:

$$S = \sqrt{R^2 + H^2}$$

*EXAMPLE:* Find the entire outside area of a cone with a diameter of 30 inches and a height of 20 inches.

Slant Height, in = $\sqrt{(\text{Radius, in})^2 + (\text{Height, in})^2}$

= $\sqrt{(15\text{ in})^2 + (20\text{ in})^2}$

= $\sqrt{225\text{ sq in} + 400\text{ sq in}}$

= $\sqrt{625\text{ sq in}}$

= 25 in

Lateral Area of Cone, sq in = $\pi$(Slant Height, in)(Radius, in)

= $3.14 \times 25$ in $\times 15$ in

= 1,177.5 sq in

Because the entire area was asked for, the area of the base must be added.

Area, sq in = (0.785)(Diameter, in)$^2$

= $0.785 \times 30$ in $\times 30$ in

= 706.5 sq in

Total Area, sq in = Area of Cone, sq in + Area of Base, sq in

= 1,177.5 sq in + 706.5 sq in

= 1,884 sq in

### A.26 Sphere

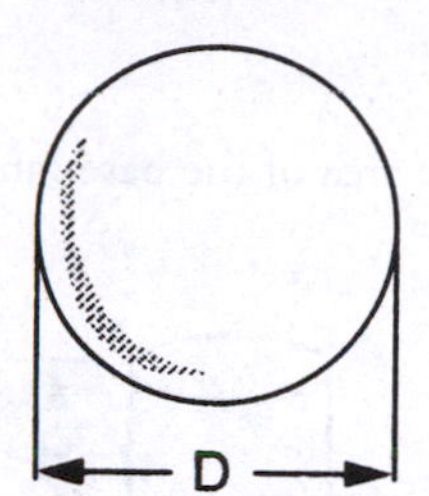


The surface area of a sphere or ball is equal to $\pi$ multiplied by the diameter squared, which is four times the cross-sectional area.

$$A_s = \pi D^2$$

If the radius is used, the formula becomes:

$$A_s = \pi D^2 = \pi \times 2R \times 2R = 4\pi R^2$$

*EXAMPLE:* What is the surface area of a sphere-shaped water tank 20 feet in diameter?

Area, sq ft = $\pi$(Diameter, ft)$^2$

= $3.14 \times 20$ ft $\times 20$ ft

= 1,256 sq ft

## A.3 VOLUMES

### A.30 Rectangle

Volumes are measured in three dimensions or in cubic units. To calculate the volume of a rectangle, the area of the base is calculated in square units and then multiplied by the height. The formula then becomes:

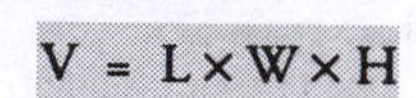

$$V = L \times W \times H$$

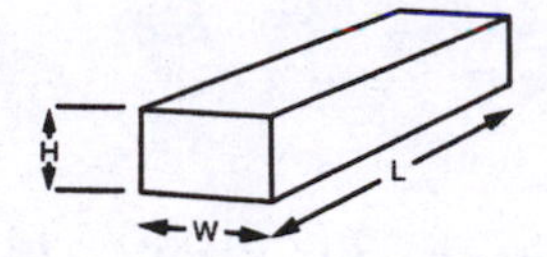


*EXAMPLE:* The length of a box is 2 feet, the width is 15 inches, and the height is 18 inches. Find its volume.

Volume, cu ft = Length, ft × Width, ft × Height, ft

= $2\text{ ft} \times \dfrac{15\text{ in}}{12\text{ in/ft}} \times \dfrac{18\text{ in}}{12\text{ in/ft}}$

= 2 ft × 1.25 ft × 1.5 ft

= 3.75 cu ft

### A.31 Prism

The same general rule that applies to the volume of a rectangle also applies to a prism.

$$\text{Volume} = \text{Area of Base} \times \text{Height}$$

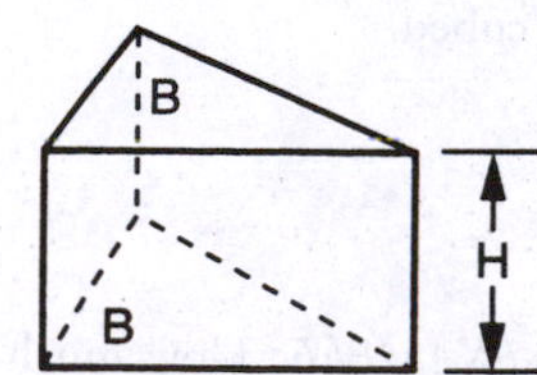

*EXAMPLE:* Find the volume of a prism with a base area of 10 square feet and a height of 5 feet. (Note that the base of a prism is triangular in shape.)

$$\begin{aligned}\text{Volume, cu ft} &= \text{Area of Base, sq ft} \times \text{Height, ft}\\ &= 10\ \text{sq ft} \times 5\ \text{ft}\\ &= 50\ \text{cu ft}\end{aligned}$$

### A.32 Cylinder

The volume of a cylinder is equal to the area of the base multiplied by the height.

$$V = \pi R^2 \times H = 0.785\ D^2 \times H$$

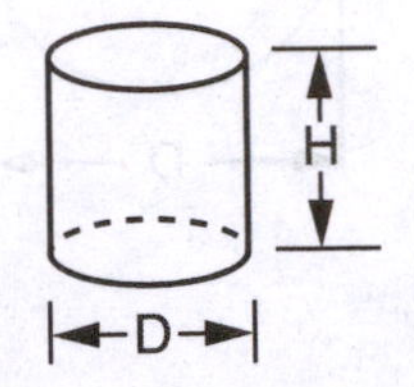


*EXAMPLE:* A tank has a diameter of 100 feet and a depth (or height) of 12 feet. Find the volume.

$$\begin{aligned}\text{Volume, cu ft} &= 0.785 \times (\text{Diameter, ft})^2 \times \text{Height, ft}\\ &= 0.785 \times 100\ \text{ft} \times 100\ \text{ft} \times 12\ \text{ft}\\ &= 94{,}200\ \text{cu ft}\end{aligned}$$

### A.33 Cone

The volume of a cone is equal to ⅓ the volume of a circular cylinder of the same height and diameter.

$$V = \frac{\pi}{3} R^2 \times H$$

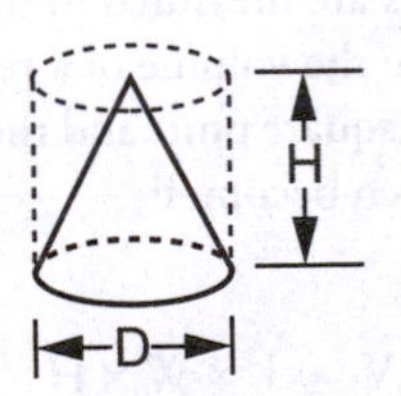


*EXAMPLE:* Calculate the volume of a cone if the height at the center is 4 feet and the diameter is 100 feet (radius is 50 feet).

$$\begin{aligned}\text{Volume, cu ft} &= \frac{\pi}{3} \times (\text{Radius})^2 \times \text{Height, ft}\\ &= \frac{\pi}{3} \times 50\ \text{ft} \times 50\ \text{ft} \times 4\ \text{ft}\\ &= 10{,}466\ \text{cu ft}\\ \text{or} \quad &= 10{,}500\ \text{cu ft}\end{aligned}$$

### A.34 Sphere

The volume of a sphere is equal to $\pi/6$ times the diameter cubed.

$$V = \frac{\pi}{6} \times D^3$$

*EXAMPLE:* How much gas can be stored in a sphere with a diameter of 12 feet?

$$\begin{aligned}\text{Volume, cu ft} &= \frac{\pi}{6} \times (\text{Diameter, ft})^3\\ &= \frac{\pi}{\not{6}} \times \overset{2}{\not{12}}\ \text{ft} \times 12\ \text{ft} \times 12\ \text{ft}\\ &= 904.3\ \text{cubic feet}\end{aligned}$$

## A.4 METRIC SYSTEM

The two most common systems of weights and measures are the English system and the metric system *(Le Système International d'Unités (SI))*. Of these two, the metric system is more popular with most of the nations of the world. The reason for this is that the metric system is based on a system of 10s and is therefore easier to remember and easier to use than the English system. Even though the basic system in the United States is the English system, the scientific community uses the metric system almost exclusively. Many organizations have urged, for good reason, that the United States switch to the metric system. Today, the metric system is gradually becoming the standard system of measurement in the United States.

As the United States changes from the English to the metric system, some confusion and controversy has developed. For example, which is the correct spelling of the following words:

1. Liter or litre?
2. Meter or metre?

The US National Bureau of Standards, the Water Environment Federation, and the American Water Works Association use litre and metre. The US Government uses liter and meter and accepts no deviations. Some people argue that METRE should be used to measure LENGTH and that METER should be used to measure FLOW RATES (like a water or electric meter). Liter and meter are used in this manual because this is most consistent with spelling in the United States.

One of the most frequent arguments heard against the US switching to the metric system was that the costs of switching manufacturing processes would be excessive. Pipe manufacturers have agreed upon the use of a soft metric conversion system during the conversion to the metric system. Past practice in the US has identified some types of pipe by external (outside) diameter while other types are classified by nominal (existing only in name, not real or actual) bore. This means that a 6-inch pipe does not have a 6-inch inside diameter. With the strict or hard metric system, a 6-inch pipe would be a 152.4-mm (6 in × 25.4 mm/in) pipe. In the soft metric system, a 6-inch pipe is a 150-mm (6 in × 25 mm/in) pipe. Typical customary and soft metric pipe-size designations are shown below:

PIPE-SIZE DESIGNATIONS

| | | | | | | | | |
|---|---|---|---|---|---|---|---|---|
| Customary, in | 2 | 4 | 6 | 8 | 10 | 12 | 15 | 18 |
| Soft Metric, mm | 50 | 100 | 150 | 200 | 250 | 300 | 375 | 450 |
| Customary, in | 24 | 30 | 36 | 42 | 48 | 60 | 72 | 84 |
| Soft Metric, mm | 600 | 750 | 900 | 1050 | 1200 | 1500 | 1800 | 2100 |

In order to study the metric system, you must know the meanings of the terminology used. Following is a list of Greek and Latin prefixes used in the metric system.

**PREFIXES USED IN THE METRIC SYSTEM**

| Prefix | Symbol | Meaning |
|---|---|---|
| Micro | μ | 1/1 000 000 or 0.000 001 |
| Milli | m | 1/1000 or 0.001 |
| Centi | c | 1/100 or 0.01 |
| Deci | d | 1/10 or 0.1 |
| Unit | | 1 |
| Deka | da | 10 |
| Hecto | h | 100 |
| Kilo | k | 1000 |
| Mega | M | 1 000 000 |

## A.40 Measures of Length

The basic measure of length is the meter.

1 kilometer (km) = 1,000 meters (m)
1 meter (m) = 100 centimeters (cm)
1 centimeter (cm) = 10 millimeters (mm)

Kilometers are usually used in place of miles, meters are used in place of feet and yards, centimeters are used in place of inches, and millimeters are used for inches and fractions of an inch.

LENGTH EQUIVALENTS

| | | | |
|---|---|---|---|
| 1 kilometer | = 0.621 mile | 1 mile | = 1.61 kilometers |
| 1 meter | = 3.28 feet | 1 foot | = 0.305 meter |
| 1 meter | = 39.37 inches | 1 inch | = 0.0254 meter |
| 1 centimeter | = 0.3937 inch | 1 inch | = 2.54 centimeters |
| 1 millimeter | = 0.0394 inch | 1 inch | = 25.4 millimeters |

*NOTE:* The above equivalents are reciprocals. If one equivalent is given, the reverse can be obtained by division. For instance, if 1 meter equals 3.28 feet, 1 foot equals 1/3.28 meter, or 0.305 meter.

## A.41 Measures of Capacity or Volume

The basic measure of capacity in the metric system is the liter. For measurement of large quantities, the cubic meter is sometimes used.

1 kiloliter (kL) = 1,000 liters (L) = 1 cu meter (cu m)

1 liter (L) = 1,000 milliliters (mL)

Kiloliters, or cubic meters, are used to measure capacity of large storage tanks or reservoirs in place of cubic feet or gallons. Liters are used in place of gallons or quarts. Milliliters are used in place of quarts, pints, or ounces.

CAPACITY EQUIVALENTS

| | | | |
|---|---|---|---|
| 1 kiloliter | = 264.2 gallons | 1 gallon | = 0.003785 kiloliter |
| 1 liter | = 1.057 quarts | 1 quart | = 0.946 liter |
| 1 liter | = 0.2642 gallon | 1 gallon | = 3.785 liters |
| 1 milliliter | = 0.0353 ounce | 1 ounce | = 29.57 milliliters |

## A.42 Measures of Weight

The basic unit of weight in the metric system is the gram. One cubic centimeter of water at maximum density weighs 1 gram, and thus there is a direct, simple relation between volume of water and weight in the metric system.

1 kilogram (kg) = 1,000 grams (gm)
1 gram (gm) = 1,000 milligrams (mg)
1 milligram (mg) = 1,000 micrograms (μg)

Grams are usually used in place of ounces, and kilograms are used in place of pounds.

WEIGHT EQUIVALENTS

| | | | |
|---|---|---|---|
| 1 kilogram | = 2.205 pounds | 1 pound | = 0.4536 kilogram |
| 1 gram | = 0.0022 pound | 1 pound | = 453.6 grams |
| 1 gram | = 0.0353 ounce | 1 ounce | = 28.35 grams |
| 1 gram | = 15.43 grains | 1 grain | = 0.0648 gram |

## A.43 Temperature

Just as you should become familiar with the metric system, it is also important to become familiar with the centigrade (Celsius) scale for measuring temperature. There is nothing magical about the centigrade scale—it is simply a different size than the Fahrenheit scale. The two scales compare as follows:

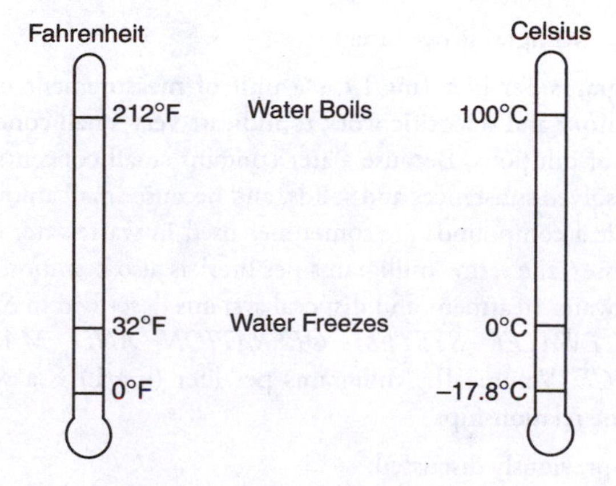


The two scales are related in the following manner:

$$\text{Fahrenheit} = \left(°\text{C} \times \frac{9}{5}\right) + 32°$$

$$\text{Celsius} = (°\text{F} - 32°) \times \frac{5}{9}$$

*EXAMPLE:* Convert 20° Celsius to degrees Fahrenheit.

$$°\text{F} = \left(°\text{C} \times \frac{9}{5}\right) + 32°$$

$$= \left(20° \times \frac{9}{5}\right) + 32°$$

$$= \frac{180°}{5} + 32°$$

$$= 36° + 32°$$

$$= 68°\text{F}$$

*EXAMPLE:* Convert –10°C to °F.

$$°F = \left(-10° \times \frac{9}{5}\right) + 32°$$

$$= \frac{-90°}{5} + 32°$$

$$= -18° + 32°$$

$$= 14°F$$

*EXAMPLE:* Convert –13°F to °C.

$$°C = (°F - 32°) \times \frac{5}{9}$$

$$= (-13° - 32°) \times \frac{5}{9}$$

$$= -45° \times \frac{5}{9}$$

$$= -5° \times 5$$

$$= -25°C$$

## A.44 Milligrams per Liter

Milligrams per liter (mg/L) is a unit of measurement used in laboratory and scientific work to indicate very small concentrations of dilutions. Because water contains small concentrations of dissolved substances and solids, and because small amounts of chemical compounds are sometimes used in wastewater system processes, the term "milligrams per liter" is also common in the wastewater treatment and disposal systems described in *SMALL WASTEWATER SYSTEM OPERATION AND MAINTENANCE,* Volume II. Milligrams per liter (mg/L) is a weight/volume relationship.

As previously discussed:

1,000 liters = 1 cubic meter = 1,000,000 cubic centimeters.

Therefore,

1 liter = 1,000 cubic centimeters.

Because 1 cubic centimeter of water weighs 1 gram,

1 liter of water = 1,000 grams or 1,000,000 milligrams.

$$\frac{1 \text{ milligram}}{\text{liter}} = \frac{1 \text{ milligram}}{1{,}000{,}000 \text{ milligrams}} = \frac{1 \text{ part}}{\text{million parts}} = \frac{1 \text{ part per}}{\text{million (ppm)}}$$

Milligrams per liter and parts per million (parts) may be used interchangeably as long as the liquid density is 1.0 gm/cu cm or 62.43 lb/cu ft. A concentration of 1 mg/L or 1 ppm means that there is 1 part of substance by weight for every 1 million parts of water. A concentration of 10 mg/L would mean 10 parts of substance per million parts of water.

To get an idea of how small 1 mg/L is, divide the numerator and denominator of the fraction by 10,000. This, of course, does not change its value because 10,000 ÷ 10,000 is equal to 1.

$$1\frac{\text{mg}}{\text{L}} = \frac{1 \text{ mg}}{1{,}000{,}000 \text{ mg}} = \frac{1/10{,}000 \text{ mg}}{1{,}000{,}000/10{,}000 \text{ mg}} = \frac{0.0001 \text{ mg}}{100 \text{ mg}} = 0.0001\%$$

Therefore, 1 mg/L is equal to one ten-thousandth of a percent, or 1% is equal to 10,000 mg/L.

To convert mg/L to %, move the decimal point 4 places or numbers to the left.

## A.45 Example Problems

Working problems using milligrams per liter or parts per million is a part of operation in wastewater systems.

*EXAMPLE:* A plant effluent flowing at a rate of 5 million pounds per day contains 15 mg/L of solids. How many pounds of solids will be discharged per day?

$$15 \text{ mg/L} = \frac{15 \text{ lbs solids}}{\text{million lbs water}}$$

$$\text{Solids Discharged, lbs/day} = \text{Concentration, lbs/M lbs} \times \text{Flow, lbs/day}$$

$$= \frac{15 \text{ lbs}}{\cancel{\text{million lbs}}} \times \frac{5 \cancel{\text{ million lbs}}}{\text{day}}$$

$$= 75 \text{ lbs/day}$$

There is one thing that is unusual about the above problem and that is the flow is reported in pounds per day. In most treatment plants, flow is reported in terms of gallons per minute or gallons per day. To convert these flow figures to weight, an additional conversion factor is needed.

One gallon of water weighs 8.34 pounds. Using this factor, it is possible to convert flow in gallons per day to flow in pounds per day.

*EXAMPLE:* A plant influent of 3.5 million gallons per day (MGD) contains 200 mg/L BOD. How many pounds of BOD enter the plant per day?

$$\text{Flow, lbs/day} = \text{Flow, } \frac{\text{M gal}}{\text{day}} \times \frac{8.34 \text{ lbs}}{\text{gal}}$$

$$= \frac{3.5 \text{ million } \cancel{\text{gal}}}{\text{day}} \times \frac{8.34 \text{ lbs}}{\cancel{\text{gal}}}$$

$$= 29.19 \text{ million lbs/day}$$

$$\text{BOD Loading, lbs/day} = \text{BOD Conc, mg/L} \times \text{Flow, M lbs/day}$$

$$= \frac{200 \cancel{\text{ mg}}^*}{\cancel{\text{million mg}}} \times \frac{29.19 \cancel{\text{ million}} \text{ lbs}}{\text{day}}$$

$$= 5{,}838 \text{ lbs/day}$$

*Remember that $1\frac{\text{mg}}{\text{L}} = \frac{1 \text{ mg}}{\text{M mg}} = \frac{1 \text{ lb}}{\text{M lbs}}$. They are identical ratios.

In solving the above problem, a relation was used that is most important to understand and commit to memory:

Feed, lbs/day = Flow, MGD × Conc, mg/L × 8.34 lbs/gal

*EXAMPLE:* A chlorinator is set to feed 50 pounds of chlorine per day to a flow of 0.8 MGD. What is the chlorine dose in mg/L?

$$\text{Conc or Dose, mg/L} = \frac{\text{Chlorine Feed, lbs/day}}{\text{Flow, MGD} \times 8.34\ \text{lbs/gal}}$$

$$= \frac{50\ \text{lbs/day}}{0.8\ \text{MG/day} \times 8.34\ \text{lbs/gal}}$$

$$= \frac{50\ \text{lbs}}{6.672\ \text{M lbs}}$$

$$= 7.5\ \text{mg/L, or } 7.5\ \text{ppm}$$

*EXAMPLE:* Treated effluent is pumped to a spray disposal field by a pump that delivers 500 gallons per minute. Suspended solids in the effluent average 10 mg/L. What is the total weight of suspended solids deposited on the spray field during a 24-hour day of continuous pumping?

$$\text{Flow, MGD} = \text{Flow, GPM} \times 60\ \text{min/hr} \times 24\ \text{hr/day}$$

$$= \frac{500\ \text{gal}}{\cancel{\text{min}}} \times \frac{60\ \cancel{\text{min}}}{\cancel{\text{hr}}} \times \frac{24\ \cancel{\text{hr}}}{\text{day}}$$

$$= 720{,}000\ \text{gal/day}$$

$$= 0.72\ \text{MGD}$$

$$\text{Weight of Solids, lbs/day} = \text{Flow, MGD} \times \text{Conc, mg/L} \times 8.34\ \text{lbs/gal}$$

$$= \frac{0.72\ \cancel{\text{M gal}}}{\text{day}} \times \frac{10\ \cancel{\text{mg}}}{\cancel{\text{M mg}}} \times \frac{8.34\ \text{lbs}}{\cancel{\text{gal}}}$$

$$= 60.048\ \text{lbs/day or about } 60\ \text{lbs/day}$$

## A.5 WEIGHT/VOLUME RELATIONS

Another factor for the operator to remember, in addition to the weight of a gallon of water, is the weight of a cubic foot of water.

One cubic foot of water weighs 62.4 lbs. If these two weights are divided, it is possible to determine the number of gallons in a cubic foot.

$$\frac{62.4\ \cancel{\text{pounds}}/\text{cu ft}}{8.34\ \cancel{\text{pounds}}/\text{gal}} = 7.48\ \text{gal/cu ft}$$

Thus, we have another very important relationship to commit to memory:

8.34 lbs/gal × 7.48 gal/cu ft = 62.4 lbs/cu ft

It is only necessary to remember two of the above items because the third may be found by calculation. For most problems, 8 ⅓ or 8.3 lbs/gal and 7½ or 7.5 gal/cu ft will provide sufficient accuracy.

*EXAMPLE:* Change 1,000 cu ft of water to gallons.

1,000 cu ft × 7.48 gal/cu ft = 7,480 gallons

*EXAMPLE:* What is the weight of 3 cubic feet of water?

62.4 lbs/cu ft × 3 cu ft = 187.2 lbs

*EXAMPLE:* The net weight of a tank of water is 750 lbs. How many gallons does it contain?

$$\frac{750\ \cancel{\text{lbs}}}{8.34\ \cancel{\text{lbs}}/\text{gal}} = 90\ \text{gal}$$

## A.6 FORCE, PRESSURE, AND HEAD

In order to study the forces and pressures involved in fluid flow, it is first necessary to define the terms used.

*FORCE:* The push exerted by water on any surface being used to confine it. Force is usually expressed in pounds, tons, grams, or kilograms.

*PRESSURE:* The force per unit area. Pressure can be expressed in many ways, but the most common term is pounds per square inch (psi).

*HEAD:* Vertical distance from the water surface to a reference point below the surface. Usually expressed in feet or meters.

An *EXAMPLE* should serve to illustrate these terms.

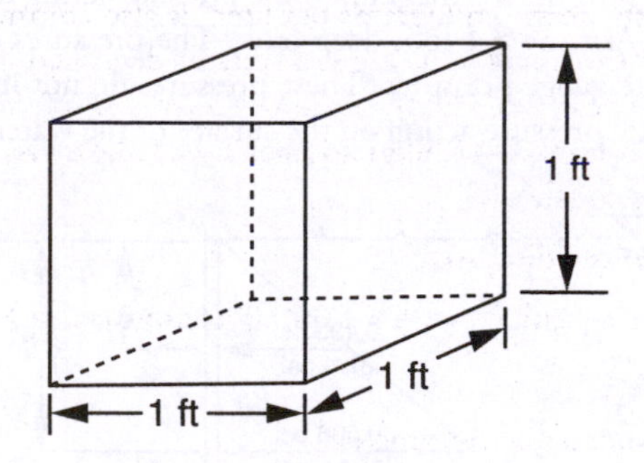


If water were poured into a 1-foot cubical container, the force acting on the bottom of the container would be 62.4 pounds.

The pressure acting on the bottom would be 62.4 pounds per square foot. The area of the bottom is also 12 in × 12 in = 144 sq in. Therefore, the pressure may also be expressed as:

$$\text{Pressure, psi} = \frac{62.4\ \text{lbs}}{\text{sq ft}} = \frac{62.4\ \text{lbs/sq ft}}{144\ \text{sq in/sq ft}}$$

$$= 0.433\ \text{lb/sq in}$$

$$= 0.433\ \text{psi}$$

Because the height of the container is 1 foot, the head would be 1 foot.

The pressure in any vessel at 1 foot of depth or 1 foot of head is 0.433 psi acting in any direction.

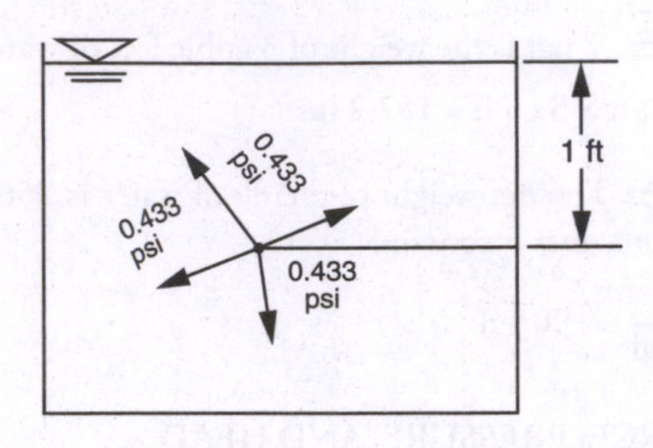


If the depth of water in the previous example were increased to 2 feet, the pressure would be:

$$p = \frac{2(62.4 \text{ lbs})}{144 \text{ sq in}} = \frac{124.8 \text{ lbs}}{144 \text{ sq in}} = 0.866 \text{ psi}$$

Therefore, we can see that for every foot of head, the pressure increases by 0.433 psi. Thus, the general formula for pressure becomes:

$$p, \text{psi} = 0.433(H, \text{ft})$$

H = feet of head
p = pounds per square inch of pressure

$$P, \text{lbs/sq ft} = 62.4(H, \text{ft})$$

H = feet of head
P = pounds per square foot of pressure

We can now draw a diagram of the pressure acting on the side of a tank. Assume a 4-foot deep tank. The pressures shown on the tank are gauge pressures. These pressures do not include the atmospheric pressure acting on the surface of the water.

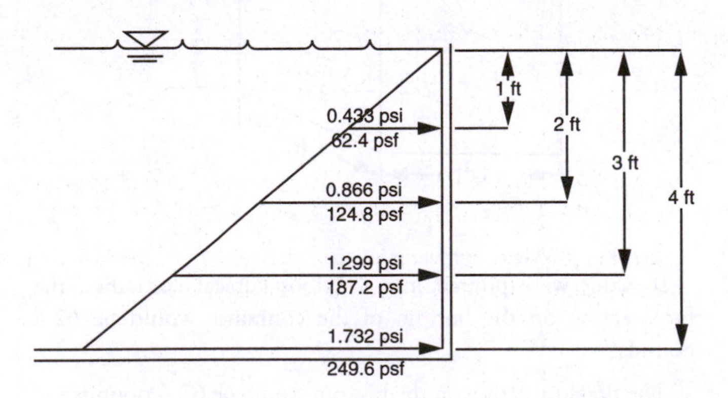


| | |
|---|---|
| $p_0 = 0.433 \times 0 = 0.0$ psi | $P_0 = 62.4 \times 0 = 0.0$ lb/sq ft |
| $p_1 = 0.433 \times 1 = 0.433$ psi | $P_1 = 62.4 \times 1 = 62.4$ lbs/sq ft |
| $p_2 = 0.433 \times 2 = 0.866$ psi | $P_2 = 62.4 \times 2 = 124.8$ lbs/sq ft |
| $p_3 = 0.433 \times 3 = 1.299$ psi | $P_3 = 62.4 \times 3 = 187.2$ lbs/sq ft |
| $p_4 = 0.433 \times 4 = 1.732$ psi | $P_4 = 62.4 \times 4 = 249.6$ lbs/sq ft |

The average pressure acting on the tank wall is 1.732 psi/2 = 0.866 psi, or 249.6 psf/2 = 124.8 psf. We divided by 2 to obtain the average pressure because there is zero pressure at the top and 1.732 psi pressure on the bottom of the wall.

If the wall were 5 feet long, the pressure would be acting over the entire 20-square-foot (5 ft × 4 ft) area of the wall. The total force acting to push the wall would be:

Force, lbs = (Pressure, lbs/sq ft)(Area, sq ft)
= 124.8 lbs/sq ft × 20 sq ft
= 2,496 lbs

If the pressure in psi were used, the problem would be similar:

Force, lbs = (Pressure, lbs/sq in)(Area, sq in)
= 0.866 psi × 48 in × 60 in
= 2,494 lbs*

*Difference in answer is due to rounding off of decimal points.

The general formula, then, for finding the total force acting on a side wall of a tank is:

$$F = 31.2 \times H^2 \times L$$

F = force in pounds
H = head in feet
L = length of wall in feet
31.2 = a constant with units of lbs/cu ft, and considers the fact that the force results from H/2 or half the depth of the water (average depth). The force is exerted at H/3 from the bottom.

*EXAMPLE:* Find the force acting on a 5-foot long wall in a 4-foot deep tank.

Force, lbs = $(31.2)(\text{Head, ft})^2(\text{Length, ft})$
= $31.2 \text{ lbs/cu ft} \times (4 \text{ ft})^2 \times 5 \text{ ft}$
= 2,496 lbs

Occasionally, an operator is warned: Never empty a tank during periods of high groundwater. Why? The pressure on the bottom of the tank caused by the water surrounding the tank will tend to float the tank like a cork if the upward force of the water is greater than the weight of the tank.

$$F = 62.4 \times H \times A$$

F = upward force in pounds
H = head of water on tank bottom in feet
A = area of bottom of tank in square feet
62.4 = a constant with units of lbs/cu ft

This formula is approximately true if the tank does not crack, leak, or start to float.

*EXAMPLE:* Find the upward force on the bottom of an empty tank caused by a groundwater depth of 8 feet above the tank bottom. The tank is 20 feet wide and 40 feet long.

$$\text{Force, lbs} = (62.4 \text{ lbs/cu ft})(\text{Head, ft})(\text{Area, sq ft})$$

$$= 62.4 \text{ lbs/cu ft} \times 8 \text{ ft} \times 20 \text{ ft} \times 40 \text{ ft}$$

$$= 399{,}360 \text{ lbs}$$

or $$= 399{,}400 \text{ lbs}$$

## A.7 VELOCITY AND FLOW RATE

### A.70 Velocity

The velocity of a particle or substance is the speed at which it is moving. It is expressed by indicating the length of travel and how long it takes to cover the distance. Velocity can be expressed in almost any distance and time units. For instance, a car may be traveling at a rate of 280 miles per 5 hours. However, it is normal to express the distance traveled per unit time. The above example would then become:

$$\text{Velocity, mi/hr} = \frac{\text{Distance, miles}}{\text{Time, hours}}$$

$$= \frac{280 \text{ miles}}{5 \text{ hours}}$$

$$= 56 \text{ miles/hour}$$

The velocity of water in a channel, pipe, or other conduit can be expressed in the same way. If the particle of water travels 600 feet in 5 minutes, the velocity is:

$$\text{Velocity, ft/min} = \frac{\text{Distance, ft}}{\text{Time, minutes}}$$

$$= \frac{600 \text{ ft}}{5 \text{ min}}$$

$$= 120 \text{ ft/min}$$

If you wish to express the velocity in feet per second, multiply by 1 min/60 seconds.

*NOTE:* Multiplying by $\frac{1 \text{ minute}}{60 \text{ seconds}}$ is like multiplying by $\frac{1}{1}$; it does not change the relative value of the answer. It only changes the form of the answer.

$$\text{Velocity, ft/sec} = (\text{Velocity, ft/min})(1 \text{ min/60 sec})$$

$$= \frac{120 \text{ ft}}{\cancel{\text{min}}} \times \frac{1 \cancel{\text{min}}}{60 \text{ sec}}$$

$$= \frac{120 \text{ ft}}{60 \text{ sec}}$$

$$= 2 \text{ ft/sec}$$

### A.71 Flow Rate

If water in a 1-foot wide channel is 1 foot deep, then the cross-sectional area of the channel is 1 ft × 1 ft = 1 sq ft.

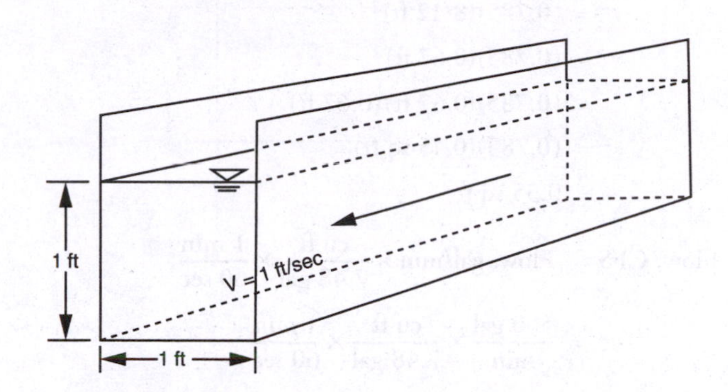


If the velocity in this channel is 1 foot per second, then each second a body of water 1 square foot in area and 1 foot long will pass a given point. The volume of this body of water would be 1 cubic foot. Since 1 cubic foot of water would pass by every second, the flow rate would be equal to 1 cubic foot per second, or 1 CFS.

To obtain the flow rate in the above example, the velocity was multiplied by the cross-sectional area. This is another important general formula.

$$Q = V \times A$$

Q = flow rate, CFS or cu ft/sec
V = velocity, ft/sec
A = area, sq ft

*EXAMPLE:* A rectangular channel 3 feet wide contains water 2 feet deep and flowing at a velocity of 1.5 feet per second. What is the flow rate in CFS?

$$Q = V \times A$$

$$\text{Flow Rate, CFS} = \text{Velocity, ft/sec} \times \text{Area, sq ft}$$

$$= 1.5 \text{ ft/sec} \times 3 \text{ ft} \times 2 \text{ ft}$$

$$= 9 \text{ cu ft/sec}$$

*EXAMPLE:* Flow in a 2.5-foot wide channel is 1.4 feet deep and measures 11.2 CFS. What is the average velocity?

In this problem, we want to find the velocity. Therefore, we must rearrange the general formula to solve for velocity.

$$V = \frac{Q}{A}$$

$$\text{Velocity, ft/sec} = \frac{\text{Flow Rate, cu ft/sec}}{\text{Area, sq ft}}$$

$$= \frac{11.2 \text{ cu ft/sec}}{2.5 \text{ ft} \times 1.4 \text{ ft}}$$

$$= \frac{11.2 \text{ cu ft/sec}}{3.5 \text{ sq ft}}$$

$$= 3.2 \text{ ft/sec}$$

*EXAMPLE:* Flow in an 8-inch pipe is 500 GPM. What is the average velocity?

$$\text{Area, sq ft} = (0.785)(\text{Diameter, ft})^2$$
$$= (0.785)(8/12\text{ ft})^2$$
$$= (0.785)(0.67\text{ ft})^2$$
$$= (0.785)(0.67\text{ ft})(0.67\text{ ft})$$
$$= (0.785)(0.45\text{ sq ft})$$
$$= 0.35\text{ sq ft}$$

$$\text{Flow, CFS} = \text{Flow, gal/min} \times \frac{\text{cu ft}}{7.48\text{ gal}} \times \frac{1\text{ min}}{60\text{ sec}}$$
$$= \frac{500\text{ gal}}{\text{min}} \times \frac{\text{cu ft}}{7.48\text{ gal}} \times \frac{1\text{ min}}{60\text{ sec}}$$
$$= \frac{500\text{ cu ft}}{448.8\text{ sec}}$$
$$= 1.114\text{ CFS}$$

$$\text{Velocity, ft/sec} = \frac{\text{Flow, cu ft/sec}}{\text{Area, sq ft}}$$
$$= \frac{1.114\text{ cu ft/sec}}{0.35\text{ sq ft}}$$
$$= 3.18\text{ ft/sec}$$

## A.8 PUMPS

### A.80 Pressure

Atmospheric pressure at sea level is approximately 14.7 psi. This pressure acts in all directions and on all objects. If a tube is placed upside down in a basin of water and a 1 psi partial vacuum is drawn on the tube, the water in the tube will rise 2.31 feet.

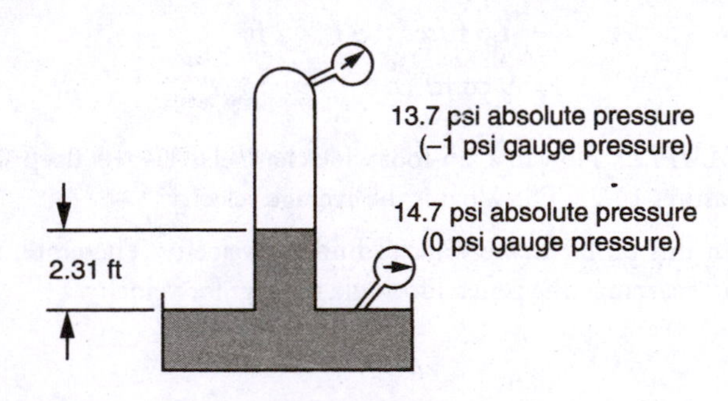


*NOTE:* 1 ft of water = 0.433 psi; therefore,

$$1\text{ psi} = \frac{1}{0.433}\text{ ft} = 2.31\text{ ft of water}$$

The action of the partial vacuum is what gets water out of a sump or well and up to a pump. It is not sucked up, but it is pushed up by atmospheric pressure on the water surface in the sump. If a complete vacuum could be drawn, the water would rise 2.31 × 14.7 = 33.9 feet; but this is impossible to achieve. The practical limit of the suction lift of a positive displacement pump is about 22 feet, and that of a centrifugal pump is 15 feet.

### A.81 Work

Work can be expressed as lifting a weight a certain vertical distance. It is usually defined in terms of foot-pounds.

*EXAMPLE:* A 165-pound man runs up a flight of stairs 20 feet high. How much work did he do?

$$\text{Work, ft-lbs} = \text{Weight, lbs} \times \text{Height, ft}$$
$$= 165\text{ lbs} \times 20\text{ ft}$$
$$= 3{,}300\text{ ft-lbs}$$

### A.82 Power

Power is a rate of doing work and is usually expressed in foot-pounds per minute.

*EXAMPLE:* If the man in the above example runs up the stairs in 3 seconds, how much power has he exerted?

$$\text{Power, ft-lbs/sec} = \frac{\text{Work, ft-lbs}}{\text{Time, sec}}$$
$$= \frac{3{,}300\text{ ft-lbs}}{3\text{ }\cancel{\text{sec}}} \times \frac{60\text{ }\cancel{\text{sec}}}{\text{minute}}$$
$$= 66{,}000\text{ ft-lbs/min}$$

### A.83 Horsepower

Horsepower is also a unit of power. One horsepower is defined as 33,000 ft-lbs per minute or 746 watts.

*EXAMPLE:* How much horsepower has the man in the previous example exerted as he climbs the stairs?

$$\text{Horsepower, HP} = (\text{Power, ft-lbs/min})\left(\frac{\text{HP}}{33{,}000\text{ ft-lbs/min}}\right)$$
$$= 66{,}000\text{ ft-lbs/min} \times \frac{\text{Horsepower}}{33{,}000\text{ ft-lbs/min}}$$
$$= 2\text{ HP}$$

Work is also done by lifting water. If the flow from a pump is converted to a weight of water and multiplied by the vertical distance it is lifted, the amount of work or power can be obtained.

$$\text{Horsepower, HP} = \frac{\text{Flow, gal}}{\text{min}} \times \text{Lift, ft} \times \frac{8.34\text{ lbs}}{\text{gal}} \times \frac{\text{Horsepower}}{33{,}000\text{ ft-lbs/min}}$$

Solving the above relation, the amount of horsepower necessary to lift the water is obtained. This is called "water horsepower."

$$\text{Water, HP} = \frac{(\text{Flow, GPM})(\text{H, ft})}{3{,}960^*}$$

$$^*\frac{8.34\text{ lbs}}{\text{gal}} \times \frac{\text{HP}}{33{,}000\text{ ft-lbs/min}} = \frac{1}{3{,}960}$$

1 gallon weighs 8.34 pounds and 1 horsepower is the same as 33,000 ft-lbs/min.

H or Head in feet is the same as Lift in feet.

However, because pumps are not 100 percent efficient (they cannot transmit all the power put into them), the horsepower supplied to a pump is greater than the water horsepower. Horsepower supplied to the pump is called "brake horsepower."

$$\text{Brake, HP} = \frac{\text{Flow, GPM} \times \text{H, ft}}{3{,}960 \times E_p}$$

$E_p$ = Efficiency of Pump (Usual range 50–85 percent, depending on type and size of pump)

Motors are also not 100 percent efficient; therefore, the power supplied to the motor is greater than the motor transmits.

$$\text{Motor, HP} = \frac{\text{Flow, GPM} \times \text{H, ft}}{3{,}960 \times E_p \times E_m}$$

$E_m$ = Efficiency of Motor (Usual range 80–95 percent, depending on type and size of motor)

The above formulas have been developed for the pumping of water and wastewater, which have a specific gravity of 1.0. If other liquids are to be pumped, the formulas must be multiplied by the specific gravity of the liquid.

*EXAMPLE:* A flow of 500 GPM of water is to be pumped against a total head of 100 feet by a pump with an efficiency of 70 percent. What is the pump horsepower?

$$\text{Brake, HP} = \frac{\text{Flow, GPM} \times \text{H, ft}}{3{,}960 \times E_p}$$

$$= \frac{500 \times 100}{3{,}960 \times 0.70}$$

$$= 18 \text{ HP}$$

*EXAMPLE:* Find the horsepower required to pump gasoline (specific gravity = 0.75) in the previous problem.

$$\text{Brake, HP} = \frac{500 \times 100 \times 0.75}{3{,}960 \times 0.70}$$

= 13.5 HP (gasoline is lighter and requires less horsepower)

## A.84 Head

Basically, the head that a pump must work against is determined by measuring the vertical distance between the two water surfaces, or the distance the water must be lifted. This is called the "static head." Two typical conditions for lifting water are shown below.

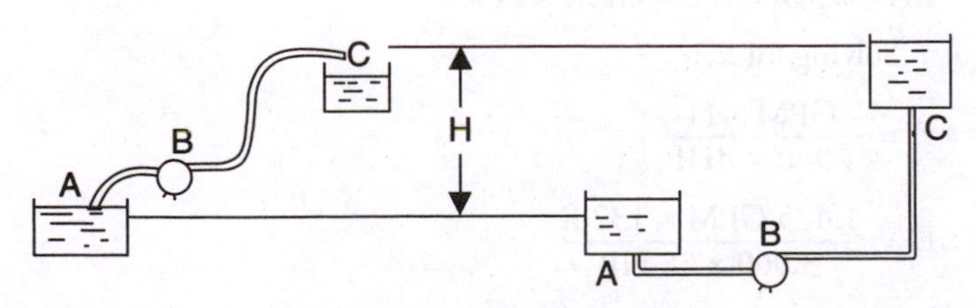


If a pump were designed in the above examples to pump only against head H, the water would never reach the intended point. The reason for this is that the water encounters friction in the pipelines. Friction depends on the roughness and length of pipe, the pipe diameter, and the flow velocity. The turbulence caused at the pipe entrance (point A); the pump (point B); the pipe exit (point C); and at each elbow, bend, or transition also adds to these friction losses. Tables and charts are available in Section A.88 for calculation of these friction losses so they may be added to the measured or static head to obtain the total head. For short runs of pipe that do not have high velocities, the friction losses are generally less than 10 percent of the static head.

*EXAMPLE:* A pump is to be located 8 feet above a wet well and must lift 1.8 MGD another 50 feet to a storage reservoir. If the pump has an efficiency of 75 percent and the motor an efficiency of 90 percent, what is the cost of the power consumed if 1 kilowatt hour costs 4 cents?

Because we are not given the length or size of pipe and the number of elbows or bends, we will assume friction to be 10 percent of static head.

Static Head, ft = Suction Lift, ft + Discharge Head, ft

= 8 ft + 50 ft

= 58 ft

Friction Losses, ft = (0.1)(Static Head, ft)

= (0.1)(58 ft)

= 5.8 ft

Total Dynamic Head, ft = Static Head, ft + Friction Losses, ft

= 58 ft + 5.8 ft

= 63.8 ft

$$\text{Flow, GPM} = \frac{1{,}800{,}000 \text{ gal}}{\cancel{\text{day}}} \times \frac{\cancel{\text{day}}}{24 \cancel{\text{hr}}} \times \frac{1 \cancel{\text{hr}}}{60 \text{ min}}$$

= 1,250 GPM (assuming pump runs 24 hours per day)

$$\text{Motor, HP} = \frac{\text{Flow, GPM} \times \text{H, ft}}{3{,}960 \times E_p \times E_m}$$

$$= \frac{1{,}250 \times 63.8}{3{,}960 \times 0.75 \times 0.9}$$

$$= 30 \text{ HP}$$

$$\text{Kilowatt-hr} = 30 \cancel{\text{HP}} \times 24 \text{ hr/day} \times 0.746 \text{ kW}/\cancel{\text{HP}}^{*}$$

= 537 kilowatt-hr/day

Cost = kWh × $0.04/kWh

= 537 × 0.04

= $21.48/day

*See Section A.10, "Basic Conversion Factors (English System)," *POWER*, page 454.

## A.85 Pump Characteristics

The discharge of a centrifugal pump, unlike a positive displacement pump, can be made to vary from zero to a maximum capacity, which depends on the speed, head, power, and specific impeller design. The interrelation of capacity, efficiency, head, and power is known as the characteristics of the pump.

The first relation normally looked at when selecting a pump is the head versus capacity. The head of a centrifugal pump normally rises as the capacity is reduced. If the values are plotted on a graph, they appear as follows:

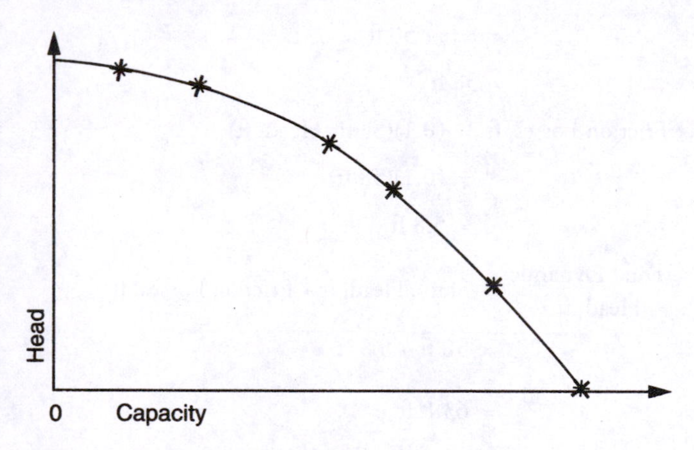


Another important characteristic is the pump efficiency. It begins from zero at no discharge, increases to a maximum, and then drops as the capacity is increased. Following is a graph of efficiency versus capacity:

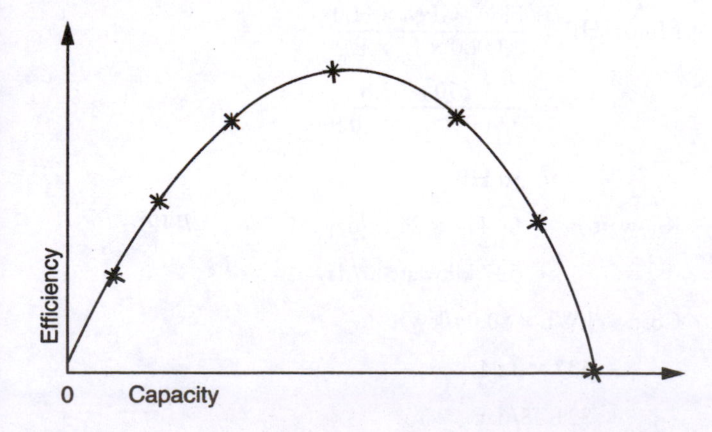


The last important characteristic is the brake horsepower or the power input to the pump. The brake horsepower usually increases with increasing capacity until it reaches a maximum, then it normally reduces slightly.

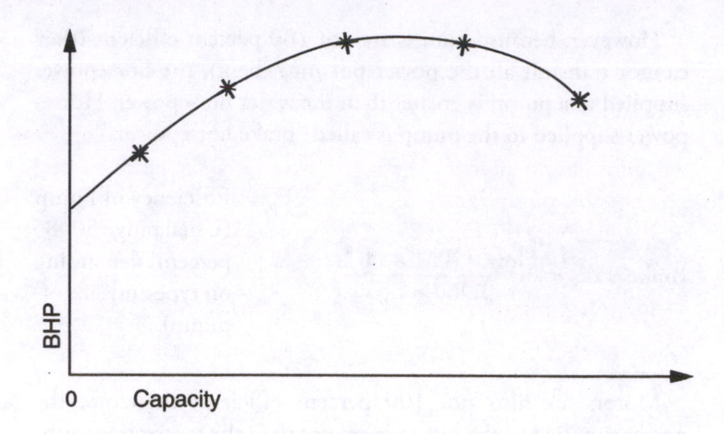


These pump characteristic curves are quite important. Pump sizes are normally picked from these curves rather than calculations. For ease of reading, the three characteristic curves are normally plotted together. A typical graph of pump characteristics is shown as follows:

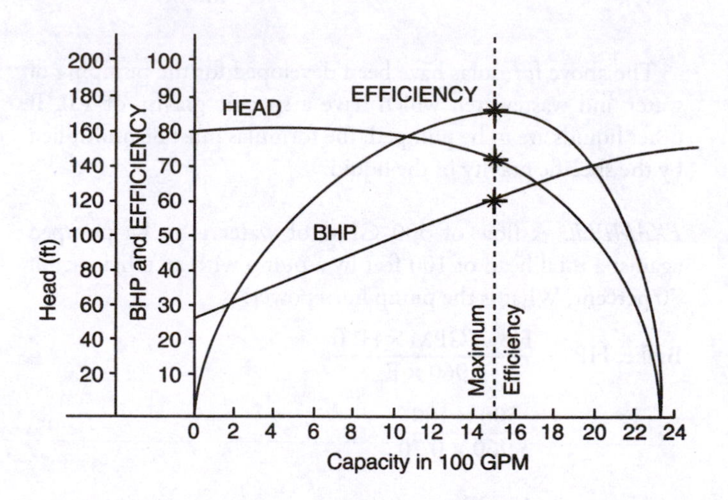


The curves show that the maximum efficiency for the particular pump in question occurs at approximately 1,475 GPM, a head of 132 feet, and a brake horsepower of 58. Operating at this point, the pump has an efficiency of approximately 85 percent. This can be verified by calculation:

$$\text{BHP} = \frac{\text{Flow, GPM} \times \text{H, ft}}{3{,}960 \times \text{E}}$$

As previously explained, a number can be written over one without changing its value:

$$\frac{\text{BHP}}{1} = \frac{\text{GPM} \times \text{H}}{3{,}960 \times \text{E}}$$

Because the formula is now in ratio form, it can be cross multiplied.

$$\text{BHP} \times 3{,}960 \times \text{E} = \text{GPM} \times \text{H} \times 1$$

Solving for E,

$$\text{E} = \frac{\text{GPM} \times \text{H}}{3{,}960 \times \text{BHP}}$$

$$\text{E} = \frac{1{,}475 \text{ GPM} \times 132 \text{ ft}}{3{,}960 \times 58 \text{ HP}}$$

$$= 0.85 \text{ or } 85\% \text{ (Check)}$$

The preceding is only a brief description of pumps to familiarize the operator with their characteristics. The operator does not normally specify the type and size of pump needed at a plant. If a pump is needed, the operator should be able to supply the information necessary for a pump supplier to provide the best possible pump for the lowest cost. Some of the information needed includes the following:

1. Flow range desired
2. Head conditions
   a. Suction head or lift
   b. Pipe and fitting friction head
   c. Discharge head
3. Type of fluid pumped and temperature
4. Pump location

### A.86 Evaluation of Pump Performance

1. Capacity

Sometimes, it is necessary to determine the capacity of a pump. This can be accomplished by determining the time it takes a pump to fill or empty a portion of a storage tank or diversion box when all inflow is blocked off.

*EXAMPLE:*

a. Measure the size of the wet well.

Length = 10 ft

Width = 10 ft

Depth = 5 ft (We will measure the time it takes to lower the well a distance of 5 feet.)

$$\text{Volume, cu ft} = \text{L, ft} \times \text{W, ft} \times \text{D, ft}$$

$$= 10 \text{ ft} \times 10 \text{ ft} \times 5 \text{ ft}$$

$$= 500 \text{ cu ft}$$

b. Record the time for the water to drop 5 feet in the wet well.

Time = 10 minutes 30 seconds

= 10.5 minutes

c. Calculate the pumping rate or capacity.

$$\text{Pumping Rate, GPM} = \frac{\text{Volume, gallons}}{\text{Time, minutes}}$$

$$= \frac{(500 \text{ cu ft})(7.5 \text{ gal/cu ft})}{10.5 \text{ min}}$$

$$= \frac{3{,}750}{10.5}$$

$$= 357 \text{ GPM}$$

If you know the total dynamic head and have the pump's performance curves, you can determine if the pump is delivering at design capacity. If not, try to determine the cause (see *OPERATION OF WASTEWATER TREATMENT PLANTS*, Volume II, Chapter 15, "Maintenance"). After a pump overhaul, the pump's actual performance (flow, head, power, and efficiency) should be compared with the pump manufacturer's performance curves. This procedure for calculating the rate of filling or emptying of a wet well or diversion box can be used to calibrate flowmeters.

2. Efficiency

To estimate the efficiency of the pump in the previous example, the total head must be known. This head may be estimated by measuring the suction and discharge pressures. Assume these were measured as follows:

2 in Mercury vacuum
20 psi
PUMP
flow
Suction side
Discharge side

No additional information is necessary if we assume the pressure gauges are at the same height and the pipe diameters are the same. Both pressure readings must be converted to feet.

$$\text{Suction Lift, ft} = 2 \text{ in Mercury} \times \frac{1.133 \text{ ft Water*}}{1 \text{ in Mercury}}$$

$$= 2.27 \text{ ft}$$

$$\text{Discharge Head, ft} = 20 \text{ psi} \times 2.31 \text{ ft/psi*}$$

$$= 46.20 \text{ ft}$$

$$\text{Total Head, ft} = \text{Suction Lift, ft} + \text{Discharge Head, ft}$$

$$= 2.27 \text{ ft} + 46.20 \text{ ft}$$

$$= 48.47 \text{ ft}$$

*See Section A.10, "Basic Conversion Factors (English System)," *PRESSURE*, page 454.

Calculate the power output of the pump or water horsepower:

$$\text{Water Horsepower, HP} = \frac{(\text{Flow, GPM})(\text{Head, ft})}{3{,}960}$$

$$= \frac{(357 \text{ GPM})(48.47 \text{ ft})}{3{,}960}$$

$$= 4.4 \text{ HP}$$

To estimate the efficiency of the pump, measure the kilowatts drawn by the pump motor. Assume the meter indicates 8,000 watts or 8 kilowatts. The manufacturer claims the electric motor is 80 percent efficient.

$$\text{Brake Horsepower, HP} = (\text{Power to Elec. Motor})(\text{Motor Eff.})$$

$$= \frac{(8 \text{ kW})(0.80)}{0.746 \text{ kW/HP}}$$

$$= 8.6 \text{ HP}$$

$$\text{Pump Efficiency, \%} = \frac{\text{Water Horsepower, HP} \times 100\%}{\text{Brake Horsepower, HP}}$$

$$= \frac{4.4 \text{ HP} \times 100\%}{8.6 \text{ HP}}$$

$$= 51\%$$

The following diagram may clarify the previous problem:

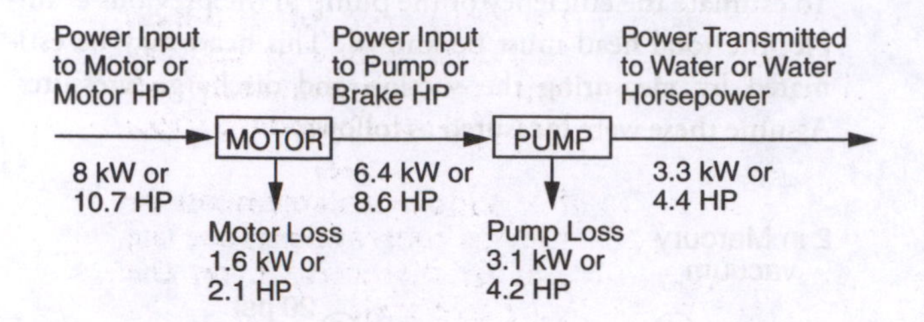


The wire-to-water efficiency is the efficiency of the power input to produce water horsepower.

$$\text{Wire-to-Water Efficiency, \%} = \frac{\text{Water Horsepower, HP}}{\text{Power Input, HP}} \times 100\%$$

$$= \frac{4.4 \text{ HP}}{10.7 \text{ HP}} \times 100\%$$

$$= 41\%$$

The wire-to-water efficiency of a pumping system (pump and electric motor) can be calculated by using the following formula:

$$\text{Efficiency, \%} = \frac{(\text{Flow, GPM})(\text{TDH, ft})(100\%)}{(\text{Voltage, volts})(\text{Current, amps})(5.308)}$$

$$= \frac{(357 \text{ GPM})(48.47 \text{ ft})(100\%)}{(220 \text{ volts})(36 \text{ amps})(5.308)}$$

$$= 41\%$$

## A.87 Pump Speed/Performance Relationships

Changing the velocity of a centrifugal pump will change its operating characteristics. If the speed of a pump is changed, the flow, head developed, and power requirements will change. The operating characteristics of the pump will change with speed approximately as follows:

$$\text{Flow, } Q_n = \left(\frac{N_n}{N_r}\right) Q_r$$

$$\text{Head, } H_n = \left(\frac{N_n}{N_r}\right)^2 H_r$$

$$\text{Power, } P_n = \left(\frac{N_n}{N_r}\right)^3 P_r$$

r = rated

n = now

N = pump speed

Actually, pump efficiency does vary with speed; therefore, these formulas are not quite correct. If speeds do not vary by more than a factor of two (if the speeds are not doubled or cut in half), the results are close enough. Other factors contributing to changes in pump characteristic curves include impeller wear and roughness in pipes.

*EXAMPLE:* To illustrate these relationships, assume a pump has a rated capacity of 600 GPM, develops 100 feet of head, and has a power requirement of 15 HP when operating at 1,500 RPM. If the efficiency remains constant, what will be the operating characteristics if the speed drops to 1,200 RPM?

Calculate the new flow rate or capacity:

$$\text{Flow, } Q_n = \left(\frac{N_n}{N_r}\right) Q_r$$

$$= \left(\frac{1{,}200 \text{ RPM}}{1{,}500 \text{ RPM}}\right)(600 \text{ GPM})$$

$$= \left(\frac{4}{5}\right)(600 \text{ GPM})$$

$$= (4)(120 \text{ GPM})$$

$$= 480 \text{ GPM}$$

Calculate the new head:

$$\text{Head, } H_n = \left(\frac{N_n}{N_r}\right)^2 H_r$$

$$= \left(\frac{1{,}200 \text{ RPM}}{1{,}500 \text{ RPM}}\right)^2 (100 \text{ ft})$$

$$= \left(\frac{4}{5}\right)^2 (100 \text{ ft})$$

$$= \left(\frac{16}{25}\right)(100 \text{ ft})$$

$$= (16)(4 \text{ ft})$$

$$= 64 \text{ ft}$$

Calculate the new power requirement:

$$\text{Power, } P_n = \left(\frac{N_n}{N_r}\right)^3 P_r$$

$$= \left(\frac{1{,}200 \text{ RPM}}{1{,}500 \text{ RPM}}\right)^3 (15 \text{ HP})$$

$$= \left(\frac{4}{5}\right)^3 (15 \text{ HP})$$

$$= \left(\frac{64}{125}\right)(15 \text{ HP})$$

$$= \left(\frac{64}{25}\right)(3 \text{ HP})$$

$$= 7.7 \text{ HP}$$

## A.88 Friction or Energy Losses

Whenever water flows through pipes, valves, and fittings, energy is lost due to pipe friction (resistance), friction in valves and fittings, and the turbulence resulting from the flowing water changing its direction. Figure A.1 can be used to convert the friction losses through valves and fittings to lengths of straight pipe that would produce the same amount of friction losses. To estimate the friction or energy losses resulting from water flowing in a pipe system, we need to know:

1. Water flow rate
2. Pipe size or diameter and length
3. Number, size, and type of valve fittings

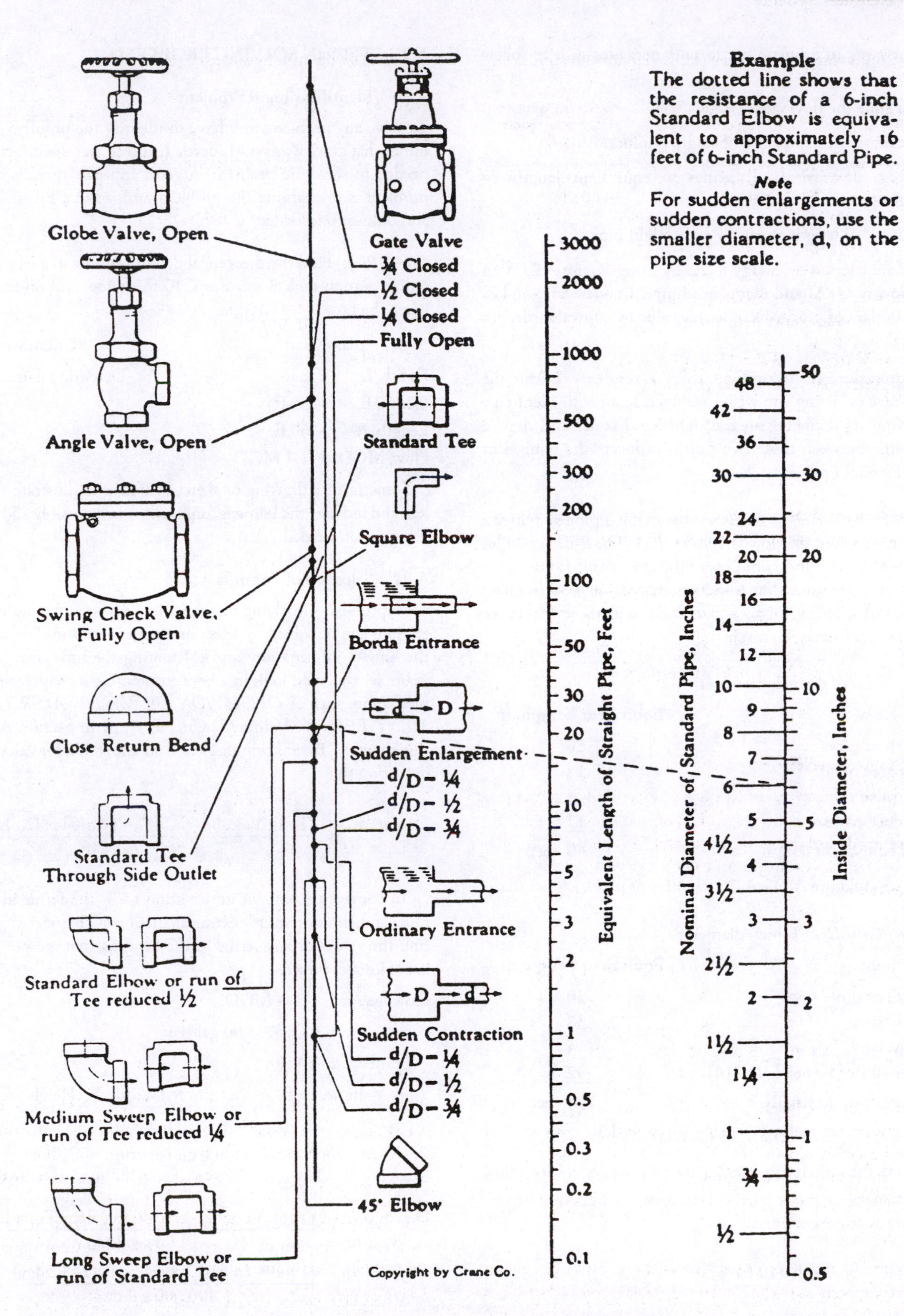


*Fig. A.1 Resistance of valves and fittings to flow of water*

(Reprinted by permission of Crane Co.)

An easy way to estimate friction or energy losses is to follow these steps:

1. Determine the flow rate.
2. Determine the diameter and length of pipe.
3. Convert all valves and fittings to equivalent lengths of straight pipe (see Figure A.1).
4. Add up total length of equivalent straight pipe.
5. Estimate friction or energy losses by using Figure A.2. With the flow in GPM and diameter of pipe, find the friction loss per 100 feet of pipe. Multiply this value by equivalent length of straight pipe.

The procedure for using Figure A.1 is very easy. Locate the type of valve or fitting you wish to convert to an equivalent pipe length; find its diameter on the right-hand scale; and draw a straight line between these two points to locate the equivalent length of straight pipe.

*EXAMPLE:* Estimate the friction losses in the piping system of a pump station when the flow is 1,000 GPM. The 8-inch suction line is 10 feet long and contains a 90-degree bend (long sweep elbow) and a gate valve. The 6-inch discharge line is 30 feet long and contains a check valve, a gate valve, and three 90-degree bends (medium sweep elbows):

*SUCTION LINE* (8-inch diameter)

| Item | Equivalent Length, ft |
|---|---|
| 1. Length of pipe | 10 |
| 2. 90-degree bend | 14 |
| 3. Gate valve | 4 |
| 4. Ordinary entrance | 12 |
| Total equivalent length | 40 feet |

Friction loss (Figure A.2) = 1.76 ft/100 ft of pipe

*DISCHARGE LINE* (6-inch diameter)

| Item | Equivalent Length, ft |
|---|---|
| 1. Length of pipe | 30 |
| 2. Check valve | 38 |
| 3. Gate valve | 4 |
| 4. Three 90-degree bends (3)(14) | 42 |
| Total equivalent length | 114 feet |

Friction loss (Figure A.2) = 7.73 ft/100 ft of pipe

Estimate the total friction losses in pumping system for a flow of 1,000 GPM.

*SUCTION*

Loss = (1.76 ft/100 ft)(40 ft) = 0.7 ft

*DISCHARGE*

Loss = (7.73 ft/100 ft)(114 ft) = 8.8 ft

Total Friction Losses, ft = 9.5 ft

## A.9 STEPS IN SOLVING PROBLEMS

### A.90 Identification of Problem

To solve any problem, you have to identify the problem, determine what kind of answer is needed, and collect the information needed to solve the problem. A good approach to this type of problem is to examine the problem and make a list of known and unknown information.

*EXAMPLE:* Find the theoretical detention time in a rectangular sedimentation tank 8 feet deep, 30 feet wide, and 60 feet long when the flow is 1.4 MGD.

| Known | | Unknown |
|---|---|---|
| Depth, ft | = 8 ft | Detention Time, hours |
| Width, ft | = 30 ft | |
| Length, ft | = 60 ft | |
| Flow, MGD | = 1.4 MGD | |

Sometimes, a drawing or sketch will help to illustrate a problem and indicate the knowns, unknowns, and possibly additional information needed.

### A.91 Selection of Formula

Most problems involving mathematics in wastewater system operations can be solved by selecting the proper formula, inserting the known information, and calculating the unknown. In our example, we could look in a book containing a chapter on sedimentation, such as *OPERATION OF WASTEWATER TREATMENT PLANTS,* Volume I, Chapter 5, or in Section A.11 of this chapter, "Basic Formulas," to find a formula for calculating detention time.

$$\text{Detention Time, hr} = \frac{(\text{Tank Volume, cu ft})(7.48\ \text{gal/cu ft})(24\ \text{hr/day})}{\text{Flow, gal/day}}$$

To convert the known information to fit the terms in a formula sometimes requires extra calculations. The next step is to find the values of any terms in the formula that are not in the list of known values.

Flow, gal/day = 1.4 MGD

= 1,400,000 gal/day

From Section A.30:

Tank Volume, cu ft = (Length, ft)(Width, ft)(Height, ft)

= 60 ft × 30 ft × 8 ft

= 14,400 cu ft

Solution of Problem:

$$\text{Detention Time, hr} = \frac{(\text{Tank Volume, cu ft})(7.48\ \text{gal/cu ft})(24\ \text{hr/day})}{\text{Flow, gal/day}}$$

$$= \frac{(14{,}400\ \text{cu ft})(7.48\ \text{gal/cu ft})(24\ \text{hr/day})}{1{,}400{,}000\ \text{gal/day}}$$

= 1.85 hr

The remainder of this section discusses the details that must be considered in solving this problem.

| US GPM | 0.5 in. | | 0.75 in. | | 1 in. | | 1.25 in. | | 1.5 in. | | 2 in. | | 2.5 in. | |
|---|---|---|---|---|---|---|---|---|---|---|---|---|---|---|
| | Vel. | Frict. | Vel. | Frict. | Vel. | Frict. | Vel. | Frict. | Vel. | Frict. | Vel. | Frict. | Vel. | Frict. |
| 10 | 10.56 | 95.9 | 6.02 | 23.0 | 3.71 | 6.86 | 2.15 | 1.77 | 1.58 | .83 | .96 | .25 | .67 | .11 |
| 20 | ...... | ...... | 12.0 | 86.1 | 7.42 | 25.1 | 4.29 | 6.34 | 3.15 | 2.94 | 1.91 | .87 | 1.34 | .36 |
| 30 | ...... | ...... | ...... | ...... | 11.1 | 54.6 | 6.44 | 13.6 | 4.73 | 6.26 | 2.87 | 1.82 | 2.01 | .75 |
| 40 | ...... | ...... | ...... | ...... | 14.8 | 95.0 | 8.58 | 23.5 | 6.30 | 10.79 | 3.82 | 3.10 | 2.68 | 1.28 |
| 50 | ...... | ...... | ...... | ...... | ...... | ...... | 10.7 | 36.0 | 7.88 | 16.4 | 4.78 | 4.67 | 3.35 | 1.94 |
| 60 | ...... | ...... | ...... | ...... | ...... | ...... | 12.9 | 51.0 | 9.46 | 23.2 | 5.74 | 6.59 | 4.02 | 2.72 |
| 70 | ...... | ...... | ...... | ...... | ...... | ...... | 15.0 | 68.8 | 11.03 | 31.3 | 6.69 | 8.86 | 4.69 | 3.63 |
| 80 | ...... | ...... | ...... | ...... | ...... | ...... | 17.2 | 89.2 | 12.6 | 40.5 | 7.65 | 11.4 | 5.36 | 4.66 |
| 90 | ...... | ...... | ...... | ...... | ...... | ...... | ...... | ...... | 14.2 | 51.0 | 8.60 | 14.2 | 6.03 | 5.82 |
| 100 | ...... | ...... | ...... | ...... | ...... | ...... | ...... | ...... | 15.8 | 62.2 | 9.56 | 17.4 | 6.70 | 7.11 |
| 120 | ...... | ...... | ...... | ...... | ...... | ...... | ...... | ...... | 18.9 | 88.3 | 11.5 | 24.7 | 8.04 | 10.0 |
| 140 | ...... | ...... | ...... | ...... | ...... | ...... | ...... | ...... | ...... | ...... | 13.4 | 33.2 | 9.38 | 13.5 |
| 160 | ...... | ...... | ...... | ...... | ...... | ...... | ...... | ...... | ...... | ...... | 15.3 | 43.0 | 10.7 | 17.4 |
| 180 | ...... | ...... | ...... | ...... | ...... | ...... | ...... | ...... | ...... | ...... | 17.2 | 54.1 | 12.1 | 21.9 |
| 200 | ...... | ...... | ...... | ...... | ...... | ...... | ...... | ...... | ...... | ...... | 19.1 | 66.3 | 13.4 | 26.7 |
| 220 | ...... | ...... | ...... | ...... | ...... | ...... | ...... | ...... | ...... | ...... | 21.0 | 80.0 | 14.7 | 32.2 |
| 240 | ...... | ...... | ...... | ...... | ...... | ...... | ...... | ...... | ...... | ...... | 22.9 | 95.0 | 16.1 | 38.1 |
| 260 | ...... | ...... | ...... | ...... | ...... | ...... | ...... | ...... | ...... | ...... | ...... | ...... | 17.4 | 44.5 |
| 280 | ...... | ...... | ...... | ...... | ...... | ...... | ...... | ...... | ...... | ...... | ...... | ...... | 18.8 | 51.3 |
| 300 | ...... | ...... | ...... | ...... | ...... | ...... | ...... | ...... | ...... | ...... | ...... | ...... | 20.1 | 58.5 |
| 350 | ...... | ...... | ...... | ...... | ...... | ...... | ...... | ...... | ...... | ...... | ...... | ...... | 23.5 | 79.2 |

| US GPM | 3 in. | | 4 in. | | 5 in. | | 6 in. | | 8 in. | | 10 in. | | 12 in. | | 14 in. | | 16 in. | | 18 in. | | 20 in. | |
|---|---|---|---|---|---|---|---|---|---|---|---|---|---|---|---|---|---|---|---|---|---|---|
| | Vel. | Frict. | Vel. | Frict. | Vel. | Frict. | Vel. | Frict. | Vel. | Frict. | Vel. | Frict. | Vel. | Frict. | Vel. | Frict. | Vel. | Frict. | Vel. | Frict. | Vel. | Frict. |
| 20 | .91 | .15 | | | | | | | | | | | | | | | | | | | | |
| 40 | 1.82 | .55 | 1.02 | .13 | | | | | | | | | | | | | | | | | | |
| 50 | 2.72 | 1.17 | 1.53 | .28 | .96 | .08 | | | | | | | | | | | | | | | | |
| 80 | 3.63 | 2.02 | 2.04 | .48 | 1.28 | .14 | .91 | .06 | | | | | | | | | | | | | | |
| 100 | 4.54 | 3.10 | 2.55 | .73 | 1.60 | .20 | 1.13 | .10 | | | | | | | | | | | | | | |
| 120 | 5.45 | 4.40 | 3.06 | 1.03 | 1.92 | .29 | 1.36 | .13 | | | | | | | | | | | | | | |
| 140 | 6.35 | 5.93 | 3.57 | 1.38 | 2.25 | .38 | 1.59 | .18 | | | | | | | | | | | | | | |
| 160 | 7.26 | 7.71 | 4.08 | 1.78 | 2.57 | .49 | 1.82 | .23 | | | | | | | | | | | | | | |
| 180 | 8.17 | 9.73 | 4.60 | 2.24 | 2.89 | .61 | 2.04 | .28 | | | | | | | | | | | | | | |
| 200 | 9.08 | 11.9 | 5.11 | 2.74 | 3.21 | .74 | 2.27 | .35 | | | | | | | | | | | | | | |
| 220 | 9.98 | 14.3 | 5.62 | 3.28 | 3.53 | .88 | 2.50 | .42 | 1.40 | .10 | | | | | | | | | | | | |
| 240 | 10.9 | 17.0 | 6.13 | 3.88 | 3.85 | 1.04 | 2.72 | .49 | 1.53 | .12 | | | | | | | | | | | | |
| 260 | 11.8 | 19.8 | 6.64 | 4.54 | 4.17 | 1.20 | 2.95 | .57 | 1.66 | .14 | | | | | | | | | | | | |
| 280 | 12.7 | 22.8 | 7.15 | 5.25 | 4.49 | 1.38 | 3.18 | .66 | 1.79 | .16 | | | | | | | | | | | | |
| 300 | 13.6 | 26.1 | 7.66 | 6.03 | 4.81 | 1.58 | 3.40 | .75 | 1.91 | .18 | | | | | | | | | | | | |
| 350 | ...... | ...... | 8.94 | 8.22 | 5.61 | 2.11 | 3.97 | 1.01 | 2.24 | .24 | | | | | | | | | | | | |
| 400 | ...... | ...... | 10.20 | 10.7 | 6.41 | 2.72 | 4.54 | 1.30 | 2.55 | .30 | | | | | | | | | | | | |
| 450 | ...... | ...... | 11.45 | 13.4 | 7.22 | 3.41 | 5.11 | 1.64 | 2.87 | .38 | 1.84 | .12 | | | | | | | | | | |
| 500 | ...... | ...... | 12.8 | 16.6 | 8.02 | 4.16 | 5.67 | 2.02 | 3.19 | .46 | 2.04 | .15 | 1.42 | .06 | | | | | | | | |
| 550 | ...... | ...... | 14.0 | 19.9 | 8.82 | 4.98 | 6.24 | 2.42 | 3.51 | .56 | 2.25 | .18 | 1.56 | .07 | | | | | | | | |
| 600 | ...... | ...... | ...... | ...... | 9.62 | 5.88 | 6.81 | 2.84 | 3.83 | .66 | 2.45 | .21 | 1.70 | .08 | 1.25 | .04 | | | | | | |
| 700 | ...... | ...... | ...... | ...... | 11.2 | 7.93 | 7.94 | 3.87 | 4.47 | .88 | 2.86 | .29 | 1.99 | .12 | 1.46 | .05 | | | | | | |
| 800 | ...... | ...... | ...... | ...... | 12.8 | 10.22 | 9.08 | 5.06 | 5.11 | 1.14 | 3.27 | .37 | 2.27 | .15 | 1.67 | .07 | | | | | | |
| 900 | ...... | ...... | ...... | ...... | 14.4 | 12.9 | 10.2 | 6.34 | 5.74 | 1.44 | 3.68 | .46 | 2.55 | .18 | 1.88 | .09 | | | | | | |
| 1000 | ...... | ...... | ...... | ...... | ...... | ...... | 11.3 | 7.73 | 6.38 | 1.76 | 4.09 | .57 | 2.84 | .22 | 2.08 | .11 | | | | | | |
| 1100 | ...... | ...... | ...... | ...... | ...... | ...... | 12.5 | 9.80 | 7.02 | 2.14 | 4.49 | .68 | 3.12 | .27 | 2.29 | .13 | | | | | | |
| 1200 | ...... | ...... | ...... | ...... | ...... | ...... | 13.6 | 11.2 | 7.66 | 2.53 | 4.90 | .81 | 3.40 | .32 | 2.50 | .15 | 1.91 | .08 | | | | |
| 1300 | ...... | ...... | ...... | ...... | ...... | ...... | 14.7 | 13.0 | 8.30 | 2.94 | 5.31 | .95 | 3.69 | .37 | 2.71 | .17 | 2.07 | .09 | | | | |
| 1400 | ...... | ...... | ...... | ...... | ...... | ...... | ...... | ...... | 8.93 | 3.40 | 5.72 | 1.09 | 3.97 | .43 | 2.92 | .20 | 2.23 | .10 | | | | |
| 1500 | ...... | ...... | ...... | ...... | ...... | ...... | ...... | ...... | 9.57 | 3.91 | 6.13 | 1.25 | 4.26 | .49 | 3.13 | .23 | 2.34 | .12 | | | | |
| 1600 | ...... | ...... | ...... | ...... | ...... | ...... | ...... | ...... | 10.2 | 4.45 | 6.54 | 1.42 | 4.54 | .55 | 3.33 | .25 | 2.55 | .13 | 2.02 | .07 | | |
| 1700 | ...... | ...... | ...... | ...... | ...... | ...... | ...... | ...... | 10.8 | 5.00 | 6.94 | 1.60 | 4.87 | .62 | 3.54 | .29 | 2.71 | .15 | 2.15 | .08 | | |
| 1800 | ...... | ...... | ...... | ...... | ...... | ...... | ...... | ...... | 11.5 | 5.58 | 7.35 | 1.78 | 5.11 | .70 | 3.75 | .32 | 2.87 | .16 | 2.27 | .09 | | |
| 1900 | ...... | ...... | ...... | ...... | ...... | ...... | ...... | ...... | 12.1 | 6.19 | 7.76 | 1.97 | 5.39 | .77 | 3.96 | .35 | 3.03 | .18 | 2.40 | .10 | | |
| 2000 | ...... | ...... | ...... | ...... | ...... | ...... | ...... | ...... | 12.8 | 6.84 | 8.17 | 2.17 | 5.67 | .86 | 4.17 | .39 | 3.19 | .20 | 2.52 | .11 | | |
| 2500 | ...... | ...... | ...... | ...... | ...... | ...... | ...... | ...... | ...... | ...... | 10.2 | 3.38 | 7.10 | 1.33 | 5.21 | .60 | 3.99 | .31 | 3.15 | .17 | | |
| 3000 | ...... | ...... | ...... | ...... | ...... | ...... | ...... | ...... | ...... | ...... | 12.3 | 4.79 | 8.51 | 1.88 | 6.25 | .86 | 4.79 | .44 | 3.78 | .24 | 3.06 | .14 |
| 3500 | ...... | ...... | ...... | ...... | ...... | ...... | ...... | ...... | ...... | ...... | 14.3 | 6.55 | 9.93 | 2.56 | 7.29 | 1.16 | 5.58 | .58 | 4.41 | .32 | 3.57 | .19 |
| 4000 | ...... | ...... | ...... | ...... | ...... | ...... | ...... | ...... | ...... | ...... | ...... | ...... | 11.3 | 3.31 | 8.34 | 1.50 | 6.38 | .75 | 5.04 | .42 | 4.08 | .24 |
| 4500 | ...... | ...... | ...... | ...... | ...... | ...... | ...... | ...... | ...... | ...... | ...... | ...... | 12.8 | 4.18 | 9.38 | 1.88 | 7.18 | .95 | 5.67 | .53 | 4.59 | .31 |
| 5000 | ...... | ...... | ...... | ...... | ...... | ...... | ...... | ...... | ...... | ...... | ...... | ...... | 14.7 | 5.13 | 10.4 | 2.30 | 7.98 | 1.17 | 6.30 | .65 | 5.11 | .38 |
| 6000 | ...... | ...... | ...... | ...... | ...... | ...... | ...... | ...... | ...... | ...... | ...... | ...... | ...... | ...... | 12.5 | 3.31 | 9.57 | 1.66 | 7.56 | .92 | 6.13 | .53 |
| 7000 | ...... | ...... | ...... | ...... | ...... | ...... | ...... | ...... | ...... | ...... | ...... | ...... | ...... | ...... | 14.6 | 4.50 | 11.2 | 2.26 | 8.83 | 1.24 | 7.15 | .72 |
| 8000 | ...... | ...... | ...... | ...... | ...... | ...... | ...... | ...... | ...... | ...... | ...... | ...... | ...... | ...... | ...... | ...... | 12.8 | 2.96 | 10.09 | 1.61 | 8.17 | .94 |
| 9000 | ...... | ...... | ...... | ...... | ...... | ...... | ...... | ...... | ...... | ...... | ...... | ...... | ...... | ...... | ...... | ...... | 14.4 | 3.73 | 11.3 | 2.02 | 9.19 | 1.18 |
| 10000 | ...... | ...... | ...... | ...... | ...... | ...... | ...... | ...... | ...... | ...... | ...... | ...... | ...... | ...... | ...... | ...... | ...... | ...... | 12.6 | 2.48 | 10.2 | 1.45 |

No allowance has been made for age, differences in diameter, or any other abnormal condition of interior surface. Any Factor of Safety must be estimated from the local conditions and the requirements of each particular installation. For general purposes, 15% is a responsible Factor of Safety.

*Fig. A.2 Friction loss for water in feet per 100 feet of pipe*

(Reprinted from the 10th Edition of the Standards of the Hydraulic Institute, 122 East 42nd Street, New York)

### A.92 Arrangement of Formula

Once the proper formula is selected, you may have to rearrange the terms to solve for the unknown term. From Section A.71, "Flow Rate," we can develop the formula:

$$\text{Velocity, ft/sec} = \frac{\text{Flow Rate, cu ft/sec}}{\text{Cross-Sectional Area, sq ft}}$$

or

$$V = \frac{Q}{A}$$

In this equation, if Q and A were given, the equation could be solved for V. If V and A were known, the equation would have to be rearranged to solve for Q. To move terms from one side of an equation to another, use the following rule:

> When moving a term or number from one side of an equation to the other, move the numerator (top) of one side to the denominator (bottom) of the other; or from the denominator (bottom) of one side to the numerator (top) of the other.

$$V = \frac{Q}{A} \text{ or } Q = AV \text{ or } A = \frac{Q}{V}$$

If the volume of a sedimentation tank and the desired detention time were given, the detention time formula could be rearranged to calculate the design flow.

$$\text{Detention Time, hr} = \frac{(\text{Tank Vol, cu ft})(7.48\text{ gal/cu ft})(24\text{ hr/day})}{\text{Flow, gal/day}}$$

By rearranging the terms,

$$\text{Flow, gal/day} = \frac{(\text{Tank Vol, cu ft})(7.48\text{ gal/cu ft})(24\text{ hr/day})}{\text{Detention Time, hr}}$$

### A.93 Unit Conversions

Each term in a formula or mathematical calculation must be of the correct units. For example, the area of a rectangular clarifier (Area, sq ft = Length, ft × Width, ft) cannot be calculated in square feet if the width is given as 246 inches or 20 feet 6 inches. The width must be converted to 20.5 feet. In the example problem, if the tank volume were given in gallons, then the 7.48 gal/cu ft would not be needed. Avoid time-consuming mistakes by always checking the units in a formula before performing any calculations.

$$\text{Detention Time, hr} = \frac{(\text{Tank Volume, cu ft})(7.48\text{ gal/cu ft})(24\text{ hr/day})}{\text{Flow, gal/day}}$$

$$= \frac{\cancel{\text{cu ft}}}{} \times \frac{\cancel{\text{gal}}}{\cancel{\text{cu ft}}} \times \frac{\text{hr}}{\cancel{\text{day}}} \times \frac{\cancel{\text{day}}}{\cancel{\text{gal}}}$$

$$= \text{hr (all other units cancel)}$$

*NOTE:* We have hours = hr. One should note that the hour unit on both sides of the equation can be cancelled out and nothing would remain. This is one more check that we have the correct units.

By rearranging the detention time formula, other unknowns could be determined.

If the design detention time and design flow were known, the required capacity of the tank could be calculated.

$$\text{Tank Volume, cu ft} = \frac{(\text{Detention Time, hr})(\text{Flow, gal/day})}{(7.48\text{ gal/cu ft})(24\text{ hr/day})}$$

If the tank volume and design detention time were known, the design flow could be calculated.

$$\text{Flow, gal/day} = \frac{(\text{Tank Volume, cu ft})(7.48\text{ gal/cu ft})(24\text{ hr/day})}{\text{Detention Time, hr}}$$

Rearrangement of the detention time formula to find other unknowns illustrates the need to always use the correct units.

### A.94 Calculations

Section A.14, "Order of Arithmetic Calculations," outlines the correct order for mathematical calculations. To summarize: (1) work all calculations within parentheses; (2) simplify any exponents or square roots, working from left to right; (3) perform all multiplications and divisions, working from left to right; (4) perform all additions and subtractions, working from left to right; and (5) if the problem includes a fraction or division bar, divide the numerator by the denominator.

$$\text{Detention Time, hr} = \frac{(\text{Tank Volume, cu ft})(7.48\text{ gal/cu ft})(24\text{ hr/day})}{\text{Flow, gal/day}}$$

$$= \frac{(14{,}400\text{ cu ft})(7.48\text{ gal/cu ft})(24\text{ hr/day})}{1{,}400{,}000\text{ gal/day}}$$

$$= \frac{2{,}585{,}088\text{ gal-hr/day}}{1{,}400{,}000\text{ gal/day}}$$

$$= 1.85\text{ hr, or}$$

$$= 1.9\text{ hr}$$

### A.95 Significant Figures

In calculating the detention time in the previous section, the answer is given as 1.9 hr. The answer could have been calculated in the following way:

$$\text{Detention Time, hr} = \frac{2{,}585{,}088\text{ gal-hr/day}}{1{,}400{,}000\text{ gal/day}}$$

$$= 1.846491429 \ldots \text{ hours}$$

The resulting string of numbers, however, is not very useful. How does one know when to stop dividing and round off the number? Common sense and the concept of "significant figures" are both helpful.

First, consider the meaning of detention time and the measurements that were taken to determine the knowns in the formula. Detention time in a tank is a theoretical value and assumes that all particles of water throughout the tank move through the tank at the same velocity. This assumption is not correct; therefore, detention time can only be a representative time for some of the water particles.

Will the flow of 1.4 MGD be constant throughout the 1.9 hours, and is the flow exactly 1.4 MGD, or could it be 1.35 MGD or 1.428 MGD? A carefully calibrated flowmeter may give a reading within 2 percent of the actual flow rate. Flows into a tank fluctuate and flowmeters do not measure flows extremely accurately; so the detention time again appears to be a representative or typical detention time.

Tank dimensions are probably precise to within 0.1 foot. A flowmeter reading of 1.4 MGD is less precise and it could be 1.3 or 1.5 MGD. A 0.1 MGD flowmeter error when the flow is 1.4 MGD is (0.1/1.4) × 100% = 7% error. A detention time of 1.9 hours, based on a flowmeter reading error of plus or minus 7 percent, also could have the same error or more, even if the flow were constant. Therefore, the detention time error could be 1.9 hours × 0.07 = ±0.13 hour.

In most of the calculations in the operation of wastewater systems, the operator uses measurements determined in the lab or read from charts, scales, or meters. The accuracy of every measurement depends on the sample being measured, the equipment doing the measuring, and the operator reading or measuring the results. Your calculation, therefore, can only be an estimate and your estimate will only be as good as the least precise measurement.

To determine how many digits or numbers mean anything in an answer, the concept of significant figures is used. Significant figures are those digits of a number that carry meaning contributing to the accuracy of the number. The concept of significant figures is often used in connection with rounding. In the calculation of wastewater system problems, results are usually rounded to a certain number of significant figures or to a particular level of precision. In order to apply the concept of significant figures, it is necessary to understand the rules for determining which numbers in a measurement are considered significant. These rules and examples of each rule are given below.

1. Every nonzero digit in a recorded measurement is significant. Example: 24.7 feet, 0.743 feet, and 714 feet all have three significant figures.
2. Zeroes appearing between nonzero digits are significant. Example: 7,003 feet, 40.79 feet, and 1.503 feet all have four significant figures.
3. Zeroes in front of all nonzero digits are *not* significant; they are merely placeholders. Example: 0.00012 has two significant figures, 1 and 2.
4. Zeroes at the end of a number containing a decimal point are significant. Example: 1,241.20 feet, 210.100 feet, and 5,600.00 all have six significant figures.

As previously discussed, most wastewater system calculations are based on measurements of varying accuracy. The number of significant figures in each measurement, as well as the accuracy of each measurement, will depend on the sample being measured, the equipment doing the measuring, and the operator reading or measuring the results. Usually, the significant figures (digits) in a measurement include all the digits that can be known precisely plus a last digit that is an estimate. To determine how many figures or numbers mean anything in the results of a calculation, the general rule to remember is:

When performing calculations, the number of significant figures in the answer should not be more than the term in the calculations with the least number of significant figures.

*EXAMPLE:* The distance between two points was divided into three sections, and each section was measured by a different group of people. What is the distance between the two points? Each group reported the distance it measured as follows:

| Group | Distance, ft | Significant Figures |
|---|---|---|
| A | 11,300 | 3 |
| B | 2,438.9 | 5 |
| C | 87.62 | 4 |
| Total Distance | 13,826.52 | |

Group A reported the length of the section it measured to three significant figures. Because this is the measurement with the least number of significant figures, the distance between the two points should also be reported to three significant figures or as 13,800 feet.

### A.96 Check Your Results

After completing your calculations, you should carefully examine your calculations and answer. Does the answer seem reasonable? If possible, have another operator check your calculations before making any operational changes.

## A.10 BASIC CONVERSION FACTORS (ENGLISH SYSTEM)

*UNITS*

| | | |
|---|---|---|
| 1,000,000 | = 1 Million | 1,000,000/1 Million |

*LENGTH*

| | | |
|---|---|---|
| 12 in | = 1 ft | 12 in/ft |
| 3 ft | = 1 yd | 3 ft/yd |
| 5,280 ft | = 1 mi | 5,280 ft/mi |

*AREA*

| | | |
|---|---|---|
| 144 sq in | = 1 sq ft | 144 sq in/sq ft |
| 43,560 sq ft | = 1 acre | 43,560 sq ft/ac |

*VOLUME*

| | | |
|---|---|---|
| 7.48 gal | = 1 cu ft | 7.48 gal/cu ft |
| 1,000 mL | = 1 liter | 1,000 mL/L |
| 3.785 L | = 1 gal | 3.785 L/gal |
| 231 cu in | = 1 gal | 231 cu in/gal |

*WEIGHT*

| | | |
|---|---|---|
| 1,000 mg | = 1 gm | 1,000 mg/gm |
| 1,000 gm | = 1 kg | 1,000 gm/kg |
| 454 gm | = 1 lb | 454 gm/lb |
| 2.2 lbs | = 1 kg | 2.2 lbs/kg |

*POWER*

| | | |
|---|---|---|
| 0.746 kW | = 1 HP | 0.746 kW/HP |

*DENSITY*

| | | |
|---|---|---|
| 8.34 lbs | = 1 gal | 8.34 lbs/gal |
| 62.4 lbs | = 1 cu ft | 62.4 lbs/cu ft |

*DOSAGE*

| | | |
|---|---|---|
| 17.1 mg/L | = 1 grain/gal | 17.1 mg/L/gpg |
| 64.7 mg | = 1 grain | 64.7 mg/grain |

*PRESSURE*

| | | |
|---|---|---|
| 2.31 ft water | = 1 psi | 2.31 ft water/psi |
| 0.433 psi | = 1 ft water | 0.433 psi/ft water |
| 1.133 ft water | = 1 in mercury | 1.133 ft water/in mercury |

*FLOW*

| | | |
|---|---|---|
| 694 GPM | = 1 MGD | 694 GPM/MGD |
| 1.55 CFS | = 1 MGD | 1.55 CFS/MGD |

*TIME*

| | | |
|---|---|---|
| 60 sec | = 1 min | 60 sec/min |
| 60 min | = 1 hr | 60 min/hr |
| 24 hr | = 1 day | 24 hr/day* |

*This may be written either as 24 hr/day or 1 day/24 hours depending on which units we wish to convert to obtain our desired results.

## A.11 BASIC FORMULAS

*SEPTIC TANKS*

1. $\text{Clear Zone, Vol, cu ft} = \dfrac{(\text{Tank Length, in})(\text{Tank Width, in})(\text{Clear Zone Depth, in})}{1{,}728 \text{ cu in/cu ft}}$

2. $\text{Clear Zone Vol, gal} = (\text{Clear Zone Volume, cu ft})(7.48 \text{ gal/cu ft})$

3. $\text{Detention Time, hr} = \dfrac{\text{Clear Zone Volume, gal}}{\text{Influent Flow, gal/day}} \times 24 \text{ hr/day}$

*SAND FILTERS*

4. $\text{Surface Area, sq ft} = (\text{Length, ft})(\text{Width, ft})$

5. $\text{Hydraulic Loading, GPD/sq ft} = \dfrac{\text{Flow, GPD}}{\text{Surface Area, sq ft}}$

*UTILITY FINANCES*

6. $\text{Operating Ratio} = \dfrac{\text{Total Revenue, \$}}{\text{Operating Expenses, \$}}$

7. $\text{Coverage Ratio} = \dfrac{\text{Total Revenue, \$} - \text{Nondebt Expenses, \$}}{\text{Debt Service Expenses, \$}}$

## A.12 HOW TO USE THE BASIC FORMULAS

One clever way of using the basic formulas is to use the pie method developed by Gerald Davidson, former manager of the Clearlake Oaks County Water District, Clearlake Oaks, California. To apply this method to the basic formula for chemical doses,

1. $\text{Chemical Feed, lbs/day} = (\text{Flow, MGD})(\text{Dose, mg/L})(8.34 \text{ lbs/gal})$

   a. Draw a circle and draw a horizontal line through the middle of the circle.

   b. Write the Chemical Feed, lbs/day in the top half.

   c. Divide the bottom half into three parts.

   d. Write Flow, MGD; Dose, mg/L; and 8.34 lbs/gal in the other three parts.

   e. The line across the middle of the circle represents the line in the equation. The items above the line stay above the line and those below the line stay below the line.

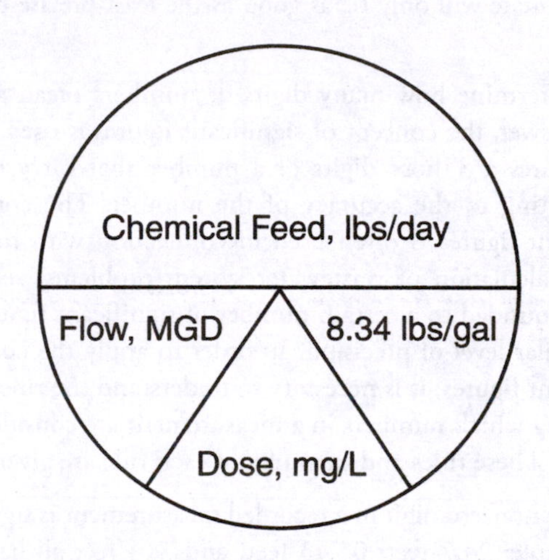


If you want to find the Chemical Feed, lbs/day, cover up the Chemical Feed, lbs/day, and what is left uncovered will give you the correct formula.

2. $\text{Chemical Feed, lbs/day} = (\text{Flow, MGD})(\text{Dose, mg/L})(8.34 \text{ lbs/gal})$

If you know the chlorinator setting in pounds per day and the flow in MGD and would like to know the dose in mg/L, cover up the Dose, mg/L, and what is left uncovered will give you the correct formula.

3. $\text{Dose, mg/L} = \dfrac{\text{Chemical Feed, lbs/day}}{(\text{Flow, MGD})(8.34 \text{ lbs/gal})}$

Another approach to using the basic formulas is to memorize the basic formula, for example the detention time formula.

4. $\text{Detention Time, hr} = \dfrac{(\text{Tank Volume, gal})(24 \text{ hr/day})}{\text{Flow, gal/day}}$

This formula works fine to solve for the detention time when the Tank Volume, gal, and Flow, gal/day, are given.

If you wish to determine the Flow, gal/day, when the Detention Time, hr, and Tank Volume, gal, are given, you must change the basic formula. You want the Flow, gal/day, on the left of the equal sign and everything else on the right of the equal sign. This is done by moving the terms diagonally (from top to bottom or from bottom to top) past the equal sign.

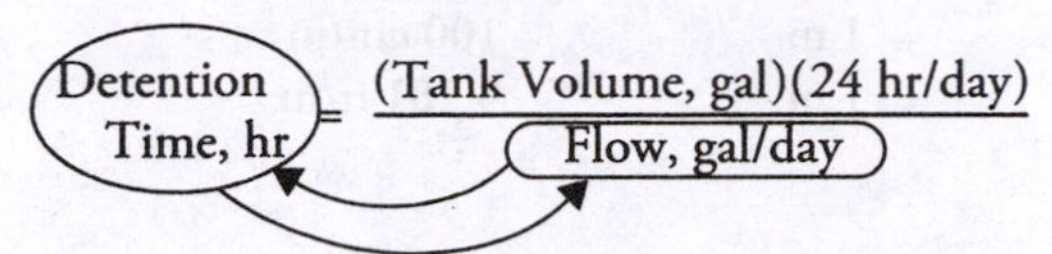


or

$$\text{Flow, gal/day} = \frac{(\text{Tank Volume, gal})(24\text{ hr/day})}{\text{Detention Time, hr}}$$

This same approach can be used if the Tank Volume, gal, is unknown and the Detention Time, hr, and Flow, gal/day, are given. We want Tank Volume, gal, on one side of the equation and everything else on the other side.

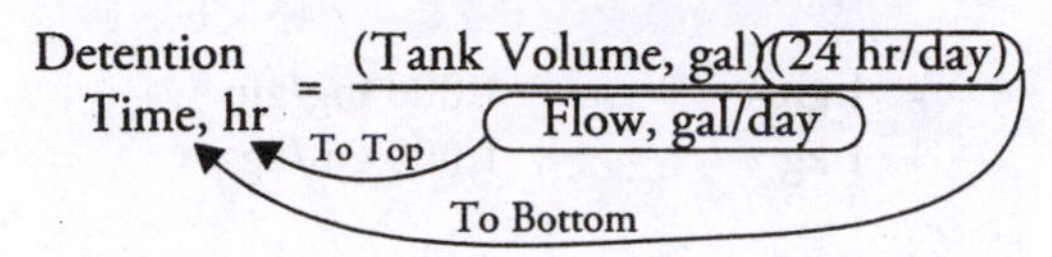


or

$$\frac{(\text{Detention Time, hr})(\text{Flow, gal/day})}{24\text{ hr/day}} = \text{Tank Volume, gal}$$

or

$$\text{Tank Volume, gal} = \frac{(\text{Detention Time, hr})(\text{Flow, gal/day})}{24\text{ hr/day}}$$

One more check you need to do is to be sure that the units in the rearranged formula cancel out correctly.

For additional information on the use of the basic formulas, refer to Sections A.91, "Selection of Formula"; A.92, "Arrangement of Formula"; A.93, "Unit Conversions"; and A.94, "Calculations."

## A.13 TYPICAL WASTEWATER SYSTEM PROBLEMS (ENGLISH SYSTEM)

### A.130 Septic Tanks

**EXAMPLES 1, 2, and 3**

In a small residential community served by individual septic tanks and subsurface wastewater infiltration systems (SWIS), the average wastewater flow per household is 150 gallons per day. The average length of a septic tank is 94 inches, the average width is 58 inches, and the clear zone depth is 11.5 inches. What is the clear zone volume in cubic feet and gallons, and what is the wastewater detention time in hours?

| Known | | Unknown |
|---|---|---|
| Influent Flow, gal/day | = 150 gal/day | 1. Clear Zone Volume, cu ft |
| Tank Length, in | = 94 in | 2. Clear Zone Volume, gal |
| Tank Width, in | = 58 in | 3. Detention Time, hr |
| Clear Zone Depth, in | = 11.5 in | |

1. Calculate the clear zone volume in cubic feet. (*NOTE:* There are 1,728 cubic inches per cubic foot.)

$$\text{Clear Zone Vol, cu ft} = \frac{(\text{Tank Length, in})(\text{Tank Width, in})(\text{Clear Zone Depth, in})}{1{,}728\text{ cu in/cu ft}}$$

$$= \frac{(94\text{ in})(58\text{ in})(11.5\text{ in})}{1{,}728\text{ cu in/cu ft}}$$

$$= \frac{62{,}698\text{ cu in}}{1{,}728\text{ cu in/cu ft}}$$

$$= 36.28\text{ cu ft}$$

2. Calculate the clear zone volume in gallons.

$$\text{Clear Zone Vol, gal} = (\text{Clear Zone Volume, cu ft})(7.48\text{ gal/cu ft})$$

$$= (36.28\text{ cu ft})(7.48\text{ gal/cu ft})$$

$$= 271.4\text{ gal, or approximately 271 gallons}$$

3. Calculate the detention time in hours.

$$\text{Detention Time, hr} = \left(\frac{\text{Clear Zone Volume, gal}}{\text{Influent Flow, gal/day}}\right)(24\text{ hr/day})$$

$$= \left(\frac{271\text{ gal}}{150\text{ gal/day}}\right)(24\text{ hr/day})$$

$$= 43.4\text{ hr}$$

### A.131 Sand Filters

**EXAMPLES 4 and 5**

A sand filter receives a flow of 1,000 gallons of wastewater per day. The filter is 10 feet long and 10 feet wide. What is the surface area of the filter in square feet, and what is the hydraulic loading on the filter in gallons per day per square foot?

| Known | | Unknown |
|---|---|---|
| Influent Flow, gal/day | = 1,000 gal/day | 1. Surface Area, sq ft |
| Tank Length, ft | = 10 ft | 2. Hydraulic Loading, GPD/sq ft |
| Tank Width, ft | = 10 ft | |

4. Calculate the surface area of the filter in square feet.

$$\text{Surface Area, sq ft} = (\text{Length, ft})(\text{Width, ft})$$

$$= (10\text{ ft})(10\text{ ft})$$

$$= 100\text{ sq ft}$$

5. Calculate the hydraulic loading on the filter in GPD/sq ft.

$$\text{Hydraulic Loading, GPD/sq ft} = \frac{\text{Flow, gal/day}}{\text{Surface Area, sq ft}}$$

$$= \frac{1{,}000\text{ gal/day}}{100\text{ sq ft}}$$

$$= 10\text{ GPD/sq ft}$$

### A.132 Utility Finances

**EXAMPLES 6 and 7**

**The total revenues for a utility are $1,237,000 and the operating expenses for the utility are $974,000. The debt service expenses are $480,000. What is the operating ratio? What is the coverage ratio?**

| Known | | Unknown |
|---|---|---|
| Total Revenue, $ | = $1,237,000 | 1. Operating Ratio |
| Operating Expenses, $ | = $974,000 | 2. Coverage Ratio |
| Debt Service Expenses, $ | = $480,000 | |

6. Calculate the utility's operating ratio.

$$\text{Operating Ratio} = \frac{\text{Total Revenue, \$}}{\text{Operating Expenses, \$}}$$

$$= \frac{\$1,237,000}{\$974,000}$$

$$= 1.27$$

7. Calculate the coverage ratio.

a. Calculate nondebt expenses for the utility.

$$\text{Nondebt Expenses, \$} = \text{Operating Expenses, \$} - \text{Debt Service Expenses, \$}$$

$$= \$974,000 - \$480,000$$

$$= \$494,000$$

b. Calculate the coverage ratio.

$$\text{Coverage Ratio} = \frac{\text{Total Revenue, \$} - \text{Nondebt Expenses, \$}}{\text{Debt Service Expenses, \$}}$$

$$= \frac{\$1,237,000 - \$494,000}{\$480,000}$$

$$= 1.55$$

## A.14 BASIC CONVERSION FACTORS (METRIC SYSTEM)

*LENGTH*

| | | |
|---|---|---|
| 100 cm | = 1 m | 100 cm/m |
| 3.281 ft | = 1 m | 3.281 ft/m |

*AREA*

| | | |
|---|---|---|
| 2.4711 ac | = 1 ha* | 2.4711 ac/ha |
| 10,000 sq m | = 1 ha | 10,000 sq m/ha |

*VOLUME*

| | | |
|---|---|---|
| 1,000 mL | = 1 liter | 1,000 mL/L |
| 1,000 L | = 1 cu m | 1,000 L/cu m |
| 3.785 L | = 1 gal | 3.785 L/gal |

*WEIGHT*

| | | |
|---|---|---|
| 1,000 mg | = 1 gm | 1,000 mg/gm |
| 1,000 gm | = 1 kg | 1,000 gm/kg |

*DENSITY*

| | | |
|---|---|---|
| 1 kg | = 1 liter | 1 kg/L |

*PRESSURE*

| | | |
|---|---|---|
| 10.015 m | = 1 kg/sq cm | 10.015 m/kg/sq cm |
| 1 Pascal | = 1 N/sq m | 1 Pa/N/sq m |
| 1 psi | = 6,895 Pa | 1 psi/6,895 Pa |

*FLOW*

| | | |
|---|---|---|
| 3,785 cu m/day | = 1 MGD | 3,785 cu m/day/MGD |
| 3.785 ML/day | = 1 MGD | 3.785 ML/day/MGD |

*hectare

## A.15 TYPICAL WASTEWATER SYSTEM PROBLEMS (METRIC SYSTEM)

### A.150 Septic Tanks

**EXAMPLES 1, 2, and 3**

**In a small residential community served by individual septic tanks and subsurface wastewater infiltration systems (SWIS), the average wastewater flow per household is 568 liters per day. The average length of a septic tank is 239 centimeters, the average width is 147 centimeters, and the clear zone depth is 29 centimeters. What is the clear zone volume in cubic meters and liters, and what is the wastewater detention time in hours?**

| Known | | Unknown |
|---|---|---|
| Influent Flow, L/day | = 568 L/day | 1. Clear Zone Volume, cu m |
| Tank Length, cm | = 239 cm | 2. Clear Zone Volume, L |
| Tank Width, cm | = 147 cm | 3. Detention Time, hr |
| Clear Zone Depth, cm | = 29 cm | |

1. Calculate the clear zone volume in cubic meters. (*NOTE:* There are 1,000,000 cubic centimeters per cubic meter.)

$$\text{Clear Zone, Vol, cu m} = \frac{(\text{Tank Length, cm})(\text{Tank Width, cm})(\text{Clear Zone Depth, cm})}{1{,}000{,}000 \text{ cu cm/cu m}}$$

$$= \frac{(239 \text{ cm})(147 \text{ cm})(29 \text{ cm})}{1{,}000{,}000 \text{ cu cm/cu m}}$$

$$= \frac{1{,}018{,}857 \text{ cu cm}}{1{,}000{,}000 \text{ cu cm/cu m}}$$

$$= 1.02 \text{ cu m}$$

2. Calculate the clear zone volume in liters.

$$\text{Clear Zone Vol, L} = (\text{Clear Zone Volume, cu m})(1{,}000 \text{ L/cu m})$$

$$= (1.02 \text{ cu m})(1{,}000 \text{ L/cu m})$$

$$= 1{,}020 \text{ liters}$$

3. Calculate the detention time in hours.

$$\text{Detention Time, hr} = \left(\frac{\text{Clear Zone Volume, L}}{\text{Influent Flow, L/day}}\right)(24 \text{ hr/day})$$

$$= \left(\frac{1{,}020 \text{ L}}{568 \text{ L/day}}\right)(24 \text{ hr/day})$$

$$= 43 \text{ hr}$$

### A.151 Sand Filters

**EXAMPLES 4 and 5**

A sand filter receives a flow of 2,500 liters of wastewater per day. The filter is 305 centimeters long and 305 centimeters wide. What is the surface area of the filter in square centimeters, and what is the hydraulic loading on the filter in centimeters per day?

| Known | | Unknown |
|---|---|---|
| Influent Flow, L/day | = 2,500 L/day | 1. Surface Area, sq cm |
| Tank Length, cm | = 305 cm | 2. Hydraulic Loading, cm/day |
| Tank Width, cm | = 305 cm | |

4. Calculate the surface area of the filter in square centimeters.

$$\text{Surface Area, sq cm} = (\text{Length, cm})(\text{Width, cm})$$

$$= (305 \text{ cm})(305 \text{ cm})$$

$$= 93{,}025 \text{ sq cm}$$

5. Calculate the hydraulic loading on the filter in centimeters per day.

$$\text{Hydraulic Loading, cm/day} = \frac{(\text{Flow, L/day})(1{,}000{,}000 \text{ cu cm/cu m})}{(\text{Surface Area, sq cm})(1{,}000 \text{ L/cu m})}$$

$$= \frac{(2{,}500 \text{ L/day})(1{,}000{,}000 \text{ cu cm/cu m})}{(93{,}025 \text{ sq cm})(1{,}000 \text{ L/cu m})}$$

$$= 26.9 \text{ cm/day}$$

## A.16 ADDITIONAL READING

1. *BASIC MATH CONCEPTS FOR WATER AND WASTEWATER PLANT OPERATORS.* Obtain from CRC Press, PO Box 409267, Atlanta, GA 30384-9267. Order No. TX8084. ISBN 978-0-87762-808-8. Price, $71.95.

# ABBREVIATIONS

| | |
|---|---|
| °C | degrees Celsius |
| °F | degrees Fahrenheit |
| μ | micron |
| μg | microgram |
| μm | micrometer |
| ac | acres |
| ac-ft | acre-feet |
| amp | amperes |
| atm | atmosphere |
| CFM | cubic feet per minute |
| CFS | cubic feet per second |
| Ci | curie |
| cm | centimeters |
| cu ft | cubic feet |
| cu in | cubic inches |
| cu m | cubic meters |
| cu yd | cubic yards |
| D | dalton |
| dB | decibel |
| ft | feet or foot |
| ft-lb/min | foot-pounds per minute |
| g | gravity |
| gal | gallons |
| gal/day | gallons per day |
| GFD | gallons of flux per square foot per day |
| gm | grams |
| GPCD | gallons per capita per day |
| GPD | gallons per day |
| gpg | grains per gallon |
| GPM | gallons per minute |
| GPY | gallons per year |
| gr | grains |
| ha | hectares |
| HP | horsepower |
| hr | hours |
| Hz | hertz |
| in | inches |
| J | joules |
| k | kilos |
| kg | kilograms |
| km | kilometers |
| kN | kilonewtons |
| kPa | kilopascals |
| kW | kilowatts |
| kWh | kilowatt-hours |
| L | liters |
| lb | pounds |
| lbs/sq in | pounds per square inch |
| M | mega |
| M | million |
| *M* | molar (or molarity) |
| m | meters |
| mA | milliampere |
| meq | milliequivalent |
| mg | milligrams |
| MGD | million gallons per day |
| mg/L | milligrams per liter |
| min | minutes |
| mL | milliliters |
| mm | millimeters |
| N | newton |
| *N* | normal (or normality) |
| nm | nanometer |
| ohm | ohm |
| Pa | pascal |
| pCi | picocurie |
| pCi/L | picocuries per liter |
| ppb | parts per billion |
| ppm | parts per million |
| psf | pounds per square foot |
| psi | pounds per square inch |
| psig | pounds per square inch gauge |
| RPM | revolutions per minute |
| SCFM | standard cubic feet per minute |
| sec | seconds |
| *SI* | *Le Système International d'Unités* |
| sq ft | square feet |
| sq in | square inches |
| W | watt |
| yd | yards |

# WASTEWATER WORDS

## A Summary of the Words Defined

## in

## SMALL WASTEWATER SYSTEM OPERATION AND MAINTENANCE

## PROJECT PRONUNCIATION KEY

by Warren L. Prentice

The Project Pronunciation Key is designed to aid you in the pronunciation of new words. While this key is based primarily on familiar sounds, it does not attempt to follow any particular pronunciation guide. This key is designed solely to aid operators in this program.

You may find it helpful to refer to other available sources for pronunciation help. Each current standard dictionary contains a guide to its own pronunciation key. Each key will be different from each other and from this key. Examples of the difference between the key used in this program and the *WEBSTER'S NEW WORLD COLLEGE DICTIONARY*[1] key are shown below.

In using this key, you should accent (say louder) the syllable that appears in capital letters. The following chart is presented to give examples of how to pronounce words using the project key.

| | SYLLABLE | | | | |
|---|---|---|---|---|---|
| WORD | 1st | 2nd | 3rd | 4th | 5th |
| aerobic | air | O | bick | | |
| bacteria | back | TEER | e | uh | |
| contamination | kun | TAM | uh | NAY | shun |

The first word, *AEROBIC,* has its second syllable accented. The second word, *BACTERIA,* has its second syllable accented. The third word, *CONTAMINATION,* has its second and fourth syllables accented.

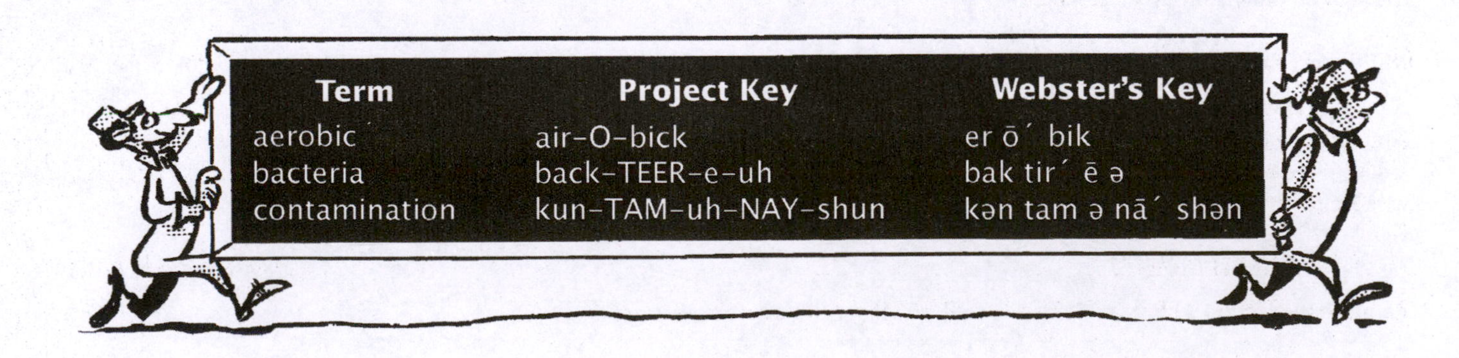

| Term | Project Key | Webster's Key |
|---|---|---|
| aerobic | air-O-bick | er ō´ bik |
| bacteria | back-TEER-e-uh | bak tir´ ē ə |
| contamination | kun-TAM-uh-NAY-shun | kən tam ə nā´ shən |

[1] The *WEBSTER'S NEW WORLD COLLEGE DICTIONARY,* Fourth Edition, 1999, was chosen rather than an unabridged dictionary because of its availability to the operator. Other editions may be slightly different.

# WORDS

**>GREATER THAN** **>GREATER THAN**

**DO >5 mg/L would be read as DO GREATER THAN 5 mg/L.**

**<LESS THAN** **<LESS THAN**

**DO <5 mg/L would be read as DO LESS THAN 5 mg/L.**

## A

**ACEOPS** **ACEOPS**

**See ALLIANCE OF CERTIFIED OPERATORS, LABORATORY ANALYSTS, INSPECTORS, AND SPECIALISTS (ACEOPS).**

**ABSORPTION (ab-SORP-shun)** **ABSORPTION**

**The taking in or soaking up of one substance into the body of another by molecular or chemical action (as tree roots absorb dissolved nutrients in the soil).**

**ACCOUNTABILITY** **ACCOUNTABILITY**

**When a manager gives power/responsibility to an employee, the employee ensures that the manager is informed of results or events.**

**ACTIVATED SLUDGE** **ACTIVATED SLUDGE**

**Sludge particles produced in raw or settled wastewater (primary effluent) by the growth of organisms (including zoogleal bacteria) in aeration tanks in the presence of dissolved oxygen. The term activated comes from the fact that the particles are teeming with bacteria, fungi, and protozoa. Activated sludge is different from primary sludge in that the sludge particles contain many living organisms that can feed on the incoming wastewater.**

**ACTIVATED SLUDGE PROCESS** **ACTIVATED SLUDGE PROCESS**

**A biological wastewater treatment process that speeds up the decomposition of wastes in the wastewater being treated. Activated sludge is added to wastewater and the mixture (mixed liquor) is aerated and agitated. After some time in the aeration tank, the activated sludge is allowed to settle out by sedimentation and is disposed of (wasted) or reused (returned to the aeration tank) as needed. The remaining wastewater then undergoes more treatment.**

**ACUTE HEALTH EFFECT** **ACUTE HEALTH EFFECT**

**An adverse effect on a human or animal body, with symptoms developing rapidly.**

**ADSORPTION (add-SORP-shun)** **ADSORPTION**

**The gathering of a gas, liquid, or dissolved substance on the surface or interface zone of another material.**

**ADVANCED WASTE TREATMENT** **ADVANCED WASTE TREATMENT**

**Any process of water renovation that upgrades treated wastewater to meet specific reuse requirements. May include general cleanup of water or removal of specific parts of wastes insufficiently removed by conventional treatment processes. Typical processes include chemical treatment and pressure filtration. Also called tertiary treatment.**

**AERATION (air-A-shun) LIQUOR** **AERATION LIQUOR**

**Mixed liquor. The contents of the aeration tank, including living organisms and material carried into the tank by either untreated wastewater or primary effluent.**

AERATION (air-A-shun) TANK — AERATION TANK

The tank where raw or settled wastewater is mixed with return sludge and aerated. The same as aeration bay, aerator, or reactor.

AEROBES — AEROBES

Bacteria that must have dissolved oxygen (DO) to survive. Aerobes are aerobic bacteria.

AEROBIC (air-O-bick) — AEROBIC

A condition in which atmospheric or dissolved oxygen is present in the aquatic (water) environment.

AEROBIC BACTERIA (air-O-bick back-TEER-e-uh) — AEROBIC BACTERIA

Bacteria that will live and reproduce only in an environment containing oxygen that is available for their respiration (breathing), namely atmospheric oxygen or oxygen dissolved in water. Oxygen combined chemically, such as in water molecules ($H_2O$), cannot be used for respiration by aerobic bacteria.

AEROBIC (air-O-bick) DIGESTION — AEROBIC DIGESTION

The breakdown of wastes by microorganisms in the presence of dissolved oxygen. This digestion process may be used to treat only waste activated sludge, or trickling filter sludge and primary (raw) sludge, or waste sludge from activated sludge treatment plants designed without primary settling. The sludge to be treated is placed in a large aerated tank where aerobic microorganisms decompose the organic matter in the sludge. This is an extension of the activated sludge process.

AESTHETIC (es-THET-ick) — AESTHETIC

Attractive or appealing.

AGGLOMERATION (uh-glom-er-A-shun) — AGGLOMERATION

The growing or coming together of small scattered particles into larger flocs or particles, which settle rapidly. Also see FLOC.

AIR BINDING — AIR BINDING

The clogging of a filter, pipe, or pump due to the presence of air released from water. Air entering the filter media is harmful to both the filtration and backwash processes. Air can prevent the passage of water during the filtration process and can cause the loss of filter media during the backwash process.

AIR GAP — AIR GAP

An open, vertical drop, or vertical empty space, between a drinking (potable) water supply and potentially contaminated water. This gap prevents the contamination of drinking water by backsiphonage because there is no way potentially contaminated water can reach the drinking water supply.

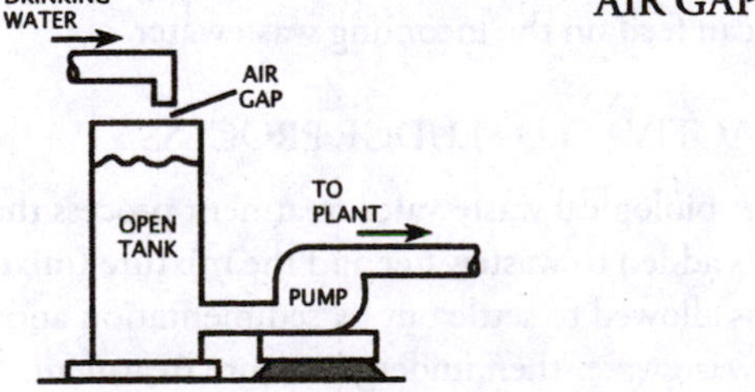


AIR LIFT PUMP — AIR LIFT PUMP

A special type of pump consisting of a vertical riser pipe submerged in the wastewater or sludge to be pumped. Compressed air is injected into a tail piece at the bottom of the pipe. Fine air bubbles mix with the wastewater or sludge to form a mixture lighter than the surrounding water, which causes the mixture to rise in the discharge pipe to the outlet.

AIR RELEASE — AIR RELEASE

A type of valve used to allow air caught in high spots in pipes to escape.

ALGAE (AL-jee) — ALGAE

Microscopic plants containing chlorophyll that live floating or suspended in water. They also may be attached to structures, rocks, or other submerged surfaces. Excess algal growths can impart tastes and odors to potable water. Algae produce oxygen during sunlight hours and use oxygen during the night hours. Their biological activities appreciably affect the pH, alkalinity, and dissolved oxygen of the water.

ALGAL (AL-gull) BLOOM — ALGAL BLOOM

Sudden, massive growths of microscopic and macroscopic plant life, such as green or blue-green algae, which can, under the proper conditions, develop in lakes, reservoirs, and ponds.

ALIQUOT (AL-uh-kwot) — ALIQUOT

Representative portion of a sample. Often, an equally divided portion of a sample.

ALLIANCE OF CERTIFIED OPERATORS, LABORATORY ANALYSTS, INSPECTORS, AND SPECIALISTS (ACEOPS) — ALLIANCE OF CERTIFIED OPERATORS, LABORATORY ANALYSTS, INSPECTORS, AND SPECIALISTS (ACEOPS)

A professional organization for operators, laboratory analysts, inspectors, and specialists dedicated to improving professionalism; expanding training, certification, and job opportunities; increasing information exchange; and advocating the importance of certified operators, lab analysts, inspectors, and specialists. For information on membership, contact ACEOPS, 3130 Pierce Street, Suite 100, Sioux City, IA 51104-3942, phone (712) 258-3464, or email: Info@aceops.org.

AMBIENT (AM-bee-ent) TEMPERATURE — AMBIENT TEMPERATURE

Temperature of the surroundings.

AMPERAGE (AM-purr-age) — AMPERAGE

The strength of an electric current measured in amperes. The amount of electric current flow, similar to the flow of water in gallons per minute.

AMPERE (AM-peer) — AMPERE

The unit used to measure current strength. The current produced by an electromotive force of one volt acting through a resistance of one ohm.

AMPEROMETRIC (am-purr-o-MET-rick) — AMPEROMETRIC

A method of measurement that records electric current flowing or generated, rather than recording voltage. Amperometric titration is a means of measuring concentrations of certain substances in water.

AMPLITUDE — AMPLITUDE

The maximum strength of an alternating current during its cycle, as distinguished from the mean or effective strength.

ANAEROBES — ANAEROBES

Bacteria that do not need dissolved oxygen (DO) to survive.

ANAEROBIC (AN-air-O-bick) — ANAEROBIC

A condition in which atmospheric or dissolved oxygen (DO) is *NOT* present in the aquatic (water) environment.

ANAEROBIC BACTERIA (AN-air-O-bick back-TEER-e-uh) — ANAEROBIC BACTERIA

Bacteria that live and reproduce in an environment containing no free or dissolved oxygen. Anaerobic bacteria obtain their oxygen supply by breaking down chemical compounds that contain oxygen, such as sulfate ($SO_4^{2-}$).

ANAEROBIC (AN-air-O-bick) DECOMPOSITION — ANAEROBIC DECOMPOSITION

The decay or breaking down of organic material in an environment containing no free or dissolved oxygen.

ANAEROBIC (AN-air-O-bick) DIGESTER — ANAEROBIC DIGESTER

A wastewater solids treatment device in which the solids and water (about 5 percent solids, 95 percent water) are placed in a large tank where bacteria decompose the solids in the absence of dissolved oxygen.

ANODE (AN-ode) — ANODE

The positive pole or electrode of an electrolytic system, such as a battery. The anode attracts negatively charged particles or ions (anions).

ANOXIC (an-OX-ick) ANOXIC

A condition in which the aquatic (water) environment does not contain dissolved oxygen (DO), which is called an oxygen deficient condition. Generally refers to an environment in which chemically bound oxygen, such as in nitrate, is present. The term is similar to ANAEROBIC.

APPURTENANCE (uh-PURR-ten-nans) APPURTENANCE

Machinery, appliances, structures, and other parts of the main structure necessary to allow it to operate as intended, but not considered part of the main structure.

AQUIFER (ACK-wi-fer) AQUIFER

A natural, underground layer of porous, water-bearing materials (sand, gravel) usually capable of yielding a large amount or supply of water.

ARTIFICIAL GROUNDWATER TABLE ARTIFICIAL GROUNDWATER TABLE

A groundwater table that is changed by artificial means. Examples of activities that artificially raise the level of a groundwater table include agricultural irrigation, dams, and excessive sewer line exfiltration. A groundwater table can be artificially lowered by sewer line infiltration, water wells, and similar drainage methods.

ASEPTIC (a-SEP-tick) ASEPTIC

Free from the living germs of disease, fermentation, or putrefaction. Sterile.

ASHING ASHING

Formation of an activated sludge floc in a clarifier effluent that is well oxidized and floats on the water surface (has the appearance of gray ash).

ASSET MANAGEMENT ASSET MANAGEMENT

The process of maintaining the functionality and value of a utility's assets through repair, rehabilitation, and replacement. Examples of utility assets include buildings, tools, equipment, pipes, and machinery used to operate a water or wastewater system. The primary goal of asset management is to provide safe, reliable, and cost-effective service to a community over the useful life of a utility's assets.

AUGER (AW-grr) AUGER

A sharp tool used to go through and break up or remove various materials that become lodged in sewers.

AUTHORITY AUTHORITY

The power and resources to do a specific job or to get that job done.

AVAILABLE EXPANSION AVAILABLE EXPANSION

The vertical distance from the sand surface to the underside of a trough in a sand filter. This distance is also called freeboard.

AVAILABLE MOISTURE CONTENT AVAILABLE MOISTURE CONTENT

The quantity of water present that is available for use by vegetation.

## B

BOD (pronounce as separate letters) BOD

Biochemical Oxygen Demand. The rate at which organisms use the oxygen in water or wastewater while stabilizing decomposable organic matter under aerobic conditions. In decomposition, organic matter serves as food for the bacteria and energy results from its oxidation. BOD measurements are used as a surrogate measure of the organic strength of wastes in water.

$BOD_5$ — $BOD_5$

$BOD_5$ refers to the five-day biochemical oxygen demand. The total amount of oxygen used by microorganisms decomposing organic matter increases each day until the ultimate BOD is reached, usually in 50 to 70 days. BOD usually refers to the five-day BOD or $BOD_5$.

BACKFLOW — BACKFLOW

A reverse flow condition, created by a difference in water pressures, that causes water to flow back into the distribution pipes of a potable water supply from any source or sources other than an intended source. Also see BACKSIPHONAGE.

BACKUP — BACKUP

An overflow or accumulation of water caused by clogging or by a stoppage.

BACTERIA (back-TEER-e-uh) — BACTERIA

Bacteria are living organisms, microscopic in size, that usually consist of a single cell. Most bacteria use organic matter for their food and produce waste products as a result of their life processes.

BACTERIAL (back-TEER-e-ul) CULTURE — BACTERIAL CULTURE

In the case of activated sludge, the bacterial culture refers to the group of bacteria classified as AEROBES and FACULTATIVE BACTERIA, which covers a wide range of organisms. Most treatment processes in the United States grow facultative bacteria that use the carbonaceous (carbon compounds) BOD. Facultative bacteria can live when oxygen resources are low. When nitrification is required, the nitrifying organisms are OBLIGATE AEROBES (require oxygen) and must have at least 0.5 mg/L of dissolved oxygen throughout the whole system to function properly.

BALANCED SCORECARD — BALANCED SCORECARD

A strategic planning and management system used extensively by water and wastewater facility managers to align business activities to the vision and strategy of the organization, improve communications, and monitor organizational performance against strategic goals. More information is available at www.balancedscorecard.org.

BASE-EXTRA CAPACITY METHOD — BASE-EXTRA CAPACITY METHOD

A cost allocation method used by utilities to determine rates for various groups. This method considers base costs (O&M expenses and capital costs), extra capacity costs (additional costs for maximum day and maximum hour demands), customer costs (meter maintenance and reading, billing, collection, accounting), and fire protection costs.

BATCH PROCESS — BATCH PROCESS

A treatment process in which a tank or reactor is filled, the wastewater (or other solution) is treated or a chemical solution is prepared, and the tank is emptied. The tank may then be filled and the process repeated. Batch processes are also used to cleanse, stabilize, or condition chemical solutions for use in industrial manufacturing and treatment processes.

BENCHMARK — BENCHMARK

A standard or point of reference used to judge or measure quality or value.

BENCHMARKING — BENCHMARKING

A process an agency uses to gather and compare information about the productivity and performance of other similar agencies with its own information. The purpose of benchmarking is to identify best practices, set improvement targets, and measure progress.

BIOASSAY (BUY-o-AS-say) — BIOASSAY

(1) A way of showing or measuring the effect of biological treatment on a particular substance or waste.

(2) A method of determining the relative toxicity of a test sample of industrial wastes or other wastes by using live test organisms, such as fish.

BIOAUGMENTATION | BIOAUGMENTATION

The addition of bacterial cultures to speed up the breakdown of grease and other organic materials. Bioaugmentation is used to clean sewers and, on a preventive basis, to remove deposits in sewers. This method is also used to prevent grease buildup in lift station wet wells.

BIOCHEMICAL OXYGEN DEMAND (BOD) | BIOCHEMICAL OXYGEN DEMAND (BOD)

See BOD.

BIODEGRADABLE (BUY-o-dee-GRADE-able) | BIODEGRADABLE

Organic matter that can be broken down by bacteria to more stable forms that will not create a nuisance or give off foul odors is considered biodegradable.

BIOFLOCCULATION (BUY-o-flock-yoo-LAY-shun) | BIOFLOCCULATION

The clumping together of fine, dispersed organic particles by the action of certain bacteria and algae. This results in faster and more complete settling of the organic solids in wastewater.

BIOMASS (BUY-o-mass) | BIOMASS

A mass or clump of organic material consisting of living organisms feeding on the wastes in wastewater, dead organisms, and other debris. Also see ZOOGLEAL MASS.

BIOMONITORING | BIOMONITORING

A term used to describe methods of evaluating or measuring the effects of toxic substances in effluents on aquatic organisms in receiving waters. There are two types of biomonitoring, the BIOASSAY and the BIOSURVEY.

BIOSURVEY | BIOSURVEY

A survey of the types and numbers of organisms naturally present in the receiving waters upstream and downstream from plant effluents. Comparisons are made between the aquatic organisms upstream and those organisms downstream of the discharge.

BLANK | BLANK

A bottle containing only dilution water or distilled water; the sample being tested is not added. Tests are frequently run on a sample and a blank and the differences are compared. The procedure helps to eliminate or reduce test result errors that could be caused when the dilution water or distilled water used is contaminated.

BLOCKAGE | BLOCKAGE

Any incident in which a sewer is partially or completely blocked, causing a backup, a service interruption, or an overflow. Also called stoppage.

BOND | BOND

(1) A written promise to pay a specified sum of money (called the face value) at a fixed time in the future (called the date of maturity). A bond also carries interest at a fixed rate, payable periodically. The difference between a note and a bond is that a bond usually runs for a longer period of time and requires greater formality. Utility agencies use bonds as a means of obtaining large amounts of money for capital improvements.

(2) A warranty by an underwriting organization, such as an insurance company, guaranteeing honesty, performance, or payment by a contractor.

BREAKOUT OF CHLORINE (CHLORINE BREAKAWAY) | BREAKOUT OF CHLORINE (CHLORINE BREAKAWAY)

A point at which chlorine leaves solution as a gas because the chlorine feed rate is too high. The solution is saturated and cannot dissolve any more chlorine. The maximum strength a chlorine solution can attain is approximately 3,500 mg/L. Beyond this concentration molecular chlorine will break out of solution and cause off-gassing at the point of application.

BREAKPOINT CHLORINATION | BREAKPOINT CHLORINATION

Addition of chlorine to water or wastewater until the chlorine demand has been satisfied. At this point, further additions of chlorine will result in a free chlorine residual that is directly proportional to the amount of chlorine added beyond the breakpoint.

BUFFER — BUFFER

A solution or liquid whose chemical makeup neutralizes acids or bases without a great change in pH.

BUFFER CAPACITY — BUFFER CAPACITY

A measure of the capacity of a solution or liquid to neutralize acids or bases. This is a measure of the capacity of water or wastewater for offering a resistance to changes in pH.

BULKING — BULKING

Clouds of billowing sludge that occur throughout secondary clarifiers and sludge thickeners when the sludge does not settle properly. In the activated sludge process, bulking is usually caused by filamentous bacteria or bound water.

BURPING — BURPING

A term used to describe what happens when billowing solids are swept by the water up and out over the effluent weirs in the discharged effluent. Billowing solids result when the settling tank sludge blanket becomes too deep (occupies too much volume in the bottom of the tank).

## C

CERCLA (SIRK-la) — CERCLA

Comprehensive Environmental Response, Compensation, and Liability Act of 1980. This act was passed primarily to correct past mistakes in industrial waste management. The focus of the act is to locate hazardous waste disposal sites that are creating problems through pollution of the environment and, by proper funding and implementation of study and corrective activities, eliminate the problem from these sites. Current users of CERCLA-identified substances must report releases of these substances to the environment when they take place (not just historic ones). This act is also called the Superfund Act. Also see SARA.

CHEMTREC (KEM-trek) — CHEMTREC

Chemical Transportation Emergency Center. A public service of the American Chemistry Council dedicated to assisting emergency responders deal with incidents involving hazardous materials. Their toll-free 24-hour emergency phone number is (800) 424-9300.

COD (pronounce as separate letters) — COD

Chemical Oxygen Demand. A measure of the oxygen-consuming capacity of organic matter present in wastewater. COD is expressed as the amount of oxygen consumed from a chemical oxidant in mg/L during a specific test. Results are not necessarily related to the biochemical oxygen demand (BOD) because the chemical oxidant may react with substances that bacteria do not stabilize.

CALL DATE — CALL DATE

First date a bond can be paid off.

CAPILLARY (KAP-uh-larry) ACTION — CAPILLARY ACTION

The movement of water through very small spaces due to molecular forces.

CAPITAL IMPROVEMENT PLAN (CIP) — CAPITAL IMPROVEMENT PLAN (CIP)

A detailed plan that identifies requirements for the repair, replacement, and rehabilitation of facility infrastructure over an extended period, often 20 years or more. A utility usually updates or prepares this plan annually. The plan consists of programs and projects to upgrade and rehabilitate wastewater collection and treatment systems and increase their capacity to allow for future growth.

CATHODIC (kath-ODD-ick) PROTECTION — CATHODIC PROTECTION

An electrical system for prevention of rust, corrosion, and pitting of metal surfaces that are in contact with water, wastewater, or soil. A low-voltage current is made to flow through a liquid (water) or a soil in contact with the metal in such a manner that the external electromotive force renders the metal structure cathodic. This concentrates corrosion on auxiliary anodic parts, which are deliberately allowed to corrode instead of letting the structure corrode.

CATION (KAT-EYE-en) EXCHANGE CAPACITY — CATION EXCHANGE CAPACITY

The ability of a soil or other solid to exchange cations (positive ions such as calcium, $Ca^{2+}$) with a liquid.

CAVITATION (kav-uh-TAY-shun) CAVITATION

The formation and collapse of a gas pocket or bubble on the blade of an impeller or the gate of a valve. The collapse of this gas pocket or bubble drives water into the impeller or gate with a terrific force that can knock metal particles off and cause pitting on the impeller or gate surface. Cavitation is accompanied by loud noises that sound like someone is pounding on the impeller or gate with a hammer.

CENTRIFUGE CENTRIFUGE

A mechanical device that uses centrifugal or rotational forces to separate solids from liquids.

CERTIFICATION EXAMINATION CERTIFICATION EXAMINATION

An examination administered by a state agency or professional association that operators take to indicate a level of professional competence. In the United States, certification of operators of water treatment plants, wastewater treatment plants, water distribution systems, and small water supply systems is mandatory. In many states, certification of wastewater collection system operators, industrial wastewater treatment plant operators, pretreatment facility inspectors, and small wastewater system operators is voluntary; however, current trends indicate that more states, provinces, and employers will require these operators to be certified in the future. Operator certification is mandatory in the United States for the Chief Operators of water treatment plants, water distribution systems, and wastewater treatment plants.

CERTIFIED OPERATOR CERTIFIED OPERATOR

A person who has the education and experience required to operate a specific class of treatment facility as indicated by possessing a certificate of professional competence given by a state agency or professional association.

CHAIN OF CUSTODY CHAIN OF CUSTODY

A record of each person involved in the handling and possession of a sample from the person who collected the sample to the person who analyzed the sample in the laboratory and to the person who witnessed disposal of the sample.

CHEMICAL OXYGEN DEMAND (COD) CHEMICAL OXYGEN DEMAND (COD)

A measure of the oxygen-consuming capacity of organic matter present in wastewater. COD is expressed as the amount of oxygen consumed from a chemical oxidant in mg/L during a specific test. Results are not necessarily related to the biochemical oxygen demand (BOD) because the chemical oxidant may react with substances that bacteria do not stabilize.

CHLORAMINES (KLOR-uh-means) CHLORAMINES

Compounds formed by the reaction of hypochlorous acid (or aqueous chlorine) with ammonia.

CHLORINATION (klor-uh-NAY-shun) CHLORINATION

The application of chlorine to water or wastewater, generally for the purpose of disinfection, but frequently for accomplishing other biological or chemical results.

CHLORINE CONTACT CHAMBER CHLORINE CONTACT CHAMBER

A baffled basin that provides sufficient detention time of chlorine contact with wastewater for disinfection to occur. The minimum contact time is usually 30 minutes. Also commonly referred to as basin or tank.

CHLORINE BREAKAWAY CHLORINE BREAKAWAY

See BREAKOUT OF CHLORINE.

CHLORINE DEMAND CHLORINE DEMAND

Chlorine demand is the difference between the amount of chlorine added to water or wastewater and the amount of chlorine residual remaining after a given contact time. Chlorine demand may change with dosage, time, temperature, pH, and nature and amount of the impurities in the water.

Chlorine Demand, mg/L = Chlorine Applied, mg/L – Chlorine Residual, mg/L

CHLORINE REQUIREMENT CHLORINE REQUIREMENT

The amount of chlorine that is needed for a particular purpose. Some reasons for adding chlorine are reducing the MPN (Most Probable Number) of coliform bacteria, obtaining a particular chlorine residual, or oxidizing some substance in the water. In each case, a definite dosage of chlorine will be necessary. This dosage is the chlorine requirement.

CHLORINE RESIDUAL — CHLORINE RESIDUAL

The concentration of chlorine present in water after the chlorine demand has been satisfied. The concentration is expressed in terms of the total chlorine residual, which includes both the free and combined or chemically bound chlorine residuals. Also called residual chlorine.

CHLORINE RESIDUAL ANALYZER — CHLORINE RESIDUAL ANALYZER

An instrument used to measure chlorine residual in water or wastewater. This instrument also can be used to control the chlorine dose rate.

CHLORORGANIC (klor-or-GAN-ick) — CHLORORGANIC

Organic compounds combined with chlorine. These compounds generally originate from, or are associated with, living or dead organic materials, including algae in water.

CHRONIC HEALTH EFFECT — CHRONIC HEALTH EFFECT

An adverse effect on a human or animal body with symptoms that develop slowly over a long period of time or that recur frequently.

CLEAR ZONE — CLEAR ZONE

See SUPERNATANT.

COAGULATION (ko-agg-yoo-LAY-shun) — COAGULATION

The clumping together of very fine particles into larger particles (floc) caused by the use of chemicals (coagulants). The chemicals neutralize the electrical charges of the fine particles, allowing them to come closer and form larger clumps. This clumping together makes it easier to separate the solids from the water by settling, skimming, draining, or filtering.

CODE OF FEDERAL REGULATIONS (CFR) — CODE OF FEDERAL REGULATIONS (CFR)

A publication of the US government that contains all of the proposed and finalized federal regulations, including safety and environmental regulations.

COLIFORM (KOAL-i-form) — COLIFORM

A group of bacteria found in the intestines of warm-blooded animals (including humans) and also in plants, soil, air, and water. The presence of coliform bacteria is an indication that the water is polluted and may contain pathogenic (disease-causing) organisms. Fecal coliforms are those coliforms found in the feces of various warm-blooded animals, whereas the term coliform also includes other environmental sources.

COLORIMETRIC MEASUREMENT — COLORIMETRIC MEASUREMENT

A means of measuring unknown chemical concentrations in water by measuring a sample's color intensity. The specific color of the sample, developed by addition of chemical reagents, is measured with a photoelectric colorimeter or is compared with color standards using, or corresponding with, known concentrations of the chemical.

COMBINED AVAILABLE CHLORINE — COMBINED AVAILABLE CHLORINE

The total chlorine, present as chloramine or other derivatives, that is present in a water and is still available for disinfection and for oxidation of organic matter. The combined chlorine compounds are more stable than free chlorine forms, but they are somewhat slower in disinfection action.

COMBINED AVAILABLE CHLORINE RESIDUAL — COMBINED AVAILABLE CHLORINE RESIDUAL

The concentration of chlorine residual that is combined with ammonia, organic nitrogen, or both in water as a chloramine (or other chloro derivative) and yet is still available to oxidize organic matter and help kill bacteria.

COMBINED CHLORINE — COMBINED CHLORINE

The sum of the chlorine species composed of free chlorine and ammonia, including monochloramine, dichloramine, and trichloramine (nitrogen trichloride). Dichloramine is the strongest disinfectant of these chlorine species, but it has less oxidative capacity than free chlorine.

COMMINUTION (kom-mih-NEW-shun) COMMINUTION

A mechanical treatment process that cuts large pieces of wastes into smaller pieces so they will not plug pipes or damage equipment. Comminution and shredding usually mean the same thing.

COMMINUTOR (kom-mih-NEW-ter) COMMINUTOR

A device used to reduce the size of the solid materials in wastewater by shredding (comminution). The shredding action is like many scissors cutting to shreds all the large solids in the wastewater.

COMMODITY-DEMAND METHOD COMMODITY-DEMAND METHOD

A cost allocation method used by water utilities to determine water rates for the various water user groups. This method considers the commodity costs (water, chemicals, power, amount of water use), demand costs (treatment, storage, distribution), customer costs (meter maintenance and reading, billing, collection, accounting), and fire protection costs.

COMMUNITY RIGHT-TO-KNOW COMMUNITY RIGHT-TO-KNOW

The Superfund Amendments and Reauthorization Act (SARA) of 1986 provides statutory authority for communities to develop right-to-know laws. The act establishes a state and local emergency planning structure, emergency notification procedures, and reporting requirements for facilities. Also see RIGHT-TO-KNOW LAWS and SARA.

COMPOSITE (PROPORTIONAL) SAMPLE COMPOSITE (PROPORTIONAL) SAMPLE

A composite sample is a collection of individual samples obtained at regular intervals, usually every one or two hours during a 24-hour time span. Each individual sample is combined with the others in proportion to the rate of flow when the sample was collected. Equal volume individual samples also may be collected at intervals after a specific volume of flow passes the sampling point or after equal time intervals and still be referred to as a composite sample. The resulting mixture (composite sample) forms a representative sample and is analyzed to determine the average conditions during the sampling period.

COMPOUND COMPOUND

A pure substance composed of two or more elements whose composition is constant. For example, table salt (sodium chloride, NaCl) is a compound.

CONFINED SPACE CONFINED SPACE

Confined space means a space that:

(1) Is large enough and so configured that an employee can bodily enter and perform assigned work; and

(2) Has limited or restricted means for entry or exit (for example, manholes, tanks, vessels, silos, storage bins, hoppers, vaults, and pits are spaces that may have limited means of entry); and

(3) Is not designed for continuous employee occupancy.

Also see DANGEROUS AIR CONTAMINATION and OXYGEN DEFICIENCY.

CONFINED SPACE, PERMIT-REQUIRED (PERMIT SPACE) CONFINED SPACE, PERMIT-REQUIRED (PERMIT SPACE)

A confined space that has one or more of the following characteristics:

(1) Contains or has a potential to contain a hazardous atmosphere,

(2) Contains a material that has the potential for engulfing an entrant,

(3) Has an internal configuration such that an entrant could be trapped or asphyxiated by inwardly converging walls or by a floor that slopes downward and tapers to a smaller cross section, or

(4) Contains any other recognized serious safety or health hazard.

CONING CONING

Development of a cone-shaped flow of liquid, like a whirlpool, through sludge. This can occur in a sludge hopper during sludge withdrawal when the sludge becomes too thick. Part of the sludge remains in place while liquid rather than sludge flows out of the hopper. Also called coring.

**CONTACT STABILIZATION** — CONTACT STABILIZATION

Contact stabilization is a modification of the conventional activated sludge process. In contact stabilization, two aeration tanks are used. One tank is for separate reaeration of the return sludge for at least four hours before it is permitted to flow into the other aeration tank to be mixed with the primary effluent requiring treatment. The process may also occur in one long tank.

**CORPORATION STOP** — CORPORATION STOP

A water service shutoff valve located at a street water main. This valve cannot be operated from the ground surface because it is buried and there is no valve box. Also called a corporation cock.

**COVERAGE RATIO** — COVERAGE RATIO

The coverage ratio is a measure of the ability of the utility to pay the principal and interest on loans and bonds (this is known as debt service) in addition to any unexpected expenses.

**CURRENT** — CURRENT

A movement or flow of electricity. Electric current is measured by the number of coulombs per second flowing past a certain point in a conductor. A coulomb is equal to about $6.25 \times 10^{18}$ electrons (6,250,000,000,000,000,000 electrons). A flow of one coulomb per second is called one ampere, the unit of the rate of flow of current.

**CURVILINEAR (KER-vuh-LYNN-e-ur)** — CURVILINEAR

In the shape of a curved line.

**CYCLE** — CYCLE

A complete alternation of voltage or current in an alternating current (AC) circuit.

## D

**DNA** — DNA

Deoxyribonucleic acid. A chemical that encodes genetic information that is transmitted between generations of cells.

**DANGEROUS AIR CONTAMINATION** — DANGEROUS AIR CONTAMINATION

An atmosphere presenting a threat of causing death, injury, acute illness, or disablement due to the presence of flammable or explosive, toxic, or otherwise injurious or incapacitating substances.

(1) Dangerous air contamination due to the flammability of a gas, vapor, or mist is defined as an atmosphere containing the gas, vapor, or mist at a concentration greater than 10 percent of its lower explosive (lower flammable) limit (LEL).

(2) Dangerous air contamination due to a combustible particulate is defined as a concentration that meets or exceeds the particulate's lower explosive limit (LEL).

(3) Dangerous air contamination due to the toxicity of a substance is defined as the atmospheric concentration that could result in employee exposure in excess of the substance's permissible exposure limit (PEL).

*NOTE:* A dangerous situation also occurs when the oxygen level is less than 19.5 percent by volume (OXYGEN DEFICIENCY) or more than 23.5 percent by volume (OXYGEN ENRICHMENT).

**DEBT SERVICE** — DEBT SERVICE

The amount of money required annually to pay the (1) interest on outstanding debts, or (2) funds due on a maturing bonded debt or the redemption of bonds.

**DECHLORINATION (DEE-klor-uh-NAY-shun)** — DECHLORINATION

The removal of chlorine from the effluent of a treatment plant. Chlorine needs to be removed because chlorine is toxic to fish and other aquatic life.

**DECIBEL (DES-uh-bull)** — DECIBEL

A unit for expressing the relative intensity of sounds on a scale from zero for the average least perceptible sound to about 130 for the average level at which sound causes pain to humans. Abbreviated dB.

DECOMPOSITION or DECAY — DECOMPOSITION or DECAY

The conversion of chemically unstable materials to more stable forms by chemical or biological action.

DELEGATION — DELEGATION

The act in which power is given to another person in the organization to accomplish a specific job.

DENITRIFICATION (dee-NYE-truh-fuh-KAY-shun) — DENITRIFICATION

(1) The anoxic biological reduction of nitrate nitrogen to nitrogen gas.

(2) The removal of some nitrogen from a system.

(3) An anoxic process that occurs when nitrite or nitrate ions are reduced to nitrogen gas and nitrogen bubbles are formed as a result of this process. The bubbles attach to the biological floc and float the floc to the surface of the secondary clarifiers. This condition is often the cause of rising sludge observed in secondary clarifiers or gravity thickeners. Also see NITRIFICATION.

DEPRECIATION — DEPRECIATION

The gradual loss in service value of a facility or piece of equipment due to all the factors causing the ultimate retirement of the facility or equipment. This loss can be caused by sudden physical damage, wearing out due to age, obsolescence, inadequacy, or availability of a newer, more efficient facility or equipment. The value cannot be restored by maintenance.

DESICCANT (DESS-uh-kant) — DESICCANT

A drying agent that is capable of removing or absorbing moisture from the atmosphere in a small enclosure.

DESICCATOR (DESS-uh-kay-tor) — DESICCATOR

A closed container into which heated weighing or drying dishes are placed to cool in a dry environment in preparation for weighing. The dishes may be empty or they may contain a sample. Desiccators contain a substance (DESICCANT), such as anhydrous calcium chloride, that absorbs moisture and keeps the relative humidity near zero so that the dish or sample will not gain weight from absorbed moisture.

DETENTION TIME — DETENTION TIME

The time required to fill a tank at a given flow or the theoretical time required for a given flow of wastewater to pass through a tank. In septic tanks, this detention time will decrease as the volumes of sludge and scum increase.

DEWATER — DEWATER

(1) To remove or separate a portion of the water present in a sludge or slurry. To dry sludge so it can be handled and disposed of.

(2) To remove or drain the water from a tank or a trench. A structure may be dewatered so that it can be inspected or repaired.

DIFFUSED-AIR AERATION — DIFFUSED-AIR AERATION

A diffused-air activated sludge plant takes air, compresses it, and then discharges the air below the water surface of the aerator through some type of air diffusion device.

DIFFUSER — DIFFUSER

A device (porous plate, tube, bag) used to break the air stream from the blower system into fine bubbles in an aeration tank or reactor.

DISCHARGE HEAD — DISCHARGE HEAD

The pressure (in pounds per square inch (psi) or kilopascals (kPa)) measured at the centerline of a pump discharge and very close to the discharge flange, converted into feet or meters. The pressure is measured from the centerline of the pump to the hydraulic grade line of the water in the discharge pipe.

$$\text{Discharge Head, ft} = (\text{Discharge Pressure, psi})(2.31\ \text{ft/psi})$$

or

$$\text{Discharge Head, m} = (\text{Discharge Pressure, kPa})(1\ \text{m}/9.8\ \text{kPa})$$

DISINFECTION (dis-in-FECT-shun) DISINFECTION

The process designed to kill or inactivate most microorganisms in water or wastewater, including essentially all pathogenic (disease-causing) bacteria. There are several ways to disinfect, with chlorination being the most frequently used in water and wastewater treatment plants. Compare with STERILIZATION.

DISSOLVED OXYGEN DISSOLVED OXYGEN

Molecular oxygen dissolved in water or wastewater, usually abbreviated DO.

DISTILLATE (DIS-tuh-late) DISTILLATE

In the distillation of a sample, a portion is collected by evaporation and recondensation; the part that is recondensed is the distillate.

DRAIN TILE SYSTEM DRAIN TILE SYSTEM

A system of tile pipes buried under agricultural fields that collect percolated waters and keep the groundwater table below the ground surface to prevent ponding.

DRAINAGE WELLS DRAINAGE WELLS

Wells that can be pumped to lower the groundwater table and prevent ponding.

DROP JOINT DROP JOINT

A sewer pipe joint where one part has dropped out of alignment. Also see VERTICAL OFFSET.

DRY PIT DRY PIT

See DRY WELL.

DRY WELL DRY WELL

A dry room or compartment in a lift station, near or below the water level, where the pumps are located, usually next to the wet well.

DUCKWEED DUCKWEED

A small, green, cloverleaf-shaped floating plant, about one-quarter inch (6 mm) across, which appears as a grainy layer on the surface of a pond.

DYNAMIC HEAD DYNAMIC HEAD

When a pump is operating, the vertical distance (in feet or meters) from a point to the energy grade line. Also see ENERGY GRADE LINE (EGL), STATIC HEAD, and TOTAL DYNAMIC HEAD (TDH).

DYNAMIC PRESSURE DYNAMIC PRESSURE

When a pump is operating, pressure resulting from the dynamic head.

$$\text{Dynamic Pressure, psi} = (\text{Dynamic Head, ft})(0.433\ \text{psi/ft})$$

or

$$\text{Dynamic Pressure, kPa} = (\text{Dynamic Head, m})(9.8\ \text{kPa/m})$$

## E

EGL EGL

See ENERGY GRADE LINE (EGL).

EASEMENT EASEMENT

Legal right to use the property of others for a specific purpose. For example, a utility company may have a five-foot (1.5 m) easement along the property line of a home. This gives the utility the legal right to install and maintain a sewer line within the easement.

EDUCTOR (e-DUCK-ter) EDUCTOR

A hydraulic device used to create a negative pressure (suction) by forcing a liquid through a restriction, such as a Venturi. An eductor or aspirator (the hydraulic device) may be used in the laboratory in place of a vacuum pump. As an injector, it is used to produce vacuum for chlorinators. Sometimes used instead of a suction pump.

EFFECTIVE SOIL DEPTH EFFECTIVE SOIL DEPTH

The depth of soil in the leach field trench that provides a satisfactory percolation area for the septic tank effluent.

EFFLUENT (EF-loo-ent) EFFLUENT

Water or other liquid—raw (untreated), partially treated, or completely treated—flowing *FROM* a reservoir, basin, treatment process, or treatment plant.

ELECTROLYSIS (ee-leck-TRAWL-uh-sis) ELECTROLYSIS

The decomposition of material by an outside electric current.

ELECTROMOTIVE FORCE (EMF) ELECTROMOTIVE FORCE (EMF)

The electrical pressure available to cause a flow of current (amperage) when an electric circuit is closed. Also called voltage.

ELECTRON ELECTRON

(1) A very small, negatively charged particle that is practically weightless. According to the electron theory, all electrical and electronic effects are caused either by the movement of electrons from place to place or because there is an excess or lack of electrons at a particular place.

(2) The part of an atom that determines its chemical properties.

ELEMENT ELEMENT

A substance that cannot be separated into its constituent parts and still retain its chemical identity. For example, sodium (Na) is an element.

ENCLOSED SPACE ENCLOSED SPACE

See CONFINED SPACE.

END POINT END POINT

The completion of a desired chemical reaction. Samples of water or wastewater are titrated to the end point. This means that a chemical is added, drop by drop, to a sample until a certain color change (blue to clear, for example) occurs. This is called the end point of the titration. In addition to a color change, an end point may be reached by the formation of a precipitate or the reaching of a specified pH. An end point may be detected by the use of an electronic device, such as a pH meter.

ENDOGENOUS (en-DODGE-en-us) RESPIRATION ENDOGENOUS RESPIRATION

A situation in which living organisms oxidize some of their own cellular mass instead of new organic matter they adsorb or absorb from their environment.

ENERGY GRADE LINE (EGL) ENERGY GRADE LINE (EGL)

A line that represents the elevation of energy head (in feet or meters) of water flowing in a pipe, conduit, or channel. The line is drawn above the hydraulic grade line (gradient) a distance equal to the velocity head ($V^2/2g$) of the water flowing at each section or point along the pipe or channel. Also see HYDRAULIC GRADE LINE (HGL).

[SEE DRAWING ON PAGE 475]

ENTRAIN ENTRAIN

To trap bubbles in water either mechanically through turbulence or chemically through a reaction.

ENZYMES (EN-zimes) ENZYMES

Organic substances (produced by living organisms) that cause or speed up chemical reactions. Organic catalysts or biochemical catalysts.

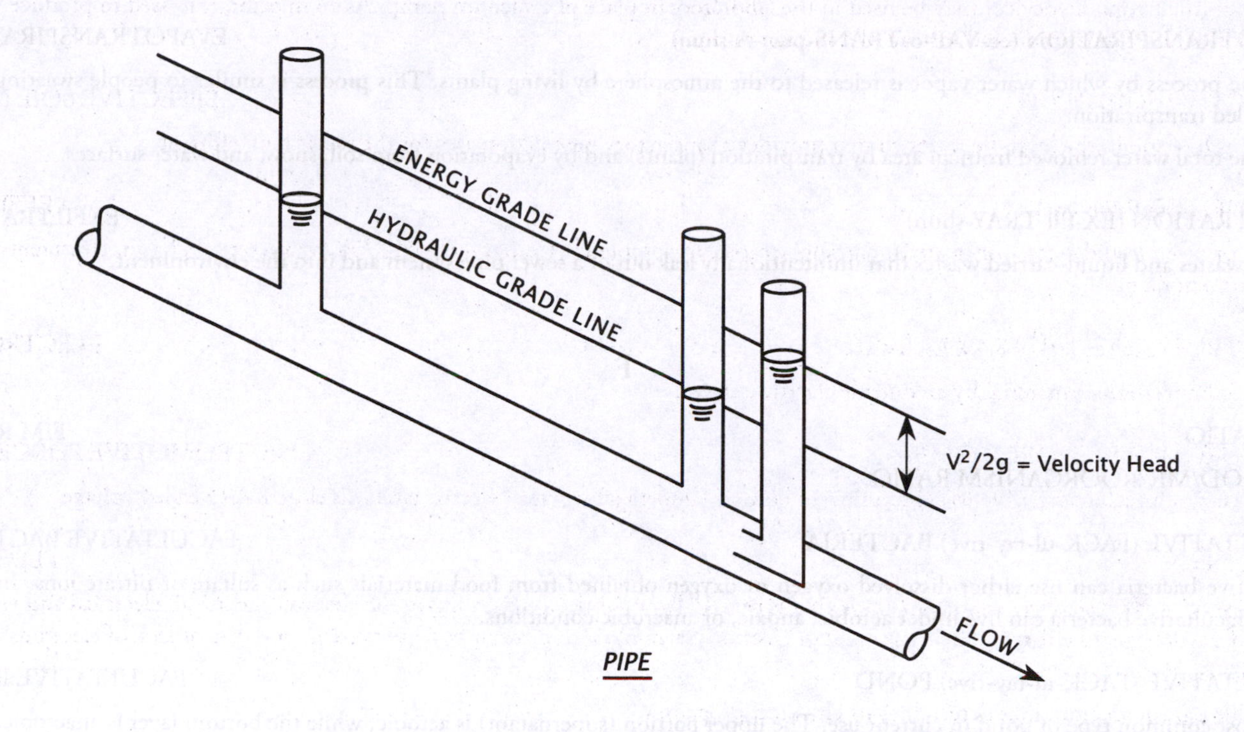


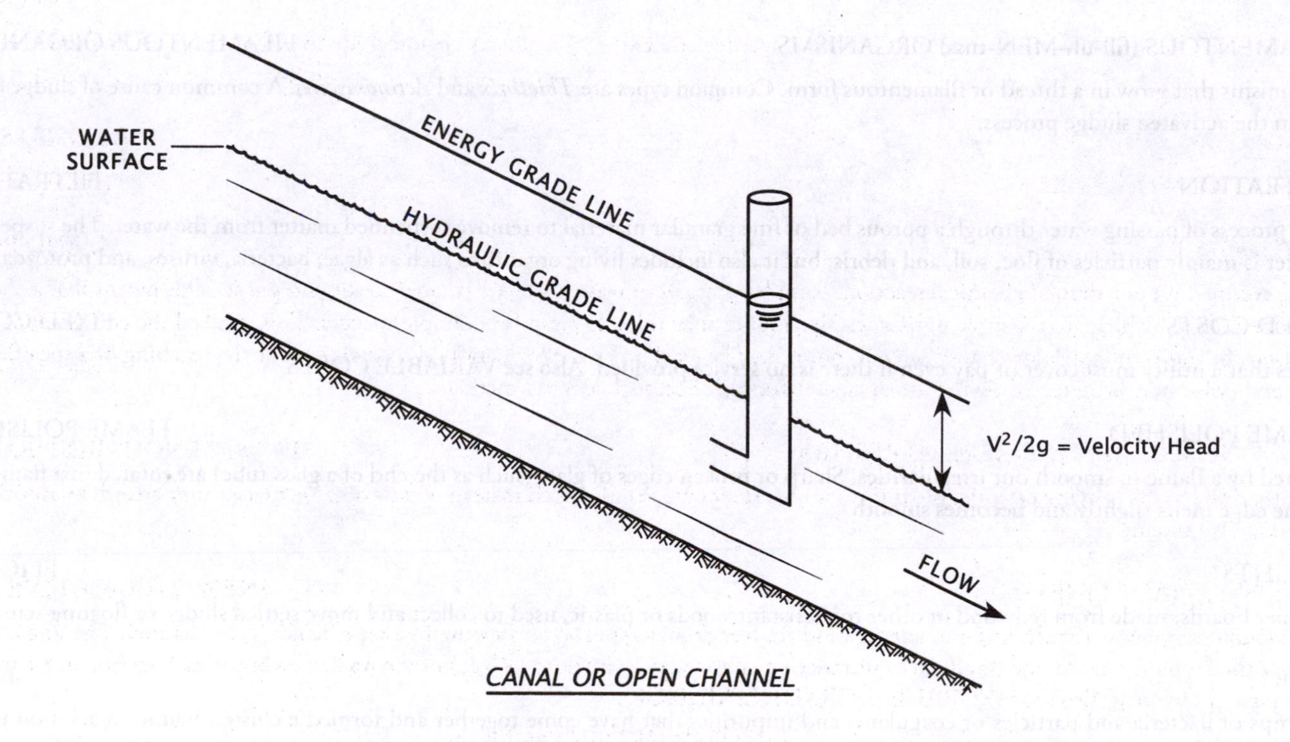


*Energy grade line and hydraulic grade line*

EQUITY EQUITY

The value of an investment in a facility.

EVAPOTRANSPIRATION (ee-VAP-o-TRANS-purr-A-shun) EVAPOTRANSPIRATION

(1) The process by which water vapor is released to the atmosphere by living plants. This process is similar to people sweating. Also called transpiration.

(2) The total water removed from an area by transpiration (plants) and by evaporation from soil, snow, and water surfaces.

EXFILTRATION (EX-fill-TRAY-shun) EXFILTRATION

Liquid wastes and liquid-carried wastes that unintentionally leak out of a sewer pipe system and into the environment.

## F

F/M RATIO F/M RATIO

See FOOD/MICROORGANISM RATIO.

FACULTATIVE (FACK-ul-tay-tive) BACTERIA FACULTATIVE BACTERIA

Facultative bacteria can use either dissolved oxygen or oxygen obtained from food materials such as sulfate or nitrate ions. In other words, facultative bacteria can live under aerobic, anoxic, or anaerobic conditions.

FACULTATIVE (FACK-ul-tay-tive) POND FACULTATIVE POND

The most common type of pond in current use. The upper portion (supernatant) is aerobic, while the bottom layer is anaerobic. Algae supply most of the oxygen to the supernatant.

FILAMENTOUS (fill-uh-MEN-tuss) ORGANISMS FILAMENTOUS ORGANISMS

Organisms that grow in a thread or filamentous form. Common types are *Thiothrix* and *Actinomycetes*. A common cause of sludge bulking in the activated sludge process.

FILTRATION FILTRATION

The process of passing water through a porous bed of fine granular material to remove suspended matter from the water. The suspended matter is mainly particles of floc, soil, and debris; but it also includes living organisms such as algae, bacteria, viruses, and protozoa.

FIXED COSTS FIXED COSTS

Costs that a utility must cover or pay even if there is no service provided. Also see VARIABLE COSTS.

FLAME POLISHED FLAME POLISHED

Melted by a flame to smooth out irregularities. Sharp or broken edges of glass (such as the end of a glass tube) are rotated in a flame until the edge melts slightly and becomes smooth.

FLIGHTS FLIGHTS

Scraper boards, made from redwood or other rot-resistant woods or plastic, used to collect and move settled sludge or floating scum.

FLOC FLOC

Clumps of bacteria and particles or coagulants and impurities that have come together and formed a cluster. Found in aeration tanks, secondary clarifiers, and chemical precipitation processes.

FLOW LINE FLOW LINE

(1) The top of the wetted line, the water surface, or the hydraulic grade line of water flowing in an open channel or partially full conduit.

(2) The lowest point of the channel inside a pipe, conduit, canal, or manhole. This term is used by some contractors, however, the preferred term for this usage is invert.

FOOD/MICROORGANISM (F/M) RATIO FOOD/MICROORGANISM (F/M) RATIO

Food to microorganism ratio. A measure of food provided to bacteria in an aeration tank.

$$\frac{\text{Food}}{\text{Microorganisms}} = \frac{\text{BOD, lbs/day}}{\text{MLVSS, lbs}}$$

$$= \frac{\text{Flow, MGD} \times \text{BOD, mg/L} \times 8.34\ \text{lbs/gal}}{\text{Volume, MG} \times \text{MLVSS, mg/L} \times 8.34\ \text{lbs/gal}}$$

or metric

$$= \frac{\text{BOD, kg/day}}{\text{MLVSS, kg}}$$

$$= \frac{\text{Flow, ML/day} \times \text{BOD, mg/L} \times 1\ \text{kg/M mg}}{\text{Volume, ML} \times \text{MLVSS, mg/L} \times 1\ \text{kg/M mg}}$$

FOOT VALVE FOOT VALVE

A special type of check valve located at the bottom end of the suction pipe on a pump. This valve opens when the pump operates to allow water to enter the suction pipe but closes when the pump shuts off to prevent water from flowing out of the suction pipe.

FREE AVAILABLE CHLORINE FREE AVAILABLE CHLORINE

The amount of chlorine available in water. This chlorine may be in the form of dissolved gas ($Cl_2$), hypochlorous acid (HOCl), or hypochlorite ion ($OCl^-$), but does not include chlorine combined with an amine (ammonia or nitrogen) or other organic compound.

FREE AVAILABLE CHLORINE RESIDUAL FREE AVAILABLE CHLORINE RESIDUAL

The amount of chlorine available in water at the end of a specified contact period. This chlorine may be in the form of dissolved gas ($Cl_2$), hypochlorous acid (HOCl), or hypochlorite ion ($OCl^-$), but does not include chlorine combined with an amine (ammonia or nitrogen) or other organic compound.

FREE OXYGEN FREE OXYGEN

Molecular oxygen available for respiration by organisms. Molecular oxygen is the oxygen molecule, $O_2$, that is not combined with another element to form a compound.

FREEBOARD FREEBOARD

(1) The vertical distance from the normal water surface to the top of the confining wall.
(2) The vertical distance from the sand surface to the underside of a trough in a sand filter. This distance is also called available expansion.

FREEBOARD
WALL HEIGHT
WATER DEPTH

FRICTION LOSS FRICTION LOSS

The head, pressure, or energy (they are the same) lost by water flowing in a pipe or channel as a result of turbulence caused by the velocity of the flowing water and the roughness of the pipe, channel walls, or restrictions caused by fittings. Water flowing in a pipe loses head, pressure, or energy as a result of friction. Also called head loss.

## G

GIS GIS

See GEOGRAPHIC INFORMATION SYSTEM (GIS).

GARNET GARNET

A group of hard, reddish, glassy, mineral sands made up of silicates of base metals (calcium, magnesium, iron, and manganese). Garnet has a higher density than sand.

GAS/LIQUID GAS/LIQUID

Gaseous/Liquid. Gaseous/liquid chlorination refers to the fact that free chlorine is delivered to small treatment plants in containers that hold liquid chlorine with a free chlorine gas above the liquid in the container. The release of chlorine gas from the liquid chlorine surface depends on the temperature of the liquid and the pressure of the chlorine gas on the liquid surface. As chlorine gas is removed from the container, the gas pressure drops and more liquid chlorine becomes chlorine gas.

GATE GATE

(1) A movable, watertight barrier for the control of a liquid in a waterway.

(2) A descriptive term used on irrigation distribution piping systems instead of the word valve. Gates cover outlet ports in the pipe segments. Water flows are regulated or distributed by opening the gates by either sliding the gate up or down or by swinging the gate to one side and uncovering an individual port to permit water flow to be discharged or regulated from the pipe at that particular point.

GEOGRAPHIC INFORMATION SYSTEM (GIS) GEOGRAPHIC INFORMATION SYSTEM (GIS)

A computer program that combines mapping with detailed information about the physical locations of structures, such as pipes, valves, and manholes, within geographic areas. The system is used to help operators and maintenance personnel locate utility system features or structures and to assist with the scheduling and performance of maintenance activities.

GRAB SAMPLE GRAB SAMPLE

A single sample of water collected at a particular time and place that represents the composition of the water only at that time and place.

GRADE GRADE

(1) The elevation of the invert (or bottom) of a pipeline, canal, culvert, sewer, or similar conduit.

(2) The inclination or slope of a pipeline, conduit, stream channel, or natural ground surface; usually expressed in terms of the ratio or percentage of number of units of vertical rise or fall per unit of horizontal distance. A 0.5 percent grade would be a drop of one-half foot per hundred feet (one-half meter per hundred meters) of pipe.

GRAVIMETRIC GRAVIMETRIC

A means of measuring unknown concentrations of water quality indicators in a sample by weighing a precipitate or residue of the sample.

GRINDER PUMP GRINDER PUMP

A small, submersible, centrifugal pump with an impeller, designed to grind solids into small pieces before they enter the collection system.

GROUND GROUND

An expression representing an electrical connection to earth or a large conductor that is at the earth's potential or neutral voltage.

GROUNDWATER GROUNDWATER

Subsurface water in the saturation zone from which wells and springs are fed. In a strict sense the term applies only to water below the water table. Also called phreatic water and plerotic water.

GROUNDWATER TABLE GROUNDWATER TABLE

The average depth or elevation of the groundwater over a selected area. Also see ARTIFICIAL GROUNDWATER TABLE, SEASONAL WATER TABLE, and TEMPORARY GROUNDWATER TABLE.

## H

**HGL** — HGL

See HYDRAULIC GRADE LINE (HGL).

**HARMFUL PHYSICAL AGENT or TOXIC SUBSTANCE** — HARMFUL PHYSICAL AGENT or TOXIC SUBSTANCE

Any chemical substance, biological agent (bacteria, virus, or fungus), or physical stress (noise, heat, cold, vibration, repetitive motion, ionizing and non-ionizing radiation, hypo- or hyperbaric pressure) that:

(1) Is regulated by any state or federal law or rule due to a hazard to health

(2) Is listed in the latest printed edition of the National Institute of Occupational Safety and Health (NIOSH) Registry of Toxic Effects of Chemical Substances (RTECS)

(3) Has yielded positive evidence of an acute or chronic health hazard in human, animal, or other biological testing conducted by, or known to, the employer

(4) Is described by a Material Safety Data Sheet (MSDS) available to the employer that indicates that the material may pose a hazard to human health

Also see ACUTE HEALTH EFFECT and CHRONIC HEALTH EFFECT.

**HEAD** — HEAD

The vertical distance, height, or energy of water above a reference point. A head of water may be measured in either height (feet or meters) or pressure (pounds per square inch or kilograms per square centimeter). Also see DISCHARGE HEAD, DYNAMIC HEAD, STATIC HEAD, SUCTION HEAD, SUCTION LIFT, and VELOCITY HEAD.

**HEAD LOSS** — HEAD LOSS

The head, pressure, or energy (they are the same) lost by water flowing in a pipe or channel as a result of turbulence caused by the velocity of the flowing water and the roughness of the pipe, channel walls, or restrictions caused by fittings. Water flowing in a pipe loses head, pressure, or energy as a result of friction. The head loss through a comminutor is due to friction caused by the cutters or shredders as the water passes through them and by the roughness of the comminutor walls conveying the flow through the comminutor. Also called friction loss.

[SEE DRAWING ON PAGE 480]

**HEADER** — HEADER

A large pipe to which the ends of a series of smaller pipes are connected. Also called a manifold.

**HEPATITIS (HEP-uh-TIE-tis)** — HEPATITIS

Hepatitis is an inflammation of the liver caused by an acute viral infection. Yellow jaundice is one symptom of hepatitis.

**HERTZ (Hz)** — HERTZ (Hz)

The number of complete electromagnetic cycles or waves in one second of an electric or electronic circuit. Also called the frequency of the current.

**HOT TAP** — HOT TAP

Tapping into a sewer line under pressure, such as a force main or a small-diameter sewer under pressure.

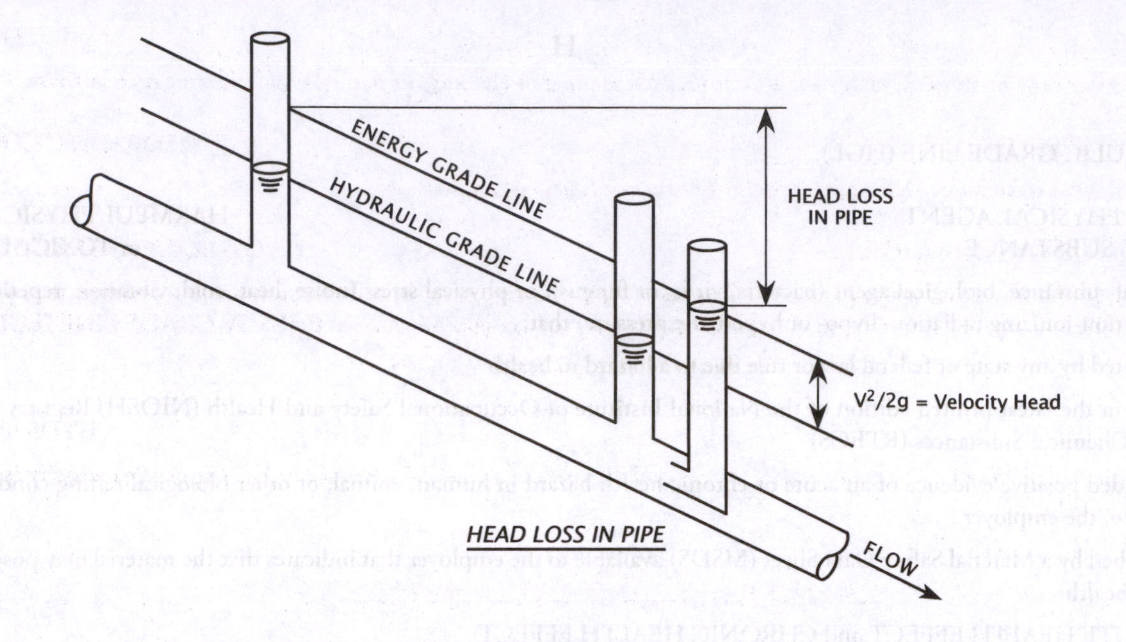
ENERGY GRADE LINE
HYDRAULIC GRADE LINE
HEAD LOSS IN PIPE
V2/2g = Velocity Head
FLOW
HEAD LOSS IN PIPE

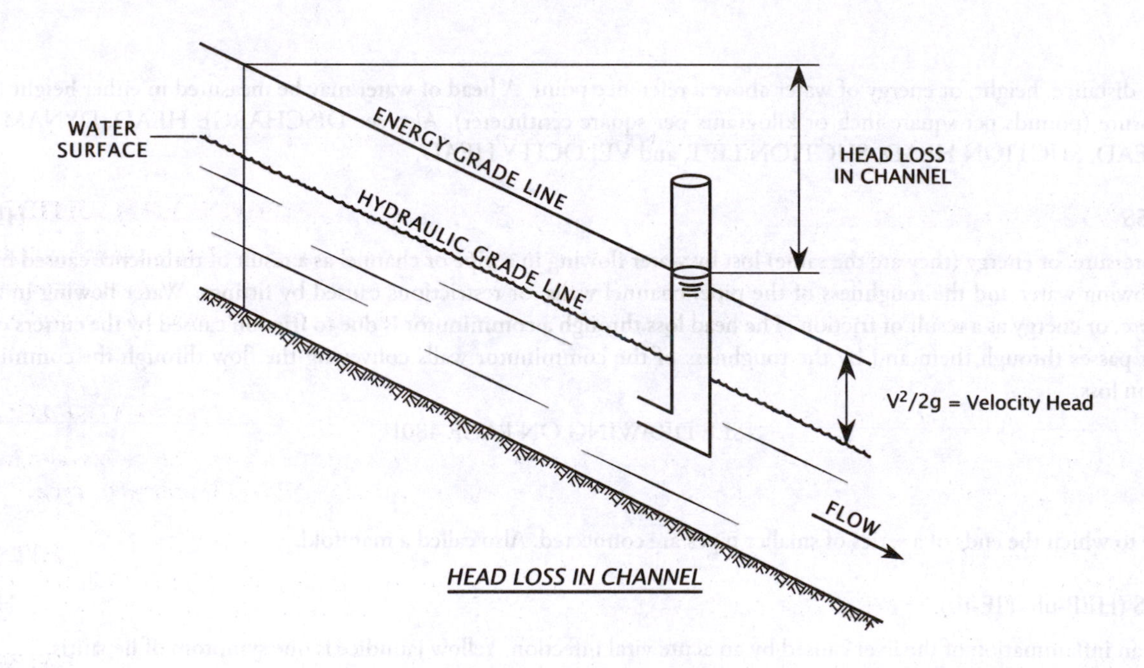
WATER SURFACE
ENERGY GRADE LINE
HEAD LOSS IN CHANNEL
HYDRAULIC GRADE LINE
V2/2g = Velocity Head
FLOW
HEAD LOSS IN CHANNEL

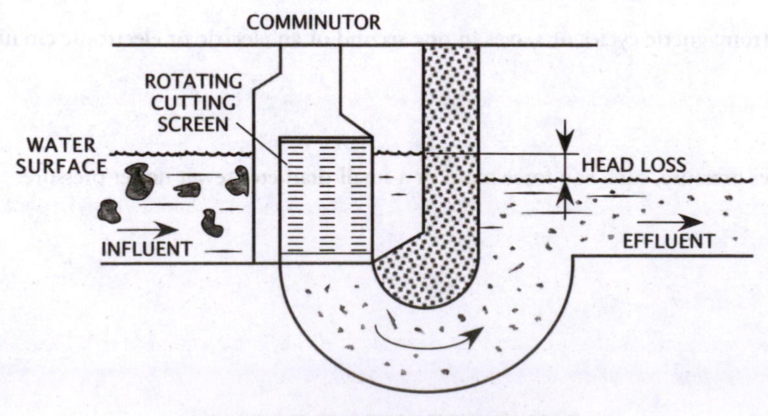
COMMINUTOR
ROTATING CUTTING SCREEN
WATER SURFACE
HEAD LOSS
INFLUENT
EFFLUENT
HEAD LOSS THROUGH COMMINUTOR

*Head loss*

HYDRAULIC HYDRAULIC

Referring to water flowing through manmade structures such as pipes or channels or natural environments such as rivers.

HYDRAULIC CONTINUITY HYDRAULIC CONTINUITY

The smooth flow of wastewater as it moves through sewers, manholes, and pipes.

HYDRAULIC GRADE LINE (HGL) HYDRAULIC GRADE LINE (HGL)

The surface or profile of water flowing in an open channel or a pipe flowing partially full. If a pipe is under pressure, the hydraulic grade line is that level water would rise to in a small, vertical tube connected to the pipe. Also see ENERGY GRADE LINE (EGL).

[SEE DRAWING ON PAGE 475]

HYDRAULIC JUMP HYDRAULIC JUMP

The sudden and usually turbulent abrupt rise in water surface in an open channel when water flowing at high velocity is suddenly retarded to a slow velocity.

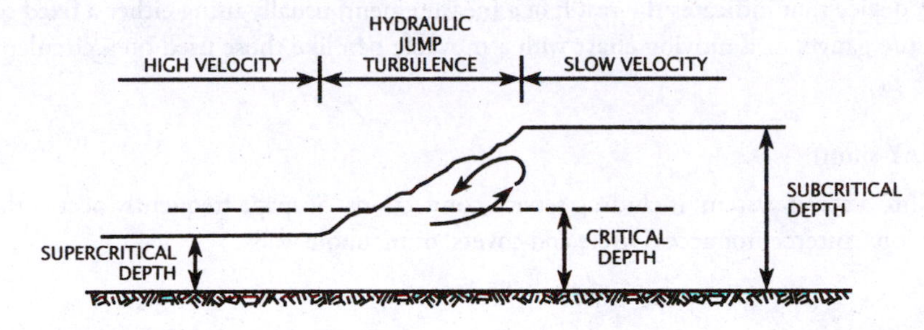


HYDROGEN SULFIDE GAS ($H_2S$) HYDROGEN SULFIDE GAS ($H_2S$)

Hydrogen sulfide is a gas with a rotten egg odor, produced under anaerobic conditions. Hydrogen sulfide gas is particularly dangerous because it dulls the sense of smell, becoming unnoticeable after you have been around it for a while; in high concentrations, it is only noticeable for a very short time before it dulls the sense of smell. The gas is very poisonous to the respiratory system, explosive, flammable, colorless, and heavier than air.

HYDROLOGIC (HI-dro-LOJ-ick) CYCLE HYDROLOGIC CYCLE

The process of evaporation of water into the air and its return to earth by precipitation (rain or snow). This process also includes transpiration from plants, groundwater movement, and runoff into rivers, streams, and the ocean. Also called the water cycle.

HYDROPHILIC (hi-dro-FILL-ick) HYDROPHILIC

Having a strong affinity (liking) for water. The opposite of HYDROPHOBIC.

HYDROPHOBIC (hi-dro-FOE-bick) HYDROPHOBIC

Having a strong aversion (dislike) for water. The opposite of HYDROPHILIC.

HYPOCHLORINATION (HI-poe-klor-uh-NAY-shun) HYPOCHLORINATION

The application of hypochlorite compounds to water or wastewater for the purpose of disinfection.

HYPOCHLORINATORS (HI-poe-KLOR-uh-nay-tors) HYPOCHLORINATORS

Chlorine pumps, chemical feed pumps, or devices used to dispense chlorine solutions made from hypochlorites, such as bleach (sodium hypochlorite) or calcium hypochlorite into the water being treated.

HYPOCHLORITE (HI-poe-KLOR-ite) HYPOCHLORITE

Chemical compounds containing available chlorine; used for disinfection. They are available as liquids (bleach) or solids (powder, granules, and pellets) in barrels, drums, and cans. Salts of hypochlorous acid.

## I

IMHOFF CONE — IMHOFF CONE

A clear, cone-shaped container marked with graduations. The cone is used to measure the volume of settleable solids in a specific volume (usually one liter) of water or wastewater.

IMPELLER — IMPELLER

A rotating set of vanes in a pump or compressor designed to pump or move water or air.

INDICATOR — INDICATOR

(1) (Chemical indicator) A substance that gives a visible change, usually of color, at a desired point in a chemical reaction, generally at a specified end point.

(2) (Instrument indicator) A device that indicates the result of a measurement, usually using either a fixed scale and movable indicator (pointer), such as a pressure gauge, or a moving chart with a movable pen like those used on a circular flow-recording chart. Also called a receiver.

INFILTRATION (in-fill-TRAY-shun) — INFILTRATION

The seepage of groundwater into a sewer system, including service connections. Seepage frequently occurs through defective or cracked pipes, pipe joints and connections, interceptor access risers and covers, or manhole walls.

INFLOW — INFLOW

Water discharged into a sewer system and service connections from sources other than regular connections. This includes flow from yard drains, foundations, and around access and manhole covers. Inflow differs from infiltration in that it is a direct discharge into the sewer rather than a leak in the sewer itself.

INFLUENT — INFLUENT

Water or other liquid—raw (untreated) or partially treated—flowing *INTO* a reservoir, basin, treatment process, or treatment plant.

INORGANIC — INORGANIC

Used to describe material such as sand, salt, iron, calcium salts, and other mineral materials. Inorganic materials are chemical substances of mineral origin, whereas organic substances are usually of animal or plant origin. Also see ORGANIC.

INORGANIC WASTE — INORGANIC WASTE

Waste material such as sand, salt, iron, calcium, and other mineral materials that are only slightly affected by the action of organisms. Inorganic wastes are chemical substances of mineral origin; whereas organic wastes are chemical substances usually of animal or plant origin. Also see NONVOLATILE MATTER, ORGANIC WASTE, and VOLATILE SOLIDS.

INSECTICIDE — INSECTICIDE

Any substance or chemical formulated to kill or control insects.

INTEGRATOR — INTEGRATOR

A device or meter that continuously measures and sums a process rate variable in cumulative fashion over a given time period. For example, total flows displayed in gallons per minute, million gallons per day, cubic feet per second, or some other unit of volume per time period. Also called a totalizer.

INTERCEPTOR — INTERCEPTOR

A septic tank or other holding tank that serves as a temporary wastewater storage reservoir for a septic tank effluent pump (STEP) system. Also see SEPTIC TANK.

INTERCEPTOR (INTERCEPTING) SEWER INTERCEPTOR (INTERCEPTING) SEWER

A large sewer that receives flow from a number of sewers and conducts the wastewater to a treatment plant. Often called an interceptor. The term interceptor is sometimes used in small communities to describe a septic tank or other holding tank that serves as a temporary wastewater storage reservoir for a septic tank effluent pump (STEP) system.

INTERFACE INTERFACE

The common boundary layer between two substances, such as water and a solid (metal); or between two fluids, such as water and a gas (air); or between a liquid (water) and another liquid (oil).

INTERSTICE (in-TUR-stuhz) INTERSTICE

A very small open space in a rock or granular material. Also called a pore, void, or void space. Also see VOID.

INVERT (IN-vert) INVERT

The lowest point of the channel inside a pipe, conduit, canal, or manhole. Also called flow line by some contractors, however, the preferred term is invert.

INVERTED SIPHON INVERTED SIPHON

A pressure pipeline used to carry wastewater flowing in a gravity collection system under a depression, such as a valley or roadway, or under a structure, such as a building. Also called a depressed sewer.

## J

(NO LISTINGS)

## K

(NO LISTINGS)

## L

LIFT LIFT

The vertical distance water is mechanically lifted (usually pumped) from a lower elevation to a higher elevation.

LINEAL (LIN-e-ul) LINEAL

The length in one direction of a line. For example, a board 12 feet (meters) long has 12 lineal feet (meters) in its length.

LOADING LOADING

Quantity of material applied to a device at one time.

LOWER EXPLOSIVE LIMIT (LEL) LOWER EXPLOSIVE LIMIT (LEL)

The lowest concentration of a gas or vapor (percent by volume in air) that explodes if an ignition source is present at ambient temperature. At temperatures above 250°F (121°C) the LEL decreases because explosibility increases with higher temperature.

# M

*M* or MOLAR | *M* or MOLAR

A molar solution consists of one gram molecular weight of a compound dissolved in enough water to make one liter of solution. A gram molecular weight is the molecular weight of a compound in grams. For example, the molecular weight of sulfuric acid ($H_2SO_4$) is 98. A one *M* solution of sulfuric acid would consist of 98 grams of $H_2SO_4$ dissolved in enough distilled water to make one liter of solution.

MCRT | MCRT

Mean Cell Residence Time. An expression of the average time (days) that a microorganism will spend in the activated sludge process.

$$\text{MCRT, days} = \frac{\text{Total Suspended Solids in Activated Sludge Process, lbs}}{\text{Total Suspended Solids Removed From Process, lbs/day}}$$

or

$$\text{MCRT, days} = \frac{\text{Total Suspended Solids in Activated Sludge Process, kg}}{\text{Total Suspended Solids Removed From Process, kg/day}}$$

*NOTE:* Operators at different plants calculate the Total Suspended Solids (TSS) in the Activated Sludge Process, lbs (kg), by three different methods:

1. TSS in the Aeration Basin or Reactor Zone, lbs (kg)
2. TSS in the Aeration Basin and Secondary Clarifier, lbs (kg)
3. TSS in the Aeration Basin and Secondary Clarifier Sludge Blanket, lbs (kg)

These three different methods make it difficult to compare MCRTs in days among different plants unless everyone uses the same method.

mg/L | mg/L

See MILLIGRAMS PER LITER, mg/L.

MLSS | MLSS

Mixed Liquor Suspended Solids. The amount (mg/L) of suspended solids in the mixed liquor of an aeration tank.

MPN | MPN

MPN is the Most Probable Number of coliform-group organisms per unit volume of sample water. Expressed as a density or population of organisms per 100 mL of sample water.

MSDS | MSDS

See MATERIAL SAFETY DATA SHEET (MSDS).

MANIFOLD | MANIFOLD

A large pipe to which the ends of a series of smaller pipes are connected. Also called a header.

MANOMETER (man-NAH-mut-ter) | MANOMETER

An instrument for measuring pressure. Usually, a manometer is a glass tube filled with a liquid that is used to measure the difference in pressure across a flow measuring device, such as an orifice or a Venturi meter. The instrument used to measure blood pressure is a type of manometer.

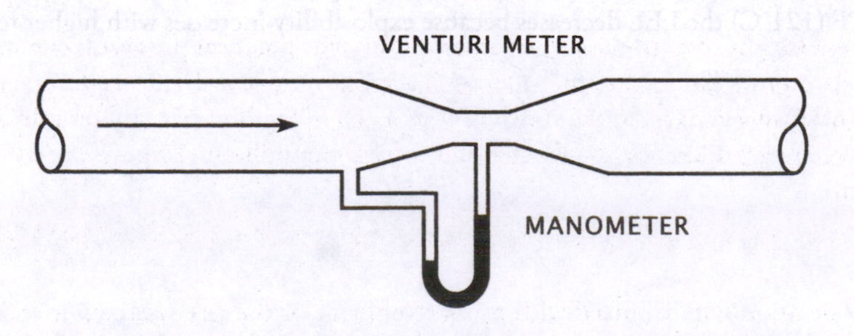

MATERIAL SAFETY DATA SHEET (MSDS) MATERIAL SAFETY DATA SHEET (MSDS)

A document that provides pertinent information and a profile of a particular hazardous substance or mixture. An MSDS is normally developed by the manufacturer or formulator of the hazardous substance or mixture. The MSDS is required to be made available to employees and operators or inspectors whenever there is the likelihood of the hazardous substance or mixture being introduced into the workplace. Some manufacturers are preparing MSDSs for products that are not considered to be hazardous to show that the product or substance is not hazardous.

MEAN CELL RESIDENCE TIME (MCRT) MEAN CELL RESIDENCE TIME (MCRT)

See MCRT.

MECHANICAL AERATION MECHANICAL AERATION

The use of machinery to mix air and water so that oxygen can be absorbed into the water. Some examples are: paddle wheels, mixers, or rotating brushes to agitate the surface of an aeration tank; pumps to create fountains; and pumps to discharge water down a series of steps forming falls or cascades.

MEDIAN MEDIAN

The middle measurement or value. When several measurements are ranked by magnitude (largest to smallest), half of the measurements will be larger and half will be smaller.

MEGOHM (MEG-ome) MEGOHM

Millions of ohms. Mega- is a prefix meaning one million, so 5 megohms means 5 million ohms.

MENISCUS (meh-NIS-cuss) MENISCUS

The curved surface of a column of liquid (water, oil, mercury) in a small tube. When the liquid wets the sides of the container (as with water), the curve forms a valley. When the confining sides are not wetted (as with mercury), the curve forms a hill or upward bulge.

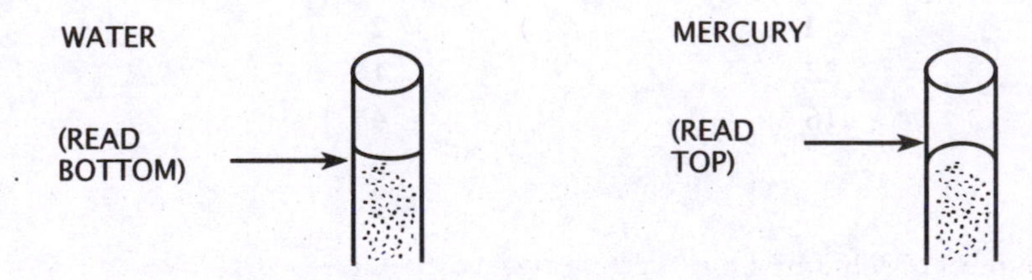


MICRON (MY-kron) MICRON

µm, Micrometer or Micron. A unit of length. One millionth of a meter or one thousandth of a millimeter. One micron equals 0.00004 of an inch.

MICROORGANISMS (MY-crow-OR-gan-is-ums) MICROORGANISMS

Very small organisms that can be seen only through a microscope. Some microorganisms use the wastes in wastewater for food and thus remove or alter much of the undesirable matter.

MIL MIL

A unit of length equal to 0.001 of an inch. The diameter of wires and tubing is measured in mils, as is the thickness of plastic sheeting.

MILLIGRAMS PER LITER, mg/L MILLIGRAMS PER LITER, mg/L

A measure of the concentration by weight of a substance per unit volume. For practical purposes, one mg/L of a substance in water is equal to one part per million parts (ppm). Thus, a liter of water with a specific gravity of 1.0 weighs one million milligrams. If one liter of water contains 10 milligrams of dissolved oxygen, the concentration is 10 milligrams per million milligrams, or 10 milligrams per liter (10 mg/L), or 10 parts of oxygen per million parts of water, or 10 parts per million (10 ppm), or 10 pounds dissolved oxygen in 1 million pounds of water (10 ppm).

MIXED LIQUOR MIXED LIQUOR

When the activated sludge in an aeration tank is mixed with primary effluent or the raw wastewater and return sludge, this mixture is then referred to as mixed liquor as long as it is in the aeration tank. Mixed liquor also may refer to the contents of mixed aerobic or anaerobic digesters.

MIXED LIQUOR SUSPENDED SOLIDS (MLSS) MIXED LIQUOR SUSPENDED SOLIDS (MLSS)

The amount (mg/L) of suspended solids in the mixed liquor of an aeration tank.

MIXED LIQUOR VOLATILE SUSPENDED SOLIDS (MLVSS) MIXED LIQUOR VOLATILE SUSPENDED SOLIDS (MLVSS)

The amount (mg/L) of organic or volatile suspended solids in the mixed liquor of an aeration tank. This volatile portion is used as a measure or indication of the microorganisms present.

MOLAR MOLAR

See *M* or MOLAR.

MOLARITY MOLARITY

A measure of concentration defined as the number of moles of solute per liter of solution. Also see *M* or MOLAR.

MOLE MOLE

The name for a quantity of any chemical substance whose mass in grams is numerically equal to its atomic weight. One mole equals $6.02 \times 10^{23}$ molecules or atoms. Also see MOLECULAR WEIGHT.

MOLECULAR OXYGEN MOLECULAR OXYGEN

The oxygen molecule, $O_2$, that is not combined with another element to form a compound.

MOLECULAR WEIGHT MOLECULAR WEIGHT

The molecular weight of a compound in grams per mole is the sum of the atomic weights of the elements in the compound. The molecular weight of sulfuric acid ($H_2SO_4$) in grams is 98.

| Element | Atomic Weight | Number of Atoms | Molecular Weight |
|---|---|---|---|
| H | 1 | 2 | 2 |
| S | 32 | 1 | 32 |
| O | 16 | 4 | 64 |
| | | | 98 |

MOLECULE MOLECULE

The smallest division of a compound that still retains or exhibits all the properties of the substance.

MOST PROBABLE NUMBER (MPN) MOST PROBABLE NUMBER (MPN)

See MPN.

## N

*N* or NORMAL *N* or NORMAL

A normal solution contains one gram equivalent weight of reactant (compound) per liter of solution. The equivalent weight of an acid is that weight that contains one gram atom of ionizable hydrogen or its chemical equivalent. For example, the equivalent weight of sulfuric acid ($H_2SO_4$) is 49 (98 divided by 2 because there are two replaceable hydrogen ions). A one *N* solution of sulfuric acid would consist of 49 grams of $H_2SO_4$ dissolved in enough water to make one liter of solution.

NPDES PERMIT NPDES PERMIT

National Pollutant Discharge Elimination System permit is the regulatory agency document issued by either a federal or state agency that is designed to control all discharges of potential pollutants from point sources and stormwater runoff into US waterways. NPDES permits regulate discharges into US waterways from all point sources of pollution, including industries, municipal wastewater treatment plants, sanitary landfills, large animal feedlots, and return irrigation flows.

NTU NTU

Nephelometric Turbidity Units. See TURBIDITY UNITS.

NAMEPLATE NAMEPLATE

A durable, metal plate found on equipment that lists critical installation and operating conditions for the equipment.

NATURAL CYCLES NATURAL CYCLES

Cycles that take place in nature, such as the water or hydrologic cycle where water is transformed or changed from one form to another until the water has returned to the original form, thus completing the cycle. Other natural cycles include the life cycles of aquatic organisms and plants, nutrient cycles, and cycles of self- or natural purification.

NEUTRALIZATION (noo-trull-uh-ZAY-shun) NEUTRALIZATION

Addition of an acid or alkali (base) to a liquid to cause the pH of the liquid to move toward a neutral pH of 7.0.

NITRIFICATION (NYE-truh-fuh-KAY-shun) NITRIFICATION

An aerobic process in which bacteria change the ammonia and organic nitrogen in wastewater into oxidized nitrogen (usually nitrate). The second-stage BOD is sometimes referred to as the nitrogenous BOD (first-stage BOD is called the carbonaceous BOD). Also see DENITRIFICATION.

NITROGENOUS (nye-TRAH-jen-us) NITROGENOUS

A term used to describe chemical compounds (usually organic) containing nitrogen in combined forms. Proteins and nitrate are nitrogenous compounds.

NONPOINT SOURCE NONPOINT SOURCE

A runoff or discharge from a field or similar source, in contrast to a point source, which refers to a discharge that comes out the end of a pipe or other clearly identifiable conveyance. Also see POINT SOURCE.

NONVOLATILE MATTER NONVOLATILE MATTER

Material such as sand, salt, iron, calcium, and other mineral materials that are only slightly affected by the actions of organisms and are not lost on ignition of the dry solids at 550°C (1,022°F). Volatile materials are chemical substances usually of animal or plant origin. Also see INORGANIC WASTE and VOLATILE SOLIDS.

NORMAL NORMAL

See *N* or NORMAL.

NORMALITY NORMALITY

The number of gram-equivalent weights of solute in one liter of solution. The equivalent weight of any material is the weight that would react with or be produced by the reaction of 8.0 grams of oxygen or 1.0 gram of hydrogen. Normality is used for certain calculations of quantitative analysis. Also see *N* or NORMAL.

NUTRIENT NUTRIENT

Any substance that is assimilated (taken in) by organisms and promotes growth. Nitrogen and phosphorus are nutrients that promote the growth of algae. There are other essential and trace elements that are also considered nutrients. Also see NUTRIENT CYCLE.

NUTRIENT CYCLE NUTRIENT CYCLE

The transformation or change of a nutrient from one form to another until the nutrient has returned to the original form, thus completing the cycle. The cycle may take place under either aerobic or anaerobic conditions.

## O

O&M MANUAL O&M MANUAL

Operation and Maintenance Manual. A manual that describes detailed procedures for operators to follow to operate and maintain a specific treatment plant and the equipment of that plant.

OSHA (O-shuh) OSHA

The Williams-Steiger Occupational Safety and Health Act of 1970 (OSHA) is a federal law designed to protect the health and safety of workers, including collection system and treatment plant operators. The Act regulates the design, construction, operation, and maintenance of industrial plants and wastewater collection systems and treatment plants. The Act does not apply directly to municipalities, *except* in those states that have approved plans and have asserted jurisdiction under Section 18 of the OSHA Act. *However, contract operators and private facilities do have to comply with OSHA requirements.* Wastewater treatment plants have come under stricter regulation in all phases of activity as a result of OSHA standards. OSHA also refers to the federal and state agencies that administer the OSHA regulations.

OBLIGATE AEROBES OBLIGATE AEROBES

Bacteria that must have atmospheric or dissolved molecular oxygen to live and reproduce.

OCCUPATIONAL SAFETY AND HEALTH ACT OF 1970 (OSHA) OCCUPATIONAL SAFETY AND HEALTH ACT OF 1970 (OSHA)

See OSHA.

ODOR THRESHOLD ODOR THRESHOLD

The minimum odor of a gas or water sample that can just be detected after successive dilutions with odorless gas or water. Also called threshold odor.

OFFSET JOINT OFFSET JOINT

A pipe joint that is not exactly in line and centered. Also see DROP JOINT and VERTICAL OFFSET.

OHM OHM

The unit of electrical resistance. The resistance of a conductor in which one volt produces a current of one ampere.

OPERATING RATIO OPERATING RATIO

The operating ratio is a measure of the total revenues divided by the total operating expenses.

ORGANIC ORGANIC

Used to describe chemical substances that come from animal or plant sources. Organic substances always contain carbon. (Inorganic materials are chemical substances of mineral origin.) Also see INORGANIC.

ORGANIC WASTE ORGANIC WASTE

Waste material that may come from animal or plant sources. Natural organic wastes generally can be consumed by bacteria and other small organisms. Manufactured or synthetic organic wastes from metal finishing, chemical manufacturing, and petroleum industries may not normally be consumed by bacteria and other organisms. Also see INORGANIC WASTE and VOLATILE SOLIDS.

ORGANIZING ORGANIZING

Deciding who does what work and delegating authority to the appropriate persons.

ORIFICE (OR-uh-fiss) ORIFICE

An opening (hole) in a plate, wall, or partition. An orifice flange or plate placed in a pipe consists of a slot or a calibrated circular hole smaller than the pipe diameter. The difference in pressure in the pipe above and at the orifice may be used to determine the flow in the pipe. In a trickling filter distributor, the wastewater passes through an orifice to the surface of the filter media.

OUTFALL OUTFALL

(1) The point, location, or structure where wastewater or drainage discharges from a sewer, drain, or other conduit.

(2) The conduit leading to the final discharge point or area. Also see OUTFALL SEWER.

OUTFALL SEWER OUTFALL SEWER

A sewer that receives wastewater from a collection system or from a wastewater treatment plant and carries it to a point of ultimate or final discharge in the environment. Also see OUTFALL.

OVERHEAD OVERHEAD

Indirect costs necessary for a utility to function properly. These costs are not related to the actual collection, treatment, and discharge of wastewater, but include the costs of rent, lights, office supplies, management, and administration.

OVERTURN OVERTURN

The almost spontaneous mixing of all layers of water in a reservoir or lake when the water temperature becomes similar from top to bottom. This may occur in the fall/winter when the surface waters cool to the same temperature as the bottom waters and also in the spring when the surface waters warm after the ice melts. This is also called turnover.

OXIDATION OXIDATION

Oxidation is the addition of oxygen, removal of hydrogen, or the removal of electrons from an element or compound; in the environment and in wastewater treatment processes, organic matter is oxidized to more stable substances. The opposite of REDUCTION.

OXIDATION DITCH OXIDATION DITCH

The oxidation ditch is a modified form of the activated sludge process. The ditch consists of two channels placed side by side and connected at the ends to produce one continuous loop of wastewater flow and a brush rotator assembly placed across the channel to provide aeration and circulation.

OXIDATION STATE/OXIDATION NUMBER OXIDATION STATE/OXIDATION NUMBER

In a chemical formula, a number accompanied by a polarity indication (+ or –) that together indicate the charge of an ion as well as the extent to which the ion has been oxidized or reduced in a REDOX REACTION.

Due to the loss of electrons, the charge of an ion that has been oxidized would go from negative toward or to neutral, from neutral to positive, or from positive to more positive. As an example, an oxidation number of 2+ would indicate that an ion has lost two electrons and that its charge has become positive (that it now has an excess of two protons).

Due to the gain of electrons, the charge of the ion that has been reduced would go from positive toward or to neutral, from neutral to negative, or from negative to more negative. As an example, an oxidation number of 2– would indicate that an ion has gained two electrons and that its charge has become negative (that it now has an excess of two electrons). As an ion gains electrons, its oxidation state (or the extent to which it is oxidized) lowers; that is, its oxidation state is reduced. Also see REDOX REACTION.

OXIDATION-REDUCTION POTENTIAL (ORP) OXIDATION-REDUCTION POTENTIAL (ORP)

The electrical potential required to transfer electrons from one compound or element (the oxidant) to another compound or element (the reductant); used as a qualitative measure of the state of oxidation in water and wastewater treatment systems. ORP is measured in millivolts, with negative values indicating a tendency to reduce compounds or elements and positive values indicating a tendency to oxidize compounds or elements.

OXIDATION-REDUCTION (REDOX) REACTION OXIDATION-REDUCTION (REDOX) REACTION

See REDOX REACTION.

OXIDIZING AGENT OXIDIZING AGENT

Any substance, such as oxygen ($O_2$) or chlorine ($Cl_2$), that will readily add (take on) electrons. When oxygen or chlorine is added to water or wastewater, organic substances are oxidized. These oxidized organic substances are more stable and less likely to give off odors or to contain disease-causing bacteria. The opposite is a REDUCING AGENT.

OXYGEN DEFICIENCY — OXYGEN DEFICIENCY

An atmosphere containing oxygen at a concentration of less than 19.5 percent by volume.

OXYGEN ENRICHMENT — OXYGEN ENRICHMENT

An atmosphere containing oxygen at a concentration of more than 23.5 percent by volume.

OZONATION (O-zoe-NAY-shun) — OZONATION

The application of ozone to water, wastewater, or air, generally for the purposes of disinfection or odor control.

## P

PEL — PEL

See PERMISSIBLE EXPOSURE LIMIT (PEL).

PARALLEL OPERATION — PARALLEL OPERATION

Wastewater being treated is split and a portion flows to one treatment unit while the remainder flows to another similar treatment unit. Also see SERIES OPERATION.

PARASITIC (pair-uh-SIT-tick) BACTERIA — PARASITIC BACTERIA

Parasitic bacteria are those bacteria that normally live off another living organism, known as the host.

PATHOGENIC (path-o-JEN-ick) ORGANISMS — PATHOGENIC ORGANISMS

Bacteria, viruses, protozoa, or internal parasites that can cause disease (such as giardiasis, cryptosporidiosis, typhoid fever, cholera, or infectious hepatitis) in a host (such as a person). There are many types of organisms that do not cause disease and are not called pathogenic. Many beneficial bacteria are found in wastewater treatment processes actively cleaning up organic wastes.

PATHOGENS (PATH-o-jens) — PATHOGENS

See PATHOGENIC ORGANISMS.

PEAKING FACTOR — PEAKING FACTOR

Ratio of a maximum flow to the average flow, such as maximum hourly flow or maximum daily flow to the average daily flow.

PERCENT SATURATION — PERCENT SATURATION

The amount of a substance that is dissolved in a solution compared with the amount dissolved in the solution at saturation, expressed as a percent.

$$\text{Percent Saturation, \%} = \frac{\text{Amount of Substance That Is Dissolved} \times 100\%}{\text{Amount Dissolved in Solution at Saturation}}$$

PERCOLATION (purr-ko-LAY-shun) — PERCOLATION

The slow passage of water through a filter medium; or, the gradual penetration of soil and rocks by water.

PERFORMANCE INDICATOR — PERFORMANCE INDICATOR

A measurable goal used to determine system performance and level of service provided. Examples of performance indicators include the number of stoppages per 100 miles of sewer per year and the number of lost time accidents per year—measurements of how *well* a utility is doing rather than how *much* a utility is doing. Also see PRODUCTION INDICATOR.

PERMEABILITY (PURR-me-uh-BILL-uh-tee) — PERMEABILITY

The property of a material or soil that permits considerable movement of water through it when it is saturated.

PERMISSIBLE EXPOSURE LIMIT (PEL) PERMISSIBLE EXPOSURE LIMIT (PEL)

The legal limit in the United States for exposure of a worker to a hazardous substance (such as chemicals, dusts, fumes, mists, gases, or vapors) or agents (such as occupational noise). OSHA sets enforceable permissible exposure limits (PELs) to protect workers against the health effects of excessive exposure. OSHA PELs are based on an 8-hour time-weighted average (TWA) exposure. Permissible exposure limits are listed in the Code of Federal Regulations (CFR) Title 29 Part 1910, Subparts G and Z. Also see TIME-WEIGHTED AVERAGE (TWA).

PERMIT-REQUIRED CONFINED SPACE (PERMIT SPACE) PERMIT-REQUIRED CONFINED SPACE (PERMIT SPACE)

See CONFINED SPACE, PERMIT-REQUIRED (PERMIT SPACE).

pH (pronounce as separate letters) pH

pH is an expression of the intensity of the basic or acidic condition of a liquid. Mathematically, pH is the logarithm (base 10) of the reciprocal of the hydrogen ion activity.

$$pH = \text{Log}\,\frac{1}{\{H^+\}}$$

If $\{H^+\} = 10^{-6.5}$, then pH = 6.5. The pH may range from 0 to 14, where 0 is most acidic, 14 most basic, and 7 neutral.

PHOTOSYNTHESIS (foe-toe-SIN-thuh-sis) PHOTOSYNTHESIS

A process in which organisms, with the aid of chlorophyll, convert carbon dioxide and inorganic substances into oxygen and additional plant material, using sunlight for energy. All green plants grow by this process.

PIG PIG

Refers to a polypig, which is a bullet-shaped device made of hard rubber or similar material. This device is used to clean pipes. It is inserted in one end of a pipe, moves through the pipe under pressure, and is removed from the other end of the pipe.

PILLOWS PILLOWS

Plastic tubes shaped like pillows that contain exact amounts of chemicals or reagents. Cut open the pillow, pour the reagents into the sample being tested, mix thoroughly, and follow test procedures.

PINPOINT FLOC PINPOINT FLOC

Very small floc (the size of a pin point) that does not settle out of the water in a sedimentation basin or clarifier. Also see FLOC.

PLAN or PLAN VIEW PLAN or PLAN VIEW

A drawing or photo showing the top view of sewers, manholes, streets, or structures.

PLANNING PLANNING

Management of utilities to build the resources and financial capability to provide for future needs.

PLUG FLOW PLUG FLOW

A type of flow that occurs in tanks, basins, or reactors when a slug of water or wastewater moves through a tank without ever dispersing or mixing with the rest of the water or wastewater flowing through the tank.

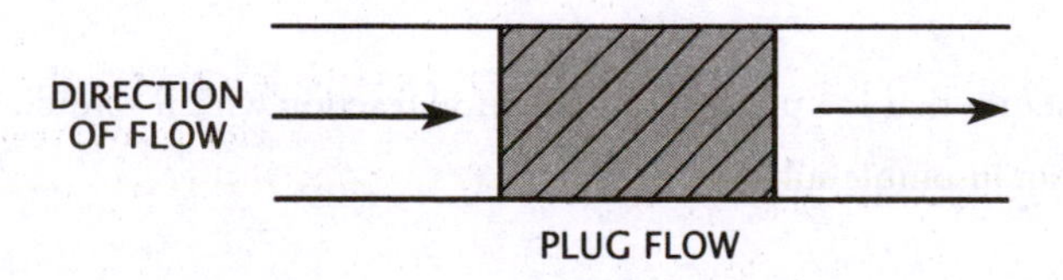

POINT SOURCE | POINT SOURCE

A discharge that comes out the end of a pipe or other clearly identifiable conveyance. Examples of point source conveyances from which pollutants may be discharged include: ditches, channels, tunnels, conduits, wells, containers, rolling stock, concentrated animal feeding operations, landfill leachate collection systems, vessels, or other floating craft. A NONPOINT SOURCE refers to runoff or a discharge from a field or similar source.

POLYELECTROLYTE (POLY-ee-LECK-tro-lite) | POLYELECTROLYTE

A high-molecular-weight (relatively heavy) substance, having points of positive or negative electrical charges, that is formed by either natural or synthetic (manmade) processes. Natural polyelectrolytes may be of biological origin or obtained from starch products or cellulose derivatives. Synthetic polyelectrolytes consist of simple substances that have been made into complex, high-molecular-weight substances. Used with other chemical coagulants to aid in binding small suspended particles to larger chemical flocs for their removal from water. Often called a polymer.

POLYMER (POLY-mer) | POLYMER

A long-chain molecule formed by the union of many monomers (molecules of lower molecular weight). Polymers are used with other chemical coagulants to aid in binding small suspended particles to larger chemical flocs for their removal from water. Also see POLYELECTROLYTE.

POPULATION EQUIVALENT | POPULATION EQUIVALENT

A means of expressing the strength of organic material in wastewater. In a domestic wastewater system, microorganisms use up about 0.2 pound (90 grams) of oxygen per day for each person using the system (as measured by the standard BOD test). May also be expressed as flow (100 gallons (378 liters)/day/person) or suspended solids (0.2 lb (90 grams) SS/day/person).

$$\text{Population Equivalent, persons} = \frac{\text{Flow, MGD} \times \text{BOD, mg/L} \times 8.34 \text{ lbs/gal}}{0.2 \text{ lb BOD/day/person}}$$

or

$$\text{Population Equivalent, persons} = \frac{\text{Flow, cu m/day} \times \text{BOD, mg/L} \times 10^6 \text{ L/cu m}}{90{,}000 \text{ mg BOD/day/person}}$$

PORE | PORE

A very small open space in a rock or granular material. Also called an interstice, void, or void space. Also see VOID.

POSTCHLORINATION | POSTCHLORINATION

The addition of chlorine to the plant discharge or effluent, following plant treatment, for disinfection purposes.

POTABLE (POE-tuh-bull) WATER | POTABLE WATER

Water that does not contain objectionable pollution, contamination, minerals, or infective agents and is considered satisfactory for drinking.

PRECHLORINATION | PRECHLORINATION

The addition of chlorine in the collection system serving the plant or at the headworks of the plant prior to other treatment processes mainly for odor and corrosion control. Also applied to aid disinfection, to reduce plant BOD load, to aid in settling, to control foaming in Imhoff units, and to help remove oil.

PRECIPITATE (pre-SIP-uh-TATE) | PRECIPITATE

(1) An insoluble, finely divided substance that is a product of a chemical reaction within a liquid.

(2) The separation from solution of an insoluble substance.

PRESENT WORTH | PRESENT WORTH

The value of a long-term project expressed in today's dollars. Present worth is calculated by converting (discounting) all future benefits and costs over the life of the project to a single economic value at the start of the project. Calculating the present worth of alternative projects makes it possible to compare them and select the one with the largest positive (beneficial) present worth or minimum present cost.

PRIMARY CLARIFIER PRIMARY CLARIFIER

A wastewater treatment device that consists of a rectangular or circular tank that allows those substances in wastewater that readily settle or float to be separated from the wastewater being treated.

PRIMARY TREATMENT PRIMARY TREATMENT

A wastewater treatment process that takes place in a rectangular or circular tank and allows those substances in wastewater that readily settle or float to be separated from the wastewater being treated. A septic tank is also considered primary treatment.

PRIME PRIME

The action of filling a pump casing with water to remove the air. Most pumps must be primed before start-up or they will not pump any water.

PRODUCTION INDICATOR PRODUCTION INDICATOR

A measure of a work activity performed by a utility's operators or crews. Examples of production indicators include miles of sewers televised per year and miles of sewers cleaned per year—measurements of how *much* a utility is doing rather than how *well* a utility is doing. Also see PERFORMANCE INDICATOR.

PROFILE PROFILE

A drawing showing elevation plotted against distance, such as the vertical section or side view of sewers, manholes, or a pipeline.

PROPORTIONAL WEIR (WEER) PROPORTIONAL WEIR

A specially shaped weir in which the flow through the weir is directly proportional to the head.

PROTOZOA (pro-toe-ZOE-ah) PROTOZOA

A group of motile, microscopic organisms (usually single-celled and aerobic) that sometimes cluster into colonies and generally consume bacteria as an energy source.

PUTREFACTION (PYOO-truh-FACK-shun) PUTREFACTION

Biological decomposition of organic matter, with the production of foul-smelling and -tasting products, associated with anaerobic (no oxygen present) conditions.

PYROMETER (pie-ROM-uh-ter) PYROMETER

An apparatus used to measure high temperatures.

## Q

(NO LISTINGS)

## R

RNA RNA

Ribonucleic acid. A chemical that provides the structure for protein synthesis (building up).

RATE OF RETURN RATE OF RETURN

A value that indicates the return of funds received on the basis of the total equity capital used to finance physical facilities. Similar to the interest rate on savings accounts or loans.

REACTIVE MAINTENANCE REACTIVE MAINTENANCE

Maintenance activities that are performed in response to problems and emergencies after they occur.

REAGENT (re-A-gent) REAGENT

A pure, chemical substance that is used to make new products or is used in chemical tests to measure, detect, or examine other substances.

RECEIVER RECEIVER

A device that indicates the result of a measurement, usually using either a fixed scale and movable indicator (pointer), such as a pressure gauge, or a moving chart with a movable pen like those used on a circular flow-recording chart. Also called an indicator.

RECEIVING WATER RECEIVING WATER

A stream, river, lake, ocean, or other surface or groundwaters into which treated or untreated wastewater is discharged.

REDOX (REE-docks) REACTION REDOX REACTION

A two-part reaction between two ions involving a transfer of electrons from one ion to the other. Oxidation is the loss of electrons by one ion, and reduction is the acceptance of electrons by the other ion. Reduction refers to the lowering of the OXIDATION STATE/OXIDATION NUMBER of the ion accepting the electrons.

In a redox reaction, the ion that gives up the electrons (that is oxidized) is called the reductant because it causes a reduction in the oxidation state or number of the ion that accepts the transferred electrons. The ion that receives the electrons (that is reduced) is called the oxidant because it causes oxidation of the other ion. Oxidation and reduction always occur simultaneously.

REDUCING AGENT REDUCING AGENT

Any substance, such as base metal (iron) or the sulfide ion ($S^{2-}$), that will readily donate (give up) electrons. The opposite is an OXIDIZING AGENT.

REDUCTION (re-DUCK-shun) REDUCTION

Reduction is the addition of hydrogen, removal of oxygen, or the addition of electrons to an element or compound. Under anaerobic conditions (no dissolved oxygen present), sulfur compounds are reduced to odor-producing hydrogen sulfide ($H_2S$) and other compounds. In the treatment of metal finishing wastewaters, hexavalent chromium ($Cr^{6+}$) is reduced to the trivalent form ($Cr^{3+}$). The opposite of OXIDATION.

REFLUX REFLUX

Flow back. A sample is heated, evaporates, cools, condenses, and flows back to the flask.

REPRESENTATIVE SAMPLE REPRESENTATIVE SAMPLE

A sample portion of material, water, or wastestream that is as nearly identical in content and consistency as possible to that in the larger body being sampled.

RESIDUAL ANALYZER, CHLORINE RESIDUAL ANALYZER, CHLORINE

See CHLORINE RESIDUAL ANALYZER.

RESIDUAL CHLORINE RESIDUAL CHLORINE

The concentration of chlorine present in water after the chlorine demand has been satisfied. The concentration is expressed in terms of the total chlorine residual, which includes both the free and combined or chemically bound chlorine residuals. Also called chlorine residual.

RESISTANCE RESISTANCE

That property of a conductor or wire that opposes the passage of a current, thus causing electric energy to be transformed into heat.

RESPONSIBILITY RESPONSIBILITY

Answering to those above in the chain of command to explain how and why you have used your authority.

RIGHT-TO-KNOW LAWS

RIGHT-TO-KNOW LAWS

Employee Right-To-Know legislation requires employers to inform employees of the possible health effects resulting from contact with hazardous substances. At locations where this legislation is in force, employers must provide employees with information regarding any hazardous substances they might be exposed to under normal work conditions or reasonably foreseeable emergency conditions resulting from workplace conditions. OSHA's Hazard Communication Standard (HCS) (Title 29 CFR Part 1910.1200) is the federal regulation and state statutes are called Worker Right-To-Know laws. Also see COMMUNITY RIGHT-TO-KNOW and SARA.

RIPRAP

RIPRAP

Broken stones, boulders, or other materials placed compactly or irregularly on levees or dikes for the protection of earth surfaces against the erosive action of waves.

RISING SLUDGE

RISING SLUDGE

Rising sludge occurs in the secondary clarifiers of activated sludge plants when the sludge settles to the bottom of the clarifier, is compacted, and then starts to rise to the surface, usually as a result of denitrification, or anaerobic biological activity that produces carbon dioxide or methane.

ROD, SEWER

ROD, SEWER

A light metal rod, three to five feet long, with a coupling at each end. Rods are joined and pushed into a sewer to dislodge obstructions.

ROTAMETER (ROTE-uh-ME-ter)

ROTAMETER

A device used to measure the flow rate of gases and liquids. The gas or liquid being measured flows vertically up a tapered, calibrated tube. Inside the tube is a small ball or bullet-shaped float (it may rotate) that rises or falls depending on the flow rate. The flow rate may be read on a scale behind or on the tube by looking at the middle of the ball or at the widest part or top of the float.

ROTATING BIOLOGICAL CONTACTOR (RBC)

ROTATING BIOLOGICAL CONTACTOR (RBC)

A secondary biological treatment process for domestic and biodegradable industrial wastes. Biological contactors have a rotating shaft surrounded by plastic discs called the media. The shaft and media are called the drum. A biological slime grows on the media when conditions are suitable and the microorganisms that make up the slime (biomass) stabilize the waste products by using the organic material for growth and reproduction.

## S

SAR

SAR

Sodium Adsorption Ratio. This ratio expresses the relative activity of sodium ions in the exchange reactions with soil. The ratio is defined as follows:

$$SAR = \frac{Na}{[\frac{1}{2}(Ca + Mg)]^{\frac{1}{2}}}$$

where Na, Ca, and Mg are concentrations of the respective ions in milliequivalents per liter of water.

$$\text{Na, meq/L} = \frac{\text{Na, mg/L}}{\text{23.0 mg/meq}} \qquad \text{Ca, meq/L} = \frac{\text{Ca, mg/L}}{\text{20.0 mg/meq}} \qquad \text{Mg, meq/L} = \frac{\text{Mg, mg/L}}{\text{12.15 mg/meq}}$$

SARA

SARA

Superfund Amendments and Reauthorization Act of 1986. The Comprehensive Environmental Response, Compensation, and Liability Act (CERCLA), commonly known as the Superfund Act, was enacted in 1980. The 1986 amendments increase CERCLA revenues to $8.5 billion and strengthen the EPA's authority to conduct short-term (removal), long-term (remedial), and enforcement actions. The amendments also strengthen state involvements in the cleanup process and the agency's commitments to research and development, training, health assessments, and public participation. A number of new statutory authorities, such as Community Right-To-Know, were also established. Also see CERCLA.

SDGS SYSTEM — SDGS SYSTEM

See SMALL-DIAMETER GRAVITY SEWER (SDGS) SYSTEM.

STEF SYSTEM — STEF SYSTEM

See SEPTIC TANK EFFLUENT FILTER (STEF) SYSTEM.

STEP SYSTEM — STEP SYSTEM

See SEPTIC TANK EFFLUENT PUMP (STEP) SYSTEM.

SANITARY SEWER — SANITARY SEWER

A pipe or conduit (sewer) intended to carry wastewater or waterborne wastes from homes, businesses, and industries to the treatment works. Stormwater runoff or unpolluted water should be collected and transported in a separate system of pipes or conduits (storm sewers) to natural watercourses.

SAPROPHYTES (SAP-row-fights) — SAPROPHYTES

Organisms living on dead or decaying organic matter. They help natural decomposition of organic matter in water or wastewater.

SCUM — SCUM

A layer or film of foreign matter (such as grease, oil) that has risen to the surface of water or wastewater.

SEASONAL WATER TABLE — SEASONAL WATER TABLE

A water table that has seasonal changes in depth or elevation.

SECCHI (SECK-key) DISK — SECCHI DISK

A flat, white disk lowered into the water by a rope until it is just barely visible. At this point, the depth of the disk from the water surface is the recorded Secchi disk transparency.

SECONDARY CLARIFIER — SECONDARY CLARIFIER

A wastewater treatment device consisting of a rectangular or circular tank that allows separation of substances that settle or float not removed by previous treatment processes.

SECONDARY TREATMENT — SECONDARY TREATMENT

A wastewater treatment process used to convert dissolved or suspended materials into a form more readily separated from the water being treated. Usually, the process follows primary treatment by sedimentation. The process commonly is a type of biological treatment followed by secondary clarifiers that allow the solids to settle out from the water being treated.

SEDIMENTATION (SED-uh-men-TAY-shun) — SEDIMENTATION

The process of settling and depositing of suspended matter carried by water or wastewater. Sedimentation usually occurs by gravity when the velocity of the liquid is reduced below the point at which it can transport the suspended material.

SEIZING or SEIZE UP — SEIZING or SEIZE UP

Seizing occurs when an engine overheats and a part expands to the point where the engine will not run. Also called freezing.

SEPTAGE (SEPT-age) — SEPTAGE

The sludge produced in septic tanks.

SEPTIC (SEP-tick) — SEPTIC

A condition produced by anaerobic bacteria. If severe, the sludge produces hydrogen sulfide, turns black, gives off foul odors, contains little or no dissolved oxygen, and the wastewater has a high oxygen demand.

SEPTIC TANK | SEPTIC TANK

A system sometimes used where wastewater collection systems and treatment plants are not available. The system is a settling tank in which settled sludge and floatable scum are in intimate contact with the wastewater flowing through the tank and the organic solids are decomposed by anaerobic bacterial action. Used to treat wastewater and produce an effluent that is usually discharged to subsurface leaching. Also referred to as an interceptor; however, the preferred term is septic tank.

SEPTIC TANK EFFLUENT FILTER (STEF) SYSTEM | SEPTIC TANK EFFLUENT FILTER (STEF) SYSTEM

A facility in which effluent flows from a septic tank into a gravity flow collection system that flows into a gravity sewer, treatment plant, or subsurface leaching system. The gravity flow pipeline is called an effluent drain.

SEPTIC TANK EFFLUENT PUMP (STEP) SYSTEM | SEPTIC TANK EFFLUENT PUMP (STEP) SYSTEM

A facility in which effluent is pumped from a septic tank into a pressurized collection system that may flow into a gravity sewer, treatment plant, or subsurface leaching system.

SEPTICITY (sep-TIS-uh-tee) | SEPTICITY

The condition in which organic matter decomposes to form foul-smelling products associated with the absence of free oxygen. If severe, the wastewater produces hydrogen sulfide, turns black, gives off foul odors, contains little or no dissolved oxygen, and the wastewater has a high oxygen demand.

SEQUENCING BATCH REACTOR (SBR) | SEQUENCING BATCH REACTOR (SBR)

A type of activated sludge system that is specifically designed and automated to mix/aerate untreated wastewater and allow solids flocculation/separation to occur as a batch treatment process.

SERIES OPERATION | SERIES OPERATION

Wastewater being treated flows through one treatment unit and then flows through another similar treatment unit. Also see PARALLEL OPERATION.

SET POINT | SET POINT

The position at which the control or controller is set. This is the same as the desired value of the process variable. For example, a thermostat is set to maintain a desired temperature.

SHOCK LOAD (ACTIVATED SLUDGE) | SHOCK LOAD

The arrival at a plant of a waste that is toxic to organisms in sufficient quantity or strength to cause operating problems. Possible problems include odors and bulking sludge, which will result in a high loss of solids from the secondary clarifiers into the plant effluent and a biological process upset that may require several days to a week to recover. Organic or hydraulic overloads also can cause a shock load.

SHOCK LOAD (TRICKLING FILTERS) | SHOCK LOAD

The arrival at a plant of a waste that is toxic to organisms in sufficient quantity or strength to cause operating problems. Possible problems include odors and sloughing off of the growth or slime on the trickling filter media. Organic or hydraulic overloads also can cause a shock load.

SHORT-CIRCUITING | SHORT-CIRCUITING

A condition that occurs in tanks or basins when some of the flowing water entering a tank or basin flows along a nearly direct pathway from the inlet to the outlet. This is usually undesirable because it may result in shorter contact, reaction, or settling times in comparison with the theoretical (calculated) or presumed detention times.

SHREDDING | SHREDDING

A mechanical treatment process that cuts large pieces of wastes into smaller pieces so they will not plug pipes or damage equipment. Shredding and comminution usually mean the same thing.

SIDESTREAM — SIDESTREAM

Wastewater flows that develop from other storage or treatment facilities. This wastewater may or may not need additional treatment.

SLOPE — SLOPE

The slope or inclination of a trench bottom or a trench side wall is the ratio of the vertical distance to the horizontal distance or rise over run. Also see GRADE (2).

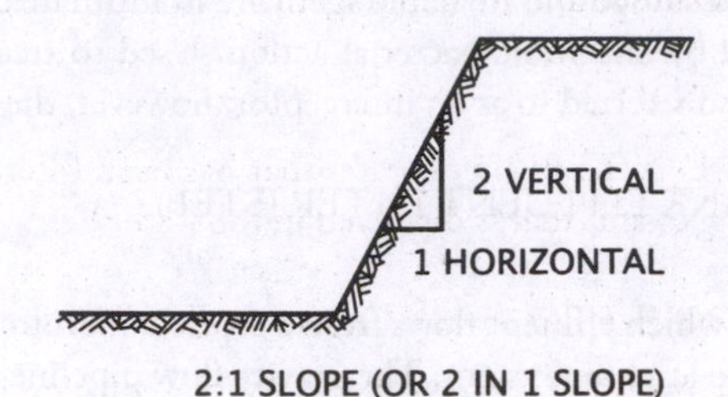


SLOUGHED or SLOUGHING (SLUFF-ing) — SLOUGHED or SLOUGHING

The breaking off of biological or biomass growths from the fixed film or rotating biological contactor (RBC) media. The sloughed growth becomes suspended in the effluent and is later removed in the secondary clarifier as sludge.

SLUDGE (SLUJ) — SLUDGE

(1) The settleable solids separated from liquids during processing.

(2) The deposits of foreign materials on the bottoms of streams or other bodies of water or on the bottoms and edges of wastewater collection lines and appurtenances.

SLUDGE AGE — SLUDGE AGE

A measure of the length of time a particle of suspended solids has been retained in the activated sludge process.

$$\text{Sludge Age, days} = \frac{\text{Suspended Solids Under Aeration, lbs or kg}}{\text{Suspended Solids Added, lbs/day or kg/day}}$$

SLUDGE DENSITY INDEX (SDI) — SLUDGE DENSITY INDEX (SDI)

This calculation is used in a way similar to the Sludge Volume Index (SVI) to indicate the settleability of a sludge in a secondary clarifier or effluent. The weight in grams of one milliliter of sludge after settling for 30 minutes. SDI = 100/SVI. Also see SLUDGE VOLUME INDEX (SVI).

SLUDGE VOLUME INDEX (SVI) — SLUDGE VOLUME INDEX (SVI)

A calculation that indicates the tendency of activated sludge solids (aerated solids) to thicken or to become concentrated during the sedimentation/thickening process. SVI is calculated in the following manner: (1) allow a mixed liquor sample from the aeration basin to settle for 30 minutes; (2) determine the suspended solids concentration for a sample of the same mixed liquor; (3) calculate SVI by dividing the measured (or observed) wet volume (mL/L) of the settled sludge by the dry weight concentration of MLSS in grams/L.

$$\text{SVI, mL/gm} = \frac{\text{Settled Sludge Volume/Sample Volume, mL/L}}{\text{Suspended Solids Concentration, mg/L}} \times \frac{1{,}000\text{ mg}}{\text{gram}}$$

SLUG — SLUG

Intermittent release or discharge of wastewater or industrial wastes.

SLURRY — SLURRY

A watery mixture or suspension of insoluble (not dissolved) matter; a thin, watery mud or any substance resembling it (such as a grit slurry or a lime slurry).

SMALL-DIAMETER GRAVITY SEWER (SDGS) SYSTEM — SMALL-DIAMETER GRAVITY SEWER (SDGS) SYSTEM

A type of collection system in which a series of septic tanks discharge effluent by gravity, pump, or siphon to a small-diameter wastewater collection main. The wastewater flows by gravity to a lift station, a manhole in a conventional gravity collection system, or directly to a wastewater treatment plant.

SOLIDS CONCENTRATION — SOLIDS CONCENTRATION

The solids in the aeration tank that carry microorganisms that feed on wastewater. Expressed as milligrams per liter of mixed liquor volatile suspended solids (MLVSS, mg/L).

SOLUBLE BOD — SOLUBLE BOD

Soluble BOD is the BOD of water that has been filtered in the standard suspended solids test. The soluble BOD is a measure of food for microorganisms that is dissolved in the water being treated.

SOLUTION — SOLUTION

A liquid mixture of dissolved substances. In a solution it is impossible to see all the separate parts.

SPECIFIC GRAVITY — SPECIFIC GRAVITY

(1) Weight of a particle, substance, or chemical solution in relation to the weight of an equal volume of water. Water has a specific gravity of 1.000 at 4°C (39°F). Wastewater particles or substances usually have a specific gravity of 0.5 to 2.5. Particulates with specific gravity less than 1.0 float to the surface and particulates with specific gravity greater than 1.0 sink.

(2) Weight of a particular gas in relation to the weight of an equal volume of air at the same temperature and pressure (air has a specific gravity of 1.0). Chlorine gas has a specific gravity of 2.5.

SPLASH PAD — SPLASH PAD

A structure made of concrete or other durable material to protect bare soil from erosion by splashing or falling water.

SPORE — SPORE

The reproductive body of certain organisms, which is capable of giving rise to a new organism either directly or indirectly. A viable (able to live and grow) body regarded as the resting stage of an organism. A spore is usually more resistant to disinfectants and heat than most organisms. Gangrene and tetanus bacteria are common spore-forming organisms.

STABILIZATION — STABILIZATION

Conversion to a form that resists change. Organic material is stabilized by bacteria that convert the material to gases and other relatively inert substances. Stabilized organic material generally will not give off obnoxious odors.

STABILIZED WASTE — STABILIZED WASTE

A waste that has been treated or decomposed to the extent that, if discharged or released, its rate and state of decomposition would be such that the waste would not cause a nuisance or odors in the receiving water.

STAFF GAUGE — STAFF GAUGE

A ruler or graduated scale used to measure the depth or elevation of water in a channel, tank, or stream.

STANDARD METHODS — STANDARD METHODS

*STANDARD METHODS FOR THE EXAMINATION OF WATER AND WASTEWATER*, 21st Edition. A joint publication of the American Public Health Association (APHA), American Water Works Association (AWWA), and the Water Environment Federation (WEF) that outlines the accepted laboratory procedures used to analyze the impurities in water and wastewater. Available from: American Water Works Association, Bookstore, 6666 West Quincy Avenue, Denver, CO 80235. Order No. 10084. Price to members, $198.50; nonmembers, $266.00; price includes cost of shipping and handling. Also available from Water Environment Federation, Publications Order Department, PO Box 18044, Merrifield, VA 22118-0045. Order No. S82011. Price to members, $203.00; nonmembers, $268.00; price includes cost of shipping and handling.

STANDARD SOLUTION — STANDARD SOLUTION

A solution in which the exact concentration of a chemical or compound is known.

STANDARDIZE | STANDARDIZE

To compare with a standard.

(1) In wet chemistry, to find out the exact strength of a solution by comparing it with a standard of known strength. This information is used to adjust the strength by adding more water or more of the substance dissolved.

(2) To set up an instrument or device to read a standard. This allows you to adjust the instrument so that it reads accurately, or enables you to apply a correction factor to the readings.

STATIC HEAD | STATIC HEAD

When water is not moving, the vertical distance (in feet or meters) from a reference point to the water surface is the static head. Also see DYNAMIC HEAD, DYNAMIC PRESSURE, and STATIC PRESSURE.

STATIC LIFT | STATIC LIFT

Vertical distance water is lifted from upstream water surface to downstream water surface (which is at a higher elevation) when no water is being pumped.

STATIC PRESSURE | STATIC PRESSURE

When water is not moving, the vertical distance (in feet or meters) from a specific point to the water surface is the static head. The static pressure in psi (or kPa) is the static head in feet times 0.433 psi/ft (or meters × 9.81 kPa/m). Also see DYNAMIC HEAD, DYNAMIC PRESSURE, and STATIC HEAD.

STATIC WATER HEAD | STATIC WATER HEAD

Elevation or surface of water that is not being pumped.

STEP-FEED AERATION | STEP-FEED AERATION

Step-feed aeration is a modification of the conventional activated sludge process. In step-feed aeration, primary effluent enters the aeration tank at several points along the length of the tank, rather than at the beginning or head of the tank and flowing through the entire tank in a plug flow mode.

STERILIZATION (STAIR-uh-luh-ZAY-shun) | STERILIZATION

The removal or destruction of all microorganisms, including pathogens and other bacteria, vegetative forms, and spores. Compare with DISINFECTION.

STOP LOG | STOP LOG

A log or board in an outlet box or device used to control the water level in ponds and also the flow from one pond to another pond or system.

STOPPAGE | STOPPAGE

Any incident in which a sewer is partially or completely blocked, causing a backup, a service interruption, or an overflow. Also called blockage.

SUBSURFACE LEACHING SYSTEM | SUBSURFACE LEACHING SYSTEM

A method of treatment and discharge of septic tank effluent, sand filter effluent, or other treated wastewater. The effluent is applied to soil below the ground surface through open-jointed pipes or drains or through perforated pipes (holes in the pipes). The effluent is treated as it passes through porous soil or rock strata (layers). Newer subsurface leaching systems include chamber and gravelless systems, and also gravel trenches without pipe the full length of the trench.

SUCTION HEAD | SUCTION HEAD

The positive pressure [in feet (meters) of water or pounds per square inch (kilograms per square centimeter) of mercury vacuum] on the suction side of a pump. The pressure can be measured from the centerline of the pump up to the elevation of the hydraulic grade line on the suction side of the pump.

SUCTION LIFT — SUCTION LIFT

The negative pressure [in feet (meters) of water or inches (centimeters) of mercury vacuum] on the suction side of a pump. The pressure can be measured from the centerline of the pump down to (lift) the elevation of the hydraulic grade line on the suction side of the pump.

SUMP — SUMP

This term refers to a facility or structure that connects an industrial discharger to a public sewer. The sump could be a sample box, a clarifier, or an intercepting sewer.

SUPERFUND ACT — SUPERFUND ACT

See CERCLA.

SUPERNATANT (soo-per-NAY-tent) — SUPERNATANT

The relatively clear water layer between the sludge on the bottom and the scum on the surface of an anaerobic digester or septic tank (interceptor).

(1) From an anaerobic digester, this water is usually returned to the influent wet well or to the primary clarifier.

(2) From a septic tank, this water is discharged by gravity or by a pump to a leaching system or a wastewater collection system.

Also called clear zone.

SURCHARGE — SURCHARGE

Sewers are surcharged when the supply of water to be carried is greater than the capacity of the pipes to carry the flow. The surface of the wastewater in manholes rises above the top of the sewer pipe, and the sewer is under pressure or a head, rather than at atmospheric pressure.

SURFACTANT (sir-FAC-tent) — SURFACTANT

Abbreviation for surface-active agent. The active agent in detergents that possesses a high cleaning ability.

SUSPENDED SOLIDS — SUSPENDED SOLIDS

(1) Solids that either float on the surface or are suspended in water, wastewater, or other liquids, and that are largely removable by laboratory filtering.

(2) The quantity of material removed from water or wastewater in a laboratory test, as prescribed in *STANDARD METHODS FOR THE EXAMINATION OF WATER AND WASTEWATER*, and referred to as Total Suspended Solids Dried at 103–105°C.

## T

TOC (pronounce as separate letters) — TOC

Total Organic Carbon. TOC measures the amount of organic carbon in water.

TWA — TWA

See TIME-WEIGHTED AVERAGE (TWA).

TAILGATE SAFETY MEETING — TAILGATE SAFETY MEETING

Brief (10 to 20 minutes) safety meetings held every 7 to 10 working days. The term comes from the safety meetings regularly held by the construction industry around the tailgate of a truck.

TELEMETERING EQUIPMENT — TELEMETERING EQUIPMENT

Equipment that translates physical measurements into electrical impulses that are transmitted to dials or recorders.

TEMPORARY GROUNDWATER TABLE — TEMPORARY GROUNDWATER TABLE

(1) During and for a period following heavy rainfall or snow melt, the soil is saturated at elevations above the normal, stabilized, or seasonal groundwater table, often from the surface of the soil downward. This is referred to as a temporary condition and thus is a temporary groundwater table.

(2) When a collection system serves agricultural areas in its vicinity, irrigation of these areas can cause a temporary rise in the elevation of the groundwater table.

TERTIARY (TER-she-air-ee) TREATMENT — TERTIARY TREATMENT

Any process of water renovation that upgrades treated wastewater to meet specific reuse requirements. May include general cleanup of water or removal of specific parts of wastes insufficiently removed by conventional treatment processes. Typical processes include chemical treatment and pressure filtration. Also called advanced waste treatment.

THIEF HOLE — THIEF HOLE

A digester sampling well that allows sampling of the digester contents without venting digester gas.

THRESHOLD ODOR — THRESHOLD ODOR

The minimum odor of a gas or water sample that can just be detected after successive dilutions with odorless gas or water. Also called odor threshold.

THRUST BLOCK — THRUST BLOCK

A mass of concrete or similar material appropriately placed around a pipe to prevent movement when the pipe is carrying water. Usually placed at bends and valve structures.

TIME-WEIGHTED AVERAGE (TWA) — TIME-WEIGHTED AVERAGE (TWA)

A time-weighted average is used to calculate a worker's daily exposure to a hazardous substance (such as chemicals, dusts, fumes, mists, gases, or vapors) or agent (such as occupational noise), averaged to an 8-hour workday, taking into account the average levels of the substance or agent and the time spent in the area. This is the guideline OSHA uses to determine permissible exposure limits (PELs) and is essential in assessing a worker's exposure and determining what protective measures should be taken. A time-weighted average is equal to the sum of the portion of each time period (as a decimal, such as 0.25 hour) multiplied by the levels of the substance or agent during the time period divided by the hours in the workday (usually 8 hours). Also see PERMISSIBLE EXPOSURE LIMIT (PEL).

TITRATE (TIE-trate) — TITRATE

To titrate a sample, a chemical solution of known strength is added drop by drop until a certain color change, precipitate, or pH change in the sample is observed (end point). Titration is the process of adding the chemical reagent in small increments (0.1–1.0 milliliter) until completion of the reaction, as signaled by the end point.

TOPOGRAPHY (toe-PAH-gruh-fee) — TOPOGRAPHY

The arrangement of hills and valleys in a geographic area.

TOTAL CHLORINE — TOTAL CHLORINE

The total concentration of chlorine in water, including the combined chlorine (such as inorganic and organic chloramines) and the free available chlorine.

TOTAL CHLORINE RESIDUAL — TOTAL CHLORINE RESIDUAL

The total amount of chlorine residual (including both free chlorine and chemically bound chlorine) present in a water sample after a given contact time.

TOTAL DYNAMIC HEAD (TDH) — TOTAL DYNAMIC HEAD (TDH)

When a pump is lifting or pumping water, the vertical distance (in feet or meters) from the elevation of the energy grade line on the suction side of the pump to the elevation of the energy grade line on the discharge side of the pump. The total dynamic head is the static head plus pipe friction losses.

TOTALIZER TOTALIZER

A device or meter that continuously measures and sums a process rate variable in cumulative fashion over a given time period. For example, total flows displayed in gallons per minute, million gallons per day, cubic feet per second, or some other unit of volume per time period. Also called an integrator.

TOXIC TOXIC

A substance that is poisonous to a living organism. Toxic substances may be classified in terms of their physiological action, such as irritants, asphyxiants, systemic poisons, and anesthetics and narcotics. Irritants are corrosive substances that attack the mucous membrane surfaces of the body. Asphyxiants interfere with breathing. Systemic poisons are hazardous substances that injure or destroy internal organs of the body. Anesthetics and narcotics are hazardous substances that depress the central nervous system and lead to unconsciousness.

TOXIC SUBSTANCE TOXIC SUBSTANCE

See HARMFUL PHYSICAL AGENT and TOXIC.

TOXICITY (tox-IS-it-tee) TOXICITY

The relative degree of being poisonous or toxic. A condition that may exist in wastes and will inhibit or destroy the growth or function of certain organisms.

TRANSPIRATION (TRAN-spur-RAY-shun) TRANSPIRATION

The process by which water vapor is released to the atmosphere by living plants. This process is similar to people sweating. Also called evapotranspiration.

TRICKLING FILTER TRICKLING FILTER

A treatment process in which wastewater trickling over media enables the formation of slimes or biomass, which contain organisms that feed upon and remove wastes from the water being treated.

TURBIDIMETER TURBIDIMETER

See TURBIDITY METER.

TURBIDITY (ter-BID-it-tee) TURBIDITY

The cloudy appearance of water caused by the presence of suspended and colloidal matter. In the waterworks field, a turbidity measurement is used to indicate the clarity of water. Technically, turbidity is an optical property of the water based on the amount of light reflected by suspended particles. Turbidity cannot be directly equated to suspended solids because white particles reflect more light than dark-colored particles and many small particles will reflect more light than an equivalent large particle.

TURBIDITY (ter-BID-it-tee) METER TURBIDITY METER

An instrument for measuring and comparing the turbidity of liquids by passing light through them and determining how much light is reflected by the particles in the liquid. The normal measuring range is 0 to 100 and is expressed as nephelometric turbidity units (NTUs). Also called a turbidimeter.

TURBIDITY (ter-BID-it-tee) UNITS (TU) TURBIDITY UNITS (TU)

Turbidity units are a measure of the cloudiness of water. If measured by a nephelometric (deflected light) instrumental procedure, turbidity units are expressed in nephelometric turbidity units (NTU) or simply TU. Those turbidity units obtained by visual methods are expressed in Jackson turbidity units (JTU), which are a measure of the cloudiness of water; they are used to indicate the clarity of water. There is no real connection between NTUs and JTUs. The Jackson turbidimeter is a visual method and the nephelometer is an instrumental method based on deflected light.

## U

### UNIFORMITY COEFFICIENT (UC)

UNIFORMITY COEFFICIENT (UC)

The ratio of (1) the diameter of a grain (particle) of a size that is barely too large to pass through a sieve that allows 60 percent of the material (by weight) to pass through, to (2) the diameter of a grain (particle) of a size that is barely too large to pass through a sieve that allows 10 percent of the material (by weight) to pass through. The resulting ratio is a measure of the degree of uniformity in a granular material, such as filter media.

$$\text{Uniformity Coefficient} = \frac{\text{Particle Diameter}_{60\%}}{\text{Particle Diameter}_{10\%}}$$

## V

### V-NOTCH WEIR

V-NOTCH WEIR

A triangular weir with a V-shaped notch calibrated in gallons (liters) per minute readings. The weir can be placed in a pipe or open channel. As the flow passes through the V-notch, the depth of water flowing over the weir can be measured and converted to a flow in gallons (liters) per minute.

### VARIABLE COSTS

VARIABLE COSTS

Costs that a utility must cover or pay that are associated with the actual collection, treatment, and discharge of wastewater. These costs vary or fluctuate on the basis of the volume of wastewater collected, treated, and discharged or reused. Also see FIXED COSTS.

### VAULT

VAULT

A small, box-like structure that contains valves used to regulate flows.

### VECTOR

VECTOR

An insect or other organism capable of transmitting germs or other agents of disease.

### VELOCITY HEAD

VELOCITY HEAD

The energy in flowing water as determined by a vertical height (in feet or meters) equal to the square of the velocity of flowing water divided by twice the acceleration due to gravity ($V^2/2g$).

### VERTICAL OFFSET

VERTICAL OFFSET

A pipe joint in which one section is connected to another at a different elevation, such as a DROP JOINT.

### VOID

VOID

A pore or open space in rock, soil, or other granular material, not occupied by solid matter. The pore or open space may be occupied by air, water, or other gaseous or liquid material. Also called an interstice, pore, or void space.

### VOLATILE (VOL-uh-tull)

VOLATILE

(1) A volatile substance is one that is capable of being evaporated or changed to a vapor at relatively low temperatures. Volatile substances can be partially removed from water or wastewater by the air stripping process.

(2) In terms of solids analysis, volatile refers to materials lost (including most organic matter) upon ignition in a muffle furnace for 60 minutes at 550°C (1,022°F). Natural volatile materials are chemical substances usually of animal or plant origin. Manufactured or synthetic volatile materials, such as plastics, ether, acetone, and carbon tetrachloride, are highly volatile and not of plant or animal origin. Also see NONVOLATILE MATTER.

### VOLATILE ACIDS

VOLATILE ACIDS

Fatty acids produced during digestion that are soluble in water and can be steam-distilled at atmospheric pressure. Also called organic acids. Volatile acids are commonly reported as equivalent to acetic acid.

VOLATILE LIQUIDS VOLATILE LIQUIDS

Liquids that easily vaporize or evaporate at room temperature.

VOLATILE SOLIDS VOLATILE SOLIDS

Those solids in water, wastewater, or other liquids that are lost on ignition of the dry solids at 550°C (1,022°F). Also called organic solids and volatile matter.

VOLTAGE VOLTAGE

The electrical pressure available to cause a flow of current (amperage) when an electric circuit is closed. Also called electromotive force (EMF).

VOLUMETRIC VOLUMETRIC

A measurement based on the volume of some factor. Volumetric titration is a means of measuring unknown concentrations of water quality indicators in a sample by determining the volume of titrant or liquid reagent needed to complete particular reactions.

VOLUTE (vol-LOOT) VOLUTE

The spiral-shaped casing that surrounds a pump, blower, or turbine impeller and collects the liquid or gas discharged by the impeller.

## W

WASTEWATER WASTEWATER

A community's used water and water-carried solids (including used water from industrial processes) that flow to a treatment plant. Stormwater, surface water, and groundwater infiltration also may be included in the wastewater that enters a wastewater treatment plant. The term sewage usually refers to household wastes, but this word is being replaced by the term wastewater.

WATER CYCLE WATER CYCLE

The process of evaporation of water into the air and its return to earth by precipitation (rain or snow). This process also includes transpiration from plants, groundwater movement, and runoff into rivers, streams, and the ocean. Also called the hydrologic cycle.

WATT WATT

A unit of power equal to one joule per second. The power of a current of one ampere flowing across a potential difference of one volt.

WEIR (WEER) WEIR

(1) A wall or plate placed in an open channel and used to measure the flow of water. The depth of the flow over the weir can be used to calculate the flow rate, or a chart or conversion table may be used to convert depth to flow. Also see PROPORTIONAL WEIR.

(2) A wall or obstruction used to control flow (from settling tanks and clarifiers) to ensure a uniform flow rate and avoid short-circuiting.

WET CHEMISTRY WET CHEMISTRY

Laboratory procedures used to analyze a sample of water using liquid chemical solutions (wet) instead of, or in addition to, laboratory instruments.

WET PIT WET PIT

See WET WELL.

WET WELL WET WELL

A compartment or tank in which wastewater is collected. The suction pipe of a pump may be connected to the wet well or a submersible pump may be located in the wet well.

## X

(NO LISTINGS)

## Y

(NO LISTINGS)

## Z

ZOOGLEAL (ZOE-uh-glee-ul) MASS — ZOOGLEAL MASS

Jelly-like masses of bacteria found in both the trickling filter and activated sludge processes. These masses may be formed for or function as the protection against predators and for storage of food supplies. Also see BIOMASS.

ZOOGLEAL (ZOE-uh-glee-al) MAT — ZOOGLEAL MAT

A complex population of organisms that form a slime growth on the sand filter media and break down the organic matter in wastewater. These slimes consist of living organisms feeding on the wastes in wastewater, dead organisms, silt, and other debris. On a properly loaded and operating sand filter, these mats are so thin as to be invisible to the naked eye. Slime growth is a more common term.

# SUBJECT INDEX

## A



## B

## J



## L



## M



## N

## O



## P

S